THREE-DIMENSIONAL MODELS OF MARINE AND ESTUARINE DYNAMICS

FURTHER TITLES IN THIS SERIES

1 J.L. MERO
THE MINERAL RESOURCES OF THE SEA
2 L.M. FOMIN
THE DYNAMIC METHOD IN OCEANOGRAPHY
3 E.J.F. WOOD
MICROBIOLOGY OF OCEANS AND ESTUARIES
4 G. NEUMANN
OCEAN CURRENTS
5 N.G. JERLOV
OPTICAL OCEANOGRAPHY
6 V. VACQUIER
GEOMAGNETISM IN MARINE GEOLOGY
7 W.J. WALLACE
THE DEVELOPMENTS OF THE CHLORINITY/SALINITY CONCEPT IN OCEANOGRAPHY
8 E. LISITZIN
SEA-LEVEL CHANGES
9 R.H. PARKER
THE STUDY OF BENTHIC COMMUNITIES
10 J.C.J. NIHOUL (Editor)
MODELLING OF MARINE SYSTEMS
11 O.I. MAMAYEV
TEMPERATURE-SALINITY ANALYSIS OF WORLD OCEAN WATERS
12 E.J. FERGUSON WOOD and R.E. JOHANNES
TROPICAL MARINE POLLUTION
13 E. STEEMANN NIELSEN
MARINE PHOTOSYNTHESIS
14 N.G. JERLOV
MARINE OPTICS
15 G.P. GLASBY
MARINE MANGANESE DEPOSITS
16 V.M. KAMENKOVICH
FUNDAMENTALS OF OCEAN DYNAMICS
17 R.A. GEYER
SUBMERSIBLES AND THEIR USE IN OCEANOGRAPHY AND OCEAN ENGINEERING
18 J.W. CARUTHERS
FUNDAMENTALS OF MARINE ACOUSTICS
19 J.C.J. NIHOUL (Editor)
BOTTOM TURBULENCE
20 P.H. LEBLOND and L.A. MYSAK
WAVES IN THE OCEAN
21 C.C. VON DER BORCH (Editor)
SYNTHESIS OF DEEP-SEA DRILLING RESULTS IN THE INDIAN OCEAN
22 P. DEHLINGER
MARINE GRAVITY
23 J.C.J. NIHOUL (Editor)
HYDRODYNAMICS OF ESTUARIES AND FJORDS
24 F.T. BANNER, M.B. COLLINS and K.S. MASSIE (Editors)
THE NORTH-WEST EUROPEAN SHELF SEAS: THE SEA BED AND THE SEA IN MOTION
25 J.C.J. NIHOUL (Editor)
MARINE FORECASTING
26 H.G. RAMMING and Z. KOWALIK
NUMERICAL MODELLING MARINE HYDRODYNAMICS
27 R.A. GEYER (Editor)
MARINE ENVIRONMENTAL POLLUTION
28 J.C.J. NIHOUL (Editor)
MARINE TURBULENCE
29 M. WALDICHUK, G.B. KULLENBERG and M.J. ORREN (Editors)
MARINE POLLUTANT TRANSFER PROCESSES
30 A. VOIPIO (Editor)
THE BALTIC SEA
31 E.K. DUURSMA and R. DAWSON (Editors)
MARINE ORGANIC CHEMISTRY
32 J.C.J. NIHOUL (Editor)
ECOHYDRODYNAMICS
33 R. HEKINIAN
PETROLOGY OF THE OCEAN FLOOR
34 J.C.J. NIHOUL (Editor)
HYDRODYNAMICS OF SEMI-ENCLOSED SEAS
35 B. JOHNS (Editor)
PHYSICAL OCEANOGRAPHY OF COASTAL AND SHELF SEAS
36 J.C.J. NIHOUL (Editor)
HYDRODYNAMICS OF THE EQUATORIAL OCEAN
37 W. LANGERAAR
SURVEYING AND CHARTING OF THE SEAS
38 J.C.J. NIHOUL (Editor)
REMOTE SENSING OF SHELF SEA HYDRODYNAMICS
39 T. ICHIYE (Editor)
OCEAN HYDRODYNAMICS OF THE JAPAN AND EAST CHINA SEAS
40 J.C.J. NIHOUL (Editor)
COUPLED OCEAN–ATMOSPHERE MODELS
41 H. KUNZENDORF (Editor)
MARINE MINERAL EXPLORATION
42 J.C.J. NIHOUL (Editor)
MARINE INTERFACES ECOHYDRODYNAMICS
43 P. LASSERRE and J.M. MARTIN (Editors)
BIOGEOCHEMICAL PROCESSES AT THE LAND–SEA BOUNDARY
44 I.P. MARTINI (Editor)
CANADIAN INLAND SEAS

Elsevier Oceanography Series, 45

THREE-DIMENSIONAL MODELS OF MARINE AND ESTUARINE DYNAMICS

Edited by

J.C.J. NIHOUL
University of Liège, B5 Sart Tilman, B-4000 Liège, Belgium

and

B.M. JAMART
MUMM, Institute of Mathematics, 15 Avenue des Tilleuls, B-4000 Liège, Belgium

ELSEVIER

Amsterdam – Oxford – New York – Tokyo 1987

ELSEVIER SCIENCE PUBLISHERS B.V.
Sara Burgerhartstraat 25
P.O. Box 211, 1000 AE Amsterdam, The Netherlands

Distributors for the United States and Canada:

ELSEVIER SCIENCE PUBLISHING COMPANY INC.
52, Vanderbilt Avenue
New York, NY 10017, U.S.A.

ISBN 0-444-42794-5 (Vol. 45)
ISBN 0-444-41623-4 (Series)

Printed in The Netherlands

FOREWORD

The International Liège Colloquium on Ocean Hydrodynamics is organized annually. The topic differs from one year to another in an attempt to address, as much as possible, recent problems and incentive new subjects in physical oceanography.

Assembling a group of active and eminent scientists from various countries and often different disciplines, the Colloquia provide a forum for discussion and foster a mutually beneficial exchange of information opening on to a survey of major recent discoveries, essential mechanisms, impelling question-marks and valuable recommendations for future research.

The Scientific Organizing Committee and the participants wish to express their gratitude to the Belgian Minister of Education, the National Science Foundation of Belgium, the University of Liège, the Intergovernmental Oceanographic Commission and the Division of Marine Sciences (UNESCO), and the Office of Naval Research for their most valuable support.

In May 1986, we learned with sadness that Dr. Norman S. Heaps would not be able to attend the Liège Colloquium as planned because of illness. Norman passed away on 26 July 1986. The modelling community has lost a pioneer, a guide, and a friend. We dedicate this volume of proceedings, which contain a small part of his large legacy, to the memory of Dr. Norman Stuart Heaps.

Jacques C. J. Nihoul | Bruno M. Jamart

TABLE OF CONTENTS

PERSPECTIVE IN THREE-DIMENSIONAL MODELLING OF THE MARINE SYSTEM
Jacques C.J. Nihoul and S. Djenidi 1

ON MODELING THREE-DIMENSIONAL ESTUARINE AND MARINE HYDRODYNAMICS
Y. Peter Sheng 35

CIRCULATION MODELLING USING ORTHOGONAL CURVILINEAR COORDINATES
Alan F. Blumberg and H. James Herring 55

PREDICTING OPEN OCEAN CURRENTS, FRONTS AND EDDIES
Allan R. Robinson 89

PREPARATION OF ESTUARY AND MARINE MODEL EQUATIONS BY GENERALIZED FILTERING METHODS
K. W. Bedford, J. S. Dingman and W. K. Yeo 113

A LIMITED AREA MODEL FOR THE GULF STREAM REGION
William R. Holland 127

STUDY OF TRANSPORT FLUCTUATIONS AND MEANDERING OF THE FLORIDA CURRENT USING AN ISOPYCNIC COORDINATE NUMERICAL MODEL
Douglas B. Boudra, Rainer Bleck and Friedrich Schott 149

DYNAMICS OF AGULHAS RETROFLECTION AND RING FORMATION IN A QUASI-ISOPYCNIC COORDINATE NUMERICAL MODEL
E. P. Chassignet and D. B. Boudra 169

MODELLING OF MESOSCALE OCEANIC INSTABILITY PROCESSES
Aike Beckmann 195

AN EDDY-RESOLVING MODEL FOR RIVER PLUME FRONTS
J. W. Dippner 211

A FINITE DIFFERENCE GENERAL CIRCULATION MODEL FOR SHELF SEAS AND ITS APPLICATION TO LOW FREQUENCY VARIABILITY ON THE NORTH EUROPEAN SHELF
J. O. Backhaus and D. Hainbucher 221

A THREE DIMENSIONAL CIRCULATION MODEL OF THE SOUTH CHINA SEA
T. Pohlmann 245

THE INFLUENCE OF BOUNDARY CONDITIONS ON THE CIRCULATION IN THE GREENLAND-NORWEGIAN SEA. A NUMERICAL INVESTIGATION
S. Legutke 269

A THREE DIMENSIONAL BAROCLINIC MODEL OF THE WESTERN BALTIC
M. J. Boehlich 285

A STUDY OF VARIOUS OPEN BOUNDARY CONDITIONS FOR WIND-FORCED BAROTROPIC NUMERICAL OCEAN MODELS
L. P. Røed and C. K. Cooper 305

COASTAL CURRENTS, INTERNAL WAVE COLLAPSES AND TURBULENCE IN THE STRAIT OF MESSINA ZONE
E. Salusti and R. Santoleri 337

A THREE-DIMENSIONAL FINITE ELEMENT MODEL FOR THE STUDY OF STEADY AND NON-STEADY NATURAL FLOWS
J.-L. Robert and Y. Ouellet 359

REAL AND SPURIOUS BOUNDARY LAYER EFFECTS IN THREE-DIMENSIONAL HYDRODYNAMICAL MODELS
Bruno M. Jamart and José Ozer 373

A TROPHIC-DIFFUSION 3D MODEL OF THE VENICE LAGOON
C. Dejak and G. Pecenik 391

THREE DIMENSIONAL CONTINENTAL SHELF HYDRODYNAMICS MODEL INCLUDING WAVE CURRENT INTERACTION
M. L. Spaulding and T. Isaji 405

THREE-DIMENSIONAL MODEL OF CURRENTS IN THE BAY OF SEINE
J. C. Salomon, B. Thouvenin and P. Le Hir 427

TIDAL STREAMS IN SHALLOW WATER
P. P. G. Dyke 441

MODELLING AND OBSERVATIONS OF THE RESIDUAL CURRENT OFF SOUTHWEST NOVA SCOTIA
K. T. Tee, P. C. Smith and D. Lefaivre 455

A THREE-DIMENSIONAL WEAKLY NONLINEAR MODEL OF TIDE-INDUCED LAGRANGIAN RESIDUAL CURRENT AND MASS-TRANSPORT, WITH AN APPLICATION TO THE BOHAI SEA

Shizuo Feng 471

THREE DIMENSIONAL NUMERICAL MODEL FOR THERMAL IMPACT STUDIES

M. Darras, P. Donnars and P. Pechon 489

ESTIMATION OF STORM-GENERATED CURRENTS

N. S. Heaps and J. E. Jones 505

A COUPLED 2-D/3-D MODELLING SYSTEM FOR COMPUTATION OF TIDAL AND WIND-INDUCED CURRENTS

J. M. Usseglio-Polatera and P. Sauvaget 539

A HIGH RESOLUTION 3D MODEL SYSTEM FOR BAROCLINIC ESTUARINE DYNAMICS AND PASSIVE POLLUTANT DISPERSION

J. Krohn, K. Duwe and K. D. Pfeiffer 555

A THREE DIMENSIONAL NUMERICAL MODEL OF SEMI-DIURNAL TIDES ON THE EUROPEAN CONTINENTAL SHELF

A. M. Davies 573

A GENERAL THREE-DIMENSIONAL EDDY-RESOLVING MODEL FOR STRATIFIED SEAS

I. D. James 591

A 3-D MODEL OF THE SEVERN ESTUARY

J. Wolf 609

THE VARIATIONAL INVERSE METHOD REVISITED (Abstract only)

C. Provost 625

THE BRANCHING OF THE GULF STREAM REVISITED USING THE VARIATIONAL INVERSE METHOD (Abstract only)

F. Martel and C. Provost 627

ABOUT A DIAGNOSTIC ANALYSIS OF THE HISTORICAL HYDROGRAPHIC DATA IN THE TROPICAL ATLANTIC (Abstract only)

C. Provost and M. S. Suk 629

LIST OF PARTICIPANTS

ADAM Y., Dr., Management Unit of the Mathematical Models (MUMM), Liège, Belgium
AIRAUDO J.L., Ing., Météorologie Nationale, Paris, France
BACKHAUS J.O., Prof. Dr., Universität Hamburg, Hamburg, Germany
BAH A., Dr., Université Laval, Québec, Canada
BECKMANN A., Mr., Universität Kiel, Kiel, Germany
BECKERS J.M., Ing., Université de Liège, Liège, Belgium
BODE L., Dr., James Cook University of North Queensland, Townsville, Australia
BOEHLICH M.J., Mr., Universität Hamburg, Hamburg, Germany
BOUDRA D.B., Prof. Dr., Rosenstiel School of Marine and Atmospheric Science, Miami, USA
CHARTIER M., Dr., Institut de Protection et de Sûreté Nucléaire, CEA, Fontenay-aux-roses, France
CHASSIGNET E., Ing., Rosenstiel School of Marine and Atmospheric Science, Miami, USA
CLEMENT F., Mr., Université de Liège, Liège, Belgium
COMELIAU B., Ing., Université de Liège, Liège, Belgium
DAHL F.E., Ing., Det norske Veritas, Høvik, Norway
DAVIES A.M., Dr., Institute of Oceanographic Sciences, Birkenhead, UK
DELECLUSE P., Dr., Museum d'Histoire Naturelle, Paris, France
DEJAK C., Prof. Dr., Enternazionale Energie Alternative, ENEA, Roma, Italy
DELEERSNIJDER E., Ing., Université de Liège, Liège, Belgium
DEMUTH Cl., Dr., Management Unit of the Mathematical Models (MUMM), Liège, Belgium
DESAUBIES Y., Prof. Dr., IFREMER, Brest, France
DINGMAN J.S., Mr., The Ohio State University, Columbus, USA
DIPPNER J.W., Dr., Universität Hamburg, Hamburg, Germany
DISTECHE A., Prof. Dr., Université de Liège, Liège, Belgium
DJENIDI S., Ing., Université de Liège, Liège, Belgium
DONNARS Ph., Ing., Laboratoire National d'Hydraulique, Chatou, France
DYKE P.P.G., Prof. Dr., Plymouth Polytechnic, Plymouth, UK
EIFLER W., Dr., Commission of the European Communities, Ispra, Italy
ELLIOTT A.J., Dr., University College of North Wales, Menai Bridge, UK
EVERBECQ E., Ing., Université de Liège, Liège, Belgium
FANDRY C., Dr., CSIRO, Hobart, Australia
FENG S., Prof. Dr., Shandong College of Oceanology, Shandong, The People's Republic of China
FLEBUS C., Mr., Université de Liège, Liège, Belgium
FRANKIGNOUL Cl., Prof. Dr., Université Pierre et Marie Curie, Paris, France
GOFFART A., Miss, Université de Liège, Liège, Belgium
GOFFART P., Ing., Université de Liège, Liège, Belgium
GOPALAKRISHNAN T.C., Dr., Kuwait Institute for Scientific Research, Safat, Kuwait
GUNST D.R.R., Ing., Ministerie van Openbare Werken, Oostende, Belgium
HAINBUCHER D., Mrs., Universität Hamburg, Hamburg, Germany
HALMES F., Mr., Université de Liège, Liège, Belgium
HAPPEL J.J., Ing., Université de Liège, Liège, Belgium
HECQ J.H., Dr., Université de Liège, Liège, Belgium
HIRES R.I., Dr., Stevens Institute of Technology, Hoboken, USA
HOLLAND W.R., Dr., National Center for Atmospheric Research, Boulder, USA
HUA B.L., Dr., IFREMER, Brest, France
HUIZINGA P., Ing., NRIO-CSIR, Stellenbosch, South Africa

JAMART B.M., Dr., Management Unit of the Mathematical Models (MUMM), Liège, Belgium
JAMES I.D., Dr., Institute of Oceanographic Sciences, Birkenhead, UK
KARAFISTAN-DENIS A., Dr., Université de Liège, Liège, Belgium
KROHN J., Dr., GKSS Research Centre, Geesthacht, Germany
LAIME A.F., Ing., Université de Liège, Liège, Belgium
LEBON G., Prof. Dr., Université de Liège, Liège, Belgium
LEFAIVRE D., Dr., Centre Champlain des Sciences de la Mer, Québec, Canada
LEGUTKE S., Mrs., Universität Hamburg, Hamburg, Germany
LI L., Dr., State Oceanic Administration, Dalian City, The People's Republic of China
LOHRMANN A., Ing., Det norske Veritas, Høvik, Norway
LYNN N.M., Mr., Royal Naval College, London, UK
MARTEL F., Miss, Université Pierre et Marie Curie, Paris, France
MILLET B., Mr., ORSTOM, Montpellier, France
MONREAL A., Dr., CONACYT, Mexico, Mexico
MOUCHET A., Miss, Université de Liège, Liège, Belgium
NEVES R., Dr., CTAMFUTL, Lisboa, Portugal
NIHOUL J.C.J., Prof. Dr., Université de Liège, Liège, Belgium
OZER J., Ing., Management Unit of the Mathematical Models (MUMM), Liège, Belgium
PECENIK G., Dr., MONTEDIPE SpA, Venezia, Italy
PECHON Ph., Ing., Laboratoire National d'Hydraulique, Chatou, France
PEDERSEN G.K., Mr., University of Oslo, Oslo, Norway
PICHOT G., Dr., Unité de Gestion Modèle Mathématique Mer du Nord et Estuaire de l'Escaut, Bruxelles, Belgium
POHLMANN Th., Mr., Universität Hamburg, Hamburg, Germany
PONTRELLI G., Dr., IRAM-CNR, Bari, Italy
POSTMA L., Mr., Delft Hydraulics Laboratory, Delft, The Netherlands
PROVOST Ch., Dr., Université Pierre et Marie Curie, Paris, France
RANDLES J., Mr., Commission of the European Communities, Ispra, Italy
ROBERT J.L., Prof. Dr., Université Laval, Québec, Canada
ROBINSON A., Prof. Dr., Harvard University, Cambridge, USA
ROCKLIFF N.J., Dr., Plymouth Polytechnic, Plymouth, UK
RODRIGUEZ I., Ing., Direction Générale des Côtes, Madrid, Spain
ROED L.P., Dr., Det norske Veritas, Høvik, Norway
ROISIN M., Mr., Université de Liège, Liège, Belgium
RONDAY F.C., Dr., Université de Liège, Liège, Belgium
RYGG O.B., Mr., University of Oslo, Oslo, Norway
SALAS DE LEON D., Dr., CONACYT, Mexico, Mexico
SALOMON J.Cl., Dr., IFREMER, Brest, France
SALUSTI S.E., Prof. Dr., Università La Sapienza, Roma, Italy
SHENG Y., Dr., Aeronautical Research Associates of Princeton, Princeton, USA
SMETS E., Ing., Waterbouwkundig Laboratorium, Borgerhout, Belgium
SMITZ J., Ing., Université de Liège, Liège, Belgium
SNYKERS Ph., Ing., Université de Liège, Liège, Belgium
SOULAIMANI A., Mr., Université de Technologie, Compiègne, France
SPAULDING M.L., Prof. Dr., Applied Science Associates Inc., Narragansett, USA
SPITZ Y., Miss, Management Unit of the Mathematical Models (MUMM), Liège, Belgium
STANLEY P., Mr., Marine Science Laboratories, Menai Bridge, UK
STEEDMAN R.K., Dr., Steedman Limited, Subiaco, Western Australia
TEE K.T., Mr., Bedford Institute of Oceanography, Dartmouth, Canada
TREGLOS Y., Mr., UNESCO, Paris, France
USSEGLIO-POLATERA J.M., SOGREAH, Grenoble, France
VALCKE A., Ing., Université de Liège, Liège, Belgium
WILLIAMS J., Mr., Branch Office of Naval Research, London, UK
WOLF J., Dr., Institute of Oceanographic Sciences, Birkenhead, UK

PERSPECTIVE IN THREE-DIMENSIONAL MODELLING OF THE MARINE SYSTEM

Jacques C.J. NIHOUL and S. DJENIDI*
School of GeoHydrodynamics and Environmental Research (GHER), University of Liège, Belgium

*Also "Unité de Modélisation de l'Environnement Marin, (MODEM), Association Université de Corse - Université de Liège, Calvi, Corse

INTRODUCTION

There is a general consensus, at least in developed countries and international institutions, that our marine environment has been dramatically deteriorating in the last decades.

The migration of industries to the coasts, the development of new harbors, the growth of vast urban centers, the use of insecticides and fongicides have led to an important pollution of coastal areas, destroying marine life and creating severe health problems for the human population. Offshore, the dumping of industrial wastes has dangerously increased since the last war. The development of maritime commercial exchanges, the exploitation of ores and hydrocarbons and other uses of the sea floor, such as dredging, continuously increase the pollution of the sea and more particularly the coastal zones.

The problems of fisheries are closely related. Not only because pollution destroys marine life or contaminates marine products or because overfishing is a form of pollution but mainly because the same physical mechanisms which govern the fate of pollutants are often responsible for creating the conditions of marine fertility.

In the same time, the cost of raw materials and energy has increased enormously, calling for a more extensive exploitation of the sea and limiting the economical resources which can be devoted to pollution control. Simultaneously, the problem of supplying the increasing world population with food has led to a more systematic harvesting of marine products.

In the next decade, the problems will become more acute and will call for a more thorough understanding and a more rational and strict control of the marine environment.

The marine system, however, is extremely complex and it is a overwhelming task to predict the intricated environmental effects of man's activities and, a fortiori, to set limits to such activities - by *international conventions and treaties,* for instance - which are not permissive or unduly severe.

Sofar, the simple collection of data and their descriptive ordering have appeared such formidable tasks that one has often ignored the need for doing more than this. One realizes that what is necessary now is the management of the marine system, the search for the necessary compromises between the requirements of increasing industrialization and affluent society and the necessity to preserve the valuable natural resources.

This is an optimal control problem which can only be approached through mathematical modelling. Mathematical models are the only alternative to zero discharge, if ecologically acceptable solutions to environmental problems are to be provided.

AN INTERTWINING OF MATHEMATICAL MODELS

The first step in modelling a marine system is the *demarcation* of the system. This includes the definition of its *support,* i.e. its extension in physical ($\underset{\sim}{x}$-t) space, and of its *scope,* i.e. its deployment in state space.

The demarcation defines the boundaries of the system and the state variables of the model and thus determines the nature, place and time of the boundary and initial conditions which will be required.

The support and the scope may differ appreciably from one study to another, depending on its particular objectives, and this generates a whole hierarchy of different mathematical models, accorded to their particular designs. These models may all be regarded as "*sub-sets*" of some - still tentative - universal model the construction of which must be pursued to keep track of all the aspects which have been sacrificed to urge conclusive, though partial, results.

The first characteristic of a model is its *object,* i.e. the geographical area, the dates and the specific events or processes one wishes to investigate. One understands easily the difference between modelling the Bering Sea or the Mediterranean, investigating ice formation, tidal fronts in mid-latitudes or trapped Kelvin waves at the equator.

The second characteristic of a model is its *span*, i.e. its dimension in physical space and in state space.

Ideally, a marine model would be a time dependent 3D-model (four-dimensional support). A reduction of the dimensions of the support can be achieved, however, by restricting attention to time and space averages, considering for instance (quasi) steady state models of low frequency residuals, depth-averaged models of shallow continental seas, cross-section averaged models of estuaries, time- and depth dependent models of surface and bottom boundary layers or primary production and finally box-models for completely space averaged ecological variables.

Ideally, also, a completely realistic marine model would have an infinite number of state variables. Computing facilities, of course, impose limitations on the number of state variables but, independently of such restrictions, there are reliability and clarity constraints on the number of state variables. A model with many state variables incorporates as many different processes and interactions and involves a correspondingly large number of parameters and boundary conditions which cannot be evaluated from existing data bases without an inevitable margin of error. The results of such a model, on the other hand, - because of its increased sophistication and fallibility - can become impossible to interpret in terms of scientific diagnosis or management recommendations.

The essence of modelling is the selection of a limited number of representative state variables. There must be sufficiently few of them for their evolution equations to be amenable to analysis but enough of them to describe adequately the system's behaviour.

The state space can be divided in several sectors corresponding to hydrodynamical, chemical, biological processes ... and one can conceive separate hydrodynamical, chemical and biological models with the necessary input-output links between them. Of these, the hydrodynamic models are by far the most advanced. In a sense, this is rather fortunate because the understanding of hydrodynamic processes is prerequisite to any form of chemical or biological modelling and, indeed, constitutes, in the present state of development of marine models, the most reliable contribution to the explanation and anticipation of ecological processes.

While *chemical and biological models* are still frequently limited to *interaction box-models* describing, by means of *differential equations*, concentrations and biomasses in a hypothetic homogeneous environment or averaged over large regions of space, *hydrodynamic models* have evolved to *transport-dispersion field-models* describing, by means of *partial differential equations* the spatial distribution and time evolution of field variables determined at all grid points.

The development of hydrodynamic models has been considerably stimulated by their direct application to coastal and off-shore engineering. It has also been made possible by the absence of any significant feed-back from chemical and biological processes on hydrodynamic phenomena : transport and dispersion in the sea are determinant factors in marine chemistry and biology but chemical and biological interactions have no appreciable effect on advection and mixing in the sea.

The state variables of hydrodynamic models are the thermo-mechanical variables, velocity, pressure, buoyancy, temperature, salinity, turbulent kinetic energy, ... depending on their degree of sophistication. They can easily be extended to include the concentrations of passive and semi-passive constituents, i.e. constituents which are simply carried along by the sea (passive) or which, while carried along, can be produced or destroyed by local reactions depending on the constituent's concentration only such as bacterial or radioactive decay (semi-passive).

The combination of hydrodynamic models and chemical-ecological interaction models leads to *active transport-dispersion models*. These however are still in an early stage of development partly because of their complexity (many coupled partial differential equations) and partly because of the lack of suitable chemical and biological data for their calibration in three-dimensions and for the determination of appropriate boundary conditions.

Limiting the scope of the model to a particular sector of the state space reduces its dimensions. This can be achieved also by restricting attention to *aggregate averages* ; considering, for instance, zooplankton biomass (with no distinction between herbivores, carnivores and omnivores and, a fortiori, between species), total organic matter (lumping together dissolved and particulate organic matter), mercury concentration in fish (with no specification of its distribution) etc... .

The third characteristic of a mathematical model is its *purview*, i.e. its *spread* in physical space, (its *"duration"* and its *"reach"*) and its *aperture* in state space (its *"frequency gamut"* and its *"wave-number arena"*).

In physical space, one thus distinguishes between *local*, *regional* and *global* models.

Although the terms are suggestive enough - a model of the near-field dispersion of a coastal discharge can be described as local, a model of the Adriatic Sea as regional, a general circulation model of the Atlantic as global - this distinction is not without some ambiguity (A model of the Mediterranean is global compared to one of the Adriatic and regional compared to a general ocean circulation model).

One can however remove the ambiguity by defining (i) *local* models as models of "small" size regions where the flow field and other hydrodynamic characteristics are given, (ii) *regional* models as models of "medium" size regions where the flow field must be determined taking into account boundary conditions imposed by larger scale circulations and (iii) *global* models as models of "large" size regions where the flow is mainly driven by internal forcing and little by open-sea boundary inflows.

Obviously, reach and duration are related and local models are naturally interested in short-term modifications of the near field while global models are more concerned with the long-term evolution of the whole system.

The marine system is characterized by fairly well-defined *"spectral windows"* i.e. domains of length-scale (inversely, wave numbers) and time scales (inversely, frequencies) associated with identified phenomena. These windows may correspond to eigenmodes of the system (internal waves, inertial oscillations, Rossby waves, El Niño ...) or external forcing (annual or daily variations of insolation, tides, storm, atmosphere climate changes ...).

The basic hydrodynamic equations contain three characteristic frequencies

(i) the Brunt-Väisälä frequency n is a measure of the stratification (n^2 is defined as the vertical gradient of buoyancy ; the maximum value of the Brunt-Väisälä frequency in the sea is of the order of $10^{-2}\ s^{-1}$) ;

(ii) the Coriolis frequency f is a measure of the Earth's rotation (f is defined as twice the vertical component of the Earth's rotation vector ; in mid-latitudes, $f \sim 10^{-4}\ s^{-1}$) ;

(iii) the Kibel frequency j is a measure of the Earth's curvature (if β denotes the gradient of f, j can be defined as $j \sim \beta r$ where

$r \sim \frac{nH}{f}$ is the Rossby radius of deformation and H a typical depth ; the maximum value of the Kibel frequency in the sea is of the order of 10^{-6} s^{-1}).

Diurnal and seasonal variations of thermal exchanges can be characterized by typical frequencies of the order of 10^{-4} s^{-1} and 10^{-7} s^{-1} respectively. A frequency of the order of 10^{-5} s^{-1} can be associated with variations in the wind field.

Finally, a frequency of the order of 10^{-8} s^{-1} may be introduced in connection with the year-to-year variations of the state of large areas of the ocean and the entire atmosphere as, for example, the self-oscillation of the Northern branch of the Gulf Stream and the El Niño Southern oscillation.

Marine processes can thus be classified according to their time scales as shown schematically in Table I.

In general, time scales and length scales are related and it is customary to associate high frequencies and high wave numbers, small frequencies and small wave numbers although the association may be different for eigenmodes and forced oscillations.

The transfer of energy between windows is effected by non-linear interactions.

Chemical and ecological interaction processes can also be characterized by specific time scales and the comparison between these time scales and those of hydrodynamic phenomena indicates which processes are actually in competition in the sea.
Obviously, at hydrodynamic scales much smaller than interaction scales, very little interaction takes place over time of significant hydrodynamic changes and basically the constituents are transported and dispersed passively by the sea. On the other hand, hydrodynamic processes with time scales much larger than interaction scales scarcely affect the dynamics of interactions over any time of interest.
The range of (time and length) scales the model can reproduce defines its *aperture*.

Time scale	Frequency (s^{-1})	Spectral windows (highlighted processes)	Smaller scale fluctuations (filtered out processes)
1 s	1	Microscale processes 3 D "eddy" turbulence (+ surface waves)	Molecular diffusion
1 m			
	10^{-2}	Mesialscale processes Internal waves Vertical microstructure Inhibited "bliny"* turbulence	Eddy turbulence
1 h			
	10^{-4}	Mesoscale processes Inertial oscillations	"Bliny turbulence"
1 d	10^{-5}	Tides, storm surges Diurnal variations	
1 w			
	10^{-6}	Synopticscale processes Frontal currents Meanders, "rossby"* turbulence	Mesoscale variability
1 month			
	10^{-7}	Seasonalscale processes	"Rossby turbulence"
1 year			
	10^{-8}	Globalscale processes Climatic processes	Seasonal variability
		(Paleo)climaticscale processes	

Table I . Schematic representation of marine variability

*A "bliny" (from the Russian "blini") is a pancake-shaped eddy contributing to an energy cascade to smaller scales via epidermic instabilities and internal waves. A "rossby" (from the scientist Rossby) is a pseudo-two-dimensional eddy column of scale of the order of the Rossby radius of deformation.

The fourth characteristic of a model is its *resolution*.
The resolution in physical space is determined by the mesh-size of the numerical grid and the time step of integration, the resolution in state space by the margins of error allowed on the state variables.

The reduction of support and scope by averaging and aggregation is, to some extent, related to the resolution of the model. A depth-integrated model for instance is equivalent to a 3D-model with a very coarse (one grid point) vertical resolution.

There is an obvious connection between the spread of a model, its aperture and its resolution. Given the complexity of the model, the limitations in computing facilities or budgets generally impose a limit on the number of grid points and time steps and thus, for a chosen reach and duration, a maximum resolution. The maximum resolution determines the largest frequencies and wave-numbers that can be resolved.
On the other hand, phenomena the length scales and time scales of which exceed the reach and duration of the support are not truly "resolved by the model" as the solution is largely determined by initial or boundary conditions (for instance, if the duration of the simulation is much smaller than the characteristic time of the process, results, at any simulation time, are completely set by their initial values).

The fifth characteristic of a model is its *accuracy*, i.e. its ability to reproduce the reality. Obviously, the accuracy of the model is not simply a question of resolution and precision of the calculations. It depends for instance to a large extent on its degree of sophistication and on the reliability of the data used for the determination of parameters and boundary conditions. A simplistic model can produce very precise results entirely disconnected from reality, an intricated model may contain too many assumptions and uncertain figures to provide a satisfactory representation of it. As pointed out before, for each problem, a compromise is inevitable.

A final distinction between models can be made here between *"process-models"* which emphasize accuracy in state space and *"engineering models"* which emphasize accuracy in physical space.

A process model is generally devised to investigate, in details, particular mechanisms, scrutinize the behaviour of specific state variables and elucidate fundamental questions. Very refined in its representation of, sometimes, rather subtle processes, it may be content with very crude approximations of the physical world (constant depths, rectilinear coasts, infinite ocean, steady two-dimensional fronts, rigid sea surface ...).

An engineering model, on the contrary, is in general called upon to tackle a practical situation and may not ignore the real field conditions (depths, coastlines, actual atmospheric forcing ...). Its aims however, are to assess the consequences of particular events and to provide the marine forecasts which will assist planning and management. The model must be sound, expeditious and efficient but is not required to provide detailed information on the delicate machinery subtending its parameterization schemes.

It is easy to imagine the excesses to which both types of models may lead if carried to extremes (ivory-tower intellectual game, on one side, foreman's rule of thumb on the other). Although purely process or engineering models may be useful for preliminary investigations, their vocation is to enlarge and progressively acquire the virtues of the other.

The final goal is a diagnostic-prognostic model providing an accurate description of all aspects of the real world. Such a model is often referred to as a *simulation model*.

The basic equations of all hydrodynamic and active dispersion models are cast in the same mould and may be regarded as different breeds of the same fundamental equations of Geophysical Fluid Dynamics and the same diffusion equations.

The differences between the models are essentially the results of their distinct aims, spans, purviews and resolutions. This diversity finds expression in the choice of state variables and related acting phenomena, the parameterization of interactions, boundary conditions and sub-grid scale processes and, to some extent, the numerical schemes.

It is easy to imagine how many different models can be conceived by considering different objectives, different places or dates, different time scales and length-scales ... even if some combinations, as pointed-out before, must be excluded.

The problems of the management of coastal waters and continental seas, coastal and off-shore engineering, pollution, eutrophication, primary production, food chain dynamics and fishing yield ... can be associated with the so-called "weather" of the sea, i.e., mesoscale and synoptic scale processes in the range of frequencies 10^{-4} - 10^{-7} s^{-1}.

The marine weather, although displaying the same features as the atmospheric weather, is characterized by time scales and length scales which are often one order of magnitude different. The most intense phenomena in continental seas are frequently found in the mesoscale range, inertial oscillations, tides, storm surges ... and the importance of synoptic features such as fronts and rossbies and macroscale flow fields which dominate atmospheric weather patterns has only been recognized, in the sea, recently, with the development of remote sensing and large scale sea-truth experiments providing unprecedented synoptic views of the ocean surface properties.

MARINE WEATHER EQUATIONS

The equations describing the weather of the sea can be obtained from the general Stratified Fluid Dynamics equations by averaging over a time of a few hours (say 10^4s). The average eliminates mesialscale and microscale processes from all the linear terms and only the effects, in the mean, of their non-linear interactions remain in the equations, in the form of turbulent or "pseudo-turbulent" diffusion terms which can be parameterized with the help of appropriate eddy diffusivities.

These diffusivities are related to global characteristics of the bliny-eddy turbulent field, viz

(i) the mean turbulent kinetic energy

$$e = \langle \frac{1}{2} \underset{\sim}{w} \cdot \underset{\sim}{w} \rangle \tag{1}$$

(ii) the turbulent kinetic energy dissipation rate

$$\varepsilon = \langle \nu \underset{\sim}{\nabla} \underset{\sim}{w} : \underset{\sim}{\nabla} \underset{\sim}{w} \rangle \tag{2}$$

where $\underset{\sim}{w}$ represents the fluctuating velocity, ν the kinematic viscosity, $:$ a double scalar product, $\underset{\sim}{\nabla}$ the vector operator

$$\underset{\sim}{\nabla} = \underset{\sim}{e}_1 \frac{\partial}{\partial x_1} + \underset{\sim}{e}_2 \frac{\partial}{\partial x_2} + \underset{\sim}{e}_3 \frac{\partial}{\partial x_3} \quad ,$$

and where angular brackets $< >$ denote an average.

Mesoscale and synoptic marine processes satisfy
(i) the Boussinesq approximation (according to which the density of sea water may be assumed constant except in the gravity term where density deviations are multiplied by the acceleration of gravity, several orders of magnitude larger than typical flow accelerations) ;
(ii) the quasi-hydrostatic approximation (according to which the vertical momentum equation reduces to a balance between vertical pressure gradient and gravity).

If ρ_0 is the constant reference ("Boussinesq") density and if one defines "buoyancy" b and "reduced pressure" q by

$$b = - g \frac{\rho - \rho_0}{\rho_0} \tag{3}$$

$$q = \frac{p}{\rho_0} + g\, x_3 + \xi \tag{4}$$

where g is the acceleration of gravity, p the pressure, x_3 the vertical coordinate (the vertical axis pointing upwards) and ξ the tidal potential, the equation of continuity and the vertical component of the equation of momentum reduce to

$$\underset{\sim}{\nabla} . \underset{\sim}{v} = \underset{\sim}{\nabla} . \underset{\sim}{u} + \frac{\partial v_3}{\partial x_3} = 0 \tag{5}$$

$$\frac{\partial q}{\partial x_3} = b \tag{6}$$

where

$$\underset{\sim}{v} = \underset{\sim}{u} + v_3 \underset{\sim}{e}_3 \tag{7}$$

is the velocity vector ($\underset{\sim}{u}$ is the horizontal velocity vector).

Eqs (5) and (6) may be regarded as defining equations for v_3 and q. The "marine weather" state variables are then
(i) the two components of the horizontal velocity vector $\underset{\sim}{u}$,
(ii) the buoyancy b,

(iii) the turbulent kinetic energy e,
(iv) the turbulent dissipation rate ε .

If y stands for any of the state variables u_1, u_2, b, e, ε, the general evolution equation can be written

$$\frac{\partial y}{\partial t} + \underset{\sim}{\nabla}\cdot(\underset{\sim}{v}y) = Q^y + \underset{\sim}{\nabla}\cdot(\tilde{\lambda}^y \underset{\sim}{\nabla} y) \tag{8}$$

where Q^y is the rate of production (destruction if negative) of y and $\tilde{\lambda}^y$ the bliny-eddy diffusivity.

The production of buoyancy is essentially due to radiation and it is generally possible to assume that radiation, absorbed in the upper few meters of the sea, can be represented by a surface source to be taken into account in the boundary conditions at the air-sea interface. (In expressing these boundary conditions for b, one must determine the buoyancy flux in terms of the fluxes of sensible and latent heat, evaporation and precipitation, taking into account the flux discontinuity due to absorbed or emitted radiation, e.g. Nihoul, 1984).

With this approximation, one can write

$$Q^b = 0 \tag{9}$$

and (e.g. Nihoul 1984, Rodi 1985)
for $y = u_j$

$$Q^j = (-\underset{\sim}{\nabla} q - f\, \underset{\sim}{e}_3 \wedge \underset{\sim}{u})_j \tag{10}$$

for $y = e$

$$Q^e = \tilde{\nu} \underset{\sim}{\nabla} \underset{\sim}{v} : \underset{\sim}{\nabla} \underset{\sim}{v} - \tilde{\lambda}^b \frac{\partial b}{\partial x_3} - \varepsilon \tag{11}$$

for $y = \varepsilon$

$$Q^\varepsilon = \gamma_1 \frac{\varepsilon}{e} [\, \tilde{\nu} \underset{\sim}{\nabla} \underset{\sim}{v} : \underset{\sim}{\nabla} \underset{\sim}{v} - \gamma_2\, \tilde{\lambda}^b \frac{\partial b}{\partial x_3} \,] - \gamma_3 \frac{\varepsilon^2}{e} \tag{12}$$

where f is the Coriolis frequency (twice the vertical component of the earth's rotation vector), $\tilde{\nu} = \tilde{\lambda}^1 = \tilde{\lambda}^2$ is the turbulent diffusivity of momentum or "turbulent viscosity", γ_1, γ_2 and γ_3 are empirical constants.

Parameterization of bliny-eddy turbulent diffusion

It can be shown (e.g. Nihoul, 1980) that bliny and eddy turbulence contributes to a pseudo Kolmogorov cascade transferring energy from the mean flow to high wave numbers (small scales) where viscous energy dissipation takes place. The viscous sink is characterized by

(i) the length scale

$$l_v \sim \varepsilon^{-1/4} \nu^{3/4} \tag{13}$$

(ii) the time scale

$$t_v \sim \nu^{1/2} \varepsilon^{-1/2} \tag{14}$$

(iii) the velocity scale

$$u_v \sim l_v t_v^{-1} \sim \varepsilon^{1/4} \nu^{1/4} \tag{15}$$

and

(iv) the Reynolds number

$$R_v = \frac{u_v l_v}{\nu} \sim 1 \tag{16}$$

Averaging and introducing an eddy viscosity to account for the effect, in the mean, of mesialscale and microscale fluctuations, amounts to replacing the energy transfer through the cascade and its ultimate dissipation at high wave numbers by a single sink at the scale of the energy containing eddies. If u_m and l_m are characteristic velocity and length scales of these eddies and $l_m u_m^{-1}$ the associated time scale, one may argue that, for the concept of eddy viscosity to be consistent, one must require

$$R_m = \frac{u_m l_m}{\tilde{\nu}} \sim 1 \tag{17}$$

$$l_m u_m^{-1} \sim \tilde{\nu}^{1/2} \varepsilon^{-1/2} \tag{18}$$

Because the turbulent energy spectrum falls off very rapidly from its peak value at scale l_m, the kinetic energy of the eddies at scale l_m, $\frac{1}{2} u_m^2$ is a very substantial fraction of the turbulent kinetic energy e and one may assume

$$u_m \sim \alpha\, e^{1/2} \tag{19}$$

Combining eqs (17), (18) and (19), one gets

$$\tilde{\nu} \sim \alpha^4 \frac{e^2}{\varepsilon} \tag{20}$$

The other turbulent diffusivities are generally expressed in the form

$$\tilde{\lambda}^s = \beta^s \tilde{\nu} \tag{21}$$

where the β^s's are new empirical functions or constants.

One generally considers that β^e and β^ε may be taken as constants of order 1 but that β^b is a function of the stratification measured by the Richardson number

$$Ri = \frac{n^2}{m^2} \tag{22}$$

or the Flux Richardson number

$$R_f = \frac{\tilde{\lambda}^b n^2}{\tilde{\nu}\; m^2} \tag{23}$$

where n is the Brunt-Väisälä frequency as before $(n^2 = |\frac{\partial b}{\partial x_3}|)$ and m is the "Prandtl frequency" given by

$$m^2 = \nabla \mathbf{v} : \nabla \mathbf{v} \tag{24}$$

This is easily understood.

In a stratified fluid, work has to be done to raise an isolated blob of fluid above its equilibrium level. In zero shear (i.e. $Ri = \infty$), the blob of fluid will fall back to its equilibrium level at a rate determined by the Brunt-Väisälä frequency n. As the shear increases (i.e. as Ri decreases), the tendency for a displaced parcel of fluid to return to its equilibrium level will decrease, but there will still be a buoyancy force acting on it to make it return. As the blob of fluid is temporarily displaced from its equilibrium position it will exchange its properties with the surrounding fluid at the new level. In the case of temperature, salinity, buoyancy and other scalar properties of the fluid, complete exchange can only be effected by small scale turbulent mixing and ultimately by molecular action. This takes a considerable time and usually the parcel of fluid will be dragged back to its equilibrium level before it can exchange more than a tiny fraction of its heat, salt, buoyancy with its new and dissimilar surroundings during its temporary residence there.

For momentum, however, the situation is different. The blob of displaced fluid has a different horizontal velocity that its new surroundings (i.e. there is a shear), and there is a drag on it. This is a bulk force which requires no molecular mixing-in : the momentum is transferred immediately by pressure.

Thus momentum exchange is likely to retain its efficiency at high Richardson number, even though the buoyancy transfer is reduced as the stratification increases. One should thus expect

$$\beta^b = f(Ri \text{ or } R_f) < 1 \tag{25}$$

<u>Parameterization of sub-grid scale diffusion</u>

The second term in the right-hand side of eq. (8) represents the mean effects of non-linear interactions of fluctuations characterized by time scales smaller than the period of averaging. Although these fluctuations are affected by the stratification, they may still be regarded as sufficiently diversified and randomly distributed to create a form of three-dimensional turbulence with rather similar efficiency in vertical and horizontal diffusion (Nihoul, 1980). In other words, if the turbulent diffusivities associated with mesialscale - microscale fluctuations are not the same as postulated in eq. (8), they may be assumed of comparable orders of magnitude.

In that case, the characteristic length scales of horizontal variations being considerably larger than the vertical length scales, one may neglect the horizontal diffusion as compared to the vertical diffusion. This does not imply that there is no horizontal diffusion in Nature. It simply means that, at this stage, the main part is still concealed in the advection term which contains irregular and variable horizontal currents responsible for a form of horizontal "pseudo turbulence" (e.g. Nihoul 1975, Monin and Ozmidov 1985).

The discrepancy between horizontal and vertical length scales however imposes, in most cases, numerical grids with much larger horizontal meshes (typically one order of magnitude larger than the lengths scale which one would associate with the time average's cut-off by similarity estimates). The discretization of the equations is then equivalent to performing a second (horizontal space) average and non-linear interactions of sub-grid scale fluctuations are responsible for an additional horizontal diffusion which it is convenient to introduce explicitly in the mathematical evolution equations, anticipating the subsequent discretization. The second term of the right-hand side of eq. (8) is then written

$$\underset{\sim}{\nabla}_h \cdot (\tilde{\kappa}^y \underset{\sim}{\nabla}_h y) + \frac{\partial}{\partial x_3}(\tilde{\lambda}^y \frac{\partial y}{\partial x_3}) \tag{26}$$

where

$$\underset{\sim}{\nabla}_h = \underset{\sim}{e}_1 \frac{\partial}{\partial x_1} + \underset{\sim}{e}_2 \frac{\partial}{\partial x_2} \tag{27}$$

The horizontal diffusivities $\tilde{\kappa}^y$ can be related to the mesh size and to the turbulent energy dissipation rate in the associated range of scales using an extension of Kolmogorov's theory developed by Ozmidov (e.g. Nihoul 1975, Monin and Ozmidov 1985). In many cases, they can be taken as constants. With eqs. (5) and (6), the evolution equations for $\underset{\sim}{u}$, b, e and ε obtained from eq. (8) (where the last term is replaced by 26 and the production rates are given by 10, 11 and 12) and eq. (20) relating $\tilde{\nu}$, e and ε, the system of marine weather equations is closed except for empirical coefficients or functions α, β, γ ... to be determined by calibration of the model.

Additional parameters (drag coefficients, albedo, ...) appear in the expression of the boundary conditions, especially at the air-sea interface, and their valuation is also part of the calibration exercices (e.g. Nihoul,1984).

THE MIXING LENGTH APPROXIMATION

The equation for the turbulent dissipation rate ε is, by common consent, the weak point of three-dimensional modelling. The fact that, through the production rate Q^ε, it introduces many empirical parameters is a demonstration of its largely empirical character and, in the same time, an indication of the amount of parameterizing which has been subjacent to the setting up of this equation.

Several authors have tried to replace the equation for ε by similar equations for different combinations of ε, e and $\tilde{\nu}$, without succeeding in decreasing the volume of parameterization and empiricism (e.g. Blumberg and Mellor, 1985).

Faced with this difficulty, one naturally tries to simplify the model and the concept of "mixing length" extended from the early work of Prandtl seems, in this respect, rather promising.

Combining eqs (18), (19) and (20), one obtains

$$\tilde{\nu} \sim \alpha\, e^{1/2}\, l_m \tag{28}$$

Predicting l_m is thus equivalent to predicting ε. Elaborating from Prandtl's early theories of turbulence, several authors have come to the conclusion that l_m, the modern version of Prandtl's "mixing length", could, in many cases, be determined - as a function of space and stratification - by simple inspection, thus sparing the analyst the solution of an additional ε (or equivalent) equation.

It is generally assumed that the mixing length l_m can be written in the form

$$l_m = l_0\, \psi \tag{29}$$

where l_0 is its value in neutral conditions ($n = 0$) and ψ is a function of the stratification. l_0 is an algebraic function of x_3 which must be such that it respects the classical logarithmic singularities in the bottom boundary layer and adjust to wind-mixed layer conditions near the surface.

The function ψ has been mostly expressed in terms of the Richardson number Ri. One of the originality of the GHER-model developed at the **G**eo**H**ydrodynamics and Environment **R**esearch Laboratory of the University of Liège is the determination of parametric relationships of ψ - as well as β^b - in terms of the flux Richardson number R_f instead of Ri (Nihoul and Djenidi, 1986).

The results of the Medalpex Experiment in the Mediterranean (Djenidi et al., 1987) suggest rather simple formulas for ψ and β^b of the type

$$\psi \sim 1 - R_f \tag{30}$$

$$\beta^b \sim (1 - R_f)^{1/2} \tag{31}$$

These expressions - however simple they appear - are not truly surprising. For instance, in the case of a stably stratified environment, eddies at scale l_m have only to transfer, to the viscous sink, via the energy cascade, an energy $\varepsilon \sim \varepsilon_0(1 - R_f)$ per unit time where ε_0 denotes the energy extracted per unit time from the mean flow. In the absence of stratification, ε_0 would be passed on to the cascade by eddies at scale l_0 .

The relation

$$l_m \sim l_0\ (1 - R_f) \tag{32}$$

amounts to requiring that, in any case, for a given energy level the characteristic time of dissipation computed with the turbulent viscosity be the characteristic time of evolution of the big eddies, i.e.

$$\varepsilon_0^{-1/2}\, l_0^{1/2} \sim \varepsilon^{-1/2}\, l_m^{1/2} \quad \rightarrow \quad l_m \sim l_0 \frac{\varepsilon}{\varepsilon_0} \sim l_0\ (1 - R_f) \tag{33}$$

THE SHALLOW WELL-MIXED SEA APPROXIMATION

If the sea is sufficiently shallow and well-mixed (for instance by intense tidal currents) as the North Sea, several simplifying hypotheses can be made, viz

(i) negligible buoyancy effects, i.e.

$$b \sim 0 \qquad Ri \sim R_f \sim 0 \qquad (34)\ (35)$$

(ii) local balance of turbulent production and destruction rates, i.e.

$$Q^e \sim \tilde{\nu} \left\| \frac{\partial \underset{\sim}{u}}{\partial x_3} \right\|^2 - \varepsilon \sim 0 \tag{36}$$

i.e. using eq. (20) and (36),

$$u_*^4 = \tilde{\nu}^2 \left\| \frac{\partial \underset{\sim}{u}}{\partial x_3} \right\|^2 \sim \alpha^4 e^2 \tag{37}$$

where u_* denotes the so-called friction velocity. Eq. (28) gives then

$$\tilde{\nu} \sim u_* \, l_m \tag{38}$$

The friction velocity u_* and the mixing length l_m can be scaled by, respectively, the square root of the bottom stress $\underset{\sim}{\tau}_b$ (per unit mass of sea water) and the total depth.
Thus

$$\tilde{\nu} \sim u_* \, l_m \sim \| \underset{\sim}{\tau}_b \|^{1/2} \, H \, \sigma(\frac{z}{H}) \tag{39}$$

where

$$H = h + \zeta \tag{40}$$

is the total depth, h is the depth, ζ the surface elevation, $z = x_3 + h$ is the altitude above the bottom (x_3=0 corresponds to the undisturbed free surface).

The fonction σ is determined empirically from experimental data. Observations and models have shown that the most important requirement on σ was its ability to take into account the logarithmic singularities in the bottom boundary layer ; the exact shape of the profile of σ being much less cogent for subsequent calculations (e.g. Nihoul 1977, Roisin 1977, Nihoul et al. 1979).

Neglecting buoyancy and using eq. (39), one can solve the continuity and momentum equations for the velocity field without determining the additional variables e and ε.

A further simplification is obtained by integrating these equations over depth and solving for the depth-mean velocity $\bar{\underset{\sim}{u}}$.

The vertically integrated equations read

$$\frac{\partial H}{\partial t} + \underset{\sim}{\nabla} . (H \bar{\underset{\sim}{u}}) = 0 \tag{41}$$

$$\frac{\partial}{\partial t} (H \bar{\underset{\sim}{u}}) + \underset{\sim}{\nabla} . (H \bar{\underset{\sim}{u}} \bar{\underset{\sim}{u}}) + f H \underset{\sim}{e}_3 \wedge \bar{\underset{\sim}{u}} = - H \underset{\sim}{\nabla} \left(\frac{p_a}{\rho_0} + g\zeta + \xi \right) + \underset{\sim}{\tau}_s - \underset{\sim}{\tau}_b + \underset{\sim}{D} \tag{42}$$

where p_a is the atmospheric pressure, $\underset{\sim}{\tau}_s$ the wind stress (per unit mass of sea water) and $\underset{\sim}{D}$ a diffusion term resulting from non-linear interactions of sub-grid scale processes and velocity fluctuations around its vertical average ("shear effect"; e.g. Nihoul, 1975).

In most cases, $\underset{\sim}{D}$ can be expressed in simple Laplacian diffusion form introducing a new (constant) viscosity coefficient. The bottom stress $\underset{\sim}{\tau}_b$ is related to the mean velocity $\bar{\underset{\sim}{u}}$ and to the wind stress. The most commonly used formula is the "quadratic friction law"

$$\underset{\sim}{\tau}_b = D \, \| \bar{\underset{\sim}{u}} \| \, \bar{\underset{\sim}{u}} - \delta \underset{\sim}{\tau}_s \tag{43}$$

where D is the "drag coefficient" ($D \sim 2\ 10^{-3}$) and δ an empirical constant ($\delta \sim 10^{-1}$).

Although eq. (43) has been very successful in marine forecasting, there are indications that it is not valid in periods of weak mean currents (at tide reversal, for instance). In such periods, in fact, the mean velocity $\bar{\underset{\sim}{u}}$ is a poor indication of the flow field : there may be a substantial veering of the velocity vector along the vertical, with quite different bottom and surface currents, and this may be important in some applications or at specific locations.

The 2D model must then be complemented by the equation for the velocity deviation $\hat{\underset{\sim}{u}} = \underset{\sim}{u} - \bar{\underset{\sim}{u}}$. The latter is easily obtained by substracting the equation for the mean (42) from the original equation for $\underset{\sim}{u}$ and reads (Nihoul et al. 1979)

$$\frac{\partial \hat{\underset{\sim}{u}}}{\partial t} + f \underset{\sim}{e}_3 \wedge \hat{\underset{\sim}{u}} + \underset{\sim}{nl} = \frac{\partial}{\partial x_3} \left(\tilde{\nu} \frac{\partial \hat{\underset{\sim}{u}}}{\partial x_3} \right) - \frac{\underset{\sim}{\tau}_s - \underset{\sim}{\tau}_b}{H} \tag{44}$$

where $\underset{\sim}{nl}$ stands in brief for all the contributions of the non-linear terms (The detailed expression is given in Nihoul et al., 1979).

Eq. (44) must be solved subject to the following boundary conditions

$$\tilde{\nu} \frac{\partial \hat{\underset{\sim}{u}}}{\partial x_3} = \underset{\sim}{\tau}_s \qquad \text{at the surface} \tag{45}$$

$$\hat{\underset{\sim}{u}} = - \bar{\underset{\sim}{u}} \quad (\underset{\sim}{u} = 0) \qquad \text{at the bottom} \tag{46}$$

In addition, one must have

$$\tilde{\nu} \frac{\partial \hat{\underset{\sim}{u}}}{\partial x_3} = \underset{\sim}{\tau}_b \qquad \text{at the bottom} \tag{47}$$

Hence $\underset{\sim}{\tau}_b$ is now determined by the model as a function of $\bar{\underset{\sim}{u}}$, $\underset{\sim}{\tau}_s$ ... .

The non-linear terms $\underset{\sim}{nl}$ are important when the current is strong but negligible when it is weak i.e., precisely, when the vertical structure of the current field may be important and when eq. (43) for $\underset{\sim}{\tau}_b$ is questionable.

Most of the information required is thus contained in the linear form of eq. (44) which one can solve, even analytically, in parallel with the 2D model at all points of interest.

Using an analytical solution based on series expansion in eigenfunctions of the vertical turbulent diffusion operator, Nihoul (1977) and Roisin (1977) have shown for instance that the bottom stress could be written, with a good approximation in the form

$$\underset{\sim}{\tau}_b = D \,\| \bar{\underset{\sim}{u}} \|\, \bar{\underset{\sim}{u}} - \delta \underset{\sim}{\tau}_s + \eta H \left[\frac{\partial \bar{\underset{\sim}{u}}}{\partial t} + f \underset{\sim}{e}_3 \wedge \bar{\underset{\sim}{u}} \right] \tag{48}$$

where η is a numerical factor.

The last term in the right hand side of eq. (48) turns out to be negligible as compared with the first one as long as the mean velocity $\bar{\underset{\sim}{u}}$ does not approach zero. It becomes important when $\bar{\underset{\sim}{u}}$ is sufficiently small and one can see that its effect, associated with the flow's inertia, may be regarded as a "memory" effect in the determination of $\underset{\sim}{\tau}_b$.

Starting from the 2D depth-averaged model and the locally 1D linear model, one can, by successive iterations, include the non-linear terms and construct a fully 3D = 2D + 1D model as shown on the sketch-plan of figure 1.

One of the advantages of this model is that the 2D submodel can be operated solely whenever one is satisfied with depth-averaged information and that the whole machinery needs only be run when and (or) where the detail of the vertical structure is required.

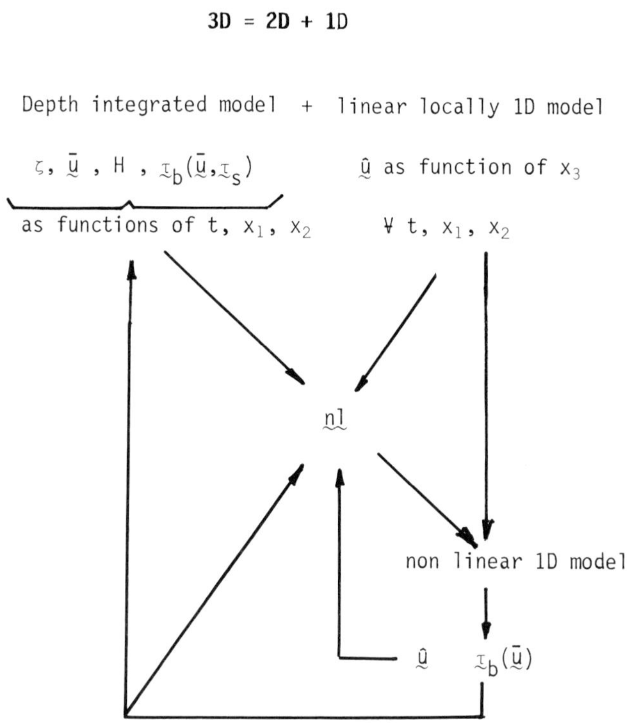

Fig. 1. Sketch of the 3D = 2D + 1D model for shallow well-mixed continental seas.

EXAMPLES OF APPLICATIONS

In the last years, the **G**eo**H**ydrodynamics and **E**nvironment **R**esearch Laboratory (GHER) of the University of Liège has developed a 3D = 2D + 1D model and a fully 3D (turbulent energy, mixing length-closure) baroclinic model. The first one was calibrated for the North-West European Continental Shelf with emphasis on the North Sea, the second one for the Mediterranean with emphasis, in a first set of simulations, on the Adriatic Sea. Some exemplary results are shown on the following figures, in illustration.

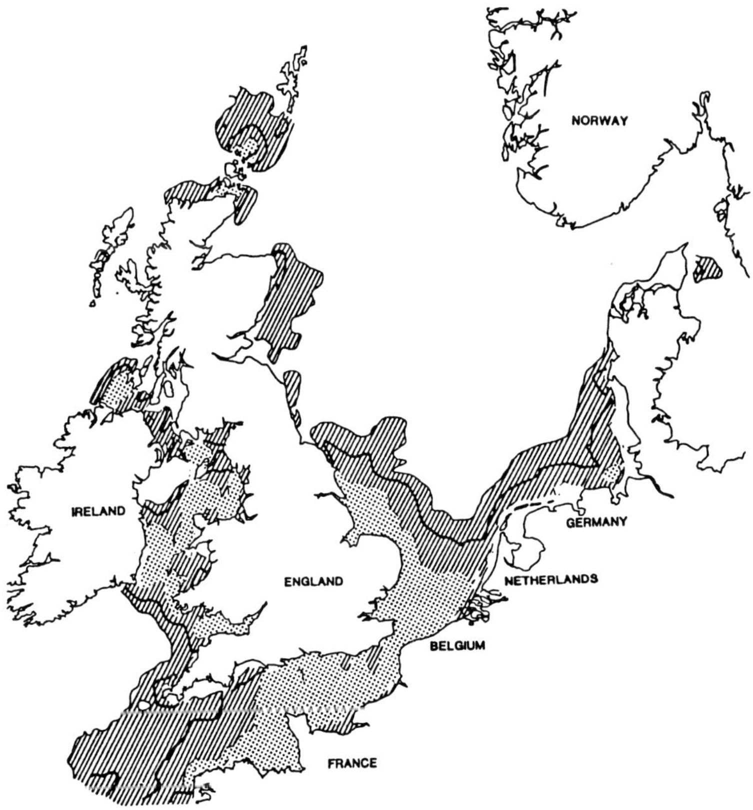

Fig. 2. Tidal fronts calculated by the 3D = 2D + 1D GHER model on the North-West European Shelf

well-mixed water

zones of transition

stratified water

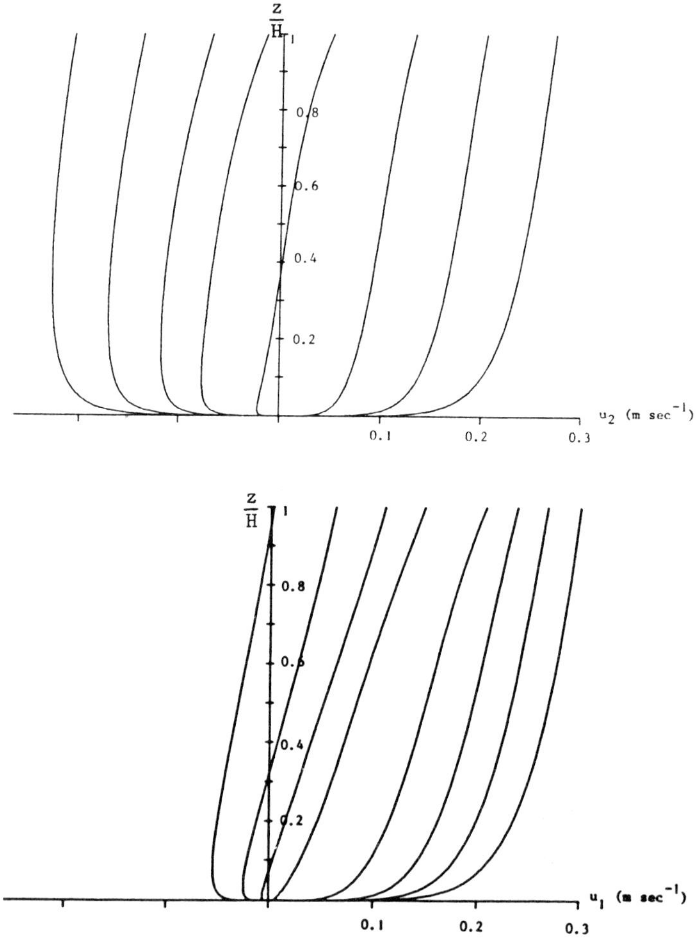

Fig. 3. Evolution with time, at tide reversal, of the two components of the horizontal velocity vector at the point 52°30'N 3°50'E in the Southern Bight of the North Sea. (Depth 22m, wind blowing to the North-East, maximum wind stress of 2 $10^{-4}m^2\ s^{-2}$). The curves from right to left are vertical profiles computed at 18' intervals. The upper curve represents the northern component, the lower curve, the eastern component (GHER 3D = 2D + 1D Model).

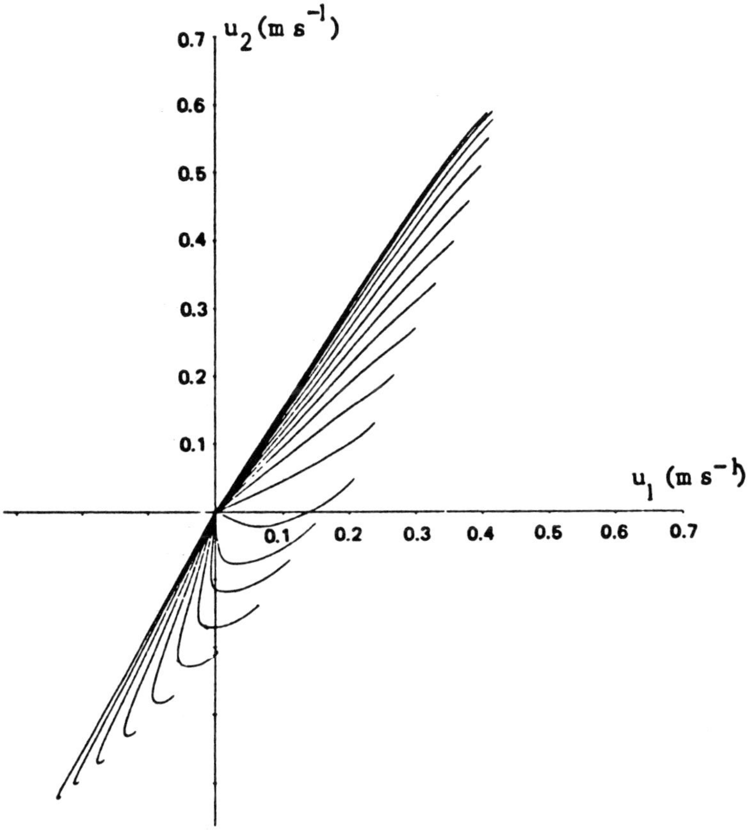

Fig. 4. Evolution with time over the first half tidal period of the Ekman diagram showing the vertical veering of the horizontal velocity vector at the point 52°30'N 3°50'E in the Southern Bight of the North Sea. The separation between two successive curves is 18'. (GHER 3D = 2D + 1D Model).

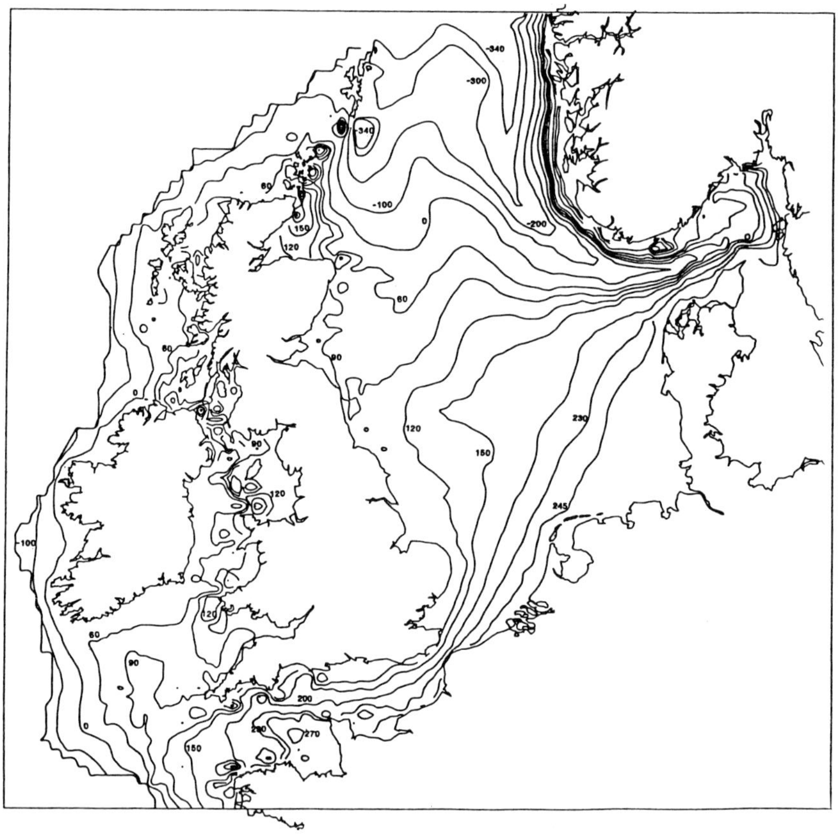

Fig. 5. Residual summer transport on the North-West European Shelf (GHER 3D = 2D + 1D Model, real winds, stream functions in $10^3 m^3 s^{-1}$).

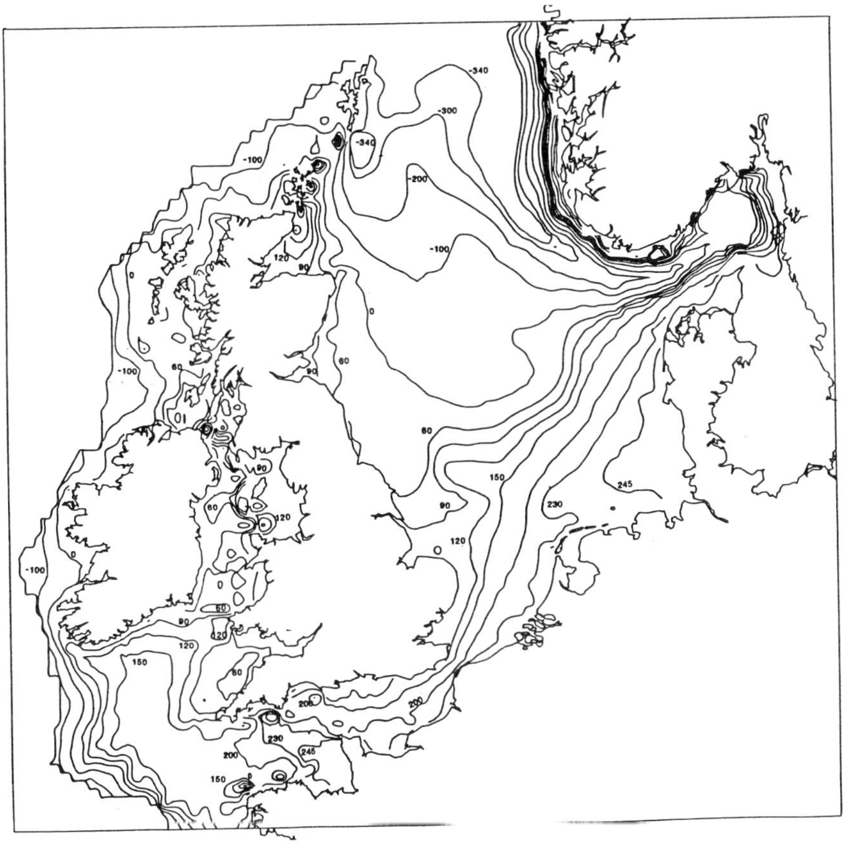

Fig. 6. Residual winter transport on the North-West European Shelf (GHER 3D = 2D + 1D Model, real winds, stream functions in $10^3 m^3 s^{-1}$).

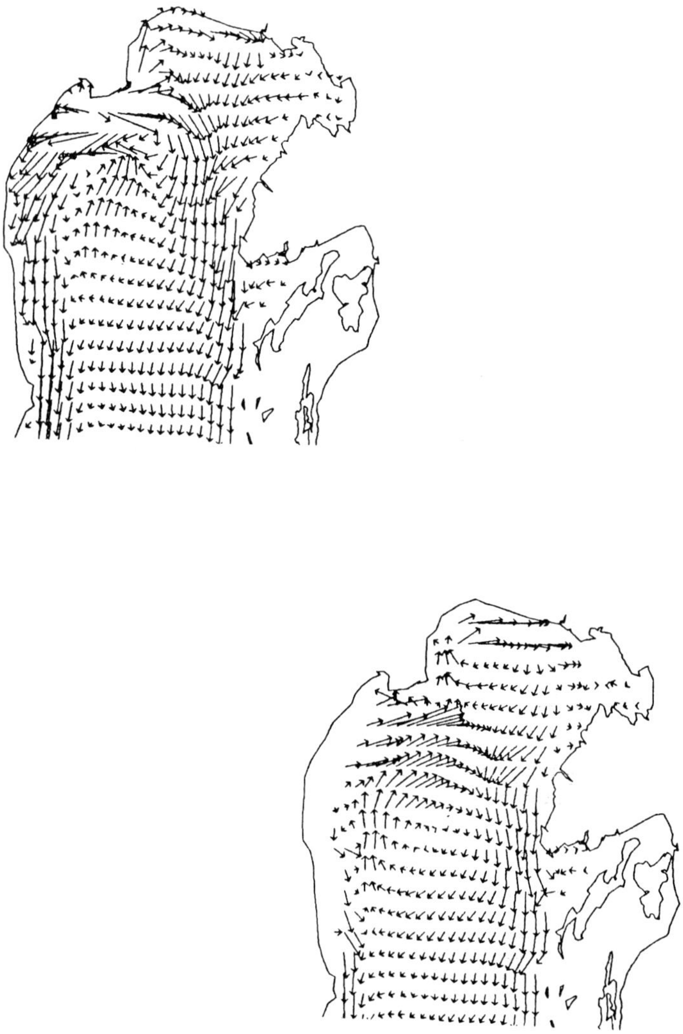

Fig. 7. Flow pattern in the Northern Adriatic Sea at 3 m (above) and 18 m(below) computed with the 3D GHER model for July 28, 1979, 4 H 12 m in negligible wind conditions. Fig. 8 and Fig. 9 which follow show that the model is able to reproduce the observed variability with the tendency to form gyres in the region of the Pô. (Djenidi et al., 1987).

a)

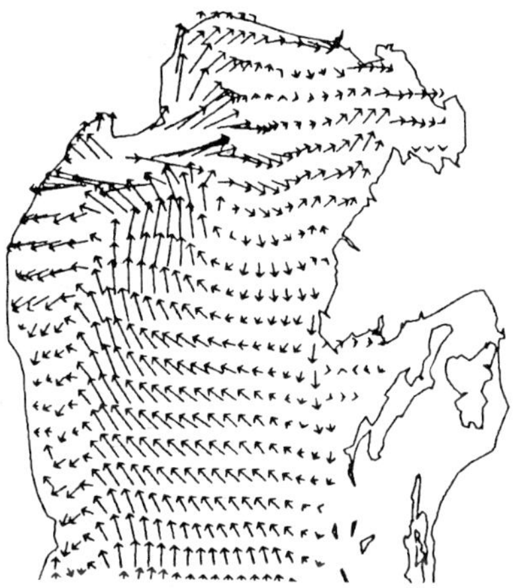

b)

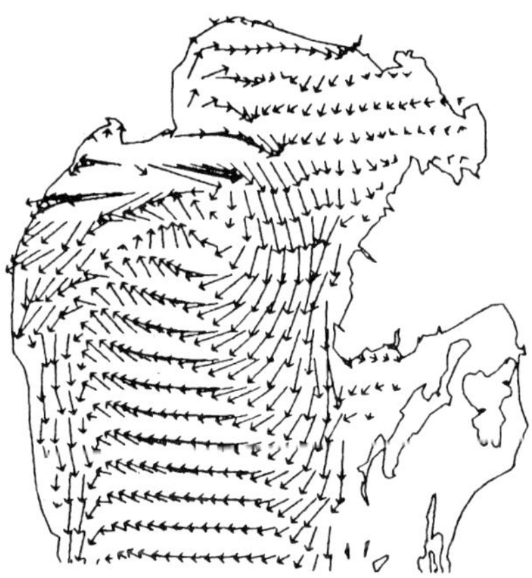

Fig. 8. Evolution of the flow pattern at 3 m, in the Northern Adriatic Sea (GHER, 3D Model, Djenidi et al., 1987).

c)

d)

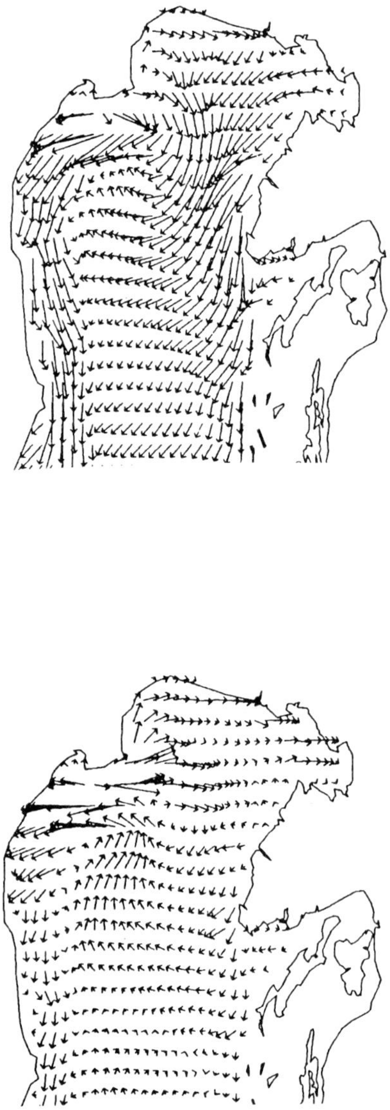

Fig. 8'. Evolution of the flow pattern at 3 m, in the Northern Adriatic Sea (GHER, 3D Model, Djenidi et al., 1987).

e)

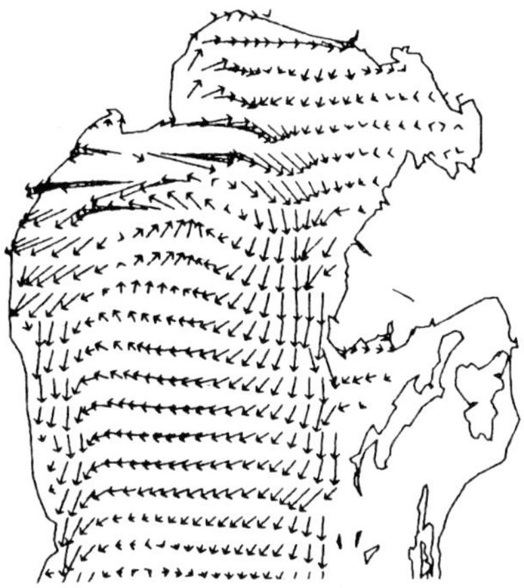

f)

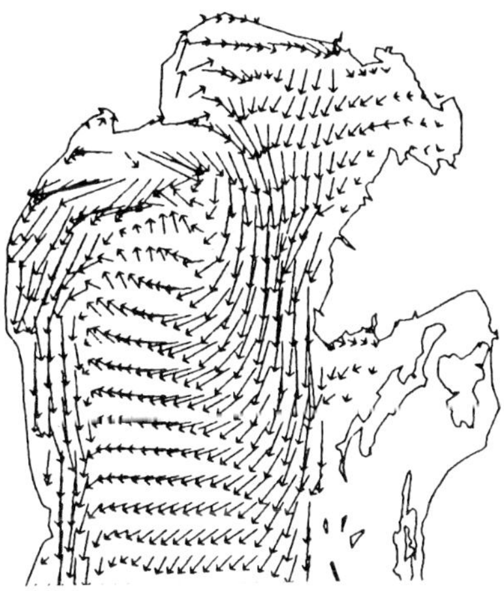

Fig. 8". Evolution of the flow pattern at 3 m, in the Northern Adriatic Sea (GHER, 3D Model, Djenidi et al., 1987).

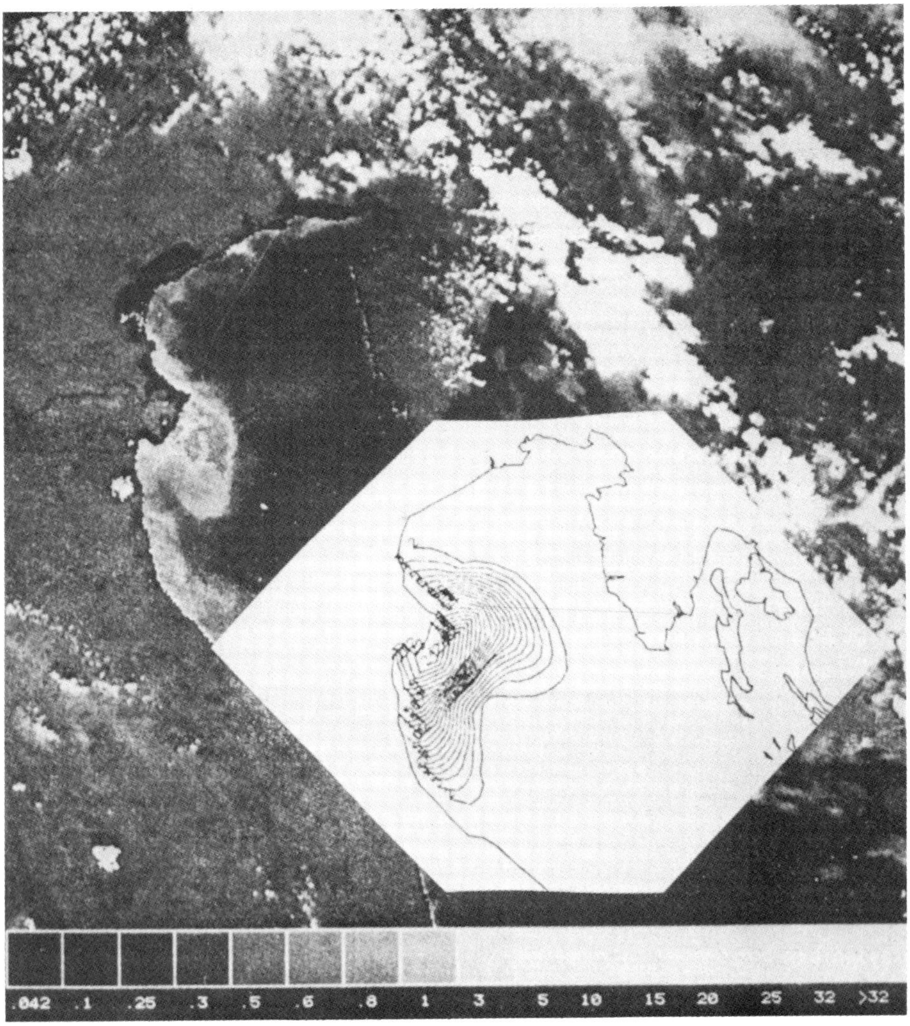

Fig. 9. Comparison between the predicted seston distribution for July 28, 79, 4 H 12 m and remote sensing observations (CZCS data, graduated in mg chlorophyll m^{-3}) (GHER 3D Model, Djenidi et al., 1987).

ACKNOWLEDGMENTS

The authors are indebted to the European Atomic Energy Community (Euratom) for partial support of this research via contracts SC-012B/BIAF/423(SD) and 2831-85/PC ISPB. They wish to express their gratitude to their colleagues of the Commissariat à l'Energie Atomique, Paris and the Joint Research Center Ispra for many fruitful discussions and authorization to reproduce some of the figures.

REFERENCES

Blumberg, A.F. and Mellor, G.L., 1985. A description of three-dimensional coastal ocean circulation model. In : N. Heaps (Editor), Three-Dimensional Shelf Models, Coastal and Estuarine Dynamics, 5, American Geophysical Union Publ..

Djenidi, S., Nihoul J.C.J., Clément, F. and Salas de Leon, D., 1987. The MODEM contribution to Medalpex. Annales Geophysicae, 5: 1-19.

Monin, A.S. and Ozmidov, R.V., 1985. Turbulence in the ocean. D. Reidel Publ. Co., Dordrecht.

Nihoul, J.C.J., 1975. Modelling of Marine Systems. Elsevier Publ. Co., Amsterdam.

Nihoul, J.C.J., 1977. Three-dimensional model of tides and storm surges in a shallow well-mixed continental seas. Dyn. Atmos. Ocean., 2: 29-47.

Nihoul, J.C.J., Runfola, Y., and Roisin, B., 1979. Non-linear three-dimensional modelling of mesoscale circulation in seas and lakes. In : J.C.J. Nihoul (Editor), Marine Forecasting, Elsevier Publ. Co., Amsterdam, chapter 15, 235-259.

Nihoul, J.C.J., 1980. The turbulent ocean. In : J.C.J. Nihoul (Editor), Marine Turbulence, Elsevier Publ. Co., Amsterdam, chapter 1, 1-19.

Nihoul, J.C.J., 1984. A three-dimensional marine circulation model in a remote sensing perspective. Annales Geophysicae, 2: 433-442.

Nihoul, J.C.J. and Djenidi, S., 1986. Three-dimensional mathematical models for "marine weather" prediction. Invited paper, Envirosoft Conference, Los Angeles, USA, November 19-21, 1986.

Rodi, W., 1985. Survey of calculation methods for flow and mixing in stratified fluids. In : Proceedgins IUTAM Symposium on Mixing in Stratified Fluids, Margaret River Western Australia, August 85, 1-51.

Roisin, B., 1977. Modèles tri-dimensionnels des courants marins. Ministry for Science Policy Brussels, Rep. ACN3, 124 pp.

ON MODELING THREE-DIMENSIONAL ESTUARINE AND MARINE HYDRODYNAMICS

Y. PETER SHENG
University of Florida, 336 Weil Hall, Gainesville, FL 32611 (U.S.A.)

ABSTRACT

Recent advances of a three-dimensional numerical model of estuarine and marine hydrodynamics are described in this paper. In particular, the parameterization of the vertical turbulent transport based on a Reynolds stress model and the adaptation of a generalized curvilinear (or "boundary-fitted") grid to the finite-difference model are highlighted. These two aspects of the three-dimensional numerical model, along with other features, allow accurate simulation of turbulent flows in estuarine and marine waters where complex geometry and bathymetry are generally present.

1. INTRODUCTION

Estuarine and marine hydrodynamic processes (e.g., tidal currents, front dynamics, and sediment dispersion) often involve three-dimensional turbulent flow in the presence of complex geometry and bathymetry. For example, tidal circulation and salinity transport in such estuaries as Suisun Bay, California (Sheng, et al., 1985) in Figure 1, and Mississippi Sound, Mississippi (Sheng and Butler, 1982) are strongly affected by complex geometry and bathymetry. Local bathymetry and geometry significantly affect flow and sediment transport around various estuarine and marine structures (e.g., bottom pipelines, break-waters, and navigation channels), and dredged material disposal mounds. Hence, to accurately simulate marine and estuarine currents due to tides, winds and density gradients, numerical models must be able to accurately and efficiently resolve (1) the turbulent transport processes, particularly the dynamics of various vertical boundary layers shown in Figure 2, and (2) the complex geometry and bathymetry (Sheng, 1986a).

Turbulent transport in the various vertical layers as shown in Figure 2 strongly affect many estuarine and marine processes. For example, oil spill trajectory is affected by the dynamics of the laminar sublayer and constant flux layer underneath the air-sea interface, while the deposition and erosion of sediments are governed by the laminar sublayer and constant flux layer near the bottom. Seeking to remove the empiricism contained in simple eddy-viscosity models, turbulence models such as the Reynolds stress model (e.g., Sheng, 1982 and 1984) and the two-equation (or k-ε) model (e.g., Rodi, 1980) have been applied to estuarine marine, and riverine environments. Two simplified versions of a Reynolds stress model (Sheng, 1984) have been incorporated into a three-dimensional model of estuarine and marine hydrodynamics by the present author. A brief description of the Reynolds stress model, and its simplified versions, along with model applications, will be presented in this paper.

Traditional multi-dimensional hydrodynamic models of marine and estuarine currents use the finite difference technique and a uniform Cartesian grid

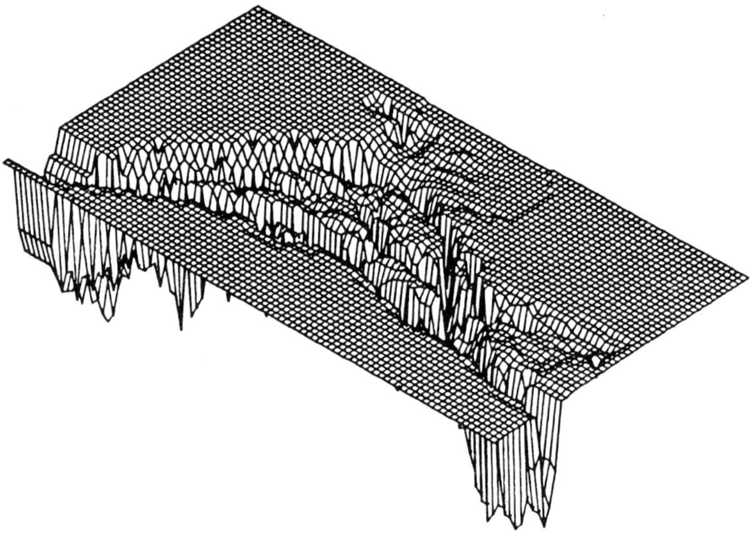

Figure 1 3-D view of Suisun Bay, California.

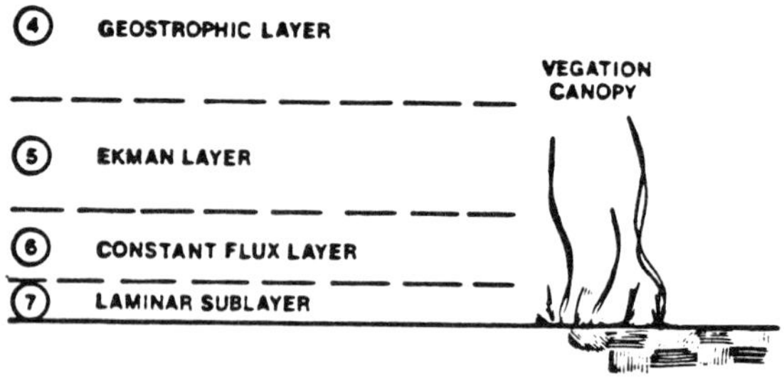

Figure 2 Vertical layers within relatively deep coastal/estuarine waters.

(e.g., Leendertse, 1967) with the shoreline and bathymetry represented by numerous stair-steps. This not only severely limits the model accuracy but also frequently dictates a prohibitively large number of grid points to resolve a complex environment. To better resolve the lateral geometry, more refined lateral grids have been used with various finite-difference models. For example, Sheng (1976) used nested and dynamically coupled Cartesian grids to study the 3-D nearshore circulation, Butler (1978) utilized stretched Cartesian grid to study coastal waves, Warnstrath (1977) employed conformal grid to study storm surge, and Waldrop and Tatom (1976) used an orthogonal curvilinear grid to study thermal plume in river. The 3-D coastal hydrodynamic model of Sheng and Butler (1982) and Sheng (1983) used a laterally exponentially-stretched Cartesian grid and a vertically σ-stretched grid, which is a special form of the "boundary-fitted grid".

Recently, Johnson (1982) employed the boundary-fitted grid technique to study the 2-D vertically-averaged riverine circulation. Non-orthogonal boundary-fitted curvilinear grids for the physical domain were first generated. Equations of motion in the Cartesian coordinates were then transformed into those in the curvilinear coordinates by performing chain-rule transformation. Although the independent variables (the coordinates) were transformed, the dependent variables (the velocity components) remained unchanged as in the Cartesian coordinates. The present model, however, solves the transformed equations of motion in terms of the "contravariant" components of velocity vectors. This technique allows for simpler equations and boundary conditions and better representation of complex geometry with relatively few number of grid points, and thus significantly improves the capability of finite difference models. Highlights and some applications of the curvilinear-grid hydrodynamic model will be presented.

2. A THREE-DIMENSIONAL CARTESIAN-GRID MODEL

2.1 Mean Equations

The basic equations describing the mean motion of coastal and estuarine waters are consisted of the continuity equation, the equations of motion, the heat equation, the salinity equation and the equation of state. Assuming (1) the hydrostatic approximation, (2) the Boussinesq approximation, and (3) the eddy-viscosity concept, the various equations can be written with respect to a right-handed Cartesian coordinate system as:

$$\nabla \cdot \bar{u} = 0 \tag{1}$$

$$\frac{\partial \overline{\underset{\sim}{u}}}{\partial t} + \nabla \cdot (\overline{\underset{\sim}{uu}}) = -f\hat{z} \times \overline{\underset{\sim}{u}} + \frac{\partial}{\partial z}\left(A_v \frac{\partial \underset{\sim}{u}}{\partial z}\right) + \underset{\sim}{\nabla} \cdot (A_H \underset{\sim}{\nabla} \overline{\underset{\sim}{u}}) \tag{2}$$

$$- \frac{1}{\rho_o} \underset{\sim}{\nabla} P_a - g \underset{\sim}{\nabla} \zeta - \frac{g}{\rho_o} \int_{\zeta}^{z} \underset{\sim}{\nabla} \rho dz$$

$$\frac{\partial T}{\partial t} + \nabla \cdot (\bar{u} T) = \frac{\partial}{\partial z} (K_v \frac{\partial T}{\partial z}) + \underset{\sim}{\nabla} \cdot (K_H \underset{\sim}{\nabla} T) \tag{3}$$

$$\frac{\partial S}{\partial t} + \nabla \cdot (\bar{u} S) = \frac{\partial}{\partial z} (D_v \frac{\partial S}{\partial z}) + \underset{\sim}{\nabla} \cdot (D_H \underset{\sim}{\nabla} S) \tag{4}$$

$$\rho = \rho\ (T,\ S) \tag{5}$$

where $\bar{u}$ represents the 3-D velocity vector (u,v,w) in the (x,y,z) directions, $\underset{\sim}{\bar{u}}$ represents the horizontal velocity vector (u,v) in the (x,y) directions, $\hat{z}$ is the unit vertical vector, t is time, f is Coriolis parameter, $\underset{\sim}{\nabla}$ represents the horizontal gradient, P_a is atmospheric pressure, g is gravitational acceleration, ζ is surface elevation, ρ is density, T is temperature, S is salinity, (A_V, K_V, D_V) represent the vertical turbulent eddy coefficients, and (A_H, K_H, D_H) represent the lateral turbulent eddy coefficients. Notice that the dynamic boundary condition at the free surface has already been incorporated into the above equations.

2.2 Boundary Conditions

Various layers shown in Figure 2 exist in the water column of relatively deep estuarine and marine waters. In relatively shallow waters, the effect of friction may be so important that the Ekman layers merge. Dynamics of the relatively thin sublayer (~1 mm) and the constant flux layer (~1 m) can affect the transport of such materials as heat, sediment, oxygen, nutrient and oil slick, which are often introduced into the water body from the surface or bottom boundaries.

The constant flux layer above the bottom is quite similar to that above the free surface. Applying the law of wall at the bottom allows one to relate the bottom stress with the velocity at some distance above the bottom as:

$$\underset{\sim}{\tau}_b = \rho_w\ c_{dw} |\underset{\sim}{\bar{u}}_+|\ \underset{\sim}{\bar{u}}_+ \tag{6}$$

where $\underset{\sim}{\tau}_b$ represents the bottom stress vector in (x,y) directions, ρ_w is water density, $\underset{\sim}{\bar{u}}_+$ is the horizontal velocity vector in (x,y) directions at some distance z_+ above the bottom, and C_{dw} is the drag coefficient determined from:

$$c_{dw} = \kappa^2 [\ln(z_+/z_o) + \phi_s(z_+/L)]^{-2} \tag{7}$$

where κ is the von-Karman constant, z_o is the physical roughness height, L is the Monin-Obhukov similarity length, and ϕ_s is a stability function. L approaches ∞ while ϕ_s approaches 0 for neutrally-stratified flows.

Equations (6) and (7) can be applied to the marine boundary layer above the air-sea interface to compute the surface wind stress from the wind velocity at some distance (z_+) above the air-sea interface (with roughness z_o), when ρ_a and c_{da} (subscript a stands for air) are used instead of ρ_w and c_{dw}. Additional constant flux layer expressions similar to (6) and (7) relate the heat flux and mass flux at bottom or free surface to local mean temperature gradient and mean concentration gradient. If one is concerned with the dispersion of non-neutrally buoyant particles such as sediments, planktons, and larvae, the boundary conditions are more complicated. For instance, erosion and deposition of sediments must be included in the bottom boundary conditions.

2.3 Turbulence Parameterization

In the present 3-D model, the vertical turbulent eddy coefficients (A_v, K_v, D_v) are determined from simplified versions of a Reynolds stress model and require no ad-hoc parameter tuning with data. The Reynolds stress model will be discussed in the next section. Since the lateral turbulent diffusion is generally much less important than the lateral advection and vertical turbulent diffusion in shallow seas, the lateral turbulent eddy coefficients are often taken to be constants to parameterize the sub-grid scale turbulence associated with the large lateral eddies.

2.4 σ-Stretching: A Boundary-Fitted Grid

Sheng et al. (1978) detailed the vertically-stretched (or "σ-stretched") version of the above equations invoking the small amplitude approximation, i.e., $\zeta \ll h$ (depth). Later, Sheng and Butler (1982) and Sheng (1983) detailed the σ-stretched equations without the small amplitude approximation. Both models treated separately the external (barotropic) modes, which are governed by the vertically integrated equations, and the internal (baroclinic) modes, which are governed by the 3-D equations, to allow for efficient numerical simulation. The recent model also uses a semi-implicit scheme to speed up the external mode simulation. As shown in Figure 3, the σ-stretched grid is a special form of "boundary-fitted" grid, and hence allows accurate resolution of complex geometry and bathymetry. The σ-stretched 3-D model has been applied to numerous water bodies, e.g., Lake Erie (Sheng et al., 1978) and Mississippi Sound (Sheng and Butler, 1982). Recent applications to two estuaries are briefly described in the following.

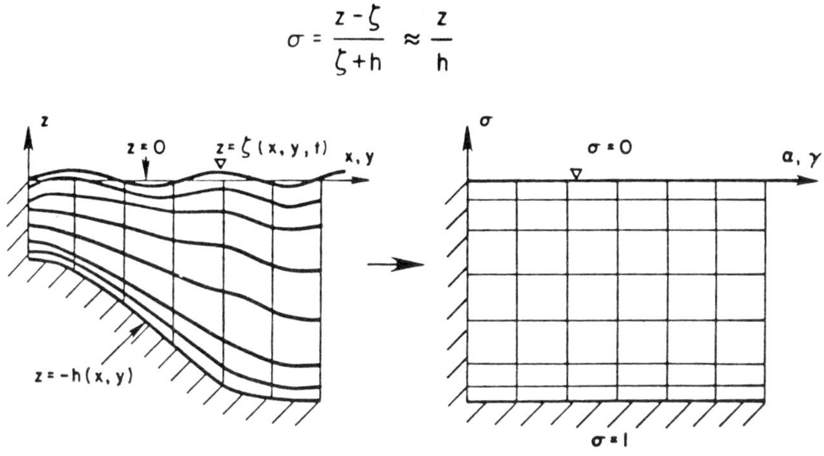

Figure 3 A vertically σ-stretched system.

2.5 Suisun Bay, California

Suisun Bay, California is part of the San Francisco Bay system. Results of tidal harmonic analysis (Cheng and Gartner, 1984) indicate M_2 and K_1 as the major constituents. Strong bi-directional tendency was observed at most tidal data stations and the basin geometry was found to significantly affect the principal tidal current direction. The tidal current speed can vary up to a factor of two between spring and neap tides. Salinity within the Suisun Bay typically varies from 10-15 ppt at the western end to 0-10 ppt at the eastern end, but exhibit significant temporal variations.

Before studying episodic events, model simulations were performed to examine some of the general features associated with the 3-D flow field within Suisun Bay (Sheng et al., 1985). A synthetic tide consisting of M_2 and K_1 constituents was used to drive the western boundary (with amplitudes of 55 cm and 31 cm for M_2 and K_1) and the eastern end (with amplitudes of 43 cm and 25 cm for M_2 and K_1), with a 90 degree phase shift. The salinity was assumed to be 18-20 ppt along the western end and 12-14 ppt along the eastern end. Boundary condition for salinity depends on the flow direction along the open boundary. During outflow, boundary salinity value is computed by considering advection from interior only. During inflow, exterior data are applied. Using an A_H of 100 m^2/sec and a bottom roughness (z_o) of 0.4 cm, a five day numerical simulation was performed. The near-surface and near-bottom currents at the end of five days are shown in Figure 4. More detailed comparison with data indicated reasonable agreement with general circulation features including stronger ebb currents, bi-directionality in tidal currents, and flow reversal at some stations. More realistic simulations are being planned.

2.6 Charlotte Harbor, Florida

The Charlotte Harbor estuarine system receives discharges from drainage of 16 percent of Florida through the Peace, Myakka and Caloosahatchee Rivers. The estuarine system is connected with the Gulf of Mexico through various inlets between the barrier islands on the west. The northern area of Charlotte Harbor, with a maximum depth of about 7 m, has been modeled recently (Sheng, et al., 1985).

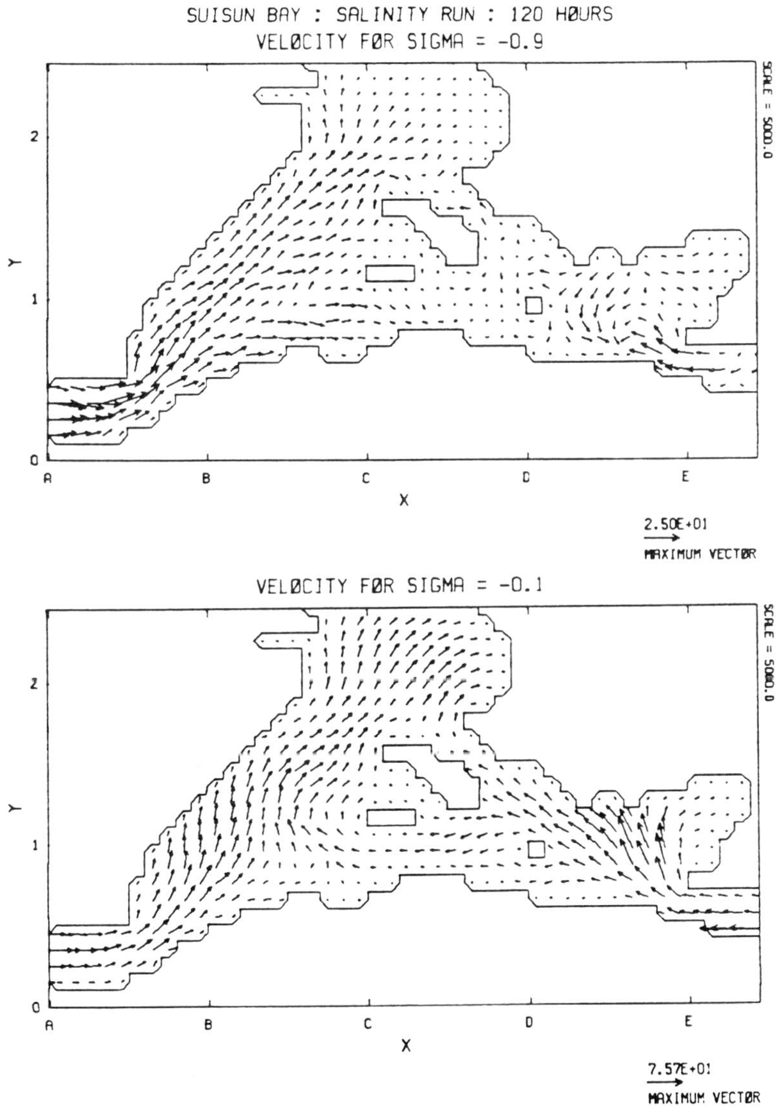

Figure 4 Tide and salinity-driven currents in Suisun Bay near the bottom (σ = -0.9) and near the surface (σ = -0.1) at 120 hours.

As shown in Figure 5, the Peace River inflow (upper right corner) was assumed to be 15,000 cfs and the tidal elevations along the southern boundary of the northern area of Charlotte Harbor was prescribed as a superposition of a strong diurnal and a weaker semi-diurnal components with the elevation varying from -17 cm (t = 0) to 28 cm (t = 15 hrs) to -17 cm (t = 24 hrs) over 24 hours (Goodwin, 1986). Again, salinity along the open boundary follows the inflow-outflow condition described before. The tidal currents and salinity

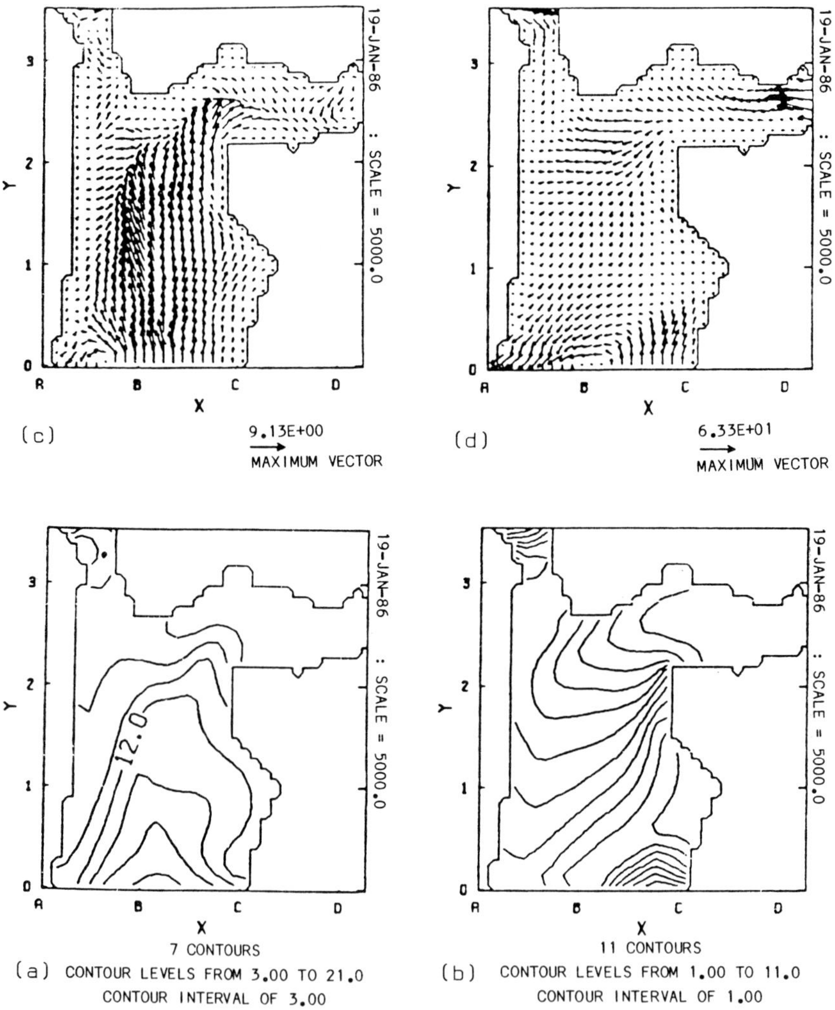

Figure 5 Computed salinity contour in Charlotte Harbor after 48 hours of model simulation with a 1/2-km grid. (a) near bottom, (b) near surface. Salinities in ppt.

Computed 3-D velocity field in Charlotte Harbor after 48 hours of model simulation with a 1/2-km grid. (c) near bottom, (d) near surface. Velocities in cm/sec.

distribution at the end of a 48-hour simulation are shown in Figure 5. A salinity wedge corresponding to the classical estuarine circulation in a partially stratified estuary is clearly exhibited. Significant salinity variation exists in the vertical direction and compares reasonably well with data.

3. VERTICAL TURBULENCE PARAMETERIZATION

Aimed at better "predicting" turbulent flows and removing the deficiencies of ad-hoc eddy viscosity models, turbulent transport modeling has been an area of active research in the last decade. While two-equation (e.g., k-ε) model, one-equation (e.g., k) model, and algebraic model have been adapted to estuarine and marine hydrodynamic models, this paper starts the discussion on vertical turbulence parameterization with "Reynolds stress model" which is the most complete "second-order closure model" and hence the origin of the various simplified turbulent transport models.

3.1 Reynolds Stress Model

The Reynolds stress model seeks closure of equations at the level of second-order correlations, i.e., through dynamic equations derived for the Reynolds stresses $\langle \underset{\sim}{v}'\underset{\sim}{v}' \rangle$, ($\underset{\sim}{v}'$ is the fluctuating velocity vector and "$\langle\ \rangle$" represents ensemble average), heat fluxes $\langle \underset{\sim}{v}'T' \rangle$, mass fluxes $\langle \underset{\sim}{v}'\omega' \rangle$, temperature variance $\langle T'T' \rangle$, concentration variance $\langle \omega'\omega' \rangle$, concentration-temperature covariance $\langle \omega'T' \rangle$, and turbulence macroscale Λ. Because of the similarity in modeling the concentration equation, salinity equation and temperature equation, we shall only include temperature as a variable for simplicity in subsequent discussion. Following the procedure originally outlined by Donaldson (1973), the second-order correlation equations in three-dimensional vector form are:

$$
\begin{aligned}
\frac{\partial \langle \underset{\sim}{v}'\underset{\sim}{v}' \rangle}{\partial t} + \underset{\sim}{v} \cdot \nabla \langle \underset{\sim}{v}'\underset{\sim}{v}' \rangle = {} & -\langle \underset{\sim}{v}'\underset{\sim}{v}' \rangle \cdot \underset{\approx}{\nabla v} - [\langle \underset{\sim}{v}'\underset{\sim}{v}' \rangle \cdot \underset{\approx}{\nabla v}]^T \\
& + \frac{\underset{\sim}{g}}{T_o} \langle \underset{\sim}{v}'T' \rangle + \langle \underset{\sim}{v}'T' \rangle \frac{\underset{\sim}{g}}{T_o} - 2\underset{\sim}{\Omega} \times \langle \underset{\sim}{v}'\underset{\sim}{v}' \rangle - 2[\underset{\sim}{\Omega} \times \langle \underset{\sim}{v}'\underset{\sim}{v}' \rangle]^T \\
& + v_c \underset{\sim}{\nabla} \cdot [q\Lambda \underset{\sim}{\nabla} \langle \underset{\sim}{v}'\underset{\sim}{v}' \rangle] - \frac{q}{\Lambda}\left(\langle \underset{\sim}{v}'\underset{\sim}{v}' \rangle - \frac{q^2}{3} \underset{\approx}{I}\right) - \frac{2bq^3}{3\Lambda} \underset{\approx}{I}
\end{aligned}
\tag{8}
$$

$$
\begin{aligned}
\frac{\partial \langle \underset{\sim}{v}'T' \rangle}{\partial t} + \underset{\sim}{v} \cdot \underset{\sim}{\nabla} \langle \underset{\sim}{v}'T' \rangle = {} & - \langle \underset{\sim}{v}'\underset{\sim}{v}' \rangle \cdot \nabla T - \underset{\approx}{\nabla v} \cdot \langle \underset{\sim}{v}'T' \rangle + \frac{\underset{\sim}{g}}{T_o} \langle T'T' \rangle \\
& - 2\Omega \times \langle \underset{\sim}{v}'T' \rangle + v_c \underset{\sim}{\nabla} \cdot (q\Lambda \underset{\sim}{\nabla} \langle \underset{\sim}{v}'T' \rangle) - \frac{Aq}{\Lambda} \langle \underset{\sim}{v}'T' \rangle
\end{aligned}
\tag{9}
$$

$$\frac{\partial\langle T'T'\rangle}{\partial t} + \underset{\sim}{v} \cdot \underset{\sim}{\nabla}\langle T'T'\rangle = -2\langle\underset{\sim}{v}'T'\rangle \cdot \underset{\sim}{\nabla}T + v_c\underset{\sim}{\nabla} \cdot (q\Lambda\underset{\sim}{\nabla}\langle T'T'\rangle) - \frac{2bsq\langle T'T'\rangle}{\Lambda} \tag{10}$$

$$\frac{\partial\Lambda}{\partial t} + \underset{\sim}{v} \cdot \underset{\sim}{\nabla}\Lambda = 0.35\frac{\Lambda}{q^2}\langle\underset{\sim}{v}'\underset{\sim}{v}'\rangle \cdot \underset{\sim\sim}{\nabla v} + 0.6bq + 0.3\underset{\sim}{\nabla} \cdot (q\Lambda\underset{\sim}{\nabla}\Lambda)$$

$$- \frac{0.375}{q}\underset{\sim}{\nabla}(q\Lambda) \cdot \underset{\sim}{\nabla}(q\Lambda) + \frac{0.8\Lambda}{q^2}\frac{\underset{\sim}{g}}{T_o} \cdot \langle\underset{\sim}{v}'T'\rangle \tag{11}$$

where q is the total turbulent intensity defined as $\langle\underset{\sim}{v}' \cdot \underset{\sim}{v}'\rangle^{1/2}$ and Λ is the turbulence macroscale representing the average turbulent eddy size. Many of the right-hand-side terms in the above equations (e.g., the production terms, the buoyancy terms, and the rotation terms) are exact, and require no parameterization. The last three terms in Equation (8) and the last two terms in Equations (9) and (10) are "modeled" terms representing the effects of third-order correlations, pressure correlations, and viscous dissipations. The model coefficients (b, A, v_c and s) have been determined from critical laboratory experiments where only one of the coefficients is important. The final coefficients (b = 0.125, A = 0.75, v_c = 0.3 and s = 2.8) thus determined have remained fixed for all model applications.

3.2 Estuarine and Marine Applications of the Reynolds Stress Model

The Reynolds stress model described above has been extensively applied to estuarine, marine and atmospheric environments. The 1-D version (the variables vary in the vertical direction only) has been used to study wave boundary layer underneath a nearly sinusoidal wave (Sheng, 1982 and 1984), wave boundary layer underneath a cnoidal wave (Sheng, 1984), current-wave interaction within bottom boundary layers (Sheng, 1983 and 1984), flow within a vegetation canopy (Sheng, 1982), ocean mixed layer dynamics (Sheng, 1984), and sediment-laden boundary layer (Sheng and Villaret, 1986). The 2-D Reynolds stress model is currently being used to study flow over a rippled bed. The fully 3-D Reynolds stress model is also being used for a dispersion study. As an example, the vertical distribution of Reynolds stress within a nearly sinusoidal wave boundary layer, a cnoidal wave boundary layer, a current-wave boundary layer, and a vegetation canopy are shown in Figure 6. While eddy viscosity models may match the measured mean velocity profiles reasonably well by adjusting the eddy viscosity coefficients, the Reynolds stress model can predict the mean flow and Reynolds stresses without ad-hoc parameter tuning. In addition, the Reynolds stress model explicitly computes the thickness of time varying logarithmic layer in wave boundary layer and clearly reproduces the effects of wave on current such as the increased Reynolds stress, turbulent intensity and apparent roughness.

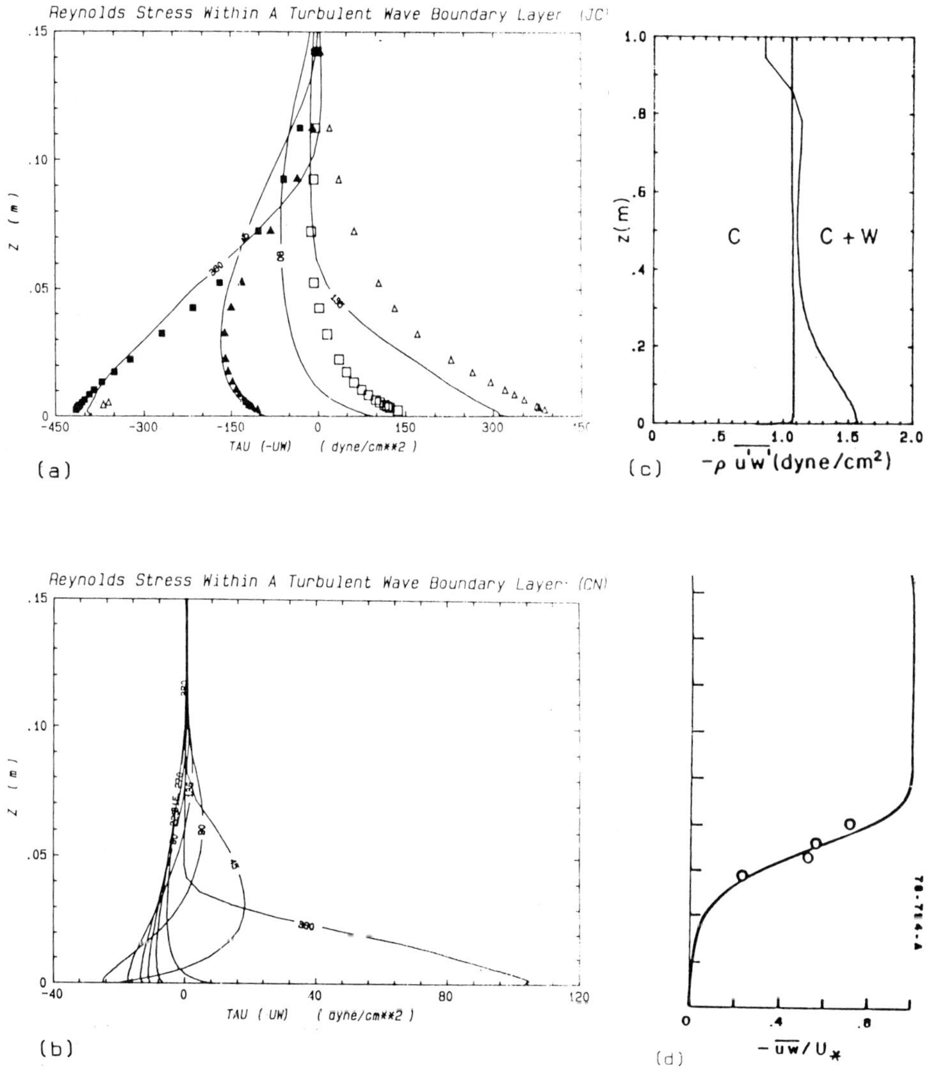

Figure 6(a) Vertical distribution of Reynolds stress within the turbulent wave boundary layer measured by Jonsson and Carlsen (1976).Comparison between model results (solid lines) and data (symbols) at 4 phase angles.

Figure 6(b) Vertical distribution of Reynolds stress within the turbulent wave boundary layer under a cnoidal wave at 8 phase angles. Model results only.

Figure 6(c) Simulated current-wave bottom boundary layer at a site in the Mississippi Sound. Vertical profiles of Reynolds stress averaged over the wave cycle. z_o = 0.1 cm, U_{100} = 10 cm/sec, and T_w = 2.5 sec.

Figure 6(d) Comparison of model predictions with in and above a corn canopy.

3.3 Simplified Versions of Reynolds Stress Model

Two simplified versions of the Reynolds stress model eliminate some of the terms in Equations (8) through (11) and are particularly useful. In the super-equilibrium version, all the second-order correlations are assumed to be in local equilibrium such that there is no time evolution or spatial diffusion of the correlations. Equations (8) to (11) are thus simplified to a closed set of algebraic equations between the second-order correlations and the gradients of mean velocity and temperature. In the quasi-equilibrium version, most second-order correlations are assumed to be in high Reynolds number local equilibrium, and algebraic relationships hold for these correlations and mean flow gradients. The dynamics of the turbulence is carried by two dynamic equations, one for $q^2 = \langle \underset{\sim}{v}' \cdot \underset{\sim}{v}' \rangle$ and one for Λ. This approximation gives reasonable results so long as the time scale of turbulence, Λ/q, is much less than the time scale of mean flow.

The two simplified versions of the Reynolds stress model have been applied to study various boundary layers in laboratory, estuarine and marine environments. The quasi-equilibrium version was able to faithfully reproduce the wave boundary layer experiment described in Sheng (1982 and 1984). In addition, it has been used to study sediment-laden boundary layers. The super-equilibrium version has been successfully used to simulate storm generated currents on continental shelf near Grand Bank (Sheng, 1986b) and tropical cyclone generated currents.

Both simplified versions have been incorporated into the 3-D hydrodynamic model developed by the present author.

4. A THREE-DIMENSIONAL CURVILINEAR-GRID MODEL

4.1 Boundary-Fitted Grid

To model flow within a rectangular domain, it is natural to use a Cartesian grid. For cylindrical or spherical domains, it is natural to use a cylindrical or spherical grid. Hence, in the presence of complex shoreline and bathymetry (see, e.g., Figure 1), it is natural to use a "boundary-fitted" grid, or generalized curvilinear grid to accurately represent the model boundaries.

Conformal grid, orthogonal grid and non-orthogonal grid are the various types of "boundary-fitted" grid with increasing complexity and generality. For relatively simple geometries, it is possible to generate conformal or orthogonal grids by rather straightforward techniques. For most estuarine and coastal applications, however, the shoreline geometries are usually quite complex and conformal or orthogonal grid cannot be generated unless the shorelines are approximated by simple curves. Hence, in general, it is essential to generate non-orthogonal grid for estuarine and coastal applications. The present 3-D hydrodynamic model employs the elliptic grid generation technique (Thompson, 1982) to generate non-orthogonal boundary-fitted grid in the horizontal directions and σ-stretched grid in the vertical direction.

As shown in Figure 7a, the basic problem is to solve:

$$\underset{\sim}{\nabla}^2 \xi = P \tag{12}$$

$$\underset{\sim}{\nabla}^2 \eta = Q \tag{13}$$

where $\underset{\sim}{\nabla}^2$ represents the Laplacian operator in x and y directions and P and Q are forcing functions for achieving desired grid resolution and alignment, with the following boundary conditions:

$$\xi = \xi(x,y),\ \eta = \text{constant} \qquad \text{on boundaries 1 and 3} \tag{14}$$

$$\xi = \text{constant},\ \eta = \eta(x,y) \qquad \text{on boundaries 2 and 4} \tag{15}$$

In practice, however, one actually solves for (x,y) in terms of (ξ,η) by interchanging the dependent and independent variables in Eqs. (12) through (15).

As indicated earlier, the present model solves the transformed equations of motion in the (ξ,η) plane in terms of the "contravariant" velocity components (Figure 7b) instead of the Cartesian velocity components. Sheng (1986c) discussed the numerous advantages of the "contravariant model" over other models that work with covariant, physical or Cartesian velocity components. The model first develops the tensor-invariant form of the equations of motion and then expands into component equations in ξ and η directions. For simplicity, we will list the tensor-invariant form of the vertically-integrated equations of motion:

$$\frac{\partial \zeta}{\partial t} + \frac{1}{\sqrt{g_o}} \frac{\partial}{\partial x_k} (\sqrt{g_o}\, U^k) = 0 \tag{16}$$

$$\frac{\partial U^k}{\partial t} + \left(\frac{U^m U^n}{H}\right)_{,m} = -g_{nj}\, \varepsilon^{kj} f U^n + \tau_s^{\ k} - c_d (g_{mn} U^n U^m)^{0.5}\, U^k + A_H U^k_{\ ,m}!^m - \frac{H}{\rho_o} P_a!^k - gH\, \zeta!^k - \frac{gH^2}{2\rho_o} \rho!^k \tag{17}$$

$$\frac{\partial T}{\partial t} + (U^k T)_{,k} = q_s - q_b + K_H\, T_{,k}!^k \tag{18}$$

where ζ is surface elevation, U is vertically-integrated velocity, T is vertically-averaged temperature, the superscript denotes a contravariant vector component, the subscript denotes a covariant vector component, H is total depth, g_o is determinant of the metric tensor g_{ij} between the prototype grid (x,y) and the transformed grid $(\xi,\eta)_{,i}$ denotes a covariant spatial derivative, $!^m$ denotes a contravariant spatial derivative, ε^{kj} is the permutation tensor, c_d is bottom friction coefficient, P_a is atmospheric pressure, ρ is density, τ_s is surface wind stress, q_s and q_b are heat fluxes at surface and bottom, and A_H and K_H are lateral turbulent eddy coefficients.

Details of the 3-D transformed equations of motion and their solutions in terms of contravariant velocity components can be found in Sheng (1986c). The

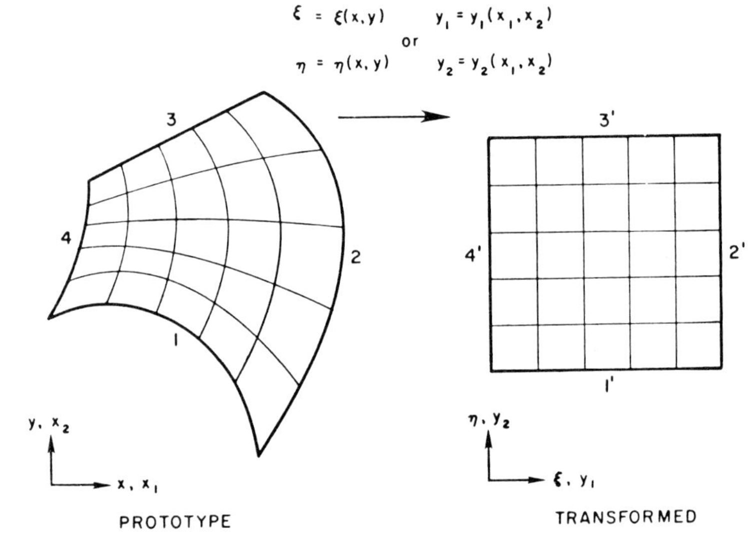

Figure 7(a) A boundary-fitted grid in the prototype and transformed system.

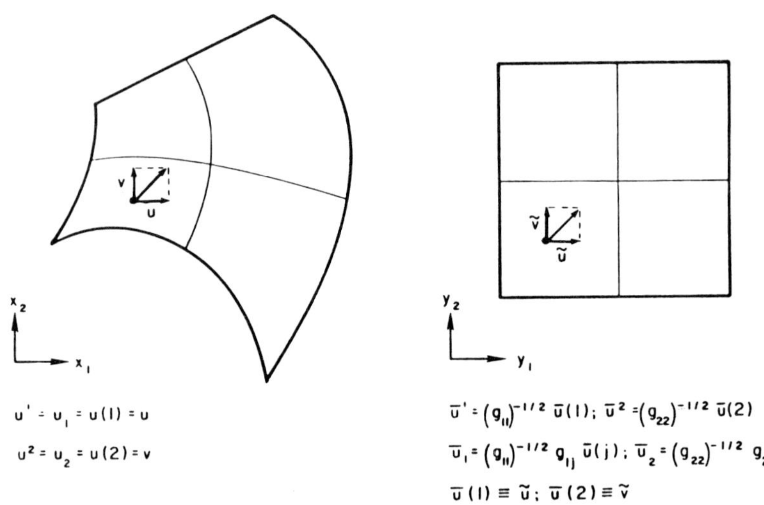

Figure 7(b) Contravariant, covariant and physical components of a vector in the prototype and transformed systems.

3-D boundary-fitted hydrodynamic model, like the finite-element model, is capable of resolving complex geometries. It is, however, more efficient than finite-element models because it only requires tridiagonal matrix inversions while finite-element models must deal with inversions of large band matrices.

It should be noted that the expansion ot the tensor-invariant equations are rather laborious and is best accomplished by means of a "symbolic manipulator" to avoid human errors. The reader is referred to Sheng (1986c) for details. It should also be pointed out that, despite the many extra terms, the transformed equations thus derived do become significantly simplified if conformal or orthogonal grid is used. The model thus is more general than those models that only work for conformal or orthogonal grids, particularly because it is extremely difficult to generate conformal or orthogonal grids for compex domains. As an example, the expanded equation of motion in the ξ-direction is:

$$\frac{\partial U}{\partial t} + \frac{\partial}{\partial \xi}\left(\frac{UV}{H}\right) + \frac{\partial}{\partial \eta}\left(\frac{UV}{H}\right) + (2D^1_{11} + D^2_{12})\frac{UV}{H} + (3D^1_{12} + D^2_{22})\frac{UV}{H} + D^1_{22}\frac{VV}{H}$$

$$= \frac{g_{12}}{\sqrt{g_o}} fU + \frac{g_{22}}{\sqrt{g_o}} fV - gH\left[\left(g^{11}\frac{\partial \zeta}{\partial \xi} + g^{12}\frac{\partial \zeta}{\partial \eta}\right) + \frac{H}{\rho_o}\left(g^{11}\frac{\partial \rho}{\partial \xi} + g^{12}\frac{\partial \rho}{\partial \eta}\right)\right]$$

$$+ \tau^1_s - c_d\left[g_{11}U^2 + 2g_{12}UV + g_{22}V^2\right]^{1/2} U$$

+ Lateral Turbulent Diffusion Terms

where (U,V) are the contravariant velocity components in (ξ,η) directions, g_{ij} and g^{ij} are metric tensor coefficients, and D^i_{jk} is Christopher symbol of the second kind.

4.2 Model Applications

The 3-D curvilinear grid model has been and is being applied to real estuaries (e.g., Chesapeake Bay and James River). For simplicity here, we will present some results obtained with the vertically-integrated version of the model.

The first example is concerned with tidal circulations in a quarter circular annular wedge driven by a simple sinusoidal tidal forcing of $\xi = \xi_o$ along the outer radius. For simplicity, depth h and gravitational acceleration g are set to unity. The domain is defined by all (r,ϕ) values that satisfy $1 < r < 2.5$ and $0 < \phi < \pi/2$. ω is equal to $\pi/3$. First of all, a simple polar coordinate was used for the linearized tidal simulation with all the metric tensor coefficients analytically computed. Because of the orthogonal grid, metric tensors are diagonal and Christopher symbols are identically zero. The model results of surface elevation and velocity agree very well with analytical results and both showed little azimuthal variation. Next, a rather skewed grid (Figure 8) by an elliptic grid generator was used.

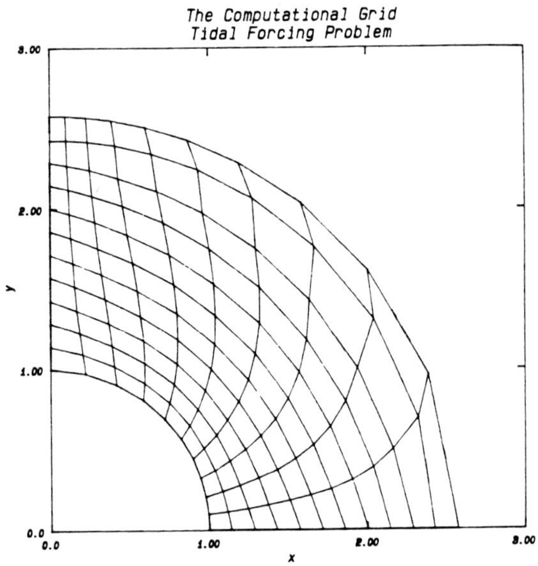

Figure 8 The computational grid for tidal forcing problem.

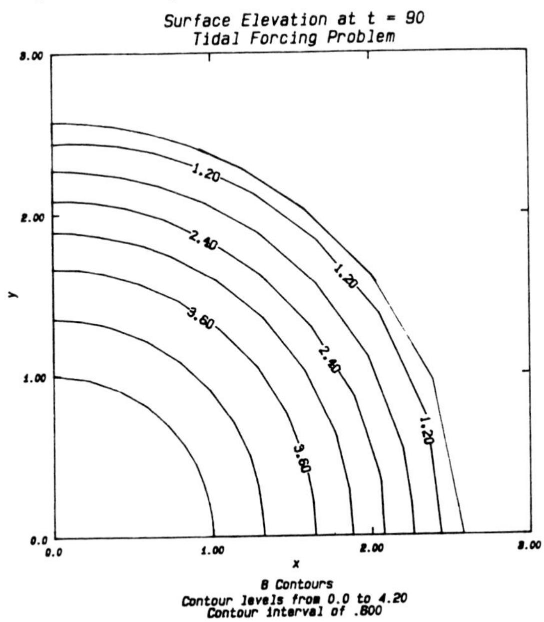

Figure 9 Contour plot of the surface elevation for the tidal forcing problem.

Solving the linearized tidal circulation in this 12 x 11 grid yields rather symmetric results for both water elevation and velocity. Figure 9 shows the water elevation over the entire domain at an instant of time. Figure 10 shows the distribution of surface elevation along two cross sections of constant ϕ (ϕ~0 and ϕ~90°). All the results compare well with the analytical results.

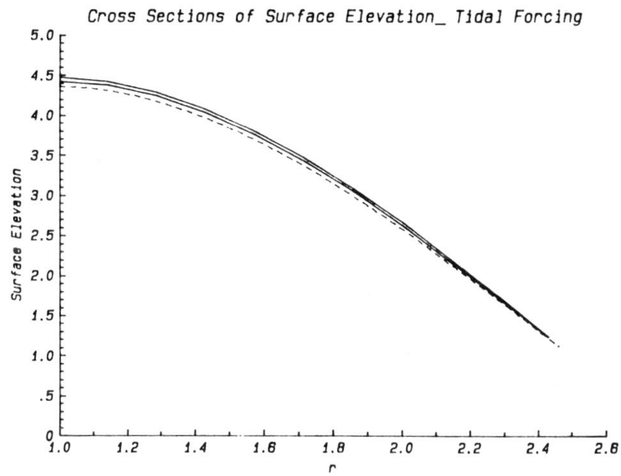

Figure 10(a) Cross sectional plot of the surface elevation for the tidal forcing problem. The solid lines are the numerical solution along the 0 and 90 degree edges, the dashed line is the analytical solution.

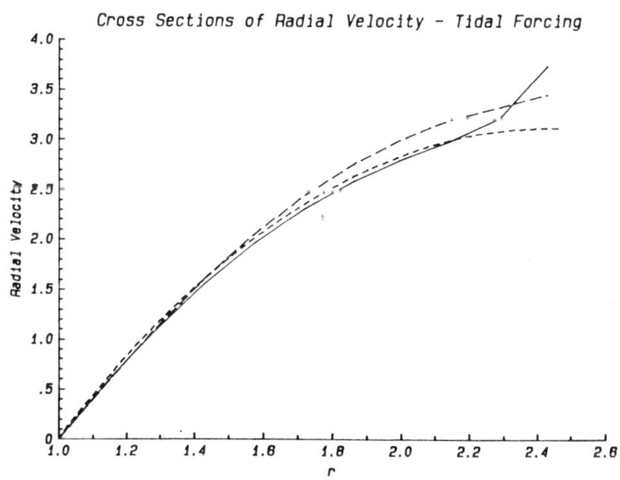

Figure 10(b) Cross sectional plot of the radial velocity for the tidal forcing problem using the numerical transformation (———) along ϕ = 0. (-- -- --) numerical solution along ϕ = 90. (--------) analytic solution.

The second example is the refraction and diffraction of a tsunami (T = 4 minutes) by a circular island (r = 10 km) and a parabolic bottom (depth = 0.444 km at shore to 4 km at r = 30 km). Considering an incident monochromatic plane wave along the positive x direction and applying radiation condition at infinity, we presented our results in terms of the wave amplitude and phase lag along the upper half of the shoreline ($\phi = 0^o$ to $\phi = 180^o$). As shown in Figure 11, the results agree well with the analytical results of Homma (1950).

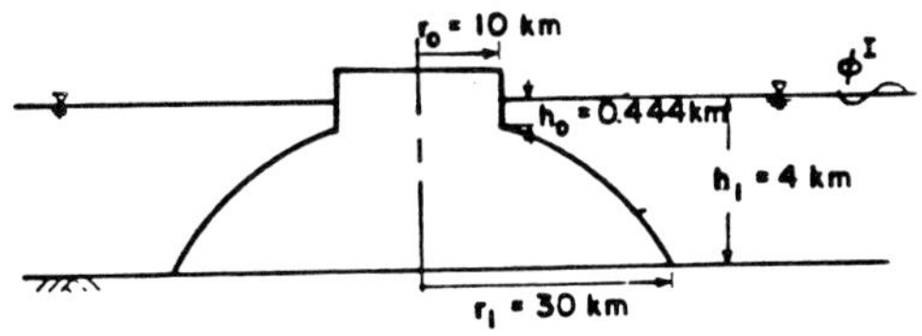

Figure 11(a) Configuration of a circular island.

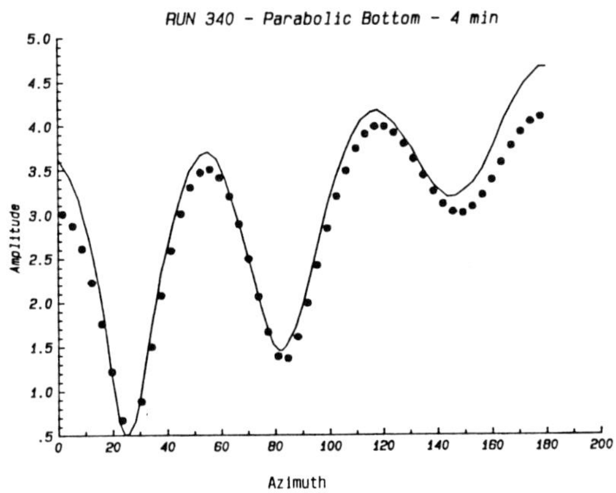

Figure 11(b) Wave amplitude along the upper shoreline of the circular island computed by the present model and Homma (1950)

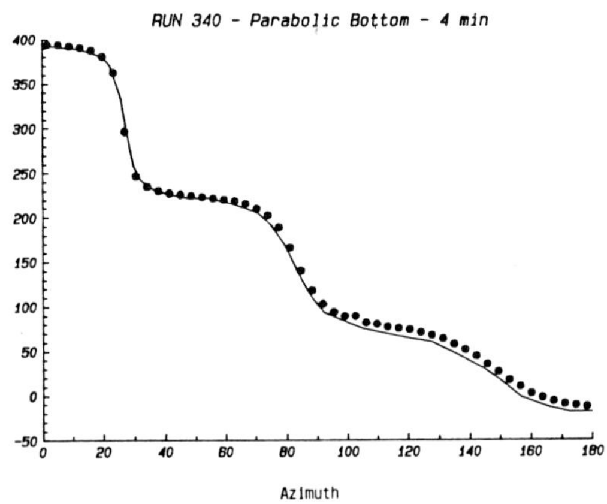

Figure 11(c) Same as 11(b), except for phase lag instead of wave amplitude.

ACKNOWLEDGEMENT

The work reported here has been partially supported by the U. S. Army Engineer Waterways Experiment Station under contract DACW 39-80-C-0087, with H. L. Butler and B. Johnson as contract monitors, and U. S. Geological Survey under contract 14-08-0001-4730, with R. T. Cheng as contract monitor.

5. REFERENCES

Butler, H. L., 1978. Coastal flood simulation in stretched coordinates. Proceedings 16th Int'l. Conf. Coastal Engineering. ASCE/Hamburg, Germany.

Cheng, R. T. and J. W. Gartner, 1984. Tides, tidal and residual currents in San Francisco Bay, results of measurements, 1979-1980. U.S.G.S. WRE Report 84-4339.

Donaldson, C. duP., 1973. Atmospheric turbulence and the dispersal of atmospheric pollutants. In: D. A. Haugen (Editor) AMS Workshop on Micrometeorology. Science Press, Boston, pp 313-390.

Goodwin, C. 1986. Personal communication.

Johnson, B. J., 1982. Numerical modeling of estuarine hydrodynamics on a boundary-fitted coordinate system. In: J. E. Thompson (Editor), Numerical Grid Generation. Elsevier Science Publishing Company, Inc., pp 409-436.

Leendertse, J. J., 1967. Aspects of a computational model for long water wave propagation. Rand Corp., Rept. RH-5299-RR

Rodi, W. 1980. Turbulence Models and Their Application in Hydraulics. IAHR.

Sheng, Y. P., 1976. Currents and contaminant dispersion in the near-shore of large lakes. J. Great Lakes Res., vol. 2, no. 2, pp 402-414.

Sheng, Y. P., 1982. Hydraulic applications of a turbulent transport model. Proceedings 1982 ASCE Hydraulic Division Specialty Conference on Applying Research to Hydraulic Practice, pp 106-119.

Sheng, Y. P., 1983. Mathematical modeling of three-dimensional coastal currents and sediment dispersion. A.R.A.P. Report 486; also Technical Report CERC-83-2, U.S. Army Engineer Waterways Experiment Station, Vicksburg, Mississippi.

Sheng, Y. P., 1984. A turbulent transport model of coastal processes. Proceedings 19th Int'l. Conf. on Coastal Engineering, ASCE, pp 2380-2396.

Sheng, Y. P., 1986a. Finite-difference models for hydrodynamics of lakes and shallow seas. In: Grey, W. G. (Editor), Physics-Based Modeling of Lakes, Reservoirs and Impoundments. ASCE.

Sheng, Y. P., 1986b. Validation of ocean current model OCM1D for storm current simulations. A.R.A.P. Report No. 580.

Sheng, Y.P., 1986c. Modeling coastal and estuarine processes using boundary-fitted grids. In: Wang, Shen, and Ding (Editors), River Sedimentation. Proceedings 3rd Int'l Symposium on River Sedimentation, pp 1416-1442.

Sheng, Y. P. and H. L. Butler, 1982. Modeling coastal currents and sediment transport. Proceedings 18th Int'l. Conf. on Coastal Engineering. ASCE, pp 1127-1148.

Sheng, Y. P. and C. Villaret, 1986. On the determination of sediment erosion relationship. In preparation.

Sheng, Y. P., S. F. Parker, D. Henn, 1985. A three-dimensional estuarine hydrodynamic model (EHSM3D), to be published as a Water Resources Investigation Report by the Water Resources Division of U.S. Geological Survey.

Sheng, Y. P., W. Lick, R. T. Gedney, and F. B. Molls, 1978. Numerical computation of the three-dimensional circulation in Lake Erie: A comparison of a free-surface and a rigid-lid model. J. Phys. Oceano., pp 72-73.

Thompson, J. E., 1982. General curvilinear coordinate systems. In: J. E. Thompson (Editor), Numerical Grid Generation. Elsevier Science Publishing Company, Inc., pp 1-30.

Waldrop, W. R. and F. B. Tatom, 1976. Analysis of the thermal effluent from the Gallatin steam plant during low river flows. Report no. 33-30, TVA.

Warnstrath, J. J., 1977. Nearshore numerical storm surge and tidal simulation. T. R. H-77-17. Army Engineer Waterways Experiment Station, Vicksburg, Mississippi.

CIRCULATION MODELLING USING ORTHOGONAL CURVILINEAR COORDINATES

ALAN F. BLUMBERG[1] and H. JAMES HERRING[2]

[1]HydroQual, Inc., 1 Lethbridge Plaza, Mahwah, New Jersey 07430

[2]Dynalysis of Princeton, 219 Wall Street, Princeton, New Jersey 08540-1512

ABSTRACT

A numerical estuarine and coastal ocean circulation model is developed in orthogonal curvilinear coordinates. The governing equations consist of the equation of continuity, the three components of momentum and conservation equations for thermal energy and salt. Other prognostic equations are solved for the turbulence kinetic energy and turbulence macroscale, both of which are part of a turbulence closure submodel representing the vertical mixing process. The circulation model employs standard finite difference techniques and accommodates irregularly shaped computational grids with no apparent difficulties. Successful two and three-dimensional simulations of a variety of problems with analytic or well established numerical solutions demonstrate that the model physics and numerics can produce meaningful results. The test cases presented here also provide a framework for objective comparison which could be used in the development of other numerical models.

1 INTRODUCTION

The tradition in ocean, estuarine and lake circulation modelling has been to use finite difference techniques on a rectangular grid. There are many instances however, when the computational resources required become excessive. This situation arises in regions where the coastline has prominent features or where boundary layer dynamics are important. If the number of points is not large enough to resolve the coastal features or boundary layer, then the numerical solution is likely to have gross errors even in the interior regions. As an example, consider the circulation in the coastal waters off California, a setting which provided the motivation for the development of the model to be described herein.

The dominant features of the circulation off the coast of California are coastal upwelling in the nearshore and the California Current offshore. Upwelling occurs in a narrow region adjacent to the coast and has a typical offshore length scale of 10-20 km (the baroclinic radius of deformation). The California Current, on the other hand, is relatively broad with a typical width of 500 km and flows southward roughly parallel to the coast. In order to resolve upwelling processes in a numerical circulation model, the spatial grid should be about 5 km, whereas a relatively coarser grid, of perhaps 25 to 50 km, suffices further offshore. The traditional uniform rectangular grid is inadequate here, because it would require an extraordinarily large number of points to cover a domain encompassing the coastal and oceanic regions of interest.

It becomes important then to have available a circulation model with the ability to resolve arbitrary topography and also the flexibility to refine the grid in regions of particular interest or in regions of large gradients. Numerous models have been developed over the years which allow for some flexibility in the grid specification. Earlier models used rectangular grids with transformed [Reid et al., 1977] or stretched [Peffley and O'Brien, 1976] coordinates. Other models such as that of Thacker [1977] employed irregular grids with triangular configurations. The explosive growth of finite element techniques [Pinder and Gray, 1977] has been motivated by the search for flexibility in the design of the computational grid. In more recent times, the use of non-orthogonal boundary-fitted coordinates (Johnson, [1982] and Spaulding, [1984]) has come into vogue. The purpose of the paper is to provide an alternative approach to those enumerated above and thus make possible long term integrations using more modest computational resources.

What follows is a detailed description of a numerical estuarine and coastal ocean circulation model that allows considerable latitude in the design of the computational grid. This additional freedom is accomplished through the use of an orthogonal curvilinear coordinate system on the horizontal coordinate surfaces. When cast in these coordinates the governing equations retain much of the analytical simplicity of the familiar Cartesian equations. One beneficial result of the simplicity is that the computational cost for the equations in orthogonal curvilinear form is only marginally increased over that for their Cartesian counterpart and is considerably less than the cost for the corresponding boundary-fitted equations.

Through a series of model test calculations, the viability and versatility of this new formulation will be demonstrated. Construction of the horizontal grid mesh used by the model is included in the discussion. The test cases which follow are examples in which analytic and other numerical solutions are available for comparison and assessment of model performance. Both two and three-dimensional cases will be considered.

2 MODEL FORMULATION

The equations which describe the circulation in estuaries and in coastal and open oceans are the equations of continuity, the three components of momentum and equations for the conservation of thermal energy and salt. These equations, together with an equation of state and a suitable form of turbulence closure, are sufficient to determine primitive variables consisting of the three components of velocity, the elevation of the free surface, the temperature, the salinity and the density.

Several simplifying assumptions have been made in formulating the equations. The hydrostatic approximation implies that the local pressure at a point is a function only of the weight of the water column above it and thus the transport of vertical momentum is negligible. The Boussinesq approximation is that the variations in the density about a mean value, ρ_0 say, are dynamically negligible except in the determination of the local hydrostatic pressure.

2.1 Equations of Motion

The equations which form the basis of the circulation model are well established in general orthogonal coordinates (see Eringen [1962] for example). Consider a system of orthogonal curvilinear coordinates with horizontal coordinates (ξ_1, ξ_2) and vertical coordinate (z) as shown in Figure 1. The metric

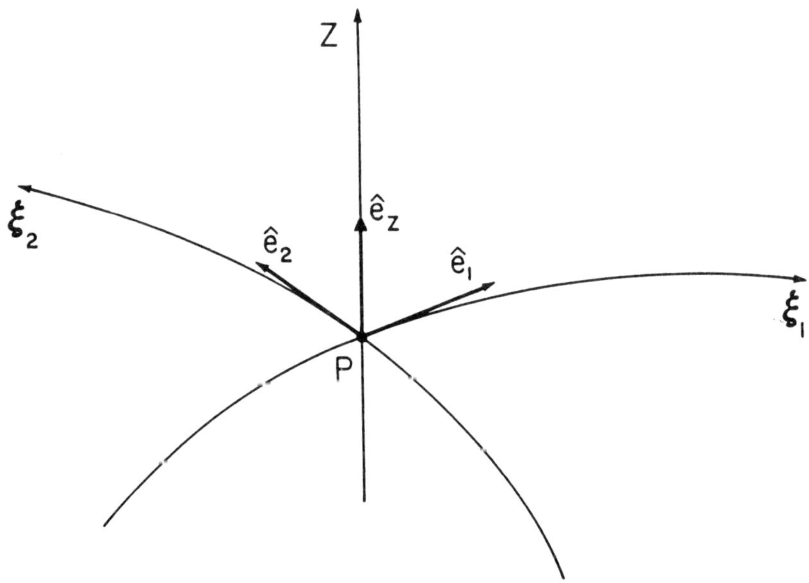

Fig. 1. The orthogonal curvilinear coordinate system used in the circulation model.

coefficients, h_1 and h_2, are defined so that a distance increment satisfies the relation

$$ds^2 = h_1^2 \, d\xi_1^2 + h_2^2 \, d\xi_2^2 \quad . \tag{1}$$

The differential arc lengths along ξ_1 and ξ_2 at point P are $ds_1 = h_1 d\xi_1$ and $ds_2 = h_2 d\xi_2$. The horizontal velocity vector has components

$$U_1 = h_1 \frac{d\xi_1}{dt} \quad \text{and} \quad U_2 = h_2 \frac{d\xi_2}{dt} \tag{2a,b}$$

in the ξ_1 and ξ_2 directions, respectively. Following Merilees [1976] and Arakawa and Lamb [1977], the continuity equation is

$$\frac{\partial}{\partial \xi_1}\left(h_2 U_1\right) + \frac{\partial}{\partial \xi_2}\left(h_1 U_2\right) + h_1 h_2 \frac{\partial w}{\partial z} = 0 \quad , \tag{3}$$

where w is the vertical velocity. The horizontal momentum equations can be written as

$$\frac{\partial U_1}{\partial t} + \frac{1}{h_1^2 h_2}\left[\frac{\partial}{\partial \xi_1}\left(U_1^2 h_1 h_2\right) + \frac{\partial}{\partial \xi_2}\left(U_1 U_2 h_1^2\right)\right] + \frac{\partial}{\partial z}\left(w U_1\right)$$

$$- \frac{U_1^2}{h_1^2}\frac{\partial h_1}{\partial \xi_1} - \frac{U_2^2}{h_1 h_2}\frac{\partial h_2}{\partial \xi_1} - fU_2 = - \frac{1}{h_1 \rho_o}\frac{\partial P}{\partial \xi_1} + \Gamma_1 \quad , \tag{4}$$

$$\frac{\partial U_2}{\partial t} + \frac{1}{h_1 h_2^2}\left[\frac{\partial}{\partial \xi_1}\left(U_1 U_2 h_2^2\right) + \frac{\partial}{\partial \xi_2}\left(U_2^2 h_1 h_2\right)\right] + \frac{\partial}{\partial z}\left(w U_2\right)$$

$$- \frac{U_1^2}{h_1 h_2}\frac{\partial h_1}{\partial \xi_2} - \frac{U_2^2}{h_2^2}\frac{\partial h_2}{\partial \xi_2} + fU_1 = - \frac{1}{h_2 \rho_o}\frac{\partial P}{\partial \xi_2} + \Gamma_2 \quad , \tag{5}$$

and the vertical momentum equation with the hydrostatic assumption is

$$\rho g = - \frac{\partial P}{\partial z} \quad , \tag{6}$$

with ρ_o, the reference density; ρ, the in situ density; g, the gravitational acceleration; f, the Coriolis parameter, and P, the pressure. The Boussinesq approximation has been used in deriving Equations (4) and (5). The pressure at depth z can be obtained by integrating Equation (6) from z to the free surface, z = (ξ_1,ξ_2,t), and is

$$P(\xi_1,\xi_2,z,t) = P_{atm} + g\rho_o \eta + g \int_z^o \rho(\xi_1,\xi_2,z',t)dz'. \tag{7}$$

In the discussions which follow the atmospheric pressure, P_{atm}, is assumed to contribute little to the pressure gradient.

The diffusion terms, represented by Γ_1 and Γ_2 in Equations (4) and (5), can be written as

$$\Gamma_1 = \frac{1}{h_1 h_2}\left[\frac{\partial}{\partial \xi_1}\left(h_2 \tau_{11}\right) + \frac{\partial}{\partial \xi_2}\left(h_1 \tau_{21}\right)\right] + \frac{\partial}{\partial z}\left(K_M \frac{\partial U_1}{\partial z}\right)$$

$$+ \frac{1}{h_1 h_2}\left[\tau_{21}\frac{\partial h_1}{\partial \xi_2} - \tau_{22}\frac{\partial h_2}{\partial \xi_1}\right] \tag{8}$$

and

$$\Gamma_2 = \frac{1}{h_1 h_2}\left[\frac{\partial}{\partial \xi_1}(h_2 \tau_{12}) + \frac{\partial}{\partial \xi_2}(h_1 \tau_{22})\right] + \frac{\partial}{\partial z}\left(K_M \frac{\partial U_2}{\partial z}\right)$$

$$+ \frac{1}{h_1 h_2}\left[\tau_{12}\frac{\partial h_2}{\partial \xi_1} - \tau_{11}\frac{\partial h_1}{\partial \xi_2}\right] . \tag{9}$$

The vertical eddy diffusivity for turbulent momentum mixing is denoted as K_M. Finally, the shear stress components in Equations (8) and (9) are

$$\tau_{11} \equiv A_M\left[\frac{h_2}{h_1}\frac{\partial U_1}{\partial \xi_1} + \frac{2U_2}{h_1 h_2}\frac{\partial h_1}{\partial \xi_2}\right] , \tag{10}$$

$$\tau_{21} \equiv A_M\left[\frac{h_1}{h_2}\frac{\partial}{\partial \xi_2}\left(\frac{U_1}{h_1}\right) + \frac{h_2}{h_1}\frac{\partial}{\partial \xi_1}\left(\frac{U_2}{h_2}\right)\right] \equiv \tau_{12} \tag{11}$$

and

$$\tau_{22} \equiv A_M\left[\frac{2U_1}{h_1 h_2}\frac{\partial h_2}{\partial \xi_1} + \frac{h_1}{h_2}\frac{\partial U_2}{\partial \xi_2}\right] . \tag{12}$$

The horizontal diffusivity is denoted as A_M. The momentum equations presented here are quite similar to the more usual Cartesian coordinate system, except for the additional terms which account for the curvature of the coordinate system itself. The Cartesian system equations (see Blumberg and Mellor, [1986]) are recovered by the transformation $h_1 d\xi_1 \rightarrow dx$ and $h_2 d\xi_2 \rightarrow dy$.

The conservation equations for temperature, θ, and salinity, S, may be written in orthogonal curvilinear coordinates as

$$\frac{\partial \theta}{\partial t} + \frac{1}{h_1 h_2}\left[\frac{\partial}{\partial \xi_1}(h_2 U_1 \theta) + \frac{\partial}{\partial \xi_2}(h_1 U_2 \theta)\right] + \frac{\partial}{\partial z}(w\theta)$$

$$= \frac{1}{h_1 h_2}\left[\frac{\partial}{\partial \xi_1}\left(\frac{h_2}{h_1} A_H \frac{\partial \theta}{\partial \xi_1}\right) + \frac{\partial}{\partial \xi_2}\left(\frac{h_1}{h_2} A_H \frac{\partial \theta}{\partial \xi_2}\right)\right]$$

$$+ \frac{\partial}{\partial z}\left(K_H \frac{\partial \theta}{\partial z}\right) \tag{13}$$

and

$$\frac{\partial S}{\partial t} + \frac{1}{h_1 h_2}\left[\frac{\partial}{\partial \xi_1}(h_2 U_1 S) + \frac{\partial}{\partial \xi_2}(h_1 U_2 S)\right] + \frac{\partial}{\partial z}(wS)$$

$$= \frac{1}{h_1 h_2}\left[\frac{\partial}{\partial \xi_1}\left(\frac{h_2}{h_1} A_H \frac{\partial S}{\partial \xi_1}\right) + \frac{\partial}{\partial \xi_2}\left(\frac{h_1}{h_2} A_H \frac{\partial S}{\partial \xi_2}\right)\right]$$

$$+ \frac{\partial}{\partial z}\left(K_H \frac{\partial S}{\partial z}\right) \quad . \tag{14}$$

The vertical eddy diffusivity for turbulent mixing of heat and salt is K_H while the horizontal diffusivity is denoted as A_H. Using the temperature and salinity, the density is computed according to the equation of state

$$\rho = \rho(\theta, S), \tag{15}$$

due to Fofonoff [1962].

The horizontal mixing coefficients A_M and A_H are used to parameterize all motion which is not resolved on the numerical grid. Since the grid is non-uniform, the horizontal mixing coefficients must vary proportionally in order to maintain an uniform grid Reynolds number. The relation employed is

$$\left(A_M, A_H\right) = \frac{A_o}{h_o}\left(h_1^2 + h_2^2\right)^{\frac{1}{2}} , \tag{16}$$

where A_o is the equivalent diffusivity for an uniform grid with a grid spacing of h_o.

The boundary conditions at the free surface, $z = \eta(\xi_1, \xi_2)$, are:

$$\rho_o K_M \left(\frac{\partial U_1}{\partial z}, \frac{\partial U_2}{\partial z}\right) = (\tau_{o1}, \tau_{o2}) \quad , \tag{17a}$$

$$\rho_o K_H \left(\frac{\partial \theta}{\partial z} , \frac{\partial S}{\partial z}\right) = (\tilde{H} , \tilde{S}) \tag{17b}$$

and

$$w = \frac{\partial \eta}{\partial t} + \frac{U_1}{h_1}\frac{\partial \eta}{\partial \xi_1} + \frac{U_2}{h_2}\frac{\partial \eta}{\partial \xi_2} , \tag{17c}$$

where (τ_{o1}, τ_{o2}) is the surface wind stress vector. The local net heat flux is $\widetilde{H}$ and the salinity flux at the surface is $\widetilde{S}$.

At the bottom of the basin, the normal gradients of θ and S are zero. Additionally, at the bottom boundary, b,

$$\rho_o K_M \left(\frac{\partial U_1}{\partial z}, \frac{\partial U_2}{\partial z}\right) = (\tau_{b1}, \tau_{b2}) \tag{18a}$$

and

$$w = -\frac{U_{b1}}{h_1}\frac{\partial H}{\partial \xi_1} - \frac{U_{b2}}{h_2}\frac{\partial H}{\partial \xi_2} , \tag{18b}$$

where $H(\xi_1, \xi_2)$ is the bottom topography and (τ_{b1}, τ_{b2}) is the bottom frictional stress. As in Blumberg and Mellor [1986] the bottom stress is determined from

$$\vec{\tau}_b = \rho_o C_D \left|\vec{V}_b\right| \vec{V}_b \quad . \tag{19a}$$

The value of the drag coefficient C_D is given by

$$C_D = \left(\frac{1}{\kappa} \ln(H + z_b)/z_o\right)^{-2}, \tag{19b}$$

where z_b and $\vec{V}_b$ are the depth of and corresponding velocity at the grid point nearest the bottom and κ is the von Karman constant. The purpose of (19) is that the calculations will yield $\vec{V}_b = (\vec{\tau}_b/\kappa u_{\tau b})\ln(z/z_o)$ in the lower boundary region if enough resolution is provided. In, for example, 100 m water the log layer is well resolved, whereas in much deeper water, it is not. In the latter instance, it is advantageous to abandon (19b) and specify $C_D = 0.0025$. Specifically, the final algorithm is to specify the larger value of the two values given by (19b) and 0.0025. The parameter z_o is the local bottom roughness. A value of 1 cm is used as suggested by Grant and Madsen [1979]. This value rather effectively parameterizes the dissipation produced by short period swells and tides when they are not explicitly included in the model forcing. A somewhat smaller value of z_o (~ 0.2 cm) should be used when the swells and/or tides are included.

Lateral boundary conditions for the curvilinear coordinate system model are formulated in a similar manner to those of a rectangular coordinate system model. On closed, land boundaries the normal component of velocity is set to zero. The normal gradients of temperature and salinity are also zero so that there are no advective and diffusive heat and salt fluxes across these boundaries. Open lateral boundary conditions must be prescribed with special care since these conditions represent a parameterization of the environment external to the domain under consideration. Considerable attention is currently being devoted to the development of robust techniques for specifying this parameterization. In most of the model simulations to follow in Section 3, the simplest possible formulation for the open boundary condition is used. That is, the sea surface elevation is specified as a function of time. There is one simulation, however, which uses a radiation condition to pass transients through the boundary and this condition is described in Section 3.4.

The "σ" coordinate system proposed by Phillips [1957] is used in the model formulation to overcome the computational problems which arise in the vicinity of large bathymetric irregularities. Following Blumberg and Mellor [1983], a new set of independent variables that transforms both the surface and bottom into coordinate surfaces is introduced. The number of vertical grid points in the transformed system is thus the same for the shallow coastal regions as for the deeper oceanic regions. The governing equations are transformed from (ξ_1, ξ_2, z, t) to (ξ_1^*, ξ_2^*, σ, t^*) coordinates where

$$\xi_1^* = \xi_1 , \quad \xi_2^* = \xi_2 , \quad \sigma = \frac{z-\eta}{H+\eta} , \quad t^* = t . \tag{20}$$

In this system σ ranges from $\sigma = 0$ at $z = \eta$ to $\sigma = -1$ at $z = -H$. A new vertical velocity can now be defined as

$$\omega \equiv w - \frac{1}{h_1 h_2}\left[h_2 U_1 \left(\sigma \frac{\partial D}{\partial \xi_1^*} + \frac{\partial \eta}{\partial \xi_1^*}\right) + h_1 U_2 \left(\sigma \frac{\partial D}{\partial \xi_2^*} + \frac{\partial \eta}{\partial \xi_2^*}\right)\right] - \left(\sigma \frac{\partial D}{\partial t^*} + \frac{\partial \eta}{\partial t^*}\right) , \tag{21}$$

which transforms the boundary conditions, Equations (17c) and (18b), into

$$\omega\left(\xi_1^*, \xi_2^*, 0, t^*\right) = 0 \text{ and } \omega\left(\xi_1^*, \xi_2^*, -1, t^*\right) = 0 . \tag{22a,b}$$

The governing equations may now be written (all * will be dropped for notational convenience) as

$$h_1 h_2 \frac{\partial \eta}{\partial t} + \frac{\partial}{\partial \xi_1}\left(h_2 U_1 D\right) + \frac{\partial}{\partial \xi_2}\left(h_1 U_2 D\right) + h_1 h_2 \frac{\partial \omega}{\partial \sigma} = 0 \tag{23}$$

$$\frac{\partial}{\partial t}\left(U_1 D\right) + \frac{1}{h_1^2 h_2}\left[\frac{\partial}{\partial \xi_1}\left(U_1^2 D\, h_1 h_2\right) + \frac{\partial}{\partial \xi_2}\left(U_1 U_2 D h_1^2\right)\right] + \frac{\partial}{\partial \sigma}\left(\omega U_1\right)$$

$$- \frac{U_1^2 D}{h_1^2}\frac{\partial h_1}{\partial \xi_1} - \frac{U_2^2 D}{h_1 h_2}\frac{\partial h_2}{\partial \xi_1} - fDU_2 = - \frac{D}{h_1 \rho_o}\frac{\partial P}{\partial \xi_1} + D\Gamma_1' \quad , \qquad (24)$$

(24)

$$\frac{\partial}{\partial t}\left(U_2 D\right) + \frac{1}{h_1 h_2^2}\left[\frac{\partial}{\partial \xi_1}\left(U_1 U_2 D h_2^2\right) + \frac{\partial}{\partial \xi_2}\left(U_2^2 D h_1 h_2\right)\right] + \frac{\partial}{\partial \sigma}\left(\omega U_2\right)$$

$$- \frac{U_1^2 D}{h_1 h_2}\frac{\partial h_1}{\partial \xi_2} - \frac{U_2^2 D}{h_2^2}\frac{\partial h_2}{\partial \xi_2} + fDU_1 = - \frac{D}{h_2 \rho_o}\frac{\partial P}{\partial \xi_2} + D\Gamma_2' \quad , \qquad (25)$$

$$\frac{\partial}{\partial t}(\theta D) + \frac{1}{h_1 h_2}\left[\frac{\partial}{\partial \xi_1}(h_2 U_1 \theta D) + \frac{\partial}{\partial \xi_2}(h_1 U_2 \theta D)\right] + \frac{\partial}{\partial \sigma}(\omega\theta)$$

$$= \frac{1}{h_1 h_2}\left[\frac{\partial}{\partial \xi_1}\left(\frac{h_2}{h_1} A_H D \frac{\partial \theta}{\partial \xi_1}\right) + \frac{\partial}{\partial \xi_2}\left(\frac{h_1}{h_2} A_H D \frac{\partial \theta}{\partial \xi_2}\right)\right] +$$

$$+ \frac{1}{D}\frac{\partial}{\partial \sigma}\left(K_H \frac{\partial \theta}{\partial \sigma}\right) \qquad (26)$$

and

$$\frac{\partial}{\partial t}(SD) + \frac{1}{h_1 h_2}\left[\frac{\partial}{\partial \xi_1}(h_2 U_1 SD) + \frac{\partial}{\partial \xi_2}(h_1 U_2 SD)\right] + \frac{\partial}{\partial \sigma}(\omega S)$$

$$= \frac{1}{h_1 h_2}\left[\frac{\partial}{\partial \xi_1}\left(\frac{h_2}{h_1} A_H D \frac{\partial S}{\partial \xi_1}\right) + \frac{\partial}{\partial \xi_2}\left(\frac{h_1}{h_2} A_H D \frac{\partial S}{\partial \xi_2}\right)\right]$$

$$+ \frac{1}{D}\frac{\partial}{\partial \sigma}\left(K_H \frac{\partial S}{\partial \sigma}\right) \quad . \qquad (27)$$

The pressure gradient terms in (24) and (25) are given, in σ coordinates, by

$$\nabla P = \rho_o g \nabla \eta + gD\nabla \int_\sigma^0 \rho\, d\sigma - g\nabla D \int_\sigma^0 \sigma \frac{\partial \rho}{\partial \sigma}\, d\sigma \quad , \qquad (28)$$

where ∇ is the gradient operator in orthogonal coordinates. The horizontal and vertical diffusion terms for momentum are defined according to:

$$D\Gamma_1' = \frac{1}{h_1 h_2}\left[\frac{\partial}{\partial \xi_1}\left(h_2 D\tau_{11}\right) + \frac{\partial}{\partial \xi_2}\left(h_1 D\tau_{21}\right)\right] + \frac{1}{D}\frac{\partial}{\partial \sigma}\left(K_M \frac{\partial U_1}{\partial \sigma}\right)$$

$$+ \frac{D}{h_1 h_2}\left[\tau_{21}\frac{\partial h_1}{\partial \xi_2} - \tau_{22}\frac{\partial h_2}{\partial \xi_1}\right] \tag{29}$$

and

$$D\Gamma_2' = \frac{1}{h_1 h_2}\left[\frac{\partial}{\partial \xi_1}(h_2 D\tau_{12}) + \frac{\partial}{\partial \xi_2}(h_1 D\tau_{22})\right] + \frac{1}{D}\frac{\partial}{\partial \sigma}\left(K_M \frac{\partial U_2}{\partial \sigma}\right)$$

$$+ \frac{D}{h_1 h_2}\left[\tau_{12}\frac{\partial h_2}{\partial \xi_1} - \tau_{11}\frac{\partial h_1}{\partial \xi_2}\right] \;. \tag{30}$$

The transformed shear stress components retain their same form as in z-coordinates, Equations (10), (11) and (12), if the ideas developed by Mellor and Blumberg [1985] are used. These ideas permit the relatively simple mathematical forms for the shear stress components and the horizontal diffusion terms in Equations (26), (27), (29) and (30), yet produce an accurate computation of bottom boundary layer dynamics even on sloping bottoms.

2.2 Turbulence Closure Model

The governing equations contain parameterized Reynolds stress and flux terms which account for the turbulent diffusion of momentum, heat and salt by small-scale processes not directly included in the model. The parameterization of turbulence in the circulation model is based on the work of Mellor and Yamada [1974].

The vertical mixing coefficients, K_M and K_H, in Equations (8), (9), (13) and (14) are obtained by appealing to a second order turbulence closure scheme which characterizes the turbulence by equations for the turbulence kinetic energy, $q^2/2$, and a turbulence macroscale, ℓ , according to, (in σ coordinates),

$$\frac{\partial(q^2 D)}{\partial t} + \frac{1}{h_1 h_2}\left[\frac{\partial}{\partial \xi_1}\left(h_2 U_1 D q^2\right) + \frac{\partial}{\partial \xi_2}\left(h_1 U_2 D q^2\right)\right] + \frac{\partial}{\partial \sigma}(\omega q^2)$$

$$= \frac{2K_M}{D}\left[\left(\frac{\partial U_1}{\partial \sigma}\right)^2 + \left(\frac{\partial U_2}{\partial \sigma}\right)^2\right] + \frac{2g}{\rho_o} K_H \frac{\partial \rho}{\partial \sigma} - \frac{2q^3 D}{\Lambda_1} +$$

$$\frac{1}{h_1 h_2}\left[\frac{\partial}{\partial \xi_1}\left(\frac{h_2}{h_1} A_H D \frac{\partial q^2}{\partial \xi_1}\right) + \frac{\partial}{\partial \xi_2}\left(\frac{h_1}{h_2} A_H D \frac{\partial q^2}{\partial \xi_2}\right)\right] + \frac{1}{D}\frac{\partial}{\partial \sigma}\left(K_q \frac{\partial q^2}{\partial \sigma}\right) \tag{31}$$

and

$$\frac{\partial(q^2\ell D)}{\partial t} + \frac{1}{h_1h_2}\left[\frac{\partial}{\partial\xi_1}\left(h_2U_1Dq^2\ell\right) + \frac{\partial}{\partial\xi_2}\left(h_1U_2Dq^2\ell\right)\right] + \frac{\partial}{\partial\sigma}(\omega q^2\ell)$$

$$= \frac{\ell E_1 K_M}{D}\left[\left(\frac{\partial U_1}{\partial\sigma}\right)^2 + \left(\frac{\partial U_2}{\partial\sigma}\right)^2\right] + \frac{\ell E_1 g}{\rho_o} K_H \frac{\partial\rho}{\partial\sigma} - \frac{q^3 D}{B_1}\tilde{W} +$$

$$\frac{1}{h_1h_2}\left[\frac{\partial}{\partial\xi_1}\left(\frac{h_2}{h_1} A_H\ D \frac{\partial(q^2\ell)}{\partial\xi_1}\right) + \frac{\partial}{\partial\xi_2}\left(\frac{h_1}{h_2} A_H\ D\frac{\partial(q^2\ell)}{\partial\xi_2}\right)\right] +$$

$$\frac{1}{D}\frac{\partial}{\partial\sigma}\left[K_q \frac{\partial}{\partial\sigma}(q^2\ell)\right] \tag{32}$$

where a wall proximity function defined as

$$\tilde{W} = 1 + E_2\left(\frac{\ell}{\kappa L}\right)^2 \quad , \tag{33}$$

with

$$(L)^{-1} = (\eta - z)^{-1} + (H + z)^{-1} \quad , \tag{34}$$

has been introduced.

With the use of the closure assumptions it is possible to reduce the prescription of the mixing coefficients K_M, K_H and K_q to the following expressions,

$$K_M \equiv \ell q S_M, \quad K_H \equiv \ell q S_H \quad \text{and} \quad K_q \equiv \ell q S_q . \tag{35a,b,c}$$

The stability functions, S_M, S_H and S_q are analytically derived, algebraic relations functionally dependent upon $\partial U_1/\partial\sigma$, $\partial U_2/\partial\sigma$, $g\rho_o^{-1}\partial\rho/\partial\sigma$, q, and ℓ . These relations follow from a closure hypothesis first described by Mellor [1973] and recently summarized by Mellor and Yamada [1982].

It is convenient to define

$$G_M \equiv \frac{\ell^2}{q^2 D}\left[\left(\frac{\partial U_1}{\partial\sigma}\right)^2 + \left(\frac{\partial U_2}{\partial\sigma}\right)^2\right]^{1/2} \tag{36a}$$

and

$$G_H \equiv \frac{\ell^2}{q^2 D}\frac{g}{\rho_o}\frac{\partial\rho}{\partial z} \quad . \tag{36b}$$

Then the stability functions become

$$S_M\left[6A_1A_2G_M\right] + S_H\left[1 - 2A_2B_2G_H - 12A_1A_2G_H\right] = A_2 \tag{37}$$

and

$$S_M\left[1 + 6A_1^2G_M - 9A_1A_2G_H\right] - S_H\left[12A_1^2G_H + 9A_1A_2G_H\right] = A_1(1 - 3C_1) \ , \tag{38}$$

which are readily solved for S_M and S_H as functions of G_M and G_H, and

$$S_q = 0.20 \quad . \tag{39}$$

A necessary closure assumption is that all lengths are proportional to each other, thus,

$$(\ell_1, \ell_2, \Lambda_1, \Lambda_2) = (A_1, A_2, B_1, B_2)\ell \tag{40}$$

By appealing to laboratory data [Mellor and Yamada, 1982] the empirical constants were assigned the values:

$$(A_1, A_2, B_1, B_2, C_1) = (0.92,\ 0.74,\ 16.6,\ 10.1,\ 0.08) \tag{41a}$$

and

$$(E_1,\ E_2) = (1.8,\ 1.33) \quad . \tag{41b}$$

The surface and bottom boundary conditions on q^2 and ℓ are

$$q^2 = B_1^{2/3}\ u_{\tau s}^2 \quad , \tag{42a}$$

$$q^2\ell = 0 \tag{42b}$$

and

$$q^2 = B_1^{2/3}\ u_{\tau b}^2 \quad , \tag{42c}$$

where $u_{\tau s}$ and $u_{\tau b}$ are the friction velocities associated with the surface wind and bottom frictional stresses, respectively.

2.3 Orthogonal Curvilinear Grid Generation

The specification of an orthogonal curvilinear grid is considerably more involved than that of a Cartesian grid. Both the shape of the domain and the relative spacing of the grid lines throughout the domain must be specified in the case of a curvilinear grid. In a Cartesian grid, on the other hand, all grid lines are straight and their spacing remains uniform from end to end. The additional freedom inherent in a curvilinear grid makes it possible to adapt the grid to the features of the flow so that many grid lines fall in regions of rapid change in flow properties.

There are also constraints which prevent an orthogonal grid from being completely arbitrary, as are finite element and boundary fitted grids, for instance. An orthogonal grid consists of families of lines intersecting orthogonally. Each grid line of one family must cross each grid line of the other family once, and only once; entering on one boundary and exiting on the opposite boundary. Also, large local curvature in one family of grid lines must be avoided, since it causes a focusing of the other, orthogonal, family of grid lines to form what is referred to as a singularity in the grid.

Many continental shelf regions, an example of which will be described in Section 3, exhibit much more rapid changes in flow properties and bathymetry on transects normal to the coast than on alongshore transects. In order to model this behavior efficiently, it is advantageous to use a calculation grid with a band of closely spaced grid lines parallel to the coast. This fine mesh will require the smallest number of lines if this grid domain is bent so that the lines essentially follow the coast.

Forcing the grid to match the exact shape of the coast is impractical, however. The actual coastline may be rough and may include bays and promontories which are not reproducible on the grid scale chosen. In any case, the coastal profile often does not reflect the shape of the topography below the surface that is more important in determining the circulation. Therefore, the inner boundary of the grid is best chosen as some mean coastline profile.

The outer boundary may be chosen approximately the required distance offshore, but may be smoother than the shore boundary. The grid spacing in the offshore direction is determined by erecting a set of curvilinear lines between the inner and outer boundaries which are spaced in the required proportion. If these lines are described as lines of constant ξ_1, they may be expressed in the form

$$f(\lambda, \varphi; \xi_1) = 0 , \tag{43}$$

where λ and φ are the longitude and latitude, respectively.

Another family of grid lines for constant ξ_2, may be erected normal to the lines of constant ξ_1. The condition of orthogonality is provided by the Cauchy-Riemann equations, inverted to relate λ and φ as functions of the grid coordinates, ξ_1 and ξ_2,

$$\frac{1}{h_1}\frac{\partial\varphi}{\partial\xi_1} = \frac{1}{h_2}\frac{\partial\lambda}{\partial\xi_2}\cos\varphi \tag{44}$$

and

$$\frac{1}{h_2}\frac{\partial\varphi}{\partial\xi_2} = -\frac{1}{h_1}\frac{\partial\lambda}{\partial\xi_1}\cos\varphi \ . \tag{45}$$

Eliminating the metrics between Equations (44) and (45) yields a relation between the angles of the ξ_1 and ξ_2 coordinate lines,

$$\frac{\partial\varphi}{\partial\xi_1}\frac{\partial\varphi}{\partial\xi_2} + \frac{\partial\lambda}{\partial\xi_1}\frac{\partial\lambda}{\partial\xi_2}\cos^2\varphi = 0 \ . \tag{46}$$

Equation (46) is both necessary and sufficient to insure orthogonality. Equations (43) and (46) may be solved for the location of lines of constant ξ_2 in physical space. Beginning with the desired grid spacing at one boundary the solution proceeds across the domain to the opposite boundary. The results, written in the form

$$\lambda = f_\lambda(\xi_1 \ , \ \xi_2) \tag{47}$$

and

$$\varphi = f_\varphi(\xi_1 \ , \ \xi_2) \quad , \tag{48}$$

represent a smooth, one to one, mapping of the computational grid onto the physical surface of the earth expressed in longitude and latitude.

Although the physical location of the grid is required for interpretation and plotting of results, only the metrics of the computational grid, h_1 and h_2, are actually required for the solution of the equations of motion. From Equations (44) and (45) it can be shown that the metrics are given by the relations

$$h_1^2 = R^2\left\{\left(\frac{\partial\varphi}{\partial\xi_1}\right)^2 + \left(\frac{\partial\lambda}{\partial\xi_1}\right)^2\cos^2\varphi\right\} \tag{49}$$

and

$$h_2^2 = R^2\left\{\left(\frac{\partial\varphi}{\partial\xi_2}\right)^2 + \left(\frac{\partial\lambda}{\partial\xi_2}\right)^2\cos^2\varphi\right\} \tag{50}$$

where R is the radius of the earth.

The construction, testing and selection of a satisfactory grid is an iterative process. It begins with the specification of several basic parameters; the region to be covered by the physical grid and the matrix size of the computational grid. All of the computational grids used for the model simulations to follow have been generated with the procedures described above.

When evaluating the grid distribution it is important to consider the role of the grid in the solution of the equations of motion. In a finite difference solution the variables at a grid point represent some appropriately defined average over that grid box. If the properties of the physical flow are relatively uniform, or at least monotonically varying over a grid box, then the numerical solution can closely represent the actual circulation. Alternatively, when the physical values vary widely within a grid box, the effects of this spatial variability are not explicitly included in the numerical solution and must be introduced using a subgrid scale parametrization of missing physics.

There is yet another consideration in choosing a grid besides its adequacy in representing the solution. The impact of grid spacing on the time step of the calculation must also be assessed. The approximate maximum allowable time step (see Section 2.4) applies for the entire grid. Therefore the smallest value anywhere in the field is controlling.

2.4 Numerical Techniques

The equations which form the circulation model together with their boundary conditions are solved by finite difference techniques. A staggered grid ("C" grid, see Arakawa and Lamb, [1977]) is used for horizontal differencing. The arrangement of points has U_1 at points $\pm\Delta\xi_1/2$ away from the point where H, η are defined and U_2 at points $\pm\Delta\xi_2/2$ away from the H and η points. The metric h_1 is defined at the U_2 point and h_2 is defined at the U_1 point. The vertical differencing remains identical to that used by Blumberg and Mellor [1983]. More details of the differencing techniques can be found in that article. It should be pointed out that the finite difference equations employed conserve energy, mass and momentum, and they introduced no artificial horizontal diffusion.

A mode splitting technique similar to that described by Simons [1974] is used for computational efficiency. An external mode derived by vertically integrating Equations (3), (4), and (5) can be written as,

$$\frac{\partial\eta}{\partial t} + \frac{1}{h_1 h_2}\left[\frac{\partial}{\partial\xi_1}\left(\bar{U}_1 D h_2\right) + \frac{\partial}{\partial\xi_2}\left(\bar{U}_2 D h_1\right)\right] = 0 \quad , \tag{51}$$

$$\frac{\partial \bar{U}_1 D}{\partial t} + \frac{gD}{h_1}\frac{\partial\eta}{\partial\xi_1} - f\bar{U}_2 D = \Lambda_1 \tag{52}$$

and

$$\frac{\partial \bar{U}_2 D}{\partial t} + \frac{gD}{h_2}\frac{\partial\eta}{\partial\xi_2} + f\bar{U}_1 D = \Lambda_2 \quad , \tag{53}$$

where the pressure has been eliminated using Equations (7) and the vertically integrated velocities are defined as

$$\left(\bar{U}_1, \bar{U}_2\right) \equiv \frac{1}{H + \eta} \int_{-H}^{\eta} \left(U_1, U_2\right) dz \ . \tag{54}$$

The terms Λ_1 and Λ_2 contain vertical integrals of the density gradient, advective and diffusive terms. The computed bottom friction and prescribed wind stresses are also incorporated into the Λ's.

The computational strategy is to solve equations for the external mode, that is, the shallow water wave equations (51), (52) and (53), with a short time step to resolve high frequency motions. The terms on the right hand side of Equations (52) and (53) are supplied from the internal mode and are held constant in time over the external mode integration period. The external mode provides $\partial\eta/\partial\xi_1$ and $\partial\eta/\partial\xi_2$ for insertion into the internal mode equations which are then solved with a much longer time step, the ratio between the external and internal wave speeds. Once the vertical structure has been determined, the terms on the right hand side of Equations (52) and (53) are updated and another external mode solution begins.

The time differencing scheme consists of centered differences (leap frog). The numerical scheme is quasi-implicit in that vertical diffusion is evaluated at the forward time level. This makes possible small vertical spacing near the surface and bottom without the need to reduce the time increment or restrict the magnitude of the mixing coefficients.

From a viewpoint of computational stability, the Courant-Friedrichs-Levy condition on the vertically integrated, external mode, transport equations governs the size of the time step. An extensive Fourier stability analysis is not possible with the present model equations because of the spatially varying metrics. The critical time step thus can no longer be given in explicit algebraic form even for a linearized subset. However, by analogy to the analysis techniques used with linear equations in a Cartesian coordinate, a rule of thumb for the critical time step is found to be

$$\Delta t_c < \frac{1}{c_e} \left[\left(h_1^{-2} + h_2^{-2}\right) \right]^{-1/2} \tag{55}$$

which holds locally. Here $c_e = 2(gH)^{\frac{1}{2}} + \bar{U}$ where $\bar{U}$ is the fastest vertically averaged velocity. The entire domain must now be examined and the smallest Δt_c used. The internal mode is restricted by a similar condition except that c_e, is replaced by c_i, a value twice the fastest internal wave speed plus the largest current velocity. For most coastal and open ocean situations, c_e/c_i is from 80 to 100, while in estuaries the ratio is somewhat smaller, perhaps 10.

2.5 Vertically Integrated Model

The external mode equations can be cast into a stand alone model by explicitly integrating the terms which appear on the right hand side of Equations (52) and (53). It will be apparent from examination of the stand-alone, vertically integrated model and the external mode equations that they are quite similar. By demonstrating the validity of the curvilinear coordinate approach with the vertically integrated model, the performance of an important component of the three-dimensional model is also being proven at computing costs far below those of the full three-dimensional model. For completeness these new, integrated, equations are presented here,

$$\frac{\partial(\bar{U}_1 D)}{\partial t} + \frac{1}{h_1^2 h_2}\left[\frac{\partial}{\partial \xi_1}\left(\bar{U}_1^2 D h_1 h_2\right) + \frac{\partial}{\partial \xi_2}\left(\bar{U}_1 \bar{U}_2 D h_1^2\right)\right]$$

$$- \frac{1}{h_1^2}\frac{\partial h_1}{\partial \xi_1}\bar{U}_1^2 D - \frac{1}{h_1 h_2}\frac{\partial h_2}{\partial \xi_1}\bar{U}_2^2 D + \frac{gD}{h_1}\frac{\partial \eta}{\partial \xi_1} - fU_2 D - D\bar{\tau}_1 = \Phi_1 \tag{63}$$

and

$$\frac{\partial(\bar{U}_2 D)}{\partial t} + \frac{1}{h_1 h_2^2}\left[\frac{\partial}{\partial \xi_1}\left(\bar{U}_1 \bar{U}_2 D h_2^2\right) + \frac{\partial}{\partial \xi_2}\left(\bar{U}_2^2 D h_1 h_2\right)\right]$$

$$- \frac{1}{h_1 h_2}\frac{\partial h_1}{\partial \xi_2}\bar{U}_1^2 D - \frac{1}{h_2^2}\frac{\partial h_2}{\partial \xi_2}\bar{U}_2^2 D$$

$$+ \frac{gD}{h_2}\frac{\partial \eta}{\partial \xi_2} + f\bar{U}_1 D - D\bar{\tau}_2 = \Phi_2 \quad . \tag{64}$$

The terms Φ_1 and Φ_2 contain the bottom friction (parameterized with a quadratic drag law) and the specified wind stress,

$$\Phi_1 = C_D\left(\bar{U}_1^2 + \bar{U}_2^2\right)^{\frac{1}{2}}\bar{U}_1 + \frac{1}{\rho_o}\tau_1^w \tag{65}$$

and

$$\Phi_2 = C_D\left(\bar{U}_1^2 + \bar{U}_2^2\right)^{\frac{1}{2}}\bar{U}_2 + \frac{1}{\rho_o}\tau_2^w \quad , \tag{66}$$

where C_D is a drag coefficient usually taken to be 0.0025 and τ_1^w and τ_2^w are the wind stress components in the ξ_1 and ξ_2 directions at the surface. The quantities $\bar{\tau}^1$ and $\bar{\tau}^2$ are vertical integrals of the horizontal momentum diffusion and are written as

$$D\bar{\tau}_1 = \frac{1}{h_1 h_2}\frac{\partial}{\partial \xi_1}\left[\frac{2A_M h_2}{h_1}\frac{\partial}{\partial \xi_1}\left(\bar{U}_1 D\right) + \frac{2A_M}{h_1}\frac{\partial h_1}{\partial \xi_2}\left(\bar{U}_2 D\right)\right]$$

$$+ \frac{1}{h_1 h_2}\frac{\partial}{\partial \xi_2}\left[\frac{A_M h_1^2}{h_2}\frac{\partial}{\partial \xi_2}\left(\frac{\bar{U}_1 D}{h_1}\right) + A_M h_2 \frac{\partial}{\partial \xi_1}\left(\frac{\bar{U}_2 D}{h_2}\right)\right]$$

$$+ \frac{A_M}{h_1 h_2}\frac{\partial h_1}{\partial \xi_2}\left[\frac{h_1}{h_2}\frac{\partial}{\partial \xi_2}\left(\frac{\bar{U}_1 D}{h_1}\right) + \frac{h_2}{h_1}\frac{\partial}{\partial \xi_1}\left(\frac{\bar{U}_2 D}{h_2}\right)\right]$$

$$- 2A_M\left[\bar{U}_1 D\left(\frac{1}{h_1 h_2}\frac{\partial h_2}{\partial \xi_1}\right)^2 + \frac{1}{h_1 h_2^2}\frac{\partial h_2}{\partial \xi_1}\frac{\partial}{\partial \xi_2}\left(\bar{U}_2 D\right)\right] \tag{67}$$

and

$$D\bar{\tau}_2 = \frac{1}{h_1 h_2}\frac{\partial}{\partial \xi_1}\left[A_M h_1 \frac{\partial}{\partial \xi_2}\left(\frac{\bar{U}_1 D}{h_1}\right) + A_M \frac{h_2^2}{h_1}\frac{\partial}{\partial \xi_1}\left(\frac{\bar{U}_2 D}{h_2}\right)\right]$$

$$+ \frac{1}{h_1 h_2}\frac{\partial}{\partial \xi_2}\left[\frac{2A_M}{h_2}\frac{\partial h_2}{\partial \xi_1}\left(\bar{U}_1 D\right) + \frac{2h_1}{h_2}A_M\frac{\partial \bar{U}_2 D}{\partial \xi_2}\right]$$

$$+ \frac{1}{h_1 h_2}\frac{\partial h_2}{\partial \xi_1}A_M\left[\frac{h_1}{h_2}\frac{\partial}{\partial \xi_2}\left(\frac{\bar{U}_1 D}{h_1}\right) + \frac{h_2}{h_1}\frac{\partial}{\partial \xi_1}\left(\frac{\bar{U}_2 D}{h_2}\right)\right]$$

$$- 2A_M\left[\frac{1}{h_1^2 h_2}\frac{\partial h_1}{\partial \xi_2}\frac{\partial}{\partial \xi_1}\left(\bar{U}_1 D\right) + \left(\frac{1}{h_1 h_2}\frac{\partial h_1}{\partial \xi_2}\right)^2 \bar{U}_2 D\right] \quad . \tag{68}$$

3 MODEL APPLICATIONS

To establish confidence in the curvilinear coordinate model physics and computer coding, a hierarchy of numerical experiments has been conducted. The numerical experiments have been designed to convincingly test the model at the same time minimizing computational costs. The initial numerical experiments involve the external mode portion as a stand alone model. Other simulations involve the full three-dimensional model used to investigate specific oceanic processes. The test cases which follow are not meant to be exhaustive but rather to illustrate the versatility and viability of the curvilinear coordinate system approach. The examples are cases in which analytic and other numerical solutions are available for comparison and assessment of model performance.

3.1 Wedge Shaped Domain

Lynch and Gray [1978] have derived analytic solutions for a variety of linearized, shallow water equations problems. Consider the physical setting and grid configuration shown in Figure 2 which involves the propagation of waves into a wedge shaped domain. At the walls, $r = r_0$, $\theta = 0$ and $\theta = \pi/2$, no flow is

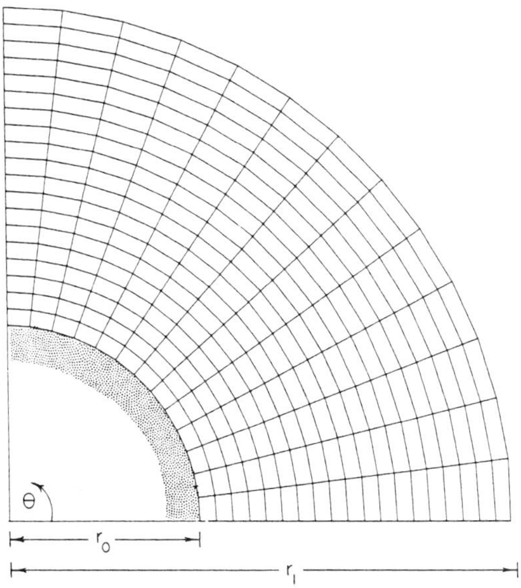

Fig. 2. The grid configuration used in the wedge shaped domain experiments.

allowed and at $r = r_1$, the open boundary, the surface elevation is prescribed to vary sinusoidally in time with uniform amplitude and phase. The Coriolis as well as non-linear and horizontal diffusion terms are neglected in this application. A linear bottom friction formulation is used in these cases instead of Equations (65) and (66) with a frictional coefficient of 10^{-4} s^{-1}. The metrics for this geometry, $h_1 = 1$ and $h_2 = r$, have been numerically computed and solutions obtained. The computed elevations and velocities together with their analytical counterparts are shown in Figures 3 for both a flat bathymetry and one that slopes in a quadratic fashion. The agreement is excellent. In addition, the computed velocities and elevations are perfectly radial at every grid point as are the analytic solutions. Other cases involving these geometries with wind produce similar, rather good, comparisons.

3.2 Island Circulation

Now, consider the circulation around a circular island with a narrow shelf investigated numerically by Wang [1982]. The grid configuration is shown in Figure 4. Two cases are considered; the steady response to a uniform wind and

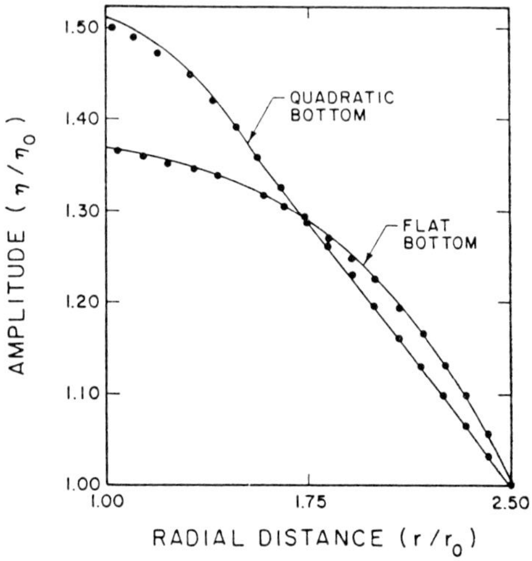

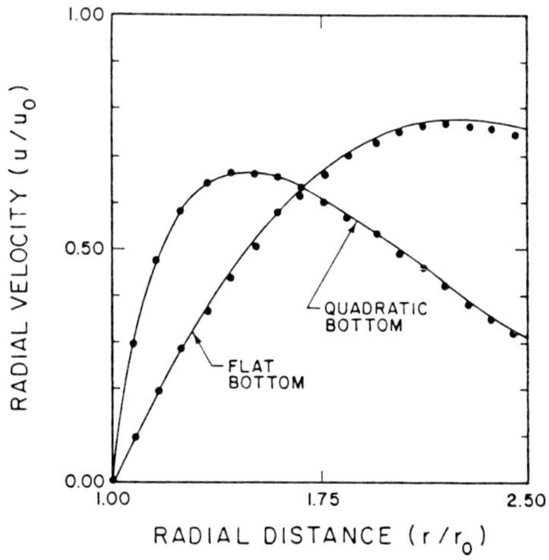

Fig. 3. A comparison of the model's computed (•) surface elevation (left) and circulation (right) with the analytical solution (——) for flat and quadratic sloping bottoms in a wedged shaped domain.

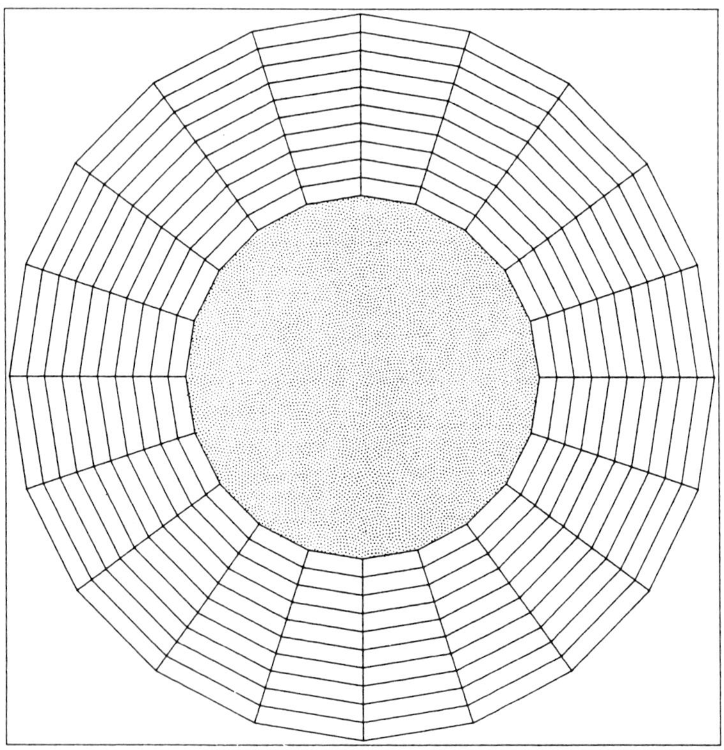

Fig. 4. The grid configuration used in the circular island domain experiments.

the generation and propagation of quasi-geostrophic, shelf waves. The island radius is 50km, the shelf width is 16km and the bottom depth is linear (40m depth at the island to 200m at the shelf) for the steady state case and flat (100m) for the wave one. The Coriolis parameter is 10^{-4} s^{-1}, the non-linear terms are included and the friction provided by Equations (65) and (66) are used. The boundary conditions on the shelf break assume the shelf motion to be completely trapped in the shelf region, that is, $\eta = 0$. In additinn, the non-linear terms in the momentum equation are also neglected at the shelf break. Figure 5 illustrates the steady response to an eastward 2 dyne cm^{-2} wind stress. The surface elevation with its antisymmetric patterns, maximum

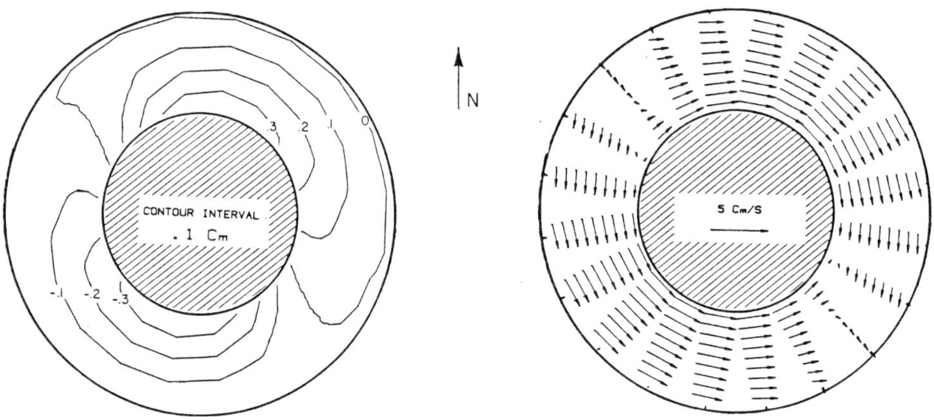

Fig. 5. The steady-state response of an island with a sloping continental shelf to a uniform eastward wind: the surface elevation (left) and the vertically averaged current (right). The island radius is 50 km and the shelf width is 16 km.

amplitudes at the northern and southern tips of the island and the velocity field agree well with Wang's numerical results.

The island configuration with a flat bottom permits quasi-geostrophic waves which propagate clockwise around the island with a phase speed of 90km day^{-1}. Figure 6 illustrates the surface elevation and velocity fields driven by a westward wind stress of 1 dyne cm^{-2} at 20 hr intervals. The computed phase and direction are correct and in agreement with Wang's results, if note is taken that there is a discrepancy between what parameters Wang used and what were reported in his article.

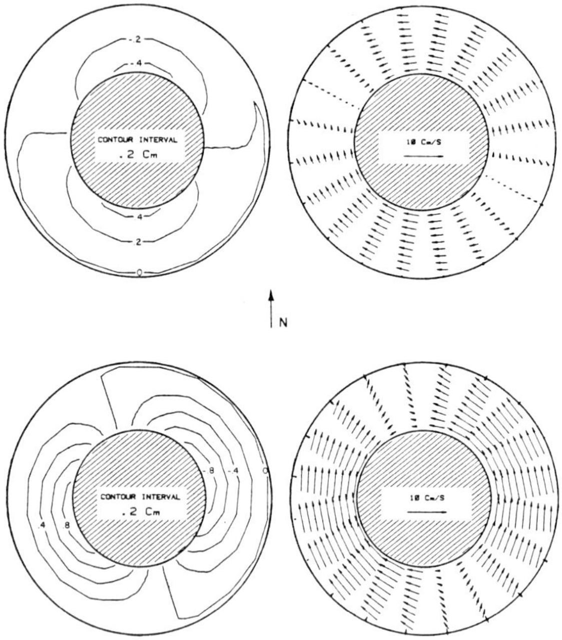

Fig. 6. The surface elevation (left) and vertically averaged circulation (right) for an island area responding to a westward wind. The top distributions are at 10 hours after the onset of the wind and the bottom ones are 20 hours later. The island radius is 50 km and the shelf width is 16 km.

3.3 Enclosed Basin

The next example is concerned with testing the curvilinear coordinate model in a situation where the horizontal grid spacings change radically. Consider the "Bow Tie" shaped closed basin illustrated in Figure 7. The orthogonality of the grid was insured by selecting the metrics based upon an elliptic cylindrical coordinate system. For this setting the North-South axis is ~100km and the width at the most narrow portion is ~17km. The basin is 10m deep and situated at a latitude of 45°N ($f = 10^{-4}\ s^{-1}$). The spatial resolution varies substantially, from 1.2 km to 17 km. The steady-state response to a uniform wind is examined. Here the simple wind setup on a complicated grid is examined. The solutions should not depend on the grid configuration but rather on the physical forces driving the system. To achieve a fast spin-up of the basin, the non-linear bottom friction coefficient is increased by a factor of 100. The model starts from rest and is driven in the first experiment by an eastward wind stress of 1 dyne cm^{-2} and in the second experiment by a wind of similar magnitude but directed to the north.

At steady state, which occurs after ~60 hours, the velocities should be zero and the elevation distribution perfectly symmetric. The results of the two

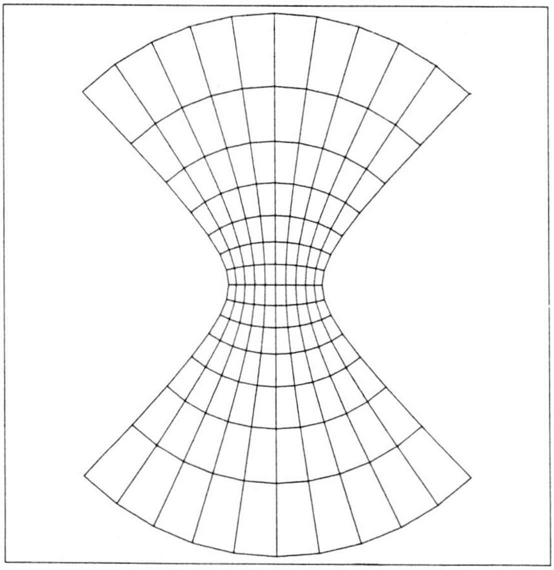

Fig. 7. The geometry and grid configuration in the "Bow Tie" experiments. The basin has a north south extent of ~ 100 km and at the narrowest portion, the width is ~ 17 km.

experiments are shown in Figure 8. The amount of setup and the direction agree almost perfectly with what one obtains analytically. Also, the velocities have vanished as expected.

3.4 Charleston Bump

As a last example of the two-dimensional cases, the vertically integrated model has been run in a reduced gravity mode to simulate the internal dynamics of the flow as it passes around a topographic feature.

The two-layer, reduced gravity model consists of two layers of stably stratified fluid with a fixed density contrast. The bottom layer is infinitely deep so that the barotropic mode is filtered out. The model describes the evolution of the first internal (baroclinic) mode of the water column (see [Gill 1982] for a derivation). The domain has a large bump in the western boundary coastline similar to the "Charleston bump", the ridge and bottom trough feature off Charleston, South Carolina at about 32°N. The Charleston bump was chosen here because it is a region which exhibits recurring meander and eddy activity.

The model domain and grid are shown in Figure 9. The latitudinal width of the domain is 800km and the longitudinal width is 300km with the region being centered about 32°N. The domain has been rotated by 45°, counterclockwise. The grid resolution of the domain is finest near the western coast where the off-

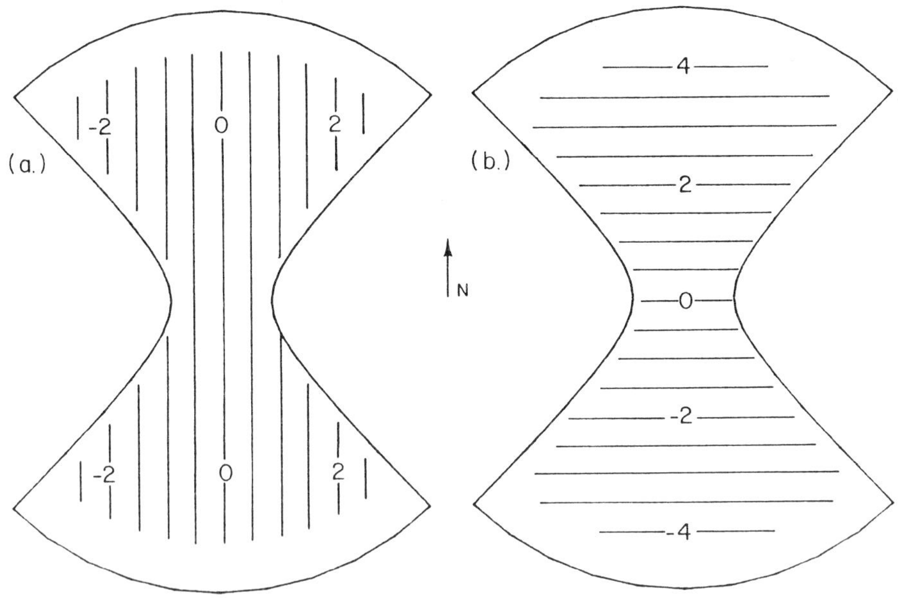

Fig. 8. The steady state elevation in response to a) an eastward 1 dyne cm^{-2} wind stress and b) a northward 1 dyne cm^{-2} wind stress. The units of the elevation are in cm.

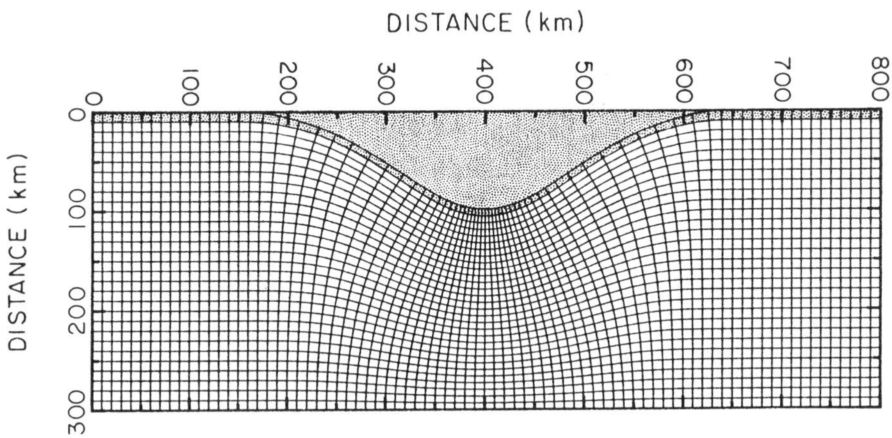

Fig. 9. The computational grid used for the idealized Charleston bump simulations.

shore spacing is ~6km and increases to ~10km away from the bump. The alongshore spacing varies from ~3km near the bump to ~15km further away. With these grid intervals, the physical processes characterized by the baroclinic radius should be well resolved and mesoscale eddy activity should be captured.

The coefficient of the horizontal diffusion is $100 m^2 s^{-1}$ in this simulation. The initial depth of the upper layer is 500m and the reduced gravity is $0.02 ms^{-2}$, so that the radius of deformation is about 40km. A depth of 500m is chosen here since it best characterizes the Charleston bump region. The model bump shown in Figure 9 has roughly the same amplitude as the perturbation found in the actual 500m isobath but is somewhat larger in latitudinal extent. This difference is of minor consequence to the results of these idealized experiments; if anything the results will be smoother than the observations since the model bump is less abrupt. The associated gravity wave speed is ~$3 ms^{-1}$, the Coriolis parameter is $f_o + \beta y$, where $f_o = 7.8 \times 10^{-5} s^{-1}$, $\beta = 1.936 \times 10^{-11} m^{-1} s^{-1}$, and y is the latitude in kilometers.

One of the major difficulties in modelling geographically limited regions, as mentioned previously, is the problem associated with the open lateral boundary conditions. Externally prescribed or gradient boundary conditions are usually employed although it can be shown that these conditions can produce misleading model results. A recently developed boundary condition by Blumberg and Kantha [1985] seems to circumvent the problems in certain physical settings. It is a modified form of the Sommerfeld radiation condition which permits transients to pass out through the boundary and at the same time allows the inflow and outflow at the boundaries to vary about some mean. Mathematically, the condition is:

$$\frac{\partial V}{\partial t} + C \frac{\partial V}{\partial n} = - \left[\frac{V - V_k}{T_f} \right] \tag{69}$$

where C is the phase speed of the gravity wave and n is the direction normal to the planar boundary. The term on the right represents damping that forces the value of V at the boundary to some known V_k with a time scale of the order to T_f. $T_f = 0$ corresponds to an externally fixed boundary where no disturbances are allowed to pass out through the boundary and $T_f \to \infty$ represents a pure radiation condition which renders a boundary transparent to waves travelling in the positive n direction with phase speed C.

A parabolic velocity profile is chosen to represent the Gulf Stream,

$$\begin{aligned} V_k &= V_o \left[1 - (x/L)^2\right], \quad 35 \text{ km} \le x \le 125 \text{ km}, \\ V_k &= 0, \qquad \text{otherwise}, \end{aligned} \tag{70}$$

where L is the half width of the Gulf Stream, ~45km, and V_o is 60cms^{-1}. The transport associated with this profile is 20 x $10^6 m^3 s^{-1}$, a realistic amount for the top 500m of the water column. The velocity profile given by (70) is used along both the northern and southern boundaries. A value of 25 days is selected for T_f; it is approximately the average advective transit time for a parcel to traverse the domain. The pure radiation condition ($T_f \to \infty$) is used along the eastern boundary. At the coast, a no normal flow condition is used. For the initial state, motion is specified in the interior of the domain according to (70). The interface is taken to have no initial displacement. During the spin-up processes, the system attains a quasi-geostrophic balance within a few inertial periods (~2 days).

A sequence of snapshots from the bump experiment of the upper layer transport and interface displacement is shown in Figure 10. Positive values for the interface displacement indicate regions where the layer thickness has increased. The snapshots taken at 20 day intervals clearly illustrate mesoscale eddies of ~120km diameter appearing spontaneously in the vicinity of the bump. In spite of the time independent specification for the boundary conditions, a time dependent field of eddies develops. By day 10 two anticyclonic eddies are present. As time progresses these eddies move to the northwest, probably the result of advection by the mean flow, and exit the domain. Figure 10b shows the formation of a cyclonic eddy just downstream of the bump and an anticyclonic eddy upstream of the bump. Both eddies also grow to a diameter of ~120km and move northwest.

In contrast, a parallel numerical experiment using all the same parameters but in a domain without a bump produced no eddies. The flow essentially reached steady state and was in geostrophic equilibrium. A simple inflow/outflow had been established. It should be noted that by increasing the flow Reynolds number ($Re \sim 2LV/A_M$) by a factor of 2, to ~450, did however produce eddies in this case.

3.5 Three-Dimensional Idealized California Shelf Model

The finite difference computational code of the curvilinear model is quite intricate. Additional complexities are introduced by the mode splitting technique. It is therefore necessary to establish the self-consistency of the computer program and to demonstrate that the governing equations are properly coded.

The first of these tests is the null case where the model (30 x 40 x 16) is not externally forced and is closed on its lateral boundaries. Any initially prescribed horizontally homogeneous (in z space) density distribution should not result in any motion for arbitrary topography. This is a rather good test to identify coding errors in a "z" coordinate model but is an even more impor-

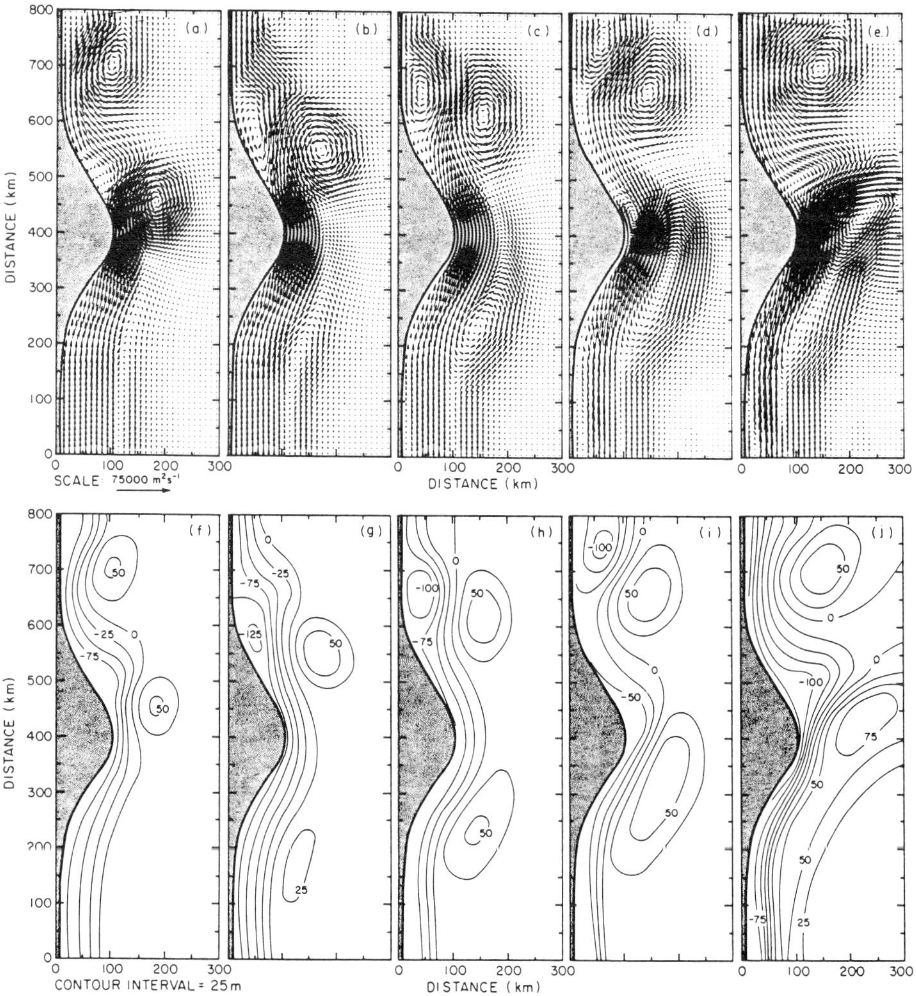

Fig. 10. A time sequence of upper layer transports and pycnocline displacements at 10 day intervals beginning at day 10.

tant test when using the σ formulation. A several hundred timestep simulation with realistic California Coast topography produced no velocity fields. Symmetry of the code was established by forcing a square basin with a northerly wind and then with the Coriolis parameter rotated, an easterly wind. The numerical simulations showed the same solutions after the second set was rotated through 90°. A host of other symmetry tests were conducted.

The use of global diagnostics also provided a means of checking the computations. The programmed finite difference equations like their differential counterparts conserve mass, volume, heat, salt and total energy. The global (over the computational domain) budgets of these quantities were monitored in each of the above mentioned experiments and in every case there was conservation to within computer round-off error.

The three-dimensional circulation model was then used to investigate the response of a region characteristic of the California Coast to wind forcing. The computational grid employed here is illustrated in Figure 11. An enlargement of the near coastal region along central California is also provided in that figure. The grid spacing varies from ~4km near the coast to ~25km near the

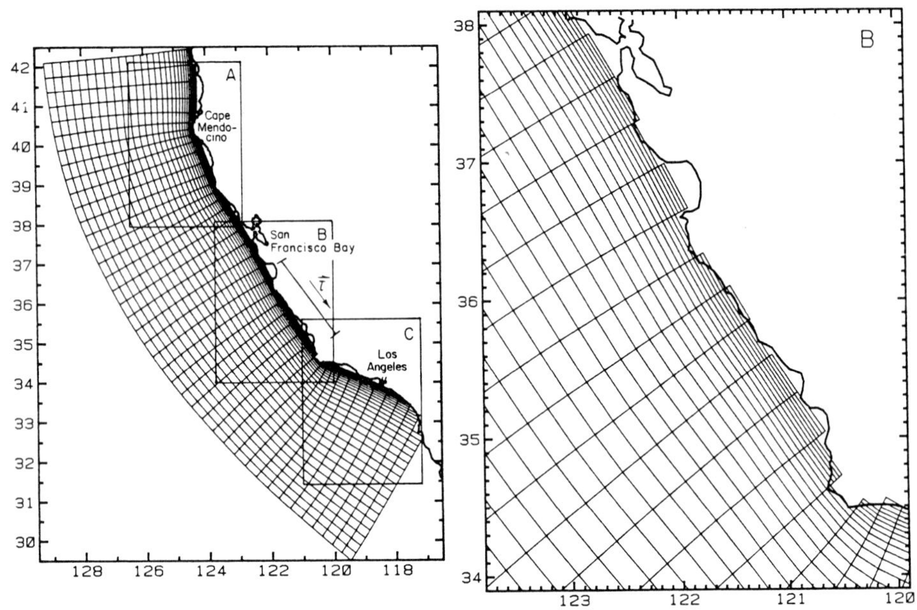

Fig. 11. The California coast grid (left) used in the three-dimensional model experiment. The forcing zone is indicated. A magnified view of the subregion B is on the right.

western portion of the domain. The alongshore resolution varies from ~16km to 50km with highest resolution occurring near the coastal promontories. The region has a continental shelf-slope with characteristics typical of central California and is

$$H(r) = \begin{cases} 5.67 \times 10^{-3}\, r + 30\text{m} & r \le r_1 \ , \\ 2.57 \times 10^{-2}(r - r_1) + 200\text{m} & r_1 \le r \le r_2 \ , \\ 2000\text{m} & r \le r_2 \ , \end{cases} \tag{71}$$

where r is the distance from the coast and r_1 = 30km, r_2 = 100km (see Figure 14 for a plot of the topography). The domain is represented by a 30 x 40 x 21 lattice of points with the vertical distribution of points having an irregular spacing as shown in Table I. The increased resolution near the surface and bottom is apparent. The horizontal mixing coefficients are chosen to be

Table I: Vertical Resolution of the California Shelf Circulation Model. Depths in meters.

k	σ	$Z_{100\ m}$	$Z_{500\ m}$	$Z_{1000\ m}$
1	0.000	.00	.0	.0
2	0.0006	.06	.3	.6
3	0.0012	.12	.6	1.2
4	0.0030	.30	1.5	3.0
5	0.0060	.60	3.0	6.0
6	0.0120	1.20	6.0	12.0
7	0.0240	2.40	12.0	24.0
8	0.0360	3.60	18.0	36.0
9	0.0600	6.00	30.0	60.0
10	0.1000	10.0	50.0	100.0
11	0.1800	18.0	90.0	180.0
12	0.2600	26.00	130.0	260.0
13	0.3400	34.00	170.0	340.0
14	0.4200	42.00	210.0	420.0
15	0.5000	50.00	250.0	500.0
16	0.6000	60.00	300.0	600.0
17	0.7000	70.00	350.0	700.0
18	0.8000	80.00	400.0	800.0
19	0.8800	88.00	440.0	880.0
20	0.9400	94.00	470.0	940.0
21	1.0000	100.00	500.0	1000.0

$100 m^2\ s^{-1}$, a value small enough not to smooth out important features. At the edges of the domain, the normal components of the velocity are set to zero. While this prescription of the velocity field is not general, it suffices here as long as only a short duration experiment is considered. The initial stratification employed is horizontally homogenous with a vertical structure as shown in Figure 12. Alongshore winds in a 300km band (see Figure 11 for the limits) are imposed with a magnitude of 1 dyne cm^{-2} . The salinity is set everywhere equal to 34 ppt and is used as a check of the computer code's conservation properties. The ocean is initially at rest.

The response of this ocean basin over an 6 day period to the sudden onset of alongshore winds in the 300km band is now examined. The evolution of the surface circulation is illustrated in Figure 13. Here the alongshore velocity on a line about 6 km from the coast is used. It is evident that a strong alongshore coastal jet develops quickly within the forcing region and spreads poleward as time progresses. By the end of the simulation the current has reached the

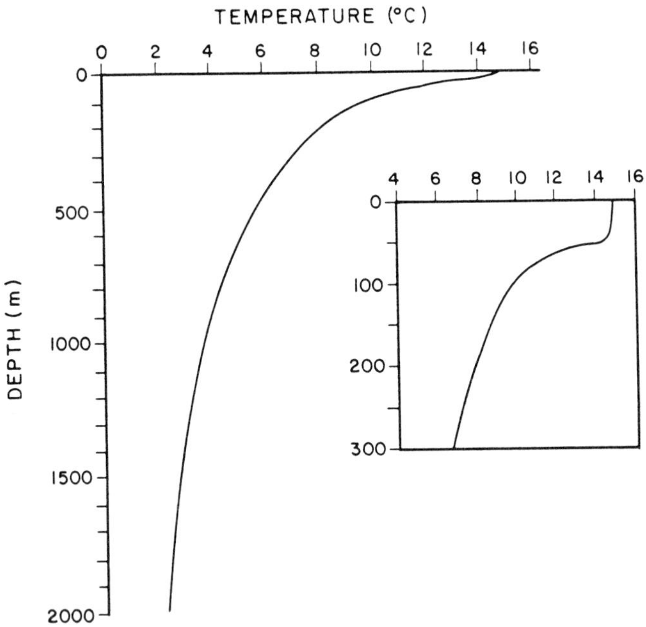

Fig. 12. The initial temperature distribution used in the prognostic model experiments. The distribution is typical of that observed off the coast of California. The insert is a detail of the upper 300 m.

northernmost boundary while not penetrating southward of the forcing zone. Thus in agreement with the recently developed ideas of coastal trapped waves [see Philander and Yoon, 1982; Suginohara, 1982 and Yoon and Philander, 1982], this model produces disturbances that take place initially at the southern edge of the forcing zone and extend poleward. As demonstrated by others and reproduced here, the equatorward velocity initially increases with time and stops increasing from the southern edge. North of the forcing zone the time evolution shows the free propagation of coastal trapped (the first mode) waves travelling with a speed of ~225km day^{-1} . The time evolution of the circulation at 200m depth also is shown in Figures 13. The development of a poleward undercurrent is evident from the alongshore velocity at this depth. The poleward undercurrent develops at a slower rate than does the coastal jet through waves travelling with phase speeds of ~75km day^{-1} . The phase speeds of the waves appearing in Figure 13 agree with to those computed using the techniques described by Brink [1982] for calculating free coastal-trapped wave modal structures and dispersion curves. Within the forcing zone the upwelling is quite intense with surface temperatures becoming ~1 1/2°C cooler by the end of day 6, as shown in Figure 14. It should be noted that Figure 14 also shows the formation of a well defined bottom boundary layer on the shelf. In addition, downwelling is present along the shelf break at depths below 300m.

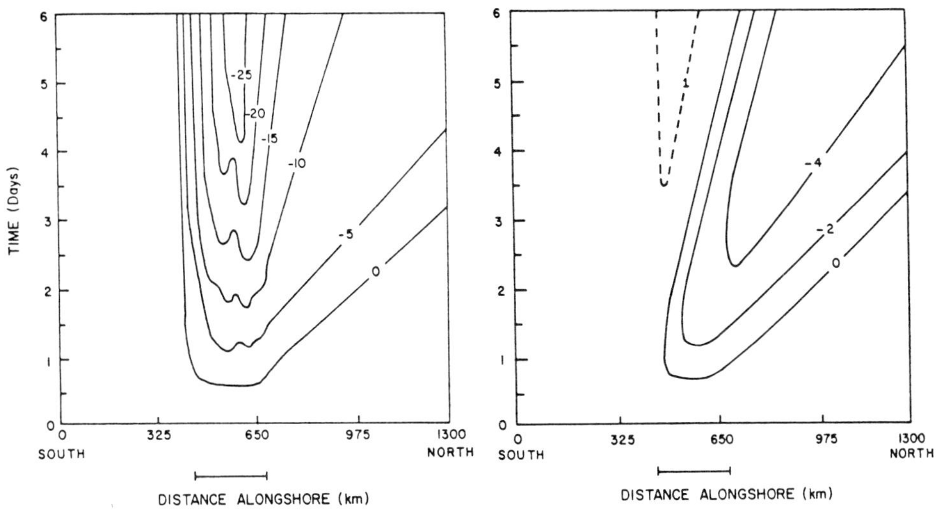

Fig. 13. The time evolution of the surface alongshore velocity (cm s^{-1}) along a line ~6 km from the coast (left) and the 200 m depth alongshore velocity (cm s^{-1}) along a line ~20 km from the coast (right).

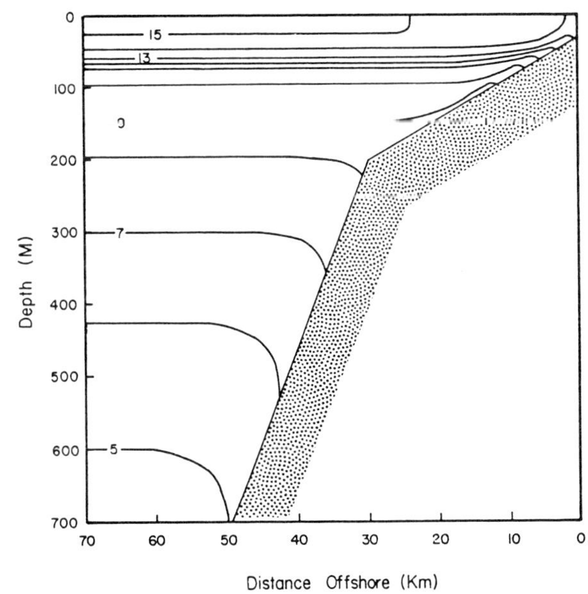

Fig. 14. The temperature distribution along a vertical section near the center of the wind forcing zone (CI = 1°C). Initially the temperature distribution is only a function of depth.

4 CONCLUSIONS

A numerical circulation model has been developed using an orthogonal curvilinear coordinate system. A series of numerical experiments have been presented for cases where analytic and numerical solutions are available for comparison. These test cases indicate that the curvilinear model is a viable option when modelling an area where details of the coastlines are important as in continental shelf and estuarine regions. No numerical problems have been encountered as far as the grid aspect ratio (values of 1:15 have been used) and design of the computational grid are concerned. Idealized experiments with a vertically integrated model, that is, the external mode component of the three-dimensional model, run in a reduced gravity mode have illustrated the dramatic impact of the Charleston bump on the internal dynamics of the Gulf Stream. The results demonstrate that when a Gulf Stream flow is forced around a topographic feature, intense eddies develop spontaneously; however, when the same flow traverses a region without such a feature no eddies are produced.

To further establish confidence in the three-dimensional model, the response of a stratified ocean with a coastline characteristic of the California Continental Shelf region to an alongshore upwelling favorable wind is examined. The resulting circulation with upwelling and the propagation of various modes of coastal trapped waves is in substantial agreement with previous studies. In a case not presented here, a year-long model simulation for the year 1981 has been conducted with this shelf model. The model was driven with synoptic wind fields derived from the National Weather Service predictions and climatological hydrography and circulation for model initialization and boundary conditions. The interested reader is referred to Blumberg et al. [1985].

In considering whether to use a curvilinear model or a Cartesian model for a particular problem, it should be noted that the curvilinear coordinate model has additional terms in the governing equations which a rectangular coordinate model does not have. The result is about a 60% increase in computer time for the same number of grid points, which tends to some extent to offset the savings from improved grid resolution. Therefore, a Cartesian coordinate model may continue to be more attractive for calculations where the computational domain has a regular shape and the desired resolution is relatively uniform throughout. The principal application of the orthogonal coordinate system model is to obtain high resolution where it is required without paying the penalty of unnecessarily high resolution in other parts of the modeled region. There appear to be no additional limitations imposed by the use of curvilinear coordinates.

Finally, it is hoped that the cases selected for examples of the application of the model in Section 3 will provide a framework for testing and comparing the results of other models as well.

5 ACKNOWLEDGEMENT

This work was completed while AFB was with Dynalysis of Princeton. The assistance of Dr Lakshmi H. Kantha in computing the analytical wave speeds, discussed in the California shelf model simulation, is gratefully acknowledged. Funding for the study was provided by the Minerals Management Service of the U. S. Department of the Interior under contract Numbers AA851-CT1-67 and 14-12-0001-29113.

6 REFERENCES

Arakawa, A. and Lamb, V.R., 1977. Computational design of the basic dynamical process of the UCLA General Circulation Model. Methods in Computational Physics, 17. Academic Press, pp. 173-265.

Blumberg, A.F. and Mellor, G.L., 1983. Diagnostic and prognostic numerical circulation studies of the South Atlantic Bight. J. Geophys. Res., 88: 4579-4592.

Blumberg, A.F. and Mellor, G.L., 1986. A description of a three-dimensional coastal ocean circulation model. In: N. Heaps (Editor), Three-Dimensional Shelf Models, Coastal and Estuarine Sciences, 5, American Geophysical Union (AGU).

Blumberg, A.F., Kantha, L.H., Herring, H.J. and Mellor, G.L., 1985. California shelf physical oceanography circulation model: Final report. Dynalysis of Princeton Report No. 88, 368 pp.

Blumberg, A.F. and Kantha, L.H., 1985. Open boundary conditions for circulation models. J. of Hydraulic Engin., 111: 237-255.

Brink, K.H., 1982. A comparison of long coastal trapped wave theory with observations off Peru. J. Phys. Oceanogr., 12: 897-913.

Eringen, A.C., 1962. Nonlinear theory of Continuous Media. McGraw-Hill Book Company, New York.

Fofonoff, N.P., 1962. Physical properties of sea-water. In: N.M. Hill, Ed., The Sea. Vol. 1, Interscience Publishers of John Wiley and Sons, New York, pp. 3-30.

Grant, W.D. and Madsen, O.S., 1979. Combined wave and current interaction with a rough bottom. J. Geophys. Res., 84: 1797-1808.

Gill, A.E., 1982. Atmosphere-Ocean Dynamics. Academic Press, New York.

Johnson, B.H., 1982. Numerical modelling of estuarine hydrodynamics on a boundary fitted coordinate system. In: J. Thompson (Editor), Numerical Grid Generation, Elsevier, pp. 409-436.

Lynch, D.R. and Gray, W.G., 1978. Analytic solutions for computer flow model testing. J. of Hydraulics Div., ASCE, 104: 1409-1428.

Mellor, G.L., 1973. Analytic prediction of the properties of stratified planetary surface layers. J. Atmos. Sci., 30: 1061-1069.

Mellor, G.L. and Blumberg, A.F., 1985. Modeling vertical and horizontal diffusivities with the sigma coordinate system. Mon. Wea. Rev., 113: 1379-1383.

Mellor, G.L. and Yamada, T., 1974. A hierarchy of turbulence closure models for planetary boundary layers. J. Atmos. Sci., 31: 1791-1806.

Mellor, G.L. and Yamada, T., 1982. Development of a turbulence closure model for geophysical fluid problems. Rev. Geophys. and Space Phys., 20: No. 4, 851-875.

Merilees, P.E., 1976. Fundamentals of numerical weather prediction. In: A. Murphy and D. Williamson (Editors), Weather Forecasting and Weather Forecasts: Models, Systems, and Users, NCAR Tech. Note NCAR/CA-5 + 1976-ASP, Boulder, Colorado, 1, 2-138.

Peffley, M.B. and O'Brien, J.J., 1976. A three-dimensional simulation of coastal upwelling off Oregon. J. Phys. Oceanogr., 6: 164-180.

Philander, S.G.H. and Yoon, J.-H., 1982. Eastern boundary currents and coastal upwelling. J. Phys. Oceanogr., 12: No. 8, 862-879.
Phillips, N.A., 1957. A coordinate system having some special advantages for numerical forecasting. J. of Meteorology, 14: 184-185.
Pinder, G.F. and Gray, W.G., 1977. Finite Element Simulation in Surface and Subsurface Hydrology. Academic Press, New York.
Reid, R.O., Vastano, A.C., Whitaker, R.E. and Wanstreth, J.J., 1977. Experiments in storm surge simulation. In: E.D. Goldberg, I.N. McCave, J.J. O'Brien and J.H. Steele (Editors), The Sea, Vol. 6, John Wiley, New York.
Spaulding, M.L., 1984. A vertically averaged circulation model using boundary-fitted coordinates. J. Phys. Oceanogr., 14: 973-982.
Suginohara, N., 1982. Coastal upwelling: Onshore-offshore circulation, equatorward coastal jet and poleward undercurrent over a continental shelf-slope. J. Phys. Oceanogr., 12: 272-284.
Thacker, W.C., 1977. Irregular grid finite-difference techniques: simulations of oscillations in shallow circular basins. J. Phys. Oceanogr., 7: 284-292.
Wang, D-P., 1982. Development of a three-dimensional, limited-area (island) shelf circulation model. J. Phys. Oceanogr., 12: 605-617.
Yoon, J-H. and Philander, S.G.H., 1982. The generation of coastal undercurrents. J. Oceanogr. Soc. of Japan, 38: 215-224.

PREDICTING OPEN OCEAN CURRENTS, FRONTS AND EDDIES

ALLAN R. ROBINSON
Division of Applied Sciences, Harvard University, Cambridge, Massachusetts

ABSTRACT

The concept of ocean prediction science, related to the forecasting of the "internal weather of the sea" is introduced as important for ocean scientific research generally and for practical application in the areas of marine operations and planetary environmental management. Systematic field estimation, melding observations obtained by remotely located and *in situ* sensors with dynamical model forecasts (4-dimensional data assimilation) is advocated. The characteristics of the oceanic mesoscale (analogous to the atmospheric synoptic scale) are reviewed. Forecasting the mesoscale (O(10-10^2 km) and O(10-10^2 days)) is feasible and has been initiated, but a regional approach is now necessary. Examples of recent progress in mesoscale prediction research are presented, including an ongoing forecast system which is nowcasting and forecasting in the Gulf-Stream region in real time (GULFCASTING).

1 INTRODUCTION — OCEAN PREDICTION SCIENCE

Knowledge of the kinematical structure and dynamical characteristics of flows in the deep ocean has accrued rapidly in recent years. The energetically dominant flow is variable in space and time, generally extending smoothly throughout the water column from surface to bottom with the most intensive flow in the main thermocline region. This so-called oceanic mesoscale variability is the oceanic analogue of the atmospheric synoptic scale weather phenomena, and is the "internal weather of the sea" (Robinson, 1983). Nowcasting and forecasting oceanic mesoscale phenomena is interesting and compelling simply because it is the dominant phenomena in most of the ocean. From a practical viewpoint, prediction is of critical importance to applied marine operations and to scientific research at sea. Although the necessity of prediction has been realized for some time (Mooers, Piacsek, and Robinson, 1981) it is only now that forecasting has become feasible and initial efforts have begun. The basic physical fields of velocity, pressure, density, temperature and salinity in the ocean are linked; forecasting any of them implies essentially the forecasting of all of them. Additionally, nowcasting and forecasting of the basic physical fields allows by relatively simple extension, the prediction both of associated fields (e.g., heat flux, sound speed) and active and passive dispersing materials (e.g., nutrients, tracers, and pollutants).

Directed and concerted utilization of the sea and its resources is increasing, and the inadvertent impact of an ongoing terrestrial technical/industrial society on the oceans is a serious concern (Malone and Roederer, 1985). Deep open ocean forecasting is a valuable aid to marine technology and to large scale environmental management. Underwater operations associated with resource exploration and exploitation (e.g., oil, minerals, hydrothermal energy, fisheries), transportation and defense, do or would benefit substantially from useful predictions from an economic viewpoint and for human safety. Navies increasingly operate underwater. For marine environmental management nowcasts and forecasts increase the effectiveness in coping with and assessing the actual or potential effects of accidental or planned releases of pollutants and foreign material. Chemical and sludge dumps are routine, accidental oil spills occur and must be dealt with, low-level nuclear waste dumps exist, and several nations are considering disposing of high level nuclear wastes in the seabed.

Predictions are important to ocean science i) for physical oceanographic research, ii) for research studies in which physical fields influence biological, chemical or geological processes, and iii) as a means for generating resources for otherwise unobtainable datasets. Consider first the impact on physical research. Mesoscale phenomena, such as bursts of midocean baroclinic instability (McWilliams *et al.*, 1983; Pinardi and Robinson, 1987) are notoriously intermittent and eventful. Pinpointing the anticipated occurrence of such events for focused research reduces resource requirements and accelerates research progress. The alternative is waiting for such events to occur at a fixed mooring site or waiting for an adequate array of floats improbably but fortuitously to enter the relevant space-time domain. Moreover, research on regional mesoscale dynamical processes in the ocean overlaps significantly regional forecasting research. Hypothesis testing associated with the theory of dynamical processes involves the identical regional initial/boundary condition problem central to forecasting and hindcasting (Robinson and Walstad, 1987). Energy and vorticity balances require the inference from measurements and observations of accurate field estimates which can be achieved by a dynamical filtering process technically akin to forecasting procedures (Pinardi and Robinson, 1987). Biological, chemical and geological oceanographers (the latter involved for example with sediment and/or bottom boundary layer processes) have become increasingly aware of the mesoscale physical environment and of the necessity for knowing the synoptic transport fields. These are of crucial importance in real time for the efficient design and execution of experiments at sea and are ultimately necessary for the interpretation of biogeochemical process measurements which generally will also exhibit variability on the mesoscale. Finally we note that since synoptic/mesoscale prediction is in its infancy, dynamical research issues and practical forecasting issues are impossible to separate now from the practicing viewpoint. Thus cooperation between fundamental researchers and marine operators can be of substantial mutual benefit. Not only can the capability of accurate and efficient forecasts be achieved relatively rapidly, but well-designed initialization, updating and verification data can also provide an extensive ongoing dynamical experimental database of a scope generally unavailable for fundamental research studies.

The elucidation of oceanic mesoscale phenomena and the initiation of deep sea predictions are occurring in a broadly based supportive intellectual and technical context. Physical oceanography has evolved rapidly during recent years. New sensors often borne on new platforms and sampling for the first time at definitive rates have revealed new phenomena and are providing a quantitative, definitive and permanent kinematic database. New ideas and theories are being brought to bear and large computer-based numerical models are a new component of the science. The comprehensive and abstract basis of geophysical fluid dynamics helps to focus dynamical oceanographic problems and relate them to analogues which arise in sister sciences such as meteorology, astrophysics, and engineering fluid dynamics. Contemporary techniques of applied mathematics, including numerical analysis and incorporating computational physics, are now available to grapple with essential nonlinearities. A general methodology for dealing with quasideterministic and quasirandom nonlinear dynamical systems is slowly evolving. New technology, new ideas and accelerated effort have evolved ocean science so rapidly that ninety percent of our "corpus" of credible knowledge today has accrued during the last decade.

A most significant element of the new methodology for nowcasting, forecasting and descriptive-synoptic oceanography generally is *systematic field estimation*. Here the term "systems" refers to a multicomponent approach, built out of ongoing and historical observations, a validated general dynamical model, and various regional statistics. All estimates have errors, but a combination of two independent estimates can be made which has an expected error lower than either of the two component errors. This technique is efficient and is utilized by engineers and astronomers under the terminology of optimal estimation. One independent estimate can be derived from observations and another from the dynamical model. Oceanographers have traditionally combined data with model constraints via geostrophic computations and more recently in terms of diagnostic numerical computations (e.g., Sarmiento and Bryan, 1982). The inverse method is a modern technique of constraining data by models which oceanographers (e.g., Wunsch, 1978) share with geophysicists. The most powerful and relevant systematic field estimation methods for application to ocean prediction occur in meteorology where they are known as four-dimensional data assimilation and are used for melding numerical forecasts and asynoptic observations. Real-time data assimilation requires real-time acquisition, telemetering, and transmission of data, the rapid movement of large data sets, the accessing of computers remotely and/or the use of powerful miniaturized cpu devices at sea. Microelectronics and networking advances make real-time deep ocean forecasting a reality and rapid technological progress in these areas is to be expected.

Overall, ocean scientists are aware that the greatest impact of contemporary technology on their discipline will come from two sources: satellites and supercomputers. A step function of progress is anticipated (ACOS-NSF, 1984) and the influences are, of course, already felt. Both of these devices specifically bear upon opportunities for progress in ocean prediction. Remote sensing from satellites provides an extensive coverage of measurements in the space-time domain not attainable previously nor now by any other means. Supercomputers, essential for the development and application of ocean current and circulation models are the fastest and most powerful machines available (e.g., the Crays and Cybers) and are still severely overstressed by the demands of ocean models. Here the frontiers of hardware development and intellectual progress go hand in hand. As in other branches of modern physics and engineering, realistic simulations of (regions of) the ocean are now possible. The use of such simulation in scientific research (numerical experimentation, sensitivity and process studies, etc.) is thought by many to represent the first major step forward in the basic *scientific method* since the seventeenth century. Science is now a tripartite endeavor with Simulation added to the two classical components. Experiment and Theory.

2 THE MESOSCALE FORECAST PROBLEM

The synoptic/mesoscale phenomena in the ocean encompasses a wide range of features including: the meandering and filamenting of currents and jets, fronts, planetary waves, midocean eddy fields, solitons, etc. The pervasive horizontal spatial scale is on the order of the internal Rossby radius of deformation (the depth scale times the ratio of the buoyancy frequency to the Coriolis frequency) with the associated time scales related to advections, planetary wave propagation rates, or possibly forcing. The vertical scales are usually associated with the full water column and the main thermocline depth, but may be as short as a few hundred meters. Aspects of the strongest mesoscale variability (e.g., Gulf Stream meanders and rings) have been known for decades (Stommel, 1965; Richardson, 1976) but it is only since the 1970's that a realistic synoptic picture of the ocean is becoming available (Robinson, 1983). Dedicated experiments defined the Northwestern Atlantic midocean eddies and provided the basis for extensive geographical exploration. Recent discoveries include Caribbean (Kinder, 1983) and Mediterranean (Robinson *et al.*, 1987a) eddies. It is important to note that new sampling and interpretative techniques are still revealing qualitatively new features. Such new features may exist in relative isolation such as sub-mesoscale "lenses" (McWilliams *et al.*, 1983) and squirts (Davis, 1985) or may be associated with major *a priori* known variability such as Gulf Stream shingles and outbreaks (Cornillon, Evans and Large, 1986).

The characteristics, distribution and statistics of the eddy field of the global ocean are emerging. The global eddy field has both homogeneous and heterogeneous aspects. First, consider the dynamically homogeneous ocean. Mesoscale eddies are ubiquitous, occurring almost everywhere they are sought and usually with dominant energy. Quiescent regions exist (e.g., the mid North East Pacific and the beta triangle region of the North Atlantic), but they are the exception rather than the rule. Moreover, types of eddy features recur in the world ocean in similar circumstances, consider e.g. the characteristics of the variabilities of the Gulf Stream, Kuroshio, and other major Western boundary current systems (meanders, rings extension regions, etc.). Dynamically analogous mesoscale phenomena may appear superficially dissimilar only because of effects due to local circumstances including stratification, topography, and interaction with other scales. There are, however, important heterogeneities. Certain features occur only in restricted regions and may be tied to local details of topography and coastal geometry. Coastal and island mesoscale features and effects are distinctive, have deep sea influences and interactions, and require more dedicated study. The kinematics and physics of equatorial mesoscale phenomena differ from that at mid and higher latitudes, and the polar ocean variability is tied to ice dynamical processes. The remainder of this discussion will be concerned with mid-latitude mesoscale variability and its forecasting. This problem is closest to the atmospheric weather prediction problem. Many phenomenological features are still implicated, and from a synoptic viewpoint, a degree of statistical heterogeneity exists related to the space-time intermittency of energetic events.

Several time and space scales relevant for the prediction problem arise: from the phenomena itself, from the space-time domain of interest, and from the techniques adopted for the prediction procedure. The spatial scales, somewhat larger than the internal deformation radius, range from tens to hundreds of kilometers. The temporal scales range from several days to weeks and months. Compared to the atmosphere (thousands of kilometers, hours-days), the problem of internal weather prediction in the ocean involves very many small features extending over vast regions but evolving slowly. There are disadvantages and advantages. On the one hand, data collection and computer resolution requirements make it impossible now for the oceanographer to adopt the global or hemispheric approach of the meteorologists. The regional approach to mesoscale ocean prediction is necessary. On the other hand, updating and the assimilation of data in "oceanic real time" is facilitated. The location, extent, and duration of regional forecasts are dictated of course by practical or scientific interest.

Homogeneity factors should be exploited as much as possible by the development of flexible and portable techniques and of regional models which can be easily relocated. The prediction domain may need to be larger than the practical interest domain, so as to include vorticity interactions which influence the evolution of the mesoscale within the domain of interest, or so as to remove boundary effects to a distance of negligible influence, etc.

There are two basic physical types of regional mesoscale forecast problems in which: (i) evolution occurs in the domain via internal oceanic dynamical processes only and (ii) evolution in the domain involves the response to local and atmospheric forcing. In the latter case (ii), a knowledge of the surface fluxes of momentum, heat, energy, and density must be known, and a forecast of the atmospheric winds is a critical input to the oceanic forecast. Thus regional atmospheric skills and atmospheric predictability time scales constrain the ocean forecaster. The type (i) problem however appears to be the significant one for most of the oceanic mesoscale. The energization of the mesoscale within the region of interest may have occurred remotely in both space and time (e.g., rings snap off the Gulf Stream which itself integrates wind energy slowly accumulated over the open region of the gyre). We will further restrict our discussion here to the type (i) problem, again the closest analogue to meteorological forecasting. Two limiting subcases are noteworthy. In the first case (a *propagation region* problem) signals essentially propagate through the domain; features enter through one boundary and exit through another. In the second case (an *interaction region* problem) features evolve under the influence of local dynamical processes in the domain, nonlinear conversion occurs resulting in birth, growth, transformation, decay, of eddies, etc.

3 MESOSCALE PREDICTION RESEARCH

Operational and research nowcasting and forecasting of the mesoscale requires running realistic deep oceanic dynamical models with real ocean data and the implementation of regional forecast systems. The regional dynamical models should be as flexible and portable as possible. We regard such models as *modular components* of an hierarchy of ocean models. Conceptually, the hierarchy has both i) *dynamical* and ii) *regional* bases. Firstly, the internal dynamics may consist of quasigeostrophic, intermediate or primitive equations; secondly, the mesoscale model may have interconnected surface and bottom boundary layer models and must be explicitly or parametrically embedded in a larger scale basin or general circulation model. The dynamical model must be calibrated, validated and ultimately verified. Model-model, as well as model-data intercomparisons are essential. Observational network research and development, including observational system simulations, is a major task. Adequate and feasible data bases must be established with an efficient mix of remotely sensed and *in situ* measurements. Data assimilation methods need adaptation from meteorology and fundamental research for special oceanic circumstances. Error sources and structures which arise from computation, physics, and observations must be determined and sensitivities ascertained, component-wise and systematically. These issues represent a challenging and enduring range of research problems but progress is occurring. A useful and comprehensive review and overview of the mid-1986 status of research from the viewpoint of the growing community of American ocean prediction scientists is provided by Mooers, Robinson and Thompson (1987), and Hurlburt has assessed aspects.

We have been pursuing a line of research involving systematic field estimation conceptualized in terms of an Oceanic Descriptive Predictive System (ODPS) schematized in Figure 1. In this context real ocean studies and forecast trials

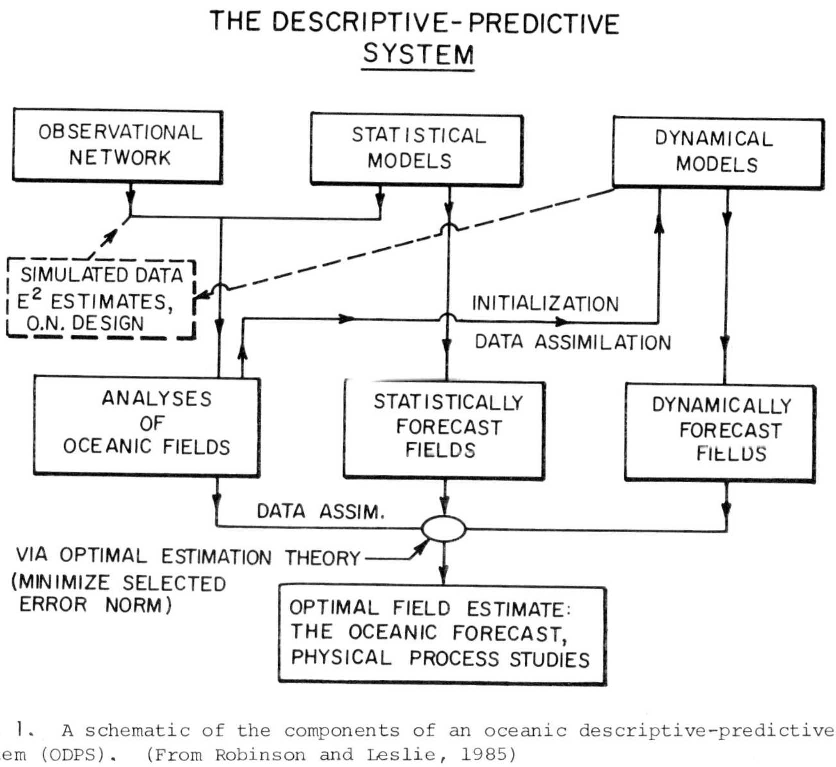

Fig. 1. A schematic of the components of an oceanic descriptive-predictive system (ODPS). (From Robinson and Leslie, 1985)

have been carried out by a number of investigators utilizing the quasigeostrophic version of the Harvard Open Ocean Model illustrated in Figure 2. The quasigeostrophic model is a powerful and robust tool. The fundamental initial/boundary condition problem for the model requires some means of specification at each

HARVARD OPEN OCEAN MODEL
-PORTABLE-

$$\frac{\partial \zeta}{\partial t} + aJ(\psi, \zeta) + \beta \psi_x = \mathcal{F}_{Pqr}$$

$$\zeta = \nabla_H^2 \psi + \Gamma^2 (\sigma \psi_z)_z$$

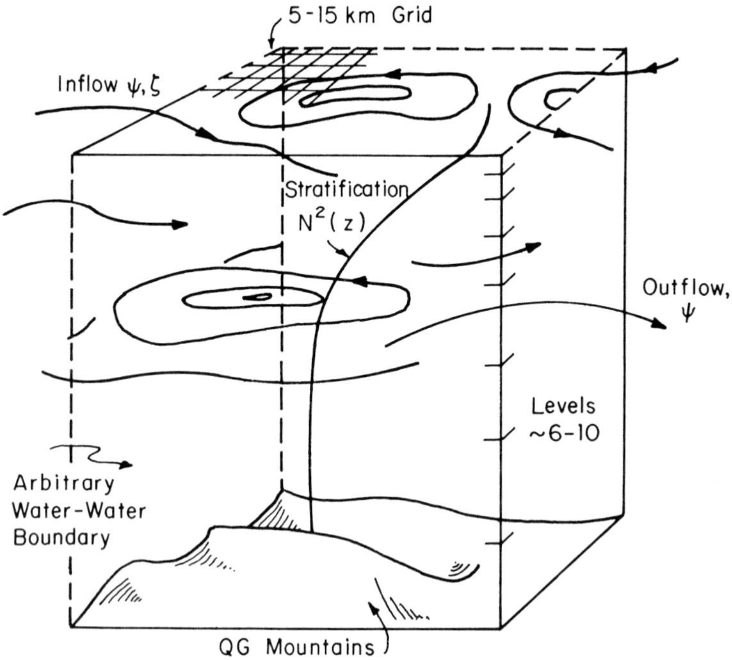

Fig. 2. A schematic of the Harvard Open Ocean Model with the governing equations included. The horizontal and vertical grids are indicated. Arrows indicate the current flow. At the base of the model domain is an idealized bottom topography. (From Robinson and Walstad, 1987)

time step of the inflow/outflow field with vorticity on the inflow (Charney, Fjortoft, and von Neumann, 1950). The model subgridscale dissipation (F_{pqr}) is a Shapiro (1971) type filtering with arbitrary strength and frequency, of the vorticity field. The calibration of the model and its application to dynamical process, forecasting and data assimilation studies is presented by Robinson and Walstad (1987) and Robinson (1986) reviews the data assimilation problem and related research progress. Here we will briefly summarize the background and then cite four recent examples of research findings related to the prediction and predictability of mesoscale eddies and to the theory and practice of data assimilation.

The POLYMODE Synoptic Dynamics Experiment (SDE) carried out in the western North Atlantic in the latter half of the 1970's cooperatively between scientists from the USA and the USSR (Robinson, 1982, 1983) provided the first continuous data set of (almost) synoptic observations in the ocean over several independent realizations in the space-time domain. The SDE lasted for more than a year and subsets of data were collected over areal regions extending from about two to six hundred kilometers on a side. This data set, the first of its kind ever, provided a unique opportunity for the development and calibration of the regional open model, including sensitivity studies with respect to initial and boundary condition errors in both flow and vorticity. Regional simulation methodologies, real data initialization procedures, and EVA, a model consistent Energy and Vorticity Analysis scheme for the inference of regional dynamical processes (Pinardi and Robinson, 1986) were initially developed in the SDE data context. During the first half of the 1980's the Harvard group scientists collaborated with Prof. CNK Mooer's group from the Naval Postgraduate School at Monterey in a program called OPTOMA (Ocean Prediction Through Observations Models and Analysis) dedicated to mesoscale regional predictive and dynamical methodological research. The setting was the region of intense jets filaments eddies and fronts in the California Current system, basically a deep water open ocean environment, but logistically convenient and extensible for coastal interaction studies. Repeated sampling of a central domain has provided an adequate set of partially connected synoptic realizations and related statistics. The OPTOMA data context continues to provide for the systematic development of the ODPS concepts, the development of data assimilation methods, and the testing and application of EVA. Successful real-time prediction experiments have been carried out and the role of dynamical model interpolation in modern synoptic/descriptive oceanography exemplified (Robinson, Carton, Pinardi and Mooers, 1986). Presently a major research effort of the Harvard group is directed towards the Gulf Stream meander and ring region extending eastward of Cape Hatteras past the Grand Banks, partly as a contribution to the ONR multi-institutional research initiative SYNOP, for which the main field program will be for two years starting in the fall of 1987. (Hogg, 1986). Nowcasting and forecasting are now performed routinely (see Section 4) and research on the vorticity dynamics and energetics on the birth and death of rings (via ring-stream interactions) (Robinson, Pinardi and Spall, 1987) is continuing.

The first example is concerned with the practical problem of producing a regional forecast in an open domain requiring, therefore, the specification of future boundary conditions (Walstad, 1987). The data employed (XBTs and current meters) is a POLYMODE-SDE subset called Mark-2 mapped over the inner 300 km^2 region via a multivariate analysis (in Fig. 1, the upper left-hand and central boxes inputting to the lower left-hand box) and used for the calibration and initialization studies mentioned above. A recently completed study of the major energetic events in the Mark-II data (Pinardi and Robinson, 1987) provides the detailed dynamical context for the interpretation of the prediction study. The technique adopted for a Regional Dynamical Forecast (RDF — the right-hand column of Fig. 1) involves a prior statistical forecast of the boundary conditions around the edge of the domain by forward extrapolating in time with a mixed space-time correlation function in an objective analysis scheme. This we call an RDF with Statistically Forecast Boundary Condition (RDF-SFBC). Comparison fields of interest are the Regional Statistical Forecast (RSF — the central column of Fig. 1), and the RDF resulting from updating the boundary conditions continually with the best available data (Benchmark Boundary Condition, RDF-BBC). The RDF-SFBC procedure has been applied to the data 3 times with a sliding starting time, with and without the effects of bottom topography. The results for four examples in terms of the domain averaged difference field between the analyzed observation and the forecast (called the normalized root mean square error) are shown in Figure 3a. In this region where the statistical model is good, the RDF-SFBC gives results after 15 days with an error level of about 45% which compares with 25% for the benchmark RDF-BBC and 62% for the purely statistical procedure. Thus model dynamics generally improves the forecasts' accuracy and the accuracy of forecasts made from only one independent synoptic data realization are useful generally for at least 15 days. Examining the results in more detail indicates: 1) a quantitative dependence on the types of features propagating through the domain, ii) a dominant error arising from boundary condition errors, and therefore iii) a correlated degradation of the RDF-SFBC and the RSF, iv) that topographic effects generally help after about three weeks but may be detrimental to shorter forecasts of the thermocline fields if the deep fields are in error. It is noteworthy that v) good predictions via the RDF-SFBC procedure can be made even in an *interaction region* problem. This is the case for days 3640-3670 shown in Fig. 3b, during which cyclone intensification accompanied by a burst of baroclinic energy conversion is occurring. (In Fig. 3b and all subsequent figures streamfunctions are non-dimensional. See primary references for details.)

Comprehensive maps of the POLYMODE region have been constructed over the 500 km sq. domain of the Synoptic Dynamics Experiment (SDE) by Carton and McGillicudy (1985) from SDE current meter and hydrographic data and also incorporating Local Dynamics Experiment (LDE) and SOFAR float data. Analysis errors are estimated which arise from gappy sampling, from extending measurements into the deep water, and from the lack of absolute velocity measurements around much of the 200 km outer rim. This analysis provides a unique set of mesoscale ocean data over a large domain with continuous synoptic realizations over several

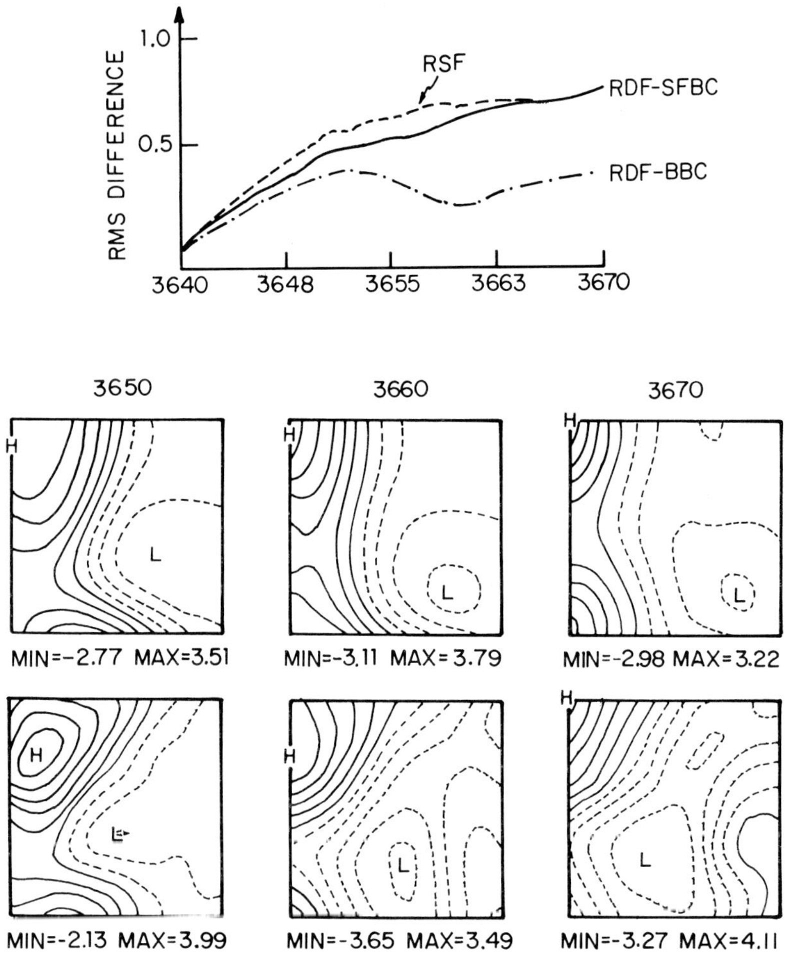

Fig. 3. a) Comparison of the domain averaged difference between analyzed observation fields and the numerically forecast fields for 30-day forecasts. RDF indicates Regional Dynamical Forecast; RSF indicates Regional Statistical Forecast; BBC indicates Benchmark Boundary Conditions and SFBC indicates Statistically Forecast Boundary Conditions. b) Comparison of the analyzed observation fields (top row) with the numerically forecast fields (bottom row) at ten-day intervals from the POLYMODE data set. Contour interval is 0.7. (From Walstad, 1987)

eddies from dedicated *in situ* data. In the second example cited Carton (1987) utilizes this analysis to provide initial conditions, boundary conditions and verification data for a set of now-, fore-, and hindcasting experiments in a first month of a dynamical forecast; iv) that benchmark forecasts can maintain a 60% error level for at least several months; and v) that in benchmark forecasts, major errors propagate in from the boundaries.

The predictability question is approached by attempting to minimize the role of errors in the objectively mapped data in a study of the comparison of forecasts which have the same boundary conditions but different initial conditions. Different model-derived initial conditions with successively greater difference are achieved by starting a new forecast every ten days into a sixty-day forecast originally initialized with mapped data. The results are summarized in Fig. 4b, which depicts the ensemble averaged model-model difference fields. Forecasts initially close diverge to an expected error level of 35% but forecasts initially far apart decrease their differences. These results are interpreted as indicating that the combination of the model, the domain and the prediction scheme contains an irreducible error level, tied to internal instabilities, of about 30-40%.

The Kalman filter is an optimal sequential estimation technique used by engineers and researched by meteorologists which is theoretically optimal but demanding of computer resources (Ghil *et al.*, 1981). Our next example is an idealized study by Miller (1986) exploring the potential of this method for regional data assimilation in quasigeostrophic models. A series of tests was run with the Kalman filter and a one-dimensional linearized barotropic quasigeostrophic model. This one-dimensional scheme was tied to the Harvard Open Ocean Model by choice of dynamical model parameters and by tying the error models to the calibrated computational error of the three-dimensional model (Miller *et al.*, 1983). An unstable numerical scheme was deliberately chosen for the one-dimensional computational model. These tests had a twofold purpose: i) to determine whether the Kalman filter using sparse data would allow a model to track an unstable process without diverging in time from a reference solution, and ii) to test the suitability of the Kalman filter for open boundary problems by means of examples designed to determine whether updating with data confined to the interior of the model region could control the error in the interior, leaving the greatest error near the boundary.

Figure 5 shows a summary of results from this pilot study. Figures 5a and 5b show results from an experiment with periodic boundary conditions which focuses on questions (i) and figure 5c, an experiment with open boundary conditions adding the complications of (ii). The reference solution was a sum of Rossby waves contaminated by noise in the form of Fourier components with Gaussian random amplitudes. The total RMS amplitude of the solution was normalized to 1; the period of the fastest wave was 4π, and the temporal resolution in the model was $4\pi/32$. Without updating, the error variances rapidly grow to unity everywhere. In this experiment, the forecast solution was updated by sampling the study slanted towards questions of regional ocean predictability. What are the limits to the potential accuracy of a prediction scheme set by internal instabilities, and how do the errors they generate compare to those arising from data inadequacies and model errors? Predictability questions are fundamental (Holloway and West, 1984) and provide a useful viewpoint even though ocean prediction is at an early stage. Six experiments are carried out each invoking 34 model integrations representing an attempt to produce ensemble statistics for a variety of forecasting examples based on independent synoptic realizations. Carton calls the RDF-BBC a "hindcast with accurate boundary conditions." The errors after both 10 days and 60 days for 34 different such hindcasts are shown in Fig. 4a, which thus illustrates the degree of variation of individual results that go into the error statistics. The *error* here is again defined as the domain averaged normalized root mean square difference between the mapped observations (containing data sampling and analysis errors) and the model runs (containing model and computational errors). Results of this study indicate: i) that a forecast by simply persisting the initial fields everywhere leads to a 60% expected error in 10 days and continues to grow rapidly beyond that time; ii) that a dynamical model forecast made with persistent boundary conditions maintains an expected error of 60% in the inner third of the domain for three weeks; iii) that the major influence of initial conditions is felt during the

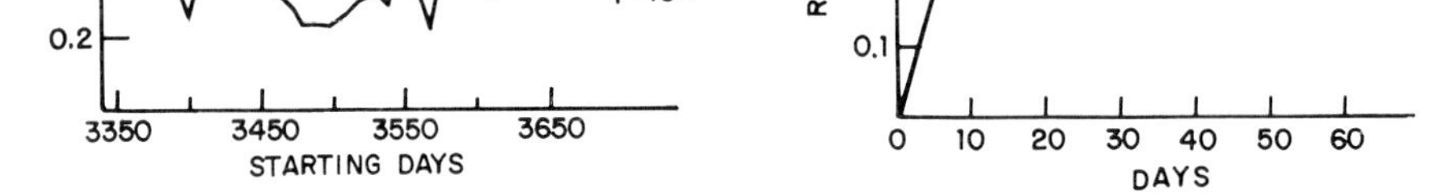

Fig. 4. a) Domain averaged difference between analyzed observation fields and the numerically forecast fields after ten days (solid line) and sixty days (dahsed line) for 34 separate forecasts; each with a different starting date. b) Domain average difference between forecasts with identical boundary conditions but differeing initial conditions. K indicates the number of days into a sixty day forecast the subsequent forecast begins. (From Carton, 1987)

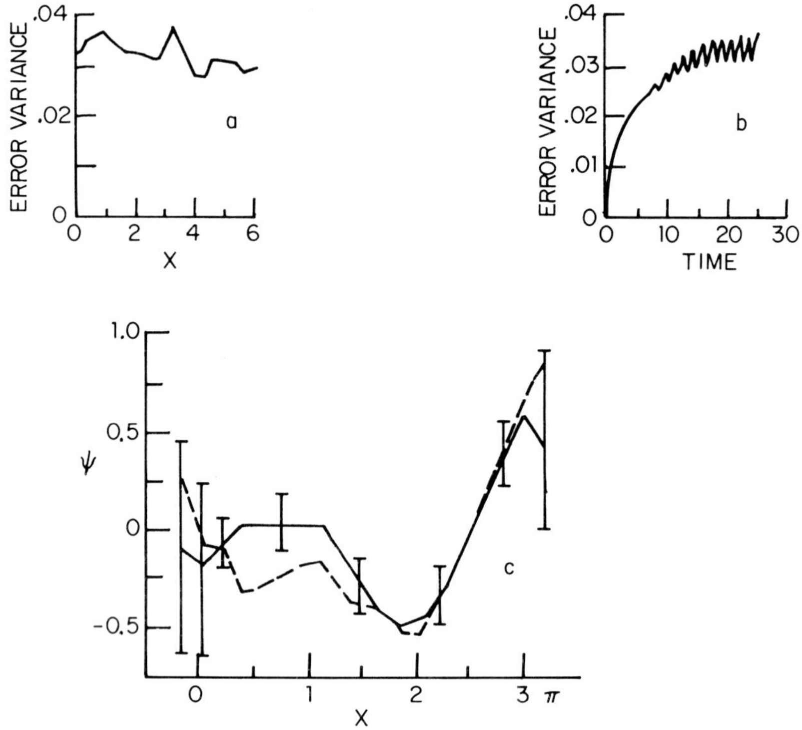

Fig. 5. a) The expected error variance as a function of space for a one-dimensional linearized barotropic quasigeostrophic model using a Kalman filter. b) The evolution of the error variance in a region upstream of the data dense region. c) A comparison of streamfunction values for a sample realization (solid line) and the reference solution (dashed line) over the model domain. The computed expected error bars are one standard deviation wide. (From Miller, 1986).

reference solution at two points in the interior every fourth time step. There are 19 grid points in the domain and points 8 and 12 are updated. The expected error variance as a function of space converged rapidly to the pattern shown in Fig. 5a; Fig. 5b shows the evolution of the error variance in a region upstream of the data dense region. The oscillations in that graph reflect the updating frequency. Note that the error is controlled around the 0.3 level; these are worst-case results since Fig. 5a is just before the next updating time step and Fig. 5b is upstream of updating.

The second experiment involves the specification of open boundary conditions; the boundary condition prediction scheme introduces an additional error source. Figure 5c shows a sample realization compared to the reference solution together with computed expected error bars. Updating was performed each step at four points in the interior. The solid curve is the forecast and the dashed curve is the reference solution. The error bars are one standard deviation wide. Note that the error in the interior is well controlled, despite the relatively large expected error near the boundaries which are far from the data locations. Ten experiments were studied and it was generally found that sparse data could result in reasonable forecasts, and that interior updating could control the interior. In addition, although Kalman filtering is notoriously demanding of computer resources, these results indicate that the resource demand is expected to be manageable on modern computers for typical midlatitude eddy prediction problems.

A variety of strategies of field estimation and prediction have been applied to the OPTOMA data and intercompared by Rienecker, Mooers, and Robinson (1987). Of particular interest is a simple and computationally efficient data assimilation scheme which melds objectively analyzed synoptic observations (± 1 day) with the model dynamical forecast, the dynamical forecast utilized for the assimilation is with statistically forecast boundary conditions (RDF-SFBC). The flow in the 150 km square study domain during the period illustrated consisted of a jet entering from the east, flowing westward and then turning southward. The standard objective mapping is shown in Fig. 6e. Figure 6a is

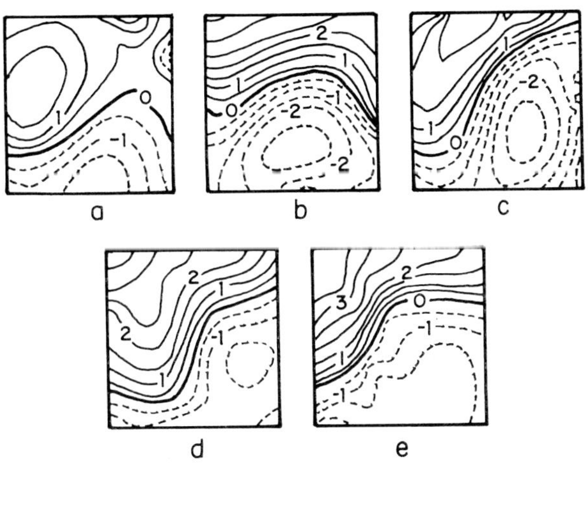

Fig. 6. A comparison of the strategies of field estimation and prediction for Julian day 5522 (6 July, 1983). a) Statistical forecast, b) dynamical forecast with persistent boundary conditions, c) dynamical forecast with benchmark boundary conditions, d) dynamical forecast with data assimilation, and e) objective analysis. Contour interval is 0.5. (From Rienecker *et al.*, 1987).

a purely statistical forecast based on a time-forward extrapolation via objective analysis, Fig. 6b is a forecast with persistent boundary conditions, and Fig. 6c is the dynamical forecast. Our focus is on Fig. 6d, which is the dynamical forecast with data assimilation. Typical synoptic objective analysis fields for assimilation are shown in Figure 7; the area shaded has, γ, the expected error estimate of the objective analysis greater than fifty percent ($\gamma > 50\%$). The melding simply weights the forecast stream function (ψ^f) and the analysis stream function (ψ^a) inversely with respect to γ, i.e., $\psi^{est} = \gamma\psi^f + (1-\gamma)\psi^a$. Barotropic simulations with this scheme were explored by Tu (1981). Although it is not theoretically optimal, it is a significant qualitative and quantitative (Rienecker *et al.*, 1987, Table 5) improvement of the estimate which utilizes readily available fields and correlations. It seems reasonable to implement such strategies while both developing ocean forecasting technology and tailoring optimal assimilation schemes to oceanic flow fields and open regional conditions. In comparing Figs. 6d and 6e it is important to recognize that the analysis of 6e is not the best estimate of the OPTOMA II fields because it does not contain the effect of dynamical interpolation, which Rienecker *et al.* (1987) believe to improve the estimate. This consideration further enhances the value of data assimilation in the interior of the forecast domain.

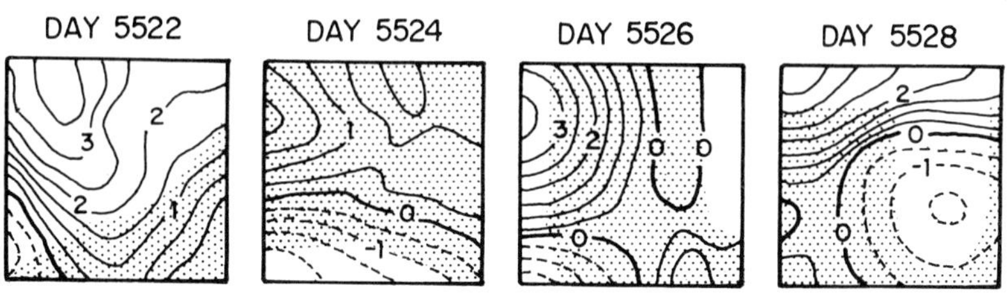

Fig. 7. Typical synoptic objective analysis fields used for data assimilation. The area shaded in each box has an expected error estimate of the objective analysis greater than fifty percent. Contour interval is 0.5. (From Rienecker *et al.*, 1987)

4 A GULF-STREAM FORECAST SYSTEM — *GULFCASTING*

We have established and are now operating a real time Gulf-Stream Descriptive Predictive System (GS-ODPS) from 50°-70° W longitude in open ocean domains one to two thousand kilometers on a side. In this region the Gulf-Stream current after breaking away from Cape Hatteras, amplifies generally eastward propagating meanders which aperiodically snap off to the north and south as warm core or cold core rings (Watts, 1983; Fofonoff, 1981). Several rings of both types typically exist there and (multiple) ring-stream, ring-ring interactions occur including recoalescence of rings into the stream and ring mergers (Richardson, 1983). These variable currents and eddies represent some of the most energetic mesoscale phenomena known to exist in the world ocean and they are accompanied by associated mesoscale features such as shingles, outbreaks, external eddies related to current looping, etc. This is a complex region both kinematically and dynamically, characterized by inhomogeneous, non-stationary and anisotropic statistics. Several factors, however, make this an attractive region for research and forecast trials including the practical and scientific importance of the region, the strength of the mesoscale signals, and historical phenomenological knowledge.

The latter two factors provide the basis for a special dynamical model initialization scheme called *feature initialization*. The major mesoscale features, i.e., the Gulf-Stream axis and the rings are identified and generally located by satellite InfraRed (IR) observations (obtained from NOAA-7). Several passes over several days or a few weeks are usually necessary to deal with the cloud obscuration problem. The model is then initialized with standard forms of features for the current jet and ring structures. Each of these feature models has a simple analytical form with a few free indices which must be set, such as the maximum swirl speed and radius of a ring, etc. This approach is applicable because experience indicates that the Gulf-Stream profiles are always remarkably similar when viewed along the local and instantaneous axis of the stream. Furthermore, rings have characteristic and common structures even as they age. Figure 8 shows schematically the features "hanging in place" in the dynamical model just prior to a run. The model will dynamically adjust the features and interact them, dynamically interpolate between the features, and then evolve the fields forward in time. To estimate future boundary conditions we construct a simple propagation model for ring positions and for meander crests and troughs based on the last few weeks observations and project forward. The model was tuned and validated for this region and scheme by a detailed dynamical study for the period November-December 1984 (Robinson, Pinardi and Spall, 1987) and by five real-time forecast research exercises carried out in the period November 1985 through June 1986 (Robinson, Spall, Walstad and Leslie, 1987). During these exercises AXBT flights were used to obtain data for improving the location of critical features and for verifications. Forecasts are carried out for a week or two, which can be characterized either by simple propagation or by major events. A striking and successfully forecast week's development is illustrated in Figure 9. The large amplitude wave grew within the region in the dynamical forecast; it was later observed in the IR, and AXBT's dropped between 65° and 67° W agreed with the forecast within aircraft navigational accuracy (within ±2 km).

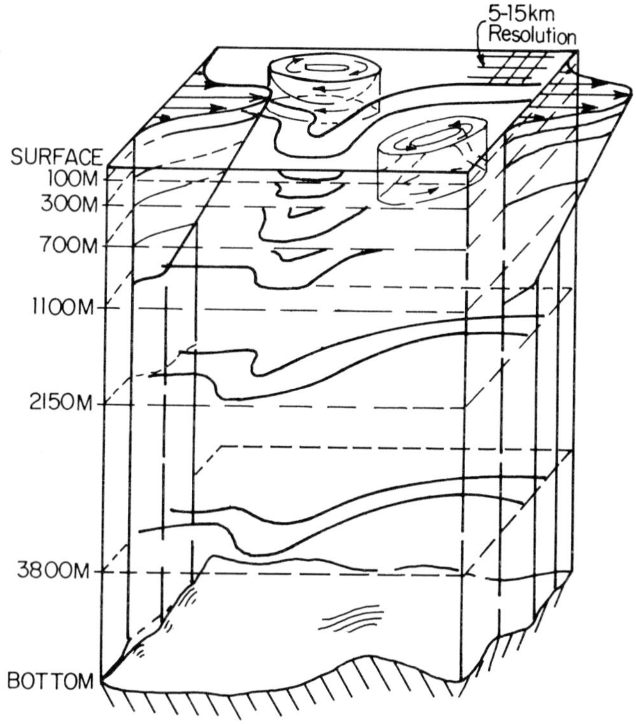

Fig. 8. Schematic of the Harvard Open Ocean Model as used in the Gulf Stream ODPS with the various feature models (stream, warm eddy, cold ring) in place. Actual model levels and horizontal resolution are also indicated. (From Robinson, Pinardi and Spall, 1987)

Since the fall of 1986 we have been maintaining the GS-ODPS, nowcasting and forecasting on a regular basis. The components are: i) Satellite derived sea surface temperature data ii) dynamical model runs on a supercomputer in a large domain, iii) subdomain model runs performed to test sensitivities to feature locations and details of major interactive events, and iv) AXBT flights to remove ambiguities in IR data, to pinpoint critical features and to provide data for updating and assimilation. Figure 10a shows a large domain supercomputer forecast for 23 May, 1986 on which is indicated a subdomain selected for sensitivity study and an AXBT flight track designed for defining four critical axis crossings and two ring locations. The major sensitivity in question involved primarily a warm core ring-stream interaction which would result in a rapid axis distortion. Figure 10b shows a vertical section of model output flow and temperature fields, which can of course be taken at any time and orientation. In summary, we have implemented a rudimentary system which has coverage via a

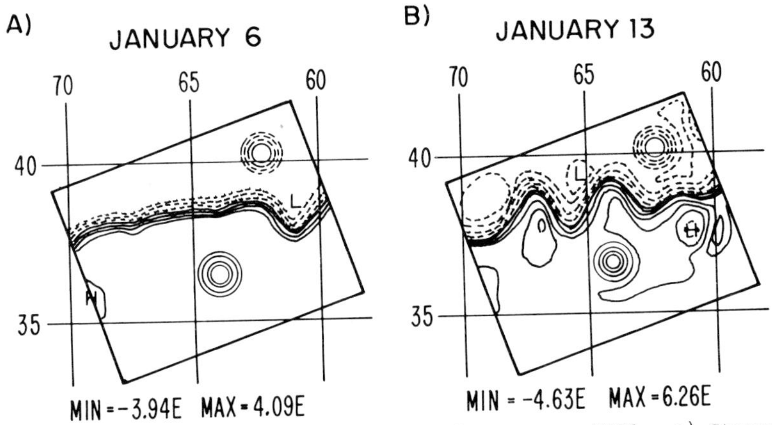

Fig 9. Results of a Gulf Stream forecast for January 1986. a) Streamfunction for January 6 at 100 m. b) As in a) but for January 13. Contour interval for both is 1.0. (From Robinson, Spall, Walstad and Leslie, 1987.)

remotely sensed data component, an interpretative model component, and a dedicated component for the acquisition of critical subsurface *in situ* data. The roles of the dynamical model are of course i) to provide subsurface fields derived from remotely sensed surface observations, ii) to interpolate sparse data dynamically, and iii) to extrapolate forward in time, i.e., to predict.

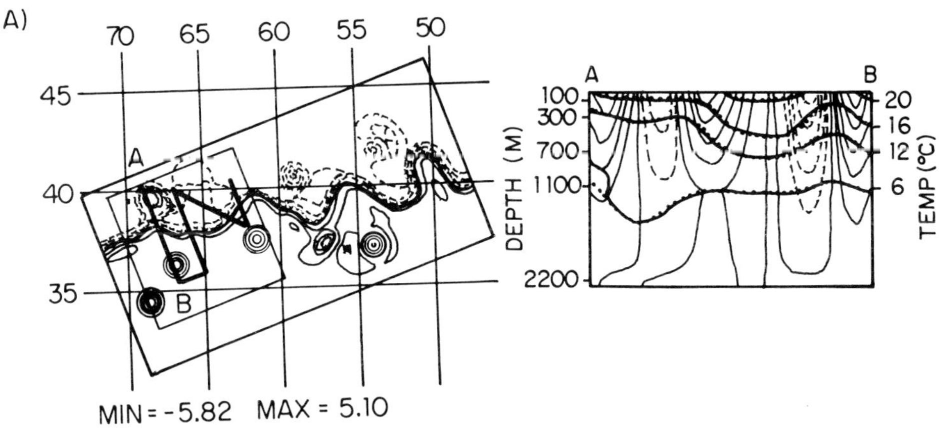

Fig. 10. a) Streamfunction at 100 m for a large domain supercomputer forecast for 23 May 1986. The smaller box defines a domain selected for sensitivity studies. An AXBT flight track is shown by the bold line. A and B indicate the end points for the first leg of the flight track. b) Vertical section of model output flow and temperature fields along the leg of the flight track in a) indicated by the A and B. Solid lines indicate positive streamfunction; dashed lines indicate negative streamfunction; bold lines are contours of temperature. Contour interval for streamfunction in a) and b) is 1.0. (From Robinson, Spall, Walstad and Leslie, 1987)

5 CONCLUSIONS

The oceanic mesoscale is the analogue of the atmospheric synoptic scale. A variety of mesoscale phenomena occur which are the energetically dominant flows over much of the ocean and are the "internal weather" phenomena of the sea. Forecasting research presents scientific and technical problems rooted in modern nonlinear mechanics and computational fluid dynamics. Systematic field estimation, especially four-dimensional data assimilation (involving remotely sensed and *in situ* data and realistic numerical dynamical models), is relevant and essential. Phenomenological scales on the order of tens to hundreds of kilometers and of several days to months make regional forecasting presently necessary but facilitate data assimilation in oceanic real time. Mesoscale forecasting is now feasible and successful real time forecasts have been carried out. Ocean prediction research is being vigorously pursued encompassing: model development and verification; data assimilation and predictability studies; and physical process and prediction experiments.

The prospectus for ocean mesoscale forecasting is excellent, provided the systematic approach is implemented with the potentials both of the individual components and of their combination exploited. Physically, the classification of regional kinematics and dynamics is required. Standard regions for operational forecasts should be identified and relevant tuned and verified regional models established for use when accurate regional forecasts are desired. On the large scale (very large regional, basin or global), observations of sea surface height and sea surface temperatures with mesoscale resolution should be continually available via satellite coverage. Also a relatively coarse resolution eddy resolving model should be kept running on a supercomputer assimilating remotely sensed and *in situ* data. (Scatterometer winds and remotely sensed air-sea fluxes will ultimately be necessary). Data from an *in situ* subsurface observational network must be telemetered in real time and should consist of an efficient mix of: time series from free floating and moored sensors, repeated sections, and remotely sensed data (e.g., acoustic tomographic). Regions of special interest will require denser data sampling and fine mesoscale resolution models for prediction and data assimilation. Accurate runs for the subregions can draw upon the coarser large regional model output for boundary and initial condition data, should be used for sensitivity runs, and can be carried out at forecasting centers, on ships at sea, or wherever a powerful but portable microcomputer can be made available. The interaction among the components of the forecast system is symbiotic and the whole is much more powerful than the sum of the parts.

ACKNOWLEDGMENTS

The general concepts of ocean prediction science and the approach to the regional mesoscale forecasting problem were first presented in Cambridge, Massachusetts at the OCEAN PREDICTION WORKSHOP in April 1986 and then further developed for the 18th International Liege Colloquium on Hydrodynamics — "Three-dimensional models of marine and estuarine dynamics." In Liege Professor J.C.J. Nihoul presented the interesting opportunity for deep sea dynamicists, long three-dimensional, to interact with shallow water dynamicists, long predictive. The interfacing of coastal and open ocean models offers challenging and important opportunities for several years to come.

It is a pleasure to acknowledge Mr. Michael A. Spall's scientific contributions to the establishment of the Gulf-Stream ODPS which were essential to its success. I am grateful for the valued assistance of Mrs. Renate D'Arcangelo, Ms. Marsha G. Cormier and Mr. Wayne G. Leslie in the production of the manuscript. This research was supported by the Office of Naval Research (N00014-84-0461) and the Institute of Naval Oceanography under contracts to Harvard University.

REFERENCES

ACOS-NSF, 1984. Emergence of a Unified Ocean Science: A Report by the Advisory Committee for Ocean Studies — The National Science Foundation.

Carton, J.A., 1987. How Predictable are the Geostrophic Currents in the Recirculation Zone of the North Atlantic? J. Phys. Oceanogr, in press.

Carton, J.A. and McGillicuddy, D.J., 1985. Comprehensive Objective Maps of POLYMODE Streamfunctions. Reports in Meteorology and Oceanography, 21, Harvard University, Cambridge, MA.

Charney, J.G., Fjortoft, R., and von Neumann, J., 1950. Numerical Integration of the Barotropic Vorticity Equation. Tellus, 2: 237-254.

Cornillon, P., Evans, D., and Large, W., 1986. Warm Outbreaks of the Gulf Stream into the Sargasso Sea. J. Geophys. Res., C5, 91: 6583-6596.

Davis, R.E., 1985. Drifter Observations of Coastal Surface Currents During CODE. J. Geophys. Res., 90: 4741-4755.

Fofonoff, N.P., 1981. The Gulf Stream System. In: B.A. Warren and C. Wunsch (Editors), The Evolution of Physical Oceanography, MIT Press, Cambridge, MA.

Ghil, M., Cohn, S., Tavantzis, J., Bube, K., and Isaacson, E., 1981. Application of Estimation Theory to Numerical Weather Prediction. In: L. Bengtsson, M. Ghil and E. Kallen (Editors), Dynamic Meteorology, Data Assimilation Methods, Springer-Verlag, New York/Heidelberg/Berlin, pp. 139-224.

Hogg, N.G. (Editor), 1986. SYNOP — Synoptic Ocean Prediction — Program Cover Document. Woods Hole Oceanographic Institution.

Holloway, G., and West, B.J., 1984. Predictability of Fluid Motions, AIP Conference Proceedings No. 106, American Institute of Physics, New York.

Hurlburt, H.E., 1984. The potential for ocean prediction and the role of altimeter data. Mar. Geodesy, 8: 17-66.

Kinder, T.H., 1983. Shallow Currents in the Caribbean Sea and Gulf of Mexico as Observed with Satellite-Tracked Drifters. Bull. Mar. Sci., 33 (2): 239-246.

Malone, T.F. and Roederer, J.G. (Editors), 1985. Global Change. ICSU Press by Cambridge University Press, Cambridge/New York.

McWilliams, J.C., Brown, E.D., Bryden, H.L., Ebbesmeyer, C.C., Elliot, B.A., Heinmiller, R.H., Lien Hua, B., Leaman, K.D., Lindstrom, E.J., Luyten, J.R., McDowell, S.E., Owens, W.B., Perkins, H., Price, J.F., Regier, L., Riser, S.C., Rossby, H.T., Sanford, T.B., Shen, C.Y., Taft, B.A., and Van Leer, J.C., 1983. The Local Dynamics of Eddies in the Western North Atlantic. In: A.R. Robinson (Editor), Eddies in Marine Science. Springer-Verlag, Berlin/Heidelberg/New York/Tokyo, pp. 92-113.
Miller, R.N., 1986. Toward the Application of the Kalman Filter to Regional Open Ocean Modeling. J. Phys. Oceanogr., 16: 72-86.
Miller, R.N., Robinson, A.R. and Haidvogel, D.B., 1983. A Baroclinic Quasigeostrophic Open Ocean Model. J. Comput. Phys., 50 (1): 38-70.
Mooers, C.N.K.M., Piacsek, S.A. and Robinson, A.R. (Editors), 1981. Ocean Prediction: The Scientific Basis and the Navy's Needs, A Status and Prospectus Report. Proceedings of the Ocean Prediction Workshop, Monterey, CA, May 1981.
Mooers, C.N.K.M., Robinson, A.R. and Thompson, J.D. (Editors), 1987. Ocean Prediction Workshop 1986, A Status and Prospectus Report on the Scientific Basis and the Navy's Needs. Proceedings of the Ocean Prediction Workshop: Phase I — Cambridge, MA, April 1986; Phase II — Long Beach, MS, November 1986.
Pinardi, N. and Robinson, A.R., 1986. Quasigeostrophic Energetics of Open Ocean Regions. Dyn. Atmos. Oceans, 10 (3): 185-221.
Pinardi, N. and Robinson, A.R., 1987. Dynamics of Deep Thermocline Jets in the POLYMODE Region. J. Phys. Oceanogr., in press.
Richardson, P.L., 1983. Gulf Stream Rings. In: A.R. Robinson (Editor), Eddies in Marine Science. Springer-Verlag, Berlin/Heidelberg/New York/ Tokyo, pp. 19-45.
Rienecker, M.M., Mooers, C.N.K.M. and Robinson, A.R., 1987. The Evolution of Mesoscale Features off Northern California: Dynamical Interpolation and Forecast Experiments. J. Phys. Oceanogr., in press.
Robinson, A.R., 1982. Dynamics of Ocean Currents and Circulation: Results of POLYMODE and Related Investigations. In: A. Osborne and P.M. Rizzoli (Editors), Topics in Ocean Physics, Society Italiana di Fisica, Bologna, Italy, pp. 3-29.
Robinson, A.R., 1983. Overview and Summary of Eddy Science. In: A.R. Robinson (Editor), Eddies in Marine Science. Springer-Verlag, Berlin/ Heidelberg/New York/Tokyo, pp. 3-15.
Robinson, A.R., 1986. Data Assimilation, Mesoscale Dynamics and Dynamical Forecasting. In: J.J. O'Brien (Editor), Advanced Physical Oceanographic Numerical Modelling, Proceedings of the NATO Advanced Studies Institute, R. Reidel, Dordrecht, The Netherlands, pp. 465-483.
Robinson, A.R. and Leslie, W.G., 1985. Estimation and Prediction of Oceanic Fields. Progr. Oceanogr., 14: 485-510.
Robinson, A.R. and Walstad, L.J., 1987. The Harvard Open Ocean Model: Calibration and Application to Dynamical Process, Forecasting, and Data Assimilation Studies. J. Appl. Numer. Math., in press.
Robinson, A.R., Carton, J.A., Pinardi, N. and Mooers, C.N.K.M., 1986. Dynamical Forecasting and Dynamical Interpolation: An Experiment in the California Current. J. Phys. Oceanogr., 16: 1561-1579.
Robinson, A.R., Pinardi, N. and Spall, M.A., 1987a. Gulf Stream Simulations and the Dynamics of Ring and Meander Process, in preparation.
Robinson, A.R., Spall, M.A., Walstad, L.J. and Leslie, W.G., 1987b. Forecasting the Internal Weather of the Sea: A Real-Time System for Gulf Stream Meanders and Rings, in preparation.
Sarmiento, J.L. and Bryan, K., 1982. An Ocean Transport Model for the North Atlantic. J. Geophys. Res., 87: 394-408.
Shapiro, R., 1971. The Use of Linear Filtering as a Parameterization for Atmospheric Diffusion. J. Atmos. Sci., 28: 523-531.
Stommel, H., 1965. The Gulf Stream. University of California Press, Berkeley/Los Angeles/London.

Swallow, J., 1976. Variable Currents in Mid-Ocean. Oceanus, 19: 18-25.

Tu, K., 1981. A Combined Dynamical and Statistical Approach to Regional Forecast Modeling of Open Ocean Currents. Ph.D. thesis, Harvard University, Reports in Meteorology and Oceanography, 13.

Walstad, L.J., 1987. The Harvard Quasigeostrophic Model: Hindcasting, Forecasting, and Development of the Coupled Quasigeostrophic – Surface Boundary Layer Model. Ph.D. thesis, Harvard University, Cambridge, MA.

Watts, D.R., 1983. Gulf Stream Variability. In: A.R. Robinson (Editor), Eddies in Marine Science. Springer-Verlag, Berlin/Heidelberg/New York/ Tokyo, pp. 114-144.

Wunsch, C., 1978. The North Atlantic Circulation West of 50 W Determined by Inverse Methods. Rev. Geophys. Space Phys., 16: 583-620.

PREPARATION OF ESTUARY AND MARINE MODEL EQUATIONS BY GENERALIZED FILTERING METHODS

K. W. BEDFORD, J. S. DINGMAN and W. K. YEO
Department of Civil Engineering
The Ohio State University, 2070 Neil Avenue
Columbus, Ohio 43210 (USA)

ABSTRACT

Higher order averaging procedures which circumvent the limitations of traditional Reynolds averaging are presented and applied to the three-dimensional equations used for marine and estuary models. These averages or filters can either be analog or digital, and a review of the classes of such filters and their characterization is presented first. Low pass filters in both single and cascaded form are then applied to the governing equations, and closures via traditional and high pass filter expansions are identified. Finally, the relationship between analog filters, digital filters and commonly used higher order numerical schemes is explored, and it is shown that certain numerical schemes are indeed digital filter forms of the analog filters.

1 INTRODUCTION

During the last thirty years of surface water flow and transport modeling (meteorological models as well), the basic Reynolds averaged form of the mean flow turbulence equations has been the unquestioned basis for the governing model equations. Accompanying this singular set of averaged equations has come an overwhelming variety of numerical representations for these equations. Additional complexity exists in the selection of which closure to use and, for many closure forms, what empirical coefficients to use. Considerable research activity is devoted to closure and discretization forms while very little effort is expended in reviewing and possibly improving the averaging and therefore structure of the basic governing equations.

Recently a small body of research suggests that indeed a cause of the multitudinous forms of computational and closure forms resides in difficulties with the method used to average and prepare the basic turbulence equations. The basis for this suggestion resides in the requirement that the marine equations must be averaged to account for time varying average flows and that consistency in the averaging must be achieved. The currently used Reynolds average is not all together adequate in meeting this requirement. Analog filters based upon a generalized averaging or filter definition are being developed to address these averaging requirements and will be used in this article to formulate new equations for the marine and estuary model problem. Additionally, and in complete analogy to digital signal processing, digital filter forms of the analog filters are available; however, their relation to the analog equation forms and to existing numerical methods is unexplored. Therefore, the objectives of this article are to: 1) present a review of filtering procedures; 2) summarize the rules or procedures by which these filters are applied to the equations; 3) review the types of analog averaging or filters used in marine models to date; 4) apply these procedures to the derivation of a new set of marine and estuary equations; and 5) contrast the analog equation forms and presently used higher order discretizations. The last objective seeks to begin determining if many of the proposed numerical methods are simply digital forms of the newly recognized analog filter terms.

2 BASIC AVERAGING DEFINITIONS

2.1 Definition of filtering operation

A continuous field variable, say $f(\underline{x},t)$, can be decomposed into its mean and fluctuating components as (Dakhoul and Bedford, 1986a):

$$f(\underline{x},t) = \overline{f}(\underline{x},t) + f'(\underline{x},t) \tag{1}$$

where

$$\overline{f}(\underline{x},t) = \int\int\int\int_{-\infty}^{+\infty} G(\underline{x}-\underline{x}', t-t')\, f(\underline{x}',t')\, d\underline{x}'dt' \tag{2}$$

In these equations: $\underline{x}$ denotes the Cartesian vector spatial coordinate ($x = x_i + y_j + z_k$); t denotes time; and $G(\underline{x},t)$ is a weight or filter function constrained such that

$$\int\int\int\int_{-\infty}^{+\infty} G(\underline{x},t)\, d\underline{x}\, dt = 1.0 \tag{3}$$

2.2 Analysis and types of filters

The distinctions between the various analog and digital filters are based upon the behavior of the filter in resolving desirable portions of the wavenumber or frequency spectrum of the problem and suppressing or eliminating undesirable portions. The filter response function, R, defined as the Fourier transform of the filter function, is a measure of this activity, e.g., $R = \overline{f}^*/f^* = G^*$; where the asterisk denotes the Fourier transform of the variable or function. The response function may be either parameterized as a function of wavenumber, k, or frequency, ω. Based upon $R_{k,\omega}$, four types of response functions and therefore filters are identified according to which portion of the wave number or frequency spectrum is resolved; these are lowpass (L), highpass (H), bandpass (BP) and bandreject (BR) response functions.

Without question, the dominant type of averaging used in model preparation is the low pass filter wherein high frequency and/or wavenumber information above a cutoff value ω_c or k_c is eliminated while the lower frequency motion is allowed or retained. An additional attribute of the low pass filter is that the resolved portion of the variable or signal not possess either amplitude or phase distortion. To date, no use has been made in surface water wave or turbulence modeling (or any other fluids modeling for that matter) of the other averaging definitions. In passing, it should be noted that the response function for the high pass filter is simply related to the low pass function, i.e., $H(k,w) = 1-L(k,\omega)$. Therefore, in principle, if a low pass filter is known, then so is an equivalent high pass filter.

2.3 Types of low pass filters

A review of low pass filter functions used in equation preparation has been presented in Dakhoul and Bedford (1986a) and can be roughly broken into two major categories; the uniform and the Gaussian filter. Two further subdivisions occur within each category in that either spatial or temporal versions of these filters are possible. A fifth filter, a generalized spatial-temporal filter, has also been designed and tested by Dakhoul and Bedford (1986a, 1986b), and thereby the previous four filters become special cases of this general filter.

i) Uniform filter. The definition for the general uniform space-time filter for n=1 to 3 spatial dimensions is:

$$G_{L,u}(\underline{x},t) = \frac{1}{\delta t}\prod_{i=1}^{n}\frac{1}{\delta i} \qquad \text{for} \quad \begin{array}{l} -\frac{\delta t}{2} \leqslant t \leqslant \frac{\delta t}{2} \\ -\frac{\delta i}{2} \leqslant x_i \leqslant \frac{\delta i}{2} \end{array} \tag{4}$$

where δt and δi are averaging scales to be selected by the analyst.

The response function of this filter is

$$L_u(k,\omega) = \frac{\sin(\omega\delta t/2)}{\omega\delta t/2}\prod_{i=1}^{n}\frac{\sin(k_i\delta i/2)}{k_i\delta i/2} \tag{5}$$

Spatial or temporal filters are easily derived from these functions. It is recognized that the fixed interval Reynolds average (Reynolds, 1895) is a specialized temporal filter, i.e.,

$$\bar{f}(\underline{x}) = \frac{1}{\delta t} \int_{t-\delta t/2}^{t+\delta t/2} f(\underline{x},t')\, dt'. \tag{6}$$

A similar fixed interval volume average has been used in atmospheric models since the work of Smagorinsky (1963) and Deardorff (1973).

ii) Gaussian filter The generalized Gaussian space-time filter is defined for n=1 to 3 spatial dimensions as:

$$G_{LG}(\underline{x},t) = \left(\frac{\gamma_t}{\pi}\right)^{1/2} \frac{1}{\delta t} \exp\left\{\frac{-\gamma_t t^2}{\delta t^2}\right\} \prod_{i=1}^{n} \left(\frac{\gamma_i}{\pi}\right)^{1/2} \frac{1}{\delta i} \exp\left\{\frac{-\gamma_i x_i^2}{\delta i^2}\right\} \tag{7}$$

with a response function defined as:

$$L_G(k,\omega) = \exp\{-\omega^2 \delta t^2/4\gamma_t\} \prod_{i=1}^{n} \exp\{-k_i \delta i^2/4\gamma_i\} \tag{8}$$

In this equation, γ is a coefficient which commonly varies between 1 and 6.

The use of moving-average higher order filters, defined by eqns. (4) and (7), was first performed with spatial filters by Leonard (1974) and in surface water flow and transport models by Bedford (1981) Babajimopolous and Bedford (1980) and Bedford and Babajimopolous (1980). In surface water flow modeling, Bedford (1981) found it necessary to use anisotropic spatial filters and further noticed that three-dimensional models based upon hydrostatic pressure created, propagated, and dissipated turbulence as a horizontal two-dimensional flow field even though a three-dimensional velocity field is calculated.

3 APPLICATION OF ANALOG FILTERING

3.1 Rules of averaging

Using the fixed interval Reynolds averaging definition now in use today, it is possible (Hinze, 1975) to define a series of averaging rules for averaging various combinations of functions and operations; the second column in Table 1 summarizes those operations. In this table the overbar stands for the averaging of either the function f or g; furthermore, t stands for time and s indicates a general spatial dimension.

The third column contains the rules permitted if one assumes a moving average interpretation for the filtering operation. It is noticed immediately that rules No. 4, 5 and 6, valid for fixed interval averaging, are no longer valid. As a result, considerable differences in the final average equation form will result.

TABLE 1

Summary of averaging rules for fixed and moving average filter definitions

Averaging rule No.	Fixed interval average	Moving average
1	$\overline{f+g} = \bar{f} + \bar{g}$	$\overline{f+g} = \bar{f} + \bar{g}$
2	$\overline{\frac{\partial f}{\partial t}} = \frac{\partial \bar{f}}{\partial t}$	$\overline{\frac{\partial f}{\partial t}} = \frac{\partial \bar{f}}{\partial t}$
3	$\overline{\frac{\partial f}{\partial s}} = \frac{\partial \bar{f}}{\partial s}$	$\overline{\frac{\partial f}{\partial s}} = \frac{\partial \bar{f}}{\partial s}$
4	$\overline{f'} = 0$	$\overline{f'} \neq 0$
5	$\bar{\bar{f}} = \bar{f}$	$\bar{\bar{f}} \neq \bar{f}$
6	$\overline{\bar{f}\,\bar{g}} = \bar{f}\,\bar{g}$	$\overline{\bar{f}\,\bar{g}} \neq \bar{f}\,\bar{g}$

3.2 The application of Reynolds averaging to the Navier-Stokes equations

The Navier-Stokes equations are written in tensor notation as:

$$\frac{\partial u_i}{\partial t} + \frac{\partial}{\partial x_j} u_i u_j = -\frac{1}{\rho}\frac{\partial p}{\partial x_i} + \nu \frac{\partial^2 u_i}{\partial x_j \partial x_j} \tag{9}$$

In this equation ρ is the density, p is the pressure, ν is the kinematic viscosity and u_i is the velocity vector, $u_i = u_1 i + u_2 j + u_3 k$, for which $u_1 = u$, $u_2 = v$ and $u_3 = w$. Repeated indices in a term imply summations over all the indices.

To apply Reynolds averaging eqn. (9) is averaged (Hinze, 1975). After using Rules No. 1, 2, and 3, the following equation results:

$$\frac{\partial \bar{u}_i}{\partial t} + \frac{\partial \overline{u_i u_j}}{\partial x_j} = -\frac{1}{\rho}\frac{\partial \bar{p}}{\partial x_i} + \nu \frac{\partial^2 \bar{u}_i}{\partial \psi_j \partial x_j} \tag{10}$$

The decomposition of the nonlinear term occurs by a further averaging, i.e.,

$$\overline{u_i u_j} = \overline{(\bar{u}_j + u'_i)(\bar{u}_j + u'_j)} = \bar{u}_i \bar{u}_j + \overline{u'_i u'_j} \tag{11}$$

Employing rule No. 6 and eqn. (11) gives the final Reynolds averaged form of the Navier-Stokes equations

$$\frac{\partial \bar{u}_i}{\partial t} + \frac{\partial \overline{ui}\ \overline{uj}}{\partial x_j} = -\frac{1}{\rho}\frac{\partial \bar{p}}{\partial x_i} + \nu \frac{\partial^2 \bar{u}_i}{\partial x_j \partial x_j} - \frac{\partial \overline{(u_i' u_j')}}{\partial x_j} \qquad (12)$$

The last terms in eqn. (12) are the Reynolds stress terms about which considerable closure discussion occurs.

3.3 Application of generalized averaging to the Navier-Stokes equations

For the case of moving or generalized averaging which doesn't restrict flows to spatially or temporally constant means, the following defects in the Reynolds procedure are identified and remedied.

i) Inertia term decomposition. As Leonard (1974) so clearly and provocatively pointed out, Rule No. 6 in Table 1 is not correct. Using a Gaussian spatial filter and the Taylor series, he was able to improve the inertia term decomposition as in the following equation:

$$\overline{\bar{u}_i\ \bar{u}_j} = \int \int_{-\infty}^{\infty} \int G\ (\underline{x} - \underline{x}')\ \overline{u_i}\ \overline{u_j}\ (\underline{x}\ ')\ d\underline{x}'$$

$$= \overline{u_i}\ \overline{u_j} + \frac{\delta i^2}{4\gamma_i}\ \overline{u_i}\ \overline{u_j} + 0\ (\delta i^4) \qquad (13)$$

In this expansion the filter coefficients δi are all assumed equal $(=\delta)$ as are the γ_i $(= \gamma)$.

ii) Cross correlation term decomposition. Due to the inapplicability of Rules No. 6 and 4, the cross correlation terms $\overline{u_i}\ u_j'$ are no longer zero. Therefore Clark et al. (1977) found that

$$\overline{\bar{u}_i u_j'} = -\frac{\delta i^2}{4\gamma_i}\ \bar{u}_i \nabla^2\ \bar{u}_j \qquad (14)$$

iii) Averaging inconsistencies and cascade filtering. Recently these authors have noted an inconsistency in the application of the moving average derivation of the basic equations in addition to those in 3.3(i) and 3.3(ii). It is noted that the decomposition suggested by Leonard (eqn. 13) for the inertia terms involves a two step or two-fold averaging technique. In digital signal processing literature this is called a cascaded filter (Rabiner and Gold, 1975). We note that because Rule 5 is not valid for the moving average approach, then not only must the nonlinear/inertia term be cascade filtered, but also the linear differential terms in the governing equations. In other words, to be consistent the entire equation must be cascade filtered. Therefore, the temporal acceleration term is cascade filtered as follows:

$$\frac{\partial \bar{u}_i}{\partial t} = \frac{\partial \overline{(\bar{u}_i + u_i')}}{\partial t} = \frac{\partial \bar{\bar{u}}_i}{\partial t} + \frac{\partial \overline{u_i'}}{\partial t} ; \tag{15}$$

and using the Taylor series expansion and spatial Gaussian filter, eqn. (15) is rewritten as:

$$\frac{\partial \bar{\bar{u}}_i}{\partial t} = \frac{\partial \bar{u}_i}{\partial t} + \frac{\delta i^2}{4\gamma} \frac{\partial}{\partial t} \left(\frac{\partial^2 \bar{u}_i}{\partial x_\ell \partial x_\ell} \right) + \frac{\partial \overline{u_i'}}{\partial t} \tag{16}$$

Similar expansions occur for the pressure gradient terms as well as any Coriolis or source/sink terms.

iv) <u>Summary Navier-Stokes equation</u>. When combining the developments in eqns. (13-16), new generalized turbulent Navier-Stokes equations emerge which now permit turbulence to be defined relative to a non-constant mean. In tensor notation, the equation becomes (dropping the viscous term)

$$\frac{\partial \bar{u}_i}{\partial t} + \frac{\partial}{\partial x_j}(\bar{u}_i \bar{u}_j) + Fi = -\frac{1}{\rho} \frac{\partial \bar{p}}{\partial x} + Si + Ri \tag{17}$$

where Fi represents the acceleration filter terms

$$Fi = \frac{\partial}{\partial t} \left\{ \frac{\delta_\ell^2}{4\gamma} \left(\frac{\partial^2 \bar{u}_i}{\partial x_\ell \partial x_\ell} \right) \right\} + \frac{\partial}{\partial x_j} \left\{ \frac{\delta_\ell^2}{2\gamma} \left(\frac{\partial \bar{u}_i}{\partial x_\ell} \right) \left(\frac{\partial \bar{u}_j}{\partial x_\ell} \right) \right\} \tag{18}$$

Si represents the pressure filter term

$$Si = \frac{\partial}{\partial x_i} \left\{ \frac{\delta_\ell^2}{4\gamma} \left(\frac{\partial^2 \bar{p}}{\partial x_\ell \partial x_\ell} \right) \right\} \tag{19}$$

and Ri represents the subgrid scale term

$$Ri = \frac{\partial \overline{u_i'}}{\partial t} + \frac{\partial \overline{(u_i' u_j')}}{\partial x_j} = -\frac{1}{\rho} \frac{\partial \overline{p'}}{\partial x_i} \tag{20}$$

Note that this derivation has been done with a spatial Gaussian filter. A derivation for a spatial/temporal filter, as discussed previously, is possible; however, in light of the new cascaded filter approach used to derive eqns. (17) - (19) and the direct realization of a temporal effect due to the spatial filter, the use of a direct temporal filter component may no longer be necessary. Such investigations are being pursued by the second and third authors of this article as part of their Doctoral Dissertation research.

4 SHALLOW-WATER MODEL EQUATIONS

It is a very simple task to extend the cascaded filter or averaging method to the derivation of shallow-water model equation for use in three-dimensional marine and estuary simulations. If it is assumed that the coordinate surface is placed at the still water level with z being positive upwards, then for very weak vertical acceleration the following model equations for continuity, x, y, and z momentum and a passive contaminant c can be written for either spatial or temporal filters as:

$$\frac{\partial \bar{u}_i}{\partial x_i} = 0 \tag{21}$$

$$\frac{\partial \bar{u}}{\partial t} + \frac{\partial(\bar{u}\bar{u}_j)}{\partial x_j} + Fu - f\bar{v} = - g \frac{\partial \bar{\eta}}{\partial x} + Nu + Ru \tag{22}$$

$$\frac{\partial \bar{v}}{\partial t} + \frac{\partial \bar{v}\ \bar{u}_j}{\partial x_j} + Fv + f\ \bar{u} = - g \frac{\partial \bar{\eta}}{\partial y} + Nv + Rv \tag{23}$$

$$\frac{\partial p}{\partial z} = - \rho g \tag{24}$$

$$\frac{\partial \bar{c}}{\partial t} + \frac{\partial(\bar{u}_j \bar{c})}{\partial x_j} + M = G \tag{25}$$

In the x and y momentum, equations (eqns. 22 and 23), f represents the earth's angular rotation frequency for a particular latitude and η is the free surface position measured from the stillwater level.

4.1 Definition of filters - spatial forms

The filter forms for a Gaussian spatial filter are:

$$Fu = F\overset{S}{u} = F(i=1) - \frac{f\delta_\ell^2}{4\gamma} \left\{ \frac{\partial^2 \bar{v}}{\partial x_\ell \partial x_\ell} \right\} \tag{26}$$

$$Fv = F\overset{S}{v} = F(i=2) + \frac{f\delta_\ell^2}{4\gamma} \left\{ \frac{\partial^2 \bar{u}}{\partial x_\ell \partial x_\ell} \right\} \tag{27}$$

$$M = M^S = \frac{\partial}{\partial t} \left\{ \frac{\delta_\ell^2}{4\gamma} \left(\frac{\partial^2 \bar{c}}{\partial x_\ell \partial x_\ell} \right) \right\} + \frac{\partial}{\partial x_j} \left\{ \frac{\delta_\ell^2}{4\gamma} \left(\frac{\partial \bar{c}}{\partial x_\ell} \right) \left(\frac{\partial \bar{u}_j}{\partial x_\ell} \right) \right\} \tag{28}$$

$$Nu = N\overset{S}{u} = - g \frac{\partial}{\partial x} \left\{ \frac{\delta_\ell^2}{4\gamma} \frac{\partial^2 \bar{\eta}}{\partial x_\ell \partial x_\ell} \right\} \tag{29}$$

$$Nv = N\overset{S}{v} = - g \frac{\partial}{\partial y} \left\{ \frac{\delta_\ell^2}{4\gamma} \frac{\partial^2 \bar{\eta}}{\partial x_\ell \partial x_\ell} \right\} \tag{30}$$

Here F(i=1) and F(i=2) are respectively Fi (eqn. 18) evaluated at i = 1 and i = 2; and $R\overset{S}{u}$, $R\overset{S}{v}$, and G^S are the appropriate residual or subgrid scale terms which must be expressed as a function of the mean flow variables.

4.2 Definition of filters - temporal forms

If only a temporal filter is desired then the filter terms become:

$$Fu = F_u^t = \frac{\delta t^2}{4\gamma}\frac{\partial}{\partial t}\left(\frac{\partial^2 \bar{u}}{\partial t^2}\right) + \frac{\partial}{\partial x_j}\left\{\frac{\delta t^2}{4\gamma}\frac{\delta^2(\bar{u}\,\bar{u}_j)}{\partial t^2}\right\} - f\,\frac{\delta t^2}{4\gamma}\frac{\partial^2 \bar{v}}{\partial t^2} \tag{31}$$

$$Fu = F_v^t = \frac{\delta t^2}{4\gamma}\frac{\partial}{\partial t}\left(\frac{\partial^2 \bar{v}}{\partial t^2}\right) + \frac{\partial}{\partial x_j}\left\{\frac{\delta t^2}{4\gamma}\frac{\partial^2(\bar{v}\bar{u}_j)}{\partial t^2}\right\} + f\,\frac{\delta t^2}{4\gamma}\frac{\partial^2 \bar{u}}{\partial t^2} \tag{32}$$

$$M = M^t = \frac{\delta t^2}{4\gamma}\frac{\partial}{\partial t}\left(\frac{\partial^2 \bar{c}}{\partial t^2}\right) + \frac{\partial}{\partial x_j}\left\{\frac{\delta t^2}{4\gamma}\frac{\partial^2(\bar{u}_j\bar{c})}{\partial t^2}\right\} \tag{33}$$

$$Nu = N_u^t = -\,g\,\frac{\partial}{\partial x}\left\{\frac{\delta t^2}{4\gamma}\frac{\partial^2 \bar{\eta}}{\partial t^2}\right\} \tag{34}$$

$$Nv = N_v^t = -g\,\frac{\partial}{\partial y}\left\{\frac{\delta t^2}{4\gamma}\frac{\partial^2 \bar{\eta}}{\partial t^2}\right\} \tag{35}$$

The appropriate forms for R_u^t, R_v^t, and G^t are also easily defined and must be expressed in terms of the averaged variables.

5 RELATIONSHIP OF ANALOG TO DIGITALLY AVERAGED OR FILTERED MODEL EQUATIONS

It is natural to ask whether all the additional terms represented by the filter terms in eqns. (21) to (35) have any digital equivalent; or is it possible that many of the higher order numerical schemes might, in an ad-hoc fashion, be representing the analog filter terms derived above. Such a comparison requires expanding all higher order procedures on the same grid and a comparison with the analog filtered terms digitized in a consistent fashion. Such a comparison is beyond this paper, however, several examples serve to illustrate that this may indeed be the case. Some three-dimensional expansions for all the governing equations are quite space consuming, therefore only simple term expansions are presented here with the extensions to full three-dimensional discretizations a matter of algebra and not conceptual difficulty.

5.1 The finite element approximation

If for example the linearized averaged x-momentum equation is taken as:

$$\frac{\partial \bar{u}}{\partial t} + g\,\frac{\partial \bar{\eta}}{\partial x} - f\bar{v} = 0 \tag{36}$$

then for a two-dimensional rectangular finite element representation approximated with bilinear basis functions the above equations becomes:

$$\frac{1}{36}\left\{\frac{\partial}{\partial t}(\bar{u}_{11}+4\bar{u}_{01}+\bar{u}_{-11})+4\left(\frac{\partial}{\partial t}(\bar{u}_{10}+4\bar{u}_{00}+\bar{u}_{-10})\right)+\frac{\partial}{\partial t}(\bar{u}_{i-1}+4\bar{u}_{0-1}+\bar{u}_{-1-1})\right\}$$

$$+\frac{g}{12\Delta x}\left\{(\bar{\eta}_{11}+4\bar{\eta}_{10}+\bar{\eta}_{1-1})-(\bar{\eta}_{-11}+4\bar{\eta}_{-10}+\bar{\eta}_{-1-1})\right\}$$

$$-\frac{f}{36}\left\{(\bar{v}_{11}+4\bar{v}_{01}+\bar{v}_{-11})+4(\bar{v}_{10}+4\bar{v}_{00}+\bar{v}_{-10})+(\bar{v}_{1-1}+4v_{0-1}+\bar{v}_{-1-1})\right\}=0 \quad (37)$$

Where the numerical subscript notation represents the evaluation of the dependent variables at grid points i+1, j+1, i, j, i-1, j-1; etc.

If eqn. 22 is expanded for this two dimensional case with either a Gaussian spatial filter (γ =6) or a uniform filter, the form becomes:

$$\frac{\partial}{\partial t}\left\{\bar{u}+\frac{\delta x^2}{6}\frac{\partial^2\bar{u}}{\partial x^2}+\frac{\delta y^2}{6}\frac{\partial^2\bar{u}}{\partial y^2}+\frac{\delta x^2\delta y^2}{36}\frac{\partial^4\bar{u}}{\partial x^2\partial y^2}\right\}$$

$$+g\frac{\partial}{\partial x}\left\{\bar{\eta}+\frac{\delta x^2}{6}\frac{\partial^2\bar{\eta}}{\partial x^2}+\frac{\delta y^2}{6}\frac{\partial^2\eta}{\partial y^2}+\frac{\delta x^2\delta y^2}{36}\frac{\partial^4\bar{\eta}}{\partial x^2\delta y^2}\right\}$$

$$-f\left\{\bar{v}+\frac{\delta x^2}{6}\frac{\partial^2\bar{v}}{\partial x^2}+\frac{\delta y^2}{6}\frac{\partial^2\bar{v}}{\partial y^2}+\frac{\delta x^2\delta y^2}{36}\frac{\partial^4\bar{v}}{\partial x^2\partial y^2}\right\}=0 \quad (38)$$

If the simplest second order centered approximations are used for the differential terms in eqn. (36), and δx and δy are set equal to $2\Delta x$ and $2\Delta y$, respectively, then the resulting expression is exactly identical to the finite element discretization.

5.2 Shuman filter factor form: linear equation

Shuman (1962) presented an advanced discretization for the pressure and Coriolis terms in the equations. Again, by way of example, eqn. (36) is used. Shuman's scheme for this equation results in the following expression:

$$\frac{\partial\bar{u}}{\partial t}+\frac{g}{8\Delta X}\left\{(\bar{\eta}_{11}+2\bar{\eta}_{10}+\bar{\eta}_{1-1})-(\bar{\eta}_{-11}+2\bar{\eta}_{-10}+\bar{\eta}_{-1-1})\right\}$$

$$-\frac{f}{16}\left\{(\bar{v}_{11}+2\bar{v}_{01}+\bar{v}_{-11})+2(\bar{v}_{10}+2\bar{v}_{00}+\bar{v}_{-10})+(\bar{v}_{1-1}+2\bar{v}_{0-1}+v_{-1-1})\right\}=0 \quad (39)$$

By using a Gaussian filter applied only on the spatial derivatives and with discretizations again only centered, second order finite difference approximations, an exact equivalence is obtained. It is interesting to note that the averaging has not been consistently applied to the time differential term; however in this technique as used by Jelensianski (1972) in the SPLASH and SLOSH storm surge models, other ad hoc time filtering was required.

5.3 Pressure averaging technique; dissipative interface scheme

As suggested by Abbott (1979) and used by Jensen (1983), the dissipative interface scheme, described on the simple x-momentum equation without the Coriolis term, the expansion is:

$$\frac{\partial \bar{u}}{\partial t} + \frac{g}{\Delta x} \{\theta (\bar{\eta}_1^{n+1} - \bar{\eta}_0^{n+1}) + (1-2\theta)(\bar{\eta}_1^{n} - \bar{\eta}_0^{n}) + \theta(\bar{\eta}_1^{n-1} - \bar{\eta}_0^{n-1})\} \tag{40}$$

Using a general Gaussian filter, the expansion becomes

$$\frac{\partial \bar{u}}{\partial t} + g \frac{\partial}{\partial x} \{\bar{\eta} + \frac{\delta t^2}{\gamma} \frac{\delta^2 \bar{\eta}}{\partial t^2} + \text{HOT} \} \tag{41}$$

and by replacing the terms in eqn. (41) with second order centered time and space finite difference approximations, the following approximation occurs:

$$\frac{\partial \bar{u}}{\partial t} + \frac{g}{\Delta x} \{ \frac{1}{\gamma} (\bar{\eta}_1^{n+1} - \bar{\eta}_0^{n+1}) + (1 - \frac{2}{\gamma})(\bar{\eta}_1^{n} - \bar{\eta}_0^{n}) + \frac{1}{\gamma} (\bar{\eta}_1^{n-1} - \bar{\eta}_0^{n-1})\} \tag{42}$$

Equivalences between analog and digital forms occur as follows: for $\theta = 1/\gamma$ and $\gamma \geqslant 2$ the filter scheme is equivalent to Abbott's (1979) method; for $\gamma = 4$ and $\theta = 0.25$, the method is identical to the method by Shuman (1971); while for $\gamma = 2.0$, and $\theta = 0.50$, the filter expansion is equivalent to the method of McCowan (1978) and Hansen (1983).

Many more comparisons of this nature are currently underway by the second author for his dissertation research. The correspondence between the analog and digital forms lends credence to the analog derivations performed in the first part of this paper. It is also the case that since the analog derivations are complete and robust, that the non-presence of various digital filter terms in a numerical model might expose a flaw in the numerical treatment of a term in the basic equations.

6 CONCLUSIONS

A generalized averaging or filtering procedure for deriving turbulent flow equations accounting for continuously varying average dependent variables has been presented and applied to the three-dimensional shallow water marine and estuary equations. In the course of these derivations, questions as to the adequacy of the Reynolds average approach are raised and a method for remedying the flaws by the filter expansion approach is recommended. Since so many models are based upon these Reynolds average equations, it is suggested that a number of higher order discretizations in these models are used to remedy the averaging defects. As a means of initial testing of this hypothesis, various numerical schemes are shown to be equivalent to the analog filter forms derived here. Further work is proceeding on this comparison.

7 ACKNOWLEDGMENTS

This work was supported by the National Science Foundation research grant No. CEE 8410522 and their support is greatly appreciated.

8 REFERENCES

Abbott, M. A., 1979. Computational Hydraulics. Pittman, London.

Babajimopoulos, C., and Bedford, K. W., 1980. Formulating lake models which preserve spectral statistics. J. Hyd. Div., ASCE, 106 (HY1): 1-19.

Bedford, K. W., 1981. Spectra preservation capabilities of Great Lakes transport models. In: Hugo B. Fischer (Editor), Transport Models for Inland and Coastal Waters. Academic Press, New York, 172-221.

Bedford, K. W., and Babajimopoulos, C., 1980. Verifying lake transport models with spectral statistics. J. Hyd. Div., ASCE, 106 (HY1): 21-38.

Clark, R. A., Ferziger, J. H., and Reynolds, W. C., 1977. Evaluation of Subgrid-scale Turbulence Models Using a Fully Simulated Turbulent Flow. Rept. TF-9, Stanford University Thermosciences Division.

Dakhoul, Y. M., and Bedford, K. W., 1986a. Improved averaging method for turbulent flow simulation. Part I; theoretical development and application to Burger's transport equation. Int. J. Numer. Meth. Fluids, 6: 49-64.

Dakhoul, Y. M., and Bedford, K. W., 1986b. Improved averaging method for turbulent flow simulation. Part II; calculations and verification. Int. J. Numer. Meth. Fluids, 6: 65-82.

Deardorff, J. W., 1973. The use of subgrid transport operations in a three-dimensional model of atmospheric turbulence. J. Fluid Eng., 429-438.

Hinze, J. O., 1975. Turbulence, 2nd ed., McGraw Hill, New York.

Jelesnianski, C. P., 1972. SPLASH 1. (Special Program to List Amplitudes of Surges from Hurricanes) Landfall Storms. NOAA Tech. Mem. NWS TDL-46, Silver Springs, Md.

Jensen, R. E., 1983. A Consistent Analysis of Boussinesq-type Water Wave Equations in Continuous and Discrete Form. Ph.D. thesis, Texas A and M University, College Station.

Leonard, A., 1974. Energy cascade in large-eddy simulation of turbulent flows. Adv. Geophys, 18A: 237-248.

McCowan, A. D., 1978. Numerical simulation of shallow water waves. Proc. Fourth Australian Conference on Coastal and Ocean Engineering, Adelaide, Australia, 132-136.

Rabiner, C. R., and Gold, B., 1975. Theory and Application of Digital Signal Processing. Prentice Hall, NJ.

Reynolds, O., 1895. On the dynamical theory of incompressible viscous fluids and the determination of the criterion. Philos. Trans. R. Soc. London, Ser. A, 1986: 123-164.

Shuman, F. G., 1962. Numerical experiments with the primitive equations. Proc. Int. Symp. Num. Weather Prediction in Tokyo, November 1960. Meteorological Society of Japan, Tokyo, 85-107.

Shuman, F. G., 1971. Resuscitation of an integration procedure. NWC Office Note 54.

Smagorinsky, J., 1963. General circulation experiments with the primitive equations: I. The basic experiment, Mon. Weather Rev., 91: 99-161.

A LIMITED AREA MODEL FOR THE GULF STREAM REGION

WILLIAM R. HOLLAND
National Center for Atmospheric Research
P.O. Box 3000, Boulder, Colorado 80307

ABSTRACT

Studies of eddy/mean flow interactions in basin-scale, eddy-resolving numerical models have been carried out for a decade or so. Recently, as a result of a need for better vertical and horizontal resolution, a new generation of models designed to accomplish calculations in limited areas of a basin, models with open boundary conditions, have begun to be developed. These models have been applied to the Gulf Stream region, to the Agulhas Current, to the Brazil/Falklands Current confluence region, and to the Tasman Sea region. These results will be discussed, both in terms of "physics" of these energetic western boundary current regions and in terms of the "numerics" associated with very fine resolution and with open boundary conditions. The most ambitious undertaking of this kind by the author and his colleagues has been a very fine resolution model study of the Gulf Stream region from Cape Hatteras to the Grand Banks and bordered on the north and south by boundaries at 50 and 25 Degrees North respectively. The Gulf Stream enters the region of interest as a western boundary current with a certain specified inflow. The flow exits the region by way of open boundaries on the east, north, and south. Various possible boundary conditions have been tried and the implications of these for interior physical behavior examined. In particular, the nature of Gulf Stream meander processes and the role of important bottom relief in the region are discussed.

1. INTRODUCTION

Studies of eddy/mean flow interactions in basin scale, eddy-resolved, numerical models have been carried out for a decade or so. The early work focused upon the origin of mesoscale eddies as a result of baroclinic and barotropic instabilities of the western boundary current and its seaward extension, and of the Recirculation nearby (Holland and Lin, 1975a,b; Holland, 1978). Recent work has begun to refine the picture and has examined various theoretical issues. These include studies of the homogenization of potential vorticity (Holland, Keffer, and Rhines, 1984), instability mechanisms (Haidvogel and Holland, 1978; Holland and Haidvogel, 1980), eddy mixing and gyre equilibration (Rhines and Holland, 1979; Holland and Rhines, 1980), and the penetration scale of the Gulf Stream (Holland and Schmitz, 1985).

In addition to these theoretical studies, comparisons of model results with observations have played an important part in model refinement and in identifying important issues regarding the physics of the Gulf Stream system

(Schmitz and Holland, 1982; Schmitz et al., 1982; Holland, 1985; Schmitz and Holland, 1986). This work is currently being extended with models of much higher vertical resolution than heretofore to examine the vertical structure of mean and eddy fields in the Gulf Stream and Kuroshio. As illustrations of this kind of comparison, Schmitz and Holland (1986) show several observational/model intercomparisons that are currently being examined. In particular meridional sections of mean zonal flow and eddy kinetic energy in eight layer numerical experiments have much in common with North Atlantic and North Pacific current meter mooring data. The correspondence is by no means exact but it is clear that both vertical and horizontal structure in the numerical experiment has many features in common with the data in both mean and eddy quantities, including approximately correct ratios of surface to deep mean and eddy currents in the intense flow, and similar meridional structure in terms of eastward and westward (recirculating) mean flows. An additional comparison between the data at 55W (in the Gulf Stream) and a similar point in the intense eastward flow of one of these eight layer model calculations indicates that the vertical structure of eddy kinetic energy is remarkably similar to observations, suggesting strongly that we are on the right track regarding the eddy/mean flow interactions that give rise to the intense eddy field in these vigorously eddying western boundary regions.

To date, most numerical studies have been highly idealized with respect to the geometry of the ocean basin in question. There are many good reasons for this. For one thing, simpler situations are a vital and necessary part of understanding the more complex ones. When confronting questions requiring detailed comparisons with observations, however, one must ask whether or where dynamically similar regions of an idealized basin can be found in the western North Atlantic. Or, turning the question around, where in an idealized basin should one seek to compare observations along 55W (or any other place)? Thus, as models more faithfully reproduce observed features, we are driven toward more faithful inclusion in our models of basin shape, bottom topography, and boundary conditions (e.g., wind stress, buoyancy flux, inflow and outflow across the boundaries of local domains, etc.) Such models, with various physical and geometrical factors successively put in or taken out, allow us to ascertain which features are key to understanding the dynamics and which are not.

In the last several years, we have begun to develop models of the North Atlantic (and other basins) that have somewhat 'realistic' geometry. For example, Holland (1983, 1986) examined the wind-driven circulation in the North Atlantic basin from 15°N to 65°N, using a three layer quasigeostrophic model with ¼ degree horizontal resolution. Studies of the role of eddies in the general circulation and studies of the oceanic response to transient

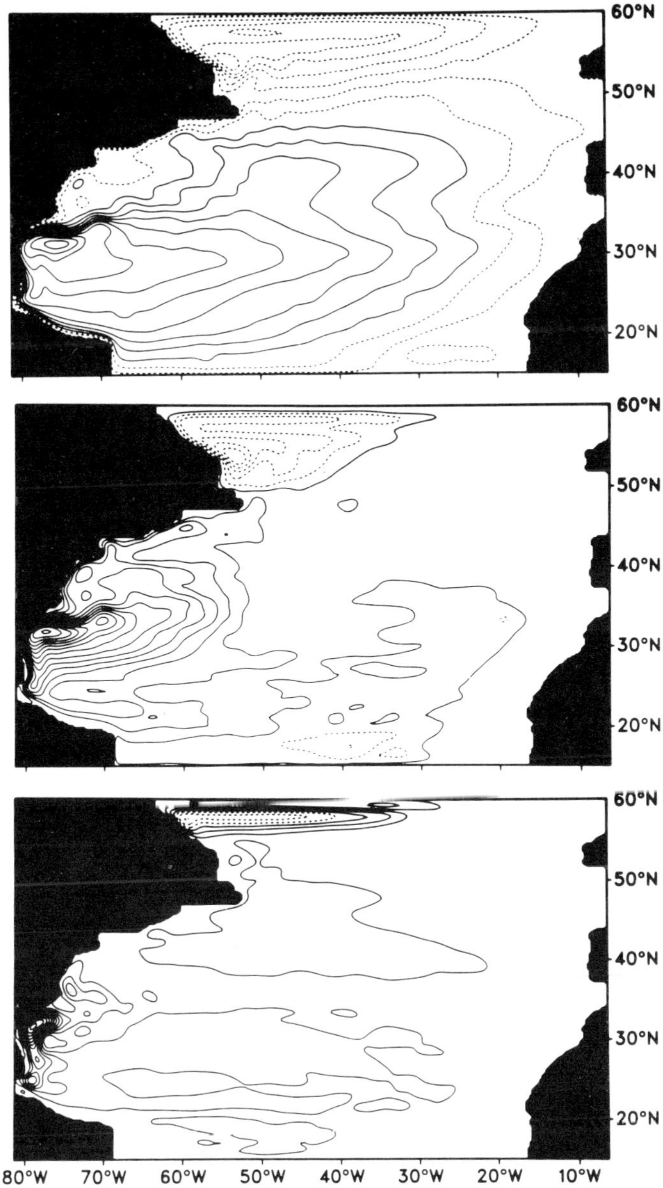

Figure 1. The time averaged circulation in the North Atlantic basin using the annual mean Hellerman wind stresses, based upon a three layer QG model. The streamfunctions at 150 m, 650 m, and 3000 m are shown.

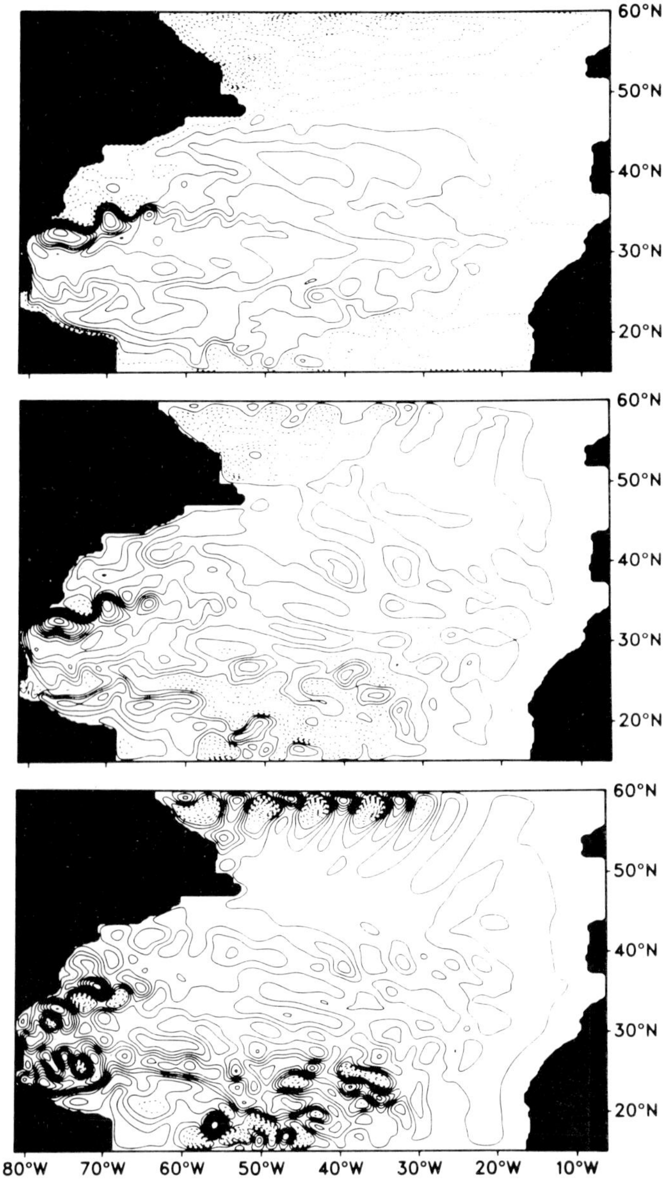

Figure 2. The instantaneous streamfunctions of the flow in the North Atlantic basin at the same levels as in figure 1. Note the strong Gulf Stream meandering in the upper layer and the important eddying circulations in the deep ocean.

wind forcing continue. Figure 1 shows, for example, the time-averaged streamfunction for a particular case with steady wind forcing. The upper layer (a) shows the mean gyre forced by the mean annual Hellerman wind stress; the middle (b) and lower (c) layers show the time-averaged eddy driven components of the flow, with a broader recirculation in the main thermocline, and a very narrow recirculation in the deep water under the mean Gulf Stream. These figures are the result of time averaging over a five year period. Figure 2 shows instantaneous views of the streamfunctions, illustrating the rich eddy field in the western North Atlantic.

Figure 3 shows the upper layer instantaneous streamfunction for a case with Hellerman's monthly forcing with the annual component removed. Thus the forcing is purely transient with periods between 2 and 12 months represented. The main response (after a 20 year spin-up time) shows westward propagating, annual period, baroclinic Rossby waves as the primary response. The amplitude of these waves is small compared to the eddy signal shown in the experiment above (figures 1 and 2) but away from the intense Gulf Stream region of instability, particularly in the eastern basin, the transient signal would dominate.

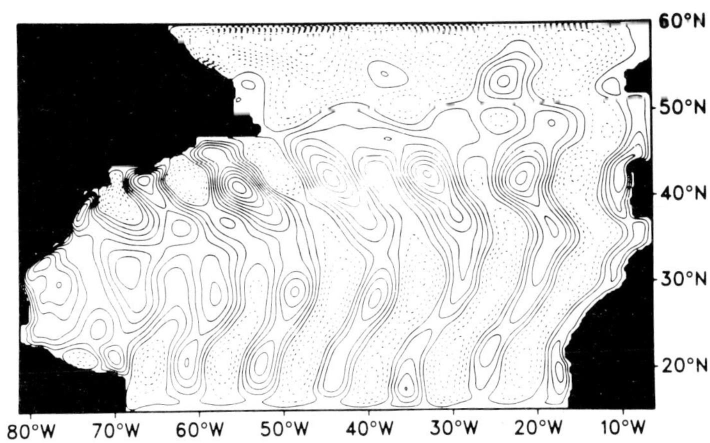

Figure 3. An instantaneous, upper layer streamfunction in the North Atlantic QG model, driven by Hellerman's seasonal winds only (no mean forcing). Westward propagating, annual period, first baroclinic mode Rossby waves dominate the solution.

Both these experiments have been run with constant depth oceans. If topography had been present and higher frequency wind forcing included, the ocean response might have shown important deep transient flows driven directly by transient forcing. Such experiments have yet to be carried out.

Even though these studies can be made with a horizontal resolution of ¼ degree in latitude and longitude, for some purposes even higher resolution will be needed. This is particularly true when higher vertical resolution (more than three layers) is used. This is due to the fact that with more vertical resolution, higher baroclinic modes, with smaller Rossby radii of deformation, are included. It should be kept in mind that such experiments will eventually be carried out using the primitive equations, with their much greater computational cost. Therefore a new generation of eddy-resolving limited area models (ELAM's) that can successfully handle open boundaries is needed, using both quasigeostrophic and primitive equation physics, that will allow us to telescope in on local regions of a larger domain.

This paper discusses some initial attempts to study a variety of eddy-rich western boundary current regions, including the Gulf Stream region, the Agulhas Current, the Brazil/Falklands Current confluence region, and the Tasman Sea. Preliminary results will be shown to illustrate the power and the limitations of the Limited Area Model approach as well as to indicate some of the interesting variety of behaviors to be found.

2. WESTERN BOUNDARY CURRENT MODELS

Regional ocean models can include very high horizontal resolution and realistic geometry at the expense of having to deal with open boundary conditions. There are many possible ways to handle the open boundaries but all must deal with two basic problems: (i) how does the external ocean influence the domain of interest, and (ii) how do features of the flow generated internally realistically cross the open boundaries, thus leaving the domain of interest? These difficulties arise of course because the open boundary conditions themselves depend upon the circulation in both the internal and external ocean regions.

Radiation boundary conditions allow, for some problems, the second of these problems to be addressed. Extrapolation techniques are used to extend to the boundary changes dictated by interior influences propagating toward that boundary. Such techniques work well for some problems and not at all for others (Orlanski, 1976; Camerlengo and O'Brien, 1980; Roed and Smedstad, 1984; Chapman, 1985).

The influence of the outer ocean on the inner one could be handled in various ways: by embedding the domain of interest in a larger, perhaps coarse resolution, domain for which numerical calculation are also done; by carrying out separately a coarser domain calculation and saving the 'appropriate' boundary conditions therefrom to be imposed on a later local calculation; by parameterizing the far field influence in some fashion, for example by imposing cross boundary flow and density information from theoretical ideas such as "Sverdrup balance" and "geostrophy". All of these techniques have important deficiences. The first technique allows feedback between the inner and outer domain but the scales of motion allowed in the coarse resolution will not adequately represent the scales in the fine resolution. The second and third techniques simplify the problem by disallowing feedback; the outer domain is not affected by the evolving solutions in the domain of interest. This may or may not be a crucial choice, depending upon the problem and upon the realism of the externally imposed boundary conditions.

Finally, both problems (i) and (ii) go away if good enough observations exist at the open boundary. At the present time, this is unlikely for many problems of interest because of the scale of effort needed observationally to fully describe the space/time behavior of the flow along a section or boundary of any length.

Western boundary regions are especially prone to a limited area approach, because the western boundary is a physical one, not requiring the approximations described above. In addition, the eastern side of the domain is often less transient and less inertial, and the beta effect causes internally created transience (due to instabilities) to propagate westward, away from the open boundary. This makes the Gulf Stream region, with open boundaries on the east at 40°W and on the south at 25°N, an especially attractive region for such a study (in addition to the more obvious reasons; that the Gulf Stream and its meandering and ring formation behavior is the best known region of the World Ocean).

Before looking at some results from this region, however, let us first examine some models under development for other western boundary regions, all from the Southern Hemisphere. These are the Agulhas Current region, the Brazil-Falklands Currents confluence region, including Circumpolar flow through Drake Passage, and the region of the Tasman Sea. In the first two of these regions, the domain is not totally blocked on the western side of the domain of interest, and in the third, the problem is complicated by the presence of the New Zealand land mass. All of these lead to special problems and situations that, in the limited area model context, require special solutions.

The Agulhas Current region is one of considerable interest, firstly because it is the only western boundary current that runs out of boundary and secondly because the region south of South Africa is a crossroads of interocean exchange (Gordon, 1986). The dynamical nature of that current thus becomes of larger significance than just the local one, and the manner in which the current carries water from the Indian Ocean into the South Altantic and the extent to which some of that water "retroflects" back into the South Indian Ocean is of considerable interest to large scale oceanographers.

Figure 4 shows results from two model calculations (Holland, 1987a) using a three layer quasigeostrophic limited area model of the Agulhas region. The upper layer streamfunctions are shown at a particular instant in each calculation to illustrate briefly three points: the technique by which the boundary regions on both east and west 'parameterize' the interaction of this local domain with the rest of the South Indian and Atlantic Oceans; the

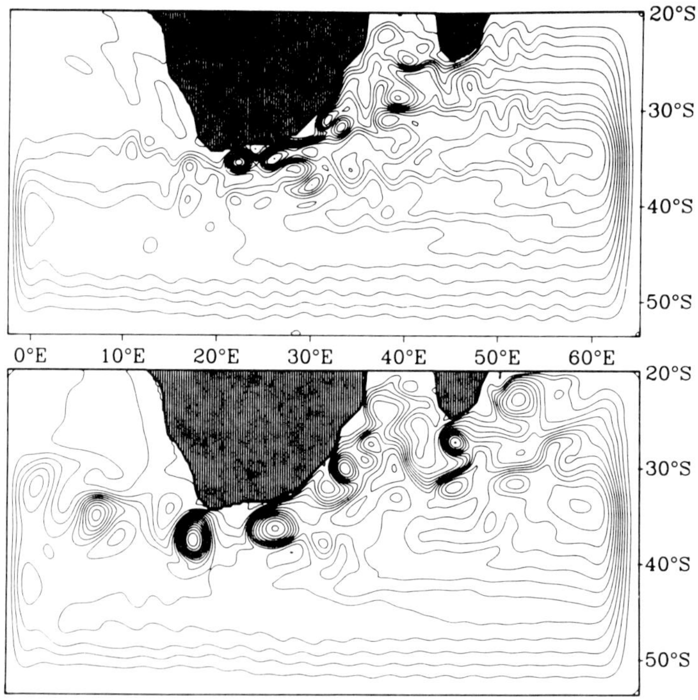

Figure 4. Instantaneous upper layer streamfunctions for two numerical experiments covering the region south of Southern Africa. The large scale, counterclockwise circulation is driven near the eastern side of the limited area to produce a given zonal flow near the eastern boundary. The land masses are Madagascar and Southern Africa.

transient nature of the flow as the Agulhas Current, flowing down the east coast of Southern Africa, turns westward and partially retroflects back to the east; and the difference between the two calculations when bottom topography is introduced.

Both calculations are the result of a long spin-up until a statistical equilibrium is reached. The local domain is actually closed but the flow near the eastern side is driven by a mass flux in and out of a narrow eastern boundary region in which special forcing functions are imposed. These conditions give a certain zonal flow whose meridional and vertical structure is known. A simple analogy to wind-stress curl forcing, acting upon each layer, is used to create the zonal flow wanted.

The same zonal flow is driven in both calculations. The westward flow toward the African continent and Madagascar in the north is unstable and creates a highly transient Agulhas Current formation region. As the boundary current moves southward and turns more and more westward, anti-cyclonic eddies form into ring-like features. The net transport by the Agulhas Current is about 60 Sverdrups, and the rings have similar transports. In the flat bottom case (a), the eddies are quite small scale and have relatively fast westward propagation speeds. In the experiment with bottom topography (b), the eddies and rings are much larger and relatively slow moving. Note that the amount of water that retroflects (returns back to the east, south of South Africa) in the topographic case is somewhat greater than that in the case without topography, suggesting the important role of eddy-topographic interactions in the retroflection process (Holland, 1987a).

The rings in both cases move westward and ultimately are absorbed in a region of enhanced friction near the western boundary. The western boundary layer acts as a passive southward recirculation zone in these cases as well as an "eddy absorbor."

This kind of local calculation, in which the dynamical behavior is determined almost entirely by the instability processes in the Agulhas formation and retroflection regions, can be carried out without complicated boundary conditions. "Pumps and baffles" can be inserted to produce the large scale circulation desired. The transient eddies are ultimately absorbed near the western edge of the domain without much reflection, so that a simple sponge layer is workable, and the forcing region near the eastern edge of the domain allows for a simple specification of the zonal flow to parameterize the gyre structure further eastward in the South Indian Ocean.

A second model study (Holland, 1987b), this time for the region of the Brazil/Falklands Current confluence, is illustrated in figure 5. The model has open boundaries on the eastern and northern sides of the domain and on the upstream side west of South America where the Circumpolar Current enters

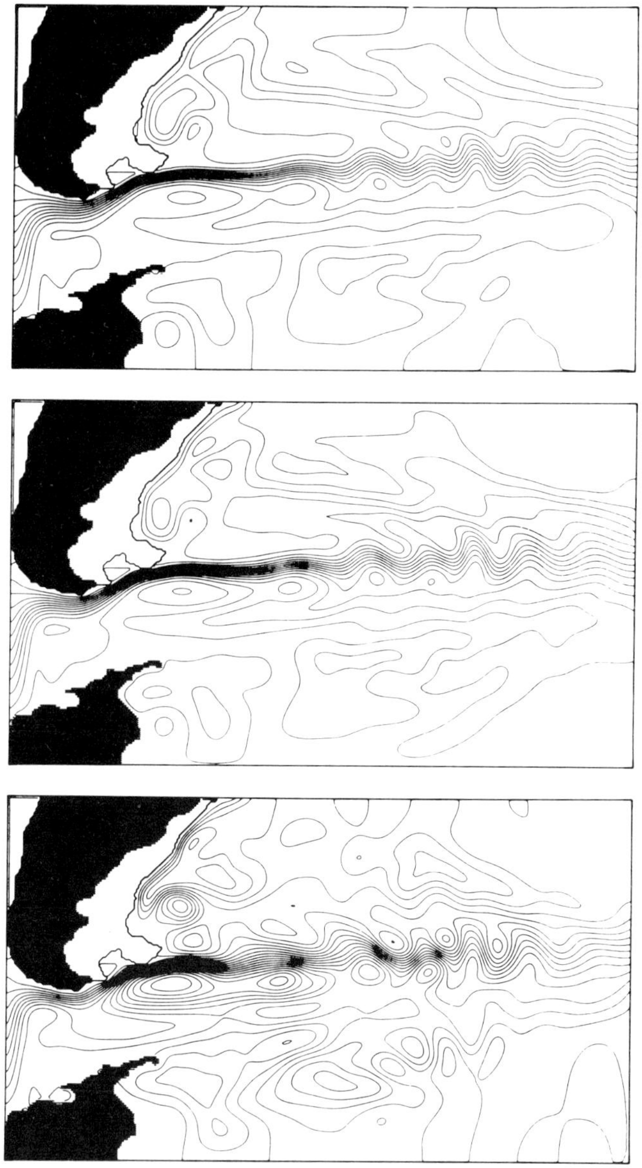

Figure 5. The quasigeostrophic streamfunctions at three levels in a limited area model of the Brazil Current/Falklands Current confluence. The Brazil Current and Circumpolar flow enter the region on the north and west boundaries respectively and the combined flow exits on the east boundary.

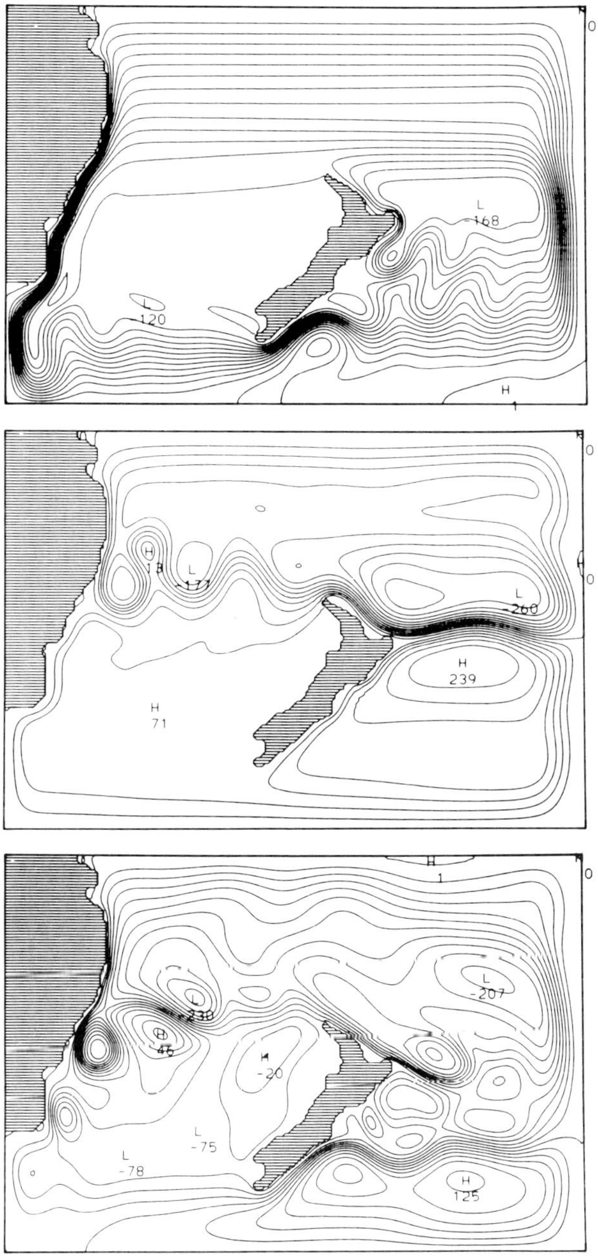

Figure 6. Three model calculations for the region of the Southwest Pacific using a barotropic model. The highly transient flow is driven near the eastern boundary by forcing terms that create multiple gyres (a parameterization of wind forcing further to the east). Top (a): Subtropical gyre dominates whole region; middle (b): the line separating the subtropical and subpolar gyres is at mid-basin; bottom (c): the line separating the gyres is at the mid-latitude of the New Zealand land mass. Note: the northernmost gyre circulates counterclockwise.

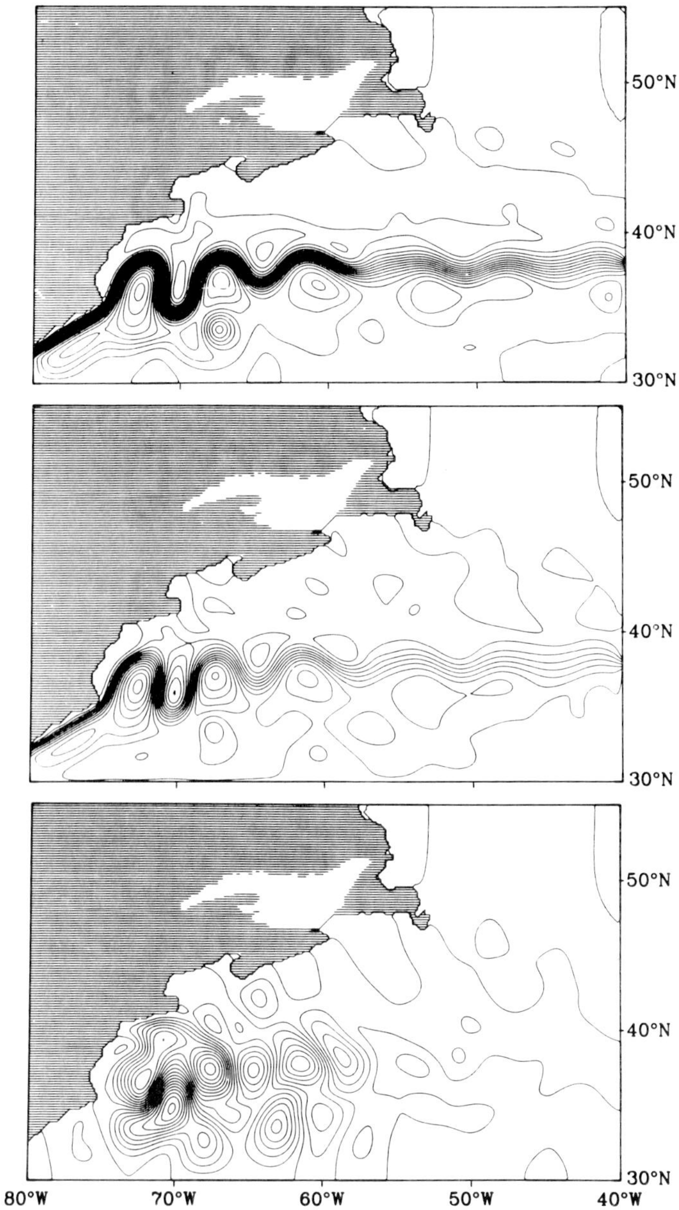

Figure 7. A regional model of the Gulf Stream region. The instantaneous streamfunctions at three levels are shown (150 m, 650 m, and 3000 m respectively). The circulation is driven by inflows and outflows only; no wind forcing present. In this experiment the ocean is of constant depth.

the region. The Brazil Current enters from the north and the combined flows exit on the east. In this three layer, constant depth calculation, boundary conditions on the streamfunction in each layer are specified so that the horizontal location and vertical structure of the inflows and outflows is fixed once and for all. The influence of the rest of the Southern Ocean circulation is "parameterized" in these boundary conditions. The location of the eastern boundary outflow is especially important to the interior solution and is chosen here to roughly coincide with the Circumpolar flow at the Prime Meridian (0°E longitude), as suggested by temperature and salinity observations and geostrophic calculations.

The vorticity at inflow points is set by the flow structure there but the vorticity at outflow points is gotten by a simple extrapolation procedure from the interior; essentially the longitudinal gradient of vorticity is set to zero. This allows the circulation some freedom to export vorticity from the region and provides an opportunity for transients to be absorbed and/or exported across this boundary. In addition, the western, northern and eastern open boundaries have adjacent narrow zones of enhanced friction that help absorb transients reaching these boundaries.

The Circumpolar Current turns northward as it rounds the southern tip of South America and brushes the southern edge of Falklands plateau. (Note that our continental boundary is chosen to be the 200 meter depth contour so that shallow shelves connect the Falklands plateau to South America). The Brazil Current coming southward turns seaward before encountering the topography of the plateau, at least at this instant of the calculation. The entire region is highly transient; the Brazil Current develops strong eddies as it turns eastward and the combined flow downstream exhibits instability-driven meandering almost to the eastern boundary. There is considerable recirculation in the deep ocean south of the Circumpolar Current.

A third set of calculations, using the limited area approach for the region of the Southwestern Pacific, shows a variety of possible behaviors for the East Australia Current (figure 6). In particular, the nature of the circulation varies enormously as the parameterized wind gyre forcing to the east of this domain of interest is changed. An examination of the wind stresses over the South Pacific shows very large variability on the seasonal time scale. The boundary between the subtropical and subpolar gyres moves over a wide range of latitudes, suggesting the three calculations shown. Here the important role of the New Zealand land mass and the relationship of the wind-stress curl distribution is explored using a barotropic, constant depth model.

The top picture (figure 6a) shows an instantaneous streamfunction when the entire domain is "forced" from the east, as in the Agulhas case above, with a single anticlockwise (subtropical type) wind forcing. The zonal flow in the north forms a western boundary current (our East Australia Current) that flows southward right through the Tasman Sea and right around New Zealand. The southern subtropical gyre boundary, as demarked by our "wind forcing" region on the east, is near the southern boundary of the domain. The middle picture (figure 6b) shows the opposite extreme when the boundary between the northern subtropical gyre and the southern subpolar gyre is much further north (near the northern tip of New Zealand). Finally the bottom picture (figure 6c) illustrates the middle ground in which the gyre boundary is about at the midlatitude of the New Zealand land mass.

The circulations are very different. In the last two cases, much of the East Australia Current separates from the boundary to flow north of New Zealand. The eastward flow is highly transient and, particularly in the final case, large counterclockwise eddies form to move southward into the relatively quiet region west of New Zealand (a Rossby Wave shadow zone that is shielded from the distant forcing to the east). This final solution has the flavor of the actual situation and is being investigated in a baroclinic model ocean (Holland, 1987c).

3. THE GULF STREAM MODEL

For the purpose of studying the meandering and ring formation processes in the Gulf Stream region, a quasigeostrophic limited area model of the region from 30°N to 55°N and 80°W to 40°W has been constructed. In the numerical experiments shown here, there is no wind forcing acting upon the region of interest; the circulation is entirely driven by the specified inflow of the Gulf Stream as a western boundary current south of Cape Hatteras and by the specified outflow of the Stream across the eastern boundary just south of 40°N. The model has three layers in the vertical; the two numerical experiments here have inflow and outflow transports of 30 Sverdrups in the upper layer (300 m thick), 32 Sverdrups in the second layer (700 m thick), and no inflow or outflow in the lowest layer (4000 m thick). The boundary conditions are similar to those in the Brazil/Falklands regional problem; streamfunction and vorticity specified on inflow, streamfunction specified on outflow, and vorticity gradient set to zero at outflow points. The location of both inflow and outflow is fixed in time, restricting the meandering process near the eastern boundary.

Figure 7 shows the streamfunction at a particular instant after a long spinup process for a case of constant depth. The meanders seen here are strongly time dependent and occasionally a warm-core or cold-core ring is shed. The deep layer, not driven by boundary inflows, is dominated by an eddy field that extends from just off Cape Hatteras to 50°W beneath the Gulf Stream. Figure 8 shows the upper layer streamfunction 20 days later, just as a cold ring is about to break off after the deep meander development.

Figure 9 shows the streamfunction at a particular instant for a case with bottom topography (shown in figure 10). The Stream has a very convoluted character in the vicinity of the New England Seamounts, but it is not clear from any instantaneous picture how important the topographic influence might be. The deep eddy field is clearly strongly influenced by bathymetry, but analyses of various statistics are required to ascertain whether the eddy field can communicate upward the location of the Seamounts.

One such statistic is shown in figure 11. The central streamline for each of these cases is shown every 20 days for a 680 day period, to indicate something of the nature of the envelope of the various paths of the Stream. Figure 11a shows the flat bottom case, figure 11b the case with bottom topography. The latter case shows the propensity for cold core ring formation just west of the New England Seamounts. The rings formed there move southwestward until they "feel" the continental slope southeast of Cape Hatteras, where they decay but also affect the Stream north of their location.

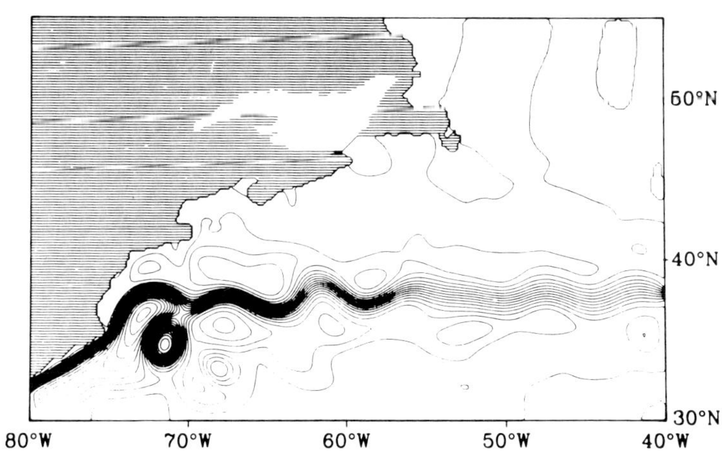

Figure 8. The upper level streamfunction 20 days following the maps shown in figure 7. Note that the meander in figure 7 has deepened and has just formed a ring-like feature south of the Stream.

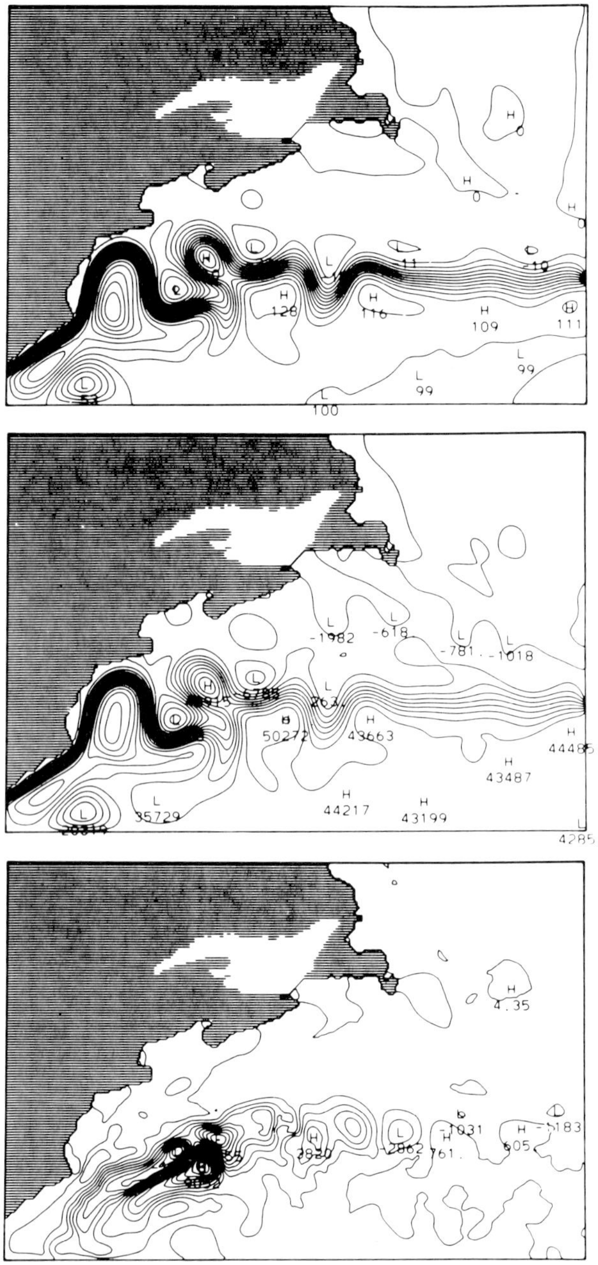

Figure 9. Instantaneous streamfunctions at three levels are shown (150 m, 650 m, and 3000 m respectively) for calculation like that in figure 7 except that bottom topography is included (see figure 10). The regional model is driven precisely the same as the flat bottom case in figure 7.

Figures 12 and 13 show another statistic for these two cases - the eddy kinetic energy patterns based upon a five year average. The maxima follow the Gulf Stream axis and are within a factor of two of observed values at all levels. The case with bottom topography (figure 13) shows clearly the effects of the vigorous cold core ring development and shedding, as a southward extension of the EKE pattern. Note also the large differences in the two abyssal patterns, presumably directly related to the presence or absence of variable depth.

These results are intended mainly to illustrate the nature of the limited area model approach. Much deeper analyses into these numerical experiments and many more experiments themselves are needed to even begin to understand the dynamical behaviors found here and to even begin to sort out the dependence upon boundary conditions, topographic effects, local wind forcing, and the other various parameters that govern the flow (friction coefficient, vertical and horizontal resolution, strength of inflow/out, etc., etc.). Studies of this kind are being vigorously pursued.

Moreover it is likely that primitive equation models with much higher vertical resolution will have to play a role in model studies of the Gulf Stream region. Such studies are also underway and new ways of coping with

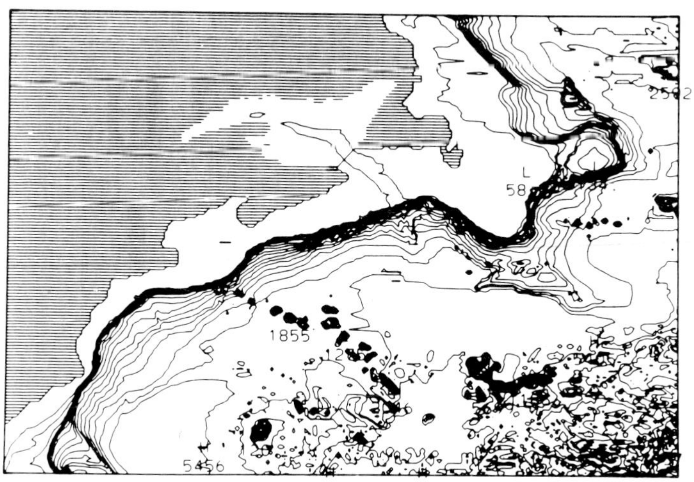

Figure 10. The bottom topography for the calculation illustrated in figure 9. The topographic variations are considered to exist only in the lowest layer. Note the line of the New England seamounts that cross the path of the Gulf Stream.

the open boundary conditions are being examined. In the end, it may be necessary to acquire an observational description of the Gulf Stream, particularly at outflow from the domain where the Stream meanders broadly over several degrees of latitude (say at 50°W or 40°W) to adequately handle the "prediction" of behavior in this region. It is likely that satellite data (AVHRR, altimetry) can very usefully help us to initialize models of this kind for dynamic calculations and to develop assimilation schemes for predictive calculations.

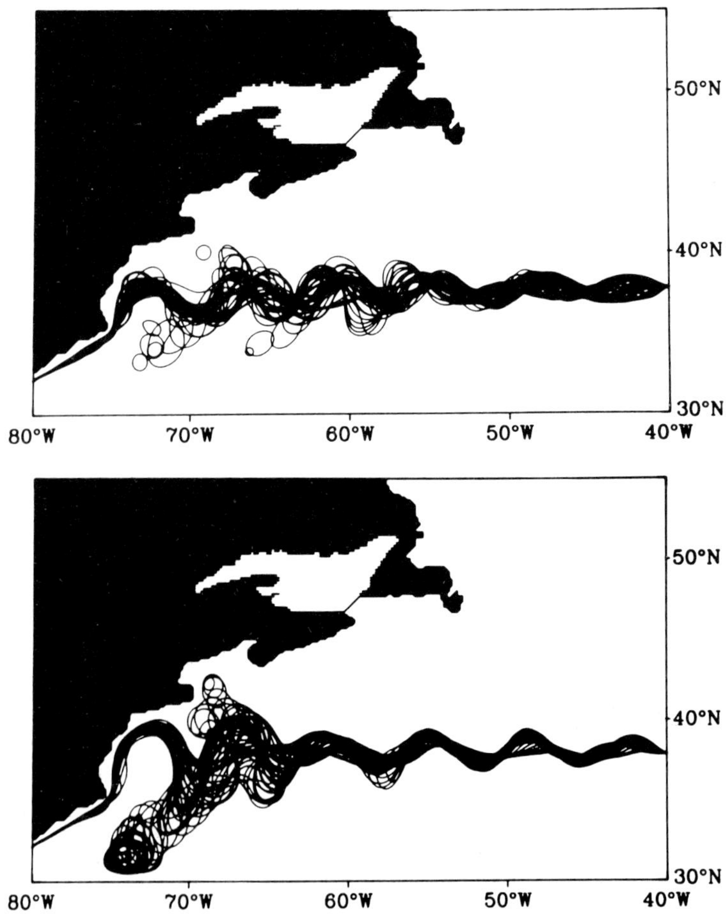

Figure 11. A time sequence of individual upper layer streamlines marking the middle of the inflowing Gulf Stream. The streamline position every 20 days (for 680 days total) is shown as an indication of the envelope of Gulf Stream paths. Upper:the constant depth case; Lower: the case with variable depth.

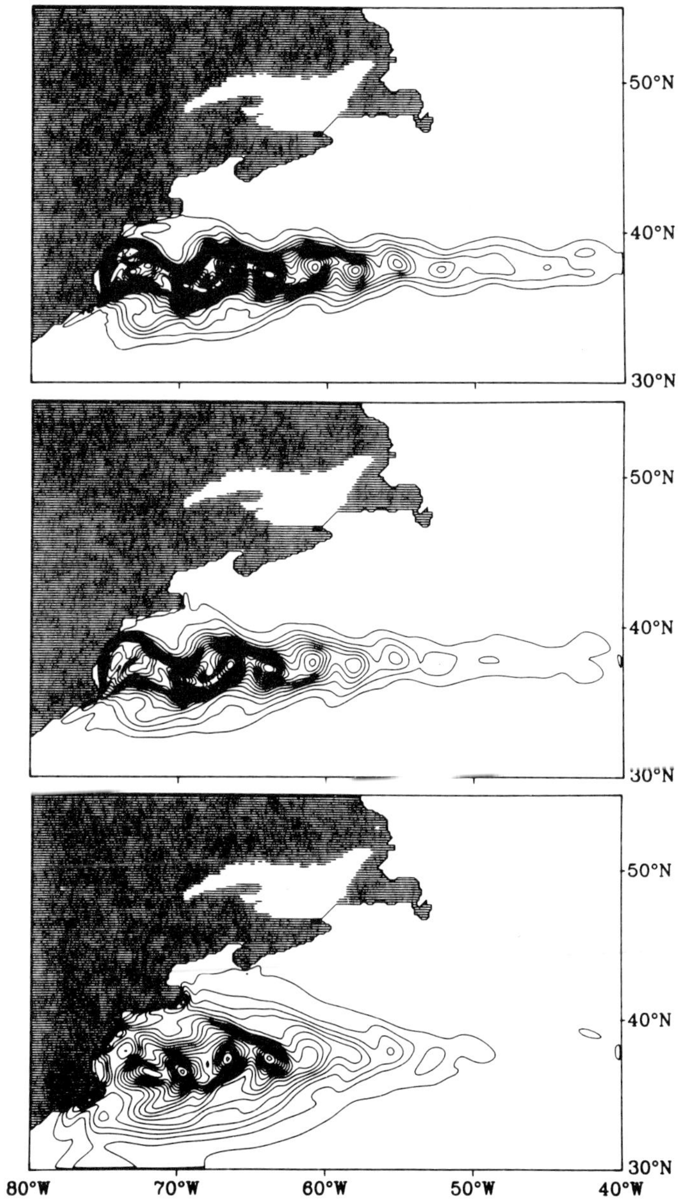

Figure 12. The patterns of eddy kinetic energy for the constant depth case, based upon five years of time averaging. The three levels are the same as in figure 7.

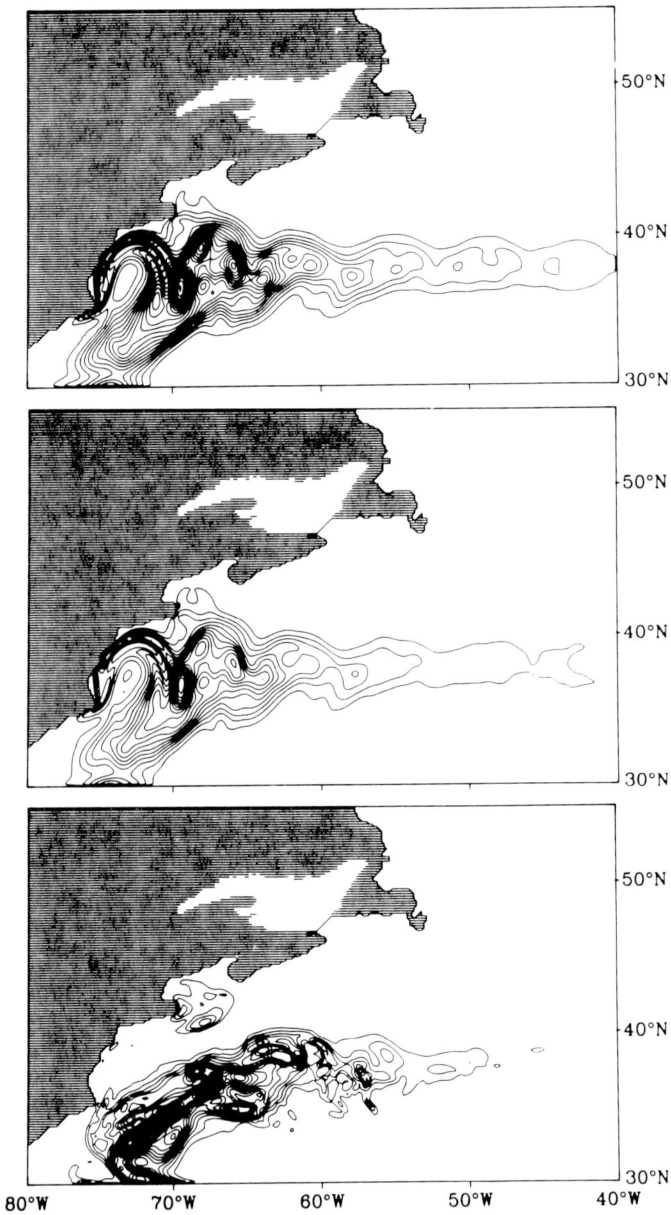

Figure 13. The patterns of eddy kinetic energy for the case with variable depth, based upon five years of time averaging. The three levels are the same as in figure 9.

4. CONCLUSION

Regional models with open (ocean) boundaries on some sides of a domain of interest look quite promising as members of a hierarchal approach to ocean modelling. Clearly, every model choice involves compromises and tradoffs. Global, basin scale and regional models, used in conjunction with each other, allow us to gain a much broader perspective upon the important influences on large scale ocean circulation. Coarse resolution interbasin models without eddies, medium resolution eddy-resolving basin studies, and high resolution limited area models all have something to contribute to our understanding of the dynamics of ocean circulation.

5. REFERENCES

Camerlengo, A.L., and J.J. O'Brien, 1980. Open boundary conditions in rotating fluids. J. Comput. Phys., 35: 12-35.

Chapman, David C., 1985. Numerical treatment of cross-shelf open boundaries in a barotropic coastal ocean model. J. Phys. Oceanogr., 15: 1050-1075.

Gordon, A.L., 1986. Interocean exchange of thermocline water. J. Geophys. Res., 91: 5037-5046.

Haidvogel, D.B., and W.R. Holland, 1978. The stability of ocean currents in eddy-resolving general circulation models. J. Phys. Oceanogr., 8: 393-413.

Holland, W.R., 1978. The role of mesoscale eddies in the general circulation of the ocean: Numerical experiments using a wind-driven quasigeostrophic model. J. Phys. Oceanogr., 8: 363-392.

Holland, W.R., 1983. Simulation of midlatitude variability. In: The Role of Eddies in the General Ocean Circulation, Proceedings Hawaiian Winter Workship, University of Hawaii, January 5-7.

Holland, W.R., 1985. Simulation of mesoscale ocean variability in midlatitude gyres. In: Atmospheric and Oceanic Modeling - Volume 28A of Advances in Geophysics, Academic Press, Orlando.

Holland, W.R., 1986. Quasigeostrophic modelling of eddy-resolved ocean circulation. In: Proceedings of the Nato Advanced Study Institute, Banyuls Sur Mer, France (in press).

Holland, W.R., 1987a. Numerical studies of the Agulhas Current region using a regional ocean model (in preparation).

Holland, W.R., 1987b. The Brazil Current/Falklands Current confluence region: Numerical model studies of a local region (in preparation).

Holland, W.R., 1987c. Geometrical influences on the western boundary current (the East Australia Current) of the South Pacific (in preparation).

Holland, W.R., and L.B. Lin, 1975a. On the origin of mesoscale eddies and their contribution to the general circulation of the ocean. I. A preliminary numerical experiment. J. Phys. Oceanogr., 5: 642-657.

Holland, W.R., and L.B. Lin, 1975b. On the origin of mesoscale eddies and their contribution to the general circulation of the ocean. II. A parameter study. J. Phys. Oceanogr., 5: 658-669.

Holland, W.R., and D.B. Haidvogel, 1980. A parameter study of the mixed instability of idealized ocean currents. Dyn. Atmos. & Oceans, 4: 185-215.

Holland, W.R., and P.B. Rhines, 1980. An example of eddy induced ocean circulation. J. Phys. Oceanogr., 10: 1010-1031.

Holland, W.R., and W.J. Schmitz, Jr., 1985. On the zonal penetration scale of model midlatitude jets. J. Phys. Oceanogr., 15: 1859-1875.

Holland, W.R., T. Keffer, and P.B. Rhines, 1984. Dynamics of the ocean general circulation: The potential vorticity field. Nature, 308: 698-705.

Orlanski, I., 1976. A simple boundary condition for unbounded hyperbolic flows. J. Comput. Phys., 21: 251-269.

Rhines, P.B., and W.R. Holland, 1979. A theoretical discussion of eddy-driven mean flows. Dyn. Atmos. & Oceans, 3: 289-325.

Roed, L.P., and O.M. Smedstad, 1984. Open boundary conditions for forced waves in a rotating fluid. SIAM, J. Sci. Stat. Comput., 5: 414-426.

Schmitz, W.J., Jr., and W.R. Holland, 1982. A preliminary comparison of selected numerical eddy-resolving general circulation experiments with observations. J. Mar. Res., 40: 75-117.

Schmitz, W.J., Jr., P.P. Niiler, R.L. Bernstein, and W.R. Holland, 1982. Recent long-term moored instrument observations in the Western North Pacific. Jour. Geophys. Res., 87: 9425-9440.

Schmitz, W.J., Jr., and W.R. Holland, 1986. Observed and modeled mesoscale variability near the Gulf Stream and Kuroshio extension. J. Geophys. Res., 91: 9624-9638.

STUDY OF TRANSPORT FLUCTUATIONS AND MEANDERING OF THE FLORIDA CURRENT USING AN ISOPYCNIC COORDINATE NUMERICAL MODEL

DOUGLAS B. BOUDRA, RAINER BLECK and FRIEDRICH SCHOTT
Rosenstiel School of Marine and Atmospheric Science, University of Miami, 4600 Rickenbacker Causeway, Miami FL 33149, (USA)

ABSTRACT

An isopycnic coordinate numerical model, using the Flux-Corrected Transport algorithm to control isopycnal outcropping and intersection with the ocean bottom, is configured in a channel with the bottom topography of the Florida Straits at 27° N. Bulk parameters determined from analysis of observations in the Subtropical Atlantic Climate Studies (STACS) program are combined with a dynamical initialization procedure, generating a Florida Current-like mass/flow pattern. Ten isopycnal layers and 2 km horizontal grid spacing resolve this cross-sectional flow. The channel is created by duplicating the cross-section in the downstream direction.

Investigation focuses on whether the transport fluctuations on time scales of a few days to several weeks and the meandering observed in the Florida Current may be either locally forced by the wind or due to inherent dynamical instabilities. When a single cross-section is forced with spatially constant but temporally fluctuating downstream wind stress, the barotropic transport response exhibits an almost 90° phase lag to the wind. The response amplitude is slightly less than directly proportional to the amplitude and period of the forcing. The baroclinic response, defined as the difference between the transport flucutations above and below 200 m depth, exhibits the almost 90° phase lag to the wind, but has an order of magnitude less amplitude than the barotropic.

In the three-dimensional channel, when the cross-stream density gradient is strong enough, perturbations with wavelengths greater than 60 km exhibit substantial amplification. The current is most unstable to wavelengths of a little more than 200 km when using the dynamically initialized fields and 150 km when using analyzed STACS Pegasus data to initialize. Events of wave amplification are of limited duration, however, and leave the basic structure of the current unchanged. In corresponding flat bottom channel experiments, the perturbation continues to amplify until the baroclinic structure has been substantially modified. From an analysis of energy conversions it is concluded that the primary mechanism of wave amplification in all cases is the release of baroclinic instability.

1 INTRODUCTION

The Florida Current is that part of the North Atlantic western boundary current system which continues on from the Gulf of Mexico Loop Current through the Florida Straits -- first between the Keys and Cuba and then, turning northward, between the mainland and the Bahamas -- and along the central and north Florida coast. The straits gradually narrow downstream from the Gulf and are at their narrowest at 27° N, after which the passageway opens up rapidly on the east side and there is no more constriction thereafter.

Fluctuations of the total transport through the straits on a wide range of time scales have been measured (Niiler and Richardson, 1973; Brooks, 1979; Lee, et al, 1985). The temporal mean transport is about equal to the annual mean Sverdrup transport at the latitude of South Florida -- 30 to 32 X 10^{6} m^{3} s^{-1}. The difference between the annual maximum, usually in June, and the minimum in October is 10-15% of the total. But larger fluctuations have been observed with periods of several days to a few weeks.

In addition to fluctuations in transport, the Florida Current is known to meander through the straits. The amplitude of the meanders seems to depend partly on the width of the straits and, thus, alongstream position (Schmitz and Richardson, 1968), owing to the above-mentioned constriction downstream from the Gulf. The period range of the meanders is, like that of some of the large transport fluctuations, from a few days to a few weeks.

Besides their possible influence on the heat budget of the North Atlantic and whatever climatic impact which that may entail, these phenomena are interesting in themselves, and various mechanisms have been invoked to explain their existence and character. The transport fluctuations with periods of a few days to two weeks are most commonly attributed to similar period fluctuations in the local or synoptic scale wind forcing (Wunsch and Wimbush, 1977; Düing et al, 1977; Lee, et al, 1985). The meandering behavior has been related to the transport variations by Düing (1975), to current instabilities by Niiler and Mysak (1971), De Soeke (1975), and Orlanski (1969), and to shelf wave modes by Schott and Düing (1976) and Brooks and Mooers (1977).

In 1982-1984, an intense 27 month observational experiment was carried out across the straits at 27° N as part of the Subtropical Atlantic Climate Studies (STACS) program. The overall goal of the experiment was to determine the most appropriate system for monitoring the transport fluctuations and meandering of the current. But, in addition, a wealth of new information was gathered, using the PEGASUS current profilers and moored current meter arrays, which allows additional documentation and analysis of the above-mentioned phenomena. Moreover, these new data sets, the most complete generated to date, may be used to initialize numerical models which may further assist in developing an understanding of Florida Current behavior. It is this last endeavor with which the current paper is concerned.

In what follows, we describe a numerical model of an idealized Florida Current and experiments in which we have studied 1) the response of the model current to fluctuations in wind forcing and 2) the stability of the current to perturbations of various wavelengths. Since the Florida Current is one small component of an enormous and complex circulation system, we cannot at this stage hope to model all of its observed behavior. But as a starting point in studying Florida Current behavior with a numerical model, and particularly because the data set recently compiled for the 27° N cross-section is the best available, we choose to focus our attention on the structure and variability of a current with the characteristics exhibited in that data set. Here we illustrate the model current behavior and give some preliminary comparison with statistics from the STACS data set. A more thorough comparison will be given in a forthcoming paper.

In Section 2, we describe the model and a theoretically-based initialization procedure. We briefly illustrate the transport response of the model current to wind stress fluctuations in Section 3. In Section 4, we show how the initialized current responds to presence of alongstream perturbations of various wavelengths, at the same time contrasting this response with that of a flat bottom version of the model. In Section 5, we first describe an initialization based more on the analyzed PEGASUS data, which possess a potential vorticity structure rather different from that given by the first procedure. Experiments testing the stability of the more realistic current are then illustrated. In Section 6, we summarize.

2 THE MODEL AND INITIALIZATION PROCEDURE

2.1 The numerical model

Since western boundary currents are characterized by substantial horizontal density gradients which are crucial to their structure, we use the isopycnic coordinate primitive equation model recently described by Bleck and Boudra (1986). Use of this model guarantees that 1) frontal structures will be optimally resolved for a given vertical resolution and 2) horizontal density gradients will not be smeared out with integration time due to lateral diffusion, which is generally the case in a z-coordinate system. The advantages of the isopycnal coordinate system can be overwhelmed by the numerical difficulties associated with the coordinate surfaces coming together with the upper surface, with each other and with bottom topography. The recent Bleck and Boudra model, in contrast to the quasi-isopycnic coordinate model of Bleck and Boudra (1981), controls collapsing of coordinate layers using a special algorithm, Flux-Corrected Transport, developed originally by Boris and Book (1973) and extended to multi-dimensional application by Zalesak (1979). While space is not available here to give the details of the FCT model, we may briefly describe the essential concept and two aspects of the model introduced here.

First of all, the algorithm is incorporated in the mass continuity equation, which predicts layer thickness. A high order scheme, in this case fourth order, is used to estimate mass fluxes into and out of a grid box provided the thickness value which would result from these estimates is bounded away from zero. When this is not the case, the estimates are combined with those made with a first order scheme, through which no values less than zero are generated.

In the model calculation, the equations are integrated in full at massless grid points as well as those with greater than zero layer thickness. This is relatively trouble-free in flat bottom calculations when the coordinate surfaces do not intersect the lower boundary. For applications such as the Florida Staits, with steeply sloping topography, a potential problem is encountered, as more than one layer comes in contact with the lower boundary, that is, unless vertical resolution is very low. If grid points in more than one massless layer coincide at the ocean bottom, the pressure force variable, the Montgomery potential, $gz + p\alpha$, which changes vertically due to the change in specific volume, will do so even in the absence of real fluid. Particularly if this configuration is just next to a region where one of those layers has greater than zero thickness, the computation of horizontal pressure force in the momentum equations will give an artificial result. Therefore, the pressure force is averaged over the bottom 30 meters of real fluid, and this value is assigned to all grid points in the column confined within this layer. This problem is not encountered at the upper boundary since pressure is zero there, and, thus, the Montgomery potential does not change from one massless layer to the next.

Velocity values in massless regions of a layer in the above model can become noisy. To prevent this from causing numerical problems in the adjacent non-zero-thickness regions, we have adopted a weighting of velocities which, at the end of each time step, replaces velocities at any grid point with less than 5 m layer thickness by a 5 m vertical average. In particular, massless grid points at the upper and lower boundary are reassigned velocities computed as a mean from the 5 m just below and above them, respectively. For additional description of the model, the reader is referred to Bleck and Boudra (1986).

2.2 Domain shape, boundary conditions, lateral friction

In its simplest form, the model is configured as a two-dimensional cross-section with the approximate bathymetry of the Florida Straits at 27°N, shown in Fig. 1. All gradients perpendicular to the cross-section are assumed zero. The lateral boundary conditions are no-slip, and bottom friction is incorporated according to a standard bulk formula in a 25 m bottom boundary layer.

Formulated as a three-dimensional channel, cyclic boundary conditions connect the 'ends' of the channel. Also in this 3-D form, the channel is assumed oriented north-south, but the variation of f is considered a higher order effect, and the constant value true for 27°N is used. Internal lateral friction is incorporated in a Laplacian-type term with constant viscosity:

$$\left(\frac{\partial p}{\partial \rho}\right)^{-1} A_m \nabla \cdot \left(\frac{\partial p}{\partial \rho} \nabla u,v\right).$$

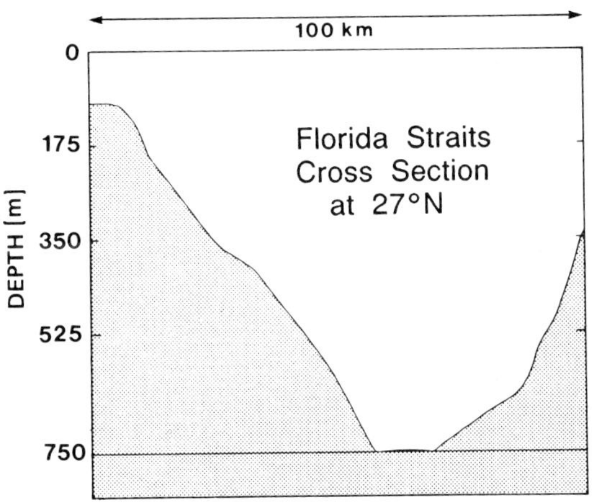

Fig.1. Illustration of the model cross-section used to represent the Florida Straits at 27°N.

The vertical is approximated in ten isopycnal layers with an increment of .4 σ_T units between consecutive layers. In the x-z cross-section the horizontal is resolved by 51 grid points with 2 km spacing and, thus, has 100 km total width. In the north-south direction, a total of 16 grid points is used to resolve a perturbation wavelength. Thus, the north-south grid spacing varies as the wavelength varies among the experiments.

2.3 Initialization

As indicated in the Introduction, two approaches have been taken to initialize the model with a Florida Current-like mass/flow field. One of these is based closely on the mean structure of the 27^{0} N cross-section compiled from the PEGASUS current profiler measurements during the STACS observations and analyzed by Leaman, et al (1987). Experiments with the model initialized in such a fashion are described here in Section 5. The other consists of initial specification of the mass field with an analytic function relating pressure and specific volume, followed by a horizontal deformation process meant to simulate oceanic frontogenesis (Hoskins and Bretherton, 1972). The function uses parameters representing the top-to-bottom and cross-stream density change, the steepness of the front, its position in the cross-channel direction, and the vertical scale of the density variation. The procedure, thus, allows great flexibility, since one may determine the stability of the current as these parameters are varied. Space is not available here to give full details. This will be done in the above mentioned forthcoming paper. The interested reader may also find the method detailed in a study of mesoscale frontogenesis by Bleck, et al (1987). The initilialized cross-section, well-balanced with the lateral and bottom boundary conditions and with a downstream transport of 31.7 X 10^6 m^3 s^{-1} is shown in Figure 2. Hereafter, we refer to this mass/flow configuration as our analytic initial fields.

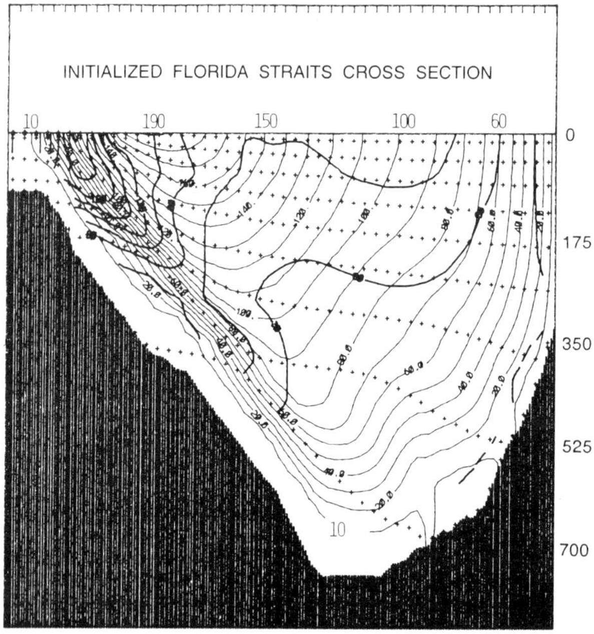

Fig.2. Illustration of the analytic initial fields. The northward velocity is contoured using thin solid lines with an interval of 10 cm s^{-1}. Selected isotachs are labelled. The position of the isopycnal coordinate surfaces is indicated by the rows of '+' signs. The potential vorticity, $(\frac{\partial v}{\partial x} + f)\frac{\partial \rho}{\partial p} + \frac{\partial v}{\partial p}\frac{\partial \rho}{\partial x}$, is contoured in heavy solid lines at intervals of 1 X 10^{-15} cm^{-2} s.

3 MODEL RESPONSE TO FLUCTUATING WIND

Recently, Lee, et al (1985) and Schott and Lee (1986) have found significant coherence between Florida Straits transport and local wind stress in certain energetic wind bands, notably periods of 3.5 days and 20 days. It can be shown that, in our model, if friction is negligible, application of a fluctuating north-south wind stress should yield a barotropic transport response which is proportional to the amplitude and period of the forcing and has a 90^0 phase lag to the wind. As friction is increased, the phase lag should gradually decrease from 90^0 and the relationship between forcing amplitude and period and the response amplitude should fall further below strict proportionality.

To determine the actual model behavior in this regard, we perform experiments with the analytic initial fields in which downstream wind stress is applied in a linearly decreasing fashion over the upper 100 m of the model, as in Bleck and Boudra (1981, 1986). In this case, the stress has no horizontal variation, but is a sinusoidal function of time. A constant eastern boundary condition on the mass transport streamfunction, used in initialization, is relaxed. Mean transport is maintained at approximately 30 X 10^6 m^3 s^{-1} by adding a constant stress of approximately .2 X 10^{-4} m^2 s^{-2} to the above fluctuating component. In these experiments, the eastern boundary condition for the streamfunction is computed in a manner analogous to that used for islands by Bryan (1969), according to the methodology developed by Kamenkovitch (1962).

The transport response of the model current to fluctuating wind is tested by successively doubling the forcing period while holding the forcing amplitude constant at .5 X 10^{-4} m^2 s^{-2} . Our experiments with free-slip lateral boundaries, an internal lateral friction coefficient of 10 m^2 s^{-1}, and zero bottom drag exhibit an approximate doubling of the barotropic transport response and a 90^0 phase lag between the wind and the response, as the above argument implies. Baroclinic response, defined as the transport above 200 m minus that below 200 m, shows a similar relationship with the wind but is an order of magnitude smaller.

When a drag coefficient of .003 for the bottom boundary layer is incorporated, along with an internal lateral viscosity of 65 m^2 s^{-1} and no-slip lateral boundaries, the response, illustrated in Table 1, is less than strictly porportional to the forcing period, as expected. However, the phase difference between the forcing and response decreases only slightly from 90^0.

TABLE 1

Model Florida Current response to sinusoidally fluctuating downstream wind stress. Forcing Amplitude - .5 X 10^{-4} m^2 s^{-2}.

Forcing Period (Days)	Response Amplitude (10^6 m^3 s^{-1})
8	.55
16	1.00
32	1.82
64	3.4
128	6.37

The forcing/response phase lag in the STACS data is 90^0 at a period of 3.5 days, but only 20^0 at the 20 day period (Lee, et al, 1985). This represents about 1 day lag in both cases. It seems likely that this phase lag is determined by factors other than friction: for instance, residence time of fluid in the Straits versus the period and spatial scale of the forcing. With our simple model geometry we are limited to describing the response of a continuous channel being forced uniformly.

4 GROWTH OF MEANDERS

Coherent, energetic meandering signals in the Florida Current at periods of approximately 5 and 12 days have been detected by Johns and Schott (1987) through frequency-domain empirical mode analysis of STACS current meter observations. The current meters were moored in an array across the straits at or near 27° N from December 1983 to June 1984. Their analysis suggests that these meanders have downstream propagation speeds and wavelengths of 36 km d^{-1}, 170 km and 28 km d^{-1}, 340 km, respectively. They find no strong correlation between meandering and total transport fluctuations, suggesting that the two are relatively unrelated. On examining the energy conversion terms associated with these modes, they conclude that the meanders are giving up energy to the mean flow through barotropic conversion. The baroclinic conversion term indicates small positive conversion - from the mean to the eddies - but its magnitude is apparently insignificant in view of the error bars. Also, the flux terms calculated from moored current meter data are deficient because the top 100 m, where much of the energy conversion takes place, is not covered by instrumentation.

In a computation of mean to eddy energy conversions using the full set of PEGASUS data, there is also little evidence that the Florida current at 27° N is unstable. In this analysis, however, Leaman, et al (1987) obtain such a small net conversion of energy between the mean flow and the eddies in both barotropic and baroclinic terms that they conclude that this conversion is of indeterminate sign. The possibility thus exists that the current in the straits is at the threshold of instability and that this is occasionally manifested in conversion of energy into meanders with wavelengths such as those reported by Johns and Schott (1987).

In this section, we wish to determine whether the current in our analytic initial fields exhibits tendencies to transfer energy into meanders through barotropic or baroclinic conversion. Because of the complexities involved in the incorporation of downstream variations in channel width and bottom topography, we have chosen to exclude those variations for the time being. As mentioned earlier, the initialized cross-section of Figure 2 is duplicated in the downstream direction to create the channel, so that there are 16 downstream grid points. The stability of this current to perturbations with wavelengths of 96 to 256 km has been tested by varying the downstream grid spacing from 6 km to 16 km at 2 km intervals. At some of the longer wavelengths, the number of downstream grid points has been doubled to verify that 16 downstream grid points is sufficient. The solutions show little sensitivity to this change in resolution.

The initial perturbation is specified only in the cross-stream velocity field and has a sinusoidal variation in the downstream direction. The amplitude of the perturbation is a velocity field obtained at some arbitrary time during the initialization procedure. It represents a cross-channel 'sloshing' due to lack of perfect geostrophic balance between the initial mass and flow fields, as well as their adaptation to the lateral and bottom boundary conditions. The perturbation is thus a function of the vertical and cross-stream directions and exhibits velocities varying approximately from -.17 m s^{-1} at the upper surface to +.23 m s^{-1} at the bottom. For these experiments, the eastern boundary condition for mass transport streamfunction is held constant at 31.7 X 10^6 m^3 s^{-1}. This can be physically related to the downstream pressure head yielded by an upper surface downstream slope of approximately 1 cm km^{-1}.

16.1 DAYS

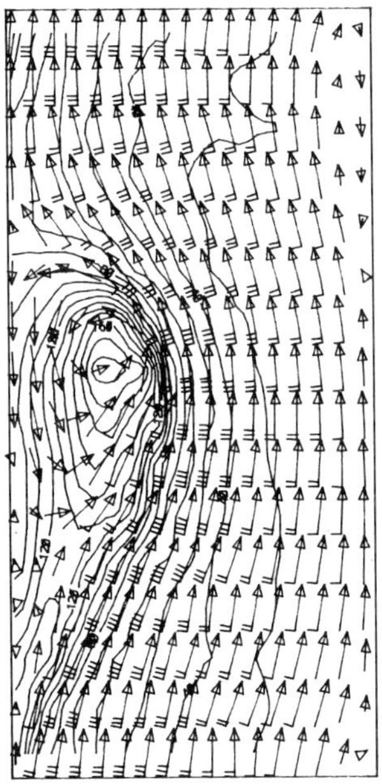

Fig.3. Top 30 m mean flow and density. Experiment with flat bottom, 100 km wide channel, and perturbation λ = 208 km. Density C.I. = .1 σ_T. Full length arrow and each additional barb = 25 cm s^{-1} speed. Arrows plotted every third point in x-direction, every y-point.

Before illustrating the results for the Florida Straits, it is instructive to point out the behavior of a current initialized in the same manner as described above in a channel of 100 km width but with a flat bottom. It is found that such a current is unstable to wavelengths of greater than approximately 60 km and that the maximum perturbation growth rate is at a wavelength of 208 km. The amplitude of this meander has increased noticeably after several days, and the upper 30 m flow pattern at 16.1 days (Figure 3) exhibits a strong cyclonic eddy to the left of the current core, which is draining energy from the available potential energy field (Figure 4a). The energy conversion terms for this model, derived in Bleck (1985) and subsequently applied to an ocean model intercomparison by Bleck and Boudra (1986) can be written as follows for conversion between potential and alongstream (i.e., zonal) mean and perturbation kinetic energies:

$$P\text{->}K_E = -\vec{v}\,' \cdot \frac{\partial p}{\partial \rho} \nabla_\rho MP$$

$$P\text{->}K_M = -\tilde{\vec{v}} \cdot \frac{\partial p}{\partial \rho} \nabla_\rho MP$$

$$K_E\text{->}K_M = \vec{v}\,' \frac{\partial p}{\partial \rho} \cdot (\vec{v}\,' \cdot \nabla_\rho)\tilde{\vec{v}}$$

MP, Montgomery Potential

$(\tilde{\ })$ along-stream mass-weighted average

$(\ )'$ departure from $(\tilde{\ })$.

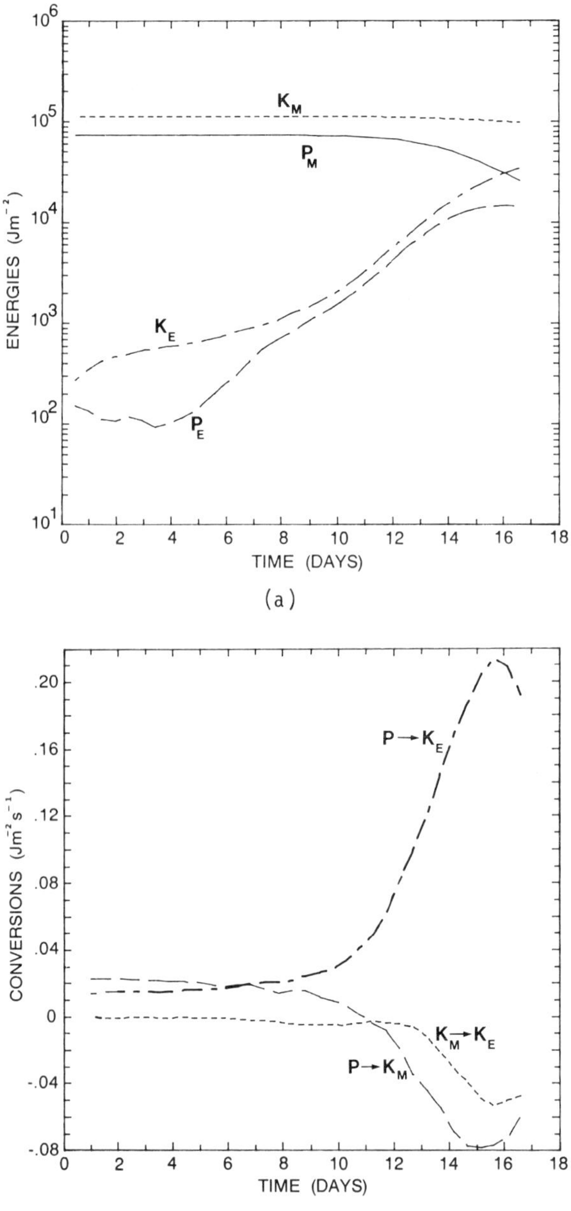

Fig.4. (a)Mean and eddy potential and kinetic energies, averaged over the channel as a function of time for the λ = 208 km flat bottom experiment. (b)Energy conversion rates as a function of time for the same experiment.

The conversion from potential to eddy kinetic is the one associated with baroclinic conversion, and that from mean kinetic to eddy kinetic with barotropic conversion. The graph of energy conversion versus time (Fig. 4b) shows that even at the start of this experiment there is a small P to K_E, which is of slightly less magnitude than P to K_M. This latter conversion is normally positive and restores mean kinetic energy lost through dissipation. As the eddy begins to grow rapidly from 8 to 12 days, baroclinic conversion climbs sharply while the P to K_M changes sign due to the loss of potential energy. Barotropic conversion likewise becomes strongly negative. This scenario signifies the release of baroclinic instability of the current.

By the end of this experiment, the energy conversion has peaked (Fig. 4b), but during the last several days, the mean potential energy has dropped from 7.2 X 10^4 to 2.5 X 10^4 J m^{-2}. Thus, the basic baroclinic structure of the channel has been substantially modified. This suggests that a current with many of the same bulk parameters as the Florida Current is baroclinically unstable in a flat bottom channel with the width of the Florida Straits. The perturbation generated in stabilizing the flow (Fig. 3) is much larger than any meander ever observed in the Florida Straits. It seems unlikely, then, that the model current with the 27° N bottom topography will exhibit such strong instability.

In fact, the model with bathymetry does exhibit meander growth, primarily through the baroclinic conversion term. But the horizontal and temporal scales of the wave growth are more limited. In addition, the growth is slower. It is also found that the growth rate is strongly dependent on the total cross-channel density change at the surface, a parameter used in the initial mass field specification which determines, to a large degree, the baroclinicity of the current. Since the surface layer of the Florida Current is horizontally well-mixed, the total east to west surface density increase in the STACS data (Leaman, et al, 1987) is but .2 σ_T units. At 50 m depth the increase from east to west is 1.2 units. The value chosen for the experiments described here is an intermediate one, .7 units, corresponding to the actual value at 30 to 40 m depth. Values of .3 or less lead to very little meander growth.

Parameters describing the instability as a function of wavelength are given in Table 2. It is found that again the most unstable wavelength is 208 km. The meander has a period of approximately 8 days. In general, more time is required for the meander to reach maximum amplitude as wavelength is increased. The maximum in P_E occurs before or at the same time as the K_E maximum. The maximum K_E to K_M conversion generally occurs at the same time or shortly after the maximum P to K_E conversion.

TABLE 2
Energy and energy conversion peaks as a function of meander wavelength (λ). Energy is in units of J m^{-2} and energy conversion in J m^{-2} s^{-1}. Time is in days.

λ(km)	K_E	K_E Time	P_E	P_E Time	P to K_E	P to K_E Time	K_E to K_M	K_E to K_M Time
96	4.3	7.3	2.6	6.3	.039	6.8	.007	8.8
128	4.9	9.2	3.0	6.8	.04	8.3	.013	7.8
160	6.6	11.2	3.25	8.8	.052	9.8	.016	9.8
192	7.7	12.2	3.9	12.2	.07	12.2	.019	13.8
208	8.4	15.1	4.4	15.1	.074	14.2	.028	14.6
224	6.3	17.0	3.4	17.0	.059	15.6	.017	15.6
256	5.8	17.1	3.3	17.1	.045	17.1	.007	17.1

17.1
DAYS

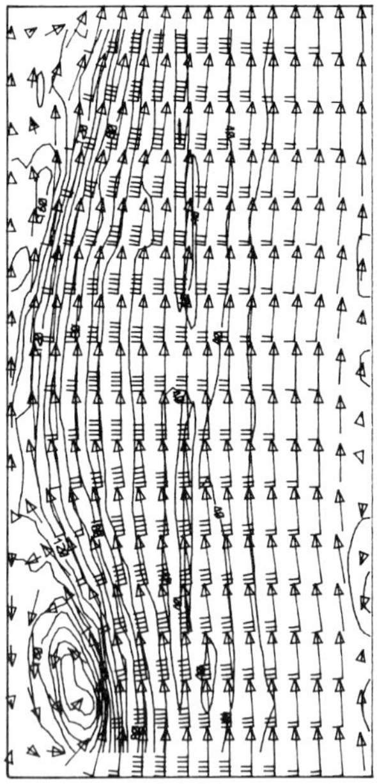

Fig.5. As in Fig.3, but for the experiment with the bottom topography illustrated in Fig.1.

The upper layer flow pattern for the 208 km wavelength shortly after the time of its maximum amplitude (Fig. 5) shows again a cyclonic eddy to the left of the current core but with considerably less spatial extent than in the flat bottom case. Further, the graphs of energy and energy conversion vs. time (Fig. 6), while they implicate release of baroclinic instability as the physical mechanism of meander growth, show that the mean potential and kinetic energy are little affected by the event of eddy growth. After the instability is released the baroclinic conversion term decreases to approximately the same value as at the initial time and the barotropic conversion term rises slightly above zero. The system is, therefore, one which is only marginally unstable. Events of meander growth leave the mean mass/flow structure relatively unchanged.

5 GROWTH OF MEANDERS USING THE STACS ANALYZED DATA

5.1 Development of initial conditions

Within the context of the analytically derived Florida Current, as described above, meander growth has been investigated as a function of the most influential physical parameters. Space limitations prevent us from detailing all of this experimentation here. The experiments described in the previous section were initialized using the parameter values which develop the most realistic looking cross-section with respect to the STACS data (Leaman, et al, 1987). A feature which could not be easily included in that initial field is the horizontally well-mixed surface layer, which is perhaps 30 m thick at the western boundary and 60 to 70 m at the eastern boundary. In addition, the eastern boundary (the steep slope of Little Bahama Bank) has a SSE to NNW tilt in the Straits, which leads to converging flow at mid-depth and a subsurface maximum in northward velocity near that boundary. Likewise, this feature cannot be easily included in our analytic initial conditions. At any rate, these special characteristics give the real Florida Current a

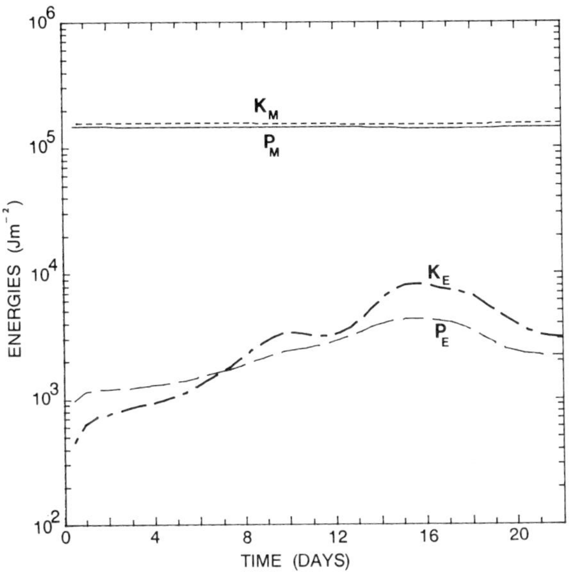

(a)

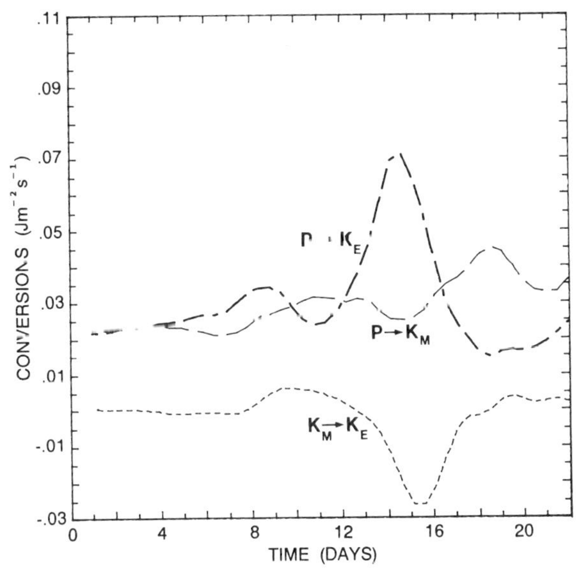

(b)

Fig.6. As in Fig.4, but for the experiment with the bottom topography of Fig.1.

rather different potential vorticity structure from that illustrated by the thick solid lines of Figure 2 for our analytical initial fields. Since the change in sign of horizontal and vertical gradients of potential vorticity are generally considered crucial factors in the stability of currents, it seems warranted to explore the stability of a model current with a more realistic potential vorticity structure.

We begin development of the initial field for these experiments with the analyzed density vs. pressure data of Leaman, et al (1987). The thickness of density layers representing increments of .4 σ_T is determined through linear interpolation, given density analyzed at 10 dbar intervals. The cross-channel spacing of their data analysis is 1.878 km, which becomes the new model grid point spacing. In this case, the grid points on the western side of the straits with less than 90 dbar bottom pressure have been discarded, so that the total straits width is 82.6 km.

The analyzed observed surface velocity minus bottom layer velocity is assumed geostrophically balanced with the surface Montgomery potential gradient. This velocity difference is also specified for the top layer velocity initially and that in the remainder of the column is computed integrating downward using the thermal wind relation. The model cross-section is then integrated for ten days of simulated time with the eastern boundary condition on the mass transport streamfunction of 31.7 X 10^6 m^3 s^{-1} and the same lateral and bottom boundary conditions as used in Section 4. An average of the downstream velocity and pressure fields over the final 200 time steps yields a well-balanced mass/flow configuration (Fig. 7), which exhibits the essential structure of the STACS mass and velocity analysis, as well as the primary potential vorticity tongue, which results from the strong stratification at the base of the surface mixed layer.

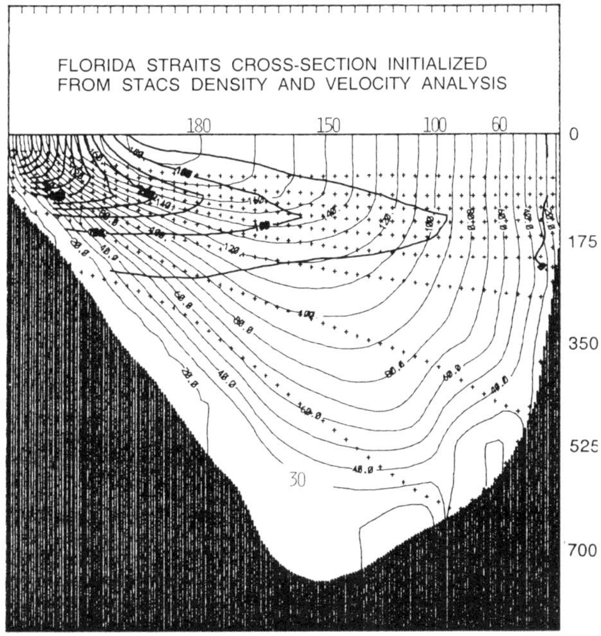

Fig.7. As in Fig.2, but for the initial fields developed from the STACS analyzed data. Potential vorticity is contoured at intervals of 5 X 10^{-15} cm^{-2} s.

5.2 Stability test

As with the analytically derived current, a cross-stream velocity field from the initial two-dimensional integration is saved to provide the amplitude for the sinusoidal perturbation used for the stability test. The maximum values in this field are somewhat less than in that used above, but are considered large enough for the current purpose. The stability characteristics of this current as a function of wavelength are summarized in TABLE 3. It is found that this more realistic current is noticeably baroclinically unstable to perturbations with wavelength greater than 60 km, as in the previous case. More than twice as much time is required for the meanders to reach peak amplitude than in the previous case. This could be an inherent characteristic of the dynamical difference between the current structures or it could be related to the amplitude and structure of the initial perturbation. More study is required before a conclusion can be reached. In Table 3, the trends with respect to wavelength are not as consistent as with the previous data set. Notably, while the meander reaching greatest amplitude in eddy potential and kinetic energy is 150 km, the maximum growth rate is at a wavelength of 120 km. Significantly, the 150 km wavelength meander has a period of approximately 5 days, giving it about the same spatial and temporal scale as the shorter of the two meanders found in the moored current meter data by Johns and Schott (1987).

TABLE 3
Energy and conversion peaks for the current initialized from the STACS analysis. Units are as in Table 2.

λ(km)	K_E	K_E Time	P_E	P_E Time	P to K_E	P to K_E Time	K_E to K_M	K_E to K_M Time
90	8.2	16.6	2.7	15.8	.07	15.8	.035	16.7
120	10.8	22.6	3.9	22.6	.081	19.2	.042	20.0
150	11.3	28.7	3.98	28.7	.046	28.0	.024	28.7
180	7.8	29.6	3.13	29.6	.035	24.4	.012	28.0
210	7.93	27.9	2.89	27.9	.032	29.4	.015	28.2

The graphs of mean and perturbation energy and energy conversion for the 150 km wavelength (Fig. 8) show that the perturbation energies initially decrease while the P to K_E and K_M to K_E conversions are near zero. At approximately 8 days, both conversions begin to rise and the perturbation energy begins to grow. The barotropic conversion to the eddies briefly rises above the baroclinic conversion and then begins to fall at 13 days. There are two peaks in baroclinic conversion corresponding to peaks in negative barotropic conversion as well, the second of which is largest. This double maximum is similar to one encountered in the previous case. Because of the proximity of the final peaks to the end of the experiment, it would be premature to state that no further peaks will follow these. However, both perturbation energies have begun to decrease at the end, for the first time since their initial fall off. For the current purposes, it is felt to be sufficient to explore the initial stability of this current. Study of its long term stability is saved for future work. As in the previous case, the P_M and K_M remain essentially unchanged during wave growth, and the initial current can be classified again as only marginally unstable.

The upper 40 m flow pattern for the experiment with meander wavelength 150 km at the time of maximum amplitude exhibits a somewhat elongated cyclonic eddy

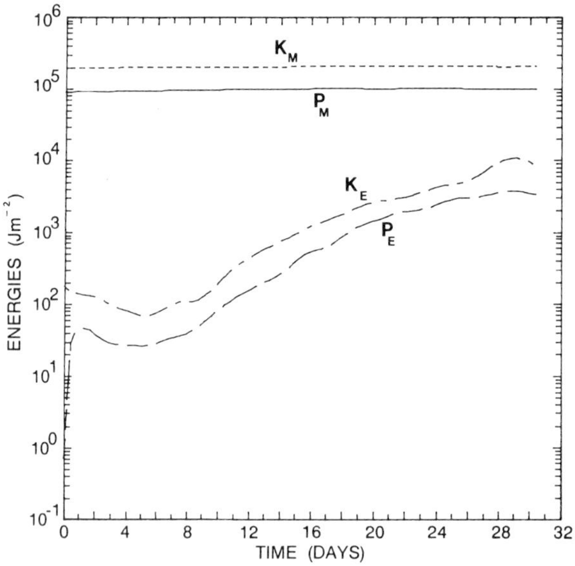

(a)

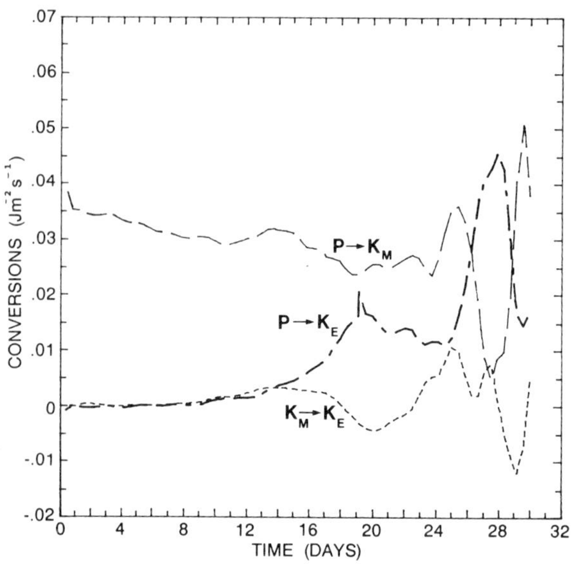

(b)

Fig.8. As in Fig.4, for the experiment initialized with analyzed STACS observations.

just west of the current core in its cyclonic turn (Fig. 9). In this pattern, the western edge of the current core varies in cross-stream position by roughly 8 km. This value is well within the range of those typically observed in association with Florida Current meanders at this latitude (Schmitz and Richardson, 1968; Johns and Schott, 1987).

An instructive way to describe the cross-sectional structure of the perturbation is through the alongstream mean of the perturbation kinetic energy (PKE). In Figure 10, we compare this field, computed at the time of maximum wave amplitude, with that computed by Leaman, et al (1987) for the full spectrum of perturbation energy analyzed from the PEGASUS data. By far, the greatest concentration of PKE in both model and observation is in the cyclonic shear zone, near and to the left of the current core, and above 350 m. The maximum in perturbation energy in the observations is about twice that in our field and is slightly below the surface, whereas ours is at the upper surface. The factor of 2 is probably accounted for in that the analysis of observations includes fluctuations on all time and spatial scales, and our model field is for a single component. Both illustrations suggest fluctuation energy near the eastern slope, especially the model. Inspection of instantaneous cross sections (not shown) shows the axis of the current in the lower two layers shifting back and forth several grid points along the meander wavelength. A bottom trapping of wave energy is suggested, and Rhines (1970) has suggested that conditions are favorable in the Florida Straits for generation of bottom trapped waves. The amplitude of the feature, however, along the eastern slope from 475-350 m depth may be exaggerated from the numerical strain of dealing with such a steeply sloping bottom.

28.7 DAYS

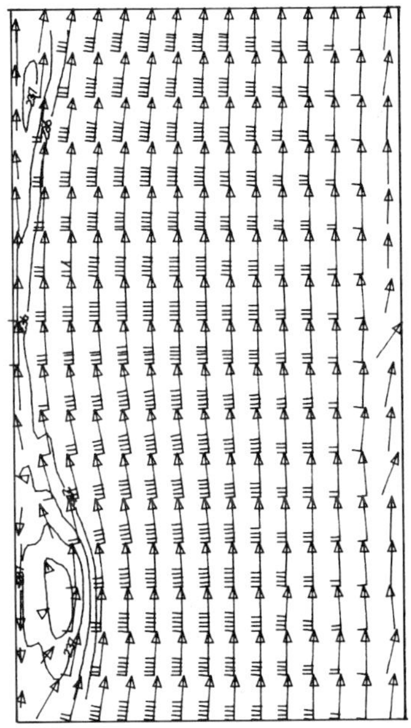

Fig.9. Instantaneous upper 40 m mean flow pattern and density for experiment initialized with analyzed STACS observations. Arrows are as in Fig.3. Density is contoured in intervals of .05 σ_T.

Notably absent from the model PKE diagram is the surface layer tongue reaching across the straits. This is most likely due to the lack of seasonal and local wind stress-induced flucutations, which are, of course, present in the analyzed observations and likely introduce this degree of variability, although the latter is much more apparent in the perturbation potential than kinetic energy (Leaman, et al, 1987).

6 SUMMARY

An isopycnic coordinate numerical model has been configured in a channel with the Florida Straits bottom topography at 27°N for study of Florida Current 1) transport fluctuations with periods of a few days to several weeks, and 2) meanders which may result from weak instabilities in the current. Initial conditions are developed, first, using an analytic function relating pressure and density, which incorporates bulk parameters from STACS observations, and which is put through a dynamical initialization, and, second, using actual grid point analyzed data from STACS.

Using the single cross-section from the first set of initial conditions, the current is subjected to horizontally uniform, but temporally fluctuating,

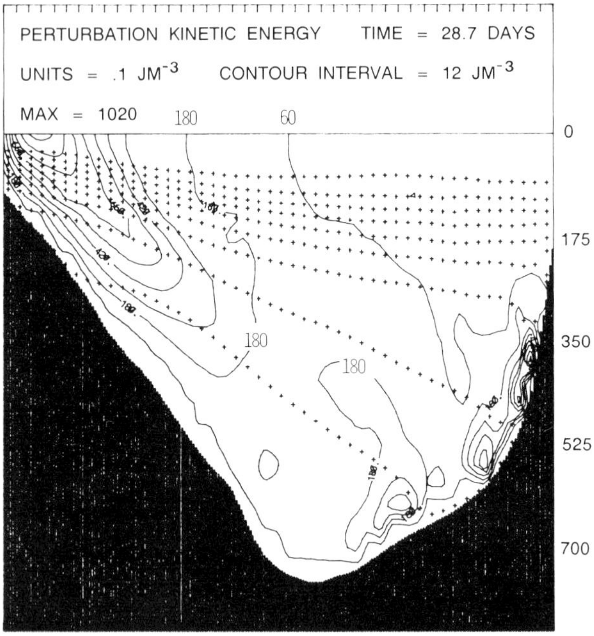

(a)

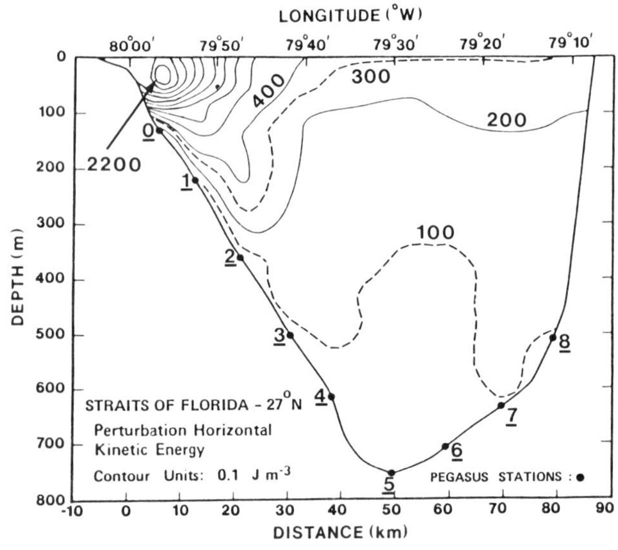

(b)

Fig.10. (a)The alongstream mean of perturbation KE (computed as the difference between the total KE and K_M) at the time when this parameter has its basin-averaged maximum. (b)Mean perturbation KE from the full spectrum of perturbations measured with the PEGASUS profilers during the STACS experiment in the Florida Straits (after Leaman, et al., 1987).

alongstream wind stress. This evokes a barotropic transport response which, with weak friction, varies linearly with forcing period and lags 90^0 in phase behind the forcing. As friction is increased, the response exhibits a noticeable decrease from linear proportionality with forcing period, but it varies only slightly from the 90^0 phase lag. This last result differs with the roughly one day observed phase lag (Lee, et al, 1985), and the discrepancy is likely due to the two-dimensional structure of the model compared with residence time of fluid in the Straits versus the temporal and spatial scale of the wind forcing.

When the cross-sectional current is duplicated 16 times in the downstream direction to form a channel, the flow is found to be baroclinically unstable to perturbations of greater than about 60 km wavelength. In the range 60 to 256 km, the most unstable wavelength is

208 km. The instability is manifested notably in development of a cyclonic eddy to the west of the current core. The wave amplification, however, is of limited horizontal and temporal scale and, when complete, the mean kinetic and potential energies are largely unchanged.

Likewise, the current developed from the analysis of STACS PEGASUS data (Leaman, et al, 1987) exhibits release of baroclinic instability for perturbations with wavelengths greater than 60 km. In this case, the maximum perturbation energy is reached for a wavelength of 150 km and period about 5 days, similar to one of two significant meandering modes detected by Johns and Schott (1987) in STACS moored current meter data. The alongstream mean of perturbation kinetic energy for this meander in the x - z plane is found to bear the same patterns as that obtained for the PEGASUS data, including all of the observed fluctuations -- that is, except for those features likely associated with seasonal and local wind event-induced fluctuations. Finally, the meander growth is again of limited horizontal and temporal scale and, when complete, the mean kinetic and potential energies have changed little.

These results, albeit with a much simplified model, suggest that the current flowing through the Florida Straits is at the threshold of baroclinic instability, and that occasional conversion of energy from potential to eddy kinetic may account for the meandering behavior of the Florida Current. The instability is apparently not strong enough to cause formation of a large eddy in the Straits. But it may, for example, lead to brief surface flow reversals along the western edge in the cyclonically curved portion of the meander.

ACKNOWLEDGEMENTS

We wish to acknowledge helpful discussions with Drs. Tom Lee, Bill Johns, and Kevin Leaman, and to thank Leaman and Peter Vertes for making their analysis of PEGASUS data available. This work has been supported by NSF grants OCE 83-11510 and OCE 86-00593. Calculations were performed on the Cray-1 computers at the National Center for Atmospheric Research, which is sponsored by the National Science Foundation.

6 REFERENCES

Bleck, R., and Boudra, D.B. 1981. Initial testing of a numerical ocean circulation model using a hybrid (quasi-isopycnic) vertical coordinate. J. Phys. Oceanogr., 11: 755-770.

Bleck, R., 1985. On the conversion between mean and eddy components of potential and kinetic energy in isentropic and isopycnic coordinates. Dyn. Atmos. and Oceans, 9: 17-37.

Bleck, R. and Boudra, D., 1986. Wind-Driven Spin-up in eddy-resolving ocean models formulated in isopycnic and isobaric coordinates. J. Geophys. Res., 91(C6): 7611-7621.

Bleck, R., Onken, R., and Woods, J.D., 1987. A two-dimensional model of mesoscale frontogenesis in the ocean. Submitted to Quart. J. Roy. Met. Soc.

Boris, J.P., and Book, D.L., 1973. Flux-corrected transport. I. SHASTA. A fluid transport algorithm that works. J. Comput. Phys., 11: 38-69.
Brooks, D.A. and Mooers, C.N.K., 1977. Free, stable continental shelf waves in a sheared barotropic boundary current. J. Phys. Oceanogr., 7: 380-388.
Brooks, I.H., 1979. Fluctuations in the transport of the Florida Current at periods between tidal and two weeks. J. Phys. Oceanogr., 9: 1048-1053.
Bryan, K., 1969. A numerical method for the study of the circulation of the World Ocean. J. Comput. Phys., 4: 347-376.
De Soeke, R.A., 1975. Some effects of bottom topography on baroclinic stability. J. Mar. Res., 33: 93-122.
Düing, W.O., 1975. Synoptic studies of transients in the Florida Current. J. Mar. Res., 33: 53-73.
Düing, W.O., Mooers, C.N.K., and Lee, T.N., 1977. Low-frequency variability in the Florida Current and relations to atmospheric forcing from 1972 to 1974. J. Mar. Res., 35: 129-161.
Hoskins, B.J. and Bretherton, F.P., 1972. Atmospheric frontogenesis models: Mathematical formulation and solution. J. Atmos. Sci., 29: 11-37.
Johns, W.E. and Schott, F., 1987. Meandering and transport variations of the Florida Current. J. Phys. Oceanogr. In press.
Kamenkovitch, V.M., 1962. Trudy Instituta Okeanologii, Akad. Nauk SSSR, 56: 241.
Leaman, K.D., Molinari, R.L., Vertes, P.S., 1987. Structure and variability of the Florida Current at 27° N: April 1982 - July 1984. J. Phys. Oceanogr. In press.
Lee, T.N., Schott, F., and Zantopp, R., 1985. Florida Current Low-frequency variability as observed with moored current meters during April 1982 to June 1983. Science, 227: 298-301.
Niiler, P.P. and Mysak, L.A., 1971. Barotropic waves along an eastern continental shelf. Geophys. Fluid Dyn., 2: 273-288.
Niiler, P.P. and Richardson, W.S., 1973. Seasonal variability of the Florida Current. J. Mar. Res., 31: 144-167.
Orlanski, I., 1969. The influence of bottom topography on the stability of jets in a baroclinic fluid. J. Atmos. Sci., 26: 1216-1232.
Rhines, P.B., 1970. Edge-, bottom, and Rossby waves in a rotating stratified fluid. Geophys. Fluid Dyn., 1: 273-302.
Schmitz, W.J., Jr., and Richardson, W.S., 1968. On the transport of the Florida Current. Deep-Sea Research, 15: 679-693.
Schott, F. and Düing, W.O., 1976. Continental shelf waves in the Florida Straits. J. Phys. Oceanogr., 6: 451-460.
Schott, F. and Lee, T.N., 1986. Variability of structure and transport of the Florida Current in the period range of days to seasonal. In Preparation.
Wunsch, C., and Wimbush, M., 1977. Simulataneous pressure, velocity, and temperature measurements in the Florida Straits. J. Mar. Res., 35: 75-104.
Zalesak, S.T., 1979. Fully dimensional flux-corrected transport algorithm for fluids. J. Comput. Phys., 31: 335-362.

DYNAMICS OF AGULHAS RETROFLECTION AND RING FORMATION IN A QUASI-ISOPYCNIC COORDINATE NUMERICAL MODEL

E.P. CHASSIGNET AND D.B. BOUDRA
Rosenstiel School of Marine and Atmospheric Science, University of Miami, 4600 Rickenbacker Causeway, Miami FL 33149-1098, (USA)

ABSTRACT

The Agulhas retroflection region of the idealized South Atlantic-Indian ocean model described by De Ruijter and Boudra (1985) and Boudra and de Ruijter (1986) is analyzed in detail. First, the physical mechanism of the model retroflection is presented through illustration of the Agulhas' vorticity balance among various experiments, and second, the ring formation process is described in terms of its vertical structure and the associated energy conversions.

Analysis of the vorticity balance in a one-layer model shows that both inertia and internal friction may account for a partial retroflection where a linear, weakly viscous system has none. In a series of one-, two-, and three-layer nonlinear, weakly viscous experiments, it is shown that retroflection is accomplished through a free inertial boundary layer, as suggested originally by De Ruijter (1982), and, furthermore, that it is the planetary vorticity advection, rather than the inertial overshooting distance, which is of greatest importance. Halving horizontal grid spacing (from 40 to 20 km) is shown to have a minor impact on the vorticity balance and the retroflection strength. The importance of a substantial viscous stress curl along the coast of Africa, as provided by the no-slip condition, is illustrated through comparison with a slippery Africa experiment. Because of the apparent importance of the planetary vorticity advection in the retroflection regime, an experiment with a more realistic South African coastal geometry is described. It is shown that the retroflection is still strong but that the associated recirculation is less intense. The primary terms of the vorticity equation have smaller magnitude but they exhibit the same basic balance as in the nonlinear, weakly viscous, rectangular Africa experiments: between planetary vorticity advection and viscous stress curl along the coast, and between planetary and relative vorticity advection on the seaward side of the coastal Agulhas and elsewhere throughout the retroflection.

The mean energetics of the experiments are examined in order to gain additional understanding of the model retroflection. Also, the signatures of barotropic and baroclinic instability in the ring formation process for three experiments are studied in detail using eddy-mean energetics. Ring formation is accompanied by development of a pronounced cyclonic circulation in the lower layer. However, both barotropic and baroclinic conversions reach a maximum at the moment of ring cutoff. Therefore, a mixed instability is suggested.

1 INTRODUCTION

One of the most intriguing features in world ocean circulation is the retroflection of the Agulhas Current south of Africa, and an important phenomena associated with it is ring formation at its western edge. The Agulhas constitutes the primary boundary current of the southwestern Indian Ocean, and as it separates from the southern tip of Africa in a southerly or southwesterly direction, the major part of it retroflects, returning eastward to rejoin the subtropical gyre. Several times a year, in retroflecting, the current cuts itself off and a ring of warm, salty Indian Ocean water is formed. These rings drift westward and, along with other leakage of Agulhas water around the tip of Africa, likely have an important impact on the heat and salt budget of the South Atlantic.

In this paper, we present some basic dynamical ideas and numerical results concerning the Agulhas retroflection and the ring formation. The dynamical topics include, 1) the retroflection vorticity balance, 2) the mean energetics and 3) the nature of the instability associated with ring formation.

2 DESCRIPTION OF THE MODEL AND THE EXPERIMENTS

The numerical models used are described in De Ruijter and Boudra (1985) and Boudra and de Ruijter (1986)(henceforth referred to as DB and BD, respectively). The authors included temporal variability, friction, inertia and the beta-effect in a study of the South Atlantic-Indian Ocean circulation, using first a one-layer simplification of, and then the full quasi-isopycnic coordinate numerical model of Bleck and Boudra (1981). Perhaps overly idealized aspects of

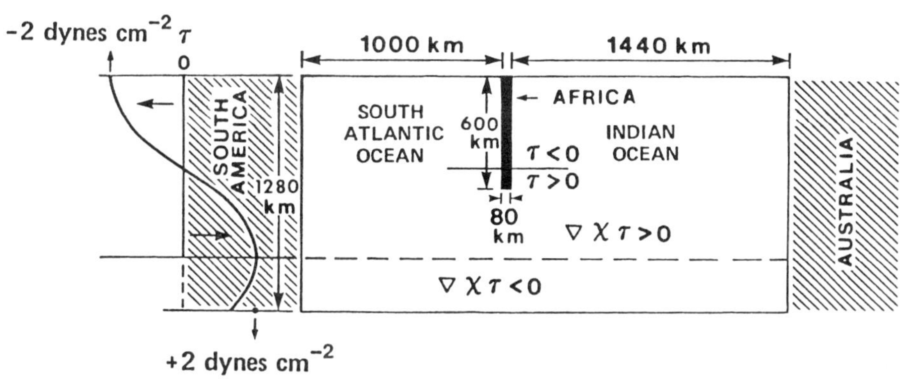

Fig.1. Geometry of the flat bottom ocean basin used in all the numerical experiments, except E11. The wind stress profile is shown at the left.

their model were the simple rectangular shape for Africa and the small total basin size. In DB and BD, the primary focus was on parameters affecting the exchange of fluid between the two basins. It was surmised that the no-slip African coast and the magnitude of the planetary vorticity advection at separation figured prominently in the retroflection vorticity balance. Influence of several other model parameters among their experiments, including internal lateral friction, horizontal resolution, stratification, incorporation of bottom drag, and interaction with the other wind-driven currents, were also examined.

The parameters for the experiments discussed in this study are presented in Table 1. Many of these are experiments discussed in DB and BD, but several new

TABLE 1

Parameters for the experiments discussed in this study.
Blanks indicate no change from the previous experiment.

Experiment	LIN1	NLIN1	LIN2	NLIN2	E1	E2	E3	E8	E9	E10	E11
Nonlinear terms	NO	YES	NO	YES							
Number of layers	1				2	3					
Thickness of the layers (m)	1000				1000 4000	600 700 3700	300 900 3800				
$g'(ms^{-2})$					.02 .005	.02 .005					
Viscosity (m^2s^{-1})	330		3300		330						
Bottom drag coeff. (s^{-1})								10^{-7}			
Boundary conditions on Africa[1]	NS								FS	NS	
Horizontal resolution (km)	40									20	
Africa's geometry[2]	A										B

[1]NS=No-slip; FS=Free-slip [2]A=Rectangular (Fig. 1); B=More realistic (Fig.5)

ones have been introduced to clarify aspects of the vorticity balance and one to give the Beta-effect more realistic importance. The basin used in the experiments is described in DB and BD, but is illustrated here for convenience (Fig.1). The steady wind stress profile is shown at the left of the figure. The direct effect of the wind stress decreases linearly from its maximum near the surface to zero at a pressure of 100 db. In the multilayer cases, the wind forcing is therefore partitioned among the top two layers if the upper layer thickness becomes less than 100 db. The conditions on meridional boundaries are generally no-slip (u=v=0) and those for zonal boundaries free-slip (v=du/dy=0), except in one experiment where the coast of Africa is treated homogeneously as a free-slip boundary. The character of retroflection and ring formation in the experiments is summarized in Table 2.

TABLE 2

Description of the upper layer flow pattern of the model retroflection region

LIN1	Small recirculation cell east of Africa. Most of the flow (35 Sv) goes into the Atlantic.
NLIN1	Substantial retroflection extending southwest of the tip of Africa. Approx. 15 Sv leakage into the Atlantic. High temporal variability.
LIN2	Establishment of a retroflection but with 20-25 Sv leakage.
NLIN2	Similar to LIN2. The retroflection is moved south and west of the tip of Africa. Steady solution.
E1	Similar to NLIN1, but with 25 Sv leakage and less variability. Rings are formed continuously at a rate of about 5 per year.
E2	Similar to E1, but with 15 Sv leakage and the position of the retroflection closer to the tip of Africa. The rate of ring formation is about 4 per year.
E3	The current retroflects very soon after it separates from the coast. Small leakage (less than 5 Sv). High eddy energy in the retroflection. No rings.
E8	Similar to E3, but the circulation is more intense. No rings.
E9	Most of the flow turns westward and then northward along the coast into the Atlantic.
E10	Similar to E3, but with less intense retroflection and a strong downstream meander. Only 4 rings are formed in 10 years, each of which requires a strong interaction with the subpolar front.
E11	The boundary current is broader with smaller velocities than E10. The retroflection is still strong, but is less intense and there is a 5-10 Sv leakage. About 3 rings are formed per year, but not in such a regular fashion as in E1.

3 THE VORTICITY BALANCE

3.1 Method of analysis

Taking an average over time, assuming statistical equilibrium, the full vorticity equation can be written

$$\underbrace{\overline{\vec{V}\cdot\nabla_s\zeta_s}}_{\text{RVA}} + \underbrace{\overline{\beta v}}_{\text{BETA}} + \underbrace{\overline{(\zeta_s + f)\,\nabla_s\cdot\vec{V}}}_{\text{STRCH}}$$

$$+ \underbrace{\overline{\vec{k}\cdot\left(\nabla_s \times \dot{s}\frac{\partial\vec{V}}{\partial s}\right)}}_{\text{VERT}} + \underbrace{\overline{\vec{k}\cdot\left(\nabla_s\alpha \times \nabla_s p\right)}}_{\text{SOL}}$$

$$+ \underbrace{\overline{A\,\mathrm{curl}_z\left(\left(\frac{\partial p}{\partial s}\right)^{-1}\nabla_s\cdot\left(\left(\frac{\partial p}{\partial s}\right)\nabla_s\vec{V}\right)\right)}}_{\text{VISC}} + \underbrace{\mathrm{curl}_z\left(\frac{\partial\tau_x}{\partial z}\right)}_{\text{WIND}} = 0$$

where s is the generalized vertical coordinate, density here; ζ_s is the relative vorticity evaluated at s=constant; curl_z is the vertical component of the curl; A is the constant horizontal eddy viscosity and the other symbols are conventional. (RVA) is the relative vorticity advection, (BETA) the planetary vorticity advection, (STRCH) the stretching term, (VISC) the viscous stress curl, (WIND) the wind stress curl, (SOL) the solenoidal term and (VERT) the curl of the vertical advection. The actual finite difference formulation of the vorticity equation which we use in computation of these terms conserves potential vorticity and potential enstrophy (see Bleck, 1979; Bleck and Boudra, 1981). Finding the mean values of the terms for the final five years of each experiment, a balance was obtained. Our computations have shown that SOL and VERT are zero or negligible. This can be expected from the use of isopycnal coordinates. In addition, STRCH is equal to zero in the one layer case and has very small temporal mean values in the multilayer case. It, therefore, is not addressed in our analysis.

3.2 The vorticity balance in a one-layer model

It is instructive to begin our discussion of the model vorticity balance with a weakly viscous, linear experiment. Removing the inertial terms from the equations of motion, the vorticity equation is simplified and the balance is among BETA, VISC and WIND. Our steady solution LIN1 (not illustrated here) has qualitatively the same characteristics as the analytical solution of De Ruijter (1982). Along the no-slip coast of Africa in the Munk layer, lateral friction balances the β-induced input of relative vorticity. The magnitude of the latter is determined by the southward velocity, which is governed by the

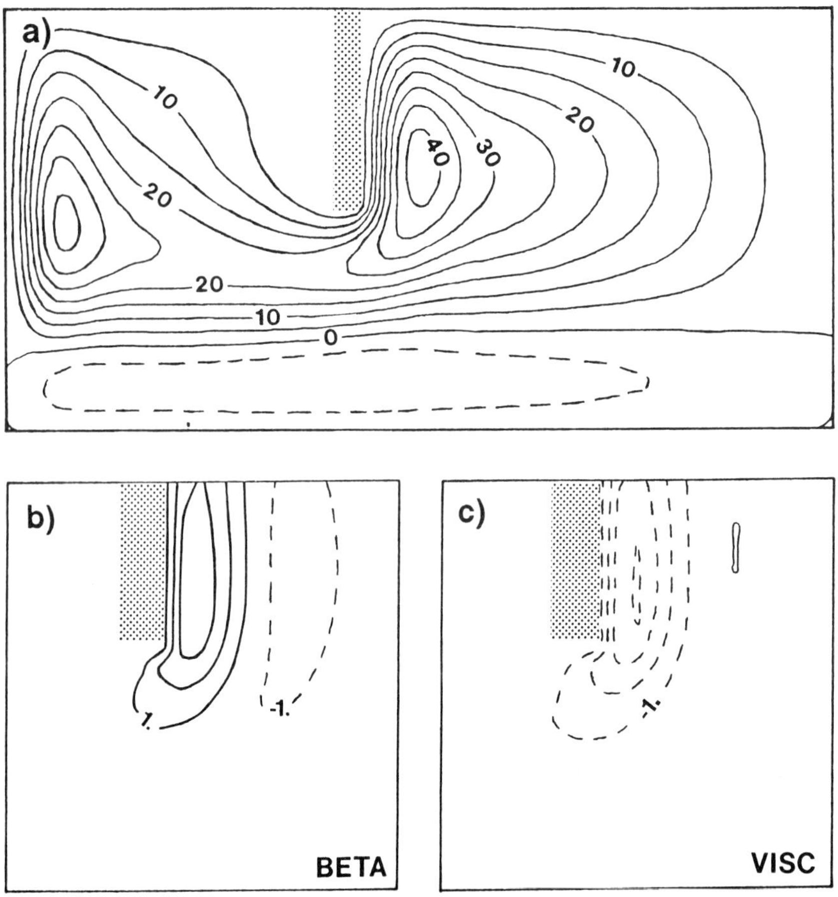

Fig.2. Experiment LIN2. (a) The mean mass transport streamfunction. The contour interval is 5 X $10^6 m^3 s^{-1}$. (b) The time mean distribution of planetary vorticity advection (BETA) around the southern tip of Africa. The contour interval is 2 x $10^{-12} s^{-2}$. (c) As in (b), except viscous stress curl (VISC).

combination of integrated wind curl and layer depth, as well as friction. In the low viscosity case (LIN1), the boundary layer is relatively thin and friction becomes very small after separation, and therefore, so does the planetary vorticity advection. The only way for the flow to satisfy this condition at the southern tip of Africa is to immediately turn westward or eastward. The wind curl maximum is only 120 km north of the latitude of Africa's tip. Therefore, there is very little linear return flow toward the Indian Ocean interior north of that latitude. Thus, most of the flow branches westward into the Atlantic.

If the viscosity is increased by an order of magnitude (LIN2) (Fig.2), friction becomes important in the area south of Africa. The current can continue southward for some distance since BETA (Fig.2b) can be balanced by VISC (Fig.2c). In other terms, the frictional boundary layer is broadened enough so that more streamlines of the separating Agulhas are able to connect with returning Sverdrup lines to the south and, thus, establish a partial retroflection. Important inertial effects are, therefore, not required in order to obtain retroflection; so that the latter may be obtained in highly diffusive, coarse-resolution models primarily because of the large lateral viscosity required by those models.

Introducing the inertial terms adds a new term to the vorticity balance, the relative vorticity advection RVA. Experiments NLIN1 (Fig.3) and NLIN2 (not illustrated here) correspond respectively to the nonlinear runs of LIN1 and LIN2. In the high viscosity case (NLIN2), most of the retroflection is moved south and west of the separation point, but the general characteristics of the flow pattern are unchanged. Inclusion of the inertial terms results in an increase in the diffusion of the negative relative vorticity. This is because the total velocity is a bit stronger and the boundary layer, now inertial-frictional, is thinner. BETA does not increase, however, since the orientation of the retroflection gains a SW-NE tilt. The increase in VISC is balanced by the new term RVA. The major conceptual difference from LIN2 is that the fluid may now flow southward from the tip of Africa due to its own inertia. Frictional effects are, however, still dominant.

In the corresponding low viscosity case (NLIN1) (Fig.3) where the boundary layer is relatively thin, friction is unimportant after separation and the effect of inertia is predominant. A substantial retroflection is still obtained as RVA now becomes the primary term to balance BETA. Inclusion of the inertial terms in the weakly viscous case substantially modifies the flow pattern and the

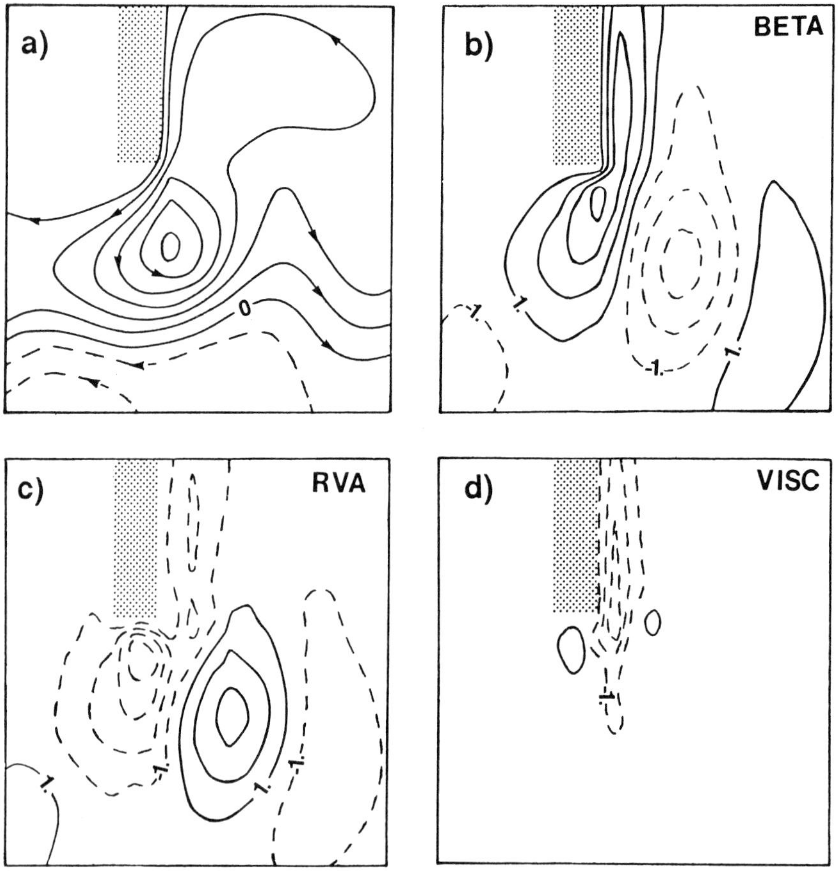

Fig.3. Experiment NLIN1. As in Fig.2 for (a) and (b). (c) Time mean distribution of relative vorticity advection (RVA). The contour interval is 2 x $10^{-12} s^{-2}$. (d) As in Fig.2c.

distribution of terms of the retroflection vorticity balance. The circulation east of Africa is southward intensified due to inertia, and weaker along the coast itself.

In each of LIN2, NLIN1 and NLIN2, there is a substantial retroflection. The similarity among all three is the establishment of a boundary layer after separation in which the beta effect is balanced by the turning eastward of a major portion of the current. In LIN2 the boundary layer is purely frictional, in NLIN2 it is inertial-frictional, and in NLIN1 it is a free inertial boundary layer. While the physical mechanism of retroflection is not the one proposed by De Ruijter (1982) -- the inertial overshooting distance -- his concept of the retroflection being accomplished within a free inertial boundary layer is apparently applicable. In the experiments remaining to be discussed, the retroflection is accomplished in this free boundary layer. NLIN1 possesses a similar distribution of the terms as the weakly nonlinear multilayer cases and will be discussed further in parallel with them.

3.3 The vorticity balance in the multilayer model

(i) Variation of inertia. As described in BD, in experiments E1, E2 and E3, the strength of the retroflection depends on the inertia of the currents, which can be quantified by the Rossby number (Ro) of the overshooting boundary current. For E1 and E2, the top layer vorticity balance is similar to that for NLIN1 above (Fig.3). There is a change in the spatial distribution of the terms, however, for E3 (Fig.4). These experiments have high temporal variability, which is not the case in LIN1, LIN2 and NLIN2. RVA, therefore, can be separated into two components, one related to the mean flow and the other one to the transients. The RVA pattern is largely dominated by the component associated with the mean flow. However, the component due to the transients can have a canceling effect on that due to the mean flow, especially in NLIN1 and E1.

Along the coast of Africa and close to the boundary, viscous effects are important. Negative relative vorticity is generated by the no-slip boundary condition and then transported eastward by diffusion. BETA and VISC increase as Ro does from E1 to E3 and they balance each other. Away from the boundary, the effect of diffusion decreases and the balance is mainly between RVA and BETA.

Just after the flow leaves the coast, VISC is still important because the flow is strongly sheared. Before the separation, the southward velocity is maximum at the first grid point east of the coast. The viscous stress curl of such a profile has a minimum at this grid point. If the boundary current were very well resolved, a positive maximum in VISC would appear near the coast. As the current overshoots the tip of Africa, the current broadens and is better resolved by the grid lattice. A positive maximum does appear on the western side of the free jet, and this feature is present in most of the nonlinear experiments discussed (NLIN1, NLIN2, E1, E2, E3 and E10).

After separation, the experimental vorticity balances differ. Changes in VISC leave BETA dominant, leading to generation of positive relative vorticity. From here on RVA balances BETA. For each case, at least a portion of the current turns eastward into the Indian Ocean. On the westernmost side of the current, inertia is not strong enough to dominate the linear tendency, and this part of the flow drifts westward into the Atlantic. This leakage is very small in the high Rossby number cases, since only a very small part of the current's western edge has weak motion.

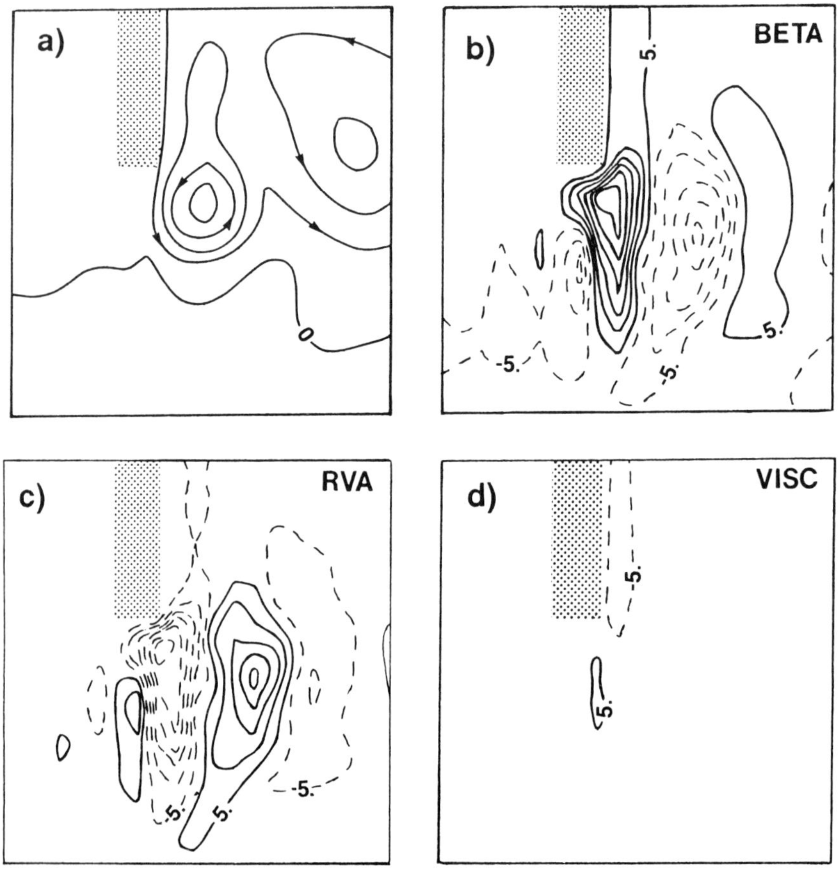

Fig.4. Experiment E3. As in Fig.3, except the contour interval in 10 X 10^{-12} s^{-2} in (b), (c) and (d).

The planetary vorticity advection (BETA), proportional to the southward component of the velocity, is of primary importance in controlling the shape and the strength of the retroflection. As the Rossby number of the separating current increases, so does BETA and the tendency to turn eastward. Therefore, this turn moves closer to the tip of Africa as southward inertia increases, in contradiction to the inertial overshooting concept, and has no direct connection with the distance to the returning Sverdrup lines.

Inclusion of bottom drag in E8 (not illustrated here) decreases the eddy-induced mean circulation in the bottom layer, but increases the intensity of the upper layer circulation, since the bottom layer flow regime in E3 is a primary mechanism for draining energy from the retroflection area. In this experiment, the terms of the model vorticity balance are slightly larger in magnitude, but their horizontal distribution remains as in E3.

(ii) Horizontal resolution. As pointed out by BD, 40 km grid resolution cannot adequately describe the details of boundary layer flows. They performed E10 with grid spacing halved in order to test the sensitivity of the retroflection to improved resolution of the boundary layer and release of baroclinic instabilty. The retroflection flow pattern of E10 (not illustrated here) is similar to that of E8. In E10, the bottom layer circulation is revived by the improved resolution of instabilities and the upper layer circulation in the retroflection area is less intense than in E3 and E8. There is a similar horizontal distribution of the terms of the vorticity equation in E10 as in E3 and E8. BETA has a maximum near the coast as in E3, but VISC is even larger because the maximum velocity is greater and closer to the coast with the improved resolution. An additional contribution from RVA balances this enhanced VISC.

Regarding the principal balance the same conclusion as in Section 3.3.a can be drawn. As the current separates at the tip, VISC decreases and the remaining dominant term, BETA, generates positive relative vorticity. As in the 40 km resolution case, RVA balances BETA as the current turns eastward rejoining the Indian Ocean subtropical gyre.

(iii) Boundary condition along the African coast. BD obtained a very different two-basin flow configuration when a free-slip boundary condition along the coast of Africa was substituted for the no-slip condition. In the corresponding no-slip case (E8) only a small amount of fluid escapes into the Atlantic, the major part of the current retroflecting into the Indian Ocean. In the free-slip experiment, most of the fluid goes around the southern tip of Africa into the Atlantic.

In the top layer vorticity balance for E9 (not illustrated here), VISC is negligible in comparison to RVA and BETA along the coast and north of the tip of Africa. The principal balance is between these last two terms. Contrary to the no-slip case, there is no cyclonic vorticity present along the coast. The boundary layer is primarily inertial as opposed to the inertial-frictional boundary layer of the no-slip case. In order to overshoot the tip (and afterward retroflect), the current must coexist with a region beyond its western edge which is largely motionless. In the no-slip case, the western boundary condition for the current along the coast is already a motionless one and, thus, provided nonlinear effects are strong enough, the current shoots past the tip in a relatively natural fashion. In contrast, the condition at the western edge of the coastal current in E9 results in the current having its maximum velocity there. In an overshooting configuration it would have to adjust to a strongly sheared condition.

3.4 South African coastal geometry

The shape and orientation of Africa in the previous experiments is one which maximizes the importance of BETA in the model retroflection. Inasmuch as this term plays a dominant role, giving it a more realistic level of importance seems warranted. After leaving the South African coast, the real Agulhas flows along the Agulhas Bank shelf break, which is oriented approximately from northeast to southwest, before flowing out into deep water. In this configuration, the component of velocity advecting planetary vorticity is about 70% of the total, rather than 100%, as with the meridionally oriented coast of BD's model. The basin geometry shown in Fig.5a has been chosen, therefore, as a next step toward model realism.

The time mean top layer flow pattern (Fig.5a) for this 20 km resolution experiment (E11) bears many of the same characteristics as those of E10 (for comparison, see Fig.13 in BD). However, there are some notable differences, especially in the Indian Ocean sector. The retroflection is still strong, but less intense. The mean Agulhas along the coast is somewhat broader, and its maximum velocity is less than in E10. With the tilted coast, the energy of the boundary region has more of a tendency to leak into the Atlantic rather than to be trapped in the Indian Ocean.

The retroflection vorticity balance of E11 (Fig.5) is mostly similar to that in the high Rossby number rectangular geometry experiments with weak friction. Along the coast BETA (Fig.5b) and VISC (Fig.5d) essentially balance each other. Again, however, at certain points, VISC is particularly strong, and RVA makes up for what BETA does not supply there. Over the remainder of the retroflection, BETA and RVA (Fig.5c) are in balance. Incorporation of the more realistic geometry has no major impact on the retroflection vorticity balance. The magnitude of BETA at separation still requires an eastward turn and thus, retroflection.

Conspicuously absent from the diagram is the feature in VISC just beyond and to the west of the tip, which is present in the other multi-layer experiments discussed here. This is apparently because the coastal turn at separation is not nearly so sharp as with rectangular geometry and there is much more motion of fluid around this turn, principally within the rings drifting into the Atlantic. In fact, the retroflection of the Agulhas is usually upstream of the tip of Africa and the depicted flow pattern of Fig.5a averages through many ring formation events. The field of RVA exhibits a larger contribution from the transients than in E3, E8 and E10 because of the much greater frequency of ring formation. More details concerning the transient aspects of E11 are provided in Section 4.

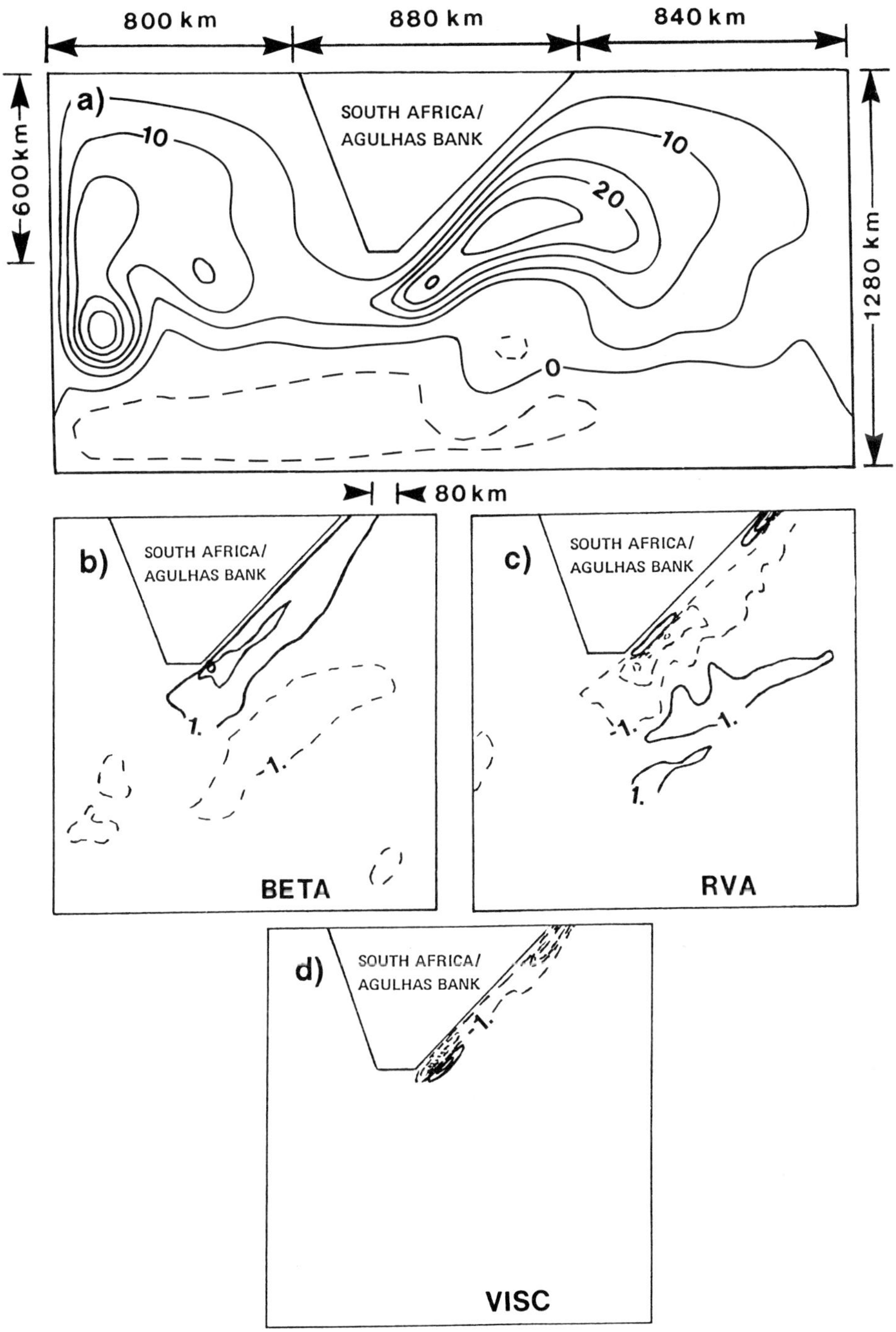

Fig.5. Experiment E11. (a) Upper layer mean mass transport streamfunction for the last five years. The C. I. is 5 x $10^6 m^3 s^{-1}$; (b), (c) and (d) as in Fig.3.

4 ENERGETICS AND RING FORMATION

4.1 Formulation of the energy conversion terms

A detailed derivation of the time mean and transient eddy energy conversion terms in isopycnic coordinates is given by Bleck (1985). Following his notation, they are as follows:

$$C(P,K_E):\quad \overline{\phi\nabla_s\cdot\left(\frac{\partial p}{\partial s}\,v'\right)}-\overline{v'\cdot\frac{\partial p}{\partial s}\,\alpha\nabla_s p}$$

$$C(P,K_M):\quad \overline{\phi\nabla_s\cdot\left(\frac{\partial p}{\partial s}\,\tilde{v}\right)}-\tilde{v}\cdot\overline{\frac{\partial p}{\partial s}\,\alpha\nabla_s p}$$

$$C(K_E,K_M):\quad \overline{v'\frac{\partial p}{\partial s}\cdot\left(v'\cdot\nabla_s+\dot{s}'\,\frac{\partial}{\partial s}\right)}\tilde{v}$$

where ϕ is the geopotential; $(\bar{\ })$ is the time average; $(\tilde{\ })$ is the mass weighted time average; $(\ )'$ the departure from the mass weighted time average; K_M is the kinetic energy associated with the mean flow and K_E, with the transient eddies; P is the total available potential energy and the other symbols are conventional.

Bleck (1985) showed that in the isopycnic framework a conversion term appears in the energetics suggesting an exchange between the eddy potential energy and K_M. To avoid confusion, therefore, P is not decomposed into mean and eddy components. It is assumed that the exchange between P and K_E is associated with baroclinic conversion and that the transfer from K_M to K_E is associated with barotropic conversion.

4.2 Mean energetics

In order to gain some additional insight into the impact of the model parameters most influential on the retroflection, we now examine the mean energetics of some of the experiments discussed above, illustrated in box diagram form in Fig.6. The time means of the energies and conversions terms are computed over the last five years of a ten year run for E1, E2, E3, E8, E9 and E11 and over selected years for E10.

Considering the influence of inertia on the basin averaged energetics, we first compare E1 and E2. The layer receiving most of the direct wind forcing is thinner in E2 than in E1, leading to increased kinetic energy in the boundary

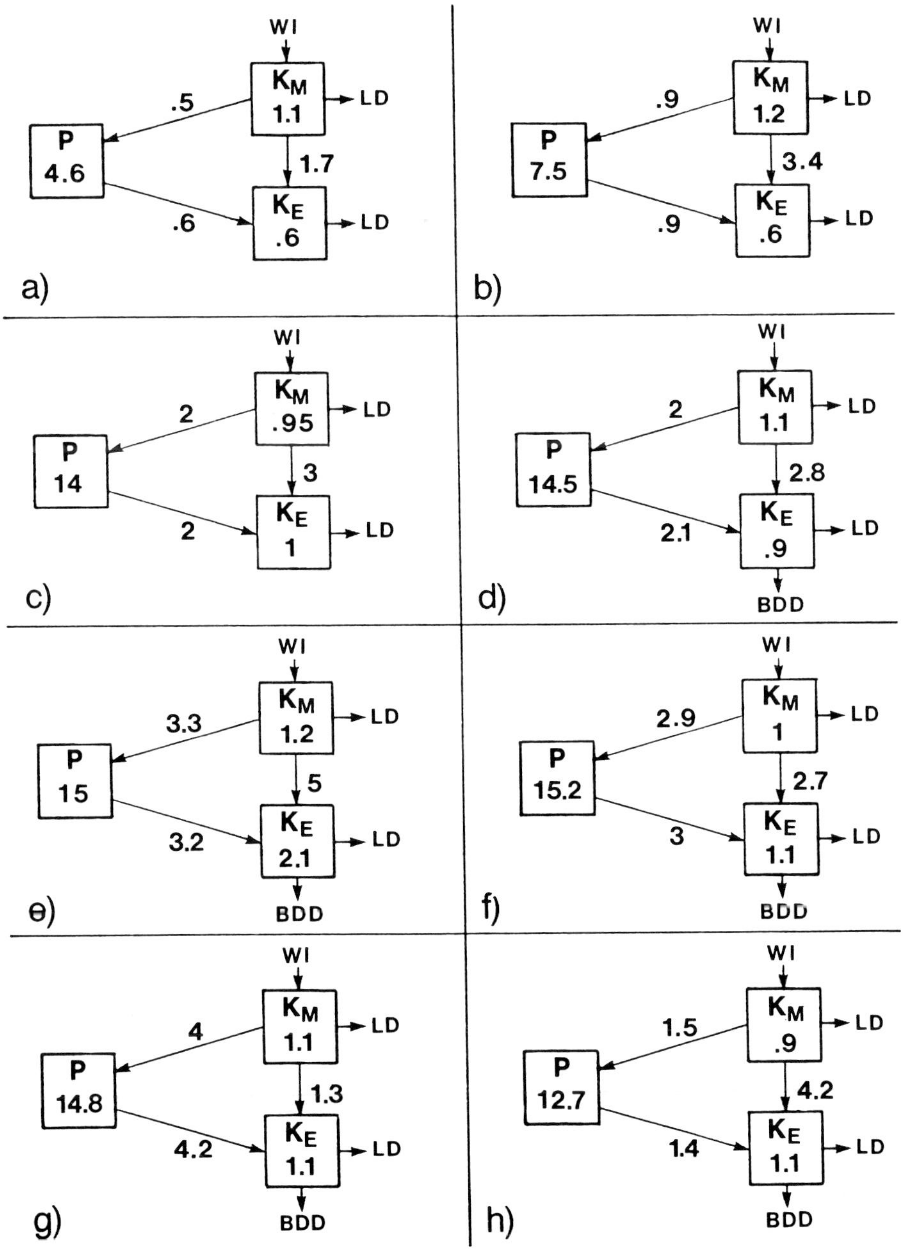

Fig.6. Eddy-mean energetics for (a) E1; (b) E2; (c) E3; (d) E8; (e) E9; (f) E10, year 4; (g) E10, year 7; (h) E11. WI is the wind input; LD, the lateral dissipation and BDD the bottom drag dissipation.

currents, even though the basin average kinetic energy does not change much (Fig.6a,b). As the top layer flow increases in speed, so does the vertical shear. From the thermal wind relation, this leads to more strongly tilted isopycnal surfaces and, thus, to more energy exchange between K_M and P. The available potential energy, however, is distributed over the whole domain, as opposed to the kinetic energy, which is concentrated mostly in the boundary currents (Fig.3 of BD). In the basin averaged sense, then, this parameter has a much larger impact on the potential energy than on the kinetic energy. Horizontal plots show that most of the conversion from K_M to K_E occurs in the retroflection area. Rings are continuously being formed there and, therefore, the increase in strength of the boundary currents from E1 to E2 apparently results in an increase in kinetic energy exchange between the mean flow and the eddies in this highly transient region.

The upper layer is thinner yet in E3 (Fig.6c), so that inertia is even greater in the boundary currents, leading to an increase of 300% in available potential energy from E1. This increase is accentuated by the fact that the circulations in the subtropical Indian and Atlantic oceans are almost separated. The boundary currents' kinetic energy increases, but no rings are formed in this experiment, and so there is less conversion between K_M and K_E than in E2, even though there is substantially more K_E in E3. Inclusion of bottom drag in E8 (Fig.6d) increases dissipation in the bottom layer. As mentioned in Section 3.3, this leads to a more stable and intense retroflection in the upper layer. Therefore, there is less K_E, but the conversions are essentially as in E3.

In the free-slip case (E9) (Fig.6e), the total kinetic energy is 50% higher than in E8, primarily because of the reduction of kinetic energy dissipation along the coast of Africa. The boundary currents and the free jet in the South Atlantic become stronger, and through geostrophic adjustment, P and the exchange between kinetic and potential energies increase also. The flow regime is particularly turbulent south of Africa in this experiment leading to large values of K_E and $C(K_M,K_E)$.

The increased horizontal resolution of E10 better represents the release of baroclinic instability. Increases are noticed in P, and in the potential to kinetic energy conversions because the boundary currents are even stronger than in E3 and E8. Since four rings were formed during the ten year experiment, it is also interesting to note the difference in the energetics for years with and without ring formation. In the latter case, the K_M-P-K_E appears to be the primary energy path (Fig.6f). During a year with a ring, the baroclinic and barotropic transfers are of the same order of magnitude (Fig.6g). This comparison strongly suggests the association between $C(K_M,K_E)$ transfer and formation of a ring. Supporting this assertion is the fact that the change in

South Africa' s geometry (E11) results in more barotropic conversion. In fact, the dominant transfer is the one from K_M to K_E (Fig.6h). The major difference between the two experiments is the much greater number of rings formed in E11. Also, P is approximatively 15% smaller in E11 than in E10 because the retroflection is less intense and the Agulhas along the boundary is somewhat broader.

The question arises as to whether basin-averaged energetics are truly relevant in discussing the dynamical processes in the retroflection region. Harrison and Robinson (1978) showed that energy transfer averaged over a model domain may not be characteristic of any important subregion. BD showed, however, that especially in the high Rossby number experiments, most of K_M and K_E are localised in the retroflection area. Our calculations have shown that almost all the $C(K_M,K_E)$ is in the Agulhas region and $C(P,K_E)$ is maximum in this region. Therefore, the energetics averaged over the whole basin give a good indication of events occuring in the retroflection area for these cases.

4.3 Ring formation

Agulhas rings have been observed and studied by Gordon (1985) and Olson and Evans (1986). Lutjeharms (1981) was able to capture the formation of one ring in an analysis of satellite imagery, but little has been established about the frequency of their formation and about the dynamics that drive the shedding, mostly because the rings are strongly nonlinear features. For example, might the formation of the rings and the associated leakage be strongly influenced by seasonal factors (wind-stress or inflow) or by nonlinear oscillations of the recirculation region off South Africa? Observational efforts to find time-dependence in the Agulhas Current transport by Pearce and Grundlingh (1982) were not conclusive and, in our model, rings form with constant forcing. In their study of the dynamics of the loop current and eddy shedding of the Gulf of Mexico, Hulburt and Thompson (1980) obtained a similar result with a quasi-annual eddy shedding. Even with realistic time variation of the upper-layer inflow, the eddy-shedding period was dominated by the quasi-annual period rather than by the forcing period, although the influence of the latter was not negligible.

In order to get some understanding of ring formation in the model, we describe a typical event of E11, illustrated here in Fig.7. In the upper layer, at day 2950 (Fig.7a), the flow pattern is in a quasi-steady configuration. An already formed ring is centered just southeast of the tip of Africa and the retroflection of the Agulhas proper is a short distance up the coast. Thirty days later, the meander east of the retroflection has begun to amplify and the center of the retroflection is moving southwest (Fig.7c). The already formed ring has begun moving toward the Atlantic. The lower layer at day 2950 shows a

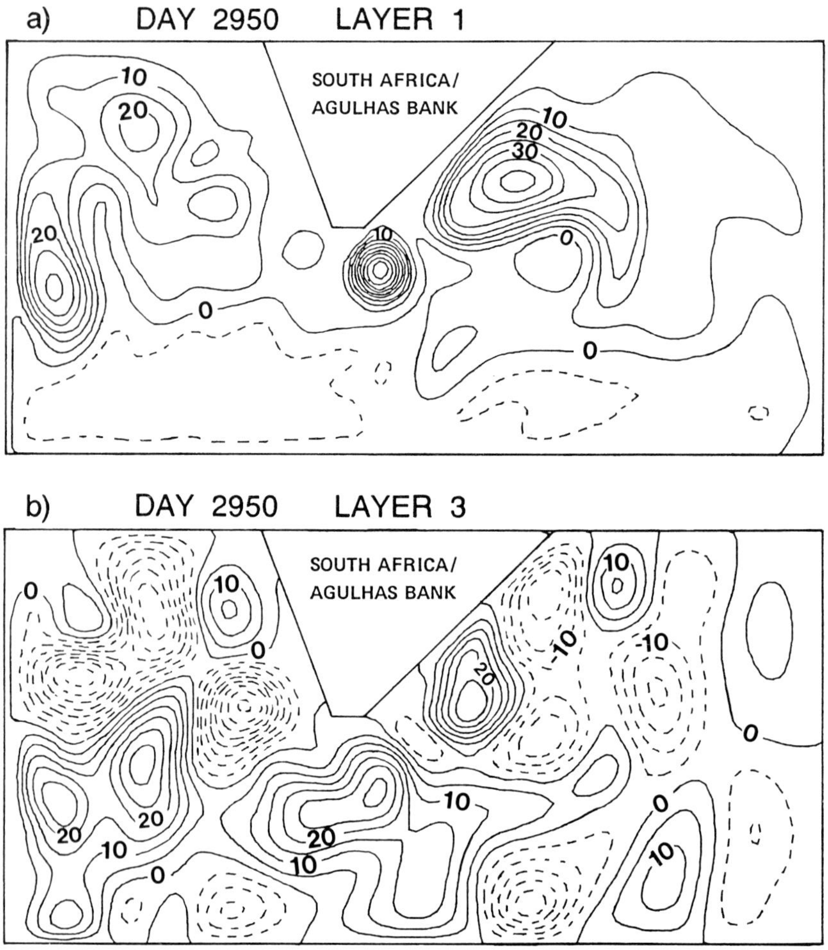

Fig.7. Ring formation in E11. Upper and lower layer mean mass transport streamfunctions. The contour interval is $5 \times 10^6 m^3 s^{-1}$.

couplet of cyclonic gyres located under the meander (Fig.7b). As the amplitude of the upper layer meander grows, they merge and intensify (Fig.7f). This intensification reaches a maximum when the ring cuts off. Its intensity weakens rapidly thereafter (Fig.7j).

The dramatic growth of this cyclonic circulation in the third layer is suggestive that baroclinic instability is being released during ring formation. In his recent model study of the Gulf of Mexico Loop Current, Hulburt (1986) also found that an intense cyclonic circulation develops in the lower layer when an eddy is shed. In order to get more insight into the instabilities involved,

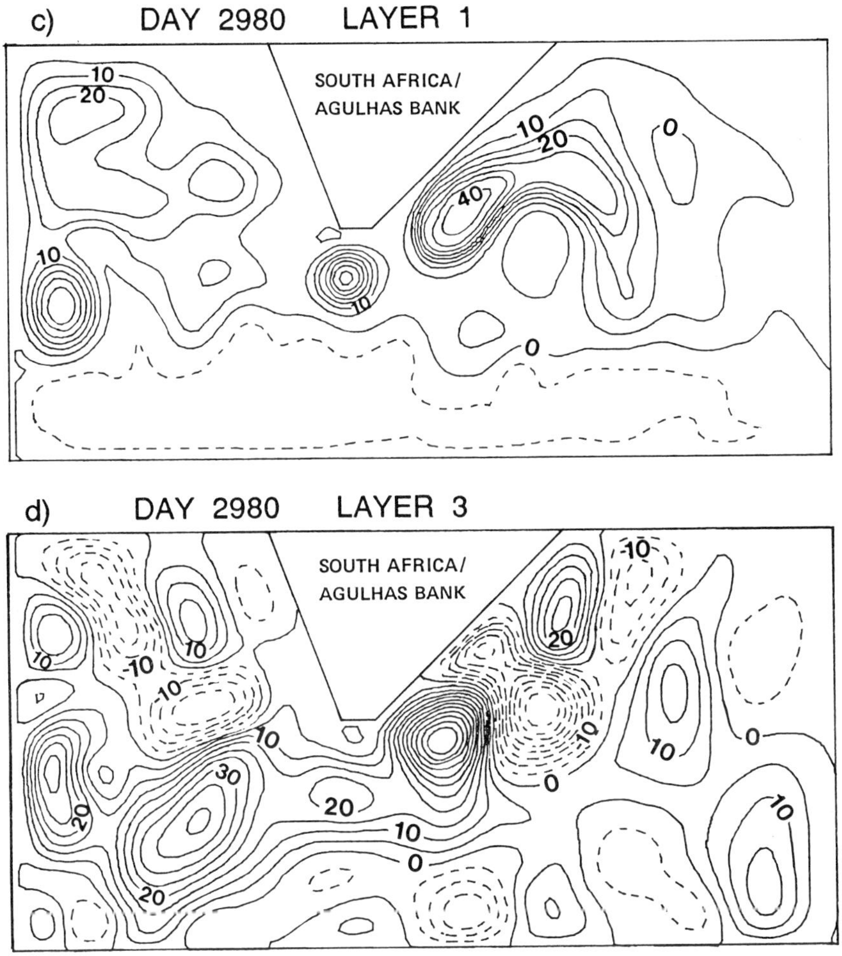

Fig.7. (Continued)

we now look at the time evolution of the energy conversion terms during this and the subsequent event (Fig.8). The mean components in the energetics are averaged over a year. The two strong signals in $C(K_M,K_E)$ in Fig.8 correspond to the two events of ring formation. In both cases, before the meander begins to grow, the baroclinic conversion term is negative. Then, as the amplitude of the meander increases, $C(P,K_E)$ rises above zero and reaches a maximum when the ring is formed. The barotropic conversion term is larger and reaches a maximum at the same time. It is not clear, however, that $C(K_M,K_E)$ is directly related to barotropic instability. Possibly, it is simply suggestive that kinetic energy which has been residing in the time mean flow field is being transfered into a

e) DAY 2990 LAYER 1

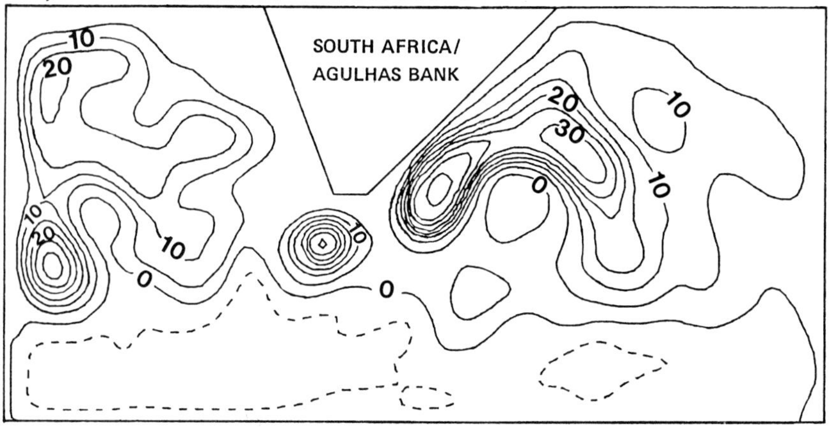

f) DAY 2990 LAYER 3

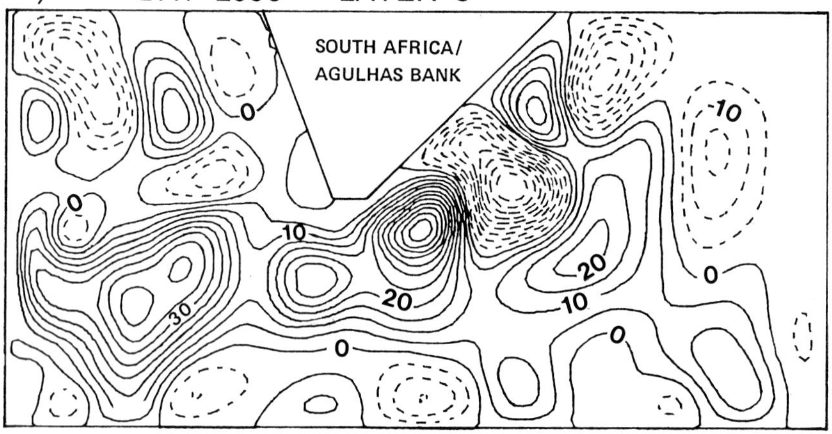

Fig.7. (Continued)

different flow configuration. Therefore, no conclusion can be reached as to whether barotropic or baroclinic instability is dominant. We can only say that both barotropic and baroclinic conversions have significant peaks during ring formation. It would appear that the instability is of the mixed type.

The character and frequency of ring formation in the experiments depend on the choice of parameters, as shown in Table 2. Analysis of the vorticity balance in Section 3 showed how important is the magnitude of the planetary vorticity advection in the strength of the retroflection. As the Rossby number of the boundary currents is increased, more fluid retroflects and fewer rings are formed. With 40km grid resolution, rings are formed in the low Rossby

g) DAY 3000 LAYER 1

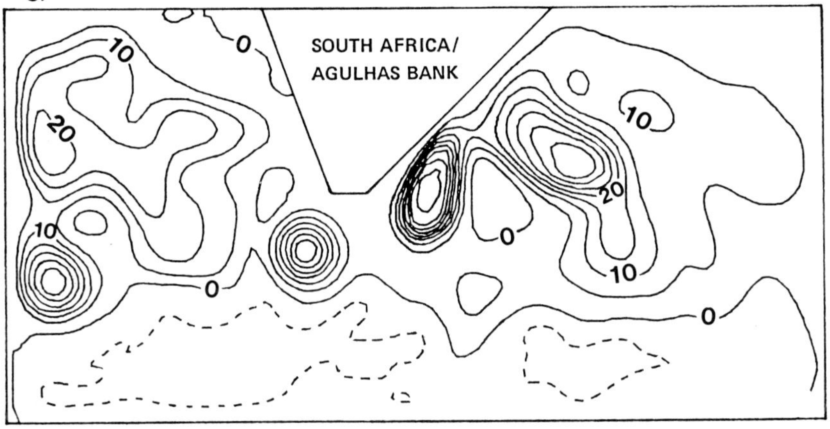

h) DAY 3000 LAYER 3

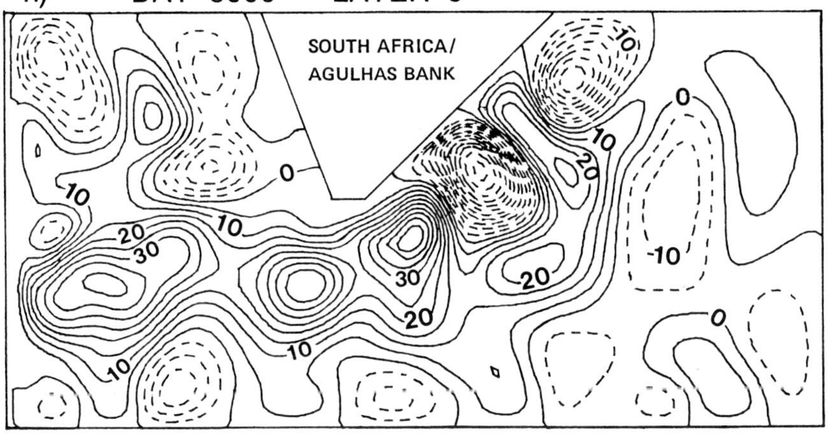

Fig.7. (Continued)

number cases (E1 and E2) and none in the high Rossby number cases (E3 and E8). But in the latter case, a change in the geometry of Africa in which the retroflection is less intense (same as E11 with 40km grid resolution, not presented here) leads to formation of rings. A few rings were also formed with an increase in horizontal resolution (E10) in which the release of baroclinic instability is improved.

The analysis of rings formed in E1 and E10 supports the result found in E11. In both experiments, as shown in Fig.9 and Fig.10 respectively, $C(K_M,K_E)$ and $C(P,K_E)$ reach maxima of the same magnitude at the moment of ring formation. As in E11, a cyclonic circulation develops in the lower layer and reaches maximum

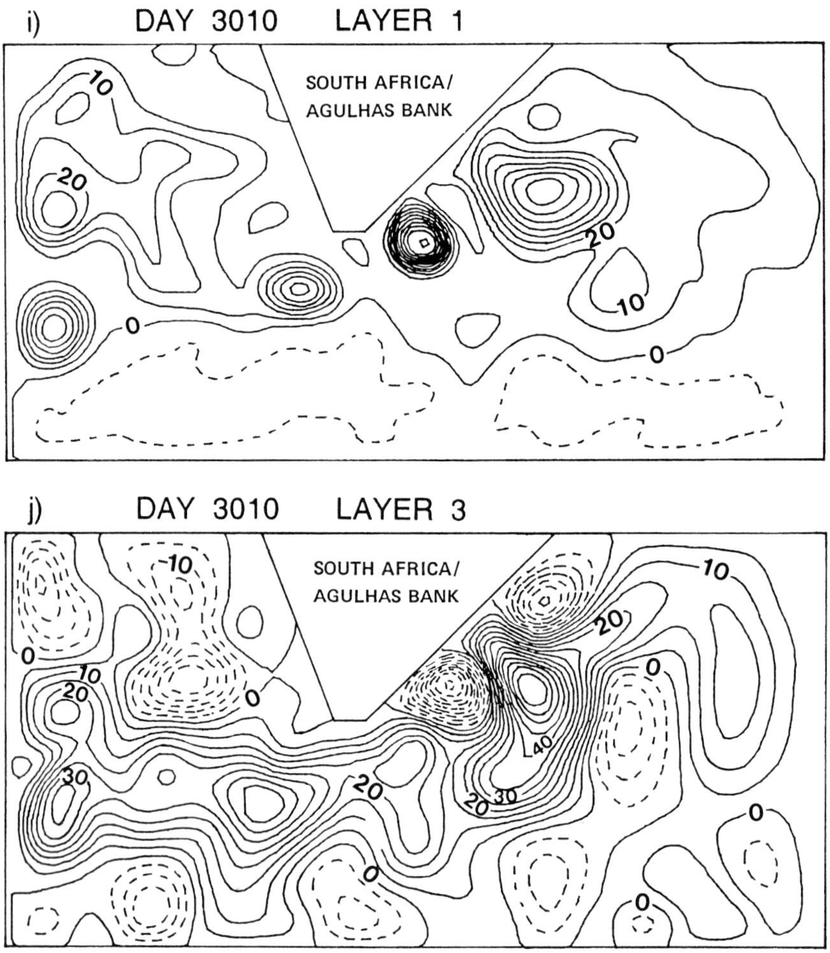

Fig.7 (Continued)

intensity when the ring separates from the gyre. To conserve space here, instantaneous streamfunctions for those two experiments are not presented. In E1, rings are formed in a continuous manner at the southern tip of Africa at a rate of about five per year. As soon as an already formed ring begins to drift toward the Atlantic, the center of the retroflection moves southward and shortly thereafter a new ring cuts off. The two peaks in Fig.9 correspond to two consecutive events of E1 in year 9. The formation process is quite different in E10, where only four rings are formed in ten years. The core of the Agulhas Current possesses more planetary vorticity advection than in E1. In E10, for a ring to cutoff, the center of the retroflection must be rather far south and a

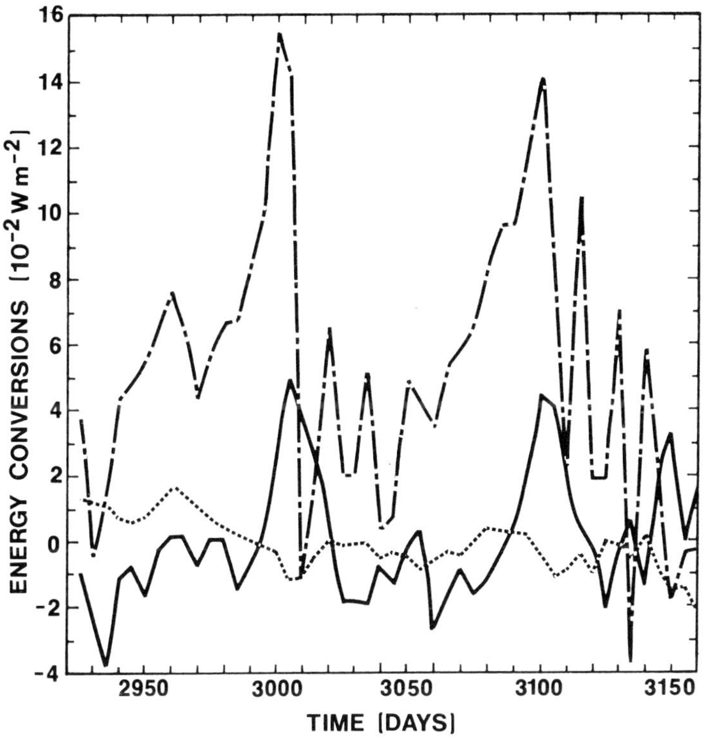

Fig.8. Experiment E11. Variations of the energy conversion terms with time. The dashed-dotted line corresponds to $C(K_M, K_E)$, the solid line to $C(P,K_E)$ and the dotted line to $C(P,K_M)$. Rings are formed at day 3000 and 3100.

dome of low potential vorticity fluid from the subpolar gyre must intrude northward along the east side of the retroflection region. In addition, the front between the gyres needs to be far enough south so that the cutoff ring can escape into the Atlantic before being recaptured by the Agulhas. For the event presented in Fig.10 occuring at the beginning of year 9 (illustrated in Fig.14 of DB), the first peak in $C(K_M,K_E)$ corresponds to the initial ring cut off, the second one to a temporary reabsorption and the third one to the final cut off of the ring. No baroclinic conversions are associated with the last pinching off.

5 SUMMARY

The vorticity balance, the mean energetics and energy conversions have been used as tools with which to describe the dynamics of Agulhas retroflection and ring formation in a quasi-isopycnic coordinate numerical model. In addition, analysis of some new experiments has increased understanding of the model

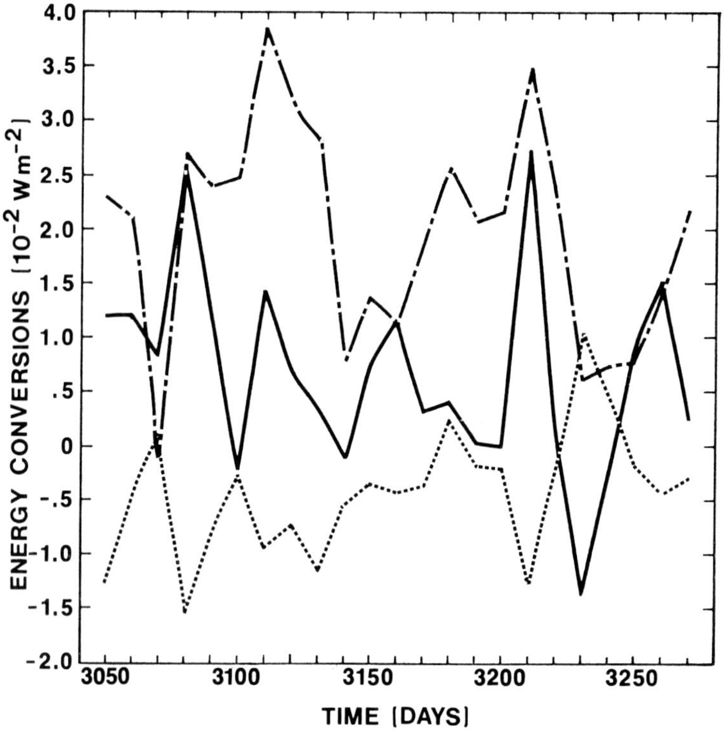

Fig.9. As in Fig.8 for experiment E1. Rings are formed at day 3080 and 3210.

parameters most influential in the retroflection.

The analysis of the vorticity balance of a one-layer simplification of the Bleck and Boudra (1981) numerical model in an idealized South Atlantic-Indian Ocean basin showed that either strong internal friction or inertia can bring about a partial retroflection where a linear model with weak friction has none. With the full quasi-isopycnic coordinate model, it was first shown how important is the magnitude of the planetary vorticity advection in the strength of the retroflection. As the Rossby number of the boundary currents is increased, more fluid retroflects and retroflects earlier after separation. When horizontal resolution is doubled to better resolve the boundary currents and instabilities, the magnitude of the terms of the vorticity balance increases as the current becomes narrower and its maximum velocity increases. Otherwise, the balance is unchanged. The importance of the frictional boundary condition was brought out through a no-slip/free-slip comparison. A new South African geometry was also adopted, in which the planetary vorticity advection is less important. The retroflection was shown to be less intense, but the primary balance is unchanged.

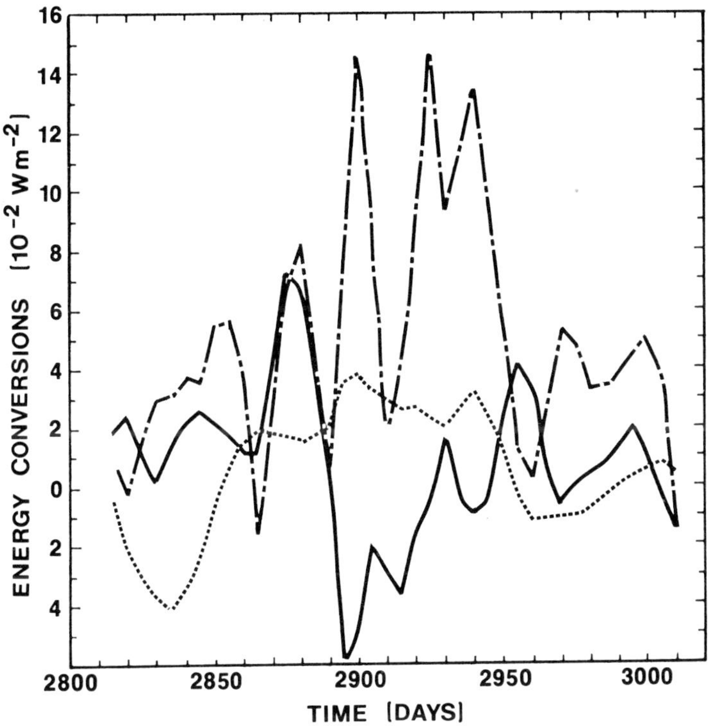

Fig.10. As in Fig.8 for experiment E10. The ring is formed at day 2880, then reabsorbed at day 2900. The final cutoff occurs at day 2930.

Additional insight on the model retroflection was gained through an analysis of the mean energetics of the experiments. It was shown that as inertia increases, more kinetic energy is present in the boundary currents, even though the basin averaged kinetic energy does not change much. This leads to more strongly tilted isopycnal surfaces and to more energy exchange between K_M and P. Energy exchange between K_M and K_E is especially large in those experiments in which rings are formed often. As the Rossby number of the boundary currents increases, more fluid retroflects and fewer rings are formed. Events of three experiments were examined in detail. Certain aspects of ring formation are different among the experiments, but each exhibits a maximum in $C(K_M,K_E)$, $C(P,K_E)$ and a pronounced cyclonic development in the lower layer when a ring cuts off. While both barotropic and baroclinic conversions have peaks, whether one principal form of instability leads to ring formation is not clear. Instability of the mixed type seems more likely.

ACKNOWLEDGEMENTS

We wish to acknowledge helpful discussions with Prof. Claes Rooth and Prof. Rainer Bleck's assistance in clarifying the finite difference form of the model vorticity equation. This work has been supported by the Office of Naval Research Contract No. N00014-85-C0020. Computations were performed using the Cray-1 computers at the National Center for Atmospheric Research, which is sponsored by the National Science Foundation.

REFERENCES

Bleck, R., 1979. Finite-difference equations in generalized vertical coordinates. Part II: Potential vorticity conservation. Contrib. Atmos. Phys., 51, 360-372.

Bleck, R., 1985. On the conversion between mean and eddy components of potential and kinetic energy in isentropic and isopycnic coordinates. Dyn. Atmos. Oceans, 9, 17-37.

Bleck, R., and Boudra, D.B., 1981. Initial testing of a numerical model ocean circulation model using a hybrid (quasi-isopycnic) vertical coordinate. J. Phy. Oceanogr., 11, 755-770.

Boudra, D.B., and de Ruijter, W., 1986. The wind-driven circulation in the South Atlantic-Indian Ocean. II. Experiments using a multi-layer numerical model. Deep-Sea Res., 33, 447-482.

de Ruijter, W., 1982. Asymptotic analysis of the Agulhas and Brazil Current system. J. Phys. Oceanogr., 12, 361-373.

de Ruijter, W., and Boudra, D.B., 1985. The wind-driven circulation in the South Atlantic-Indian Ocean. I. Numerical experiments in a one-layer model. Deep-Sea Res., 32, 557-574.

Gordon, A.L., 1985. Indian-Atlantic transfer of thermocline water at the Agulhas Retroflection. Science, 227, 1030-1033.

Harrison, D.E., and Robinson, A.R., 1978. Energy analysis of open regions of turbulent flows - mean eddy energetics of a numerical ocean circulation experiment. Dyn. Atmos. Oceans, 2, 185-211.

Hulburt, H.E., and Thompson, J.D., 1980. A numerical study of Loop Current intrusions and eddy shedding. J. Phys. Oceanogr., 10, 1611-1651.

Hulburt, H.E., and Thompson, J.D., 1982. The dynamics of the Loop Current and shed eddies in a numerical model of the Gulf of Mexico. Hydrodynamics of semi-enclosed seas. By J.C.J. Nihoul (ed.), 243-298.

Lutjeharms, J.R.E., 1981. Spatial scales and intensities of circulation of the ocean areas adjacent to South Africa. Deep-Sea Res., 28a, 1289-1302.

Olson, D.B., and Evans, R.H., 1986. Rings of the Agulhas. Deep-Sea Res., 33, 27-42.

Pearce, A.F., and Grunlingh, M.L., 1982. Is there a seasonal variation in the Agulhas Current? J. Mar. Res., 40, 177-184.

MODELLING OF MESOSCALE OCEANIC INSTABILITY PROCESSES

Aike Beckmann
Institut für Meereskunde an der Universität Kiel
Düsternbrooker Weg 20
D-2300 Kiel 1 (FRG)

ABSTRACT

Using a quasigeostrophic multi-level model on the β-plane for a limited area of order 1000 * 1000 km the influence of the vertical discretization on the evolution of mesoscale instability structures is studied. A special way of chosing levels with optimal representation of shear-mode growth-rates is presented. Two cases (local perturbation instability and large scale meander instability) are taken as a test, comparing the solutions with observations in the Canary basin and with other numerical model results.

1 INTRODUCTION

An important part of the research-project "Warmwatersphere of the Atlantic" at the Institut für Meereskunde Kiel is the observation of large scale phenomena and mesoscale variability in the eastern North Atlantic. Several cruises to various areas took place in the last few years. As a result from the spring 1982 survey detailed measurements of the hydrography of the Canary basin are available. Since then, repeated cruises revealed the picture of a permanent Azores-current with 20 cm/s maximum jet speed and a cross-jet scale of order 100 km. Fig 1. shows the quasi-synoptic dynamic height field from Käse et al. (1985), illustrating the meandering structure of the front.

Recent studies in connection with the TOPOGULF experiment also show meandering jet behaviour west of the Mid Atlantic Ridge and a more Rossby-wave dominated regime eastward of the ridge (M. Arhan, pers. comm.). There is evidence that the frontal band can be followed back to the branching of the Gulf-stream in the Newfoundland basin (e.g. in Gould, 1985).

A first glance at the involved scales yields U = 20 cm/s, L = 50 km, H = 500 m and $W = 2\cdot10^{-4}$ m/s; these values resulting in the principal parameters:

Rossby-number	:	Ro = 0.04
β-Rossby-number	:	R_β = 4.0
Burger-number	:	Bu = 1.0
ratio divergence/rotation	:	α = 0.02

Thus the flow is clearly linear with respect to momentum advection, only slightly divergent, but strongly nonlinear and baroclinic in the vorticity balance.

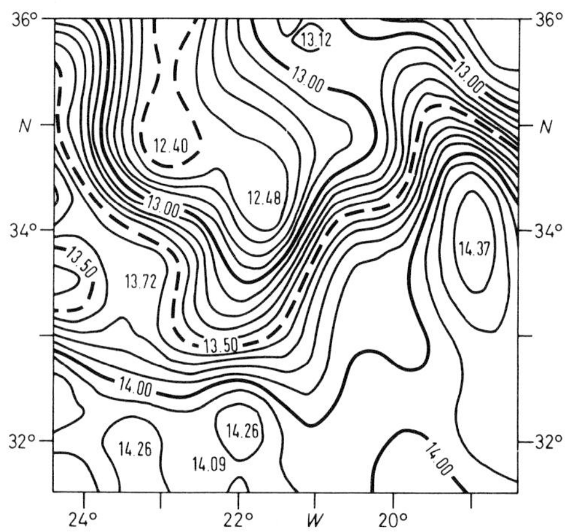

Fig. 1. Objective analysis of the geopotential anomaly field 25/1500 dbar (m^2/s^2) with the approximate centre of the frontal band marked by the 13.5 m^2/s^2 isoline (after Käse et al., 1985).

For the modelling of such a current regime and the dominant dynamical processes (like Rossby wave motion, nonlinear vorticity advection and barotropic and baroclinic instability) a quasigeostrophic model with high resolution in the horizontal as well as the vertical direction was developed.

2 THE QUASIGEOSTROPHIC MULTI-LEVEL MODEL

2.1 Model design

As we have seen we are concerned with spatial scales of 10 to 1000 km and associated time-scales of order days to months. According to this we treat any larger scale phenomenon as a background field, climatologically generated and influenced. This background field can be described by the thermal wind relation alone.

$$f_0 \cdot u_{0_z} = g/\rho_* \cdot \rho_{0_y} \qquad (1)$$

In this part of the subtropical gyre the mean current is directed eastward, resulting from the increase of density to the north.

In the quasigeostrophic approximation the equations to deal with are the nonlinear vorticity and density equation, which can be combined to the well known baroclinic vorticity-equation; however we use a diagnostic relation for the vertical velocity resulting from elimination of the time-derivative in the vorticity and density equation.

Together they form our set of equations:

$$\nabla^2\Psi_t + J(\Psi,\nabla^2\Psi) + \beta\Psi_x - f_0 w_z = D(\Psi) \quad (2)$$

$$\nabla^2 w - (f_0^2/N_0^2)w_{zz} = -(f_0/N_0^2)\cdot\left[\beta\Psi_{xz} + \nabla^2 J(\Psi,\Psi_z) - J(\Psi,\nabla^2\Psi)_z + D_z(\Psi)\right] \quad (3)$$

A suitable parametrization of sub-grid-scale processes $D(\Psi)$ is included using the horizontal Austausch-concept (biharmonic friction) for the relative vorticity $A\nabla^4\xi$. The choice of the friction parameter is due to numerical reasons only, i.e. to avoid unacceptable growth of the smallest scales a value of $2\cdot10^9\,m^4/s$ is necessary. This corresponds to a biharmonic friction for the streamfunction of $A_h = 50\,m^2/s$ for the 40 km-wave. For the present study vertical diffusion is neglected. This is possible even in a numerical model if the horizontal projection of any vertical process can be resolved by the horizontal grid.

In the horizontal directions the pseudospectral approach is introduced to minimize the phase errors and to asure a good representation of the nonlinear advection terms (Merilees & Orszag, 1979 for a detailed description). The associated boundary-conditions are periodicity in both x and y direction; 32 * 32 wavenumbers are used.

The product terms are evaluated energy-conserving on the physical grid using FFT for the transformations. Enstrophy-conservation was dropped because it appears to have only a relatively small influence on the horizontal fields within 50 days of simulation. In the frame of this numerical concept mass-conservation is automatically fulfilled.

A totally periodic model is not quite what is desired, since there has to be a net transport through the region under consideration. To incorporate this a separation of the mean zonal velocity from the Fourier-expansion was performed according to.

$$\Psi(x,y,z,t) = -u_0(z,t)\cdot y + \Psi'(x,y,z,t) \quad (4)$$

where u_0 is the meridionally averaged zonal velocity.

The vertical discretization is done using a level-concept with computation of and w at staggered level depths (following for example Bengtsson & Temperton, 1979). The boundary-conditions are vanishing vertical velocities together with $\Psi_z = 0$ at the top and at the bottom (thus omitting any effects of bottom-topography and wind-induced forcing).

Another more detailed look at the model applies to the energy budget: the separation of the mean zonal transport from the purely periodic part has to be performed at each Ψ-level. This results in a mean zonal transport profile to be treated separately. Writing down the energy-balance-relations for this modified quasigeostrophic system (friction neglected) we find that there is a net gain of energy from the climatological background flow field.

$$dE/dt = \int \frac{1}{2} (f_0^2/N_0^2)\, u_{0_z} \iint \Psi'_x \Psi'_z \, dx\, dy\, dz \tag{5}$$

Unfortunately there is no additional prognostic equation for the time dependence of the mean flow velocity in the frame of the quasigeostrophic concept; so this system has still to be called a linearized one. In order to take into account even the influence of the periodic part on the horizontally integrated part we change the vertical shear of the mean flow according to the loss of its available potential energy, keeping the vertically integrated transport through the basin constant. This results in a set of equations for each w-level:

$$\frac{d}{dt} (u_{0_z})^2 = - \frac{3}{XL \cdot YL^2} u_{0_z} \iint \Psi'_x \Psi'_z \, dx\, dy \tag{6}$$

It is easily seen from eqn. (6) that if the initial shear is zero it remains unchanged. This additional equation leads to an energetically consistent model of local mesoscale dynamics including arbitrary mean zonal flow profiles.

2.2 Vertical Calibration

Since the quasigeostrophic approximation refers to the mean stratification of the region under consideration the measured mean density-profile from the Canary-basin and the corresponding Brunt-Väisälä-frequencies are the basis for this numerical model (fig. 2a). For these given profiles the vertical structure of the first and second baroclinic mode are shown in fig. 2b. The first mode zero-crossing happens at roughly 1000 m depth. The first and second baroclinic Rossby-radius of deformation are found to be 26.3 and 8.9 km respectively.

The initial flow-profile is chosen to be dominated by the first and second baroclinic mode. For the maximum velocity of 20 cm/s at the surface the total transport through the basin reaches 8 Sverdrups.

For a mean depth of approximately 4400 m and a typical value of $2 \cdot 10^{-3}\, s^{-1}$ for N_0 we find a reasonable vertical resolution of order 500 m for horizontal grid-spacing of $\Delta x = \Delta y = 10$ km (in order to fix the Burger-number at unity) so there will be 9 level in the vertical. A higher vertical resolution requires vertical diffusion for numerical stability reasons.

The first approach in chosing the vertical grid would be to fix the levels equidistant, but for a better representation of the vertical structure found in the ocean (e.g. fig. 2a) it seems more convenient to concentrate the layers in the upper part of the water-column; for instance a grid spacing of 100, 5 * 200, 300, 500 and 2500 m depth with Ψ-level depths at 50, 200, 400, 600, 800, 1000, 1250, 1650 and 3150 m respectively.

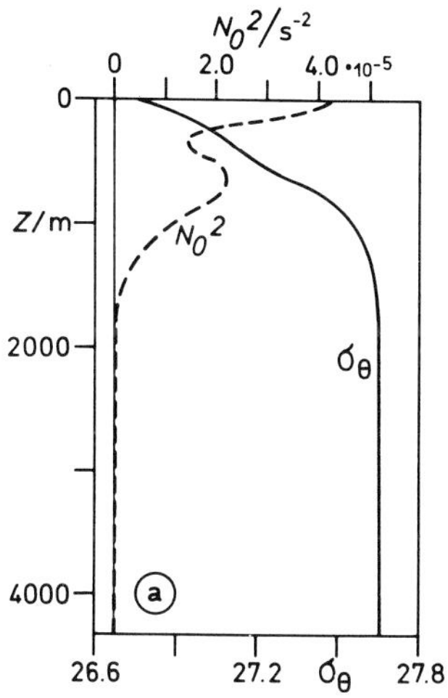

Fig 2. a) Measured mean density-profile (σ_θ) and corresponding N_0^2 for the Canary Basin in spring 1982,

b) computed vertical structure of first and second baroclinic mode,

c) 8th baroclinic eigenfunction used for the vertical discretization,

d) configuration of staggered Ψ- and w-level.

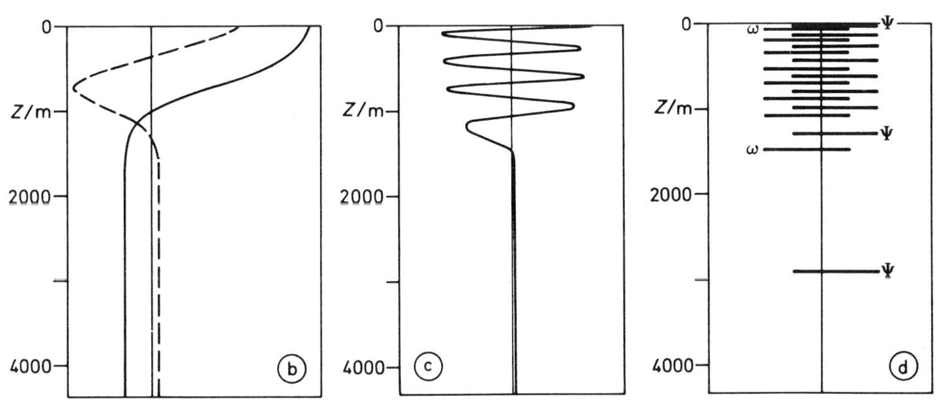

Instead, it is not a priori clear how such a vertical discretization represents the vertical structure of dynamical processes in a numerical model.

So we make use of the measured density-profile in form of the higher order vertical eigenfunction (fig 2c). For a 9-degrees-of-freedom model the 8th baroclinic vertical mode yields the desired 8 zero-crossings where the layer-boundaries are fixed. The streamfunction-levels are located in the center of the layers (fig. 2d).

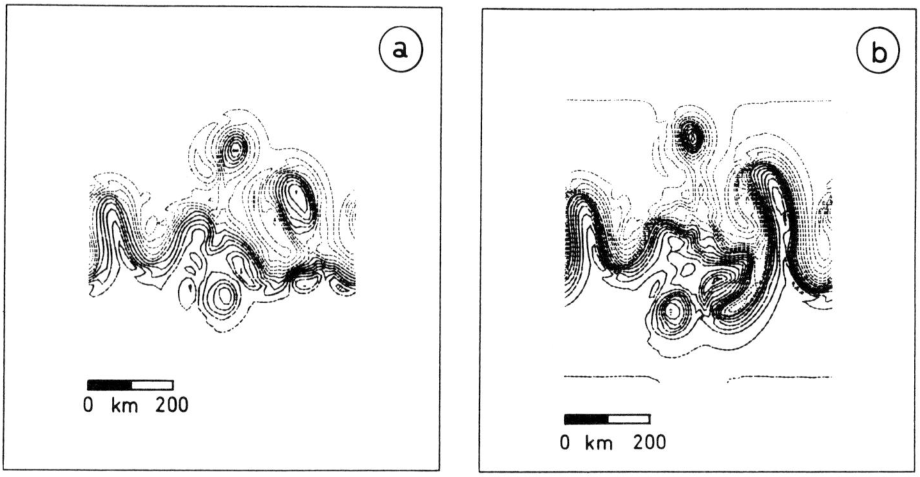

Fig. 3. An example for the different results of models with a) conventional vertical discretization and b) choice of levels according to 8th eigenfunction of the density profile. Shown is the density in the main thermocline (ca. 700 m depth) after 50 days of integration.

A first comparison of these two distinct level-choices is shown in fig. 3 for the density in the main thermocline (∿700 m) after 50 days of integration. Several features can be recognized (the detached eddies and enhanced fronts) in both versions. The dominant processes seem to be unchanged but the details are rather different.

3 LINEAR INSTABILITY PROCESSES

3.1 Unstable shear-modes

To test the sensitivity of the solutions (e.g. instabilities) on different ways of vertical discretization in a more quantitatively way we first look at the linearized system to determine the shear-mode growth-rates of the system; solving an eigenvalue problem for the baroclinic vorticity-equation neglecting every meridional variation and using the following approach:

$$\Psi'(x,y,z,t) = F(z)\cdot\exp(i(kx - \omega t)) \quad ; \qquad \omega = \omega_r + i\omega_i \qquad (7)$$

The computation of the continuous (that means 20 m vertical resolution) shear modes for the maximum velocities yields a growth-rate maximum at wavenumber 8 (80 km wavelength) with 4.5 days exponential time scale. The conventionally truncated system shows a significant different curve for waves shorter than that maximum, while the eigenfunction level choice compensates the effect of reduction of vertical degrees of freedom to a large amount (fig. 4a).

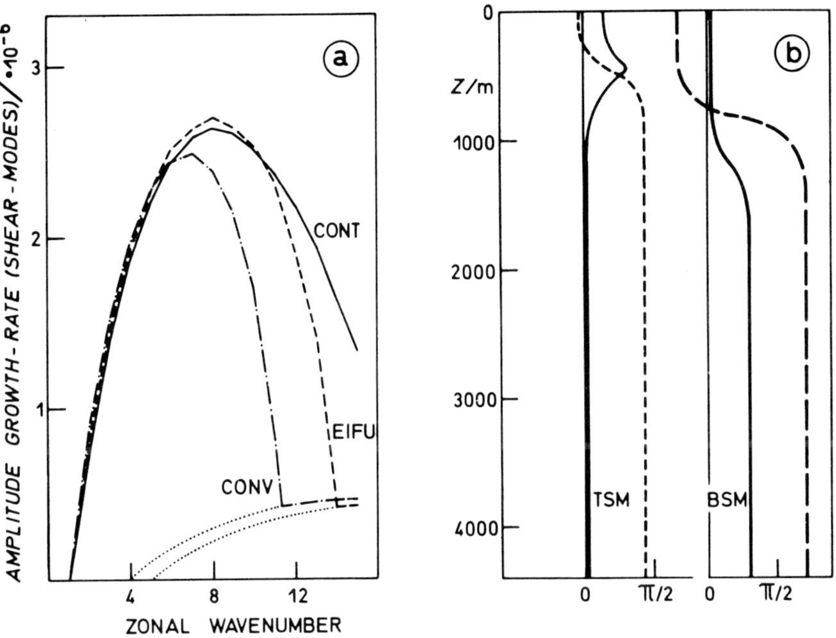

Fig. 4. a) Shear-mode growth-rates for continuous (CONT), conventional (CONV) and eigenfunction (EIFU) vertical discretization; b) amplitudes of the most unstable thermocline shear-mode (TSM) and bottom-intensified shear-mode (BSM) of the continuous model. The dashed lines show the phase.

In the 9-level model of the Canary basin there are two different unstable modes to distinguish: the first is the thermocline shear-mode (TSM), the second the bottom-intensified shear-mode (BSM); each belonging to another depth where the necessary condition for baroclinic instability ($\beta_{eff} = \beta + ((f_0^2/N_0^2)u_{0_z})_z$ changes sign) is fulfilled. Of course, critical levels cannot be sufficiently resolved by any of these discrete models.

The dominating growth rates correspond to the thermocline shear-mode (TSM). For wavelenghts shorter than 50 km there is no unstable TSM possbile, the dotted curves represent the bottom-trapped or bottom-intensified shear-mode (BSM).

Flierl (1978) pointed out that a modal representation of the vertical in a quasigeo strophic model yields better results compared with level/layer approach in case of low vertical resolution; this can also be concluded for these simulations: refering to the modal structure of the vertical problem the discrete model seems to be closer to the continuous case.

3.2 Growth-rates of unstable waves

As an extension we consider the linearized system with included y-dependency consisting of a zonal baroclinic jet and superimposed small scale perturbations in its 9-level-version.

$$\Psi'(x,y,z,t) = F(z)\cdot M(y)\cdot \exp(i(kx - \omega t)) \quad ; \qquad \omega = \omega_r + i\omega_i \tag{8}$$

$$u_0(y,z) = U(z)\cdot \exp(-\mu^2 y^2) \quad ; \qquad \mu = 1/50\ \text{km} \tag{9}$$

The analysis follows Holland & Haidvogel (1980) with three differences: nine levels are used instead of two, no friction is included and periodicity in the meridional direction is introduced to get close to the nonlinear model.

Examples of the vertical structure of the eigenfunctions are given in fig. 5. The most unstable mode is shown in fig. 5a with large amplitudes in those depths, where the necessary condition for baroclinic instability is fulfilled (predominantly in the thermocline); the most unstable bottom-intensified wave is depicted in fig. 5b. There are meridionally higher order unstable solutions but these have reduced growth-rates.

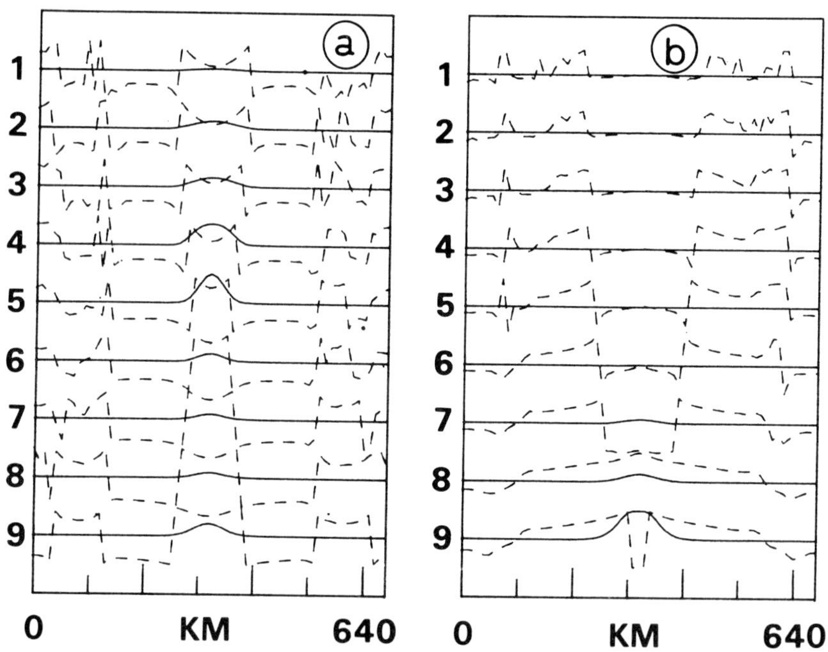

Fig. 5. a) Meridional and vertical structure of the most unstable eigenfunction from the linearized 9-level version,
b) another unstable mode with bottom-intensified amplitude. Solid lines show the amplitude, broken lines denote the phase.

Results on the sensitivity of instability processes in quasigeostrophic level models may be summarized as follows:

- a moderate change of the assumed deep sea gradient (beyond 2000 m depth) has strong effects on the growth of the bottom-trapped mode. Smaller gradients enable the system to be unstable even at smaller wavelengths leaving the growth-rates of the unstable thermocline-wave unaltered.
- the eigenfunction-discretization has strong effects in certain spectral bands: the growth-rates are larger and the maximum is shifted towards smaller wavelengths. Again it seems that first-choice-levelling cuts off certain waves just within the range of interest for simulations of the mesoscale.

Even in the nonlinear model we can prescribe a sinusoidal perturbation of one partial wave in zonal direction on a purely zonal current of an $\exp(-\mu^2 y^2)$ shape and look for the growth of this spectral component within the first 10 days (without friciton) to determine the "growth-rates".

Obviously there will be an energy-transfer through the wavenumber spectrum to smaller scales due to the energy cascade and also the reservoir of the available potential energy at the particular wavenumber is limited, so these growth-rates are expected to be smaller than the linearized ones. But as fig. 6 shows the shape of the curve is in good agreement and the maximum lies at wavenumber 8 (80 km) in both cases.

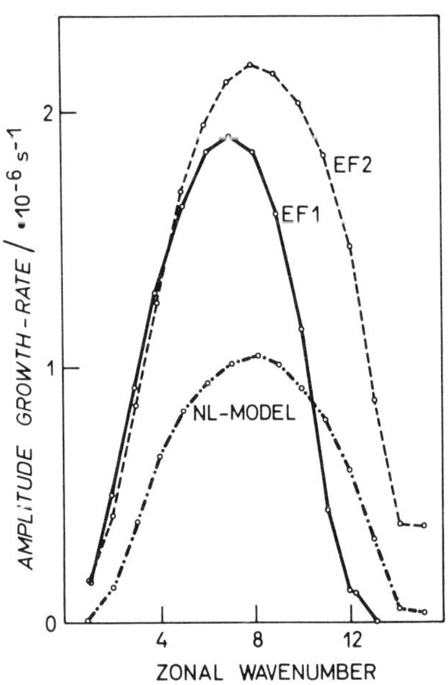

Fig. 6. Amplitude growth-rates of different configurations of linearized systems compared to the nonlinear model as a function of zonal wavenumber. EF1 denotes the conventional levelling, EF2 our eigenfunction levelling and NL-MODEL the nonlinear model with EF2 discretization. Wavenumber 1 means 640 km wavelength, wavenumber 16 means 40 km wavelength.

4 NONLINEAR QUASIGEOSTROPHIC INSTABILITY OF A JET

4.1 Local perturbation instability

As an example of the instability resulting from a local perturbation of 100 km meridional extent fig 8. shows the evolution of the external mode streamfunction at 10 days interval. Note the growth of the meridional amplitude of the perturbation, the downstream meandering and the mushroom-type instability at day 30 (fig. 8d), forcing the eddy-pairs to detach by means of enhanced advection.

Within the 50 days-simulation the maximum local transport has raised from 8 to 36 Sverdrups; the maximum velocity at the surface runs up to about 45 cm/s. As expected the dominant meander scale is roughly of order 2π times the Rossby-radius (i.e. 150 km). The density-flux directed southward due to baroclinic instability (with a northward heatflux) has a maximum about day 30; exchanging water-mass properties across the frontal jet. The momentum transport has a large southward component.

Effects of periodicity , i.e. the penetration of downstream meanders into the model region from the west does not change the structures until day 50 significantly. An experiment using an extended region was performed and no fundamental differences were found.

A short look at the change of the mean transport profile through the first 50 days (fig. 7) shows a net decrease of vertical shear as expected, especially in the upper layers, and moreover a mean-current generation in the deep layer.

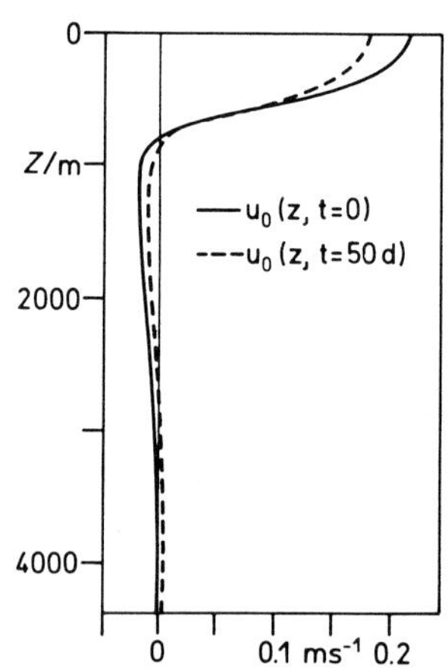

Fig. 7. Change of the mean zonal transport profile by reduction of the mean potential energy within 50 days integration.

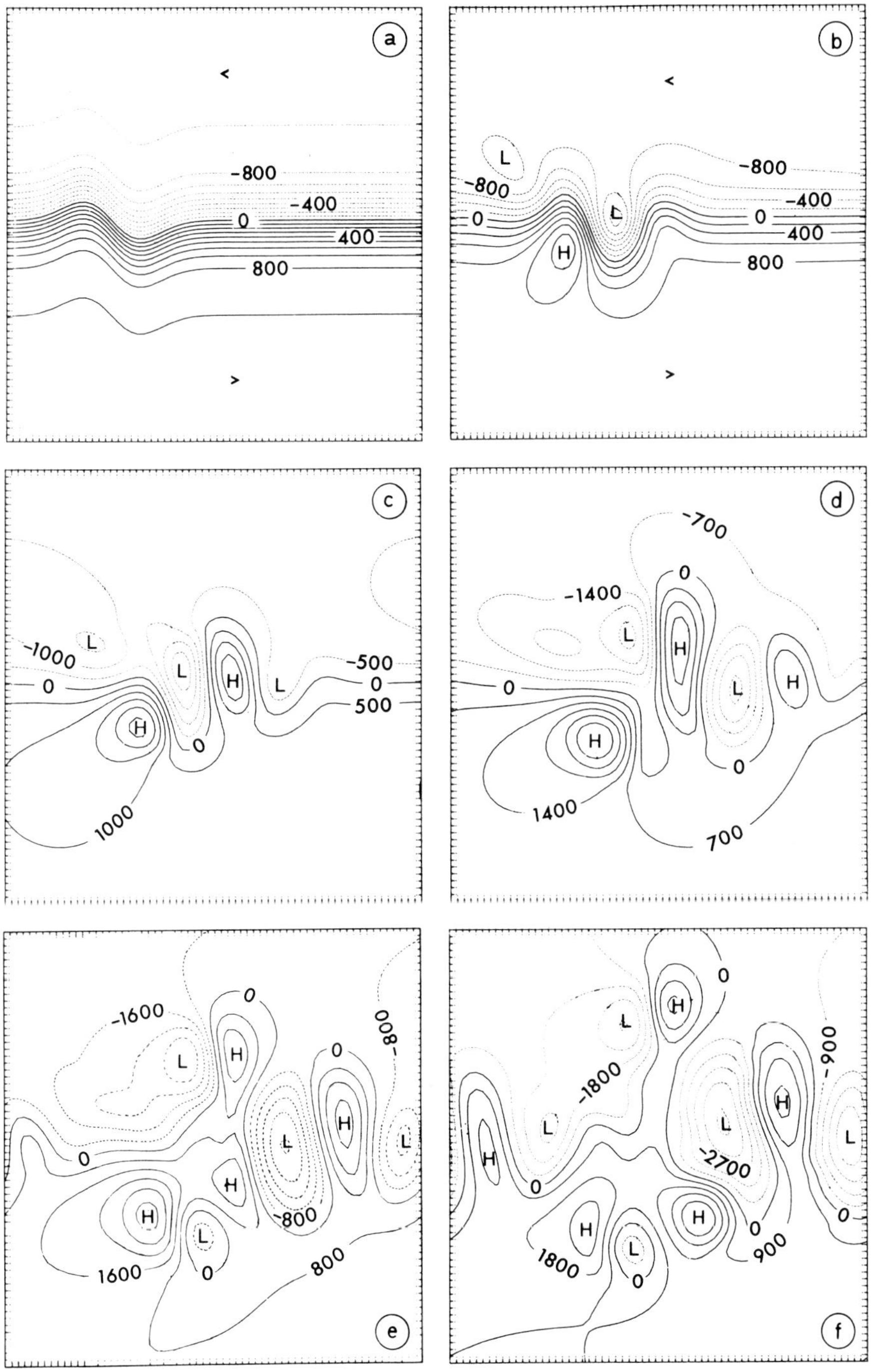

Fig. 8. Evolution of external mode streamfunction at 10 days interval in a 640 * 640 km box (units m^2/s).

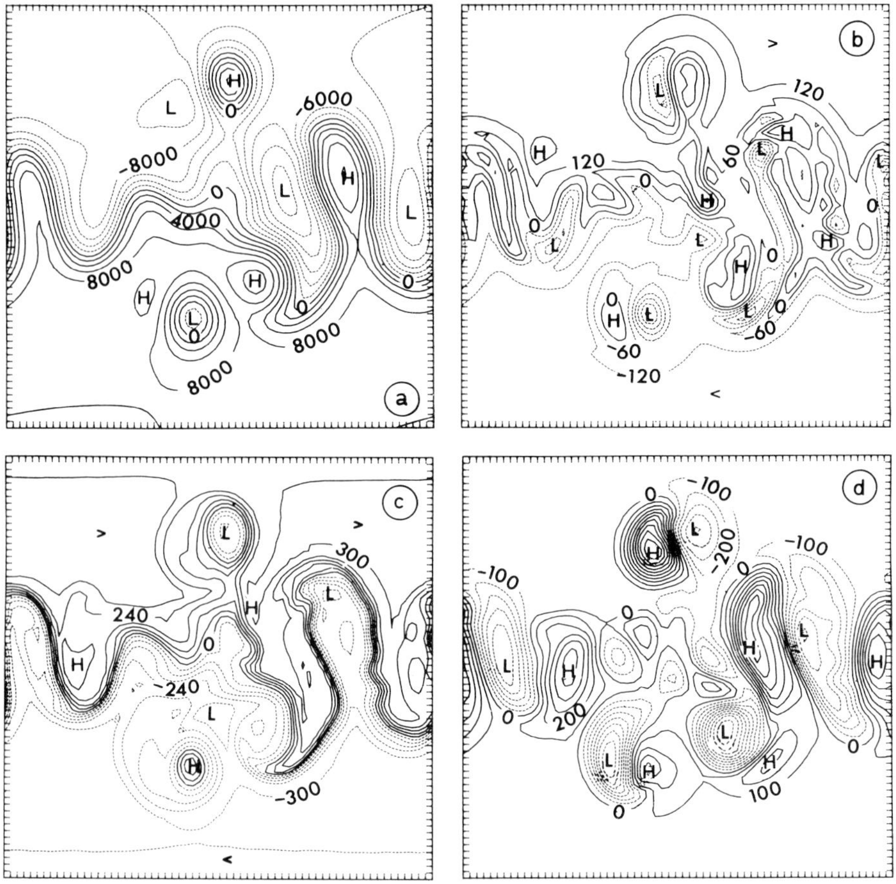

Fig. 9. a) surface streamfunction, b) surface density, c) thermocline density and d) deep sea density after 50 days of integration.

The flow field at the surface exhibits an axisymmetric ring formation (fig. 9a) but the density-field is diffuse (fig. 9b); in mid-thermocline depths the character of the front is much more obvious and a first mode vertical structure of the eddies is indicated from a comparison of fig. 9b and 9c.

The double eddy-structure evolved in the barotropic mode is mainly due to the deep sea rectification due to the bottom-intensified unstable mode (fig. 9d).

Results of Ikeda & Apel (1981) using a two-layer quasigeostrophic model of the Gulf-stream (dealing with the same lengthscales but 1 m/s maximum velocity) shows an analogous eddy-detachment-process within 36 days.
Moreover primitive-equation-model results from Kielmann & Käse (1986) applied to the same region and observed quantities do not propose additional processes to be important.

4.2 Large-scale meander instability

Käse et al. (1985) decomposed the dynamic height field of fig. 1 into a Rossby-wave field and linear trend and a residual perturbation field (fig. 10a,b).

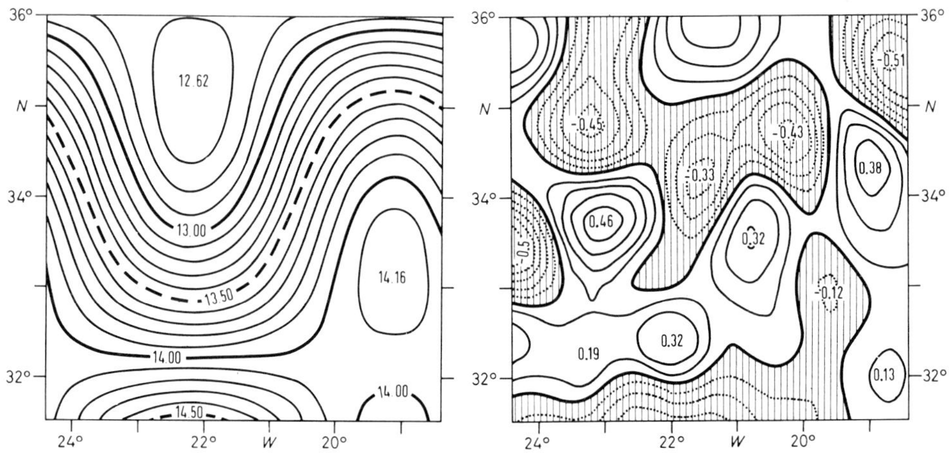

Fig. 10. a) Superposition of the linear spatial trend and the Rossby-wave fit for the geopotential topography 25/1500 dbar of fig. 1. b) Objective analysis of the residual mesoscale perturbation field 25/1500 dbar after removal of the composite mean field (after Käse et al., 1985).

In order to simulate such a flow-pattern, dominated by strong nonzonal flows we initialized the model with a large scale meander of the same vertical structure as in section 4.1 (fig. 11a).

This idealized meander of 640 km zonal wavelength and 200 km meridional amplitude splits within the 50 days integration to waves of zonal wavenumber 3 (213 km wavelength), but even after the initial structure is recognizable from the arrangement of the eddies and indicates an average propagation of 2 cm/s to the east, i.e. it is nearly stationary (fig. 11). The local transports and surface velocities have raised by a factor of three. Note the radiating Rossby-waves outside the jet-regime.

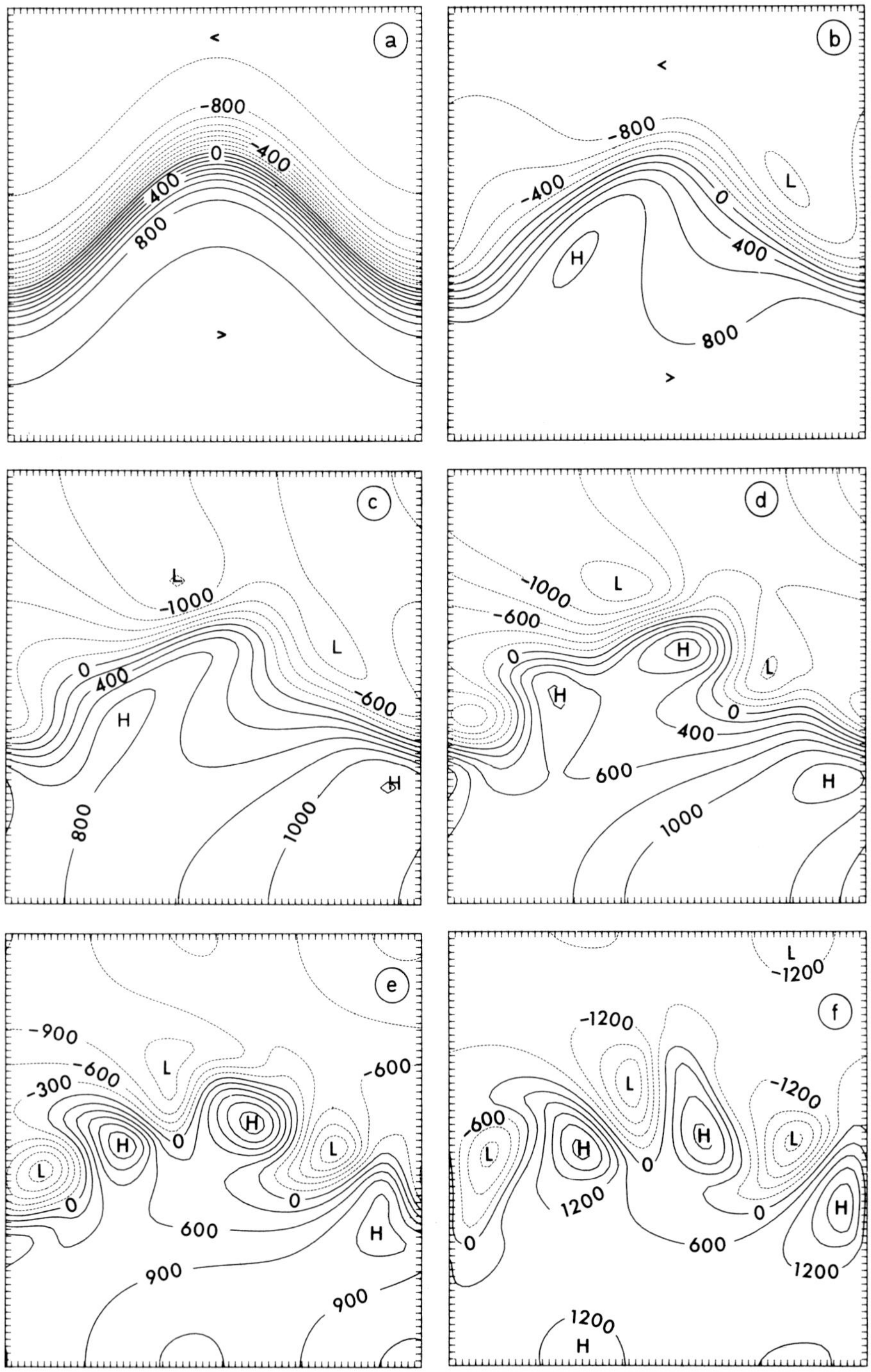

Fig. 11. Evolution of external mode streamfunction at 10 days interval in a 640 * 640 km box (units m^2/s).

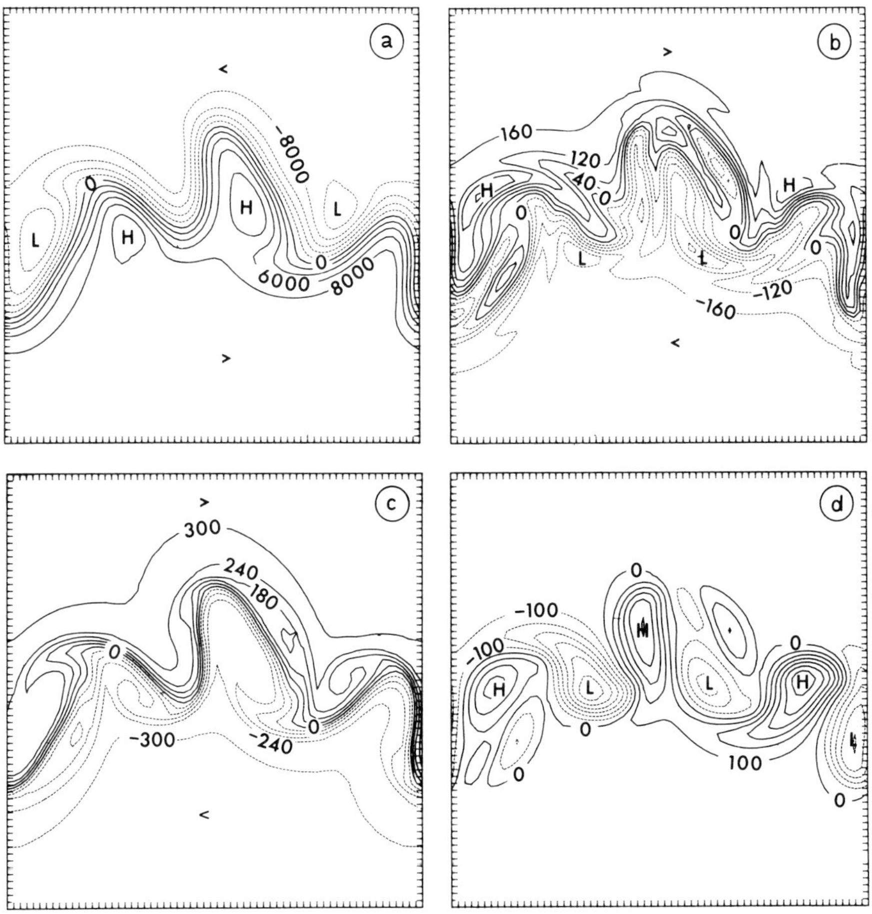

Fig. 12. a) surface streamfunction, b) surface density, c) thermocline density and d) deep sea density after 50 days of integration.

The flow at the surface shows totally meridionally arms of a continuous flow band; strong frontal zones can be seen in the mid-depth density (fig. 12 c). The eddy field is driving the deep sea currents (fig. 12d).

Comparisons between day 30 (fig. 11d) of the simulations and the observed field (fig. 10b) show structural similarities. From this we may conclude that the hydrographic survey took a snapshot of an emerging instability of such a large-scale meander. The reason for the occurrence of the observed large-scale feature (possibly an intrusion of dense water from the north or a result of previous instability processes) cannot be explained with our model.

5 CONCLUSIONS

Quasigeostrophic model-results concerning mesoscale instability processes depend critically on the used vertical discretization for a given horizontal grid-spacing. Best results are obtained from an eigenfunction-based choice of the level-depths as may be seen from shear-mode growth-rates (section 3).

A suitably calibrated quasigeostrophic multi-level model (using vertical modes for the choice of the levels) of high horizontal resolution is able to simulate frontal dynamics (neglecting all thermodynamic processes) very accurate and efficient. Test-computations with the 9-level Canary basin model show reasonable time-scales for the growth and propagation of frontal structures. A corresponding two-level model is principally overestimating the vertical shear and amplifies the baroclinic instablity process.

Another important result is that the detachment of eddies is not restricted to Gulf-stream-parameters with large horizontal velocities. This eddy-pair-separation process requires about 50 days (section 4.1). Yet no ring-formation like in the boundary current regions could be observed in the model. This is possibly due to the different vertical structure or the larger surface kinetic energy of the Gulf-stream.

The quasi-synoptic measurements in the Canary-basin carried through by Käse et al. (1985) may be explained as an early state of an evolving instability (section 4.2).

6 REFERENCES

Bengtsson, L. Temperton, C., 1979. Difference Approximations to Quasi-geostrophic Models. In: Numerical Methods Used in Atmospheric Models. GARP Publication Series, No 17, Vol 2, 340-380.

Flierl, G.R., 1978. Models of Vertical Structure and the Calibration of Two-Layer Models. Dyn. Atmos. Oceans, 2, 341-381.

Gould W.J., 1985. Physical Oceanography of the Azores Front. Progr. Oceanogr., 14, 167-190.

Holland W.R., Haidvogel, D.B., 1980. A Parameter-Study of the Mixed Instability of Idealized Ocean Currents. Dyn. Atmos. Oceans, 4, 185-215.

Ikeda M., Apel, J.R., 1981. Mesoscale Eddies Detached from Spatially Growing Meanders in an Eastward Flowing Oceanic Jet Using a Two-Layer Quasigeostrophic Model. Journ. Phys. Ocean., 11, 1638-1661.

Käse, R.H., Zenk, W., Sanford, T.B., Hiller, W., 1985. Currents, Fronts and Eddy-Fluxes in the Canary Basin. Progr. Oceanogr., 14, 231-257.

Kielmann, J., Käse, R.H., 1986. Numerical Modelling of meander and eddy formation in the Azores-Current frontal zone. Submitted to Journ. Phys. Ocean.

Merilees, P.E., Orszag, S.A., 1979. The Pseudospectral Method. In: Numerical Methods Used in Atmospheric Models. GARP Publication Series, No 17, Vol 2, 278-301.

AN EDDY-RESOLVING MODEL FOR RIVER PLUME FRONTS

J.W. DIPPNER
Deutsches Hydrographisches Institut, Bernhard-Nocht-Straße 78,
D-2000 Hamburg 4, FRG

ABSTRACT

A baroclinic eddy-resolving model of a river plume front is developed. The numerical framework is carried out with a semi-implicit two step difference scheme and the transport equation of density is simulated with an active Lagrangian tracer technique. Preliminary results of frontal and eddy formation are presented.

1 INTRODUCTION

In the German Bight exist three different types of fronts which have been classified by Krause et al.(1986). The first is a seasonal thermal front between the stratified and well mixed area in the German Bight; the second is an upwelling front in the region of the Old Elbe Valley occurring during easterly winds; and the third is a river plume front due to the river runoff of the rivers Elbe and Weser, which is the subject of this paper.

The dynamics and the developement of river plume fronts depend strongly upon the characteristics of the estuary. The main features are the amount of river runoff, the velocity of tidal currents, and geometrical constraints such as width and depth of the estuary which are responsible in forming a highly stratified, moderately stratified, or vertically homogeneous estuary (Pritchard 1955). Recently, a lot of plume models have been developed (Kao et al. 1977, Bork 1978, Stronach 1981), but these models are two-dimensional numerical models. Only few models of shallow water fronts are three-dimensional (Harashima and Oonishi 1981, James 1984).

Recent measurements, by the use of modern profiling equipment, remote sensing techniques, and experiments with dye patches of rhodamine B at the plume front, have revealed a rather complicated structure of eddies and meanders (Becker et al. 1983, Franz and Klein 1986). In general, fronts are characterized by two different length scales. In the crossfront direction, the span is of the order of a few baroclinic radii of deformation which is of the order of 5 to 20 km in the German Bight. In the alongfront direction, the span is determined by the wavelength of meander, which is determined by bathymetric scales, atmospheric forcing scales, wavelength of coastally trapped waves, and baroclinic instability scale (Bowman 1978). To get a better understanding of the complicated structure of the river plume front and the involved processes in the German Bight, a three-dimensional eddy-resolving model is developed. This process model and the model philosophy will be outlined in the following, and preliminary results are shown.

2 EQUATIONS AND BOUNDARY CONDITIONS

The model is based upon the well known primitive equations using hydrostatic and Boussinesq approximation:

$$u_t + L(u) - fv = - 1/\rho_0 \, p_x + (\kappa_M \, u_z)_z + A_M \, \nabla^2 u \qquad (1)$$

$$v_t + L(v) + fu = - 1/\rho_0 \, p_y + (\kappa_M \, v_z)_z + A_M \, \nabla^2 v \qquad (2)$$

$$p_z = - g \, \rho \qquad (3)$$

$$L(1) = 0 \qquad (4)$$

$$\rho_t + L(\rho) = (\kappa_D \, \rho_z)_z + A_D \, \nabla^2 \rho \qquad (5)$$

Here u, v, p and ρ represent the horizontal velocities, pressure, and and density, f the Coriolis parameter, ρ_0 a reference density and g the the gravity acceleration. The operator L is defined by:

$$L(a) = (ua)_x + (va)_y + (wa)_z \qquad (6)$$

The governing equations are solved with a semi-implicit two-step method in a schematic funnel-shaped estuary model with a linear bottom topography, shown in Fig. 1.

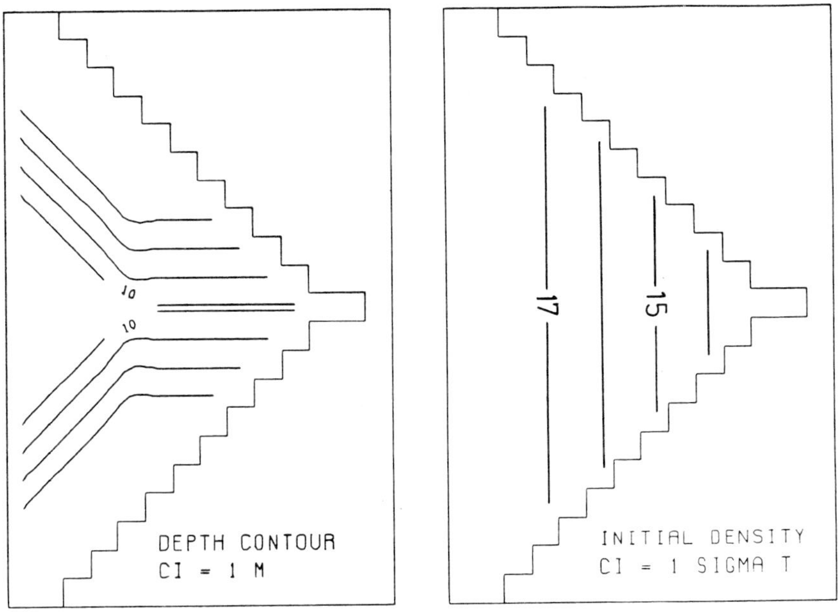

Fig. 1 Model area and topography (left), contour interval (CI) 1 m, and initial density (right) CI = $1\sigma_t$

The boundary conditions are as follows:

1) No flow through the lower boundary and no flow through the lateral coastal boundaries.
2) Free surface kinematic boundary condition.
3) A tidal signal of an M_2 tide at the open boundary.
4) A quadratic bottom stress.
5) Half slip condition at the lateral boundaries.
6) A river runoff of 1000 m^3/s, which is in the order of the River Elbe.

3 DESCRIPTION OF THE NUMERICAL MODEL

Equations 1 to 4 are vertically integrated over the layer thickness of the model and discretized forwards in time and central in space in a staggered C-grid. The vertical derivative is handled with an implicit difference scheme. The barotropic pressure gradient in the equations of motion and the horizontal divergence of the mass transport in the continuity equation are discretized with the Crank-Nicolson method. Substituting the equations of motion in the continuity equation, integrating the continuity equation over the entire water column, and using the free surface kinematic boundary condition, results in an elliptic equation for the water elevation, which is solved economically with successive overrelaxation technique. A detailed description of this semi-implicit two-step method is given by Duwe et al. (1983) or by Backhaus (1985). According to Arakawa and Lamb (1977), the non-linear terms in the equations of motion are discretized with an energy and enstrophy conserving scheme. The horizontal grid resolution is two kilometres and the vertical resolution is five metres. The model works with a time step of 745 seconds. The model parameters are a constant horizontal eddy viscosity of 2 m^2/s, a constant vertical eddy viscosity of 30 cm^2/s, and a coefficient of bottom friction of 0.0015.

The transport of density (Equation 5) is computed with an active Lagrangian tracer technique, which is explained in the following. From the numerical point of view, the main disadvantage of the Eulerian system is the advective part of the transport equation yielding the well known numerical problems, especially in the presence of strong gradients. By transformation onto Lagrangian co-ordinates it simply vanishes. For the diffusion part of the equation which is a priori assumed to be small in relevance and unknown in detail, a simple procedure which reproduces quantities of statistical significance only, may be regarded as sufficient.

A water body is interpreted not as a continuum but as a finite set of water particles with definite physical properties. For that reasons, the property of an Eulerian grid cell is the ensemble averaged property of all particles in the cell. Each particle can be identified by its actual co-ordinate $\mathbf{x}=\mathbf{x}(\mathbf{a},t)$, which depends upon the initial co-ordinate $\mathbf{a}$ and the time, where $\mathbf{x}(\mathbf{a},0)=\mathbf{a}$. The Lagrangian velocity is $d/dt\ \mathbf{x}(\mathbf{a},t)=\mathbf{u}_L(\mathbf{a},t)$ and the relation to the Eulerian velocity is $\mathbf{u}_L(\mathbf{a},t)=\mathbf{u}_E(\mathbf{x}[\mathbf{a},t],t)$. Hence, the transport and diffusion of the particles are calculated by:

$$\mathbf{x}_i^{t+\Delta t} = \mathbf{x}_i^{t} + \Delta t\ [\ \mathbf{u}_L + (\mu - 0.5)\ \mathbf{P}] \tag{7}$$

where i is the i-th particle, $\mathbf{u}_L$ the advection velocity obtained by transformation onto Lagrangian co-ordinates, $\mathbf{P}$ is the "band-width" of turbulent fluctuation, and μ is a random number uniformly distributed between 0 and 1. A reflecting boundary condition for the random walk is used. This technique has often been successfully applied to different particle properties, such as radioactivity (Maier-Reimer 1975), heat (Bork and Maier-Reimer 1978), zooplankton (Dippner 1980), or crude oil (Dippner 1983,1984).

In these applications, the particles are dynamically passive and represent a certain amount of substance. For dynamically active properties, this kind of method will not be sufficient. For example, if a particle represents a certain amount of salt, an unrealistically high number of particles would be needed to ensure a sufficient degree of accuracy. In this case, a redefinition of the particle property is necessary. To overcome this problem, the particle should be defined as a water blob with a given salinity. In the present case, each particle has a density, and the density of a grid cell of the model, which drives the circulation model, is the ensemble average of the densities of the particles in the cell.

In equation 7, only two processes are considered: the advection by the velocity field and turbulent transport by a stochastical change of the co-ordinates. Additionally, two processes should be considered; namely, the buoyancy and irreversible mixing. Since the density of a cell is the ensemble average of the density of the particles in the cell, the individual particles are the subject of buoyant forces. Taking friction into account, in analogy to Stokes' Law, a constant velocity is assumed, balancing buoyancy and friction. Therefore, the vertical particle co-ordinate yields:

$$z_i^{t+\Delta t} = z_i^{t} + \Delta t \, \gamma (\rho^t - \rho_i^t) / \rho^t \tag{8}$$

where ρ is the density of the cell, ρ_i is the density of the particle, and γ is a threshold value of a sinking or rising velocity.
Irreversible mixing in a cell is simulated by changing the individuality of the particle. This yields an adjustment of the property of the particle to the property of the cell:

$$\rho_i^{t+\Delta t} = \rho_i^{t} + \Delta t \, \kappa \, (2\pi/h)^2 \, (\rho^t - \rho_i^t) \tag{9}$$

where h is the grid size and κ is the coefficient of eddy diffusivity. A detailed description of this technique is given by Maier-Reimer and Sündermann (1981).

How many particles are necessary? There are two criteria: An empty cell should be avoided. In this case the density of the cell is averaged from the density of the surrounding cells. The second criterion is the desired accuracy. The model starts with ten particles in each grid box, which guaranties both criteria,and at the river mouth there is a continuous inflow of particles with a density of $8\sigma_t$. The initial condition of the density field is shown in Fig. 1. If a particle is leaving the model through the open boundary, it is extrapolated with the velocity on the open boundary. Thus, the particles can move in and out through the open boundary.

The parameters of the tracer model are a horizontal diffusion velocity of 5 cm/s, which is proportional to an eddy viscosity of 0.56 m²/s, no vertical diffusion velocity, a Stokes' velocity of 1 mm/s, and an eddy diffusivity of 6 m²/s.

4 RESULTS AND DISCUSSION

The model actually runs on a Cyber 180-840 and needs 100 seconds CPU-time per tidal period. The model reaches a steady state after three tidal periods, and after ten tidal periods a front is formed. The spin-up process of the kinetic energy shows a statistically steady state (Fig. 2).

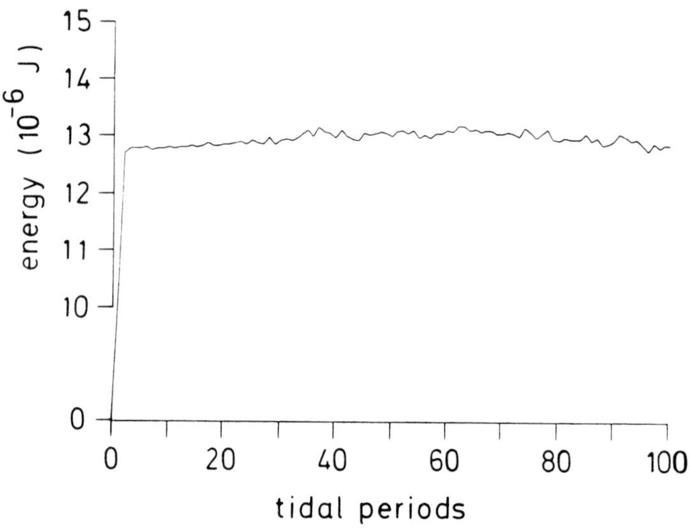

Fig. 2 Spin-up of the kinetic energy over 100 tidal cycles

Two different experiments are presented. The first version has a flat bottom with a depth of ten metres; in this case, the density adjustment (Equation 9) is neglected. The results show a formation of a front with very strong surface density gradients (Fig. 3), and the streamfunction shows - beside the typical circulation pattern - an enhancement of the velocities in the frontal zone and a typical meandering (Fig. 3). The second run is a calculation with a linear funnel-shaped topography (Fig. 1), and the density adjustment is simulated with an eddy diffusivity of 6 m²/s. That means, if a particle has a density difference of 1 σ_t to the density of the cell, the particle is adjusted after 23.8 hours. In this case a front is also formed with much weaker gradients of the surface density and parallel to the depth contour (Fig. 4). In this calculation there ist no relation between the streamfunction patterns (Fig. 4) and the distribution of the surface density.

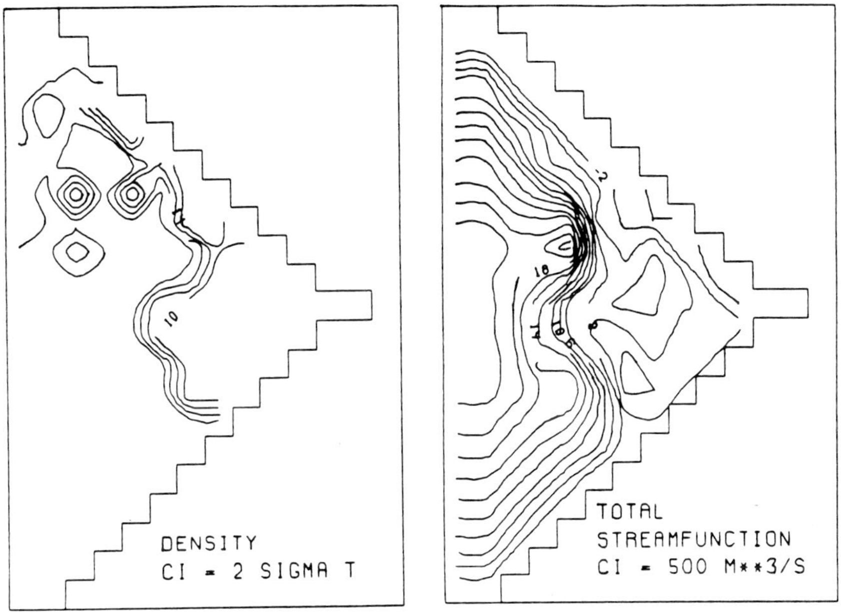

Fig. 3 Surface density (left) CI = $2\sigma_t$, and total streamfunction (right) CI = 500 m^3/s after 19 tidal periods (flat bottom)

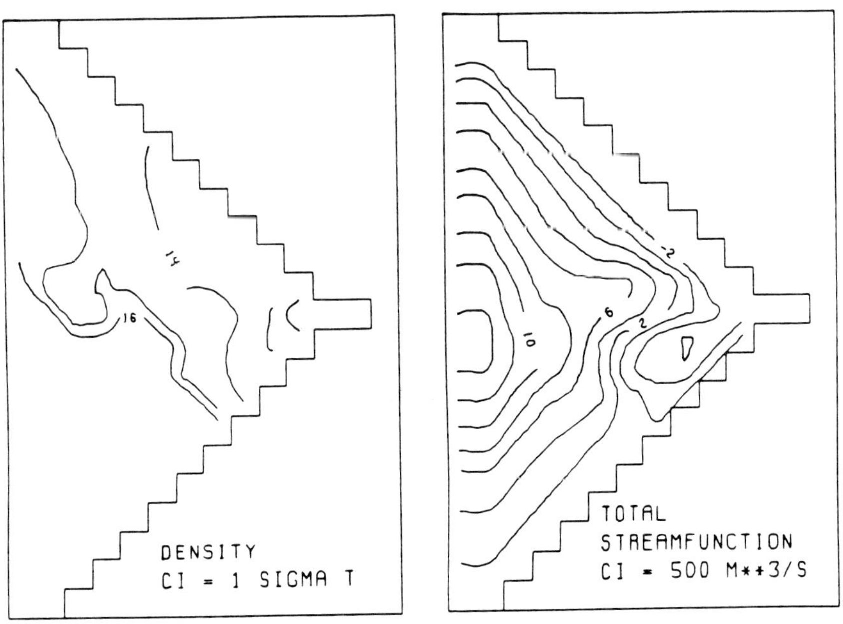

Fig. 4 Surface density (left) CI = $1\sigma_t$, and total streamfunction (right) CI = 500 m^3/s after 25 tidal periods (linear topography)

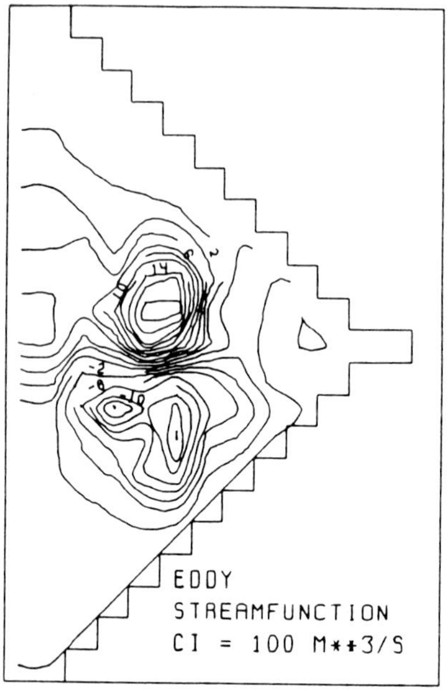

Fig. 5 Eddy streamfunction after 45 tidal periods CI = 100 m^3/s

The reason for this, is that the velocity field is controlled much more strongly by the bottom topography than by the density field.

Defining an eddy streamfunction $\psi' = \psi - \overline{\psi}$, where ψ is the total streamfunction and $\overline{\psi}$ is the mean streamfunction averaged over 97 tidal periods after the model has reached a steady state. A typical result (Fig. 5) is the development of an anticyclonic and a cyclonic eddy. These eddies are more or less persistent over the whole simulation, and they seem to be produced by the topography.

5 CONCLUSION

It is clearly demonstrated that it is possible to calculate eddies and fronts with the active Lagrangian tracer technique. For the future, an intensive parameter test is planned to obtain better knowledge of the model response. After the test the model will be extended to the dimensions of the German Bight with realistic topography, density fields, and wind input.

ACKNOWLEDGEMENT

This work is supported by the Deutsche Forschungsgemeinschaft. The author would like to thank his colleagues I. Bork, K. Duwe, E. Maier-Reimer and R. Wais for fruitful discussions.

REFERENCES

Arakawa, A. and Lamb, V.R., 1977. Computational design of the basic dynamical processes of the UCLA general circulation model. Meth. Computat. Phys. 16: 173-263.

Backhaus, J.O., 1985. A three-dimensional model for the simulation of shelf sea dynamics. Dt. hydrogr. Z. 38(4): 165-187.

Becker, G.A., Fiuza, A.F.G. and James, I.D., 1983. Water mass analysis in the German Bight during MARSEN, Phase I.J. Geoph. Res. 88 (C14): 9865-9870.

Bork, I., 1978. Preliminary model studies of sinking plumes, SMHI Rapp No. RHO 14.

Bork, I. and Maier-Reimer, E., 1978. On the spreading of power plant cooling water in a tidal river applied to the river Elbe. Adv. Wat. Res. 1(3): 161-168.

Bowman, M.J., 1978. Introduction and historical perspective. in: M.J. Bowman and W.E. Esaias (Editors), Oceanic fronts in coastal processes, Springer Verlag, Berlin/Heidelberg/New York/Tokyo.

Dippner, J.W., 1980. Numerische Simulation horizontaler Verdriftung vertikal wandernder Zooplankter. Mitt. Inst. Meeresk. Univ. Hamburg, 23: 63-113.

Dippner, J.W., 1983. A hindcast of the Bravo Ekofisk blow-out. Veröff. Inst. Meeresforsch. Bremerh. 19: 245-257.

Dippner, J.W., 1984. Ein Strömungs- und Öldriftmodell für die Deutsche Bucht. Veröff. Inst. Meeresforsch. Bremerhaven, Suppl. B. 8.

Duwe, K.C., Hewer, R.R. and Backhaus, J.O., 1983. Results of a semi-implicit two step method for the simulation of markedly nonlinear flow in coastal seas, Continental Shelf Res. 2(4): 255-274.

Franz, H. and Klein, H., 1986. Some results of a diffusion experiment at a river plume front. Dt. hydrogr. Z, 39(3): 91-112.

Harashima, A. and Oonishi, Y., 1981. The Coriolis effect against frontogenesis in steady state buoyancy driven circulation. J. Oceanogr. Soc. Japan, 37: 49-59.

James, I.D., 1984. A three-dimensional numerical shelf-sea front model with variable eddy viscosity and diffusivity. Continental Shelf Res. 3(1): 69-98.

Kao, T.W., Park, C. and Pao, H.S., 1977. Buoyant surface discharge and small scale oceanic front: a numerical study. J. Geoph. Res. 82(12): 1747-1752.

Krause, G., Budeus, G., Gerdes, D., Schaumann, K., and Hesse, K., 1986. Frontal systems in the German Bight and their physical and biological effects. In: J.C.J. Nihoul (Editor), Marine Interfaces Ecohydrodynamics. Elsevier Oceanography Ser. 42, Elsevier Amsterdam/Oxford/New York/Tokyo.

Maier-Reimer, E., 1975. Zum Einfluß eines mittleren Windschubes auf die Restströme der Nordsee. Dt. hydrogr. Z., 28(6): 253-262.

Maier-Reimer, E. and Sündermann, J., 1981. On tracer method in computational hydrodynamics, in: M.B. Abott and J.A. Cunge (Editors), Engineering Application of computation hydraulics, Pitman London.

Pritchard, D.W., 1955. Estuarine circulation patterns. Proc. Amer. Soc. Civ. Eng. 81(717): 1-11.

Stronach, J.A., 1981. The Frazer river plume front, Strait of Georgia. Oc. Manag. 6: 201-221.

A FINITE DIFFERENCE GENERAL CIRCULATION MODEL FOR SHELF SEAS AND ITS APPLICATION TO LOW FREQUENCY VARIABILITY ON THE NORTH EUROPEAN SHELF

J. O. BACKHAUS and D. HAINBUCHER
Institut für Meereskunde der Universität Hamburg,
Heimhuder Straße 71, 2000 Hamburg 13, F.R.G.

ABSTRACT

The basic ideas behind the shelf sea circulation model of the Institut für Meereskunde Hamburg (IFMH) are outlined. The stability and the computational performance of the numerical scheme are essentially based upon implicit algorithms or second order approximations introduced for those terms in the primitive equations which are likely to produce instability. As a result the scheme turns out to be rather stable and much faster than conventional, solely explicit numerical schemes. The application of the IFMH-model on the determination of the low frequency variability of the North European shelf sea provides insight into a frequency domain where information from observational data is rather scarce.

1. INTRODUCTION

The typical grid size of present shelf sea models (tidal-, storm surge- and circulation-models) is of the order of ten kilometers. These models allow for a decent resolution of the external Rossby radius of deformation. However, the internal deformation radius at mid-latitudes is well below ten kilometers. This scale is not resolved by most of the presently available shelf sea models.

Similiar as in ocean modelling we can introduce a separation or classification of baroclinic shelf sea models according to the internal deformation scale: general circulation models (GCMs) and eddy-resolving models (ERMs). In the following we assume that both are three-dimensional models.

Parallel to the developments in deep sea modelling there is a trend towards the development and application of ERMs in shelf seas in order to study small scale (local) phenomena (see for example the work of I.James, J.Dippner and M.Boehlich, 1987 all this volume) and in order to develop suitable parameterizations of these processes in GCMs. On the other hand GCMs are necessary to provide the ERMs with model-consistent boundary conditions, which contain the dynamics of the far field flow. It is anticipated that the future development of shelf sea models as a whole will profit from this bi-focal approach. Moreover, the environmental problems in shelf seas constitute a serious demand for accurate and quantitative model estimates on various scales. In the field which is often called "water-quality modelling" GCMs and ERMs are absolutely essential constituents.

The model and its application described in this contribution belongs to the class of the GCMs. However, we believe that the numerical scheme serves also as a sound basis for the development of an ERM since the numerical problems, which are delt with here, do also appear for an ERM. This statement is verified by the course of the advanced work of I. James (this volume, own citations therein) and by Boehlich´s first results on a nested system of models focussing on the Baltic (this volume), where the IFMH-scheme was used for his ERM and his GCM.

Since a detailed description of the IFMH-model and its development is given in Backhaus (1983a, 1985) we here only briefly outline the basic numerical components which differ from commonly used schemes. This will be the content of part A. In part B a time-dependent simulation of the circulation of the North European shelf sea for a period of 14 years is described and analysed with regard to intra- and inter-annual variability of the flow.

2. PART A, OUTLINE OF THE IFMH-MODEL

The following description of our model does not rely upon a specific discretisation of the space domain, the model grid, because the stabilising algorithms and operations are solely applied to the time-domain. We consider a two time-level scheme based upon finite differences applied to all three space coordinates. The scheme, a grid-box model, is applied on the primitive shallow water equations (1.1 and 1.2), which are denoted below in momentum-form for a fluid layer with a thickness h.

The equation of motion:

$$\begin{pmatrix} U \\ V \end{pmatrix}_t + \begin{pmatrix} 0 & -f \\ f & 0 \end{pmatrix} \begin{pmatrix} U \\ V \end{pmatrix} + h/\rho \begin{pmatrix} (E+I)_x \\ (E+I)_y \end{pmatrix} + \begin{pmatrix} N(U) \\ N(V) \end{pmatrix} = \begin{pmatrix} D(U) \\ D(V) \end{pmatrix} + \begin{pmatrix} \Delta\tau^x \\ \Delta\tau^y \end{pmatrix} \tag{1.1}$$

The equation of continuity:

$$\Delta w = -(U_x + V_y) \quad ; \quad \zeta_t = -(\bar{U}_x + \bar{V}_y) \tag{1.2}$$

The shear stress at the upper and lower boundaries of the depth range h is given by: $\tau^{x,y} = Av(\,(U,V)/h)_z$

In the above equations (U,V) are the components of the transport vertically integrated over the depth range h; $E=g\rho\zeta$ is the external (barotropic) and I the internal (baroclinic) pressure, separated according to the Boussinesq-approximation; N(U,V) and D(U,V) represent the nonlinear advective and the horizontal momentum diffusion terms, respectively; $\Delta\tau$ the vertical difference of the vertical shear stress τ ; Av the vertical eddy viscosity; ρ the density; f the Coriolis parameter; g the acceleration due to gravity; spatial or temporal derivatives are indicated by respective indices. In the equation of continuity (1.2) Δw is the vertical difference of the vertical current velocity; ζ is the elevation of the free surface; an overbar indicates a (total) depth mean.

In a two time-level scheme, applied to the above equations there are essentially three sources for linear numerical instability arising from different terms and combinations of terms within the equations. These are:

A) The oscillation equation, for example: $U_t = fV$. A forward in time approximation of this component of the primitive equations is a priori instable, as pointed out already by Fischer (1959). It is likely that this latent instability was damped out in a great number of models of the past by an increase of the horizontal momentum diffusion. Also in shallow water the necessary damping might have been accounted for by the effect of the bottom friction. However, once the local water depth exceeded (say) shelf values, the instability became apparent.

We cure this weakness by introducing a second order approximation for the Coriolis rotation according to Wais (1985). This approximation includes the (external and internal) pressure gradient terms in order to maintain geostrophy.

B) The well known stability limitation due to the propagation of external gravity waves in a free-surface model; the Courant-Friedrichs-Lewy (CFL) criterion. This criterion requires a Courant-number, which should be smaller than unity, in order to maintain linear computational stability. The resulting limitation of the time-step becomes particularly stringent when deep ocean regions are to be included in a shelf sea model.

We cure this weakness by introducing an implicit approximation for the external gravity waves. This requires the solution of an elliptic problem for the free surface elevation at each time-step. The extra computational effort, which is required for this solution (of a scalar quantity (the surface elevation) in the horizontal plane) is easily compensated by the advantage of the almost free choice of the time-step. That is to say the time consuming three-dimensional equations of motion need to be solved less often than in an explicit scheme. For details see Backhaus (loc. cit.).

C) The solution of the vertical shear stress terms in a grid-box model is also limited by a stability criterion, which states a quadratic dependency between the vertical increment, the time-step, and the vertical eddy viscosity coefficient: $h^2 \geqq 2\Delta t\, Av$.

Once the space- and time-increments are set, especially in view of the possibility of using a large time-step due to B), the criterion limits the choice of the vertical eddy viscosity coefficient Av. There is good reason, based upon the available literature on the vertical momentum transfer within the sea, to assume that this coefficient is space- and time-dependent. If under extreme wind forcing the coefficient is limited by an upper value to maintain stability, a feature like ´cavitation´ may occur. As a result of the unrealistic description of the momentum transfer too much of the wind induced momentum might be retained in the upper model layers, which then may "slip" away from the layers beneath. As a whole, any numerical limitation of the vertical eddy viscosity means an unwanted restriction of the physics of the vertical momentum transfer.

We cure this weakness by introducing an additional implicit scheme in the vertical for the shear stress terms. This allows for the option to include a turbulence closure scheme in the model, or any other algorithm that describes the spatial and temporal changes of the vertical momentum transfer parameter Av.

We circumvent the three numerical obstacles described above under A to C by introducing a two time-level scheme, which has the following general form:

$$\begin{pmatrix} U \\ V \end{pmatrix}^{n+1} = T_1 \begin{pmatrix} U \\ V \end{pmatrix}^{n} - T_2 h/\rho \left(\begin{pmatrix} \underline{E}_x \\ \underline{E}_y \end{pmatrix}^{n+1/2} + \begin{pmatrix} \underline{I}_x \\ \underline{I}_y \end{pmatrix}^{n} \right) + \Delta t \left(\begin{pmatrix} D(U) \\ D(V) \end{pmatrix} - \begin{pmatrix} N(U) \\ N(V) \end{pmatrix} \right)^{n} + \Delta t \begin{pmatrix} \Delta\tau^x \\ \Delta\tau^y \end{pmatrix}^{n+1/2} \tag{2.1}$$

$$\zeta^{n+1} = \zeta^{n} - \Delta t (\bar{\underline{U}}_x + \bar{\underline{V}}_y)^{n+1/2} \tag{2.2}$$

The time-levels are indicated by the upper index n, where n+1 indicates the future level. Time averages between these levels are indicated by the index n+1/2. The introduction of this centered time-level leads to the implicit system components for the external gravity waves (in 2.1 and 2.2) and for the vertical shear stress terms in (2.1). It should be noted that the centering in the time-domain leads to a scheme, which is essentially neutral with regard to the damping of amplitudes. In the equations (2.1) and (2.2) spatial difference operators (on any grid) are symbolized by an under-bar indicating the respective coordinate.

In (2.1) two operators T1,T2 appear. These are rotation matrices, applied on the Coriolis and the pressure gradient terms. They have the form:

$$T1 = \begin{pmatrix} \alpha & \beta \\ -\beta & \alpha \end{pmatrix} \quad ; \quad T2 = 1/f \begin{pmatrix} \beta & \gamma \\ -\gamma & \beta \end{pmatrix} \quad ,$$

where $\alpha = \cos f\Delta t$; $\beta = \sin f\Delta t$; $\gamma = 1 - \alpha$.

The application of the operator T2 on the pressure gradient terms in (2.1) leads to additional, ´secondary´ pressure gradient terms. For this second order approximation of the Coriolis-rotation stability will be maintained, when $f\Delta t \leqq \pi$, where π is the circle number.

The elliptic system for the free surface is obtained by replacing the momentum divergence terms in the vertically integrated equation of continuity (1.2) by the equations of motion (1.1). To demonstrate this analytically, we differentiate (1.2) with respect to time and arrive at:

$$\zeta_{tt} + \bar{U}_{tx} + \bar{V}_{ty} = 0$$

In this equation the time-derivatives of the total depth mean momentum are replaced by the vertically integrated equation of motion (1.1). The result is the elliptic system, which has the general form:

$$\zeta_{tt} + a\zeta_{xx} + b\zeta_{yy} + c\zeta_{xy} = d$$

Here the cross-derivative of the free surface elevation was added, in order to demonstrate the effect of the rotation operator T2 in the numerical approximation. Boundary conditions enter the system via the coefficients a to d. It should be noted, however, that the secondary pressure gradient terms may require additional boundary conditions, depending upon the choice of the model-grid. The numerical solution of the above system is equivalent to the solution of a linear system of equations for the free surface. We obtain this solution by means of an iteration, where the method of ´successive over-relaxation´ (SOR) was applied with good success. Matrix inversion seems to be not appropriate for this application in view of the time-dependent coefficients.

The above sketched procedure to obtain the scalar equation for the free surface is much more straight forward in the numerical approximation. There the solution at time-level n+1 of the vertical integral of the equation of motion (2.1) is directly inserted into the approximation for the momentum divergence in (2.2).

This is only possible, because the not yet known shear stress terms at level n+1/2 cancel in the vertical integral. Having obtained a solution for the free surface at level n+1, an interim solution for (2.2) can be computed for the pressure gradient terms and for all other explicitly treated terms. The final solution for (2.1) is then obtained by solving the vertical implicit system for the shear stress terms.

The succession of operations, i.e. first the solution for the free surface, and then the interim and the final solution, is the only way to integrate the entire scheme in view of the implicit algorithms which involve all three space coordinates. The scheme as a whole is semi-implicit, because the internal baroclinic pressure gradients (Ix,Iy) and the terms D(U,V) for horizontal momentum diffusion and the nonlinear advective terms N(U,V) are solved explicitly. The latter are approximated by a vector-upstream algorithm (Haltiner 1971), in order to avoid instability due to central differencing in a two time-level scheme.

As pointed out already above, we require a space- and time-dependent eddy viscosity coefficient Av for realistic simulations of the circulation. Attempts to include a two-equation turbulence-closure scheme (reviewed by Rodi 1980) within the model have failed for a number of reasons. The main reason might have been a lacking vertical resolution, which caused too large discretization errors (the vertical increments were of the order of 10 m, typical for shelf sea GCMs). Further, (time-dependent) boundary conditions for the prognostic variables in these schemes are not uniquely defined and they depend upon parameters, which have not yet been tested thoroughly under shelf sea conditions (however, the turbulence modellers pretend that they are universal). Most of the presently published applications concern simple problems, like open channel flow (Rodi loc.cit.) or tidal flow over topography (Johns 1983). Very seldom a wind stress at the sea surface is considered (Blumberg and Mellor 1980). A simplified one-equation (equilibrium) turbulence model was applied by James (1986) within an ERM. At any rate the inclusion of a complete and really and reliable working turbulence model within a shelf sea model needs future research. But already now we can estimate that it would mean a doubling of the computational effort.

Presently we circumvent this disadvantage by incorporating a simple algorithm based upon the Richardson number, which allows for an (almost nonlinear) feedback between the momentum transfer coefficient Av and the actual and local state of the flow and the stratification. Hainbucher (1985) has tested a number of parameterizations that appeared in the literature and found the best matching with observations for a stably stratified North Sea for the parameter-set, proposed by Bowden and Hamilton (1975). The main feature of this parameterization is the separation of the eddy viscosity coefficient in a portion due to neutral (Avo) and one due to stabil stratification (Avs), respectively:

Av = Avmin + Avo * Avs, whereby:

Avo = K*H*/Uj/ and Avs = $(1+\sigma Ri_j)^{-p}$

Avmin is a minimum limit of the eddy viscosity coefficient (in our case 25 cm^2/s), K a constant, H the total water depth, Uj the current velocity in the model layer j, Ri_j the Richardson number, σ and p are constants, which are chosen accordingly to Bowden and Hamilton (σ=7, p =0.25). This separation allows on the one hand to vary the eddy viscosity, i.e. the vertical momentum transfer, due to varying atmospheric forcing by the portion Avo or on the other hand to determine the rate of momentum transfer due to vertical stratification by the portion Avs.

3. PART B, AN APPLICATION

The concept of the model experiments carried out in this investigation is based upon the assumption that the variability of the circulation in a shallow shelf sea is to a very high extent induced by the action of the atmosphere. In this study the energy transfer between atmosphere and ocean will be restricted to the transfer of momentum. Suitable data which would also allow to include thermodynamics is as yet not available.

In order to obtain estimates of the flow for longer time scales, a number of diagnostic simulations were carried out for the North Sea and adjacent shelf regions by prescribing a monthly or a seasonal or even an annual climatological mean of the atmospheric forcing (Prandle 1978, Maier-Reimer 1977, Davies 1982, Backhaus and Maier-Reimer 1983, Backhaus 1983b). The resulting circulation obtained under the prescribed stationary forcing is then assumed to represent the respective climatological mean. This diagnostic approach, frequently applied also for estimates of the oceanic circulation, relies upon the assumption that the momentum transfer between atmosphere and ocean is linear. However, contrary to the deep ocean where it is justified to assume linearity (Willebrandt 1978), in a shallow shelf sea nonlinear bottom friction and nonlinear tidal interactions are dominant features. Thus, there is reason for doubt on the validity of the assumption of linearity as was demonstrated by Backhaus et. al. 1986.

Further, the diagnostic simulations do not provide information on the fluctuations of the flow for time scales smaller than the period of the respective mean. As a result practically no information exists which concerns low frequency variability of the flow induced by the transient activity of the atmosphere.

We define the low frequency variability by fluctuations of the flow with time scales which are well above the average response time of the system to atmospheric forcing. For the North European shelf sea the response time ranges between one and about five days, depending upon the rate of frictional dissipation and thus, mainly upon the local amplitude of the tides and the water depth. Since this frequency band coincides with that of synoptic scale atmospheric disturbances one may conclude that a single storm will not contribute significantly to the low frequency flow. We shall therefore call fluctuations of the flow with time scales between weeks and years "subsynoptic" fluctuations.

The obvious lack of information on the low frequency variability of the circulation and the doubt concerning the existing diagnostic means provides the main motivation for this investigation; Therefore we will place our focus on the determination of a mean circulation by considering nonlinear processes and on the determination of the order of magnitude and the time scales of atmospherically induced sub-synoptic fluctuations of the circulation, i.e. deviations from the mean.

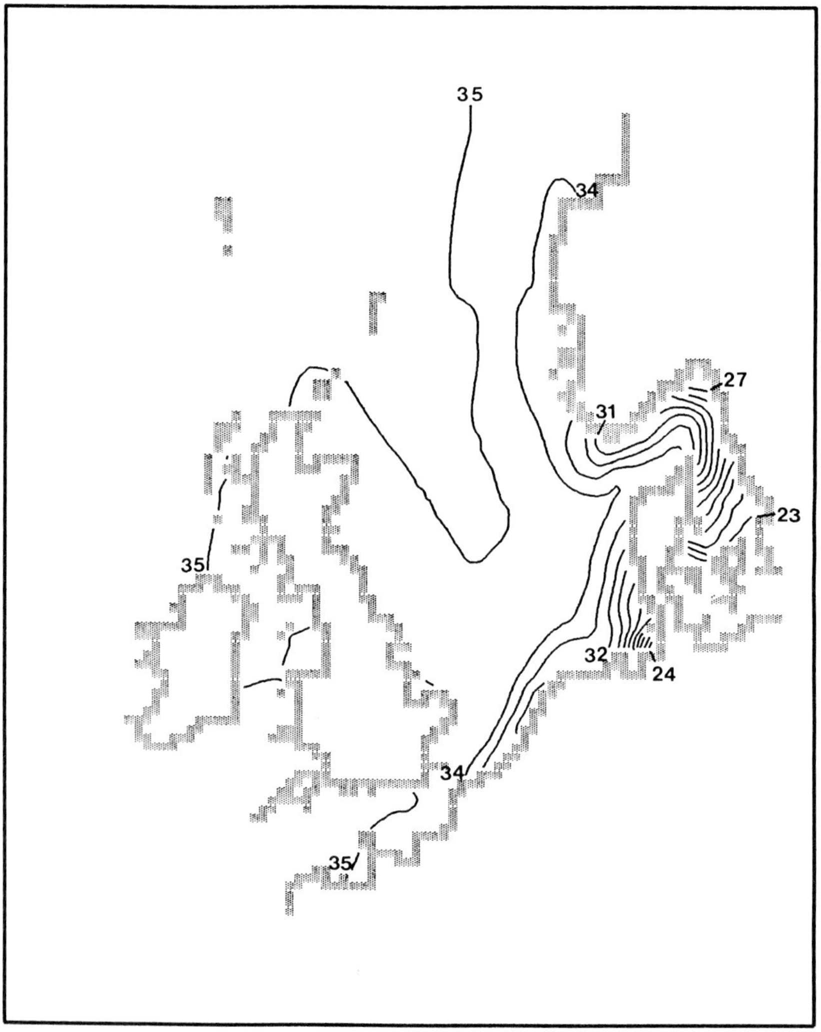

fig. 1: Mean winterly surface salinity distribution

3.1 THE FORCING

In the model simulations the effects of a time dependent atmospheric forcing, of the tide and of stratification were incorporated. The circulation will result from a superposition of these three forcing components.

The data which was used to determine the atmospheric forcing was kindly provided by the Norwegian Meteorological Institute. The record of 28 years (1955-1982) of six-hourly sea surface air-pressure distributions on a 150x150 km grid covers the north-eastern North Atlantic Ocean, the adjacent European shelf and some continental margins. The air-pressure fields were used to estimate the wind stress from the geostrophic wind by means of the nonlinear relationship by Luthardt and Hasse (1981, 1983). These parameters (air-pressure and wind stress) constitute the atmospheric forcing for our model (Backhaus et. al. 1985).

The considerable horizontal density gradient which is present throughout the year is demonstrated by the climatological mean of the winterly surface salinity displayed in figure 1 for the first model layer. With regard to the simulations it would be desirable to have at least an individual mean of the actual density distribution for each season within the simulation. However, the data coverage of presently available observations is too insufficient for this purpose which in particular applies for the shelf and ocean regions apart from the North Sea. Thus, presently we prescribe a climatological summer respectively winter mean of the stratification.

For the North Sea region the data for temperature and salinity was digitized and interpolated from the charts of climatological monthly means published by Tomczak and Goedecke (1962) and by Goedecke et. al. (1967). For the rest of the model area the climatological means on a 1x1 degree grid prepared for general ocean circulation models by Levitus (1982) were taken. Substantial spatial interpolation was necessary due to the coarse grid in order to obtain a first order approximation of the three-dimensional density field in the model domain. As a consequence the structure of the baroclinic jet at the continental shelf edge (Gould et. al. 1985) is not adequately resolved by the data.

A final dynamical interpolation was carried out by means of the baroclinic circulation model in order to eliminate inconsistencies caused by the extensive interpolation. A prognostic advection equation for the baroclinic fields was incorporated in the model scheme and an approximate geostrophic and hydrostatic balance of the density field was obtained by prescribing the forcing of the tide and of a winter respectively summer mean of the wind forcing. The baroclinic pressure fields derived from the dynamically balanced climatological means of the density distributions were prescribed as a constant forcing in all simulations in order to include at least the mean influence of the stratification on the circulation.

The M2- tide is prescribed as sea surface elevation at the open boundaries of the model domain. The resolution of the tide with a rather coarse time step of 40 minutes turned out to be sufficient enough in regard to the formulation of the question. In particular the nonlinear interactions between the tide, the density and the wind induced flow which strongly influence the low frequency (residual) flow field were taken into account.

Furthermore, another time dependent boundary condition is given by the inverse barometric effect due to air-pressure variations. A baroclinic component of the boundary condition reflects the effect of the geostrophically balanced mean winterly respectively summerly density field. This component is obtained from a line-integral of the vertically integrated pressure gradients along the open boundary, where the integration constant which fixes the mean sea level was empirically determined by adjusting the mean sea level at the open boundary so that the mean level at tide gauges at the continental coast agreed within an overall relative error of 20%.

3.2 MODEL EXPERIMENTS

An ensemble of 14 years was simulated (1969-1982) in order to obtain a representative data basis for the calculation of climatological seasonal means and for the analysis of low frequency variations.

The meridional distance of the spherical model grid is 12 minutes, the longitudinal distance 20 minutes. In the vertical the model is resolved by 7 layers for the simulations of the winter seasons (October to April) and by 12 layers for the summer simulations. The layer thickness increases with depth from 10m (summer)/20m (winter) in the upper layers of the model area to 2500m in the deeper layers of the model area outside the shelf region.

The simulation were carried out with a time step of 40 minutes, whereas the model output was reduced to daily values by integrating over two tidal periods, removing tidal and inertia oscillations.

A statistical analysis of the data was carried out due to a separation of low frequency time scales. The data was low-pass filtered with a cut-off period of 6,5 days and subsequently with a cut-off period of 90 days. The difference of both filtered data sets contains the intra-annual variability in the period range from about 10 to 90 days. Statistical analysis of transport rates in the North Sea (Backhaus et. al. 1985) has shown that evident energetic fluctuations exist in this period range. Fluctuations mainly due to the annual cycle or greater periods are included in the 90-days-low-pass filtered data. This cut-off period is somewhat arbritrary and cannot be linked to any low energy level in spectral fluctuation estimates. However, we use this data basis to evaluate the inter-annual variability.

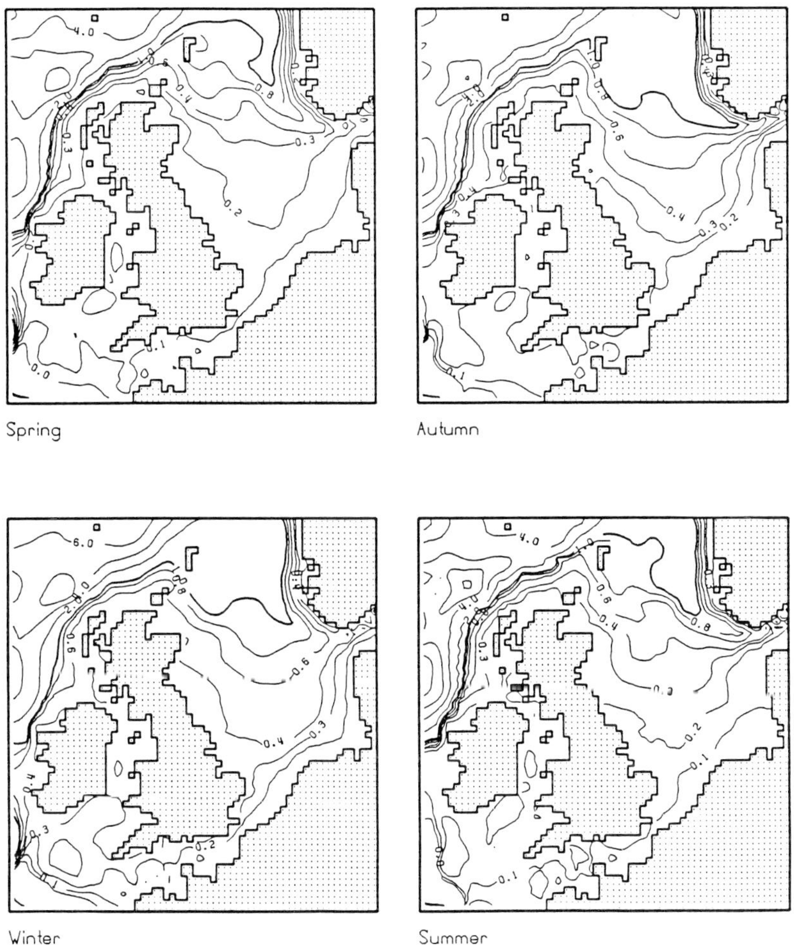

fig.2: Seasonal climatological means of the circulation represented as streamfunctions in Sverdrup ($10^6 m^3/s$).

3.3 CLIMATOLOGICAL MEANS

By averaging the 90-days-low-pass filtered data over three months (December-February, March-May, June-August, September-November) climatological means were calculated for each season. Figure 2 shows the vertically integrated seasonal means represented as stream-functions in Sverdrup ($10^6 m^3/s$).

For the entire shelf region we obtain transport rates which are well below one Sverdrup, in winter and autumn they reach higher values than in spring and summer. The 200m contour, which is the shelf edge, roughly coincides with the 1 Sv contour.

About 75% of the Atlantic water masses entering the North Sea basin in the north, especially through the Orkney-Shetland passage, are flowing back to the Atlantic via the Norwegian coastal current at the eastern boundary of the basin. This open re-circulation cell, which only weakly influences the exchange of water masses in the southern part of the North Sea, can be visualized by the 0.6 Sverdrup contour. In summer and spring the re-circulation cell is removed far to the north and the eastward going flow of the cell is much more pronounced than in the two other seasons. This result agrees well with measurements of the Fair Isle current (Dooley 1983).

The western approaches of the North European shelf sea are characterized by rather weak transport values in the Celtic Sea and in the Irish Sea, whereas the flow obtained in the area around the Hebridies is of similar magnitude as in the southern North Sea. Beyond the shelf a well organized flow of about 1-5 Sverdrup, a part of the North Atlantic eastern boundary current system, closely follows the shelf edge from Ireland towards the northern boundary of the model domain. These water masses feed the re-circulation cell within the North Sea.

3.4 THE DETERMINATION OF VARIABILITY

The determination of low frequency fluctuations is not only an academical effort. The distribution of pollutants in the sea as well as the transport of marine organismns and thereby their further fate are extensively dependent on the long periodical variability of the circulation. The abnormal distribution of herring larvae and sprat stocks in the seventies is an example of this dependency (Corten 1986).

A general view about inter- and intra-annual variability of the circulation is given by two time series of transport rates (figure 3: section through Dover Strait, figure 4: section through the Faeroe-Shetland passage). The solid line shows the inter-annual fluctuations (filter cut-off 90 days) during the years 1969 to 1981. The shaded area demonstrates the intra-annual fluctuations (band-pass: 6.5 to 90 days) during the same period. These fluctuations can be regarded as the standard deviation of the inter-annual variability. It should, however,

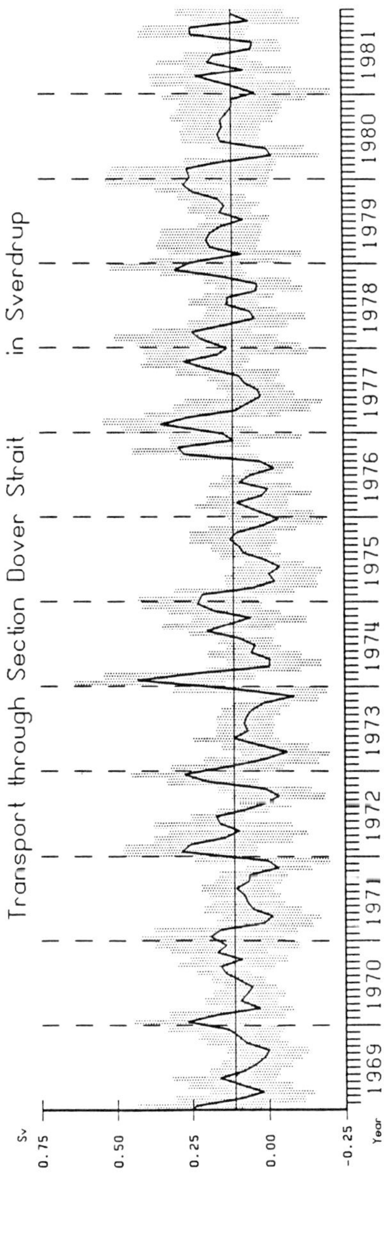

fig. 3: Time series of transport rates through section Dover (1969 - 1981). Solid line: Inter-annual variability, shaded area: Intra-annual variability

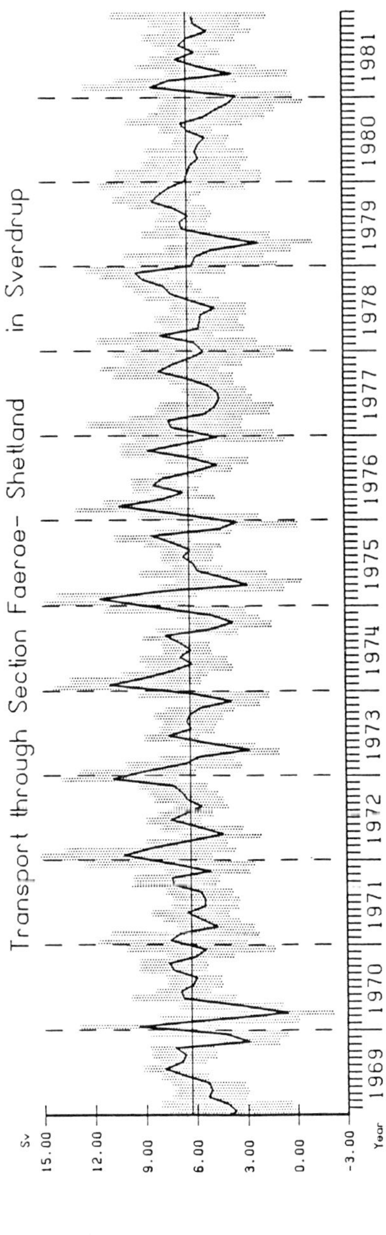

fig. 4: Time series of transport rates through section Faeroe-Shetland (1969 - 1981). Solid line: Inter-annual variability, shaded area: Intra-annual variability

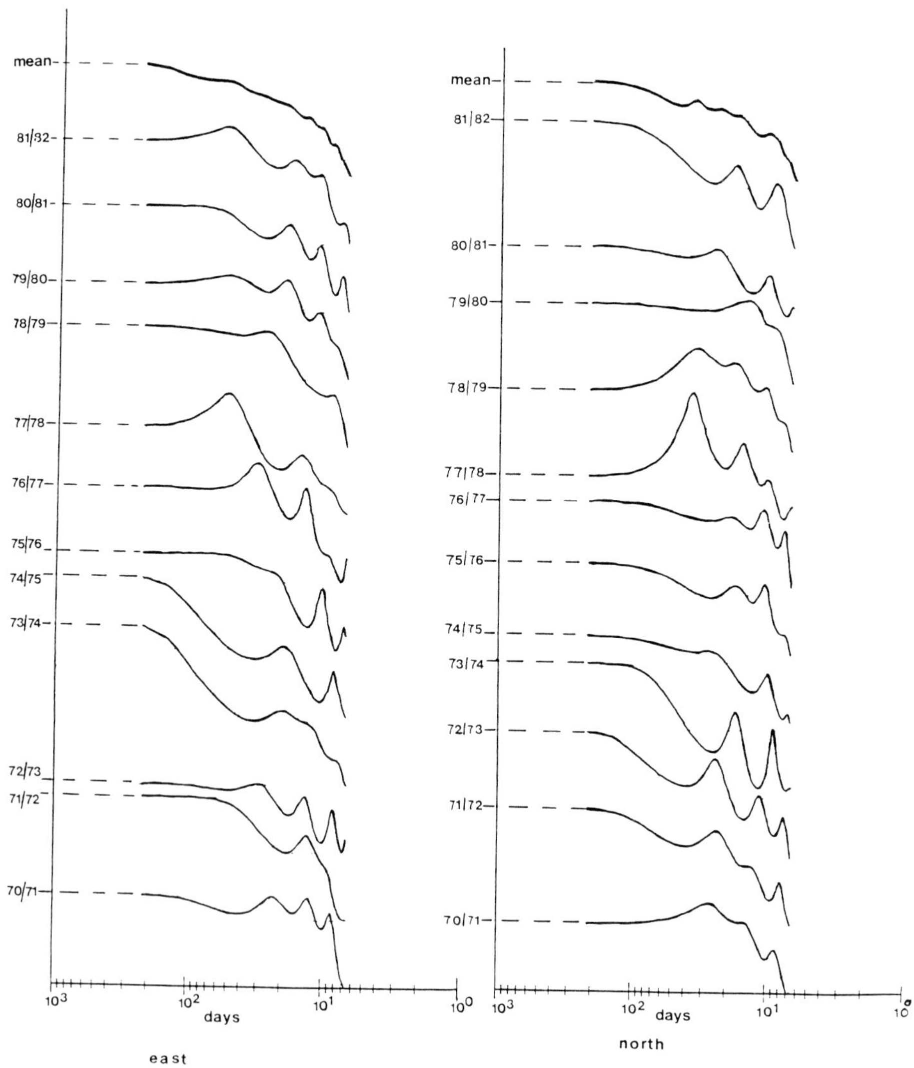

fig. 5: Spectra of the east and north wind stress components over the central North Sea. Considered winter seasons: 1970/71 - 1981/82

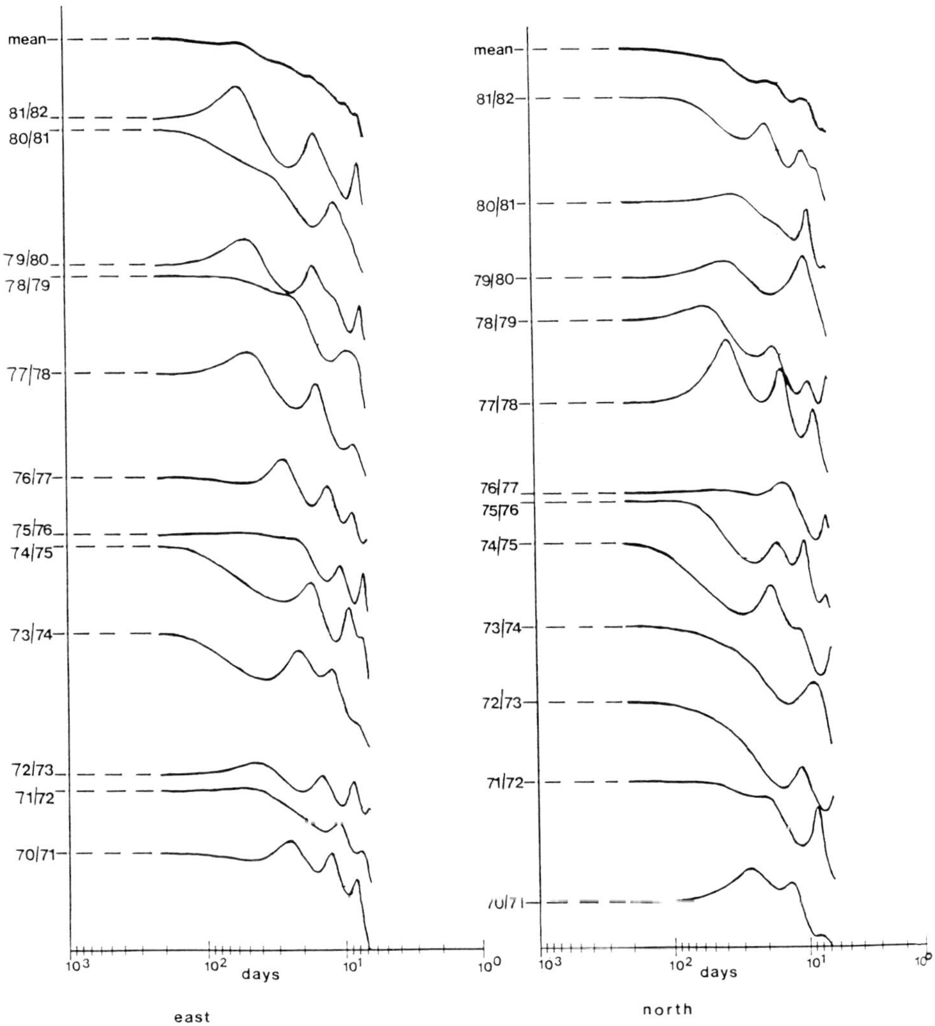

fig. 6: Spectra of the east and north wind stress components over the model area west of Ireland. Considered winter seasons: 1970/71 - 1981/82

be noticed that in this case the standard deviation is determined deterministically due to the prescribed forcing and subsequent filtering and that it should not be understood in the solely statistical sense.

3.4.1 INTRA-ANNUAL VARIABILITY

Statistical analysis of air-pressure and wind stress fields in mid and low latitudes has shown that energetic fluctuations are evident in the period range between 20 and 80 days (Madden and Julian 1972). Figure 5 and 6, representing spectra of the wind stress components at a point in the central North Sea and west of Ireland, give an example of these energetic fluctuations for the winter months (October-April). The overall mean spectrum, however, does not exhibit these signals. It is almost white (Willebrandt 1978).

The assumption that similar fluctuations may exist for the flow field is near at hand. The spectra of the transport through a section at Dover and at Faeroe-Shetland corroborate this conjecture (figure 7 and 8). They show distinct peaks in the same period range. The mean spectra averaged over all winter seasons do not exhibit any variability in the low frequency domain as in the case of the wind stress spectra.

3.4.2 INTER-ANNUAL VARIABILITY

An analysis of the inter-annual variability was carried out by determining seasonal means of the flow field (direction and magnitude) and seasonal anomalies of the kinetic energy distribution (actual mean - climatological mean) for each of the 14 years. The kinetic energy is here given as $0.5(u^2+ v^2)$. The distribution of mass was neglected in order to avoid the description of spatial variances which are only caused by varying water depths.

Figure 9 and 10 show examples of actual circulation patterns, demonstrated by the direction of the flow field and the kinetic energy anomaly. Figure 11 demonstrates the corresponding climatological means with regard to a comparison.

During spring 1974 (figure 9) the circulation was exactly reverse to the "normal" direction in the southern and central North Sea. The anomaly of the kinetic energy exhibits low values within the whole North Sea, mainly negative in the eastern parts and the Orkney- Shetland region and positive in the western parts. The altogether low values of the anomaly give rise to the conclusion that the magnitude of the "abnormal" reverse circulation in spring 74 was approximately of the same order as the climatological mean. This result is a distinct example that the effect of wind and air-pressure can by all means superpose the (density and tidal induced) residual flow and become the dominant forcing of the circulation.

The spring 1979 (figure 10) shows an enhancement of the mean circulation. Po-

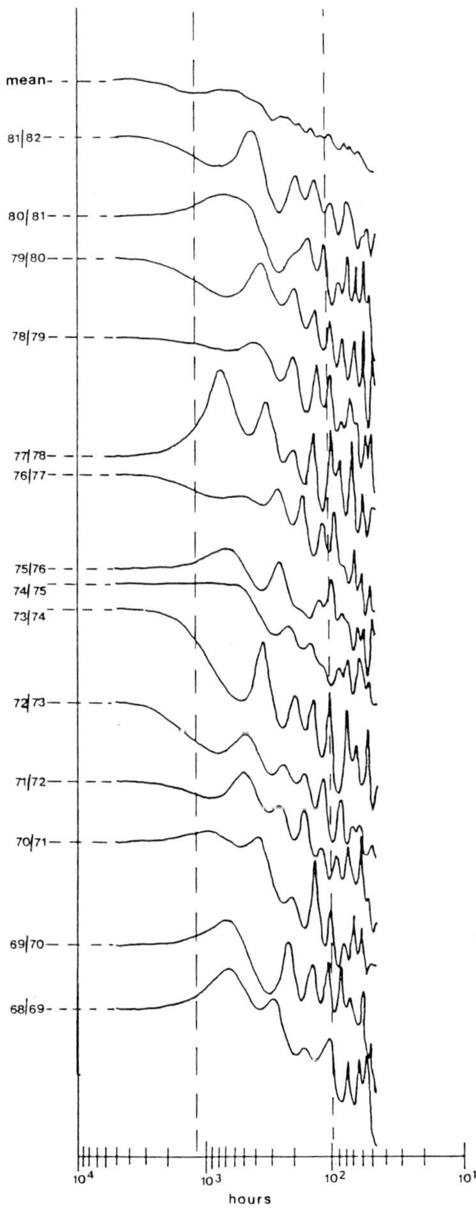

fig. 7: Spectra of the transport rates through section Dover. Considered winter seasons: 1968/69-1981/82

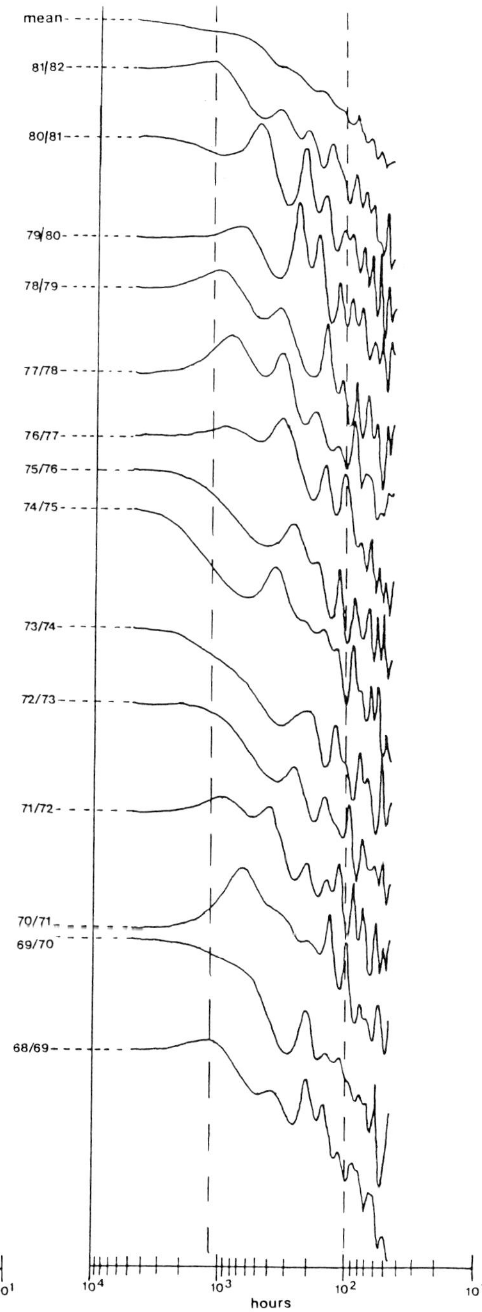

fig. 8: Spectra of the transport rates through section Faeroe-Shetland. Considered winter seasons: 1968/69-1981/82

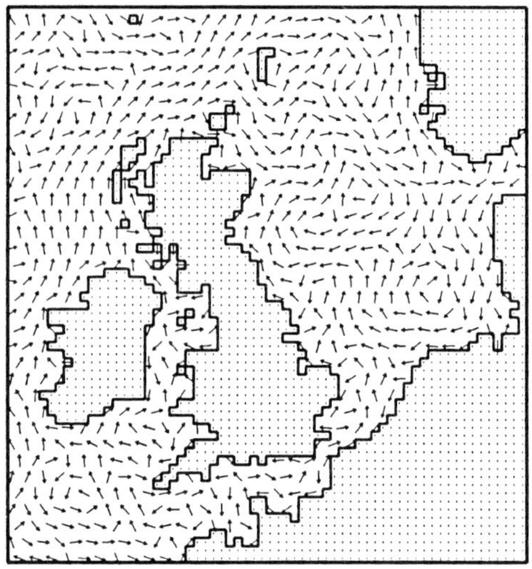

a)

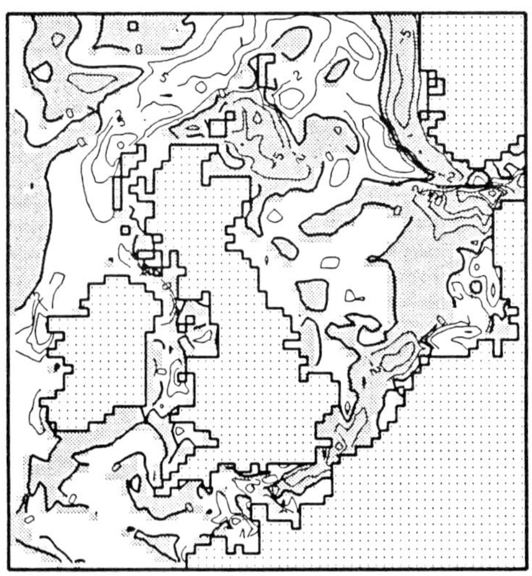

b)

fig. 9: a) Direction of the flow field in spring 1974 (depth mean flow).
b) Anomaly of the kinetic energy in spring 1974 (depth mean flow); units: $cm^2 s^{-2}$

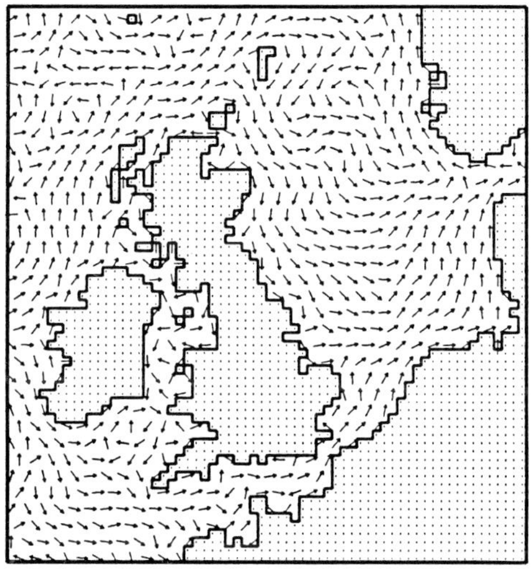

a)

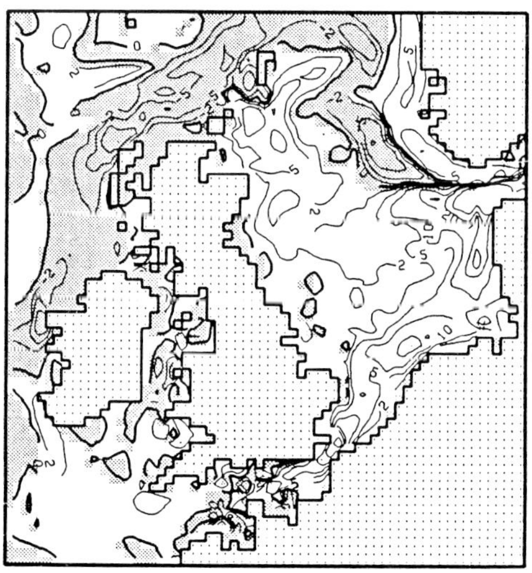

b)

fig. 10: a) Direction of the flow field in spring 1979 (depth mean flow).
b) Anomaly of the kinetic energy in spring 1979 (depth mean flow); units: $cm^2 s^{-2}$

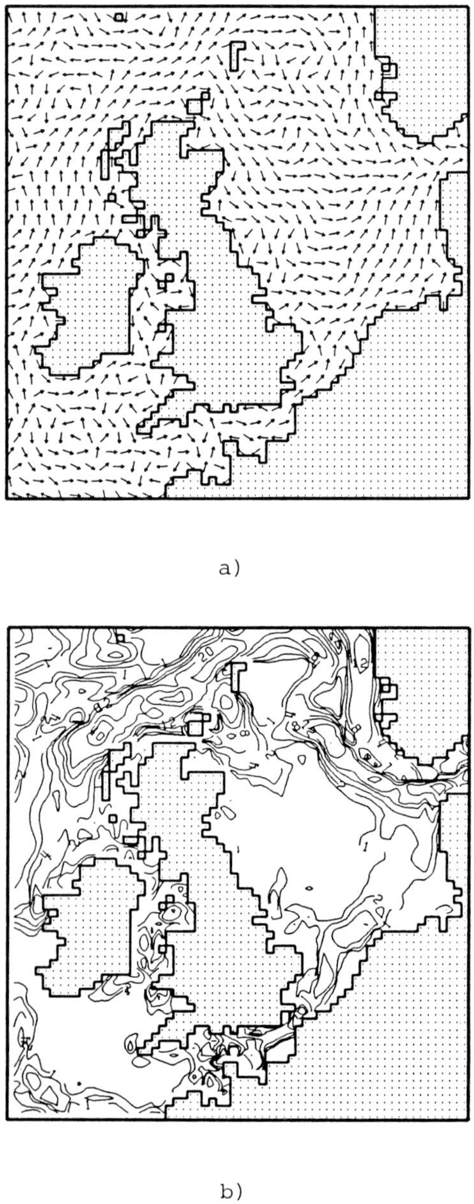

a)

b)

fig. 11: Climatological mean: spring (depth mean flow).
a) Direction of the flow field
b) Kinetic energy

sitive values of the kinetic energy anomaly can be almost found in the whole North Sea. In particular in the eastern parts the anomaly reaches very high values.

The spring 74 as well as the spring 79 are only two examples of abnormal circulation patterns. However during all considered years one can find similar anomalies which are more or less pronounced. A detailed description of these patterns is given in Hainbucher et. al. (1986).

4. FINAL REMARKS

The good stability properties of the scheme allow to apply the horizontal momentum diffusion terms in their physical meaning instead of a numerical emergency brake. We anticipate that this is an important advantage, especially with regard to the ERMs, since a realistic description of horizontal momentum shear can be crucial for the simulation of eddies.

The stability properties together with the possibility of the choice of a large, but still physically realistic time-step were necessary pre-requisites for the long term simulation presented above. To the first time an insight into the low frequency variability of the North European shelf sea was provided by the model estimates. Whereas past simulations almost solely considered (climatological) mean conditions of the circulation the here presented results suggest that deviations from the mean on various time scales are at least of similiar importance.

In the international discussion about the ´water-quality´ of the marine environment of the North Sea again primarily mean values are considered. However, Hainbucher et.al. (1987) have shown by using the above results in a Lagrangian trajectory-model that the water-quality of the North Sea may undergo similiar fluctuations as the circulation. Up to now the inadequate sampling of present environmental monitoring activities prevented the detection of these fluctuations from observations (typical sampling rates are 1 to max. 3 times per year). Hence the extensive model experiment not only served to increase our knowledge about the circulation but it initiated a discussion about important questions concerning the (better) protection/monitoring of the marine environment.

REFERENCES

BACKHAUS, J.O. and E. Maier-Reimer, 1983: On seasonal circulation patterns in the North Sea. In: North Sea dynamics (Ed.: J. Sündermann et.al.) Heidelberg: Springer, pp. 63-84

BACKHAUS, J.O., 1983a: A semi-implicit scheme for the shallow water equations for application to shelf sea modelling. Continent. Shelf Res. 2, pp. 243-254

BACKHAUS, J.O., 1983b: On the circulation of the stratified North Sea. In: Coastal Oceanography (Ed.: Gade et.al.), Plenum Publishing Corporation, pp. 93-111

BACKHAUS, J.O., 1985: A three-dimensional model for the simulation of shelf sea dynamics. Dt. hydrogr. Z. 38, pp. 165-187

BACKHAUS, J.O., J. Bartsch, D. Quadfasel and J. Guddal, 1985: Atlas of monthly surface fields of air pressure, wind stress and wind stress curl over the North Eastern Atlantic Ocean: 1955-1982. Technical Report 3-85, Institut für Meereskunde der Universität Hamburg (unpublished manuscript)

BACKHAUS, J.O., D. Hainbucher, D. Quadfasel and J. Bartsch, 1985: North Sea circulation anomalies in response to varying atmospheric forcing. ICES C.M. 1985/C:29

BACKHAUS, J.O., T. Pohlmann and D. Hainbucher, 1986: Regional aspects of the circulation on the North European shelf. ICES C.M. 1986/C:38

BLUMBERG, A.F. and G.L. Mellor, 1980: A coastal ocean numerical model. In: Mathematical Modelling of Estuarine Physics (Ed.: J. Sündermann et. al.) Heidelberg: Springer

BOEHLICH, M., 1987: A three dimensional baroclinic model of the western Baltic. (this volume)

BOWDEN, K.F. and P. Hamilton, 1975: Some experiments with a numerical model of circulation and mixing in tidal estuaries. Estuarine and Coastal Marine Science 3, pp. 281-301

CORTEN, A., 1986: Biological indications of a change in North Sea circulation in the 70´s. ICES C.M. 1986/L:10

DAVIES, A.M., 1982: Three dimensional models of North Sea circulation. In: Coastal Oceanography (Ed. Gade et.al.) Plenum Press, pp. 65-92

DIPPNER, J.D., 1987: Preliminary results of a three dimensional eddy resolving model for river plume fronts. (this volume)

DOOLEY, H.D., 1983: Seasonal variability in the position and strength of the Fair Isle current. In: Proceedings of the Symposium on North Sea Dynamics, Hamburg September 1981, (Ed.: J. Sündermann et.al.), Heidelberg: Springer, pp. 108-119

FISCHER, G., 1959: Ein numerisches Verfahren zur Errechnung von Windstau und

Gezeiten in Randmeeren. Tellus.11, pp. 60-76

GOEDECKE, E., J. Smed and G. Tomczak, 1967: Monatskarten des Salzgehaltes der Nordsee, dargestellt für verschiedene Tiefenhorizonte. Dt. hydrogr. Z., Erg.-H. B.9

GOULD, W.J., J. Loynes and J.O. Backhaus, 1985: Seasonality in slope current transport n.w. of Shetland. ICES C.M. 1985/C:7

HAINBUCHER, D., 1985: Parametrisierung des vertikalen Impulstransfers in einem dreidimensionalen Zirkulationsmodell der Nordsee. Diploma Thesis, Universität Hamburg, Institut für Meereskunde

HAINBUCHER D., J.O. Backhaus and T. Pohlmann, 1986: Atlas of climatological and actual seasonal circulation patterns in the North Sea and adjacent shelf regions: 1969-1981. Technical Report 1-86, Institut für Meereskunde der Universität Hamburg (unpublished manuscript)

HAINBUCHER, D., T. Pohlmann and J.O. Backhaus, 1987: Transport of conservative passive tracers in the North Sea: first results of a circulation and transport model. Continental Shelf Research (in press)

HALTINER, G.J., 1971: Numerical Weather Prediction. John Wiley & Sons, INC. New York, London, Sydney, Toronto

JAMES, I.D., 1986: A front-resolving sigma coordinate sea model with a simple hybrid advection scheme. Applied Mathematical Modelling, Vol.10, pp. 87-92

JAMES, I.D., 1987: The design of a general three dimensional eddy resolving model for stratified seas. (this volume)

JOHNS, B., 1983: Physical Oceanography of Coastal and Shelf Seas. Elsevier Oceanography Series 35

LEVITUS, S., 1982: Climatological atlas of the world ocean. NOAA professional paper No.13, US Government Printing Office Washington DC

LUTHARDT, H. and L. Hasse, 1981: On the relationship between surface and geostrophic wind in the German Bight area. Beitr. Phys. Atm. 54, pp. 222-232

LUTHARDT, H. and L. Hasse, 1983: The relationship between pressure field and surface wind in the German Bight area at high wind speeds. In: North Sea Dynamics (Ed.: J. Sündermann et.al.), Heidelberg: Springer, pp. 341-348

MADDEN, R.A. and P.R. Julian, 1972: Description of global-scale circulation cells in the tropics with a 40-50 day period. J. Atm. Sc., 29, pp. 1109-1123

MAIER-REIMER, E., 1977: Zum Einfluß eines mittleren Windschubes auf die Restströme. Dt. hydrogr. Z., B.28, H.6

PRANDLE, D., 1978: Residual flows and elevations in the southern North Sea. Proc. R. Soc. London (A) 359: 189-288

RODI, W., 1980: Mathematical modelling of turbulence in estuaries. In Mathematical Modelling of Estuarine Physics (Ed.: J. Sündermann et.al.), Heidelberg: Springer, pp 14-26

TOMCZAK, G. and E. Goedecke, 1962: Monatskarten der Temperatur der Nordsee, dargestellt für verschiedene Tiefenhorizonte. Dt. hydrogr. Z., Erg. H. B,7

WAIS, R., 1985: On the relation of linear stability and the representation of Coriolis terms in the numerical solution of the shallow water equations. Universität Hamburg (unpublished manuscript)

WILLEBRAND, J., 1978: Temporal and spatial scales of the wind field over North Pacific and North Atlantic. J. Phys. Oceanogr., 8, pp. 1080-1094

A THREE DIMENSIONAL CIRCULATION MODEL OF THE SOUTH CHINA SEA

T. POHLMANN
Institut für Meereskunde der Universität Hamburg
Heimhuder Straße 71, 2000 Hamburg 13, FRG

ABSTRACT

Up to now there is a great lack of observational data in the South China Sea. The best available information about the general hydrography of the region was the Naga Report compiled by Wyrtki already in 1961.

The South China Sea is an equatorial region with a complex topography. It is a regime dominated by the monsoon and stratification is of enormous importance.

A prognostic baroclinic circulation model was applied in order to increase our present knowledge and our understanding of this region. The gridsize of this 12- layer model is about 50 km in the horizontal. The layers have a thickness of 10 m to 3000 m, increasing with depth.

Simulations were carried out for the winter- and for the summer monsoon, respectively. The calculation of temperature and salinity distributions which are consistent with the circulation provide insight into new features like deep-reaching up- and downwelling phenomena. A first validation of the model results was carried out in comparison with the observational data compiled by Klaus Wyrtki.

1 INTRODUCTION

The South China Sea with an extention of 36 Mio kmxkm is the largest marginal sea in the Southeast Asian Waters (Figure 1). It is situated between the Asian continent, Borneo, the Philippines and Formosa (Taiwan). Its topography which is divided into two parts is typical for marginal seas of the Western Pacific. The northern part is a deep sea basin where depths exceed 5000 m. This basin is seperated from the main water body of the Pacific by a string of islands of volcanic origin (including the Philippines). The southern part is a shelf sea, where depths range between 50 m and 100 m.

One of the main specifications of the South China Sea is its location in tropical low latitudes, which causes two important effects on the circulation.

a) By the reduction of the Coriolis parameter near the equator nonlinear and frictional terms get an increasing importance.

b) The South China Sea is situated within the monsoon regime and is thereby strongly influenced by the, periodically semi- anual reversing, circulation of the atmosphere. Figure 2a and 2b (the windstress distribution in January and in July) reflect the fully developed winter, respectively summer monsoon situation. In winter northeasterly winds prevail over the whole region with an average magnitude of 9 m/s. In summer the wind distribution totally reverses. Weaker southwesterly winds dominate over most parts of the South China Sea (6 m/s) and only in the northern parts the direction changes to more southerly winds.

The large scale distribution of mass which has a considerable influence on the circulation of the South China Sea could be summarized as follows:

Light tropical surface water with low salinity and high temperature forms a strong contrast to the cold and salty deep water. The transition between these two water masses takes place in an extremely strong discontinuity layer situated in a depth of about 120 m.

The renewal of the deep water by water masses from the Pacific takes place through narrow passages in the string of islands which form the western boundary of the South China Sea.

Figure 3a and 3b show the surface temperature distribution in winter and in summer. The lateral gradients are much stronger in winter. They increase from 23° C in the north to 28° C in the southern parts, whereas during summer the increase is much weaker (from 28° C in the north to 29° C in the south).

The surface salinity (figure 4a and 4b) decreases from 34.0 in the northern to 32.4 in the southern parts of the South China Sea. In winter the maximum gradient is located in the central part while in summer it is situated in the southern part of the South China Sea.

Since 1961, when Wyrtki had prepared his report, no significant improvement of the general understanding of the dynamics of the South China Sea has been achieved. The sea as an important protein- source for mankind becomes particularly important for regions where the growth- rate of the population reaches maximum values. This is the case for a number of nations, which share the South China Sea. Also marine pollution as a consequence of the growing population and/ or industry has become a serious problem.

The complex topography of the region might have been the reason why up to now no attempts for model simulations have been made. In fact, the topography requires both, a shelf sea and a deep ocean model. Apart from this (numerical) challenge there seems to be reason enough to apply a numerical model on the South China Sea in order to improve our understanding of its dynamics.

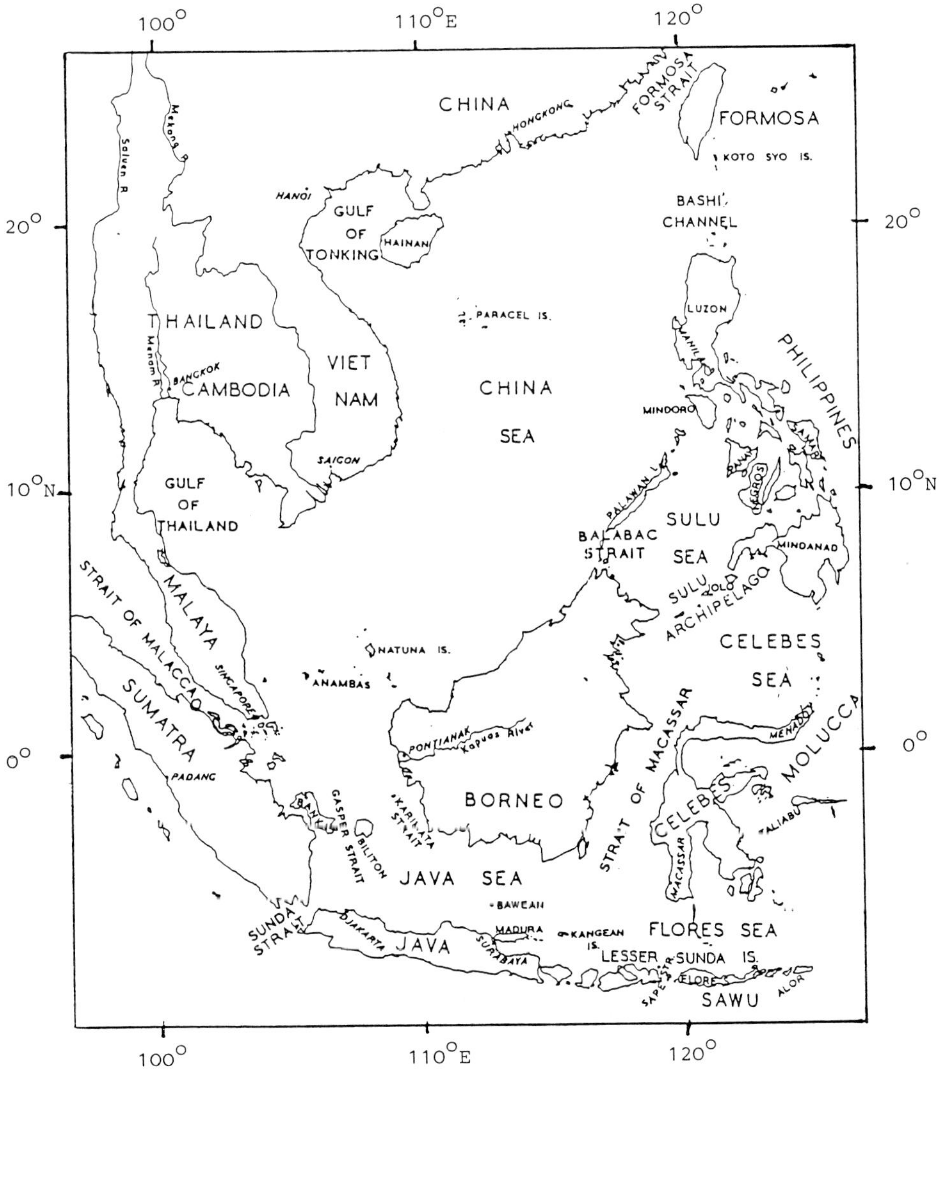

Fig. 1. Chart showing the location of the South China Sea

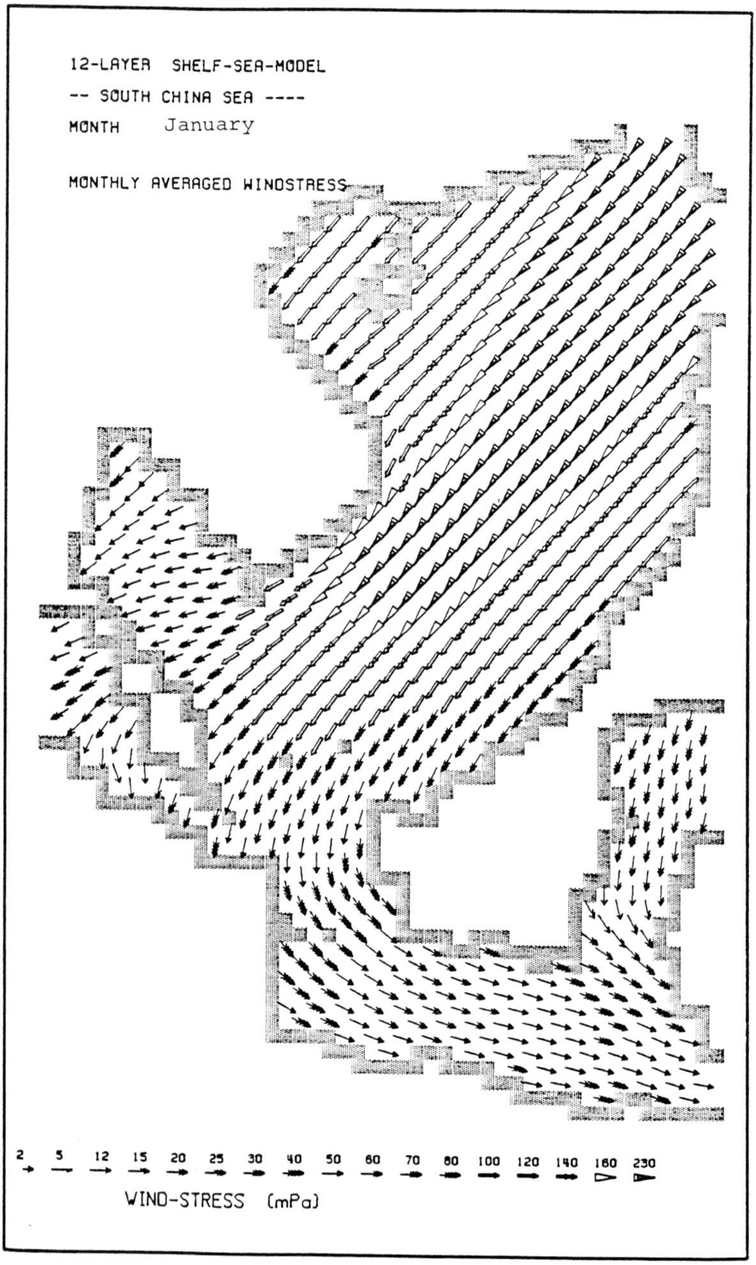

Fig. 2a. Windstress distribution in January
(after Hellerman, 1968)

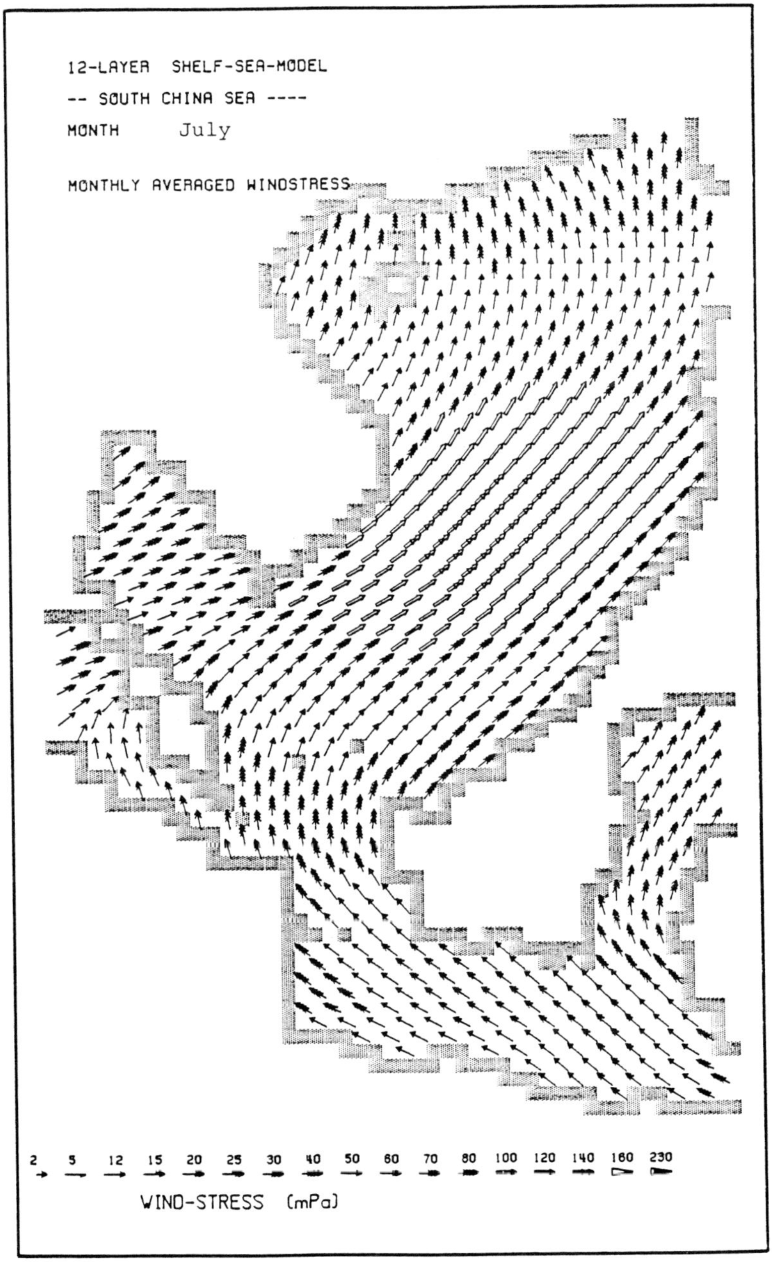

Fig. 2b. Windstress distribution in July
(after Hellerman, 1968)

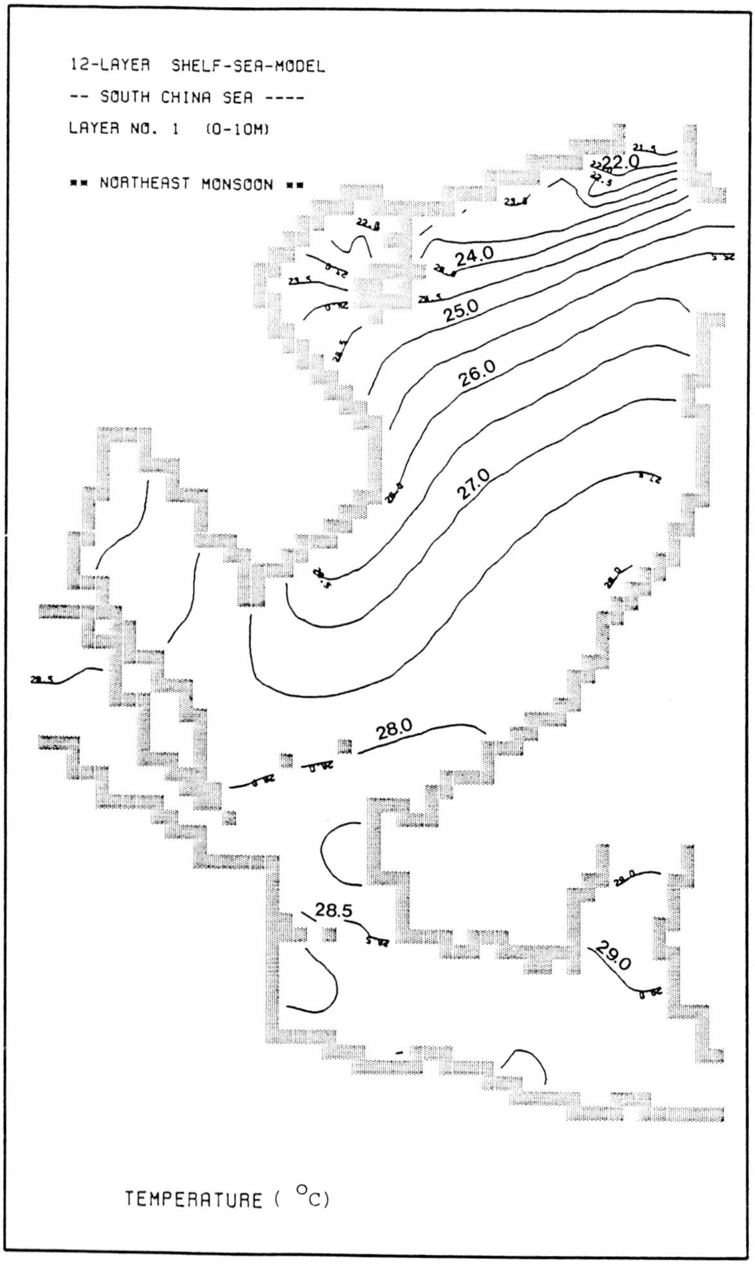

Fig. 3a. Surface temperature distribution in winter (after Levitus, 1982)

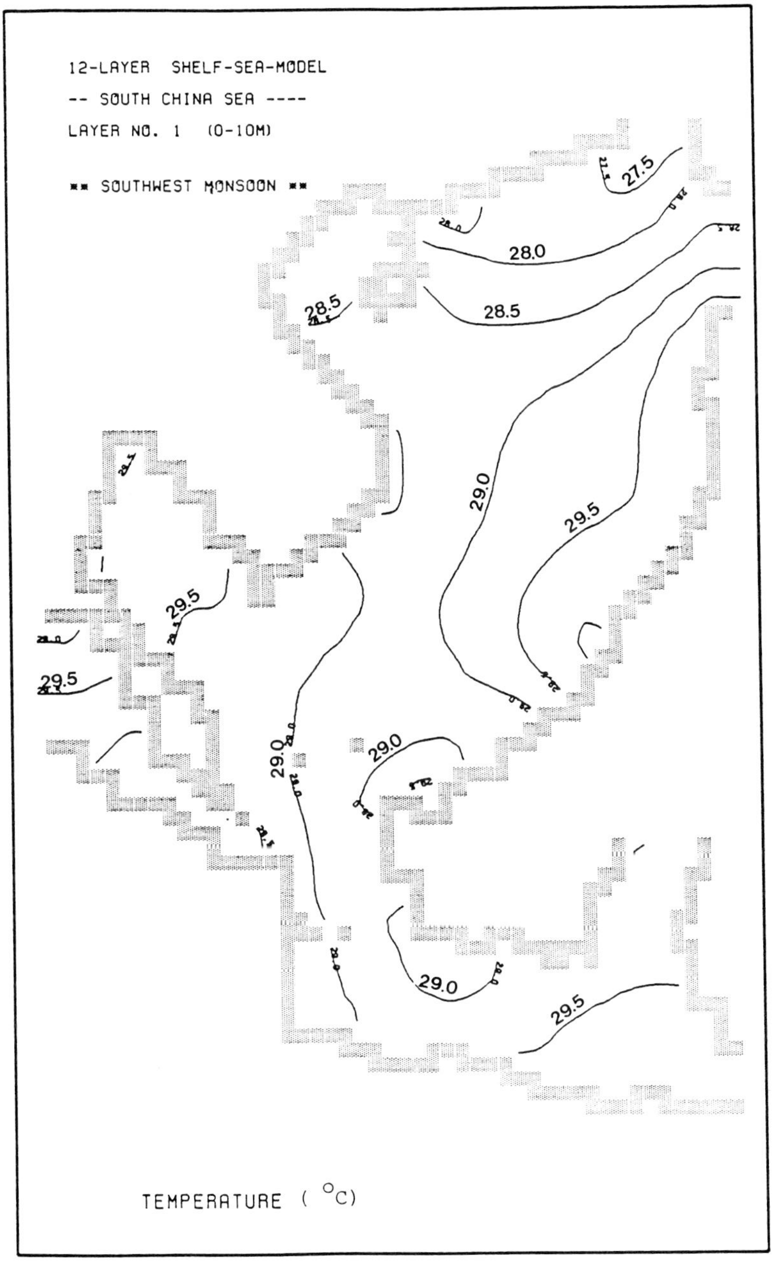

Fig. 3b. Surface temperature distribution in summer (after Levitus, 1982)

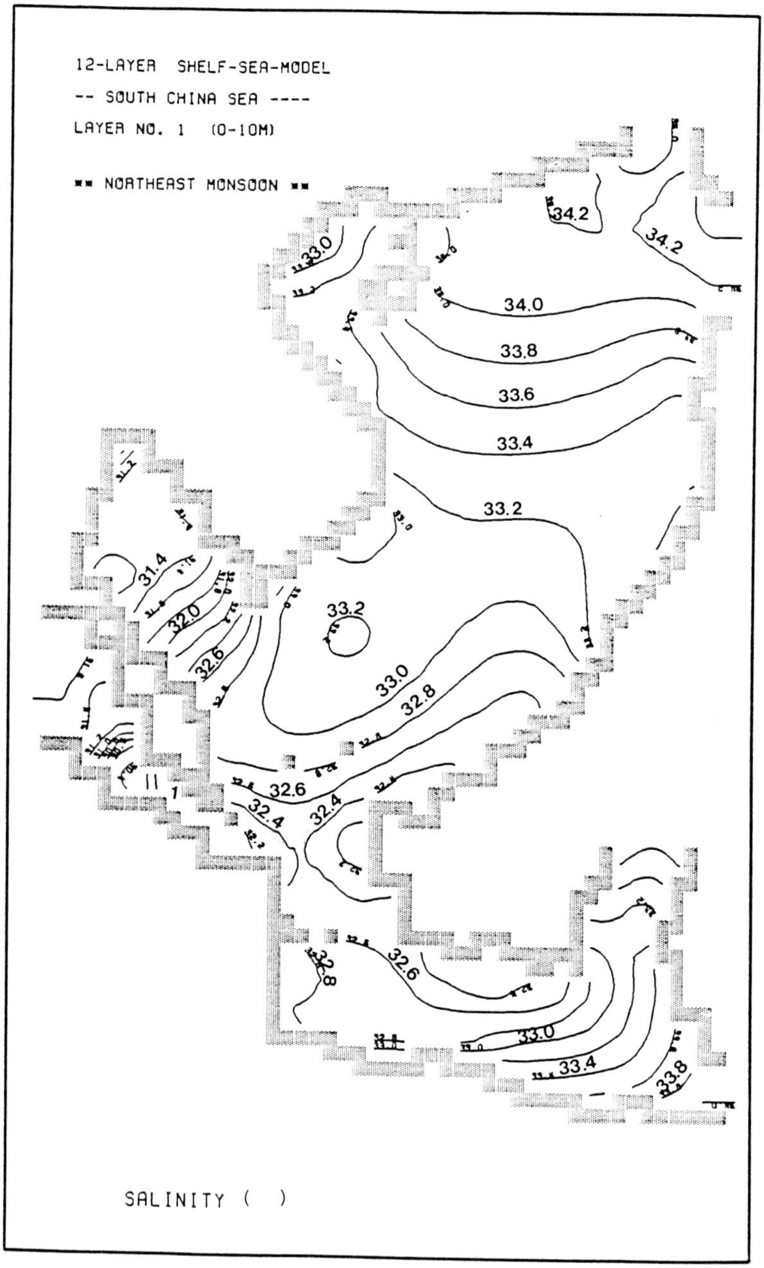

Fig. 4a. Surface salinity distribution in winter
(after Levitus, 1982)

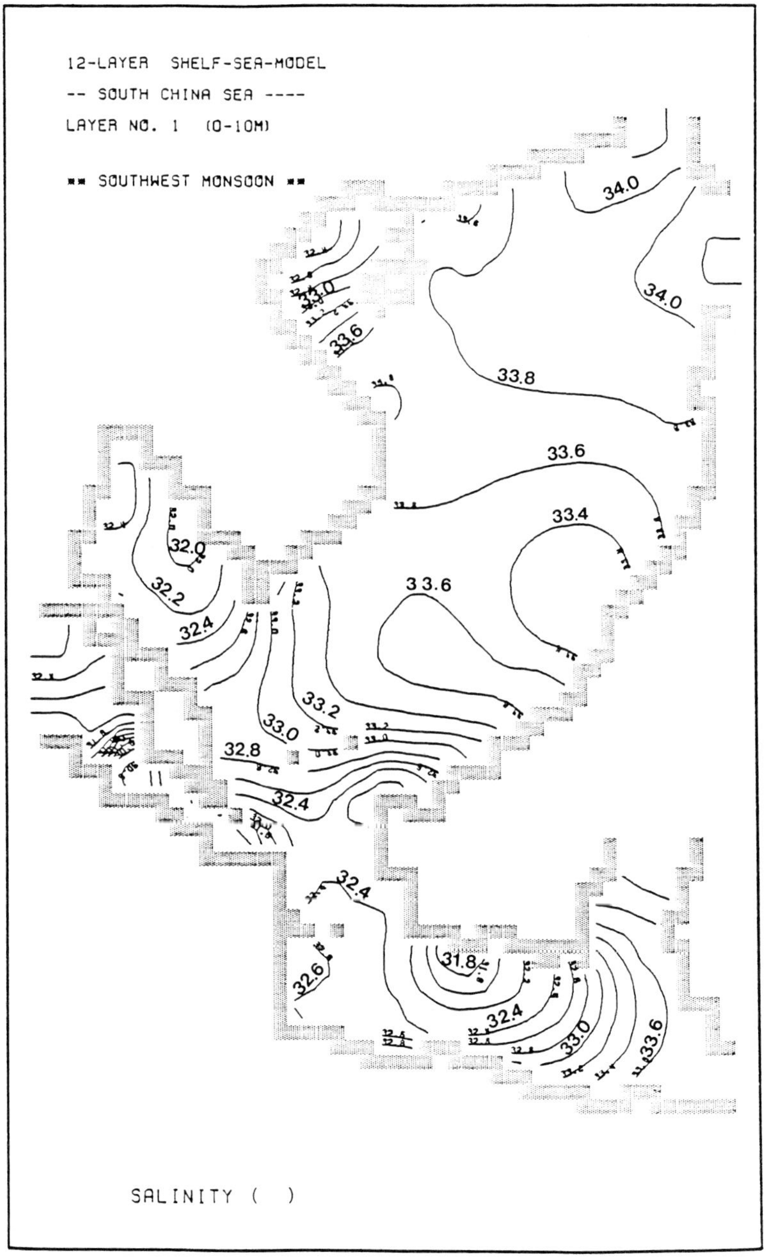

Fig. 4b. Surface salinity distribution in summer
(after Levitus, 1982)

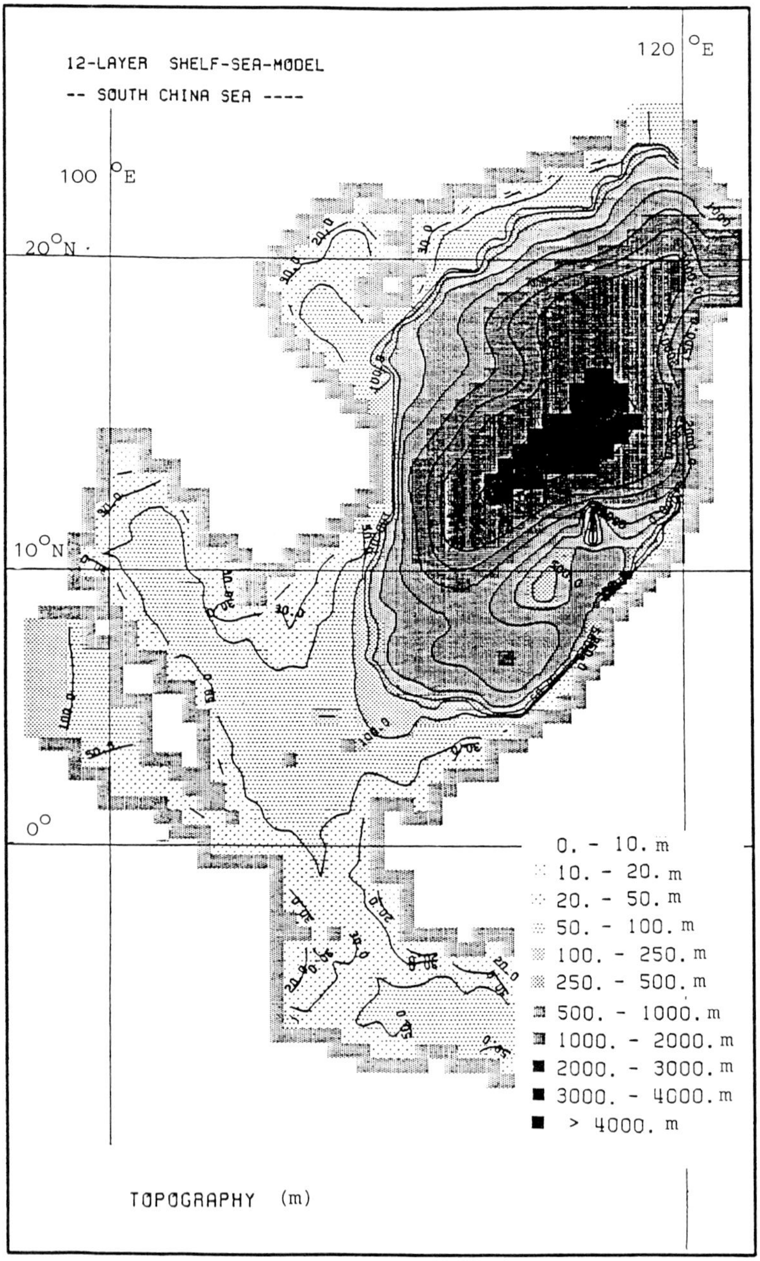

Fig. 5. Topography of the model region

2 THE MODEL

The investigations were carried out with the aid of a three dimensional baroclinic model (Backhaus, 1983). This model has already been applied to several shelf seas. (North Sea, North European Shelf, Baltic) (Backhaus, 1985, Boehlich, 1987, this issue). So, only a general description of the main features of the model is given here.

The governing equations are:

1. The equation of continuity of mass
2. The shallow water equation
3. The equation continuity of temperature and salinity
4. The equation of state for seawater

In equation 1. hydrostatic equilibrium was assumed and the Boussinesq-approximation was incorporated. Equation 3. was simplified by neglecting source and diffusion terms, as a consequence of the insufficient data amount which would permit an accurate determination of these coefficients in this region.

A semi- implicit numerical scheme was used to solve the governing equations. More detailed information about this scheme is given in Backhaus and Hainbucher (1987, this issue). The gridsize of the model is about 50 km in the horizontal. The 12 layers have a thickness of 10 m to 3000 m increasing with depth.

The location of the model region is shown in figure 5. A numerical problem arises from the fact that the South China Sea crosses the equator. The applied numerical scheme requires division by f, the Coriolis parameter. This leads to senseless results directly at the equator where f reaches zero. The problem was solved by placing the grid in a way that no gridpoint lies directly on the equator. Thereby the smallest value of the Coriolis parameter reaches still a reasonable value of 3.2×10^{-7} 1/s

The boundary conditions are the usual ones for primitive equation models. At closed lateral boundaries a no- flux and a semi- slip condition was applied. At open lateral boundaries the water elavation is prescribed and the gradients of the transport normal to the boundary are set equal to zero. For temperature and salinity a modified Sommerfeld radiation- condition (Orlanski, 1976) has been used. The wind stress at the sea surface and the bottom stress are represented by quadratic stress laws.

3 THE SIMULATION

Several simulations were carried out in order to get representative descriptions of the winter and summer monsoon circulation respectively. A simulation period of 15 days for each of the phases turned out to be sufficient to approximately reach a quasi stationary state.

The circulation in the baroclinic South China Sea model is essentially determined by:

1. Monthly averaged wind stress data for January and July extracted from the Hellerman (1980) data set. (figure 2a, 2b)
2. Seasonal averaged temperature and salinity data for the winter and the summer season. These data are taken from Levitus (1968) and interpolated into the model grid, (figure 3a, 3b, 4a, 4b) as initial fields.
3. The topography of the South China Sea. (figure 5)

The tides were neglected in these simulations because they mainly cause the high frequency variability of the circulation and are therefore of minor importance for the mean monsoon circulation. Furthermore, another simulation (Pohlmann, 1985) revealed that nonlinear interactions between wind- and tidal induced currents only reach significant values in some of the shallow southern shelf regions where high velocities appear. In most of the other parts of the model area this is not the case.

3.1 Mean winter and summer circulation patterns

The figures (6- 7) show the results for the mean winter respectively summer circulation. The first and the third model layer will be presented, whereby the first layer has an extention from the surface to 10 m depth and the latter one from 20 m to 30 m depth.

During winter (figure 6a and 6b) an inflow from the Pacific into the South China Sea through the Luzon Strait is well distinguished in the upper layers. In the northern and central parts westerly respectively northerly currents predominate, turning southward when they reach the Vietnam coast. In this area the current intensifies to a narrow band of approximately 100 km. When it has left the Vietnam coast it is widening again and later leaves the South China Sea through the Java Sea.

As it would be expected from an inspection of the wind field for July the current pattern (figure 7a and 7b) totally reverses in summer. There is an inflow from the Java Sea into the South China Sea. In the southern part a recirculation cell has developed as well as a deep- reaching cyclonic gyre in the northern part. In the central parts easterly respectively northeasterly currents predominate. The south- going flow along the Vietnam coast is much weaker than in winter. Here the flow does not reverse with the change of the monsoon.

3.2 Mean winter and summer temperature and salinity distributions

For winter and summer, respectively the differences between the initial temperature distribution and the final dynamically balanced distribution (after about 15 days) are shown in figure 8a, 8b, 9a and 9b for the first and for the

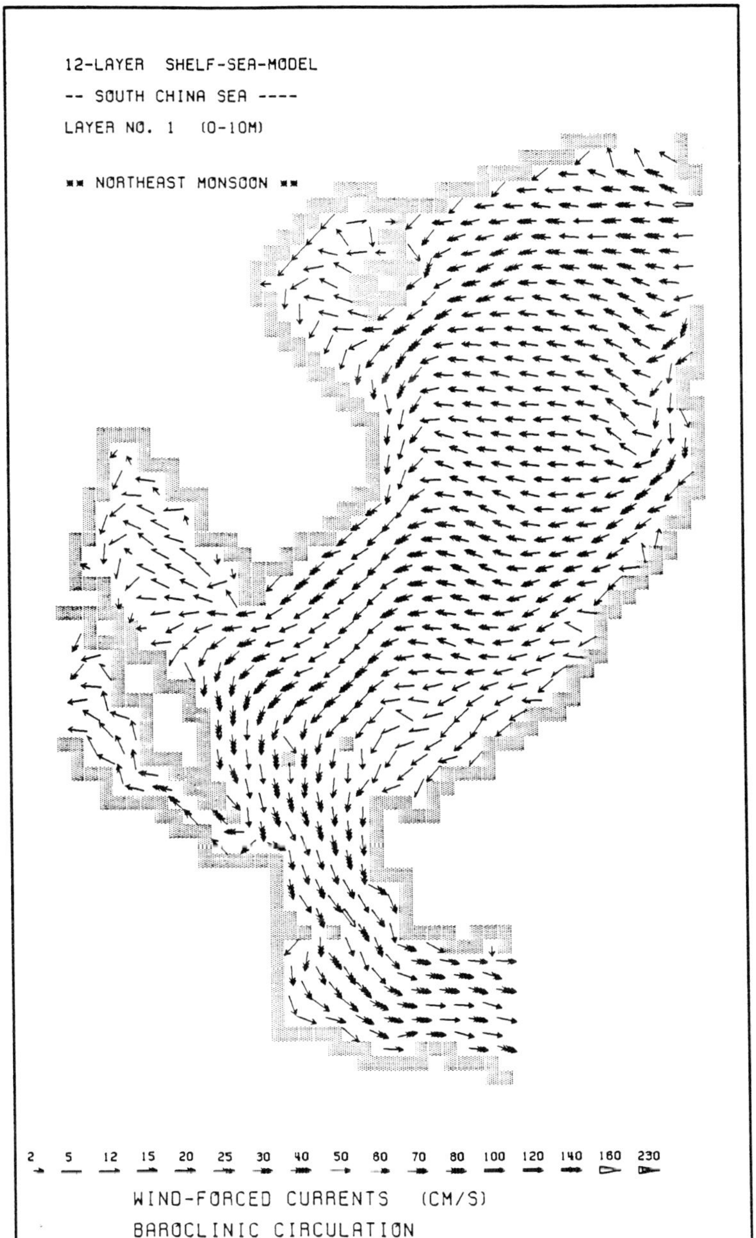

Fig. 6a. Mean winter circulation (0-10 m)

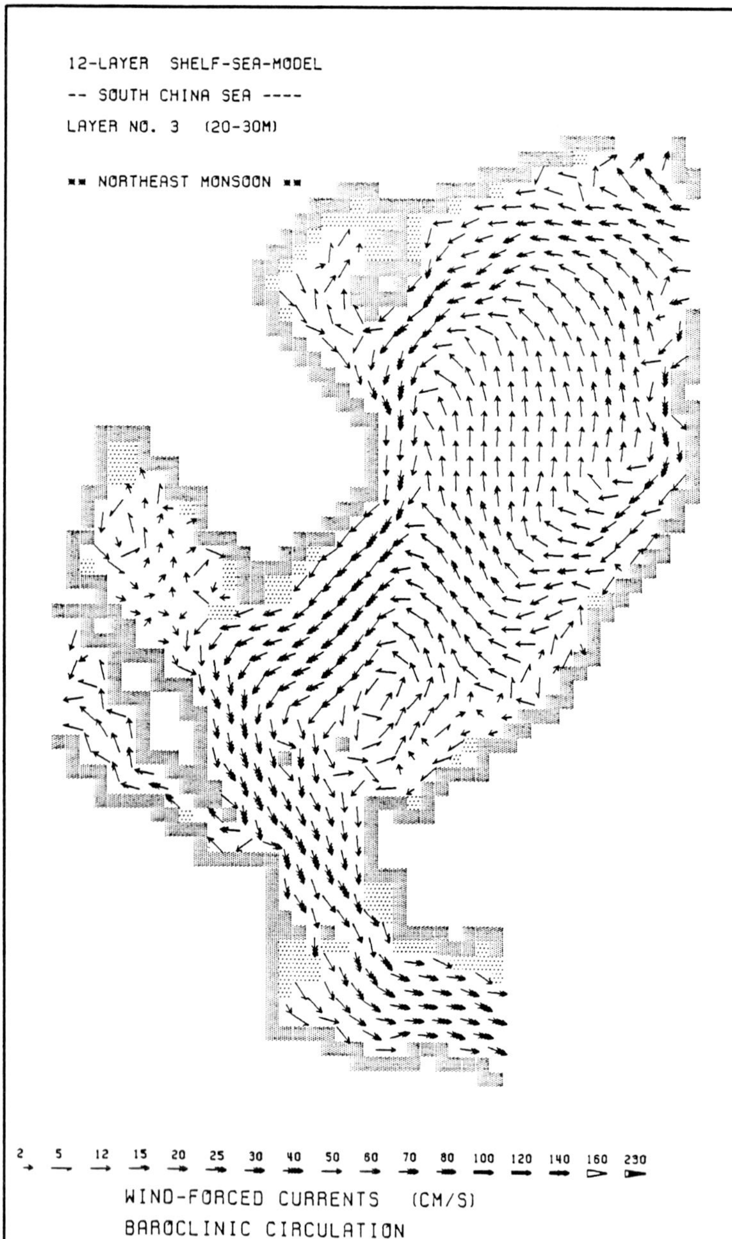

Fig. 6b. Mean winter circulation (20-30 m)

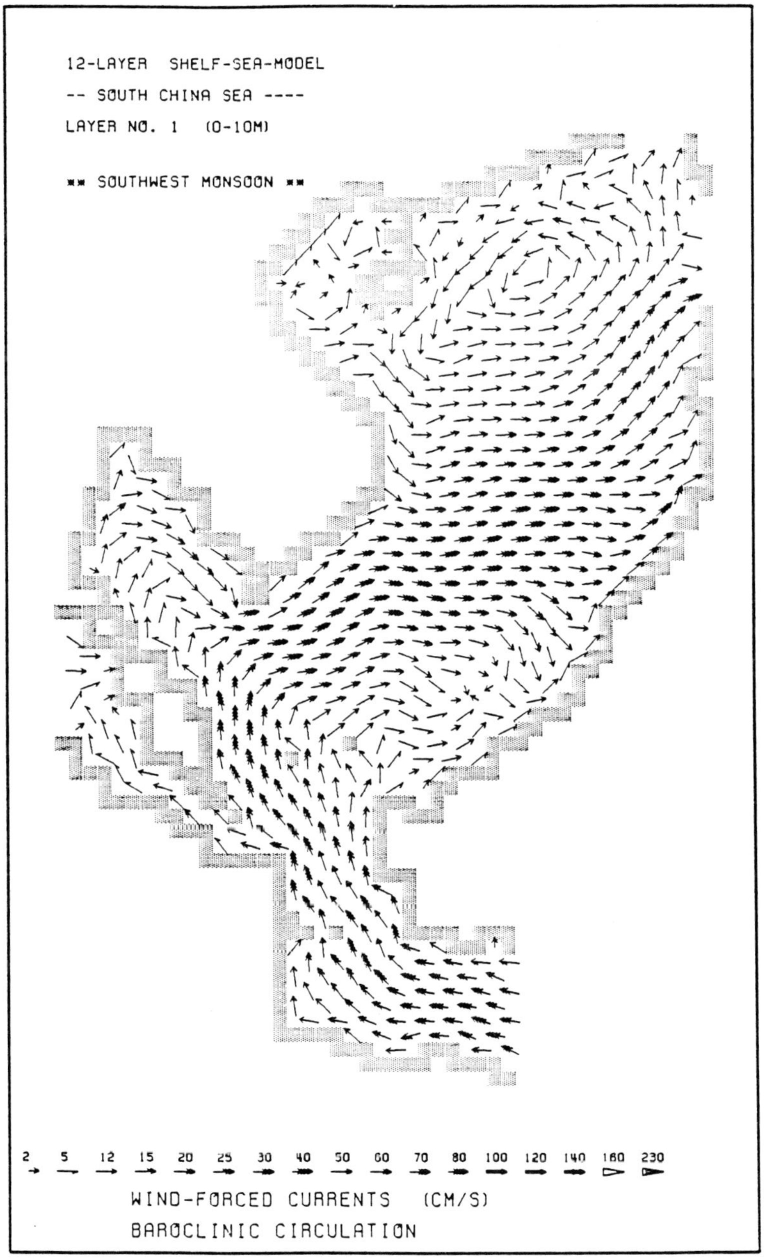

Fig. 7a. Mean summer circulation (0-10 m)

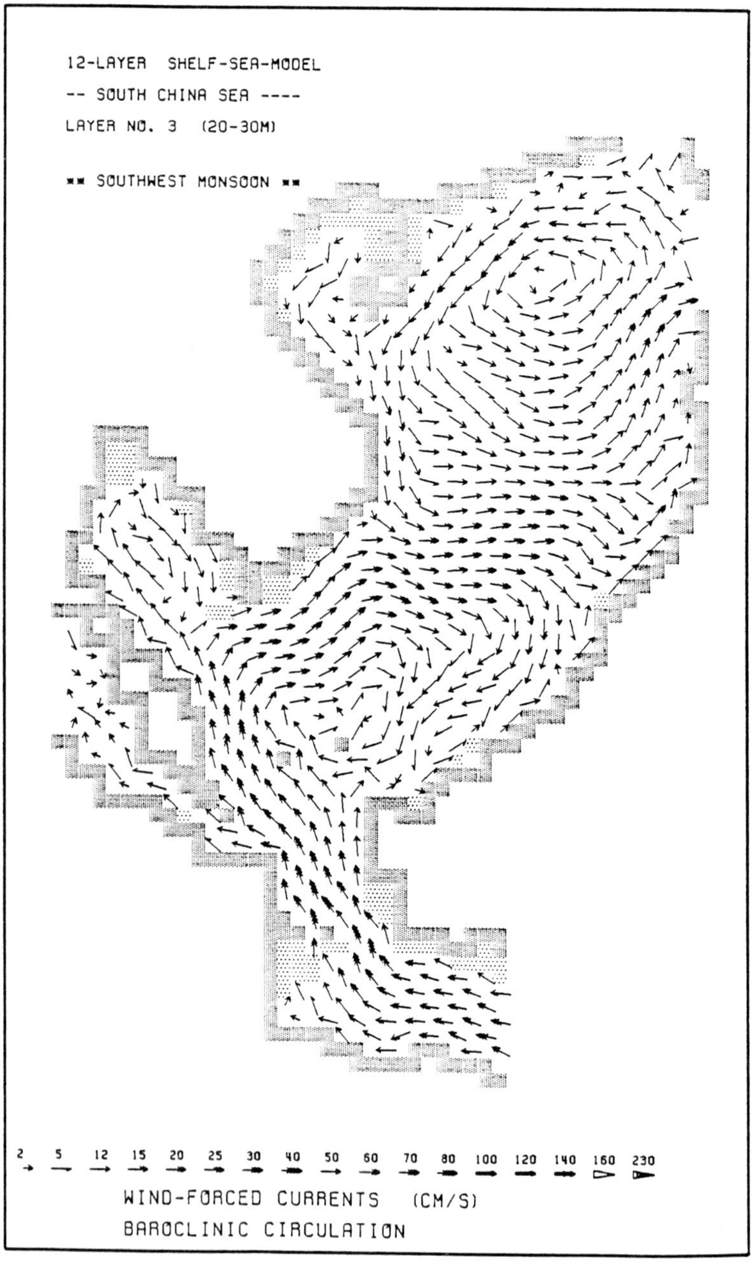

Fig. 7b. Mean summer circulation (20-30 m)

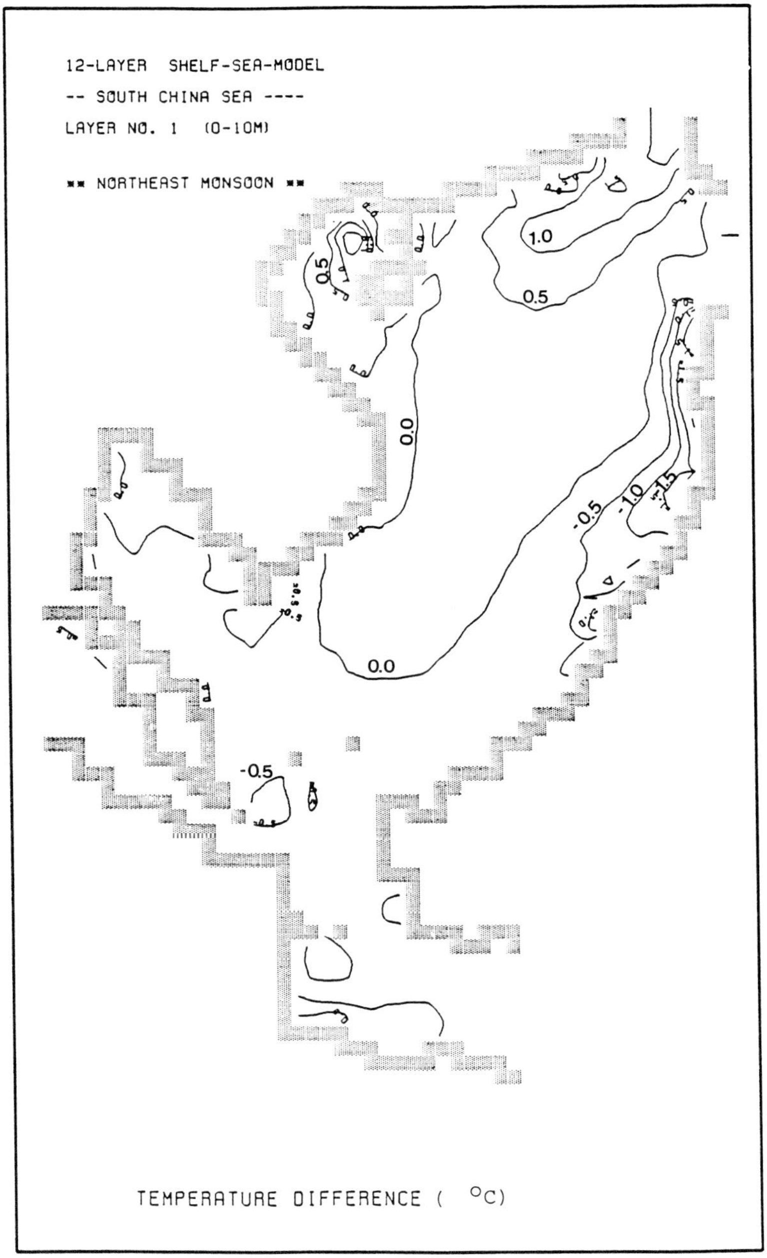

Fig. 8a. Difference between simulated and initial temperature distribution in winter (0-10 m)

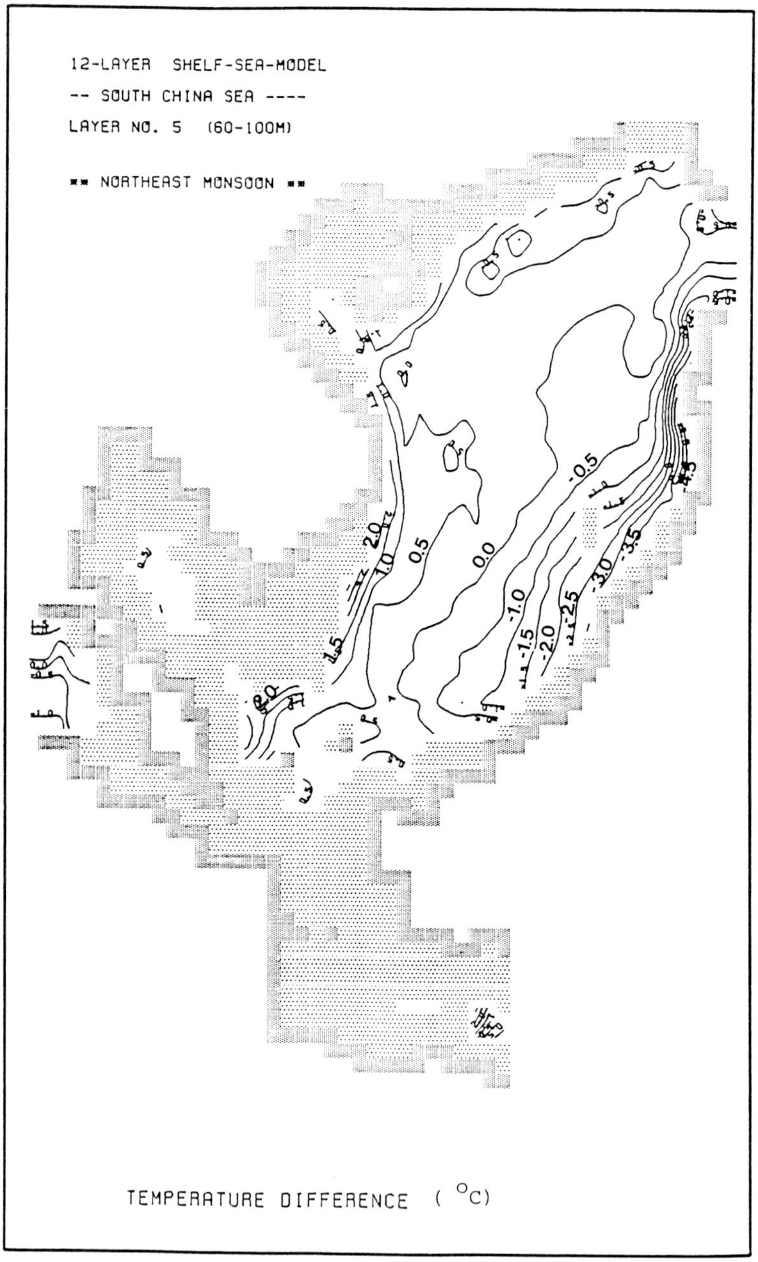

Fig. 8b. Difference between simulated and initial temperature distribution in winter (60-100 m)

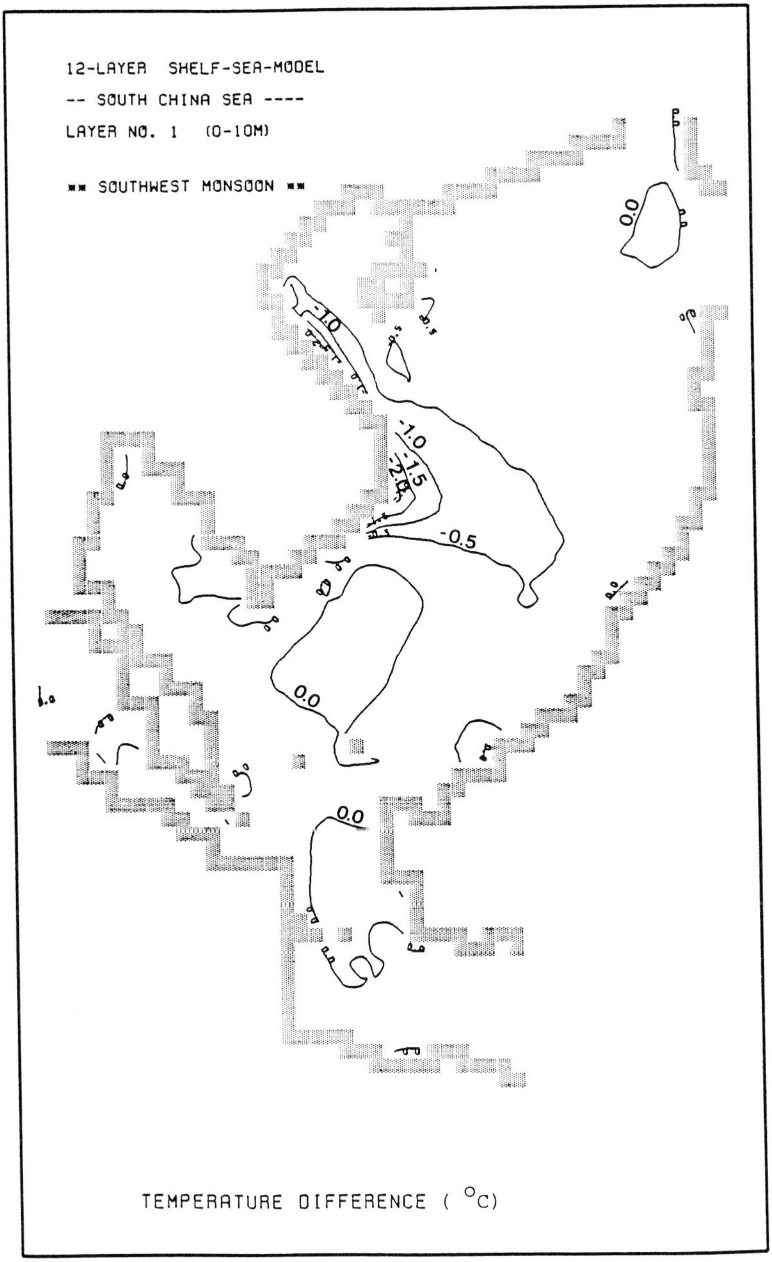

Fig. 9a. Difference between simulated and initial temperature distribution in summer (0-10 m)

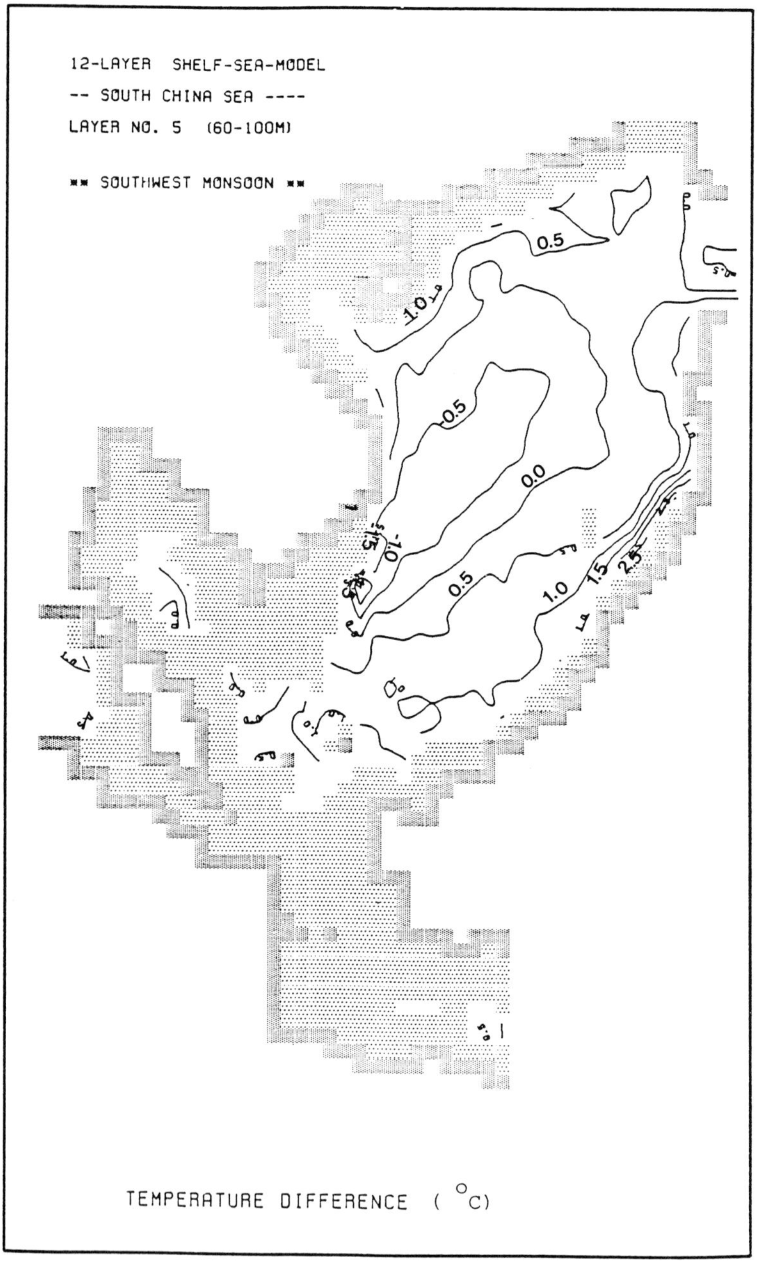

Fig. 9b. Difference between simulated and initial temperature distribution in summer (60-100 m)

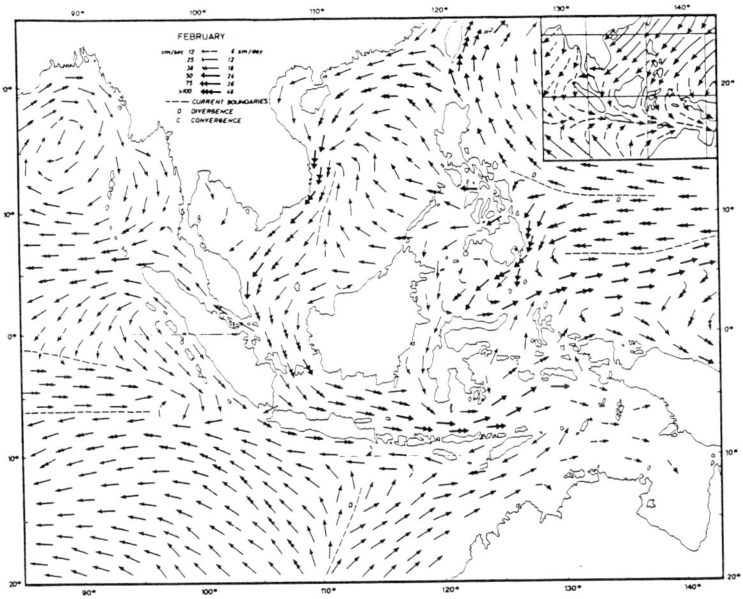

Fig. 10a. Surface currents in February (from ship drifts, Wyrtki, 1961)

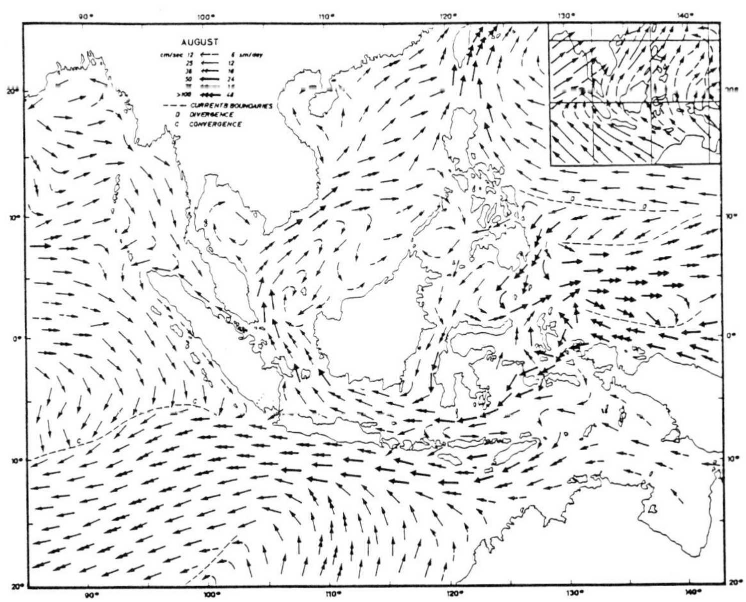

Fig. 10b. Surface currents in August (from ship drifts, Wyrtki, 1961)

fifth model layer. The fifth layer extends from 60 m to 100 m depth. During winter downwelling off the Vietnam and pronounced upwelling off the Philippine coast becomes evident (figure 8a and 8b). In summer the situation is reversed (figure 9a and 9b). In both seasons these phenomena are obviously much stronger in the fifth than in the first layer. So far deep- reaching up- and downwelling has not been observed in this region. Wyrtki´s results are based on surface measurements only, and therefore he was not able to detect this phenomenon. However, this result could be regarded as a stimulation for oceanographers measuring in the South China Sea.

4 VERIFICATION

A verification of the model results, as far as presently possible, was carried out by comparing observed and simulated transport rates through two sections. Wyrtki has calculated transport rates from observational data through the Java Sea and through a section which runs from the Vietnam coast in southwesterly direction into the South China Sea. Table 1. gives a comparision of these values calculated from observations with those transport rates calculated by the model. For the Java Sea, a shelf region, the simulations agree obviously well with the observations in both seasons. The simulated transports off the Vietnam coast are about 30 percent smaller than the observed values. This disagreement might result from the incomplete information about the location and the extent of this section, not given by Wyrtki.

TABLE 1

Comparison between observed and simulated transport rates ($x10^6$ m^3/s) through two sections.

	Off Vietnam (northeastwards positive)	Java Sea (eastwards positive)
Winter:		
Observation	-6.9	4.3
Simulation	-5.0	4.2
Summer:		
Observation	4.5	-3.2
Simulation	3.2	-3.0

The chart of the surface currents drawn by Wyrtki from ship drifts (figure 10a, 10b) allow another qualitative verification of the model results. The main features of the observed circulation are well reproduced by the simulations for both seasons. This is especially valid for the south- going current along the Vietnam coast. It is very strong in winter and forms a weak counter current in summer. The only conspicious devitation between observation and calculation is the simulated but not observed northern cyclonic gyre during the summer monsoon. By comparing the results of a baroclinic simulation against those of a barotropic simulation it is substantiated (Pohlmann, 1985) that this gyre is a baroclinic feature.

5 CONCLUSION

The simulations have shown that the model developed for the South China Sea satisfies the requirements which are formulated in chapter 2. As far as a comparision is possible the results of the model agree qualitatively and quantitatively well with Wyrtki´s compilation. The simulations have by all means qualitatively improved the knowledge about the vertical structure of the mean monsoon circulation. The possible existence of up- and downwelling at the Vietnam and Philippine coast has been pointed out for the first time. This result can be of highly economical interest for the fishery industry in this region.

The reaction, the sin- up, of the current system on the monsoon onset, presently investigated by several oceanographers (Lighthill, 1969, Cox, 1969, Quadfasel, 1982) could also be simulated with this model. However, time dependent meteorological data, momentary not available, must be supplied in a reasonable resolution in order to run such a model.

6 ACKNOWLEDGEMENTS

I am indebted to Prof. Dr. Backhaus for his valuable advice and assistance throughout this work. Also, I thank my colleage D. Hainbucher for making very helpful comments on the manuscript.

7 REFERENCES

Backhaus, J.O., 1983. A semi- implicit scheme for the shallow water equations for application to shelf sea modellinng. Continent. Shelf Res. 2: 243-254.

Backhaus, J.O., 1985. A Three- Dimensional Model for the Simulation of Shelf Sea Dynamics. Dt. hydrogr. Z. 38: 165-187.

Backhaus, J.O., Pohlmann, T., Hainbucher, D., 1986. Regional aspects of the circulation on the North European Shelf. ICES Report C.M. 1986/ C: 38 (unpubl.).

Backhaus, J.O. and Hainbucher, D., 1987. A finite difference general circulation model for shelf seas. This issue.

Boehlich, M., 1987. A three dimensional baroclinic model of the western Baltic. This issue.

Cox, M. D., 1970. A mathematical model of the Indian Ocean. Deep Sea Research 17.

Hainbucher, D. Backhaus, J.O., Pohlmann, T., 1986. Atlas of climatological and actual seasonal circulation patterns in the North Sea and adjacent shelf regions: 1969-1981. Technical Report No. 1, Institut für Meereskunde, Universität Hamburg.

Hellerman, S., 1968. An update estimate of the wind stress on the world ocean. Monthly weather review, 96.

Levitus, 1982. Climatological atlas of the world ocean. NOAA. Professional Paper No. 13. U.S. Goverment Printing office, Washington D.C.

Lighthill, M.J., 1969. Dynamical response of the Indian Ocean to onset of the Southwest Monsoon. Philos. Trans. R. Soc. 265.

Orlanski, I., 1976. A Simple Boundary Condition for unbounded Hyperbolic Flows. Journal of Computational Physics 21: 251-269.

Pohlmann, T., 1985. Simulation von Bewegungsvorgängen im Südchinesischen Meer. Diploma Thesis, Institut für Meereskunde, Universität Hamburg.

Quadfasel, D.R., Wilson, D., Leetmaa, A., 1982. Development of the flow field during onset of the Somali Currrent. Journal of Physical Oceanography. Vol 12, No. 12.

Wyrtki, K., 1961. Scientific results of marine investigations of the South China Sea and the Gulf of Thailand 1959-1961. Naga Report. Volume 2.

THE INFLUENCE OF BOUNDARY CONDITIONS ON THE CIRCULATION IN THE GREENLAND-NORWEGIAN SEA. A NUMERICAL INVESTIGATION.

S.LEGUTKE
Institut für Meereskunde, Universität Hamburg, Troplowitzstr. 7, 2000 Hamburg, F.R.G.

ABSTRACT

The dynamics of the Greenland-Norwegian Sea are investigated, using a numerical model extending from the Greenland-Scotland Ridge to the Fram Strait and including part of the Barents Shelf. The model is based on a finite difference discretisation of the primitive equations with 12 levels and a horizontal grid size of about 20 km. It is driven by windstress and buoyancy fluxes at the surface; at open boundaries volume, salt, and heat fluxes are specified. Quasi diagnostic computations have been performed using climatological seasonal mean data at the boundaries and as initial stratification. The response of the system to various situations at the inflow boundaries is investigated. The current fields produced are in good agreement with existing observations. It is found, that the bottom pressure torque is the dominant term in the vertically integrated vorticity equation almost everywhere. It causes the deep and vertically integrated flow to separate into several gyres.

1 INTRODUCTION

The Greenland-Norwegian Sea (GNS), connecting the Arctic Ocean with the North Atlantic, plays a key role in climate processes on the Northern Hemisphere. By far the major part of the heat transfer between the Arctic Ocean and its neighbouring seas occurs through the Fram Strait (Aagaard and Greisman 1975). On the other hand, the deep and bottom water formed in the Greenland Sea is the major source of bottom water in the North Atlantic (Swift 1984).

The rate of these processes is influenced by various factors. Among them are the circulation and characteristics of the water masses involved. The prevailing cyclonic wind stress distribution drives a northward flow of warm saline Atlantic water entering mainly through the Faeroe-Shetland Channel. Part of the Atlantic water leaves the GNS through the Fram Strait, thus providing a heat source for the Arctic Basin. Part of it returns to the south closing, together with the Polar water of the Eastgreenland Current, a cyclonic circulation in this region. The associated doming of isopycnals in the Greenland Basin, together with winter cooling, results in low stability water masses and bottom water formation.

Two models of the GNS have been published so far. Creegan (1976) used a model with two layers of constant densities of the region deeper than 500 m between the Fram Strait and the Greenland-Scotland Ridge to investigate the influence of wind stress distribution and inflow through the Faeroe-Shetland Channel. Obviously such a model has some shortcomings such as the exclusion of shelf areas and of the Eastgreenland Current. The 2-layer structure prohibits a proper consideration of the topographic and thermohaline influences on the circulation.

A three dimensional model of the Arctic Ocean and the GNS has been presented by Semtner (1976). Using mean forcing functions he obtained a cyclonic stream function covering the whole GNS. But with the relatively coarse horizontal grid-size of 110 km in a region of strong topographic variations the smaller scale features (see for example figures 2,3,and 4 of Metcalf (1960)) of the current system were not resolved. It has been pointed out by various authors, that the joint effect of topography and baroclinicity has a large influence on the flow field (Sarkisyan and Ivanov 1971, Holland 1973, Holland and Hirschman 1972).

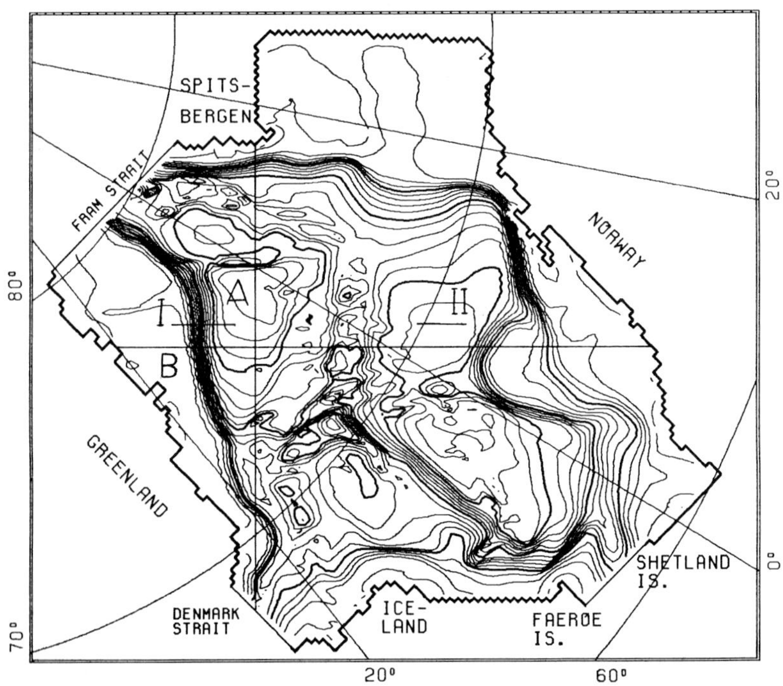

Fig.1. Model bathymetry. Heavy contour interval is 1000 m. Open passages are indicated by thin lines. Sections A,B,I,II are referred to in the text.

In the present paper a model is described, that was developed in an attempt to take into account the influence of topographic variations down to scales of 100 km. The model presented is designed to run long term prognostic calculations with variable boundary conditions both at the lateral open boundaries and at the surface. In the experiments described it is initialized with climatological seasonal mean hydrographic and wind stress data. Prognostic calculations are run for one month in order to establish a circulation field consistent with the input data. The results of these quasi diagnostic calculations will be compared with observed currents. In addition the response to varying inflow situations and anomalous wind stress is discussed.

2 DESCRIPTION OF THE MODEL

2.1 The model equations

A model similar to that described below has already been used in a study of equatorial dynamics (Latif et al. 1985). It consists of a finite difference discretisation of the primitive equations using the Boussinesq and hydrostatic approximations :

$$\underline{v}_t+\underline{v}\cdot\nabla_h\underline{v}+f(\underline{k}\times\underline{v})=-1/\rho_0\,\nabla_h p+F(\underline{v})+(A_v\underline{v}_z)_z \quad (1)$$

$$g\cdot\rho=p_z \quad (2)$$

$$\nabla_h\underline{v}+w_z=0 \quad (3)$$

$$(T,S)_t+\underline{v}\cdot\nabla_h(T,S)+w\cdot(T,S)_z=Q_{(T,S)} \quad (4)$$

The notation is as usual: $\underline{v}$ denotes the horizontal velocity vector, p is pressure, ρ density, ρ_0 a constant reference density, f the Coriolis parameter, $\underline{k}$ an unit vector upwards, g the gravitational constant, and ∇_h the horizontal gradient operator. $F(\underline{v})$ represents a parameterization of the eddy viscosity effects.

An E-grid (Arakawa and Lamb 1977) in spherical coordinates λ,φ,z is used with a zonal grid size of .25° lat. The choice of coordinates is motivated by future plans to embed the GNS-model in a model of the Atlantic and Arctic Oceans (H.Friedrich, in preparation): The poles are placed on South America and South East Asia to avoid singularities inside the world ocean. In particular, this coordinate system reduces the grid deformation due to the convergence of meridians within the area of interest. In what follows the terms zonal and meridional will be used to denote the direction of increasing λ and φ in model coordinates, while north and south is used for geographical coordinate directions.

The eddy viscosity parameterization is a simple Laplacian diffusion $F(\underline{v})=A_h\nabla_h^2(\underline{U})/DZ$ with $A_h=10^3 m^2/sec$. $\underline{U}$ denotes the velocity vector integrated over the layer depth DZ. This formulation ensures conservation of momentum away from lateral closed boundaries. At the sea surface source terms $Q_{(T,S)}$ resulting from a Newtonian coupling of observed and computed surface fields with a relaxation time scale of 16 days are specified. No explicit diffusion is needed in the prognostic equation for temperature and salt since an upwind scheme is used. The sea surface elevation Z is computed from the linearized kinematic boundary condition. The numerical scheme effectively damps the external gravity mode and inertial oscillations. The time step is then restricted by internal gravity waves. A value of 3 h is used.

The level depths of the model are 7,21,37,57,85,126,209,341,551,851,1501,and 2701 m. Since the lowest box of each column has a variable depth, the resolution of the bathymetric field does not depend on the number of boxes. In this way the levels can be concentrated in the upper 800 m where the hydrographic fields are variable. Below the Atlantic water, the water is quite homogenous and a larger spacing can be chosen.

2.2 Boundary conditions

At lateral closed boundaries no-flux and no-slip conditions are used. The system is forced at the surface by a wind stress field $\underline{T}=(Tx,Ty)=\rho_0 \cdot A_V \cdot \underline{v}_z$ and by the buoyancy flux described above.

At $z=-H(\lambda,\varphi)$, a quadratic stress law is applied ($\rho_0 \cdot A_V \underline{v}_z = e \cdot /\underline{v}/ \cdot \underline{v} \cdot Da$). The turning matrix Da and the drag coeffient e have been specified according to Weatherly (1972). With these values 2-5% of the energy dissipation is due to bottom friction.

At open lateral boundaries gradients of Z are needed to compute the barotropic pressure term. These are derived by geostrophy from specified velocity profiles, whose barotropic parts are taken from observation while the vertical shear is computed from hydrographic sections using the thermal wind relation. The same sections are used to compute the heat and salt flux by advection into the basin. Zero advective flux through the bottom and closed boundaries is ensured by the upwind formulation of the advection terms in (4).

2.3 The input data

The model domain extends from the Greenland-Scotland Ridge to the Fram Strait including the Barents Shelf west of 30°E (Fig.1).

The bathymetric field has been digitized from the U.S.Naval Research Labratory Charts, Washington D.C. (1980). It shows large gradients up to 10^{-2} anticipating strong topographic influence on the circulation pattern.

Open boundaries are assumed in the Denmark Strait, the Fram Strait and the Faeroe-Shetland Channel. No exchange with neighbouring seas is allowed for over the Iceland-Faeroe Ridge, where the transports are extremely variable but low in the average (Meincke 1983). They can be neglected in longer term calculations. On the Barents Shelf transports are low and estimates unreliable (Aagaard and Greisman 1975). For the passage between Norway and Scotland model computations indicate a transport of about 1 SV being only 1/7 of the simulated transport through the Faeroe-Shetland Channel (Backhaus et al. 1985). These boundaries are treated as closed too.

As an initial stratification and for the boundary conditions the climatological seasonal and annual mean hydrographic data of the atlas published by Levitus (1982) have been used. At the open boundaries a barotropic velocity taken from observation (Aagaard et al. 1973, Hanzlick 1984) or other model simulations (Backhaus et al. 1985) is added. The total transport through the open boundaries is 7 SV in the Westspitsbergen- and Eastgreenland Current , 5.3 SV inflow through the Faeroe Shetland Channel , an inflow of 0.6 SV of Atlantic water west of Iceland and 5.9 SV outflow in the Denmark Strait.

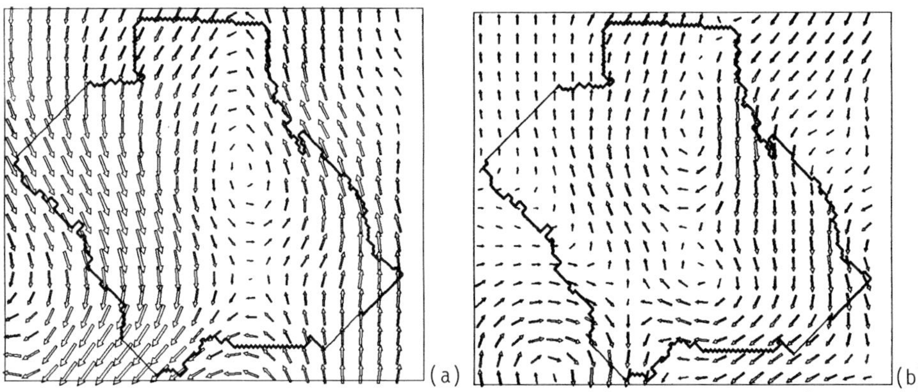

Fig.2. (a)Climatological annual mean wind stress. Maximum : 0.14 N/m^2.
(b)Monthly mean wind stress for July 1980. Maximum : 0.07 N/m^2.

In the present computations seasonal climatological mean wind stress fields computed from daily mean wind stress data for the period 1955-1982 on a 1° grid (Backhaus et al. 1985) are used. The data show a pronounced seasonal cycle with strong positive wind stress curl for the summer to winter months. During these months the wind stress is similar to the annual mean (Fig.2a). In spring it is reduced by one order of magnitude and in some years even reverses. An example is the mean wind stress of July 1980 with the largest negative anomaly of wind stress curl for the whole period (Fig.2b). Additionally, a number of runs have been performed with other data sets to test the response to inflow situations varying in the range of observation. These runs are described below.

3 THE SIMULATION RESULTS

Anderson and Gill (1975) have shown, that the response of a stratified flat ocean to a change in wind stress can be described in terms of planetary Rossby waves which are generated at the coasts in order to satisfy boundary conditions.

The time to establish a steady state at an interior point is the time required by the long wave to travel from the western boundary to that point. Taking into account the high latitude and horizontal extent of the basin in question this results in a time scale of some months for the barotropic mode and many years for the baroclinic mode. Topographic features, which divide the region into several sub-basins, add basin mode time scales (Anderson and Killworth 1977).

Previous work has shown that much of the information on the long term mean forcing is stored in the mean density field. For example Holland and Hirschman (1972) and Backhaus and Maier-Reimer (1983) have shown, that switching off the wind field results in only minor changes of the circulation pattern in diagnostic calculations. Thus an investigation of seasonal variations should extend over several cycles, using fully varying forcing functions, thereby requiring a great amount of computer time. Trying to spin up a representative monthly circulation with monthly mean wind stress data from a state of rest will lead to unrealistic results (Creegan 1976).

In order to investigate the possible influence of varying boundary conditions on the general circulation the following experiments have been made. Except for one simulation with a homogeneous model the initial stratification is always taken from the climatological atlas (Levitus 1982), i.e. annual and seasonal climatological mean. The model is then forced with the corresponding wind stress and buoyancy flux at the surface and velocity profiles at the open boundaries computed as described above. It is then allowed to adjust to these forcing functions in a prognostic calculation for one month. This is about the time needed to establish the continental shelf currents that are initiated at the open boundaries, as has been verified by direct comparison of the results with different inflow conditions. The problem remains whether the density data used are suitable for these diagnostic calculations. Climatological data tend to be smooth by averaging moving fronts.

Anyhow, once a stratification has been accepted, the response to variable forcing functions lasting for one month can be tested.

TABLE 1 Initial and boundary conditions of the experiments.

	Stratification	Wind	Open Boundaries
SOW5CL	homogenous	a.m.,climat.	closed
S5W5CL	a.m.,climat.	"	"
S5W5OP	"	"	open,climat.
S3JUOP	summer, "	July 1980	"
S5W3SY	a.m., "	summer,climat.	open,synoptic
S1W1OP	winter,climat.	winter,climat.	open,climat.
S2W2OP	spring, "	spring, "	"
S3W3OP	summer, "	summer, "	"
S4W4OP	fall, "	fall, "	"

Other test conditions except those derived from the climatological mean are : closed lateral boundaries; open boundary conditions derived from synoptic hydrographic summer sections with larger velocity shear and (keeping the total transport constant) larger surface elevation gradients; wind stress of July 1980 with a large negative anomaly. These experiments are listed in Table 1. Except for the respective changes mentioned they all use the annual mean conditions and can be compared with this case.

3.1 The annual mean case

The general picture of the surface circulation in the GNS has been known since long (Metcalf 1960) and most of its features are reproduced by the model.

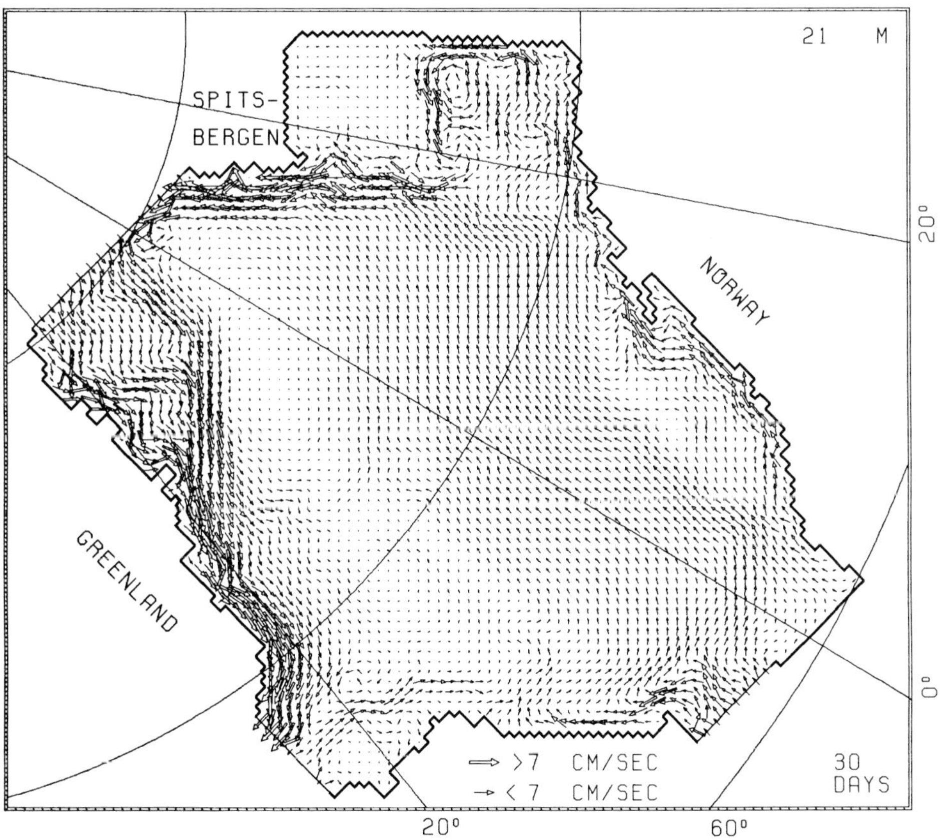

Fig.3. Current field at 21 m for the annual mean case. Maximum : 32 cm/sec.

Fig. 3 and 4 show the horizontal circulation pattern at 21 m (1/2 of the grid points are shown) below the wind driven surface circulation and at 1500 m in the deep water. A broad northward drift of Atlantic water appears in the eastern half of the basin and turns to the east when it encounters the Jan-Mayen Mohn Ridge. Velocities drop from 3 cm/sec at the surface to 0.5 cm/sec at 500 m. Helland-Hansen and Nansen (1909) have reported the advective time scale of temperature anomalies from the Sognefjord to the Barents Shelf to be 2 years. This corresponds to a velocity of 2 cm/sec. The same drift velocity of 2-3 cm/sec has been observed by Dickson and Blindheim (1984) from measurements of the large salinity minimum in the Faeroe-Shetland region in 1976 and near Bear Island in 1978/79.

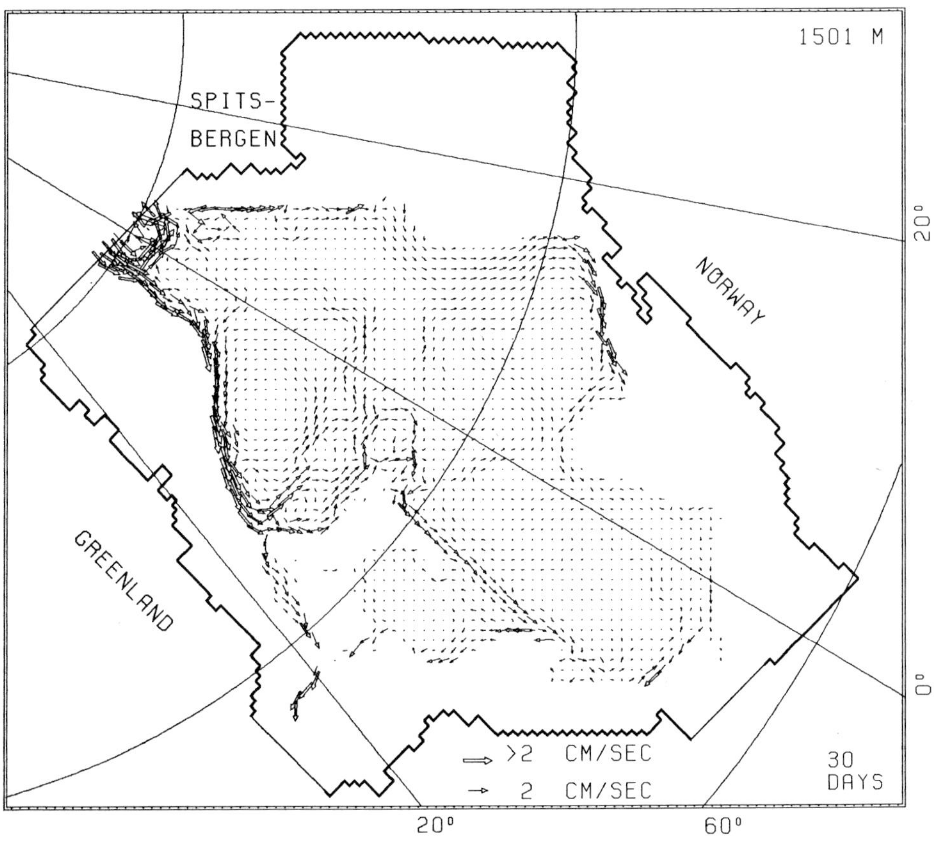

Fig.4. Current field at 1501 m for the annual mean case.

At the Barents Shelf break the greater part again turns to the north forming the Westspitsbergen Current (WSC) while one branch flows onto the Barents Shelf. A meridional section of zonal velocities (Fig.5) shows the vertical structure of the WSC. Current speeds drop from 20-30 cm/sec at the surface to 12 cm/sec in 550 m depth. This compares well with the vertical velocity shear reported by Hanzlick (1984), derived from year long current measurements in 1976-78. The total simulated transport is 6 SV, wich is comparable with the mean value of 5.6 SV of Hanzlick (1984) and 7 SV of Aagaard et al. (1973).

Coastal currents have developed at the Norwegian and the Greenland coast. Both have speeds up to 10 cm/sec. It should be kept in mind that most of the Greenland Shelf is covered by ice all year long and no allowance is made for its influence on the surface boundary conditions in the model.

On the Greenland side polar water flows southward along the shelf break in the East Greenland Current (EGC). One branch of it turns to the east at the Jan-Mayen Ridge, the Jan-Mayen Current (JMC) with velocities of 2-3 cm/sec down to the bottom (Fig.5) but most of it leaves the basin through the Denmark Strait. Thus the circulation is divided into two large gyres with their center in the Greenland and Icelandic basins. The surface velocities of the EGC over the shelf break increases from 10 cm/sec at 79° N to about 20 cm/sec at 73° N. This might be compared with direct current measurements from ice islands and drifting buoys. Reported velocities are 4 to 12 cm/sec at 80° N and increase to 14 to 24 cm/sec at 70° N (Einarsson 1972, Aagaard and Coachman 1968). The simulated transport is 7.6 SV. In both currents the transport values are mainly influenced by the downstream inflow conditions.

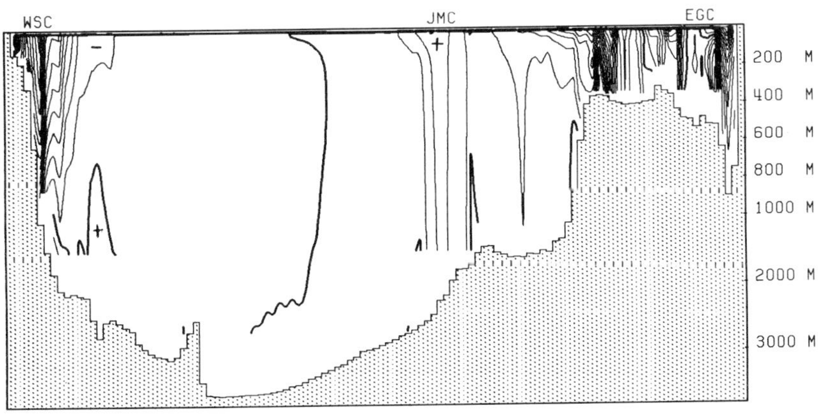

Fig.5. Vertical section of zonal velocity. Section A. The positions of the sections are given in Fig.1. Contour interval is 1 cm/sec.

The circulation of the deep water masses at 1500 m beneath the Atlantic layer is divided into several larger gyres. In the Greenland Basin the rotation is cyclonic as in the upper layers, but in the north-eastern part of the Norwegian basin it has reversed. This has already been reported by Eggvin (1961). Velocities are less than 2 cm/sec except at steep topographic features. A vertical section of meridional velocities extending from the Greenland coast at 78° N to the Norwegian coast at 65° N shows the position of the EGC, the Norwegian Current (NSC), and the coastal current (NCC) (Fig.6).

The formation of gyres in the deep flow can be discussed by means of the vertically integrated vorticity equation. It gives a balance between the time derivative of relative vorticity Z, advection of relative vorticity A, advection of planetary vorticity B, dissipation of relative vorticity V, bottom pressure torque P, dissipation of vorticity by bottom friction R, and wind stress curl W (Holland 1973). Two sections of the 5 largest terms A,B,V,P, and W are shown in

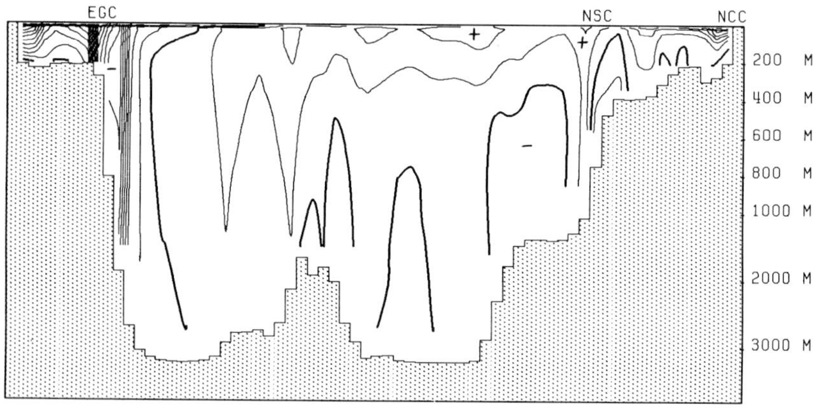

Fig.6. Same as Fig.5 but for section B.

Fig.7. Section I, perpendicular to the Greenland shelf break, shows a balance between the bottom pressure torque and viscosity effect modified by advection of planetary and relative vorticity. This picture is typical for regions with large topographic gradients, i.e. almost everywhere. Even with the much smaller slope of the inner part of the Lofoten basin (section II) the bottom pressure torque

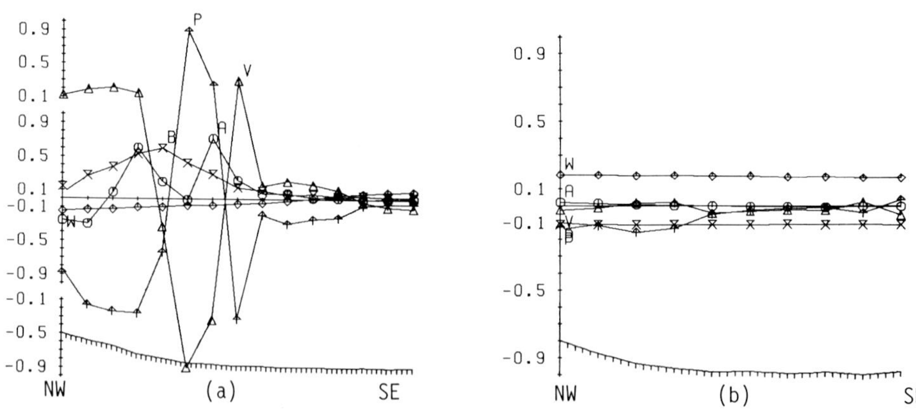

Fig.7. Main terms of the vertically integrated vorticity equation.(a) Section I (b) Section II. The positions are given in Fig.1. Symbols are explained in the text.

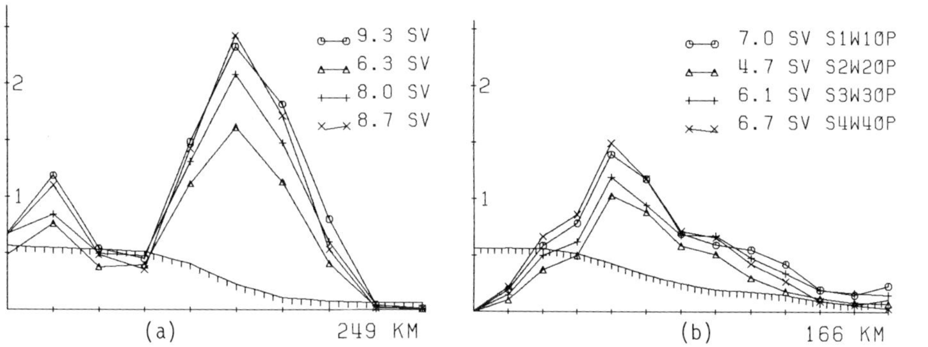

Fig.8. Transport in the EGC (a) and WSC (b) for seasonal experiments of Table 1. The coordinate units are $10^6 m^3/sec$ (velocity integrated over the area represented by each column).

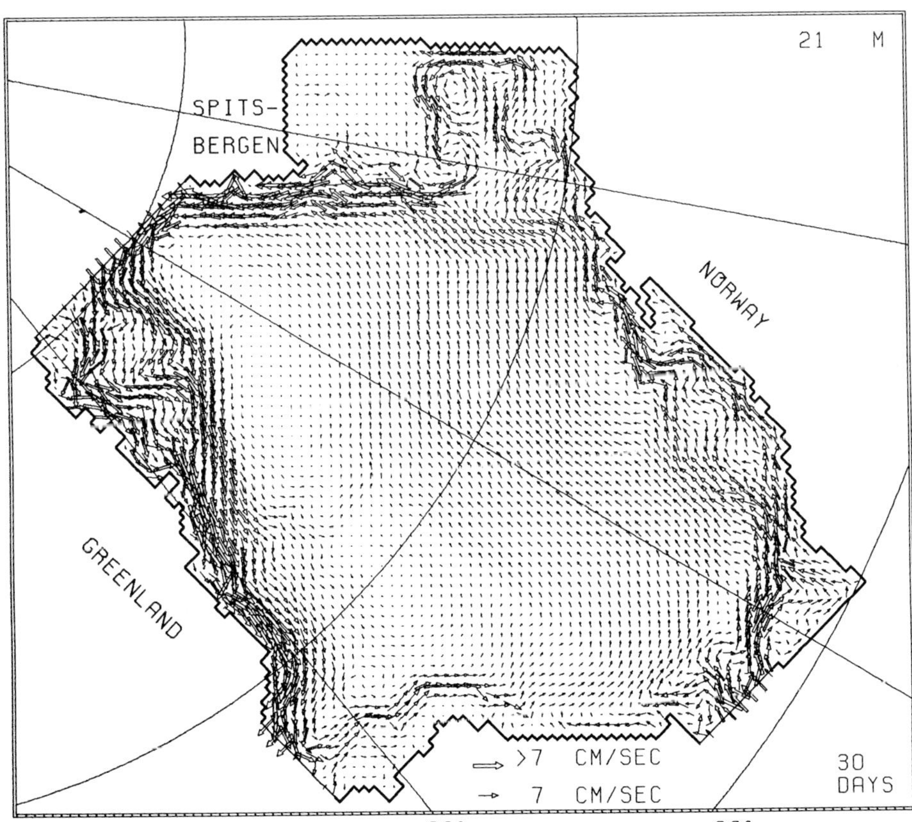

Fig.9. Circulation field at 21m depth for experiment S5W3SY. Maximum : 39cm/sec.

is not negligible. Only in the southern part of the section, where the bottom slope is practically zero, an approximate Sverdrup balance holds (Fig.7b). The relative importance of the planetary and topographic terms is β/f - H_y/H. With the small value of β and the large topographic gradients a balance of the above mentioned kind might be expected as long as the response to variable forcing is barotropic. Willebrand et al.(1980) have suggested that the response to wind varying at a scale larger than 100 km is indeed to a large extent barotropic. A longer term run of two years shows, that this kind of balance remains valid, though to a somewhat lesser extent.

3.2 The seasonal runs

The seasonal hydrographic and wind stress data give a similar circulation pattern. The main difference is in the strength of the currents. The transports on a section through the WSC and EGC are shown in Fig.8. No significant difference in the vertical velocity shear can be detected.

3.3 Influence of prescribed inflow

Except for the homogenous case, the stratification of the following experiments is the annual mean. The homogenous run shows low transports in the WSC and EGC (Fig.10), compared with experiment S5W5CL, in agreement with the result of Holland and Hirschman (1972) concerning the influence of baroclinicity and topography on the transport values. Fig.9 shows the 2nd level circulation pattern for experiment S5W3SY. The main difference lies in the strength of the continental shelf current emanating from the open boundaries and travelling along the shelf break around the basin. This is true for the closed boundary case too, which results in reduced currents. It can be seen that the variance induced by the conditions at the open boundaries, which lie within the range of observation, is at least as large as that induced by the seasonal wind stress pattern and hydrography (Fig.8 and 10).

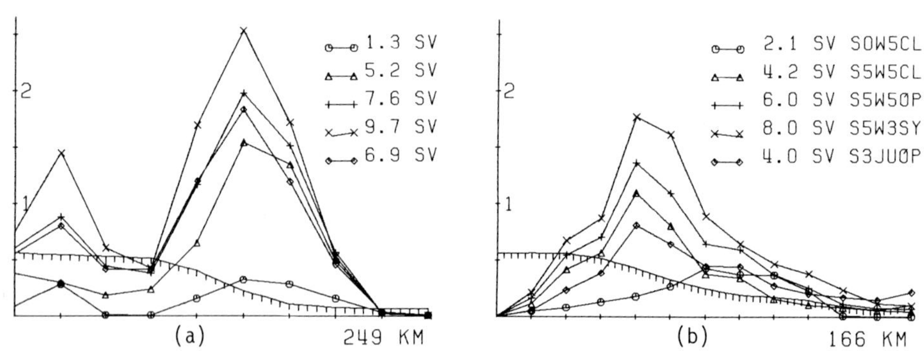

Fig.10. Same as Fig.8 but for the first 5 experiments of Table 1.

The influence of a one month anomalous wind stress is shown in Fig.11. The largest difference as compared with the annual mean case is found on the Norwegian shelf where the direction of the coastal current has reversed. The southward undercurrent off the Lofoten Islands extends now to the surface.

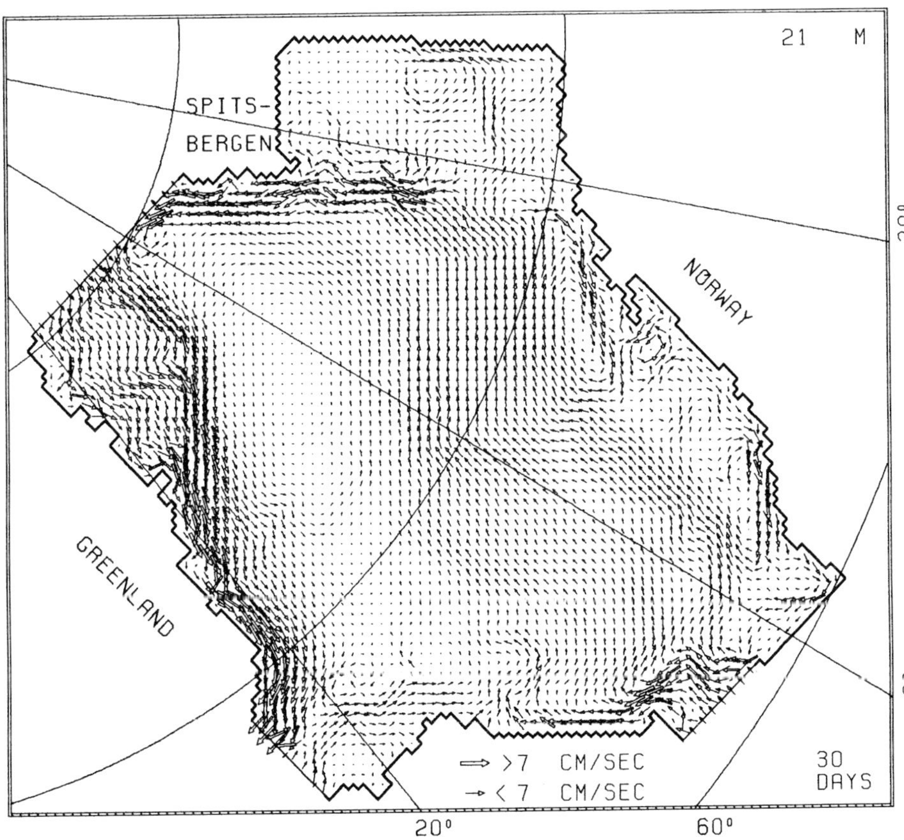

Fig.11. Same as Fig.10 for experiment S3JUOP. The maximum value is 24 cm/sec.

4 CONCLUSIONS

The results of the experiments described show a strong influence of topography on the general circulation. This can be seen by the circulation fields of a simulation of one month duration starting with observed density data. A calculation of the terms of the vertically integrated vorticity equation shows the dominance of the bottom pressure torque in regions of varying topography, i.e. almost everywhere. The circulation fields are in good agreement with observations.

The influence of various surface forcing and inflow situations at the passages connecting the GNS with the Artic and North Atlantic Oceans has been investigated. The changes in the transport values of the EGC and the WSC induced by inflow situations varying in the range of observations are comparable to those caused by seasonal mean and anomalous windstress conditions lasting for one month.

5 REFERENCES

Aagaard, K. and Coachman, L.K., 1968. The East Greenland Current North of Denmark Strait:I&II. Arctic, 21 : 181-200,267-290.

Aagaard, K. and Greisman, P., 1975. Towards new mass and heat budgets for the Arctic Ocean. J. of Geophys. Res., 80(27) : 3821-3827.

Aagaard, K., Darnell, C. and Greisman, P., 1973. Year-long current measurements in the Greenland-Spitsbergen Passage. Deep-Sea Res., 20 : 743-746.

Anderson, D.L.T. and Gill, A.E., 1975. Spin-up of a stratified ocean, with applications to upwelling. Deep-Sea Res., 22 : 583-596.

Anderson, D.L.T. and Killworth, P.D., 1977. Spin-up of a stratified ocean, with topography. Deep-Sea Res., 24 : 709-732.

Arakawa, A. and Lamb, V.R., 1977. Computational design of the basic dynamical processes of the UCLA general circulation model. Methods in Computational Physics, 17 : 173-265.

Backhaus, J.O. and Maier-Reimer, E., 1983. On seasonal circulation patterns in the North Sea. In: North Sea Dynamics, Sündermann,Lenz (Editors), Springer.

Backhaus J., Hainbucher D., Quadfasel, D. and Bartsch, J., 1985. North Sea Circulation anomalies in response to varying atmospheric forcing. I.C.E.S., C.M./C: 29, Hydrography Committee.

Backhaus, J., Bartsch, J., Quadfasel, D. and Gudall, J., 1985. Atlas of monthly surface fields of air pressure, wind stress and wind stress curl over the North Eastern Atlantic Ocean: 1955-1982. Technical Report 3-85, Inst. of Oceanography, University of Hamburg, FRG.

Carmack, E. and Aagaard, K., 1973. On the deep water of the Greenland Sea. Deep-Sea Res., 20 : 687-715.

Creegan, A., 1976. A numerical investigation of the circulation in the Norwegian Sea. Tellus, 28(5) : 451-459.

Dickson, H.D. and Blindheim, J., 1984. On the abnormal hydrographic conditions in the European Arctic during the 1970s. Rapp. P.-v. Reun. Int. Explor. Mer, 185 : 201-213.

Eggvin, J., 1960. Some results of the Norwegian hydrographical investigation in the Norwegian Sea during the IGY. Rapp. et Proc.-Verb.149, Cons. Internat. Explor. de la Mer.

Einarsson, T., 1972. Sea currents, ice drift, and ice composition in the East Greenland Current. In: Sea Ice, Karlsson (Editor), Nat. Res. Counc. of Iceland, Reykjavik, 23-32.

Hanzlick, D.J., 1983. The West Spitsbergen Currents: Transport, Forcing, and Variability. Ph.D.Thesis, University of Washington.

Helland-Hansen, B. and Nansen, F., 1909. The Norwegian Sea. Its physical oceanography based upon the Norwegian researches 1900-1904. Rept. Norw. Fish. Mar. Invest. 2(1,2), 390pp.

Holland, W.R., 1973. Baroclinic and topographic influences on the transport in western boundary currents. Geophys. Fluid Dynamics, 4 : 187-210.

Holland, W.R. and Hirschman, A.D., 1972. A numerical calculation of the circulation in the North Atlantic Ocean. J. of Phys. Oceanogr., 2 : 336-354.

Latif, M., Maier-Reimer, E. and Olbers, D.J., 1985. Climate variability studies with a primitive equation model of the Equatorial Pacific. In: J.C.J.Nihoul (Editor),Coupled ocean-atmosphere models.pp 63-81.

LeBlond, P.H. and Mysak, L.A., 1978. Waves in the Ocean. Elsevier Scientific Publ.Co.

Levitus, S., 1982. Climatological atlas of the world ocean. NOAA, Prof.Paper 13.

Meincke, J., 1983. The modern current regime across the Greenland-Scotland Ridge. In : Structure and development of the Greenland-Scotland Ridge, Bott, Saxov, Talwani, and Thiede (Editors),pp. 637-649.

Metcalf, W.G., 1960. A note on water movement in the Greenland-Norwegian Sea. Deep-Sea Res., 7 : 190-200.

Sarkisyan, A.S. and Ivanov, V.F., 1971. Joint effect of baroclinicity and bottom relief as an important factor in the dynamics of sea currents. Izv., Atmospheric and Oceanic Physics, 7(2) : 173-188.

Semtner, A.J., 1976. A numerical simulation of the Artic Ocean circulation. J. of Phys. Oceanogr., 6 : 409-425.

Swift, J.H., 1984. The circulation of the Denmark Strait and Iceland-Scotland overflow waters in the North Atlantic. Deep-Sea Res., 31(11) : 1339-1355.

Weatherly, G.L., 1972. A study of the bottom boundary layer of the Florida Current. J. Phys. Oceanogr., 2 : 54-72.

Willebrand, J., Philander, S.G.H. and Pacanowski, R.C., 1980. The oceanic response to large-scale atmospheric disturbances. J.Phys.Oceanogr., 10 : 411-429.

A THREE DIMENSIONAL BAROCLINIC MODEL OF THE WESTERN BALTIC

M.J. BOEHLICH
Institut für Meereskunde der Universität Hamburg,
Heimhuder Strasse 71, 2000 Hamburg 13, FRG

ABSTRACT

The circulation in the western Baltic is simulated by use of a three dimensional baroclinic numerical model. Eddies and other small scale flow features are resolved and can be related to the bottom topography or to baroclinic effects. Boundary conditions are provided by two coarser grid models which also take remote forcing from the North Sea and the Baltic proper into account.

1 THE PROBLEM

The depletion of oxygen in the deep layers of the Kiel Bight is a well known problem (Erhardt and Wenck, 1982, Gerlach, 1984, Miljøstyrelsen, 1984). So far it is not clear whether it was caused natural (e.g., by anomalies of the atmospheric forcing or anomalies of temperature and salinity) or man-made (by discharge of nutrients and oxygen consuming substances from land). If the oxygen depletion has a natural cause, it may be connected with the physical conditions of the area (i.e. water exchange with the North Sea, turbulent mixing, up- and downwelling). There are essentially two hypotheses for physically induced oxygen depletion. First, weak currents and the resulting strong stratification during calm weather conditions may lead to the formation of oxygen depletion when microbiological degradation of sinking organic matter consumes all the dissolved oxygen in the bottom water. Two studies of this hypothesis (Grassl and Stengel, 1985, Frey and Becker, 1986), were not able to establish a clear causal link between the physical conditions and the oxygen depletion. The second hypothesis is based on the assumption that not calm conditions but moderate and varying winds (episodes) may lead to coastal and open sea upwelling and small scale eddies. Both processes may lead to an entrainment of nutrient rich water from the bottom layer into the surface layer. Hence bottom water with high nutrient content but low oxygen concentration is pumped into the euphotic zone. This results in an increase of the primary production. The degradation of this organic matter lowers even further the oxygen concentration within the water column and especially at the sea bed.

The purpose of this paper is to investigate the second hypothesis just outlined by means of a numerical circulation model of the Kiel Bight. The area of interest is shown in fig. 1. The design of the model is based on the following known features of the system.

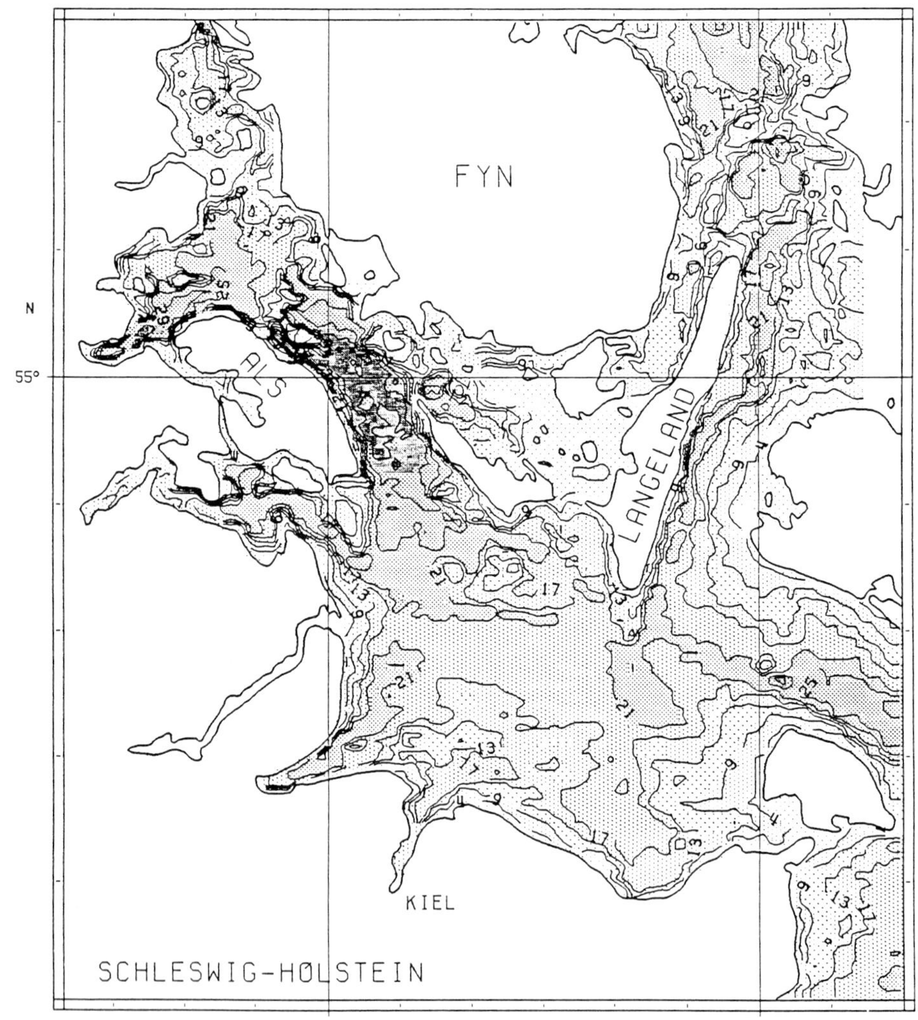

Fig. 1. Topography of the western Baltic model, (model C). Depth in meters.

The Kiel Bight is part of the transition area between the North Sea and the Baltic. The North Sea is a shelf sea with high salinity and strong tidal currents directly influenced by the ocean. The Baltic is a continental, semi-enclosed sea which has relatively low salinity and nearly no tidal currents. The system North Sea - Baltic may be understood as two large basins connected by a strait. The mean currents in the transition area are a result of the water budget of the Baltic.

Evaporation and precipation over the Baltic balance each other (Dietrich and Schott, 1974), so that the fresh water surplus due to river runoff is the driving force for the estuarine circulation in the transition area. This circulation is two layered with an outgoing current of low salinity water at the surface and an ingoing current of high salinity water at the bottom. Due to the Coriolis force the outgoing current is deflected through the Sound along the Swedish west coast, whereas the ingoing current flows along the Danish east coast through the Great Belt. Deviations from the mean flow field in the transition area are caused by deviations from the mean sea level gradient and from the mean density gradient between the two large basins. High air pressure over Scandinavia induces an outflow into the North Sea. The most favorable atmospheric situation for inflow occurs with high pressure over Jutland and low pressure over Sweden. Then, both the atmospheric pressure and the resulting wind stress lead to a negative anomaly of the sea level in the western Baltic, followed by a transport of water from the Kattegat into the Baltic.

Another reason for deviations from the mean flow field is the local wind forcing and the occurence of seiches in the entire Baltic generated by sudden changes of the strength or the direction of the windstress.

The main cause for the limitation of water exchange between the two basins and thus for the oxygen problem of the Baltic is its special topography. The Baltic is divided into a series of basins separated by shallows and sills. Inflowing water with high density creeps along the bottom, following the deepest connections between the basins. Moreover, each basin must be filled up to the height of the sill before the water of the next basin can be renewed. If the inflowing water is not heavier than the water situated in the deep basins, it is injected at a level above the bottom determined by its density. In this case, the bottom water is not exchanged until molecular and turbulent diffusions have lowered its density. Unfortunately, the connection between the Kattegat and the central Baltic is shallow and narrow so that strong baroclinic or barotropic pressure gradients are needed to exchange the bottom water. The topography of the western Baltic consists of narrow trenches, partly connected with each other and of submarine depressions. This results in a complicated struture of the flow field.

To summarize, we see that the currents whithin the Kiel Bight not only depend on the local windforcing but also on the state of the system North Sea - Baltic. The local phenomena and the remote forcing have different scales in space and time.

The flow field within the western Baltic has small time scales (local windforcing) of some hours and small spatial scales (some 100m). Due to the small scale topography the system North Sea - Baltic has large spatial scales, some 10 km and because the sytem is inert, time scales of the order of days.

These local characteristics must be taken into account when designing a numerical model of the Kiel Bight. Obviously the model requires a high resolution in space and time to cover all essential local phenomena. It also needs boundary values of satisfactory qualitiy and resolution (in space and time). Since observed boundary values are lacking, they will be provided by a larger scale circulation model.

2 MODEL DESIGN

The model system consists of three components. The first two components are needed to compute boundary values for the last, local component (referred to as "model C"). They cover the large scale remote forcing (model A), and the forcing modified by the topography of the transition area between North Sea and Baltic (model B). All components compute prognostic values of the currents, sea surface elevation, salinity and temperature. The first model (model A) is a 12 - layer baroclinic model with a horizontal resolution of 12 nm. A detailed description of the model concept is given by Backhaus (1985) and by Backhaus and Hainbucher (1987). Fig. 2 shows the areal extent of model A and the currents (depth mean flow) calculated after 50 days real time.It is forced by

a) the tidal elevation of the sea surface at the open boundaries;
b) the river runoff (mean summer conditions, 610 km^3/year (Jacobsen, 1980))into the Baltic;
c) the atmospheric wind and the pressure field (mean summer conditions (Backhaus et al., 1985)) and
d) the mean salinity and temperature fields for the summer season (May - September), taken from Lenz (1971) and from Bock (1971).

Fig. 3 shows the computed mean sea surface elevation in the Baltic in the quasi-steady state. They compare well with observations shown as dotted lines, Dietrich and Schott (1974).

The grid resolution of model A is too coarse to resolve the circulation within the transition area between the North Sea and the Baltic in a realistic way. However, model A is a good tool to compute the boundary and initial values,

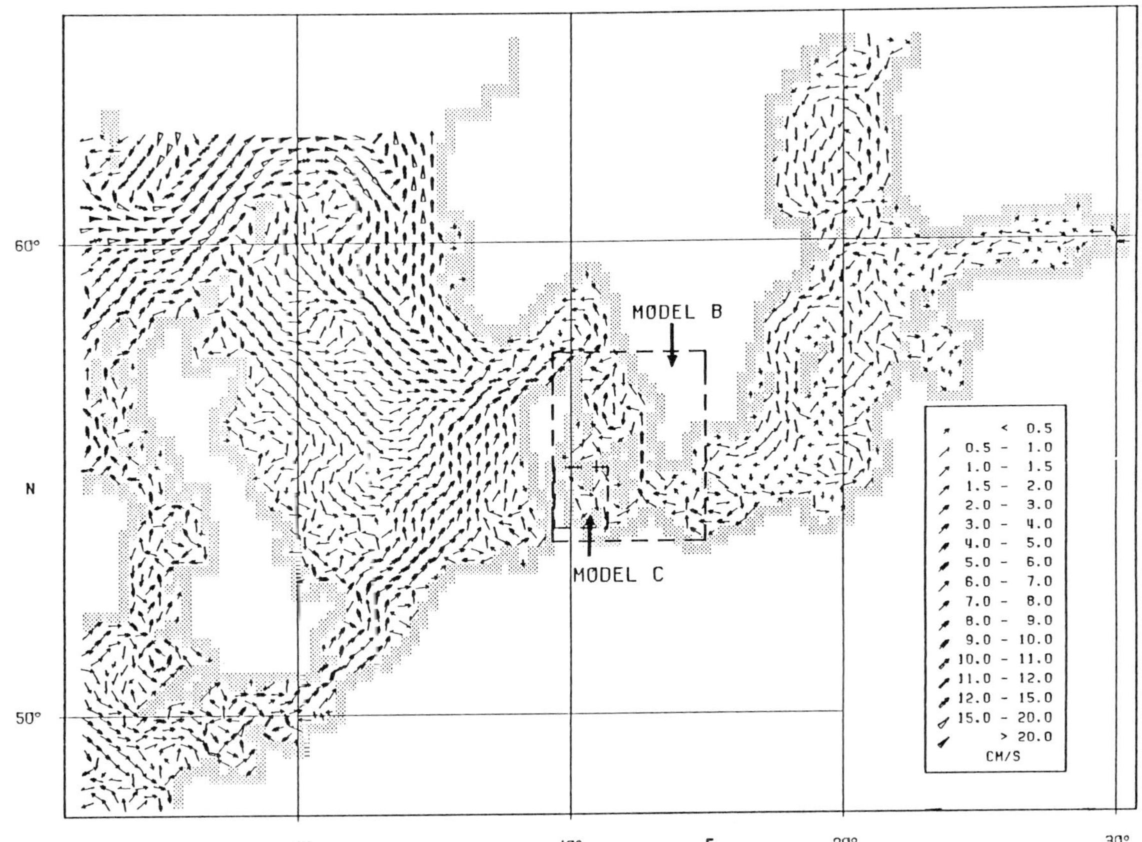

Fig. 2. Computed depth mean flow of model A with the areas of models B and C.

needed for the the medium scale model B.

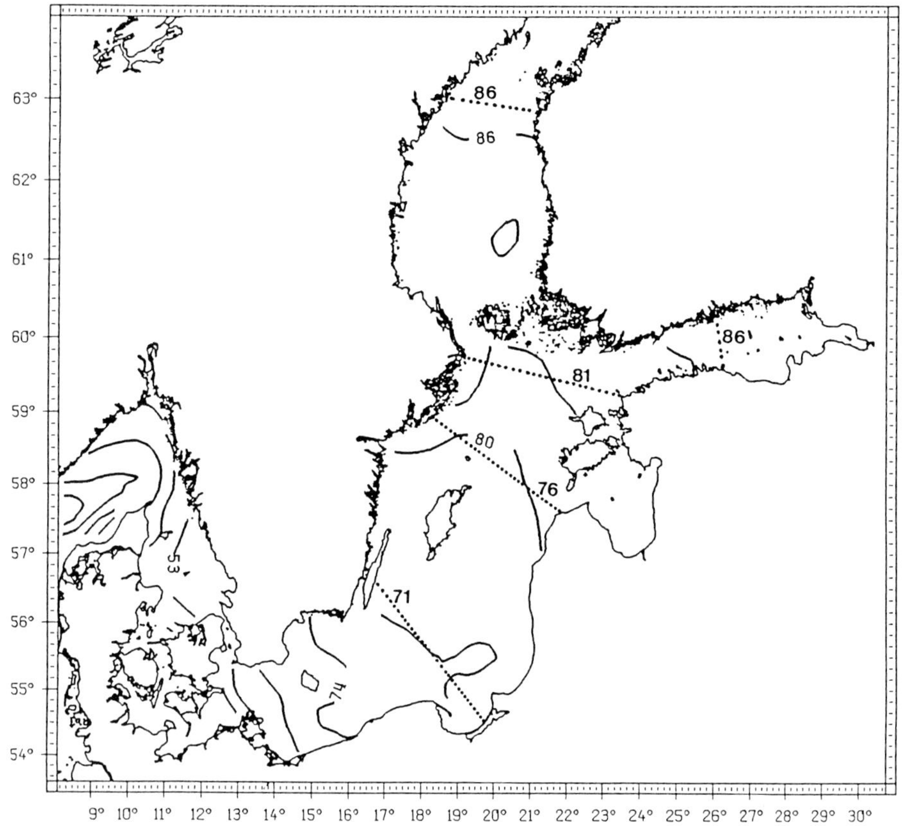

Fig. 3. Computed mean sea surface elevation of model A in cm. The dotted lines show the sea surface elevation after Dietrich & Schott, 1974.

Model B covers the transition area between the North Sea and the Baltic, the region between Skagerrak and Bornholm with a grid resolution of 3 nm. The conception of model B is the same as that of model A. It also has 12 layers, but since the deepest part of the region is only 80 m, it has a better vertical resolution (5-20m). Model B is used to describe the special features of the transition area which were not sufficiently resolved by model A (narrow trenches, sills) but which strongly influence the flow and thus the distribution of salinity and temperature. The task of model B is to interpolate the results of model A in a physical way.

To initialize model B, we use the results of model A for the region of model B by interpolating the results of the sea surface elevation, the salinity and the temperature. The advantage of this technique is, that unrealistic barotropic and baroclinic disturbances caused by the interpolation are removed after a short time of computation. Thus model B provides a dynamical interpolation.

Model B is forced by

a) The sea surface elevation, and by profiles of salinity and temperature at the open boundaries (Skagerrak and Bornholm sea) which are obtained by interpolating the results of model A;

b) the same atmospheric wind and pressure field as in model A.

Fig. 4 shows the horizontal initial field of the salinity of model B.

After 40 days of real time computation under the influence of the forcing mentioned above the salinity has advected to the pattern shown in fig. 5. The currents and the sea surface elevation corresponding to this situation are shown in figs. 6 and 7. At this stage of development the model system is already suitable to describe the water exchange between the North Sea and the Baltic. But the representation of small scale processes (e.g., eddies, up- and downwelling) is not satisfactory. To overcome this drawback, we complete the model system by appending the fine scale model C.

Model C covers the region between Fyn and Fehmarn (see fig. 1) with a grid resolution of 0.5 nm in the horizontal. The vertical resolution is between 2 and 10 m with 10 layers. The boundary values for model C (sea surface elevation, salinity and temperature) are obtained from the results of model B and the same windstress as in the preceeding steps acts at the sea surface. Fig. 8 shows the currents within the area at steady state (mean summer conditions). The advantage of model C is that it is now possible to reproduce fine scale structures (down to 3 nm) such as those observed by remote sensing techniques from satellites (Horstmann, 1983).

Looking at the several stages of the model system, it becomes clear that the reduction of the grid size not only works like a magnifying-glass, but it also allows the introduction of new physics through the better approximation of the topography. The results of the fine mesh model clearly show the dominant influence of the topography. The "large scale" current pattern does not change considerably compared to the pattern of the coarser mesh model B but below that "large scale" flow, structures appear, that were not resolved by the preceeding steps.

The benefit of this technique of a connected model system is that it is possible to study local small scale processes without neglecting the remote forcing.

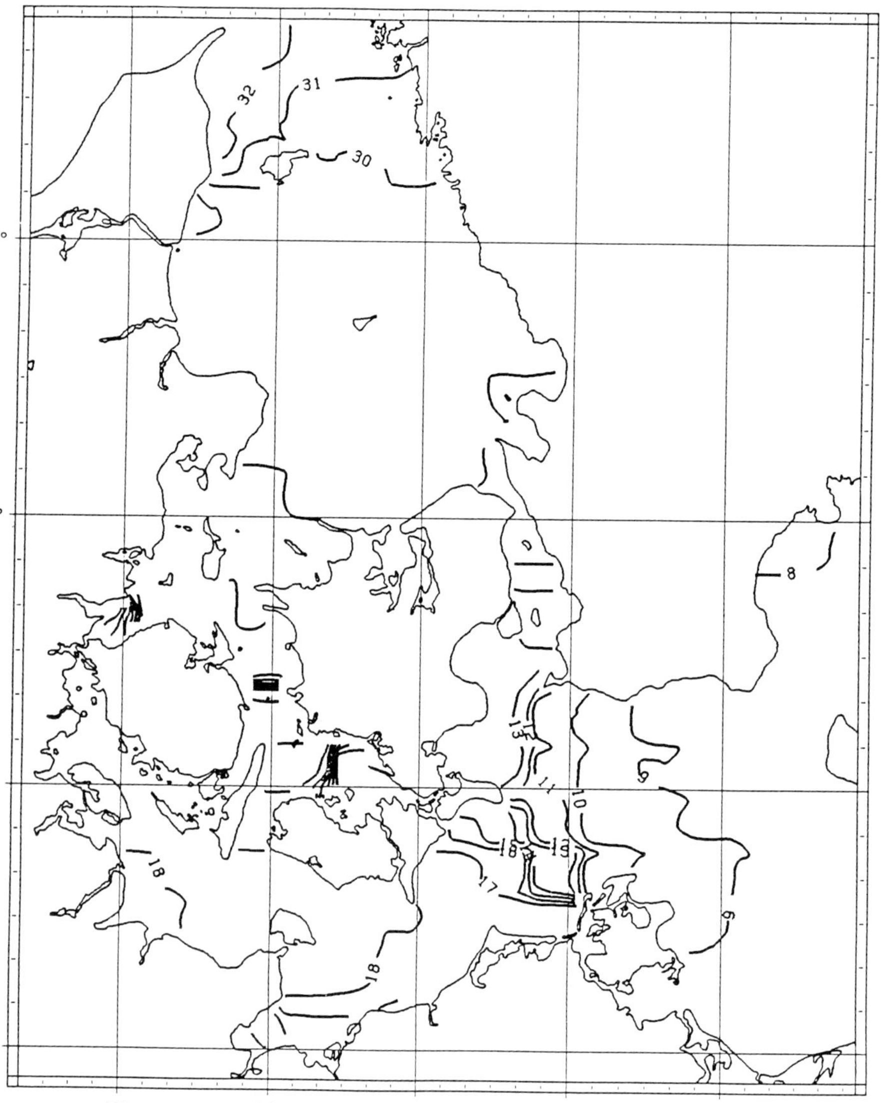

Fig. 4a. Initial salinity field of model B in the 1st layer (0 - 5 m) in ‰.

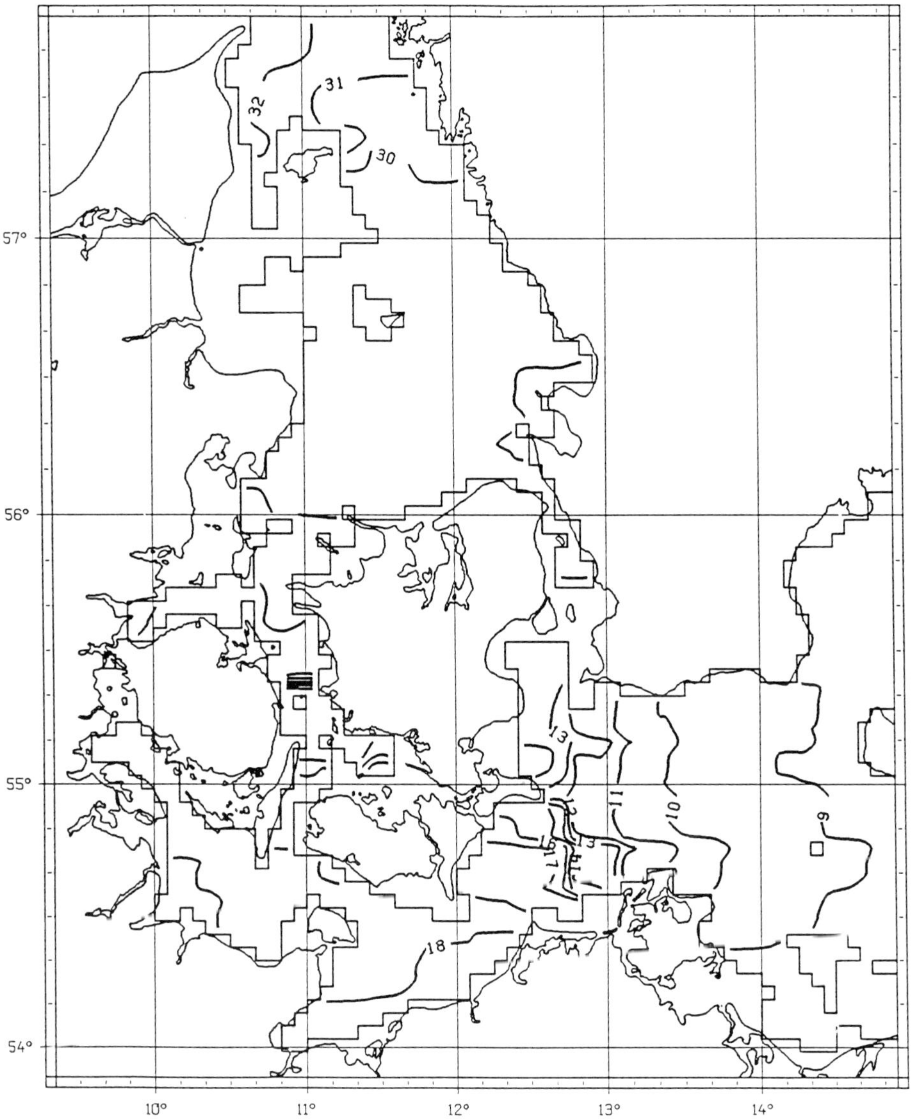

Fig. 4b. Initial salinity field of model B in the 3rd layer (10 - 15 m) in ‰.

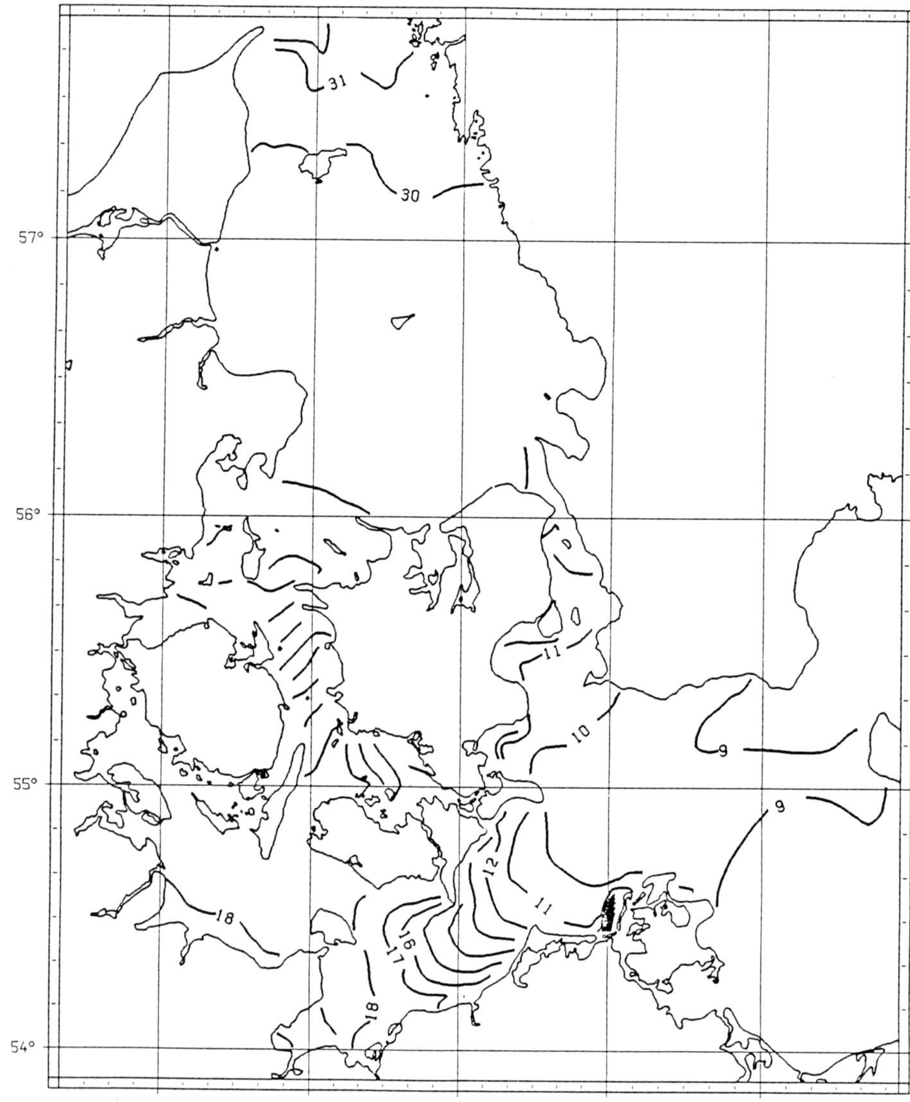

Fig. 5a. Computed mean salinity field of model B in the 1st layer (0 - 5 m) in ‰.

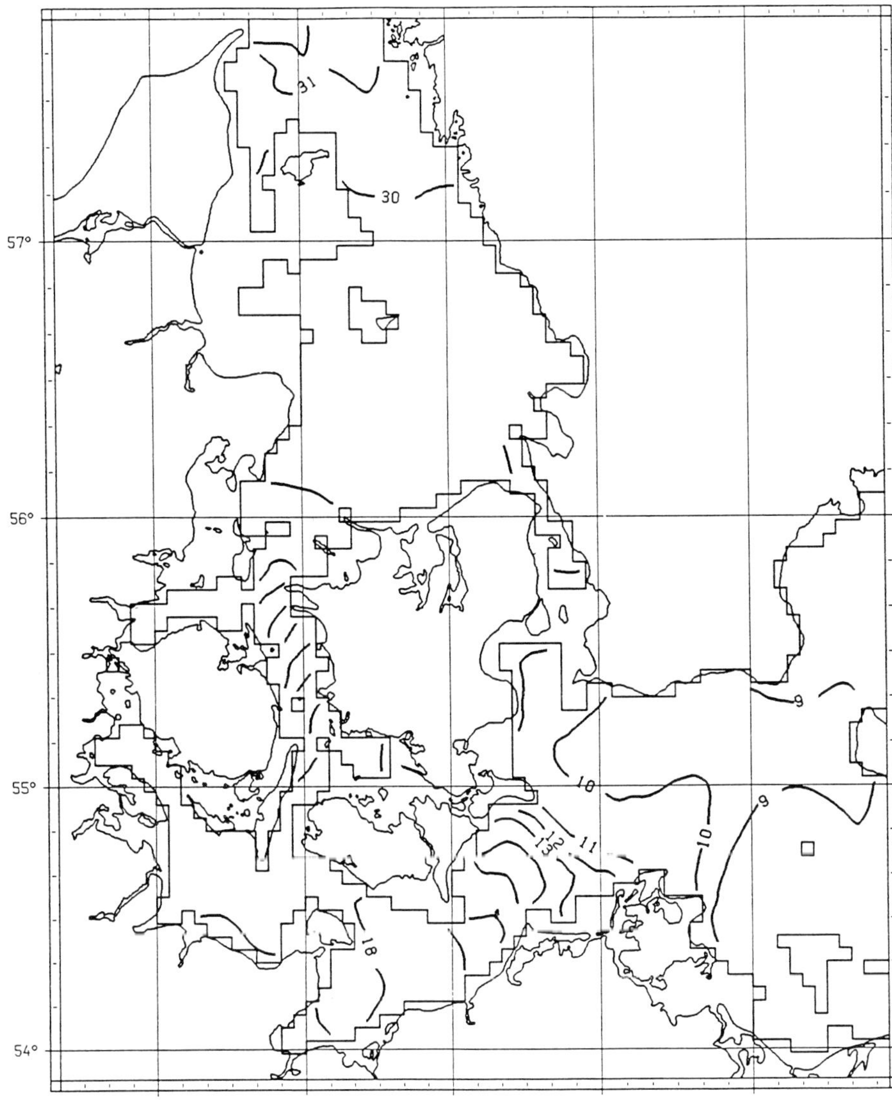

Fig. 5b. Computed mean salinity field of model B in the 3rd layer (10 - 15 m) in ‰.

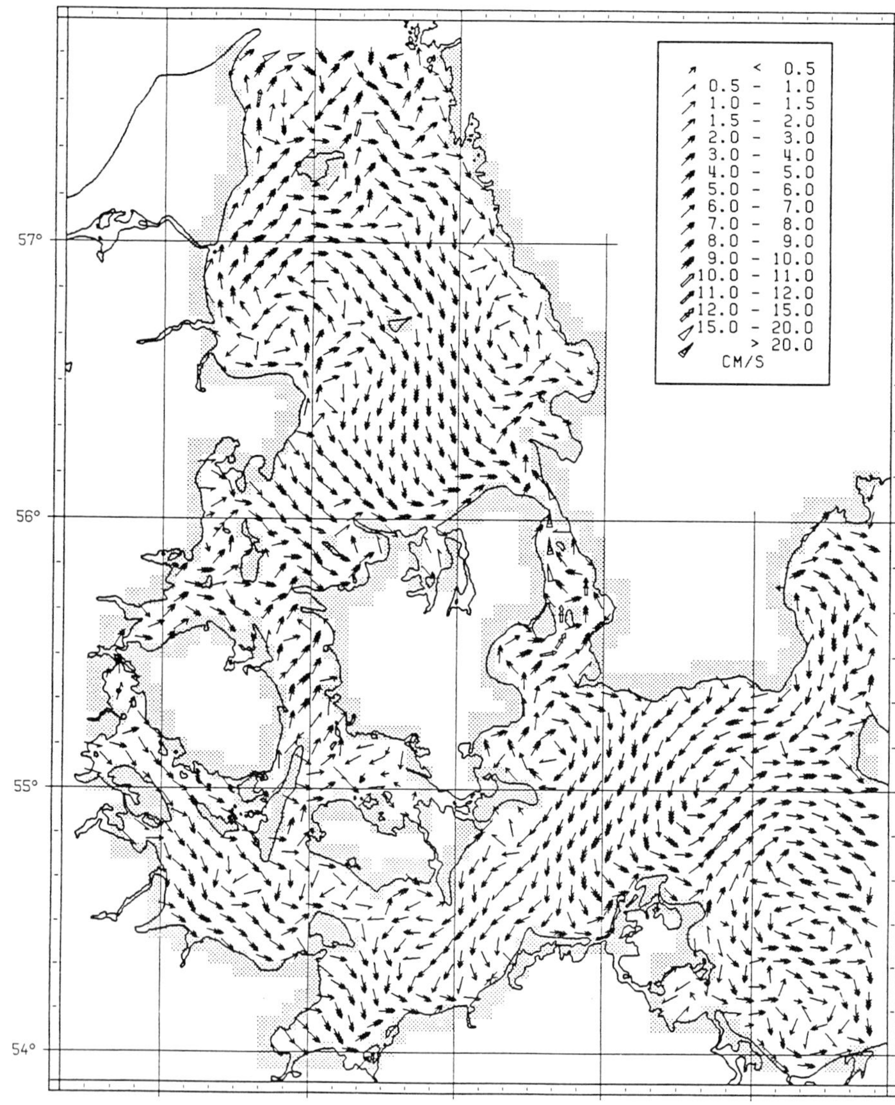

Fig. 6a. Computed mean currents of model B in the 1st layer (0 - 5 m).

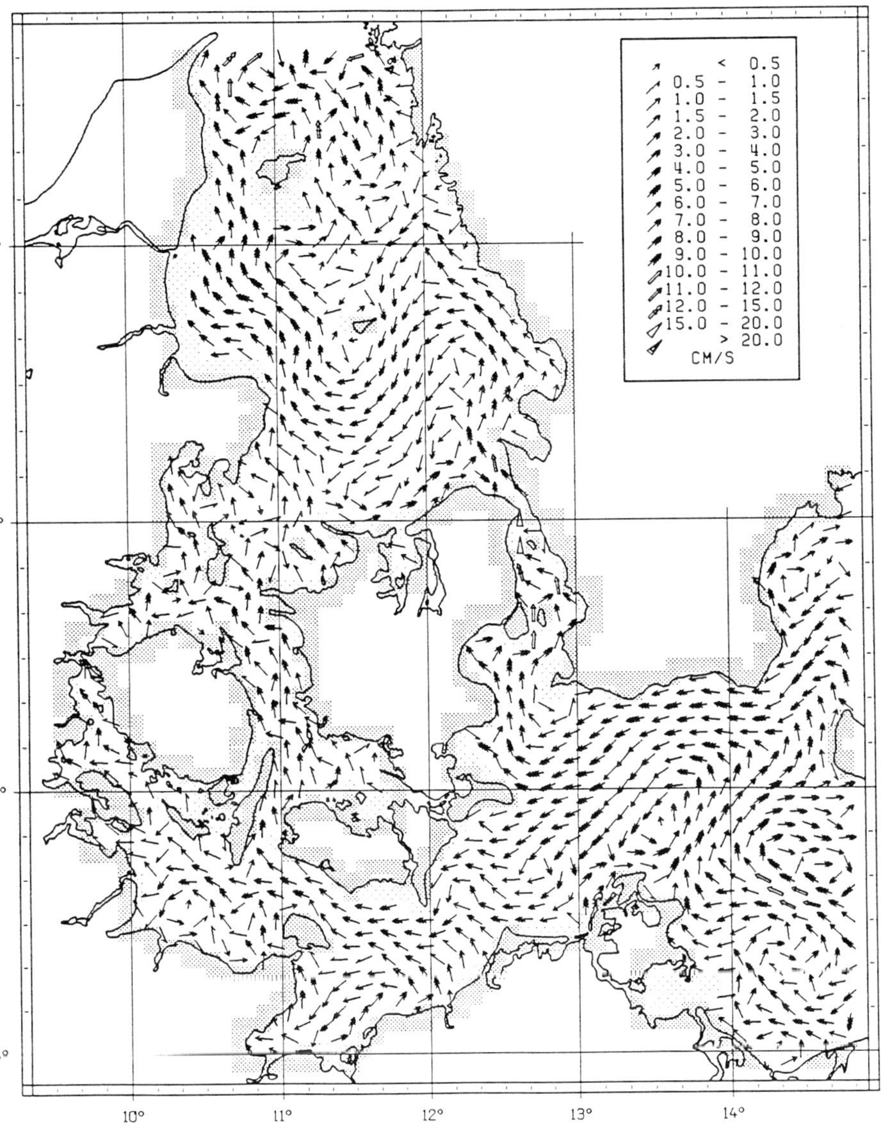

Fig. 6b. Computed mean currents of model B in the 2nd layer (5 - 10 m).

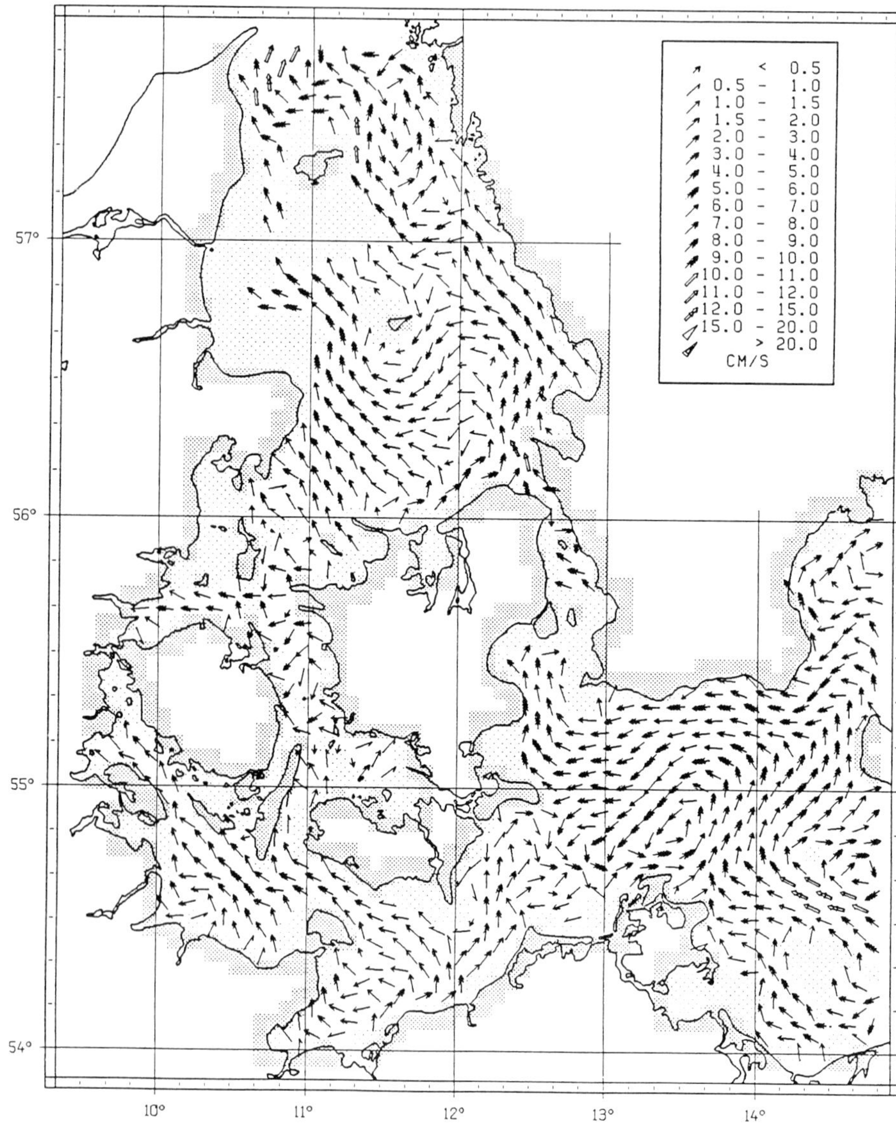

Fig. 6c. Computed mean currents of model B in the 3rd layer (10 - 15 m).

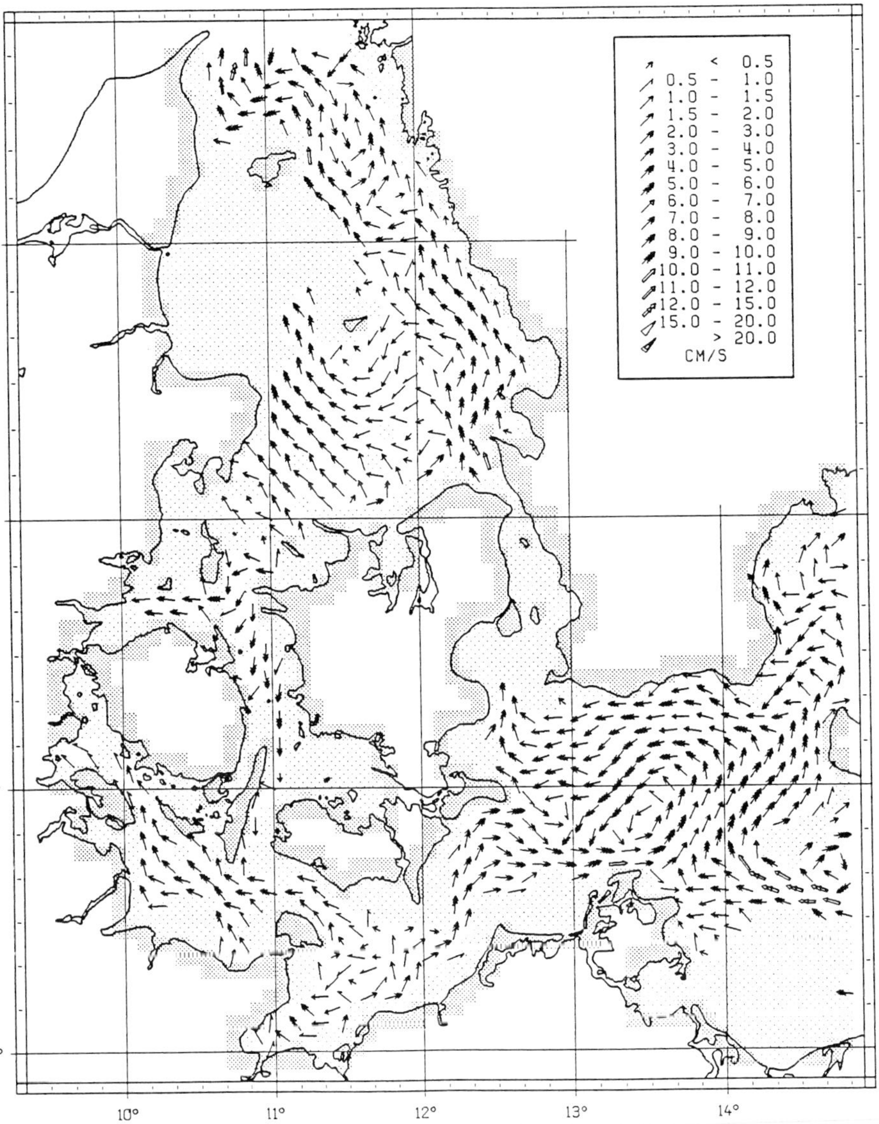

Fig. 6d. Computed mean currents of model B in the 4th layer (15 - 20 m).

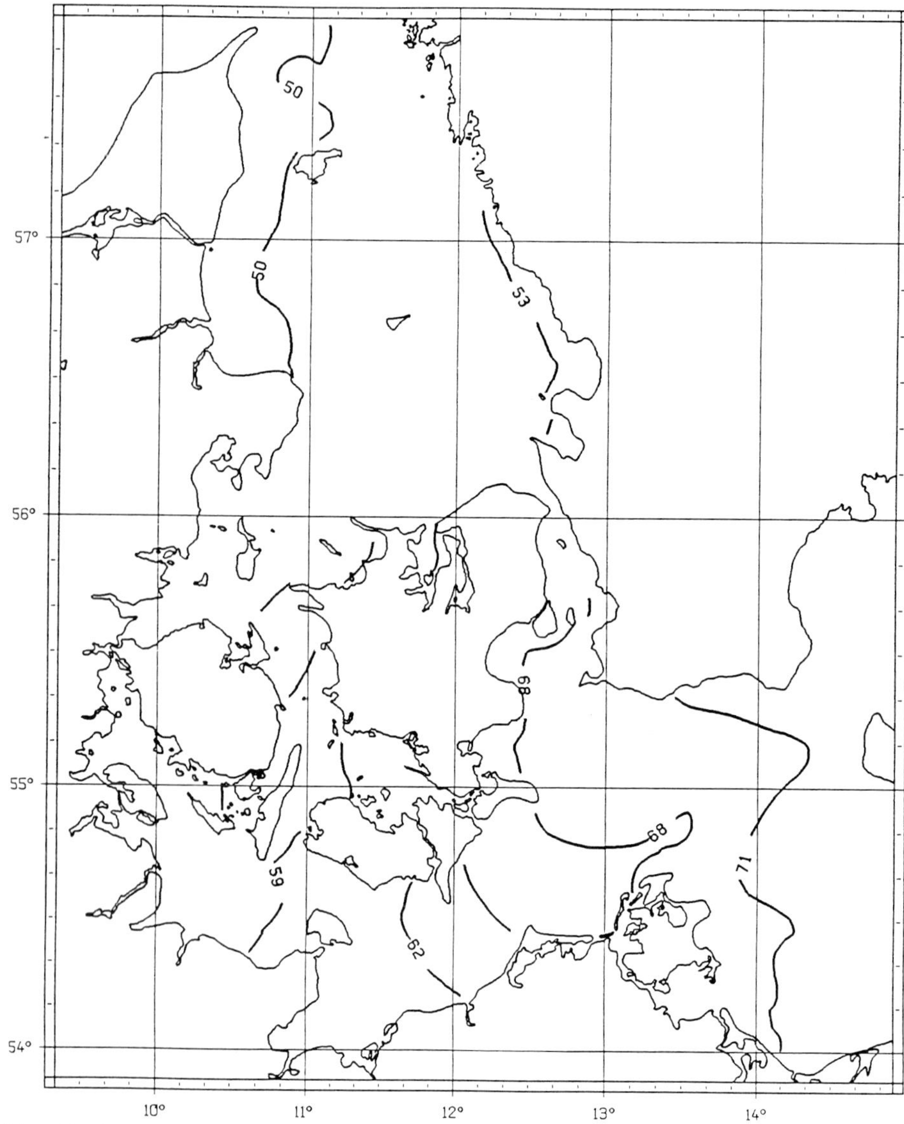

Fig. 7. Computed mean sea surface elevation of model B in cm.

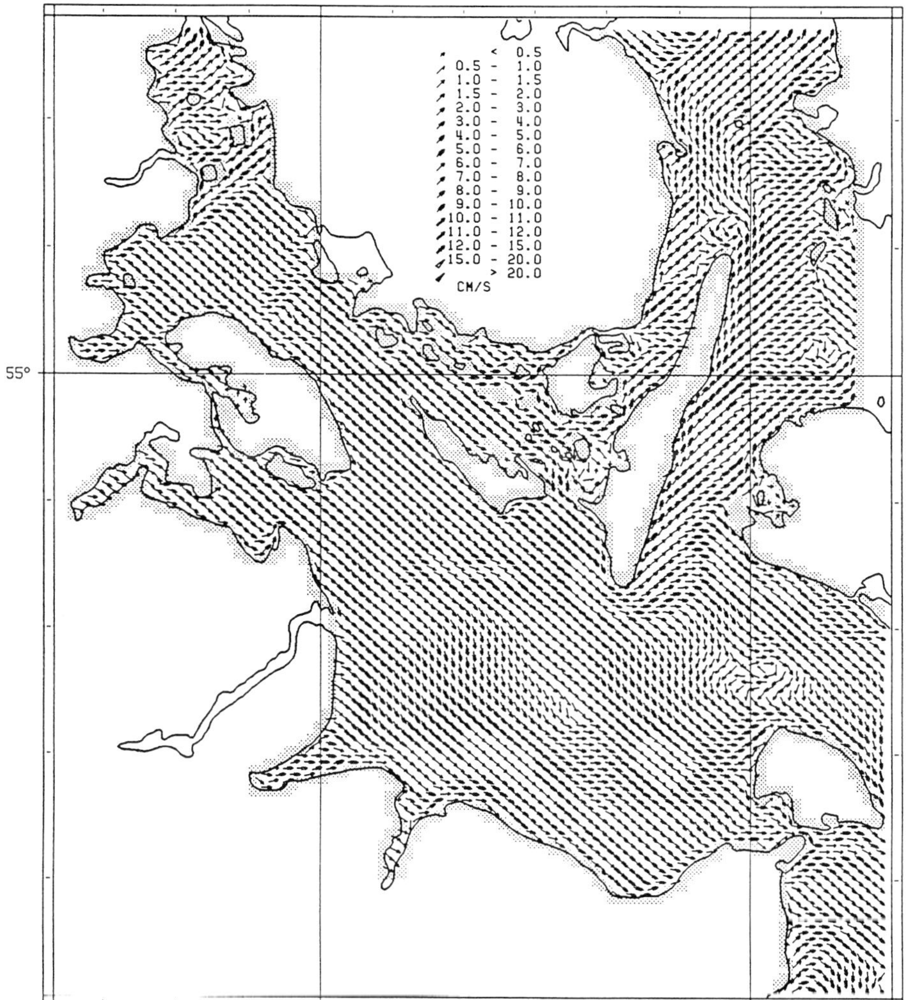

Fig. 8a. Computed mean currents of model C in the 1st layer (0 - 2 m).

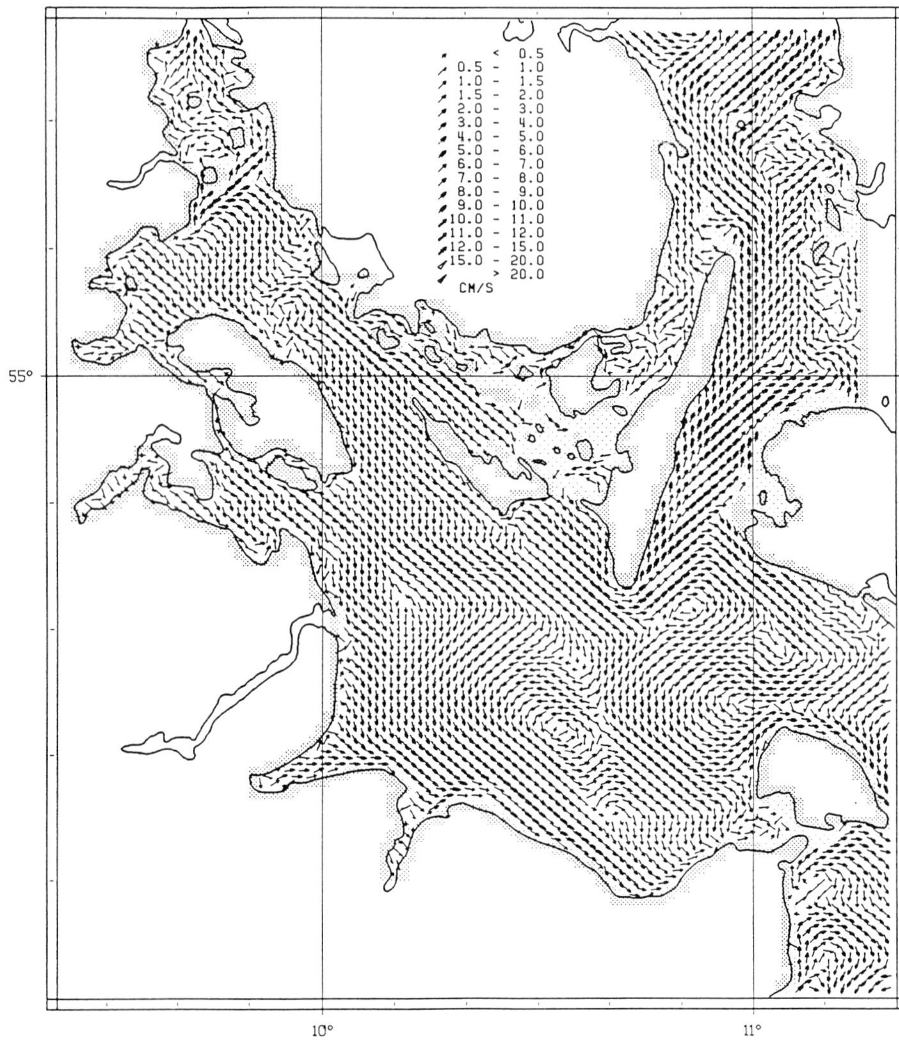

Fig. 8b. Computed mean currents of model C in the 2nd layer (2 - 4 m).

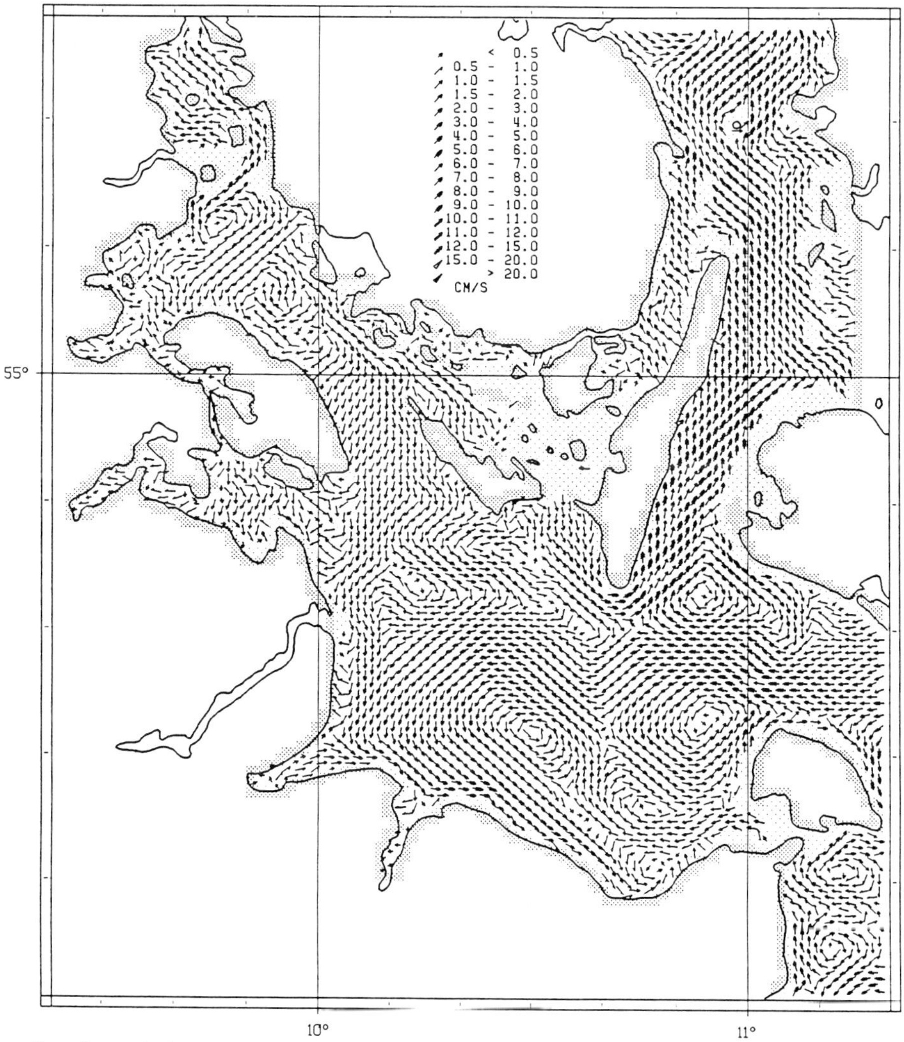

Fig. 8c. Computed mean currents of model C in the 3rd layer (4 - 6 m).

3 OUTLOOK

The next step that will be done is to investigate the time dependent flow and stratification within the Kiel Bight area with regard to frequently occuring oxygen depletions caused by telluric discharge of nutrients.

4 ACKNOWLEDGEMENTS

I am indebted to Prof. Dr. J. Backhaus for his valuable advice and assistance throughout this work.

This work was funded by the Umweltbundesamt of the Federal Republic of Germany.

5 REFERENCES

Backhaus, J. O., 1985. A three-dimensional model for the simulation of shelf sea dynamics. Deutsche Hydrographische Zeitschrift, 38: 165-187.

Backhaus, J. O. and Hainbucher, D., 1987. A finite difference general circulation model for shelf seas. This issue.

Backhaus, J., Bartsch, J., Quadfasel, D., Guddal, J., 1985. Atlas of monthly surface fields of air pressure, wind stress and wind stress curl over the North Eastern Atlantic Ocean: 1955 - 1982. Technical Report 3-85. Institute of Oceanography, University of Hamburg.

Bock, K.-H., 1971. Monatskarten des Salzgehaltes der Ostsee, dargestellt für verschiedene Tiefenhorizonte. Deutsche Hydrographische Zeitschrift. Ergänzungsheft Reihe B, Nr.12.

Dietrich, G. and Schott F., 1974. Wasserhaushalt und Strömungen. In: L. Magaard and G. Rheinheimer (Editors), Meereskunde der Ostsee. Springer, Berlin/Heidelberg/New York, pp. 33-41.

Erhardt, M. and Wenck, A.,1982. Wind pattern and hydrogen sulfide in shallow waters of the western Baltic, a cause effect relationship? 13th Conference of Baltic Oceanographers, Helsinki, 1982: 221-234

Frey, H. and Becker, G., 1986. Untersuchung der langzeitigen Variation der hydrographischen Schichtung in der Deutschen Bucht. Forschungsbericht 102 04 215/15 im Auftrage des Umweltbundesamtes.

Gerlach, S.A.,1984. Oxygen depletion 1980-1983 in coastal waters of the Federal Republic of Germany. First report of the working group "Eutrophication of the North Sea and the Baltic". Berichte aus dem Institut für Meereskunde an der Christian Albrechts Universität Kiel, 130.

Grassl, H. and Stengel ,M., 1985. Für chemisch-biologische Prozesse in deutschen Küstengewässern wichtige Wetterlagen. Forschungsbericht 102 04 215/25 im Auftrag des Umweltbundesamtes.

Horstmann, U., 1983. Distribution patterns of temperature and water colour in the Baltic sea as recorded in satellite images: Indicators for phytoplankton growth. Berichte aus dem Institut für Meereskunde an der Christian Albrechts Universität Kiel, 106.

Jacobsen, T.S.,1980. The Belt project. Sea water exchange of the Baltic, measurements and methods. The National Agency of Environmental Protection, Denmark.

Lenz, W., 1971. Monatskarten der Temperatur der Ostsee, dargestellt für verschiedene Tiefenhorizonte. Deutsche Hydrographische Zeitschrift. Ergänzungsheft Reihe B, Nr. 11.

Miljøstyrelsen, 1984. Iltsvind og Fiskedød i 1981. Omfang og årsager. Kopenhagen.

A STUDY OF VARIOUS OPEN BOUNDARY CONDITIONS FOR WIND-FORCED BAROTROPIC NUMERICAL OCEAN MODELS

L.P. RØED and C.K. COOPER
Det norske Veritas, Section for Oceanography, Høvik (Norway) and Conoco Inc., Production Research Dept., Ponca City, OK 74603 (USA)

ABSTRACT

This study focuses on the sensitivity of the interior wind-forced response of a numerical barotropic ocean model to changes in the open boundary conditions (OBCs) where the term "open" implies a sea boundary where the solution is unknown and must be assumed or extrapolated from the interior solution. Seven different OBCs are applied along the two lateral boundaries of a rectangular basin with a flat bottom. One of the longer sides of the basin is a straight coast line, while the other is clamped (temporal derivative of the sea surface is set to zero). Each OBC is studied using three wind forcing schemes which test the the boundary under strong local forcing, under weak local forcing, and with a realistic wind field (moving cyclone). The different OBCs are compared with each other as well as with a "correct" case based on either an analytic solution (when available) or model results from an expanded model domain. Comparisons are shown in terms of sea level elevation, volume fluxes and excess mass. It is found that the interior response is often highly sensitive to the implemented OBC, that the performance of some OBCs is a strong function of the local wind forcing at the boundary, and that a method based on an integration along the characteristics performs well, and is generally superior to the others. The discussion also identifies the weaknesses and strengths of the OBCs.

1 INTRODUCTION

Open boundary conditions are imposed along the edge of the model domain where the solution is unknown. Open boundaries contrast to other boundaries where the solution can be specified from data, models, or safely assumed (i.e. land). Since the solution is unknown along open boundaries an assumption must be made or the interior solution extrapolated. The importance of the OBC on the interior solution is clearly shown in previous work, e.g. Røed and Cooper (1986). OBCs are receiving increasing attention within the modelling community as suggested by a wave of recent publications: Beardsley and Haidvogel (1981), Harper and Sobey (1983), Chapman (1985), Hayashi et al. (1986), and Martinsen and Engedahl (1986).

The general approach of this work is to study the detailed response of several OBCs in a simple ocean basin subject to wind forcing. A finite difference, barotropic numerical model is used. Seven OBCs are considered: (1) a clamped sea surface (dynamic sea surface is set to zero), (2) a zero slope sea surface, (3) a combination of sponge and radiation condition (Israeli and Orszag, 1981), (4) a Sommerfeld radiation condition, (5) a forced/free wave condition (Røed and Smedstad, 1984), (6) an oblique radiation condition after Raymond and Kuo (1984), and (7) a condition constructed by integrating the compatibility equations along the characteristics (originally proposed by Hedstrom, 1979).

This study follows the same general approach of Chapman (1985), Røed and Cooper (1986), and Hayashi et al. (1986). However, this work significantly extends the earlier work in two ways. First, it considers several additional OBCs including: the forced/free wave condition by Røed and Smedstad (1984), the oblique radiation condition by Raymond and Kuo (1984), and the characteristic method by Hedstrom (1979). Second, the present study considers three wind fields: a uniform wind with strong winds on the open boundaries, a bell-shaped wind with weak forcing on the open boundaries, and a moving cyclone. The first and second test the OBCs with strong and weak local forcing, respectively. The third tests them for a realistic case with time varying wind forcing on the open boundaries.

The geometry of the model domain is chosen to simulate a shallow continental shelf. Thus, the integration area consists of straight coast, two lateral open boundaries and an offshore open sea boundary running parallel to the coast. The seven OBCs are imposed on the the two lateral boundaries. Along the offshore boundary the dynamic sea surface is set to zero (clamped). This is done because: (1) it is frequently found to be justified along the edge of the continental shelf (Csanady, 1982), and (2) the condition insures that the problem is mathematically well-posed for all seven OBCs.

The model and the OBCs are described in Section 2. Section 3 develops the "standard" results or bench marks for the three alternative wind cases. The bench marks are developed from analytic solutions, where possible, or numerical solutions derived from an extended model domain. In the latter case the cross-shore boundaries are moved so far away that they do not affect the original domain for the integration period (4 days).

The most important part of this study is provided in Section 4 where the results from the test cases are compared with the bench marks. Comparisons are made in terms of snapshots of sea surface elevation and time series of the sea surface elevation, alongshore depth mean currents and excess mass. A discussion and some final remarks are provided in Section 5.

2 FORMULATION

2.2 The model

The model employed is one commonly used for simulation of shelf circulation and storm surge predictions, i.e. the wind-forced shallow water equations. Let U,V be the volume flux components along the (x,y) axes and η the sea surface elevation above the equilibrium depth. Then the governing equations are

$$U_t - fV = -gH\eta_x + (T_s^x - T_b^x)/\rho, \tag{1}$$

$$V_t + fU = -gH\eta_y + (T_s^y - T_b^y)/\rho, \tag{2}$$

and

$$\eta_t + U_x = -V_y, \tag{3}$$

where t denotes time, g the gravitational acceleration, ρ the density of sea water, f the Coriolis parameter, H the constant equilibrium depth, $T_s^{x,y}$ the components of the wind stress at the sea surface and $T_b^{x,y}$ the similar components of the bottom stress. Subscripts x,y and t indicate differentiation with respect to subscript. A linear bottom stress formulation will be used in which the bottom stress is proportional to the depth mean current, viz.,

$$T_b^x = \rho RU/H, \text{ and } \quad T_b^y = \rho RV/H \tag{4}$$

where R is the bottom friction coefficient. This simple bottom stress parameterization is preferred because it allows analytic solutions to be derived as bench marks. Also the emphasis is on the sensitivity of the response to the changes in the imposed OBC and not on the ensuing circulation itself.

TABLE 1

Parameters and physical constants used in the numerical experiments

Symbol	Parameter	Value	Unit
f	Coriolis parameter	$1.2 \cdot 10^{-4}$	m^{-1}
ρ	Density of sea water	1025.0	kg/m^3
g	Gravitational acceleration	9.81	m/s^2
R	Bottom friction coefficient	$2.4 \cdot 10^{-3}$	m/s
L	Alongshore length of basin	1000.0	km
Δs	Grid size (between adjacent elevation points)	20.0	km
Δt	Time increment	180.0	s
H	Equilibrium depth	50.0	m

The numerical approach closely follows that of Martinsen at al. (1979). It is an explicit, finite difference foreward-backward scheme using a staggered grid (Arakawa C-grid). The various parameters used in the simulation arc given in Table 1, and the model grid is shown in Fig. 1. The grid resolution and time step chosen is well within the CFL constraint, and typical shelf waves are well resolved (Wajsowicz, 1986). The boundary condition applied at the coast (no flux) and at the outer offshore boundary (clamped) are common to all cases and may be expressed as follows,

$$V=0 \text{ at } y=0, \tag{5}$$

$$\eta=0 \text{ at } y=L/2. \tag{6}$$

Here L is the length of the regular grid (Fig. 1 and Table 1).

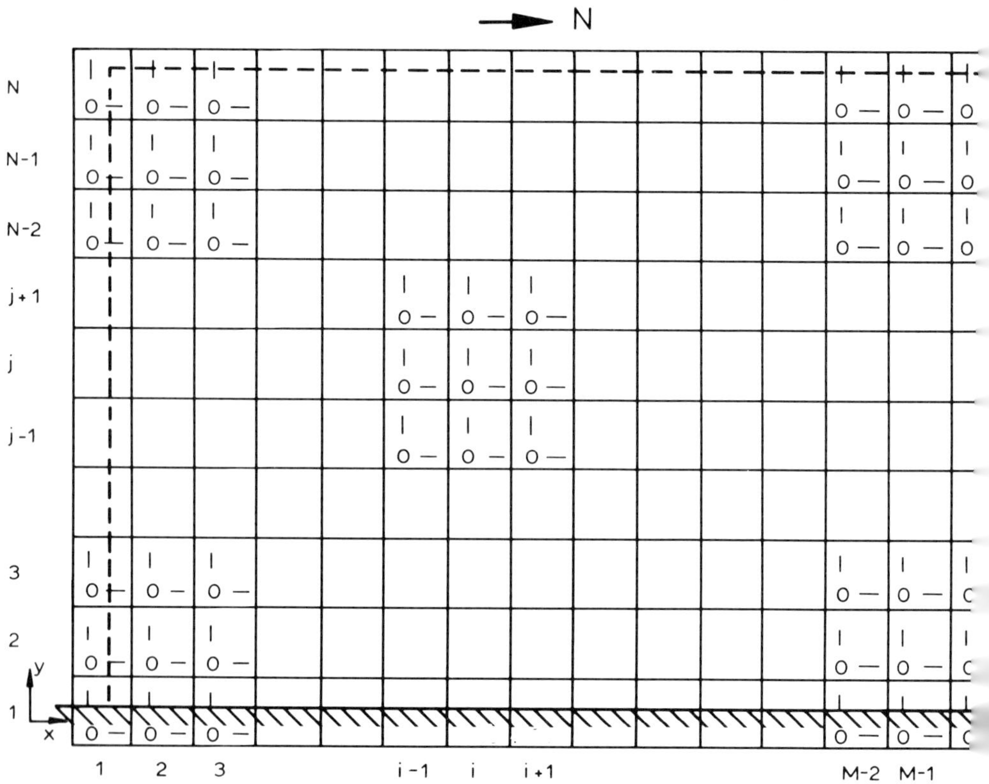

Fig. 1. The regular grid used in the numerical experiments. The dimension of the rectangle is 1000 by 500 km. The applied grid is an Arakawa C-grid. o denotes an elevation or η-point, – an alongshore flux or U-point, and | a cross-shore or V-point. Note that the boundaries located to the left (southern) and to the right (northern) are along U-points, while the offshore boundary and the coastline falls along V-points.

2.2 <u>The OBCs</u>

Seven OBCs have been considered in this study as summarized in Table 2. The notation used by Chapman (1985) has been followed. The first four (i.e. CLP, GRD, MOE and SPO) were part of his study, but different numerical analogues have been used here. The CLP and GRD were also studied by Hayashi et al. (1986). Including these two in this study provides little in the way of new findings, but does provide a common grounds for comparison with the previous studies. The details of the numerical analogues of the various OBCs are given Appendix A, while their analytic expressions are given below.

First, note that CLP, GRD, MOE, FOE, and ORC are essentially only variations of the Sommerfeld radiation condition expressed in the form

$$Q_t+c^xQ_x+c^yQ_y=0 \quad \text{at} \quad x=0,L. \tag{7}$$

Here Q is any of the three dependent variables n,U,V and $c^{x,y}$ are the projections of the wave celerity along the x,y axes.

TABLE 2

OBC types

OBC type	Abbreviation	Source
Clamped	CLP	Various authors
Gradient	GRD	Various authors
Free wave radiation	MOE	Camerlengo and O'Brien (1980)
Mixed sponge/free wave	SPO	Israeli and Orszag (1981)
Forced wave radiation	FOE	Røed and Smedstad (1984)
Oblique radiation	ORC	Raymond and Kuo (1984)
Characteristics	HOC	Hedstrom (1979)
Analytic	ANA	Various authors
Extended grid	EXT	This paper

CLP and GRD have historically been the most popular OBCs in ocean models. For CLP, both c^x and c^y are zero, so that Q (in this case the surface elevation) does not change in time. The analytic expression becomes

$$\eta=0 \quad \text{at} \quad x=0,L. \tag{8}$$

For GRD the sea surface slope is set to zero at the boundary, which is tantamount to let $c^y=0$ and $c^x \rightarrow \infty$ in (7) (with $Q=\eta$), so that

$$\eta_x=0 \quad \text{at} \quad x=0,L. \tag{9}$$

MOE (suggested by Orlanski, 1976 and later modified by Camerlengo and O'Brien, 1980) is the closest to the classical Sommerfeld radiation condition. It is assumed that the wave celerity in the along-boundary direction, c^y, is zero in which case the expression for c^x may be computed from (7), or

$$c^x=-Q_t/Q_x. \tag{10}$$

In the implementation of MOE, Q is replaced by the alongshore volume flux, and so the analytic expression for MOE becomes

$$U_t+c^uU_x=0 \quad \text{at} \quad x=0,L, \tag{11}$$

with the wave celerity computed by solving (11) with respect to c^u, viz.,

$$c^u=-U_t/U_x. \tag{12}$$

It is important to realize that in order to derive a nontrivial OBC, interior values of U has to be used in (12). Following the approach of Camerlengo and O'Brien (1980), (11) is only used when c^u signifies an outgoing disturbance (i.e. when $c^u>0$ at the northern boundary). If this is not true U is not updated.

The expression (11) is also used in FOE with U replaced by its global part, i.e. that part of U which represents the free wave part. The global part is defined so that

$$U=U^l+U^g \tag{13}$$

where U^l and U^g are the local (forced) and global (free) part, respectively. The procedure closely follows that described in Røed and Smedstad (1984) with the bottom friction as an additional term and will therefore not be repeated here.

The SPO combines the use of radiation conditions and sponge layers. The use of sponge layers in order to dampen out waves and other disturbances generated in the interior are fairly common in atmospheric models, but Israeli and Orszag (1981) was probably the first to suggest to combine a sponge layer and a radiation condition. This had been reported to be fairly successful in many cases (cf. Chapman, 1985). However, the addition of extra grid elements makes the use of sponges very costly. In the present implementation of SPO, a sponge layer 200 km wide is added to the grid in Fig. 1 at both the southern and northern boundaries. Within these layers the bottom friction coefficient is increased exponentially from its interior value to four times that value at the outer edge of the sponge by use of the formula

$$R_S=R\begin{cases}\exp(-bx) & ;\ x<0\\ \exp(b(x-L)) & ;\ x\geq L,\end{cases} \tag{14}$$

where R_S is the bottom friction coefficient within the sponges and b is a factor determined so as to give $R_S=4R$ at the outer edge of the sponges. Any disturbances left at the edge of the sponges are subject to MOE. Note that any OBC may be used for this purpose.

For ORC the projections of the wave celerity are computed from the expressions (cf. Raymond and Kuo, 1984),

$$c^{x,y}=-Q_tQ_{x,y}/(Q_x^2+Q_y^2) \tag{15}$$

ORC was suggested by Raymond and Kuo (1984) in order to handle waves hitting the boundary at an oblique angle. It differs from the other radiation conditions in that both c^x and c^y are computed. The rational for ORC is based on the fact that a wave hitting the boundary at an angle may give values for c^x by means of (12) far in excess of the numeric constraint that $c^x \leq \Delta s/\Delta t$. The analytic expression for ORC is therefore given by (7) with Q replaced by U, viz.,

$$U_t + c^x U_x + c^y U_y = 0 \text{ at } x=0,L, \tag{16}$$

with the wave celerities computed from

$$c^{x,y} = -U_t U_{x,y}/(U_x^2 + U_y^2). \tag{17}$$

HOC differs drastically from the above OBCs in that it is based on an integration of the compatibility equations derived from the governing equations. Integration of the compatibility equations rather than the primitive equations are usually referred to as the method of characteristics. This method was explored in the fifties by Hartree (1953) and Freeman and Baer (1957a,b) for use in atmospheric models and was employed by O'Brien and Reid (1967) to find the oceanic response due to a stationary hurricane. Hedstrom (1979) used this method to derive a nonreflecting OBC for nonlinear, one-dimensional and nonrotational hyperbolic problems. His condition has been modified here to include two dimensions, rotation and wind forcing. This modified version of the Hedstrom condition appears to resemble the weaklyreflective boundary condition discussed by Verboom and Slob (1984).

The compatibility equations for the system (1)-(3) are

$$D_1(U + c_o\eta)/dt = F_1 + c_o F_3 \quad \text{along } D_1x/dt = +c_o \tag{18}$$

$$D_2(U - c_o\eta)/dt = F_1 - c_o F_3 \quad \text{along } D_2x/dt = -c_o \tag{19}$$

$$D_3V/dt = F_2 \quad \text{along } D_3x/dt = 0 \tag{20}$$

where,

$$F_1 = fV + T_s{}^x/\rho - RU/H, \tag{21}$$

$$F_2 = -fU - c_o{}^2\eta_y + T_s{}^y/\rho - RV/H, \tag{22}$$

$$F_3 = -V_y \tag{23}$$

Here $c_o{}^2 = gH$ and the operators $D_j()/dt$ are defined by

$$D_j()/dt = ()_t + (D_jx/dt)()_x \;;\; j=1,2,3. \tag{24}$$

In order to minimize reflections, the slope of the incoming characteristic (defined by D_jx/dt) are forced to zero at the boundary. For instance, at the northern boundary (x=L) the incoming characteristic is given by D_2x/dt. Thus integrating (19) along $D_2x/dt=0$ and using (18) and (20) to solve for η_t, U_t and V_t it follows that,

$$\eta_t=F_3-(1/2)(U+c_o\eta)_x, \tag{25}$$

$$U_t=F_1-(1/2)c_o(U+c_o\eta)_x, \tag{26}$$

$$V_t=F_2, \tag{27}$$

to be obeyed at the northern (x=L) boundary. A similar procedure gives

$$\eta_t=F_3-(1/2)(U-c_o\eta)_x, \tag{28}$$

$$U_t=F_1+(1/2)c_o(U-c_o\eta)_x, \tag{29}$$

$$V_t=F_2, \tag{30}$$

to be satisfied at the southern (x=0) boundary. Expressions (26) and (29), which give the time rate of change of the alongshore flux, form the analytic basis for HOC.

3 STANDARD CASES

The section is broken into three subsection describing the bench marks for the three wind fields. For the first case it is possible to derive an analytic steady-state solution as well as an approximate analytic transient solution. The other two cases can not be solved analytically so bench marks are derived from grid systems extended so that the boundaries of the extended grid does not affect the solution within the regular grid during the integration period.

It should be noted that in the numerical implementation of the wind stress, the staggering of the grid is taken into account by specifying T_s^x at U-points and T_s^y at V-points (Fig. 1).

3.1 Uniform alongshore wind

The goal of this case is to test the OBCs with strong persistent wind forcing on the boundaries. The wind stress consists of a positive alongshore wind stress which decays offshore. It is instantaneously applied at model start up, or:

$$T_s^x=T=T_o\exp(-ay), \tag{31}$$

$$T_s^y=0, \tag{32}$$

where the maximum wind stress, T_o, is 0.1 Pa, and the decay factor, a, is chosen to give an e-folding of 200 km. The latter means that the offshore boundary experience only 8% of the coastal wind forcing.

In this simple case it is possible to derive analytic or at least some approximate analytic solutions to the governing eqs. (1)-(3). Initial conditions are assumed quiescent, i.e. U=V=0 and $\eta=0$ for all x,y at t=0. Along the coast, the normal flux is set to zero, i.e. V=0 at y=0. Along the offshore boundary, y=L/2, the elevation is clamped ($\eta=0$) as will be required in the numerical simulations. Details on the derivations of the analytic solution is given in Appendix B. Because of bottom friction a steady-state solution exists in the limit as $t \rightarrow \infty$ and is given by,

$$U=HT/\rho R, \tag{33}$$

$$\eta=(fT_o/a\rho gR)\{\exp(-ay)-\exp(-aL/2)\}. \tag{34}$$

Thus the steady-state solution is characterized by a wind driven alongshore flux as given by (33). This in turn is in geostrophic balance with the cross-shore pressure gradient determined by the sea surface elevation slope. Using the typical values given in Table 1 a maximum elevation of 9.1 cm is reached at the coast with a maximum depth mean current, (U/H), of 4.1 cm/s.

No general transient solution exists for this problem, but a limited solution can be found using Laplace transforms (cf. Appendix B). For instance, the alongshore flux at the coast (y=0) is given by

$$U(y=0,t)=(HT_o/\rho R)\{1-\exp(-Rt/H)\}. \tag{35}$$

Only an approximate solution for the sea surface elevation is found and is given by

$$\eta(y,t)=(fT_o/a\rho gR)\{1-\exp(-Rt/H)\}\{\exp(-ay)-\exp(-aL/2)\}. \tag{36}$$

Thus the temporal scale has an e-folding scale equal (H/R) at coastal locations. With the values used this implies that essentially U already has reached its steady-state, to within a few percent, after 20 hrs. However, a somewhat longer time scale has to be expected at offshore locations, because it will take some time for the coast to signal its presence. The approximate solution (36) clearly satisfies both the steady-state solution dictated by (34) and the initial conditions. It does ignore inertial oscillations but experience show these to be of minimal importance in the present cases. The temporal response of the alongshore depth mean current is shown in Fig. 6(a) (look at the dashed curves). A synoptic picture of the steady-state solution is shown in Fig. 5 (ANA), which clearly shows the exponential offshore decay.

3.2 Bell-shaped wind

The goal of this case is to test the OBCs in the case of a strong wind forcing in the interior and a weak wind forcing close to the boundaries. For this case, the wind stress is bell-shaped in the alongshore direction, and decreases exponentially offshore. The wind is switched on at 0 hrs then shut off after 48 hours or for $0<t<48h$

$$T_s^x=T_o\exp(-a^2x^2)\exp(-ay), \tag{37}$$

$$T_s^y=0, \tag{38}$$

and for $t>48h$

$$T_s^x=T_s^y=0. \tag{39}$$

Values for T_o and a are the same as before. This means both the alongshore and cross-shelf e-folding scales are 200 km. Thus at the offshore boundary the value is again 8% of the value at the similar coastal location. However, at the lateral boundaries it is only 0.2% of the maximum wind forcing.

No analytic solution is possible for this case, so a numerical solution was derived by extending the model domain by 4000 km to the north and south. Since the propagation speed of the fastest wave in the basin is about 22 m/s, this means that the solution within the regular grid can not be affected by any lateral boundary reflections within the 96 hr integration period. The response of this extended grid solution is shown in Fig. 2, which shows the time series of the alongshore (north-south) depth mean current. As expected the solution for the spin-down period is a mirror image of the spin-up period. Kelvin waves pass through the northern boundary at roughly 9 and 57 hr (Fig. 2, site C). These are generated by the impulsive wind forcing at 0 and 48 hr. The forced solution (Fig. 2, site B), is seen to have a temporal response exponential in nature toward a steady-state solution. The solution is characterized by a depression in the southern part and a crest in the northern part as depicted by Fig. 7 after 48 hrs of integration (look at the frame marked EXT).

3.3 Moving storm

The goal of this case is to test the OBCs for a practical real-world wind field. The wind stress is generated by a cyclone translating at 15 m/s diagonally across the grid from southwest (SW) to northeast (NE) as depicted in Fig. 3. The translation speed is slow enough for Kelvin waves to propagate ahead of the storm. The cyclone center passes the SW corner approximately 38 hrs into the integration. At 48 hrs it is located in the middle of the regular grid and leaves it approximately 10 hrs later. The wind stress components are given by

$$T_s^x=-T_o(Y/R_c)\exp(\psi/2), \tag{40}$$

$$T_s^y=T_o(X/R_c)\exp(\psi/2), \tag{41}$$

where the argument is given by

$$\psi = 1-(r/R_c)^2, \quad r^2 = X^2+Y^2, \tag{42}$$

T_o=3.0 Pa, and R_c is the distance from the center of the storm to maximum wind stress (R_c=200 km). X and Y are given by

$$X = x - x_o - u_o t, \tag{43}$$

$$Y = y - y_o - v_o t. \tag{44}$$

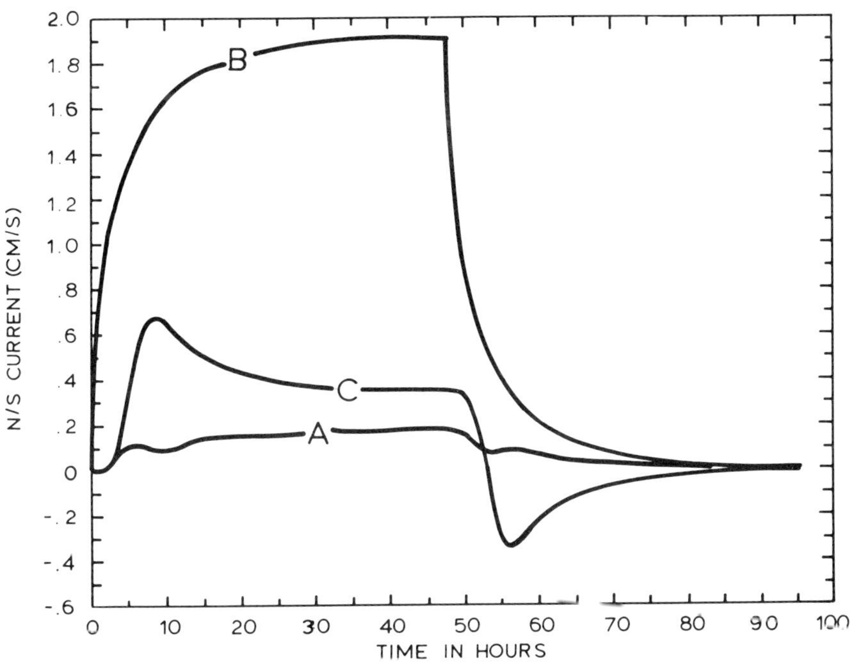

Fig. 2. Solid curves depict the temporal response of the alongshore (north-south) depth mean current (U/H) for the bell-shaped wind forcing using the extended grid. The letter attached to each curve corresponds to time series at the following grid elements: A=(2,2), B=(25,2), and C=(50,2), where (i,j) denotes grid element location as depicted by Fig. 1.

Here x_o, y_o denote the initial position of the storm center and u_o, v_o the velocity components of the center.

No analytic solution is possible for this case, so a numerical solution was derived by extending the model domain as in the bell-shaped wind case. At 36 hr the presence of the storm is becoming visible in the form of a Kelvin wave propagating into the regular grid (Fig. 3). At 48 hr the storm center is in the middle of the regular grid as indicated by the depression following in the wake of the storm. The Kelvin wave has just reached the

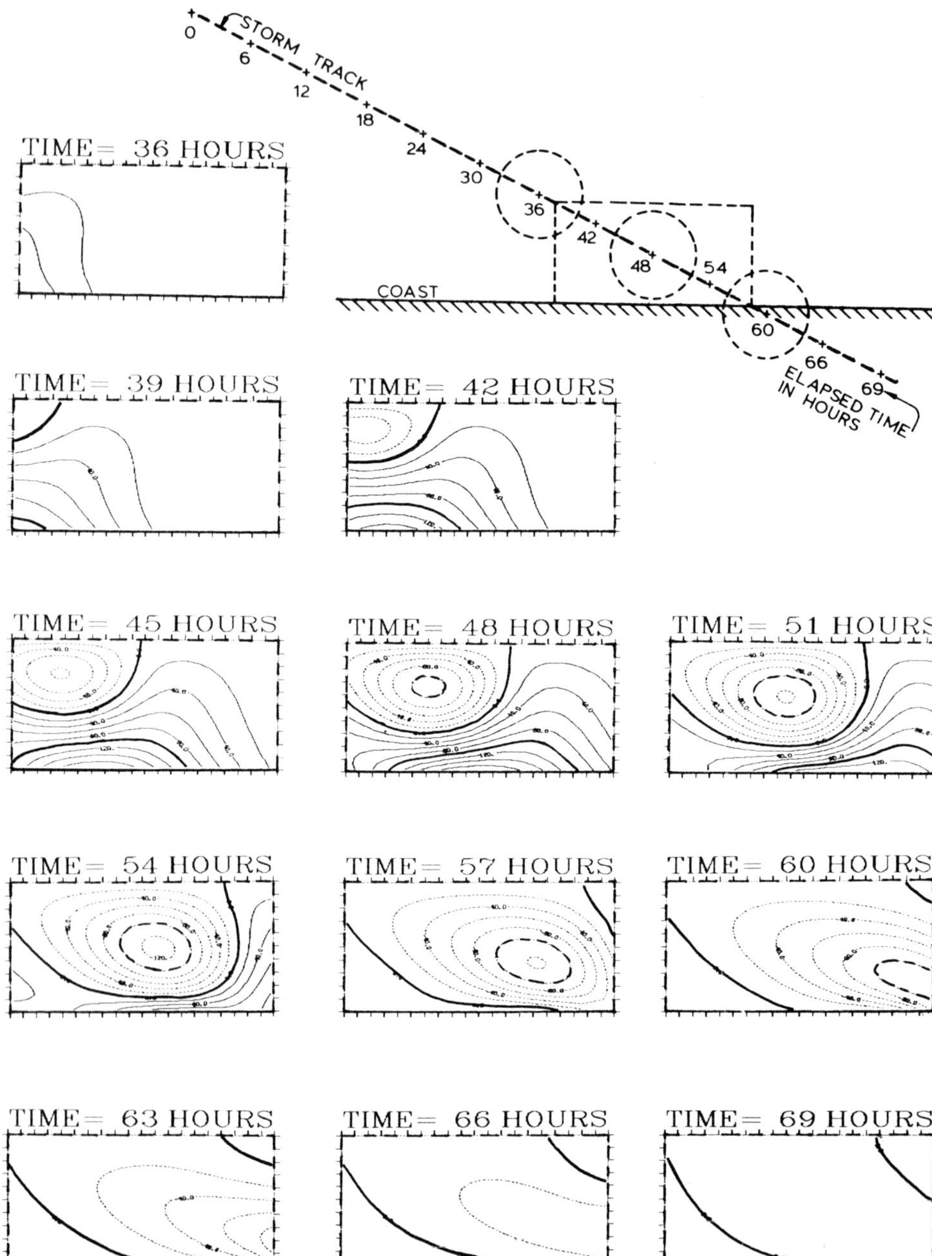

Fig. 3. Snapshots of sea surface elevation using the extended grid. Only the subdomain corresponding to the regular grid is shown. Contour interval is 20 cm between lines with 100 cm between heavy lines. The storm track of the moving cyclone is shown in the upper right panel. Numbers along track (in hours) indicate location of storm center after start of simulation. The dashed circle indicate the radius to maximum wind stress (200 km). Maximum wind stress is 3.0 Pa. The location of the regular grid with respect to the storm track is indicated by the dashed rectangle.

northern boundary at this time and starts to propagate out of the regular grid. By 54 hr the center is located close to the NE corner and the Kelvin wave is nearly gone from the domain. By 72 hr the disturbances created by the storm have essentially disappeared. Fig. 4 shows the situation from another vantage point - the time series of surface elevation at sites A, B and C.

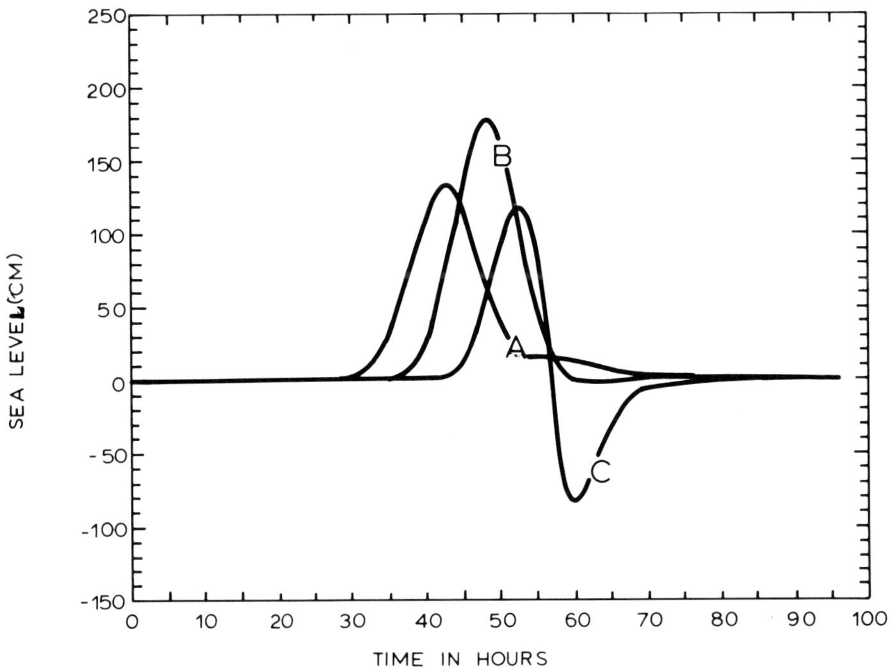

Fig. 4. The temporal response in terms of the sea surface elevation for the moving storm case at the sites A, B, and C. Solution is based on the extended grid simulation.

4 SIMULATIONS WITH OBCS

Considered in the present section is the performance of the seven OBCs for the different wind-forced cases. The solutions are generated with the wind forcing described in Section 3 and with the seven OBCs imposed at the two lateral boundaries of the regular grid. The performance is studied by comparing the numerical results with the standard cases described in Section 3 thought to represent the solution with a perfect OBC. The criteria whereby the performance of each OBC may be evaluated is based on the definition of an open boundary suggested by Røed and Cooper (1986). Thus an open boundary is one which allows disturbances originating in the regular grid to leave it without disturbing or deteriorating the solution within the regular grid.

Comparisons are made using time series of sea surface elevations, depth mean currents and snapshots of the sea surface elevation. In addition the term excess mass is often used in the comparisons. The excess mass is defined as the difference between the initial mass and the total mass at any instant. By dividing this difference with the area covered by the regular grid and the density the excess mass may be computed from the expression,

$$EM=(1/A)\iint_A \eta dxdy, \tag{45}$$

where A is the area covered by the regular grid. Thus the excess mass is a measure of the mean sea level deviation away from the equilibrium depth.

4.1 Uniform alongshore wind

The solutions after 96 hours of integration are displayed in Fig. 5, which also includes the analytic solution derived in Section 3. GRD, MOE, FOE and HOC all provide reasonable responses, while CLP, SPO and ORC perform poorly.

CLP performs poorly mainly because it is unable to produce the correct wind-forced alongshore currents. The analytic steady-state solution (33) and (34) shows this current to be in geostrophic balance with the cross-shore pressure gradient. The condition imposed by CLP effectively prohibits this wind-forced balance at the boundaries. The consequences of this failure extend to the interior of the regular grid as may be seen from the time series depicted in Fig. 6 at about 7 hours. At this time the coastal elevation levels off and reach a steady-state solution which is only 30% of the analytic solution giben by eq. (34).

The poor performance of SPO in this case may be explained by the incorrect mass fluxes generated in the sponge when strong forcing is present at the boundaries. By inspection of the steady-state solution (33), the cross boundary flux is decreased in the sponge due to the increased bottom drag coefficient there. Hence it fails to balance the wind forced flux within the regular grid. This creates a false divergence in the alongshore flux which in turn is responsible for the alongshore gradient depicted in Fig. 5 (look at the frame marked SPO).

ORC performs poorly in this case because it carries information in the cross-shelf direction by means of free waves. Thus the information about the no-flux condition at the coast propagates along the open boundary in the form of a free wave rather than a forced wave. This information is thus passed along with the speed given by c^y of (17), and this is not fast enough to give the correct cross-boundary fluxes. Divergences and convergencies are created at the boundaries which in turn create the false alongshore gradient in the sea surface elevation indicated after 96 hrs in Fig. 5.

Curiously enough MOE offers an acceptable response in this case despite its apparent similarity to ORC. This is because MOE by virtue of the Camerlengo and O'Brien (1980) modified form lacks

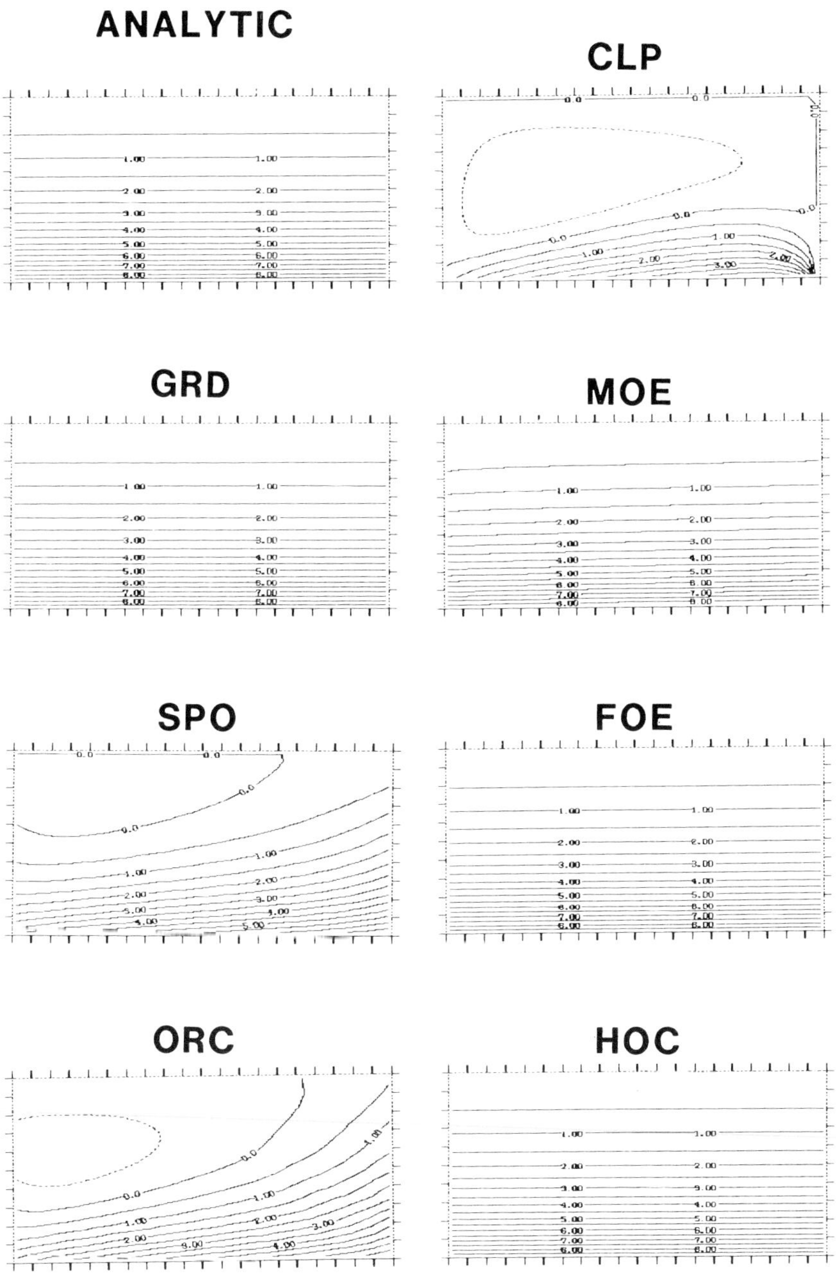

Fig. 5. The solution after 96 hours of integration in terms of contours of surface elevation for the uniform alongshore wind case using the seven OBCs. Dashed curves indicate depressions. Contour interval is 0.5 cm. The domain corresponds to the regular grid. The OBC notation follows that of Table 2.

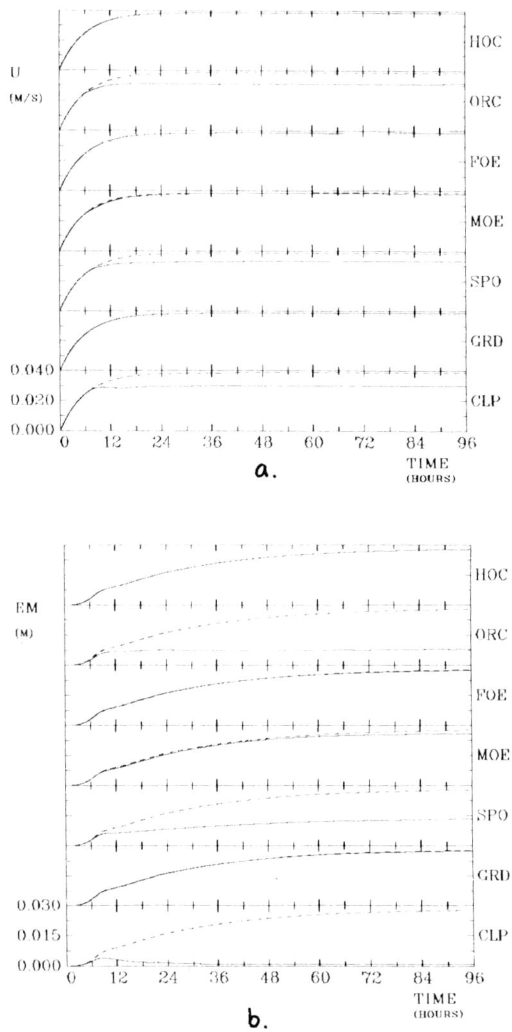

Fig. 6. Time evolution of the depth mean current (upper panel) and excess mass (lower panel) at site B for the uniform alongshore wind case. The dashed curve is the analytic solution. OBC used is indicated to the right of the time series. Scales are indicated by the bottom left numbers along the vertical axis.

the radiative nature of ORC. Thus MOE updates the alongshore flux more rapidly.

4.2 Bell-shaped wind

The solution at 48 hrs is shown in Fig. 7 just as the wind has been shut off. Comparing the different OBCs with the extended grid solution indicates that CLP, SPO, ORC, and to some degree also FOE and HOC provide reasonable responses. GRD and MOE are clearly inferior to the others. These conclusions are supported by the temporal responses shown in Fig. 8.

ORC is superior to the other radiation conditions in this case. As discussed in Section 3, Kelvin waves are created in the interior by the impulsive wind forcing. These waves are essentially free waves for this case. As noted by Raymond and Kuo (1984) ORC is carefully constructed so as to radiate free waves. Hence it is expected to be superior to most other radiation conditions for this case.

Inspection of the CLP temporal response as depicted in Fig. 8 reveals relatively small reflections despite the fact that CLP is purely reflective. This is because the bottom friction effectively dampens out these reflections. Hence the energy of the reflected waves are small in the middle of the basin. The impact of CLP is therefore controlled largely by the dissipation in the system. So for shallow shelf regions with large bottom friction, CLP may be acceptable if the site of interest is far from the boundaries. This is supported by the response shown in Fig. 7 (look at the frame marked CLP).

Like CLP, GRD is also purely reflective. But in this case this plays a minor role. More importantly, GRD causes major changes in mass as shown in the lower panel of Fig. 8. This is most apparent as the Kelvin waves hits the lateral boundaries; mass is extracted from the regular grid at a constant rate. This may be explained by GRD's inherent ability to maintain (or create) a cross-shore geostrophic balance due to the imposed zero alongshore slope. This ability was crucial for the success of GRD in the previous case, but seems to be responsible for its poor performance in this case.

The poor performance of MOE is also linked to its excess mass response. Again the difference in response between ORC and MOE is obvious and strongly suggests that ORC is superior to MOE when there is little or no forcing at the boundary.

In this particular case one would expect FOE to give a response similar to that of MOE because the local part is close to zero. Certainly, several of the features present in the MOE response are reflected in the FOE response (cf. Figs. 7 and 8). There is, however, one important difference. As the Kelvin wave leaves the boundary to the north, FOE recovers and stabilizes in contrast to MOE. This is especially true in terms of the excess mass behavior (cf. Fig. 8(b)), but is also apparent in the alongshore depth mean current (Fig. 8(a)).

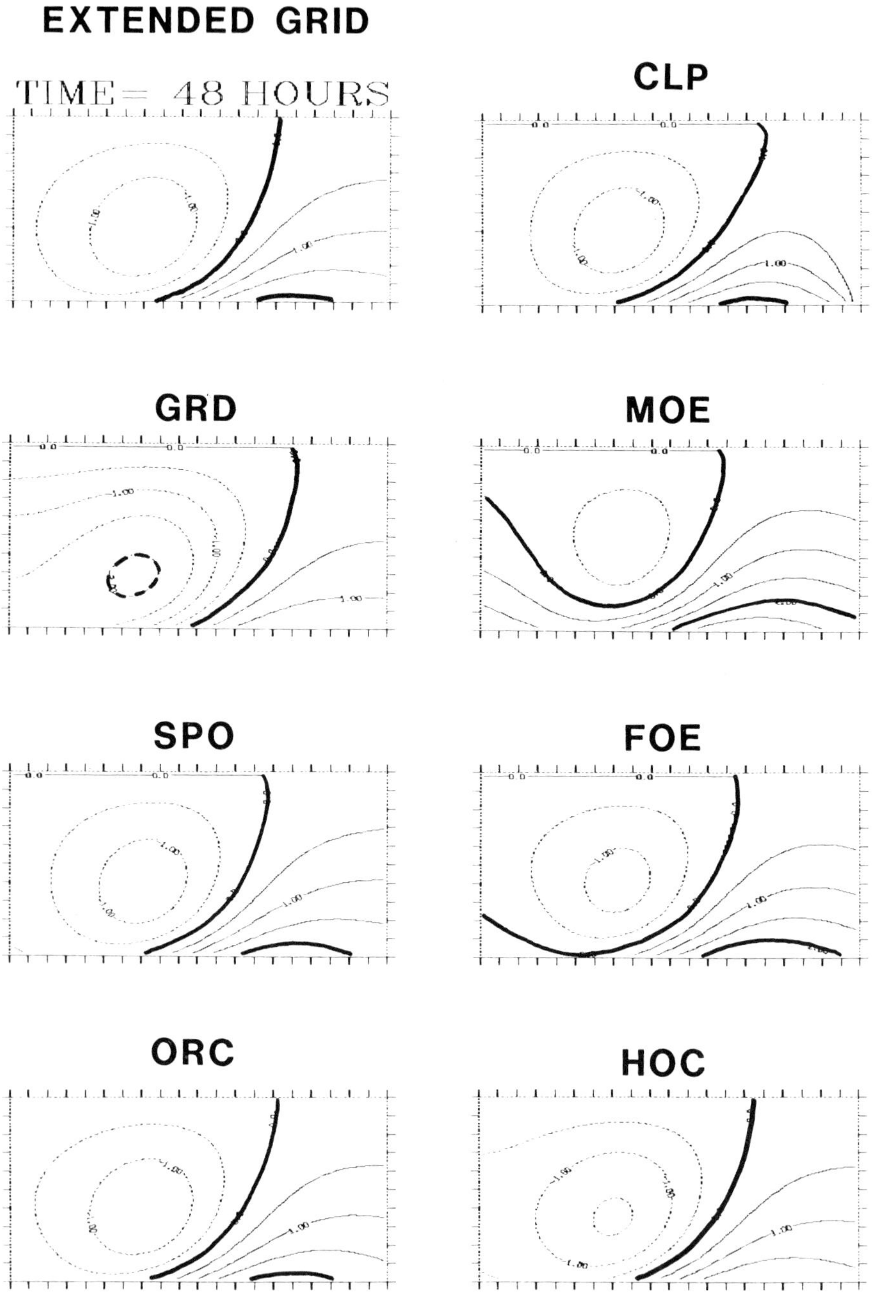

Fig. 7. Snapshots of the sea surface elevation at 48 hrs for the bell-shaped wind case. Notation otherwise as in Fig. 5. Contour interval is 0.5 cm between lines, with 2 cm between heavy lines.

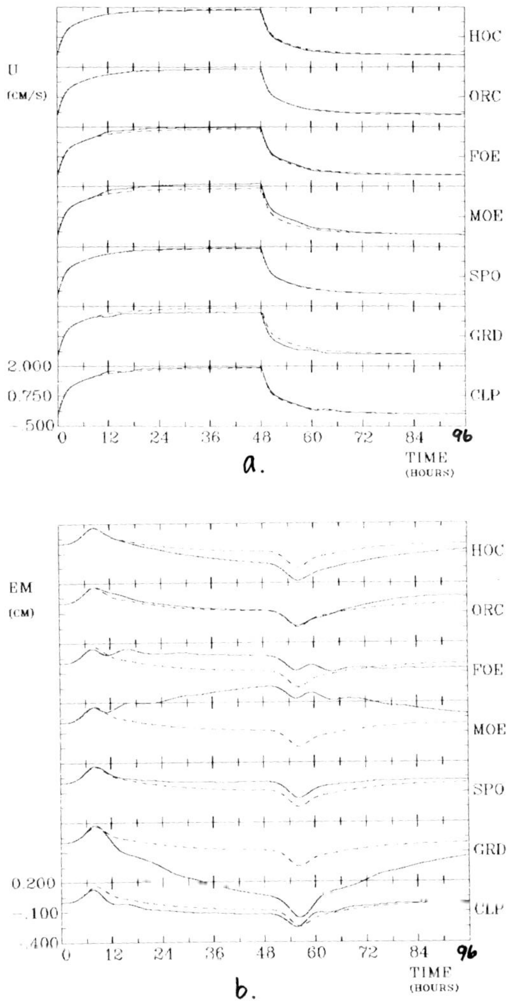

Fig. 8. Same as Fig. 6 for the bell-shaped wind case.

HOC reproduces the alongshore depth mean current well (Fig. 8(a)), but not the excess mass. This is supported by the snapshot in Fig. 7 which shows the problem to be linked to the response at the southern boundary. Indeed, looking at the time series of the alongshore depth mean current at this boundary (Fig. 9) shows that it consistently deviates from the extended grid solution in such a way as to provide less flux. This is not the case at the northern boundary.

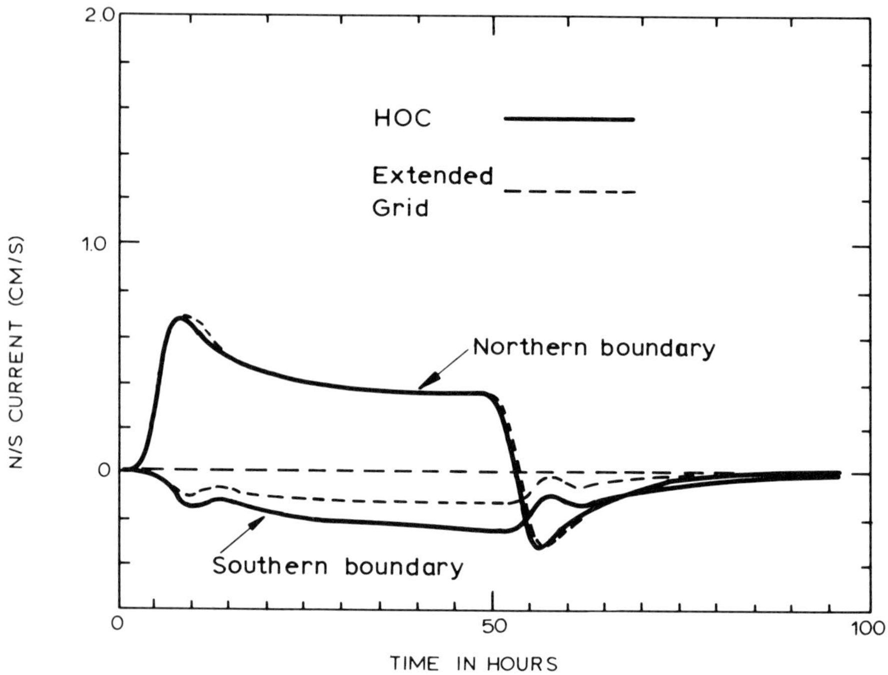

Fig. 9. The time evolution of the depth mean current at sites A and C, respectively, for the bell-shaped wind case. Dashed line indicates the response for the extended grid simulation while the solid curve corresponds to the response using HOC.

4.3 Moving storm

The response of the different OBCs for this case is shown in Figs. 10-12. CLP is not presented for this simulation because of its proved poor behavior both in the previous cases and in other cases including realistic wind fields (cf. Beardsley and Haidvogel, 1981; Chapman, 1985; Hayashi et al., 1986). All the OBCs are less than perfect in this case, but GRD, SPO and HOC seem to provide reasonable responses.

In order to evaluate the OBCs it is important to realize that some of the features in the extended grid simulation may not be included in the regular grid simulation. For example, a Kelvin wave is generated when the winds are initially started. This wave

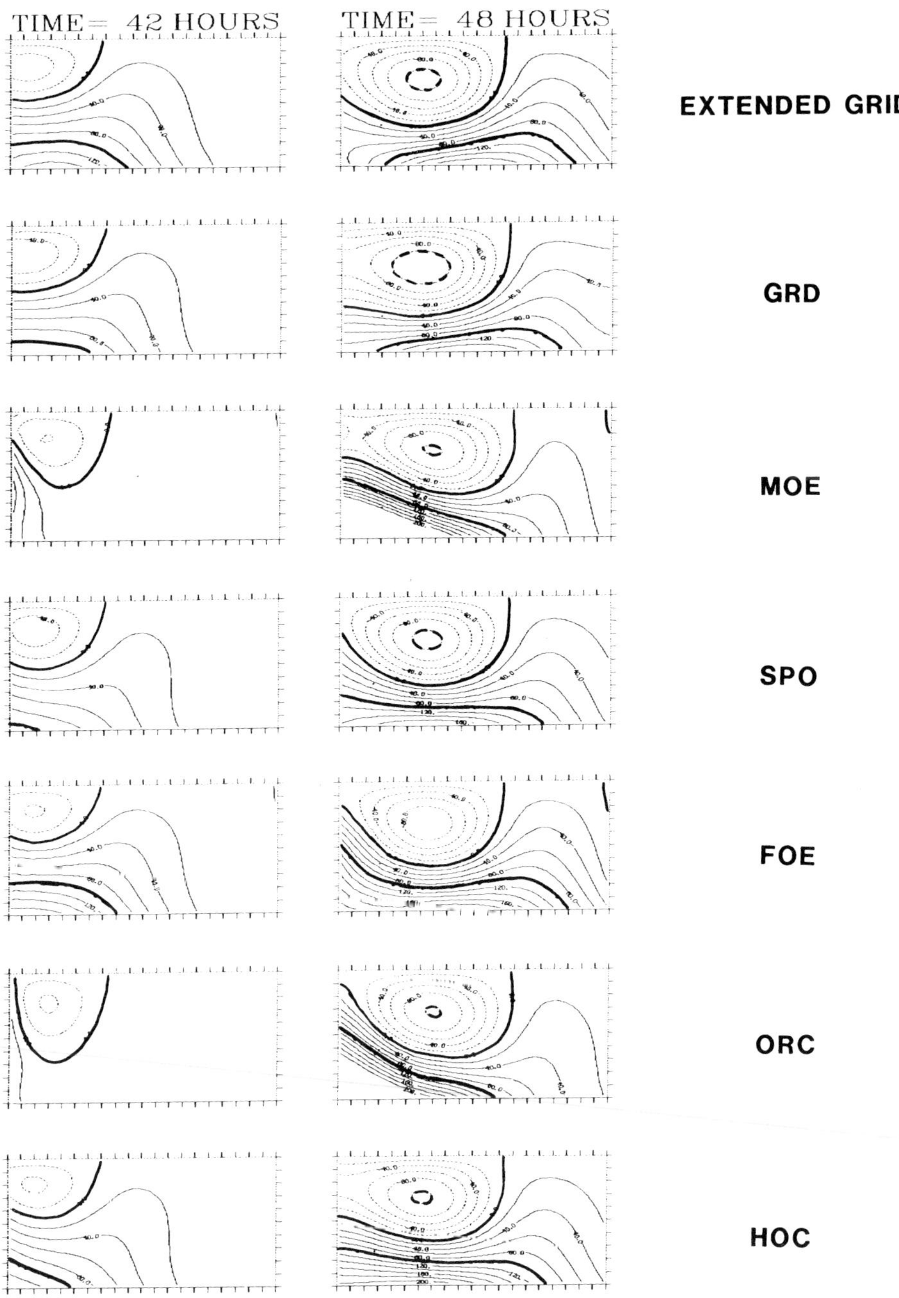

Fig. 10. Snapshots of the sea surface elevation at 42 and 48 hrs for the moving storm case using the different OBCs. Notation otherwise as in Fig. 5. Contour interval and parameters as in Fig. 3.

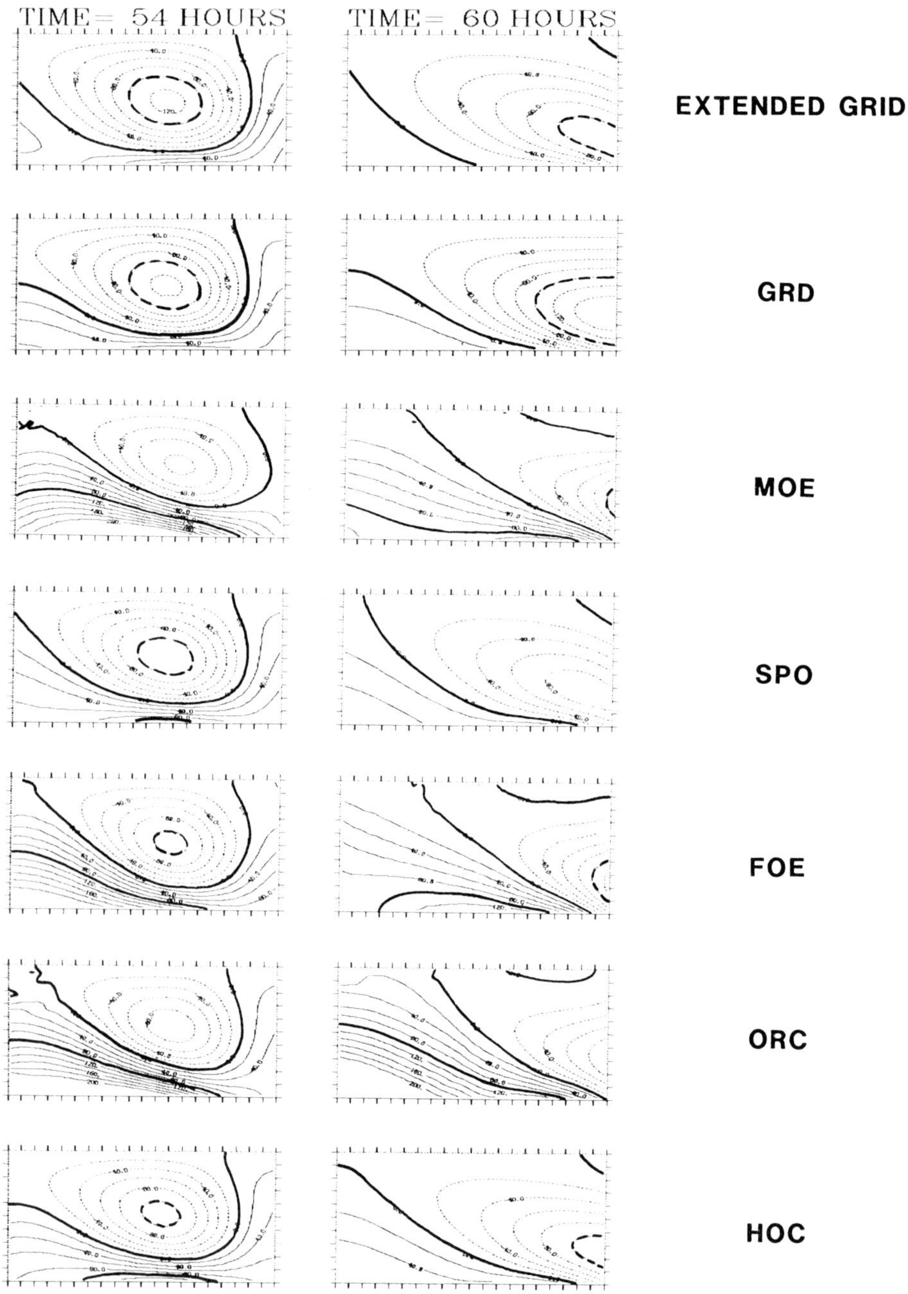

Fig. 11. Same as Fig. 10 only snapshots are at 54 and 60 hrs.

propagates in front of the storm and enters the regular grid at about 32 hrs (Fig. 12(b)). As is evident by inspection of Fig. 12(b) all the regular grid simulations are unable, except FOE, to provide this Kelvin wave. This delay is caused by the inherent inability of the regular grid simulations to form the correct Kelvin wave outside of the grid domain.

It is interesting to note that GRD does fairly well in this case. This may be explained by the conditions inherent ability to create and maintain a cross-shore geostrophic balance. It is able to, somewhat belatedly, to pick up the incoming Kelvin wave due to its strong cross-shelf geostrophic balance (cf. Figs. 10 and 12). The same is true for the depression which follows in the wake of the storm which hits the northern boundary at about 56 hr. GRD, however, is clearly reflective as depicted by Fig. 12(b). In this case, with a shallow sea the energy of the reflected wave is quickly eroded by bottom friction and is far less obvious in the middle of the basin. This is supported by the temporal response of the alongshore depth mean current as depicted in Fig. 12(a).

SPO also gives reasonable results. As revealed by Figs. 10-12 the interior response is only slightly distorted by the strong forcing at the boundaries. This is explained by the short time span that forcing is present at the boundaries, and by the fact that the forcing varies strongly in time. The presence of the sponge layer does impact the response somewhat and adversely affects the mass balance. Despite this deviation from the extended grid simulation, especially close to the boundaries, the response in terms of the depth mean current is satisfactorily.

HOC also performs reasonably well. Consistent with the previous simulation with weak forcing at the boundary, the deviations are most pronounced at the southern boundary when the storm has left the regular grid domain (Figs. 10 and 11).

MOE, FOE and ORC performs poorly for this case. For instance, both MOE and ORC do not provide the incoming Kelvin wave at all. This explains the exceptional long delay in the excess mass response (Fig. 12(b)). Due to the forced constituent of FOE this condition is able to create the incoming Kelvin wave faster than any of the other OBCs, but otherwise it behaves like MOE. Their respective responses are consistent with their responses in the previous two simulations. For instance, as the Kelvin wave hits the boundary to the north it is a free wave. Hence ORC initially transmits the wave almost perfectly, but as the forcing increases ORC deteriorates (Figs. 10 and 11).

5 DISCUSSION AND FINAL REMARKS

The response of a barotropic, numerical ocean model to different open boundary conditions (OBCs) is studied. Emphasis is on the sensitivity of the models interior wind-forced response to changes in the OBC. Seven OBCs are compared with each other as well as with benchmark simulations. These are either analytical solutions or numerical solutions derived from simulations with an

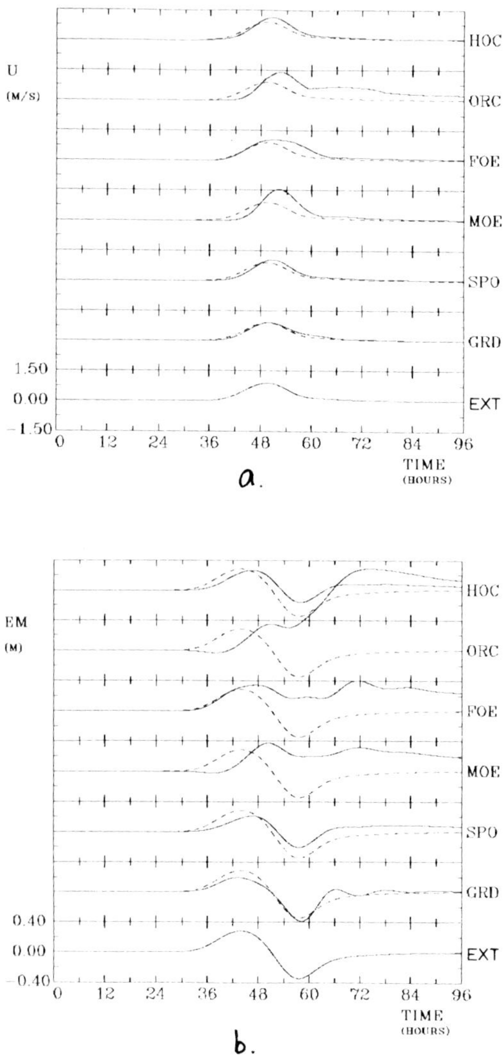

Fig. 12. Same as Fig. 6 for the moving storm case.

extended grid. The OBCs are summarized in Table 2. Three wind-forced cases have been studied with: 1) strong, 2) weak and 3) time varying wind forcing at the models open boundaries.

The results are summarized in Table 3. These evaluations are based upon an intercomparison of the response in terms of integrated quantities such as: 1) excess mass, and 2) snapshots of the sea surface elevation as well as time series of: 1) elevation and 2) depth mean current. Evaluation of performance of each OBC is based on the definition of an open boundary as a computationally boundary at which disturbances originating in the interior of the regular domain is free to leave it without disturbing or deteriorating the interior solution (Røed and Cooper, 1986).

TABLE 3

Summary of results. Y(es) indicates reasonable results while N(o) indicates a less satisfactory response. The analytic or extended grid solution is used as a comparison. Comparisons are made at the southern boundary (S) and northern boundary (N) and in terms of time series of the alongshore depth mean current at site B (U) and excess mass (EM). Finally an overall indicator (T) is provided.

OBC type	Uniform wind, strong forcing					Bell-shaped wind, weak forcing					Moving storm, varying forcing				
	S	N	U	EM	T	S	N	U	EM	T	S	N	U	EM	T
CLP	N	N	N	N	N	Y	Y	Y	Y	Y	-	-	-	-	-
GRD	Y	Y	Y	Y	Y	N	Y	N	N	N	Y	N	Y	Y	Y
MOE	N	Y	Y	Y	Y	N	N	N	N	N	N	N	N	N	N
SPO	N	Y	N	N	N	Y	Y	Y	Y	Y	Y	N	Y	Y	Y
FOE	Y	Y	Y	Y	Y	N	Y	Y	Y	Y	N	N	N	N	N
ORC	N	N	N	N	N	Y	Y	Y	Y	Y	N	N	N	N	N
HOC	Y	Y	Y	Y	Y	N	Y	Y	Y	Y	Y	Y	Y	Y	Y

In general the response is highly sensitive to changes in the imposed OBC. There is no "best" solution. For instance, OBCs that provide reasonable responses with strong forcing at the boundaries, are found to perform poorly in cases with weak forcing. This indicates that the choice of OBC depends upon the particular application that the modeler has in mind. The results also indicate that incorrect responses are most likely to show up at an upstream boundary first. Upstream is here with respect to the wind direction (here the southern boundary).

The following are the main conclusions:

(1) HOC is the only OBC which provides a reasonable response in all cases studied.

(2) SPO may prove viable, especially in cases when weak or variable forcing is applied. In cases with strong and persistent forcing, problems are likely to arise and is due to flux constraints imposed by the sponge.

(3) GRD, one of the most widely used OBCs, provides reasonably responses away from the boundary if the domain is shallow, frictionally dominated and relatively large.

(4) ORC is superior to the other radiation conditions when it comes to handling free waves at the boundaries. The results indicate that use of ORC rather than MOE may improve the FOE response. Exploratory experiments (not shown here) support this conclusion.

(5) CLP should be avoided in most applications. However, it is easy to implement and may be used in short term simulations in which disturbances originating at the boundary do not have time to travel back and adversely affect the interior solution.

(6) MOE and FOE should also be avoided in most cases. However, FOE could be improved by incorporating a better radiation condition for its global part.

The above conclusions are limited to a study of OBCs for a barotropic, linear model. Nonlinearities, stratification and topography have been neglected in this study. Most of the above OBCs are easily extended to nonlinear models, and more complicated topography. Further studies with a shelf break within the regular grid (not shown) suggest that our conclusions are unchanged. Problems will arise in extending the radiation conditions to baroclinic models. In this case the disturbance will be a mixture of baroclinic and barotropic waves and this complicates the numerical computation of the wave celerity of the disturbance. A decomposition into normal modes may possibly resolve this problem. None of the above considerations seem to constrain the applicability of HOC as long as a set of compatibility equations may be derived.

There are probably several ways whereby the response due to a particular OBC may be improved. The obvious is to experiment with combinations of different OBCs. Specifically, exploratory experiments (not shown here) with FOE in which MOE was replaced by ORC for the global part was promising. Another suitable combination is to substitute HOC for MOE in SPO. Moreover, the flow relaxation scheme suggested by Martinsen and Engedahl (1986) which combines SPO and the philosophy of FOE, is also reported to have some success.

APPENDIX A: NUMERICAL FORM OF THE SEVEN OPEN BOUNDARY CONDITIONS

All the OBCs are applied at alongshore flux or U-points (cf. Fig. 1). Only the implementation at the northern (x=L) boundary are given in detail.

1. CLP

This is the most cost efficient OBC. Application of (8) at a U-point gives (linear interpolation),

$$\eta_{M+1,j}^{n}=-\eta_{M,j}^{n} \;;\; j=2,3,..,N, \tag{A1}$$

for all time levels n.

2. GRD

This OBC is based on (9) with the numerical analogue,

$$\eta_{M+1,j}{}^{n}=\eta_{M,j}{}^{n} \; ; \; j=2,3,..,N, \tag{A2}$$

for all time levels n.

3. MOE

An upstream differencing as outlined by Miller and Thorpe (1981) has been used in order to compute c^u from (12), viz.,

$$c^u=-(\Delta s/\Delta t)\frac{U_{M-1,j}{}^{n}-U_{M-1,j}{}^{n-1}}{U_{M-1,j}{}^{n-1}-U_{M-2,j}{}^{n-1}} \quad ; \; j=2,3,..,N. \tag{A3}$$

Once c^u is computed (11) may be used to find U^{n+1} (Orlanski, 1976). In the present application, however, the modified form suggested by Camerlengo and O'Brien (1980) has been used. Hence

$$U_{M,j}{}^{n+1}=\begin{cases} U_{M-1,j}{}^{n} & ; \; c^u > 0 \\ U_{M,j}{}^{n} & ; \; c^u \leq 0 \end{cases} \quad ; \; j=2,3,..,N. \tag{A4}$$

4. SPO

As explained in Section 3 the main feature of SPO is to add a sponge outside the southern and northern boundaries. Within the sponge the governing eqs. (1)-(3) are solved as within the regular grid with due regard to the increasing bottom friction coefficient as given by eq. (14). At the outer edge of the sponge MOE is used. The sponge in the present implementation is 200 km wide which corresponds to adding 10x26 grid elements to the regular grid at the northern and southern ends. This increases the computational effort by 20%.

5. FOE

The local part of U is computed from eqs. (1)-(3) by neglecting the terms containing differentials with respect to x. It is important to realize that this subset of equations may be solved without any reference to an OBC. It provides the locally generated Ekman fluxes due to the wind and the local fluxes due to a cross shelf geostrophic current. The "free" or global part at points adjacent to the boundary is found by subtracting the local solution from the total solution. To find the global part at the boundary for time level n+1 MOE is used. Finally U^{n+1} at the boundary is found by adding the global and local parts.

6. ORC

The numerical form chosen is based on an upstream differencing of (16), viz.,

$$U_{M,j}^{n+1}=(1-r^{x})U_{M,j}^{n}+r^{x}U_{M-1,j}^{n}-r^{y}\underline{D}_{y}U_{M,j}^{n}. \tag{A5}$$

Here the coefficients r^x and r^y are given by,

$$r^{y}=\begin{cases}-1 & ; \quad c^{y}\leq-(\Delta s/\Delta t)\\ (\Delta t/\Delta s)c^{y} & ; \quad -(\Delta s/\Delta t)<c^{y}<(\Delta s/\Delta t)\\ 1 & ; \quad c^{y}\geq(\Delta s/\Delta t),\end{cases} \tag{A6}$$

$$r^{x}=\begin{cases}0 & ; \quad c^{x}\leq 0\\ (\Delta t/\Delta s)c^{x} & ; \quad 0<c^{x}<(\Delta s/\Delta t)\\ 1 & ; \quad c^{x}\geq(\Delta s/\Delta t),\end{cases} \tag{A7}$$

and

$$\underline{D}_{y}U_{M,j}^{n}=\begin{cases}U_{M,j+1}^{n}-U_{M,j}^{n} & ; \quad c^{y}\leq 0\\ U_{M,j}^{n}-U_{M,j-1}^{n} & ; \quad c^{y}>0.\end{cases} \tag{A8}$$

The celerities c^x and c^y are computed by a numerical analogue of (17) in a fashion similar to (A3), e.g.

$$c^{x}=-(\Delta s/\Delta t)\frac{(D_{t}U_{M-1,j}^{n})(D_{x}U_{M-1,j}^{n-1})}{(D_{x}U_{M-1,j}^{n-1})^{2}+(D_{y}U_{M-1,j}^{n-1})^{2}} \tag{A9}$$

where,

$$D_{t}U_{i,j}^{n}=U_{i,j}^{n}-U_{i,j}^{n-1}, \tag{A10}$$

$$D_{x}U_{i,j}^{n}=U_{i,j}^{n}-U_{i-1,j}^{n}, \tag{A11}$$

$$D_{y}U_{i,j}^{n}=(U_{i,j+1}^{n}-U_{i,j-1}^{n})/2. \tag{A12}$$

The expressions (A5)-(A12) are all evaluated for j=2,3,.,N and one-sided differences are used for j=2 and j=N whenever necessary.

7. HOC

In order to find $U_{M,j}^{n+1}$ a finite difference form of (26) has been implemented. The terms containing derivatives with respect to x have been replaced by one-sided differences. Thus,

$$U_{M,j}^{n+1}=(U_{M,j}^{n}+U_{c}+U_{f}+U_{p})/(1+R\Delta t/H_{M,j}), \tag{A13}$$

where

$$U_c=f\Delta t(V_{M,j}{}^n+V_{M,j-1}{}^n)/2, \qquad \text{(A14)}$$

$$U_f=\Delta t(T^x{}_s/\varrho)_{M,j}{}^{n+1}, \qquad \text{(A15)}$$

$$U_p=-\gamma c_{o1,j}(U_{M,j}{}^n-U_{M-1,j}{}^n+c_{o1,j}(\eta_{M,j}{}^n-\eta_{M-1,j}{}^n))$$

and $\gamma=\Delta t/2\Delta s$. Note that all the expression (A13) and (A16) are evaluated for $j=2,3,..,N$.

APPENDIX B: THE ANALYTIC SOLUTION

1. Steady-state solution

Since bottom friction is included, it is reasonable to look for a steady-state solution in the limit as $t\rightarrow\infty$. Furthermore, since the wind stress only depends on y, the dependent variables will have no x-dependence. With these observations the governing equations become,

$$-fV=T/\varrho-RU/H, \qquad \text{(B1)}$$

$$fU=-gH\eta_y-RV/H, \qquad \text{(B2)}$$

$$V_y=0. \qquad \text{(B3)}$$

Integration of the last equation in conjunction with the boundary condition at the coast yields V=0 everywhere. Using this along with the clamped condition at the offshore boundary then yields the closed-form steady-state solution (33) and (34).

2. Transient solution

No general transient solution exists for this problem. However, a limited solution can be found by transforming the governing equations by invoking Laplace transforms. Let a circumflex denote a transformed quantity. Applying the transforms in the (y,s) space to eqs. (1)-(3) then gives

$$s\hat{U}-f\hat{V}-(T/\varrho s)\ R\hat{U}/H, \qquad \text{(B4)}$$

$$s\hat{V}+f\hat{U}=-gH\hat{\eta}_y-R\hat{V}/H, \qquad \text{(B5)}$$

$$s\hat{\eta}+\hat{V}_y=0. \qquad \text{(B6)}$$

The first of these equations may be solved with respect to $\hat{U}$ to give

$$\hat{U}=[f\hat{V}+(T/\varrho s)]/[s+(R/H)]. \qquad \text{(B7)}$$

Since the transformed $\hat{V}$ satisfies the no flux condition at the coast (y=0) (35) follows by inversion of (B7).

Now, solving (B5) and (B6) with respect to $\hat{\eta}$ and $\hat{V}$ gives

$$\hat{\eta}=(\lambda^2\beta/s(a^2-\lambda^2))(a\exp(-ay)-\lambda\exp(-\lambda y)), \quad (B8)$$

$$\hat{V}=(\lambda^2\beta/(a^2-\lambda^2))(\exp(-ay)-\exp(-\lambda y)), \quad (B9)$$

where

$$\lambda=(s\{[s+(R/H)]^2+f^2\}/gH[s+(R/H)])^{1/2}, \quad (B10)$$

$$\beta=fT_o/\rho s([s+(R/H)]^2+f^2). \quad (B11)$$

Formally the solutions for U, V, and η may now be found by inverting the transformed solutions (B7)-(B9) back to (y,t) space. In general this is not possible because of the essential singularity at the point s=-R/H. Moreover, the approach taken to find U at the coast does not avoid this singularity in the inversion of $\hat{\eta}$. However, it is possible to obtain a good approximation for η everywhere by assuming the coastal elevation to be in geostrophic balance or,

$$\eta_y(y=0,t)=-(f/gH)U(y=0,t), \quad (B12)$$

where (35) specifies U(y=0,t). Assuming the offshore decay to be exponential in nature and (B12) to hold for all y, (36) follows by integration of (B12) with respect to y.

ACKNOWLEDGMENTS

The authors would like to thank Dr. R.B. Gordon for stimulating and encouraging discussions. This research was supported by Conoco Norway Inc. through a contract with Det norske Veritas. The support is gratefully acknowledged.

REFERENCES

Beardsley, R.C., and D.B. Haidvogel, 1981. Model studies of the wind-driven transient circulation in the Middle Atlantic Bight. Part 1: Adiabatic boundary conditions, J. Phys. Oceanogr., 11: 355-375.

Chapman, D.C., 1985. Numerical treatment of cross-shelf open boundaries in a barotropic coastal ocean model, J. Phys. Oceanogr., 15: 1060-1075.

Csanady, G.T., 1982. Circulation in the coastal ocean. D. Reidel Publishing Co., 279.

Freeman, J.C. Jr., and L. Baer, 1957a. Pseudo-characteristics. Trans. Amer. Geophys. Union, 38: 65-67.

Freeman, J.C. Jr., and L. Baer, 1957b. The method of wave derivatives, Trans. Amer. Geophys. Union, 38: 483-494.

Harper, B.A., and Sobey, R.J., 1983. Open boundary conditions for open-coast hurricane storm surge, Coastal Engineering, 7: 41-60.

Hartree, D.R., 1953. Some practical methods of using characteristics in the calculation of non-steady compressible flow, Rep.AECU-2713, U.S.Atomic Energy Comm., Washington, D.C.

Hayashi, T., Greenberg, D.A., and Garrett, C.J.R., 1986. Open boundary conditions for numerical models of shelf sea circulation. Cont. Shelf Res., 5: 487-497.

Hedstrom, G.W., 1979. Nonreflecting Boundary Conditions for Nonlinear Hyperbolic Systems. J. Comp. Phys., 30: 222-237.

Martinsen, E.A., and Engedahl, H., 1986. The flow relaxation scheme as an open boundary condition in a barotropic model, Submitted to Coastal Engineering.

Martinsen, E.A., Gjevik, B., and Røed, L.P., 1979. A numerical model for long barotropic waves and storm surges along the western coast of Norway. J. Phys. Oceanogr., 9: 1126-1138.

O'Brien, J.J., and R.O. Reid, 1967. The non-linear response of a two-layer, baroclinic ocean due to a stationary, axially symmetric hurricane, 1: Upwelling induced by momentum transfer, J. Atmos. Sci., 24: 197-207.

Orlanski, I., 1976. A simple boundary condition for unbounded hyperbolic flows, J. Comp. Phys., 21: 251-269.

Raymond, W.H., and Kuo, H.L., 1984. A radiation boundary condition for multi-dimensional flows, Quart. J. Roy. Met. Soc., 110: 535-551.

Røed, L.P., and Cooper, C.K., 1986. Open boundary conditions in numerical ocean models. In: J.J.O'Brien (Editor), Advanced Physical Oceanographic Numerical Modelling, NATO ASI Series C, Vol. 186, D. Reidel Publ. Co.: 411-436.

Røed, L.P., and Smedstad, O.M., 1984. Open boundary conditions for forced waves in a rotating fluid, SIAM J. Sci. Stat. Comput., 5: 414-426.

Sommerfeld, A., 1949. Partial differential equations: Lectures in Theoretical Physics, Vol. 6, Academic Press, 1949.

Verboom, G.K., and Slob, A., 1984. Weakly-reflective Boundary Conditions for Two-Dimensional Shallow Water Flow Problems. Paper presented at the 5th International Conference on Finite Elements in Water Resources, Burlington (VT), U.S.A., June 1984.

Wajsowicz, R.C., 1986. Free planetary waves in finite-difference numerical models. J. Phys. Oceanogr., 16: 773-789.

COASTAL CURRENTS, INTERNAL WAVE COLLAPSES AND TURBULENCE IN THE STRAIT OF MESSINA ZONE

E.SALUSTI
INFN, Dept. of Physics, Universita' "La Sapienza"
P.le A.Moro 2, 00185 Rome, Italy

R.SANTOLERI
Istituto di Fisica dell'Atmosfera, CNR
P.le Sturzo 31, 00144 Rome, Italy

ABSTRACT

Physical phenomena occurring in the area of the Strait of Messina, related to the current and interface rising on the sill are analyzed on the basis of recent experimental observations. The measurements of the interface time evolution on the sill have revealed for the first time the presence of turbulent columnar disturbances. The formation of a peculiar front in the Gulf of Gioia, in the southern Tyrrhenian Sea, is described as the result of turbulent mixing due to the breaking of nonlinear internal waves on the shelf of this Gulf, directly oriented toward the Strait of Messina where these waves are generated.

1. INTRODUCTION

In this note we outline some physical phenomena occurring in the area of the Strait of Messina, on the basis of new experimental observations, made by the Department of Physics of the University of Rome in the years 1980-1985, and of historical data.

The strait of Messina separates the Italian peninsula from Sicily (Fig.1) and is the natural connection between the Tyrrhenian Sea and the Ionian Sea. The width of the Strait at the narrowest point is approximately 3 km where the average depth is 75 m. The sea bottom slopes down to approximately 1300 m towards the Ionian Sea and to 300-600 m on the Tyrrhenian side. Thus the Strait constitutes a submarine barrier or sill to the marine water flowing through it.

From Vercelli's 1922 and 1923 measurements, the following

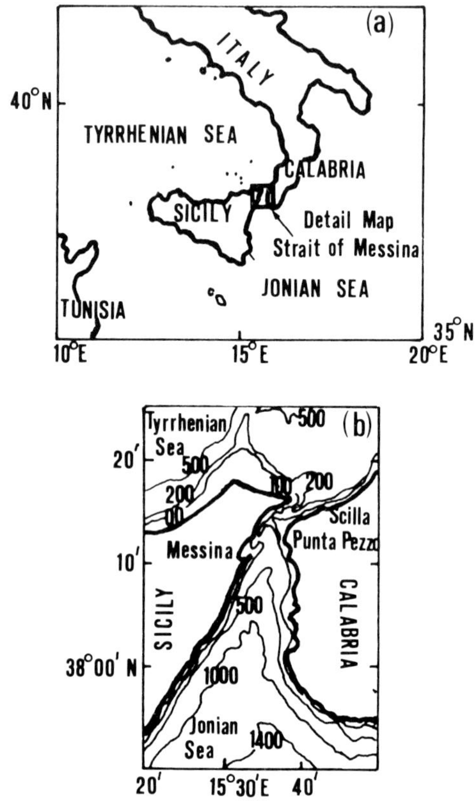

Fig.1. Bathymetric map of the Strait of Messina (from Guardiani et al., 1986).

picture of water mass distribution in the Strait of Messina has emerged (Vercelli, 1925; Hopkins et al., 1984). At the narrowest point there are two layers of marine water: surface Tyrrhenian water of Atlantic origin (T = 16.6°C, S=38.0 ‰, σ_t = 27.93) and deeper Levantine Ionian water (T=14.2°C, S=38.6 ‰,

σ_t=28.94). The time-averaged velocity profile (over a period of 15 days) shows a steady surface current flowing southward (v=10 cm/s) while a steady bottom current flows northward (v=13 cm/s).

A semidiurnal tidal current, originating from the tides of adjacent seas, is superimposed on these steady currents. In fact this channel is an amphidromic point for the semidiurnal tides of the Ionian and Tyrrhenian Seas. Tides in these basins have rather small amplitudes (10 cm south of Messina and 16 cm north of Messina) but their phase change is about five hours along the 10 km length of the Strait (as observed by Vercelli, 1925). This abrupt change is the cause of large sea surface slopes, of 1-2 cm/km, which set up strong currents, up to 2-3 m/s (Vercelli, 1925 and Defant, 1940).

The time-averaged interface depth is ~ 150 m in the Ionian and Tyrrhenian Seas, but rises to a mean depth of ~ 30 m at the sill. A strong oscillation of the interface with semidiurnal periodicity is superimposed on the steady pattern. The interface on the sill alternatively reaches the bottom and the sea surface during the two opposite tidal phases (Del Ricco, 1982; Hopkins et al., 1984). Therefore when high tide is present in the Tyrrhenian sea the current flows southward ("rema scendente") and only Tyrrhenian surface water is found on the sill. On the contrary, during the Tyrrhenian low tide, the current moves northward from the Ionian to the Tyrrhenian Sea ("rema montante") and on the sill only Ionian Intermediate Levantine Water is present. This water mass has the thermohaline characteristics of Levantine Ionian water at a depth of approximately 1000 m. For this reason the temperature difference between the waters present on the sill during the two tidal phases can even be as high as 10 °C in summer, while in winter it amounts only to several tenths of a degree (Vercelli and Picotti, 1925). Large-amplitude internal waves due to the strong tidal action give rise to extensive mixing of surface and Levantine waters: the hydrological characteristics of the Ionian and Tyrrhenian Seas in the neighbourhood of the Strait of Messina are affected by this water exchange.

During the period 1980-1985 a number of oceanographic measurements were carried out by the Department of Physics of the University of Rome in the area of the Strait of Messina. These new data revealed some peculiar characteristics of this zone related to the behavior of the current and of the interface over the sill.

Measurements of the interface time-evolution on the sill due to semidiurnal tide are analyzed in Section 2. These measurements revealed for the first time the presence of a columnar disturbance downstream from the sill, which can be related to the presence of a subcritical flow on the sill.

The above tidal mixing generates a new water mass, called C water in the following, whose hydrological characteristics are intermediate between LIW and water of Atlantic origin. This water forms a long narrow cold surface water strip flowing south of the Strait along the east coast of Sicily. This strip was detected by using both hydrological and satellite data (Bohm et al., 1983). The southward flow of the C water is analyzed in Section 3. A deep sea vein of warm salty water along the Calabrian coast was also observed. This water, coming from the Strait during the "rema montante" phase, flows northward at an average depth of 130-400 m. The water follows the bathymetry (Marullo and Santoleri, 1986) and can be considered the anolog of the C water (#3).

Another interesting phenomenon observed during these cruises was the existence of large-amplitude internal waves (Alpers and Salusti, 1983; Griffa et al., 1986; Sapia et al., 1986). They consist in packets of internal solitary waves characterized by an amplitude of 10-20 m, a phase velocity of approximately 0.9 m/s, a wavelength of 100-400 m and a total horizontal wave thickness of $\sim$2000 m. These waves are very stable and often travel long distances, before running into a solid obstacle. In our case the first obstacle met by the northward-travelling waves is the Cape Vaticano promontory. In Section 4 we show that the breaking of these internal waves in the vicinity of this cape is responsible for the formation of the thermal front ($\sim$10 km large) observable in this area.

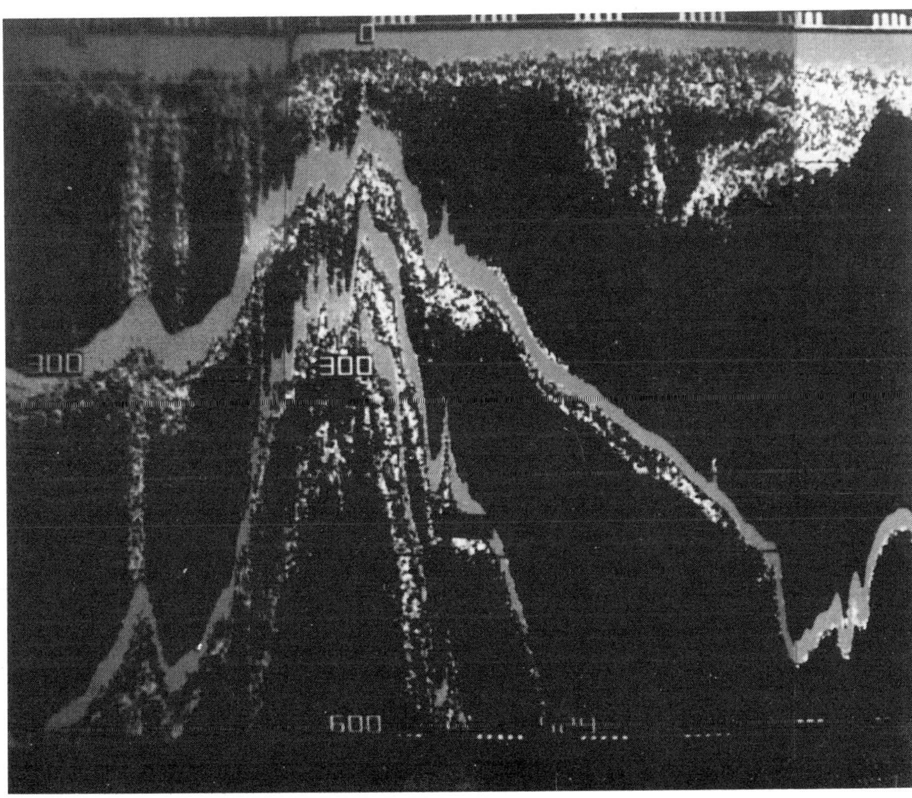

Fig.2. A photograph of a Koden image during the "rema montante" in which some columnar disturbances are visible downstream from the region where the flow is subcritical (D$\sim$250 m, from Di Sarra et al., 1986).

2. LONG INTERNAL WAVES AND TURBULENT COLUMNAR DISTURBANCES IN THE STRAIT OF MESSINA

The tidal currents in the Strait of Messina and their effect on the displacement of the water-mass interface have been discussed by several authors using Vercelli's data. Defant (1961)

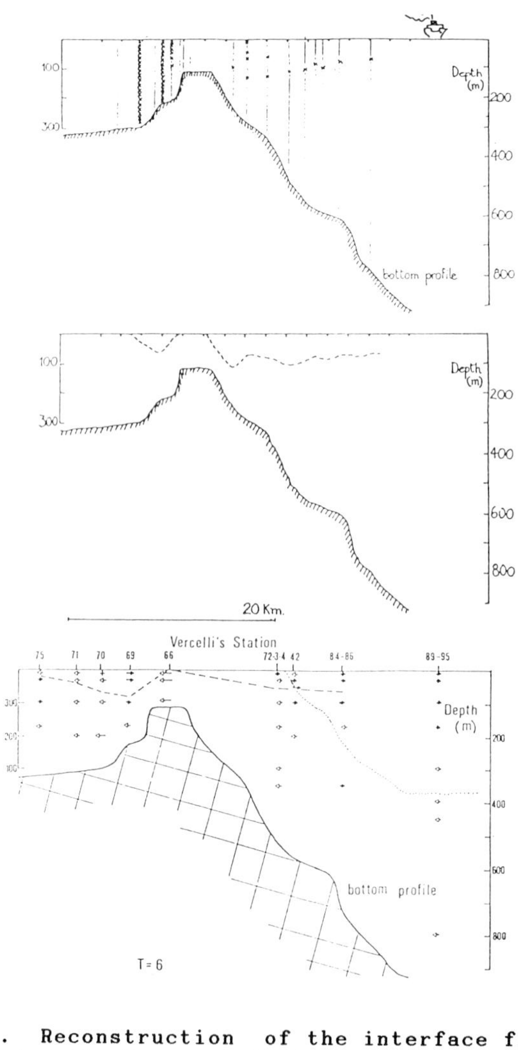

Fig. 3a. Reconstruction of the interface from the Koden images (xxx) during the "rema montante" compared with the corresponding interface position, as computed by Del Ricco (1982), and by Hopkins et al. (1984). Columnar disturbances are clearly visible downstream from the sill in both flows (from Di Sarra et al., 1986).

studied the tides inside the strait using a barotropic model; in more recent times Del Ricco (1982) used a numerical baroclinic nonlinear model to determine the time evolution of the interface between these two water layers for Vercelli's March data. He

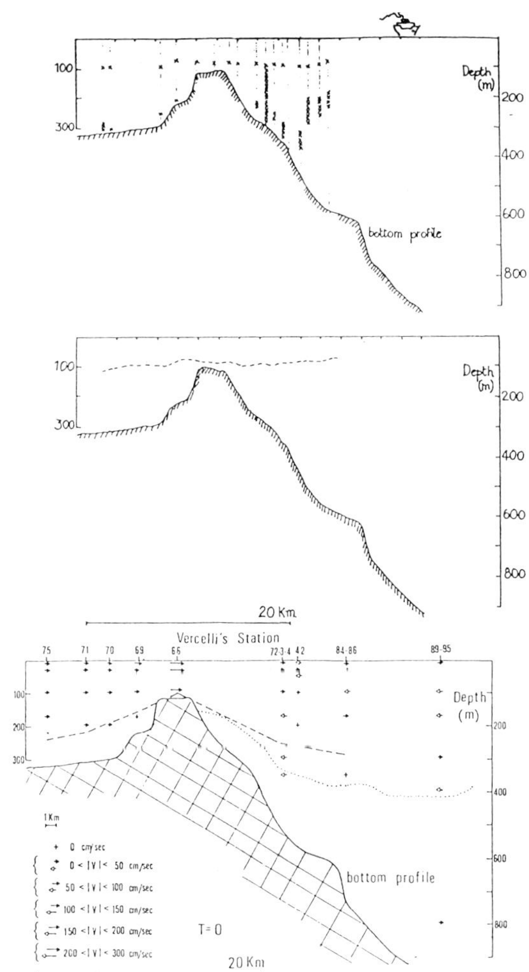

Fig. 3b. Reconstruction of the interface from the Koden images (xxx) during the "rema scendente" compared with the corresponding interface position as computed by Del Ricco (1982), and by Hopkins et al. (1984). Columnar disturbances are clearly visible downstream from the sill in both flows (from Di Sarra et al., 1986).

found an internal wave of$\sim$400 m amplitude in a zone immediately south of the sill. Hopkins et al. (1984) used Vercelli's original data to reconstruct the interface time-evolution using a simple balance between the interface slope and the acceleration of the baroclinic component. These two models show that the interface has oscillations of$\sim$100 m with a semidiurnal tidal periodicity and it alternatively touches the bottom and the air-sea surface. Moreover, these models can be used to predict the existence of a large amplitude internal wave immediately south of the sill during the rema montante. The amplitude of this wave was as high as 400 m for Del Ricco (1982) while Hopkins found a value of 100 m.

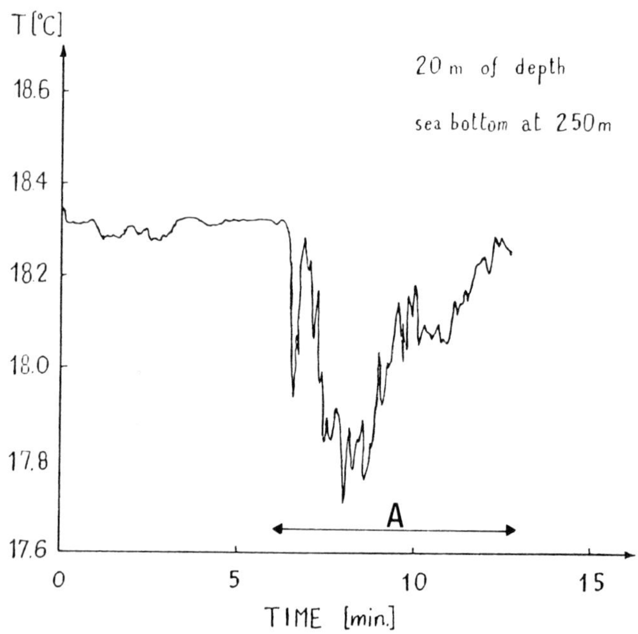

Fig. 4. Temperature as a function of time at 20 m of depth during the passage of the columnar disturbance (from Di Sarra et al., 1986).

A pilot cruise, JANE 84 (October 1984), was organized to study the interface time-evolution due to semidiurnal tide, using casts and a KODEN fish finder (28 and 200 Khz). KODEN fish finder images (Fig.2) were used to infer the explicit time evolution of the interface (Farmer et al., 1983) whilst the ship moved at $\sim$ 8 knots along the midline of the Strait. In figure 3 the reconstruction of the interface position from the Koden images is shown in comparison with the results of the Del Ricco and Hopkins models for the two different tidal periods. It can be seen that the behavior of the interface is in general agreement with the model results (Di Sarra et al., 1986).

Moreover, some of these Koden images showed an interesting phenomenon; the presence of columnar disturbances, called Bignami columns, in general visible downstream from the sill (Figs. 2-3). These disturbances ran from the surface to the bottom (D=250 m)

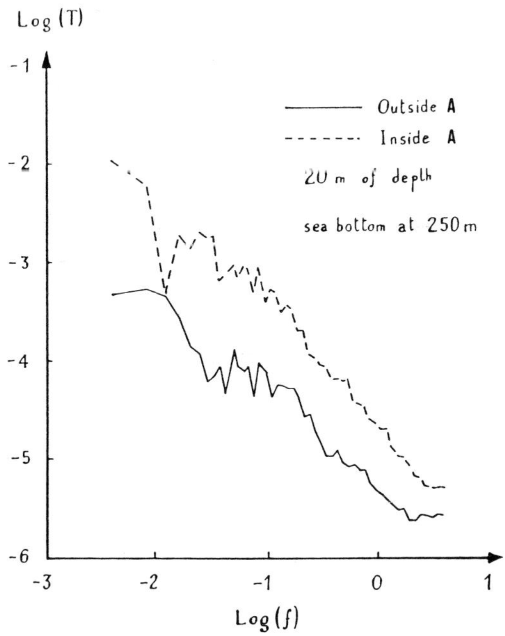

Fig. 5. Spectrum of the segments of temperature moored data shown in figure 4 (from Di Sarra et al., 1986).

and were characterized by a diameter of 400 m and a temperature of $\sim$ 0.7 °C less than the surroundings, as revealed by moored temperature data (Fig.4). The Fourier spectrum of the temperature versus time inside and outside the Bignami column showed the same behavior but the energy of the disturbance was $\sim$ 10 times greater (Fig.5).

Flow over a submerged mountain has been studied theoretically for a rather long time in time-independent formulations and compared with tank experiments. Its behaviour depends on many parameters: stratification, tidal currents and geometrical factors such as sill shape and size. However, it is possible to use a simple but useful descriptive parameter: the Froude number (Long, 1955). This parameter can be defined as the ratio of the horizontal velocity, U, to the phase velocity of the internal wave, C. Some general features of Long's results are of interest. For instance, supercritical flow ($\pi F>1$) is characterized by an interface rising above the obstacle; while in the case of subcritical flow ($\pi F<1$) the interface is lowered. In the critical case ($\pi Fn=1$, with n=1,2,...) steady wavelike disturbances appear on the sill. Recent studies have revealed other interesting features: Smith (1976) found that if the obstacle height reaches 1/9 of the lower layer depth, nonlinear effects become very important and turbulent phenomena occur. Mc Intyre (1972) noted that for $2\pi F<1$ some columnar like disturbance appeared downstream from the obstacle and was propagated upstream for some time.

The formation of the Bignami column can be explained in the light of the Mc Intyre's results. For Messina, the Froude number over the sill is greater than 1 and the flow is supercritical, but becomes subcritical immediately south and north of the sill, where the mean tidal velocity decreases. This could explain both the existence of the Bignami column and why this phenomenon was never observed on the sill.

3. DETECTION OF COASTAL CURRENTS NEAR THE STRAIT OF MESSINA

A coastal water, called C water, flowing over the Sicilian shelf south of Messina was detected in satellite imagery and by CTD casts carried out during the prime oceanographic cruise (Bohm et al., 1986). 16 AVHRR cloud-free images obtained from TIROS N and NOAA 6 were examined for different periods of the year. At the same time a cold water spot on the sill and a cold water strip which extended southward from the sill following the oriental shelf of Sicily were also observed (Fig.6). This strip was usually $\sim$ 100 km long but sometimes seemed to extend as far as Syracuse. Its width was between 4 and 10 km. This cold water strip is not always visible in satellite data, which suggests that C water may sometimes disappear or flow below the sea

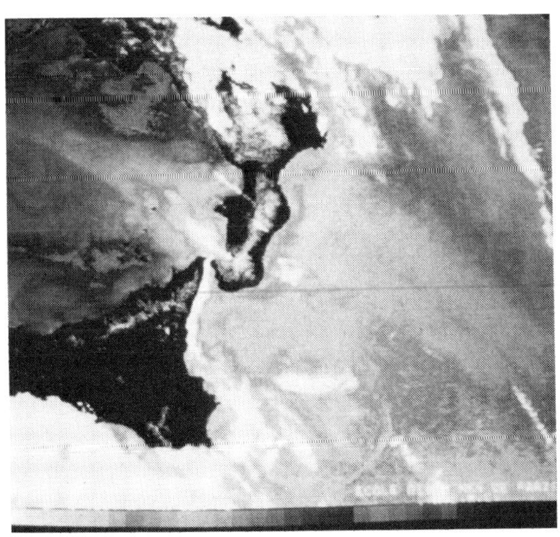

Fig. 6. Satellite image (Ch. 4 AVHRR of NOAA 6. June 2 1981 7:32 GMT) with the cold strip along the Sicilian shelf (from Bohm et al. 1986).

surface. Bohm et al. analyzed also the hydrological data collected during the PRIME oceanographic cruise of May 1982 in this zone. From the analysis of the vertical T-S diagrams they found the presence of two different water masses, C water (closer to the Sicilian coast) and Ionian water (further offshore). The main difference between these two waters is that the C water is stratified in both temperature and salinity while the latter is stratified only in temperature. The border between these two types of waters was found at 5-10 km offshore. On the basis of the similarity between T-S diagrams of the C water and Tyrrhenian

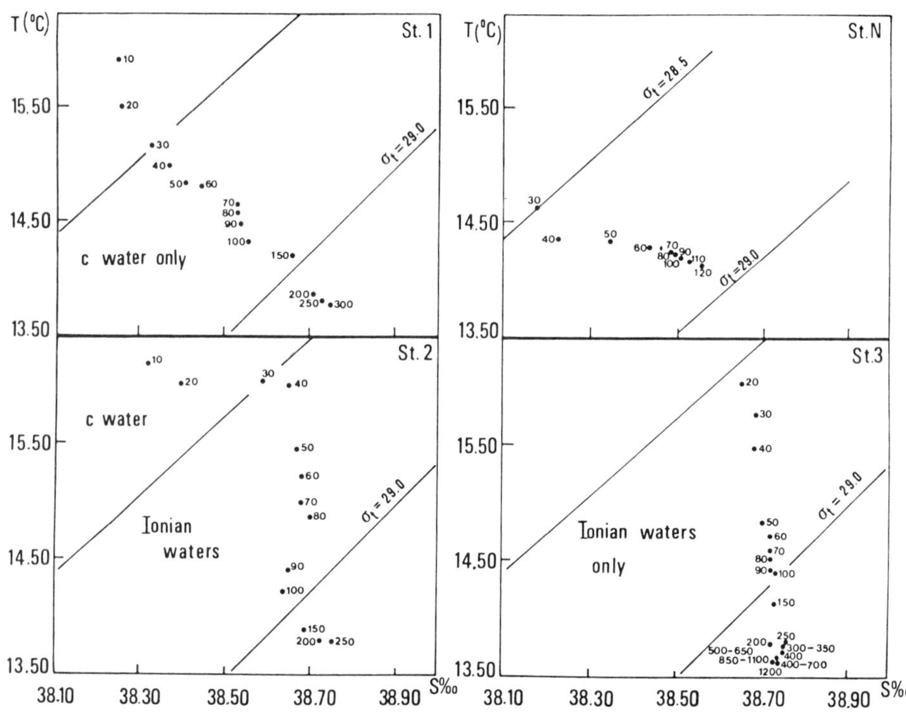

Fig. 7. Typical T-S diagrams observed during Prime cruise. Station N shows a typical T-S diagrams at Tyrrhenian Sea. Station 1, 2 and 3 concern the Ionian Sea. Station 1 shows the typical C water T-S diagram; Station 3 shows The Ionian water T-S diagram and Station 2 shows the mixing of Cwater and Ionian water masses.

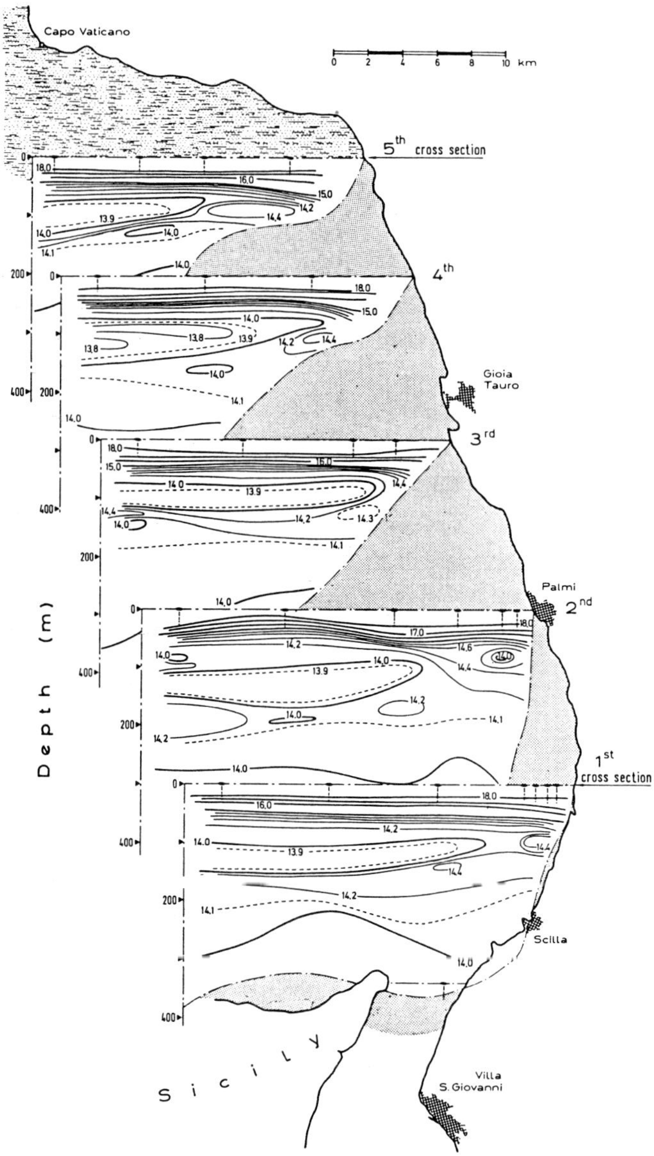

Fig. 8. Map of the isotherms in the vertical cross-sections between Messina and Capo Vaticano, (from Marullo et al., 1986).

water masses observable in the range of 30 - 100 m (see upper panels of Fig.7) Bohm et al. (1986) indicate Tyrrhenian water as possibly being the main component of C water. They consider that this water could be the result of the mixing that occurs during the rema scendente near the sill between the Tyrrhenian water and the local water.

A similar analysis for the data collected during the Vera cruise (July 1983) north of the strait of Messina in the zone between the sill and Cape Vaticano has been made by Marullo and Santoleri (1986). Their temperature cross-sections show the existence of a warm water mass between 130-400 m of depth. Above and below this deep layer the temperature decreases fairly regularly (Fig.8). This layer of deep water is characterized by an increase in temperature from 13.9 °C to 14.12-14.33 °C and in salinity from 37.8‰ to 38.3-38.6 ‰. Analysis of the T-S diagrams revealed that this internal vein has the hydrological characteristics of the Ionian Levantine water entering the Tyrrhenian sea during the rema montante. Marullo and Santoleri have shown that this water mass moves northward, following the topography, and in the vicinity of the Calabrian coast seems to mix with superficial and Levantine Tyrrhenian water.

4. TURBULENCE DUE TO THE COLLAPSE OF LARGE AMPLITUDE INTERNAL WAVES: GENERATION OF THE CAPO VATICANO FRONT

As is well known, large amplitude nonlinear internal waves are generated in the Strait of Messina by the interaction of tides with the bottom topography of the sill (Alpers and Salusti, 1983; Griffa et al., 1986; Sapia et al., 1985). Their amplitude and therefore the energy that they can transport is strongly dependent on tide intensity. All these waves propagate both northward and southward as a sequence of depressions of the interface organized in packets of solitary waves and conserving their energy and shape until they run into an obstacle. When this happens, they break, generating turbulent mixing of the adjacent stratified waters (Knickerbocker et al., 1980; Kao et al., 1985). Recently Marullo and Santoleri (1986) suggested that breaking of these internal waves could occur in the Gulf of Gioia, and could be responsible for the formation and disappearance of the thermal front observed south of Capo Vaticano both from satellite and in situ measurements.

The first experimental evidence of the existence of anomalies in the zone was given by a synthetic aperture radar (SAR) image taken on September 15, 1978, 0817 GMT by SEASAT SAR satellite from a height of 800 km. This image revealed a dark area (area of low backscattering properties) $\sim$10 km thick south of the Capo Vaticano Promontory (Alpers and Salusti, 1983) which can be interpreted as the signature of the deep water spun up by mixing.

To extend knowledge of this phenomenon, Marullo and Santoleri (1986) examined thermal satellite images from NOAA 7 covering a period of approximately two years (1982-1983, by courtesy of Centre de Meteorologie Spatiale, Lannion, France). It resulted that, out of 181 clear-sky days satellite images, 56% showed the presence of a cold front, 28% the presence of a warm front while in 16% of the cases no front was observable (Table 1). In the same period two additional oceanographic cruises were performed in the zone: the 1982 Prime cruise and the 1983 Vera cruise. During the Prime cruise, measurements around Capo Vaticano were made using a towed Guildne CTD. The primary aim of these measurements was to detect packets of nonlinear internal waves, although analysis of the data recorded also revealed the existence of a front at a distance of 10 km from Capo Vaticano. The observed thermal gradient was almost 4 °C over 7 km, and positive in the North direction (Sapia, 1986). On the contrary there were no signs of this front in the Vera cruise. To obtain more details of the type of front, two additional cruises were performed in the zone: the 1984 Jane cruise and the 1985 Greta cruise. During the Jane cruise (October 1984) CTD casts and moored temperature measurements were made. The analysis of TS profiles showed that between 40 and 100 m of depth the behavior of temperature and salinity at the innermost stations of the Gulf of Gioia is different from the surroundings, indicating a patch of mixed water (Fig.9). The observed front has an extension of ~10 Km with a

TABLE 1
Summary of observed satellite images relative to the Capo Vaticano zone for the period 1982-1983. Column TOT indicates numbers of clear-sky observed images for every month. Columns COLD and WARM indicate the number of these monthly observations in which a warm or a cold front, respectively, was observed in the zone south of Capo Vaticano, while column ABS contains the number of cases in which no presence of fronts was observed.

YEAR	MONTH	TOT	COLD	WARM	ABS
1982	January	3	1	0	2
	February	7	3	3	1
	March	11	3	6	2
	April	9	4	3	2
	May	8	8	0	0
	June	13	10	3	0
	July	17	10	6	1
	Augst	14	7	6	1
	September	9	7	1	1
1983	January	2	2	0	0
	February	10	0	7	3
	March	9	2	1	6
	April	8	5	1	2
	May	12	8	3	1
	June	6	4	1	1
	July	11	5	5	1
	August	9	7	1	1
	September	13	9	3	1
	October	10	7	1	2

gradient of 3 °C/10 km.

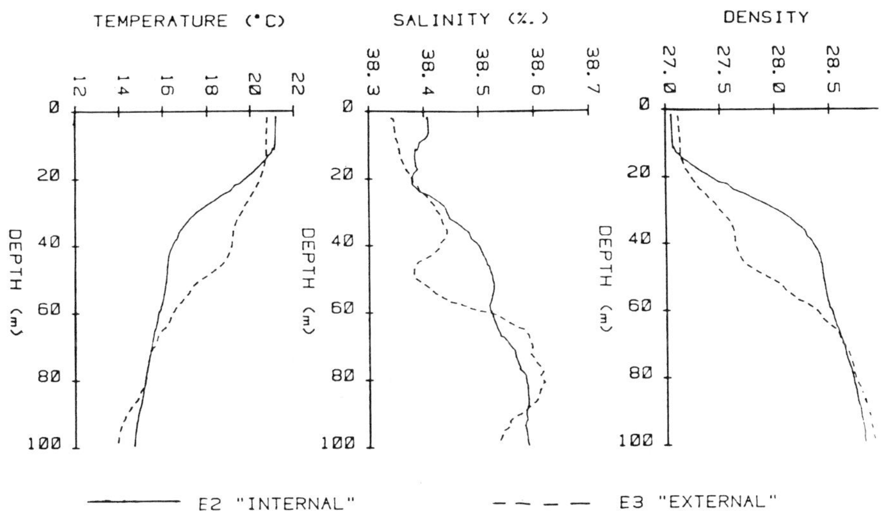

Fig. 9. Temperature, salinity and density profiles as observed at an internal station (E2) and an external station (E3) during the Jane cruise in the Gulf of Gioia (from Guardiani et al., 1986).

Moored temperature measurements were made at 30 and 80 m of depth both outside and inside the front area. The resulting temperature time series show a similar behaviour at 80 m while at 30 m the data taken inside the front show greater variation in a few minutes (1 °C in 10') than the records of the stations external to the front (Fig.10). The spectral analysis of the temperature time series showed that the spectral function S(f) has a decrease like $f^{-5/3}$, typical of a Kolmogorov turbulent spectrum (Fig.11). The Jane field measurements were repeated during the Greta cruise (2 4 August 1985). During this cruise the analysis of the CTD casts revealed that in the neighbourhood of a rising sea bottom the density profile displays an inversion at a depth of 60-90 m (Fig.12). The power spectra of the moored measurements show the same behaviour as in the Jane cruise, but no front was observable

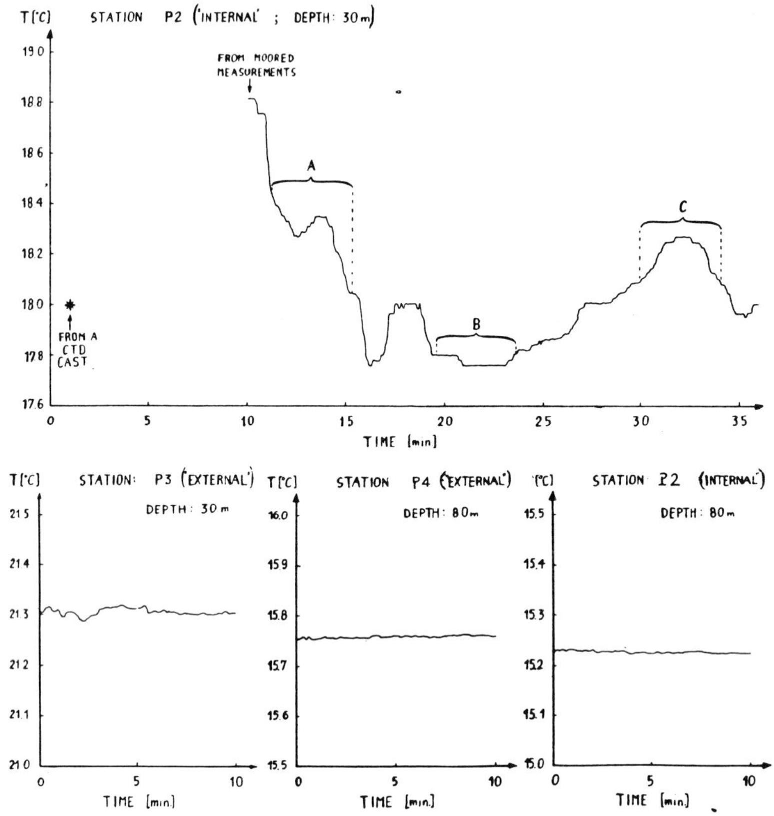

Fig. 10. Temperature time-series at different depths and different stations during Jane 1984 cruise (from Guardiani et al., 1986).

(Guardiani et al., 1986).

These new data support the idea of Marullo and Santoleri (1985) that the front observed at the Southern coast of Capo Vaticano is due to a mixing process in which energy is received from the breaking of nonlinear internal waves generated by semi-diurnal tides in the Strait of Messina. The most natural elementary mechanism of breaking occurs when the two layers become equal; the solitary waves change polarity. On approaching the shore, the height of the internal solitary wave increases until it breaks generating turbulence (Knickerboker et al., 1980;

Helfrich et al., 1984). On the contrary, tank experiments (Kao et al., 1985) seem to suggest that, when the depression internal wave intrudes over the shelf, its shape is deformed and Kelvin Helmholtz instabilities are responsible for the energy loss. Unfortunately in the light of our data, it is impossible to identify the explicit elementary breaking mechanism.

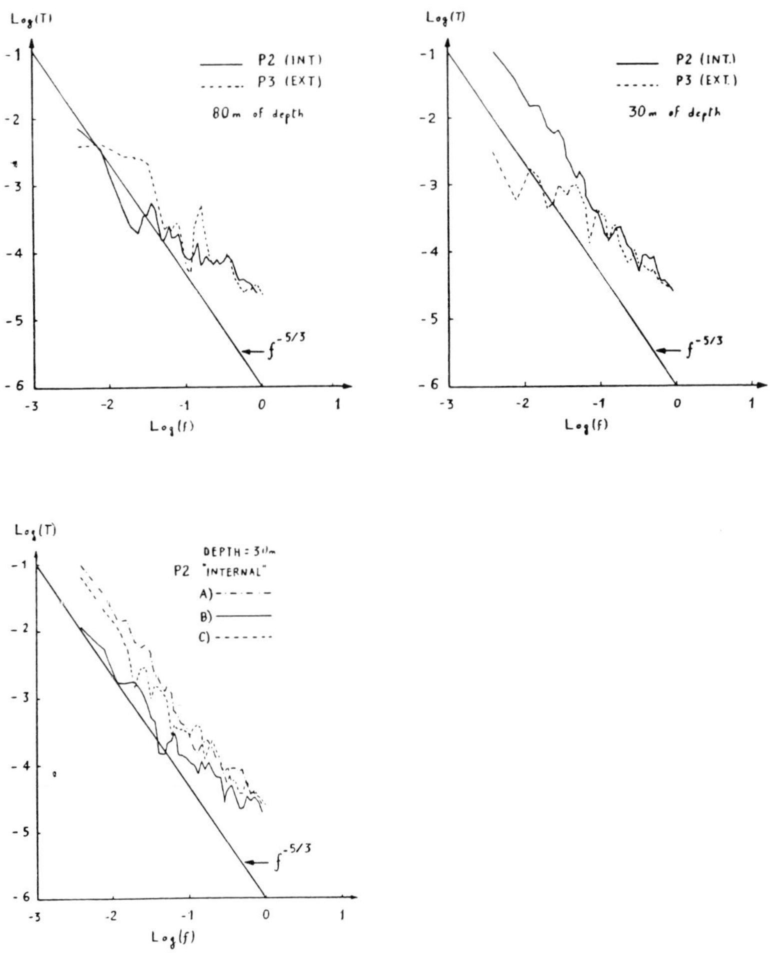

Fig. 11. Discrete Fourier Transform of temperature at various Stations and at various depths. The last graph shows the time-intervals A,B,C in Fig.10, relative to station P2, at 30 m of depth. The spectrum is larger for the time-interval "A", when a high amplitude wave is present. This fact is in agreement with the considerations of Kao et al., 1985 (from Guardiani et al., 1986).

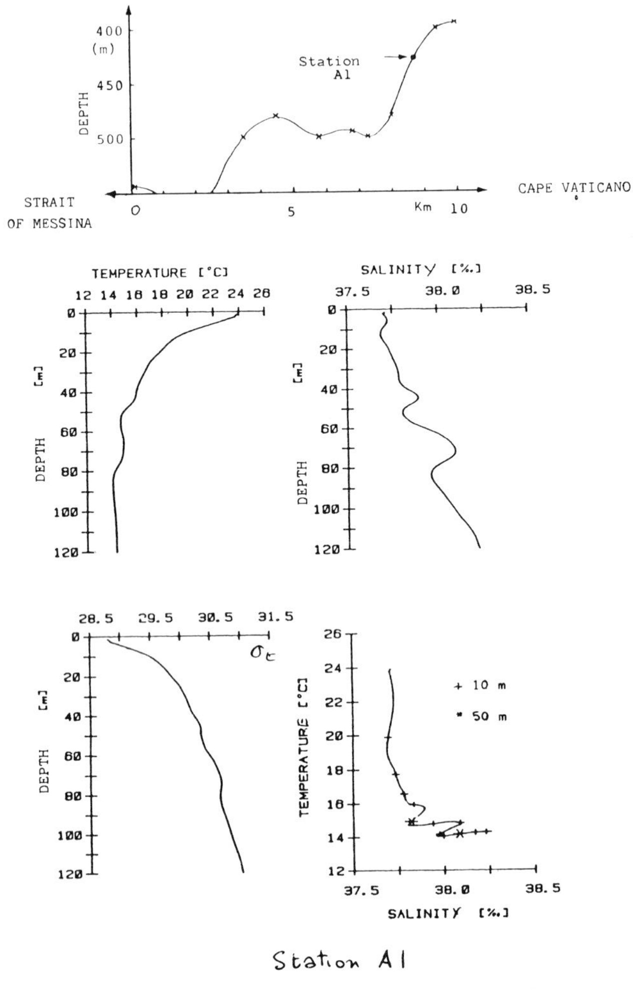

Fig. 12. Typical Temperature, Salinity and Density profiles and T-S diagram during Greta cruise at the stations located on the shelf break showing an inversion in the T-S diagram. In the upper panel the bottom depth behaviour between Messina and Capo Vaticano is also shown (from Guardiani et al., 1986).

6. DISCUSSION

To summarize, we show that a small marine strait like that of Messina can be a remarkable geophysical laboratory of general interest. In this Strait the sill is shallower than the interface of Atlantic and Levantine Waters layers; moreover it is an amphidromic point for the tides of the two basins. Consequently the tidal phenomena give rise to a peculiar process that is rather easy to observe since, in the neighbourhood of the sill, there are no strong tides. These phenomena are surface and deep currents originating on the sill and flowing to a considerable distance from the Strait. The study of the time evolution of the interface around the sill can explain most of the peculiar phenomena observed in the Messina zone in terms of the presence of nonlinear internal waves, columnar disturbances and turbulence that have been examined in this paper. These phenomena are known on theoretical grounds and their observations are of considerable interest. Of course, these phenomena are novel and would call for much more observation and speculation. However, we feel that their interest is so great that even these pilot data can be of use as a stimulus.

7. REFERENCES

Alpers, W., Salusti E., 1983. Scylla and Charybdis Observed from Space, J. Geophys. Res., 88, 1800-1808.

Bohm, E., Salusti E., 1984. Satellite and Field Observations of Currents on the Eastern Sicilian Shelf, Remote Sensing of Shelf Sea Hydrodynamics, Elsevier Sciences publisher Amsterdam, pp.51-68.

Bohm, E., Magazu', G., Wald, L., Zoccolotti, M.L., 1986. Coastal Currents on the Sicilian Shelf South of Messina. Oceanologica Acta (in press).

Defant, A., 1940. Scilla e Cariddi, e le correnti di marea nello Stretto di Messina, Geof. Pura Appl., 2, 93.

Defant, A., 1961. Physical Oceanography, Vol.II. Pergamon Press, 589 pp.

Del Ricco, R., 1982. A Numerical Model of the Vertical Circulation of Tidal Strait and its Application to Messina Strait. Nuovo Cimento, Vol. 5C, n.1, 21-45.

Di Sarra, A., Pace, A., Salusti, E., 1986. Long internal waves and turbulent columnar disturbances. (Preprint, Phys. Dept., Rome University).

Farmer, D.M., Freeland, H.J., 1983. The Physical Oceanography of Fjords; Progress in Oceanogr., Vol. 12, 147-219.

Guardiani, G., Pace, A., Salusti, E., 1986. Turbulence due to the collapse of internal solitary waves (Preprint, Phys. Dept., Roma University).

Gargett, A., 1980. Turbulence Measurements through a Train of

Breaking Internal Waves in Knight Inlet, B.C. In: Fjord Oceanography, H.J. Freeland, D.M. Farmer and C.D. Levings, eds., Plenum Press, New York.

Griffa, A., Marullo, S., Santoleri, R., Viola, A., 1986. Note on Internal Tidal Waves Generated at the Strait of Messina. Continental Shelf Res. (in press).

Helfrich, K.R., Melville, W.K., Miles, J., 1984. On the Solitary Waves over Slowly Varying Topography, J. Fluid Mech., Vol. 149, 305-317

Hopkins, T., Salusti, E., Settimi, D., 1984. Tidal Forcing of the Water Mass Interface in the Strait of Messina, J. Geophys. Res., 89, 2013-2024.

Kao, T.W., Fu-Shing, P., Renouard, D., 1985. Internal Solitons on the Pycnocline Generation, Propagation, Shoaling and breaking over a Slope. J. Fluid Mech., Vol. 159, 19-53

Knickerbocker, C.J., Newell, A.C., 1980. Internal Solitary Waves near a Turning Point. Phys. Lett., 75 A, 326-330

Long, R.R., 1955. Same Aspect of the Flow of the Stratified Fluids, III. Continuous Density Gradients. Tuellus, 7, 341-367.

Marullo, S., Santoleri, R., 1986. Fronts and Internal Currents at the Northern Mouth of the Strait of Messina. Nuovo Cimento (in press).

Mc Intyre, M.F., 1972.On Long's Ipothesis of no Upstream Influence in uniformly Stratified Fluid. J. Fluid Mech., 52, 209-243.

Osborn, A.R., Burch, T.L., 1980. Internal Solitons in the Adamann Sea, Science, 208, n.4443, 451-460.

Sapia, A., 1986. Observation of Nonlinear Internal Solitary Waves Train at the Nourthern and Southern Mouth of the Strait of Messina, submitted to Deep Sea Res..

Smith, J.D., Farmer, D.M., 1977. Nonlinear Internal Waves and Internal Hydraulic Jumps in a Fjord. In Geofluidynamical Wave Mathematics, 42, University of Washington, Seattle.

Vercelli, F., 1925. Crociere per lo Studio di Fenomeni nello Stretto di Messina (R.N. Marsigli, 1922 - 1923). Vol.1: Il Regime delle Correnti e delle Maree nello Stretto di Messina. Commissione Internazionale del Mediterraneo, Venezia, 205 pp.

Vercelli, F., Picotti, M., 1925. Crociere per lo Studio dei Fenomeni nello Stretto di Messina (R.N. Marsigli, 1922 - 1923). Vol.2: Il Regime Chimico Fisico delle Acque dello Stretto di Messina. Commissione Internazionale del Mediterraneo, Venezia, 103 pp.

A THREE-DIMENSIONAL FINITE ELEMENT MODEL FOR THE STUDY OF STEADY AND NON-STEADY NATURAL FLOWS

J.-L. ROBERT and Y. OUELLET
Civil Engineering department, Research Center on Computer application in Engineering, Laval University, Sainte-Foy, Québec, CANADA, G1K 7P4

ABSTRACT
This paper will present the basic concepts of a tridimensional finite element model for the simulation of natural flows. In a first step, the fundamentals of the formulation will be described with an emphasis on initial assumptions, mobile layer concept, type of element, discrete variables of the system. Then, special attention will be taken to expose the original technique developed to include, on the entire depth of water, the effect of the bottom friction and some justifying examples will be shown. Afterwards, a brief discussion on the strategy of resolution for such non linear system will point the specific importance of friction and inertial convective terms. This will be followed by the presentation of some examples showing the good general behaviour of the model, particularly its capability to well simulate the vertical circulation of a flow.

1 INTRODUCTION

After working since many years with finite difference and more recently with finite element depth average two dimensional models, some specific environmental problems in coastal and estuarine zones, involving the effect of density and wind on the hydrodynamic circulation, raise the need of a three dimensional model. The finite element approach was retained for mainly two reasons. The first one was its ability to discretize with variable density of computational points an irregular domain, as it can be found in the nature. This is an important advantage in a way that it allows us to choose the position of artificial boundary conditions far enough to minimize their influence and without using a prohibitive number of computational points. The second reason was linked with the possible future development of the model, specially the possibility of its coupling with existing 2-D F.E. hydrodynamic model. The major goal we kept during the development of the model was its capability to simulate the largest possible range of free surface natural flows, large or small scale, steady or non-steady, tide or wind driven. The domain of application itself has a direct influence on the formulation, in fact natural flows have a small vertical-horizontal aspect ratio. Then, the discretization technique, as it will be shown later, has been chosen differently depending on the vertical and the horizontal directions.

2 FORMULATION OF THE 3-D MODEL

2.1 Governing Equations

The physical behaviour of a fluid may be expressed by a set of equations based on the forces equilibrium and the principle of mass conservation. In a three dimension (x, y, z) domain the momentum equations are:

$$\frac{du}{dt} + \frac{1}{\rho}\frac{\partial p}{\partial x} - X - \frac{1}{\rho}\left(\frac{\partial \tau_{xx}}{\partial x} + \frac{\partial \tau_{xy}}{\partial y} + \frac{\partial \tau_{xz}}{\partial z}\right) = 0 \tag{1}$$

$$\frac{dv}{dt} + \frac{1}{\rho}\frac{\partial p}{\partial y} - Y - \frac{1}{\rho}\left(\frac{\partial \tau_{yx}}{\partial x} + \frac{\partial \tau_{yy}}{\partial y} + \frac{\partial \tau_{yz}}{\partial z}\right) = 0 \tag{2}$$

$$\frac{dw}{dt} + \frac{1}{\rho}\frac{\partial p}{\partial x} - Z - \frac{1}{\rho}\left(\frac{\partial \tau_{zx}}{\partial x} + \frac{\partial \tau_{zy}}{\partial y} + \frac{\partial \tau_{zz}}{\partial z}\right) = 0 \tag{3}$$

where:

u, v, w : the velocity components;

p : the pressure;

ρ : the density;

X, Y, Z : the external forces (Coriolis, gravity)

τ_{xx}... : the Reynolds stresses

The Reynolds stresses express the viscous similarity of the effect of the turbulence. In the model, they are included using the Boussinesq assumption using an eddy viscosity coefficient A:

$$\tau_{x_i x_i} = 2A \frac{\partial u_i}{\partial x_i} \tag{4}$$

$$\tau_{x_i x_j} = A\left(\frac{\partial u_j}{\partial x_i} + \frac{\partial u_i}{\partial x_j}\right) \tag{5}$$

Note must be taken that in the vertical direction, those terms will receive special treatments.

The mass conservation in an incompressible fluid gives the continuity equation:

$$\frac{\partial u}{\partial x} + \frac{\partial v}{\partial y} + \frac{\partial w}{\partial z} = 0 \tag{6}$$

2.2 Vertical Integration

Since the model is specially designed to simulate natural domains, some characteristics of them can be observed and then stated:

- small vertical dimension in respect of horizontal ones;
- vertical acceleration of the flow negligible;
- small vertical velocity component.

Those statements eliminate from the range of applications high slope super critical flows, closed conduits and turbomachinery flows. With this restriction the assumption of hydrostatic pressure can be taken and the acceleration and stresses term of the third momentum equation vanish keeping only the pressure and the gravity terms. In this form the set of governing equations (1), (2), (3) and (4) is integrated in the vertical z direction from the bottom position -h to the water level η.

But, in order to get some information about the vertical structure of the flow, the integration process is split into several terms:

$$\int_{-h}^{\eta} \dots \, dz = \int_{-h}^{L_{b-1}} \dots \, dz + \dots + \int_{L_2}^{L_1} \dots \, dz + \int_{L_1}^{\eta} \dots \, dz \tag{7}$$

by the introduction of intermediate integration limits (fig. 1) which have their position located by:

$$L_k = \eta - \left(\frac{\eta - h}{b}\right) k \qquad k = 1, 2, \dots b \tag{8}$$

The presence of η in the equation (8) means that the position of those intermediate level must move proportionnally with the water level, consequently the relative thickness of a layer is constant. An uniform distribution of those level has been chosen here; however, it is possible to change those relative position by any other relationship including the water level to get another distribution.

It is important to note that the intermediate integration surfaces do not have any physical meaning and allow the mass exchange between them. This is specifically prescribed by the intermediate boundary condition (Robert 1983):

$$w(L_k) = - \sum_{n=k+1}^{b} \left(\frac{\partial h_n U_n}{\partial_x} + \frac{\partial h_n V_n}{\partial_y} \right) + u(L_k) \frac{\partial L_k}{\partial_x} + v(L_k) \frac{\partial L_k}{\partial_y} \tag{9}$$

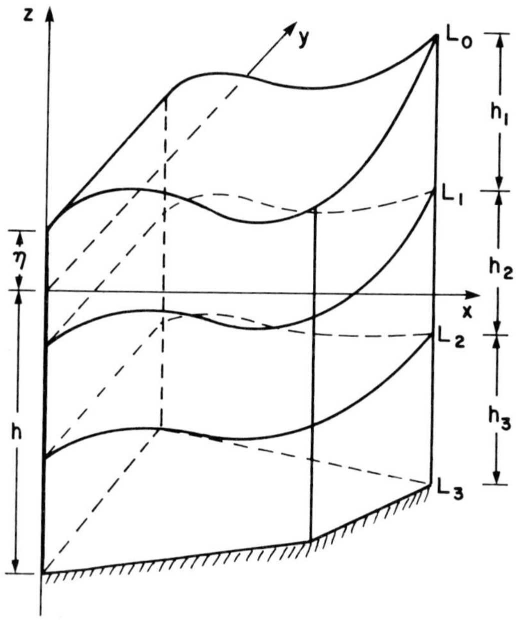

Fig. 1 Definition of the vertical integration boundaries.

where U_n and V_n are mean velocity components in a layer n. At the extremity of the vertical domain boundary conditions must be defined using a streamline condition:

$$w(\eta) = \frac{\partial \eta}{\partial t} + u(\eta)\frac{\partial \eta}{\partial x} + v(\eta)\frac{\partial \eta}{\partial y} \tag{10}$$

$$w(-h) = -u(-h)\frac{\partial h}{\partial x} - v(-h)\frac{\partial h}{\partial y} \tag{11}$$

The introduction of those conditions in the integration process using the Leibniz rule, does not eliminate all boundary velocity terms at the intermediate level, the assumption that those terms are included in the internal stresses term is made (Robert 1983).

Then the model is defined as:

- Continuity equation

$$\frac{\partial \eta}{\partial t} + \sum_{n=1}^{b}\left(\frac{\partial h_n U_n}{\partial x} + \frac{\partial h_n U_n}{\partial y}\right) = 0 \tag{12}$$

- Movement equations

$$\sum_{k=1}^{b} \left[\frac{\partial U_k}{\partial t} + U_k \frac{\partial U_k}{\partial x} + V_k \frac{\partial U_k}{\partial y} - fV_k + g \frac{\partial \eta}{\partial x} - \frac{1}{\rho h_k} \left(\frac{\partial h_k \tau_{xxk}}{\partial x} + \frac{\partial h_k \tau_{xyk}}{\partial y} + \left(\tau_{xL_{k-1}} - \tau_{xL_k} \right) \right) \right] = 0 \qquad (13)$$

$$\sum_{k=1}^{b} \left[\frac{\partial V_k}{\partial t} + U_k \frac{\partial V_k}{\partial x} + V_k \frac{\partial V_k}{\partial_y} + fU_k + g \frac{\partial \eta}{\partial y} - \frac{1}{\rho h_k} \left(\frac{\partial h_k \tau_{yxk}}{\partial x} + \frac{\partial h_k \tau_{yyk}}{\partial y} + \left(\tau_{yL_{k-1}} - \tau_{yL_k} \right) \right) \right] = 0 \qquad (14)$$

The vertical velocity component is computed from equation (9). The formulation of the horizontal stresses will be discussed in section 3.

2.3 Finite Element Formulation

This short overview of the finite element formulation will point out the specific choices made to obtain the discrete form of the model.

The integral form of equations (12), (13) and (14) is obtained by a Galerkin weighted residual method.

As one can see from the model equations, two dimensional elements can be used by associating at each node the water level η and the two horizontal components of the mean velocity in each layer U_k and V_k. Two types of interpolation functions were then considered. Based on the work of Cochet 1979 on 2-D hydrodynamic model, a quadratic function for the velocities and a linear function for the water level were tested. The result shown that in a multi-layer configuration some instabilities appeared fig. 2. Then a full quadratic function was used and improved considerably the performance of the model. Two full quadratic elements have been developed, a six nodes triangular and a nine nodes quadrilateral elements.

Special attention must be kept in the discretization of movement equations (13) and (14). Since the interpolation functions applie only on discrete parameter associated with one layer and are zero for any other parameters, the assembly process is equivalent to the sommation present in those equations. Then the resulting set of algebraic equation will have as much equations as discrete variables.

The resulting non linear system in non steady form can be expressed by:

$$[M]\{\dot{\phi}\} + [K(\phi)]\{\phi\} = 0 \qquad (15)$$

The solution method will be presented in section 4.

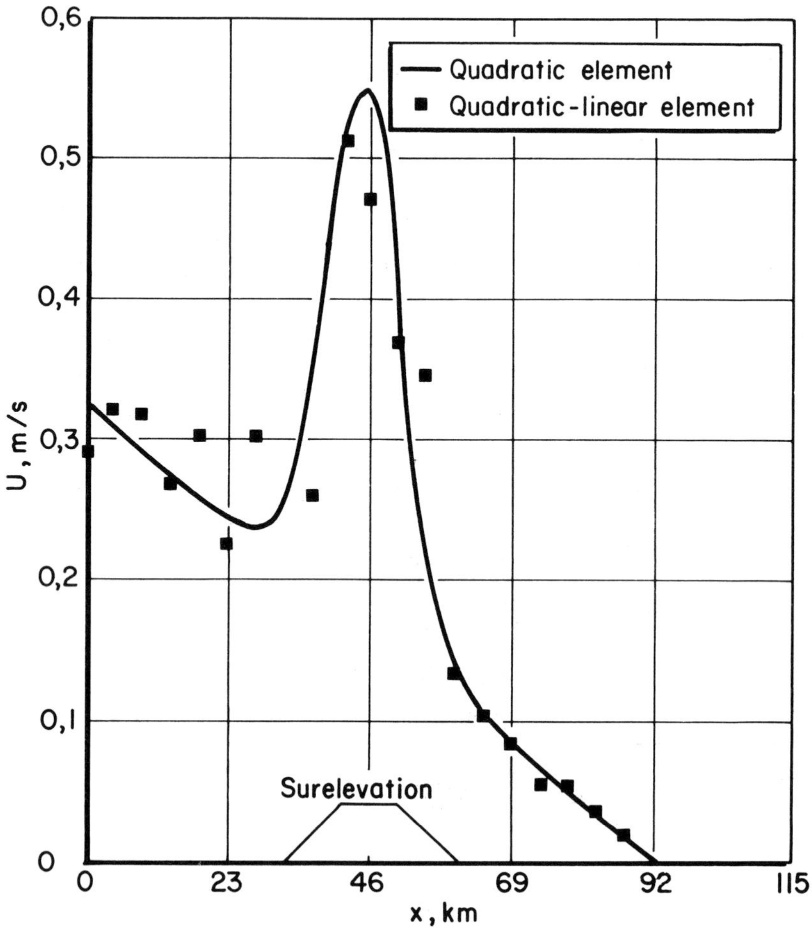

Fig. 2 Computed velocities in the center of a schematic estuary presented at Fig. 5.

3 HORIZONTAL STRESSES

3.1 Formulation

One of the most important factor in the general behaviour of a natural mass of water is the horizontal shear stress. The predominant one is the effect of the bottom roughness over the entire depth of water and in some cases, the effect of the wind at the free surface may be determinant. In such case, this specific stress component may be expressed by:

$$\tau_{x_i L_0} = C_t \rho_a W_{x_i} |W| \tag{16}$$

where C_t is a drag coefficient, ρ_a the air density and W_{x_i} the component of wind velocity W in the x_i direction. For the bottom friction term, Chézy formulation has been chosen:

$$\tau_{x_i L_b} = \frac{\rho g}{C^2} V_{x_i} |V| \tag{17}$$

Here, in the case of a two dimensional model the Chézy coefficient C, takes in account the overall effect of the bottom roughness on the flow over all the thickness of the water layer. However, when this depth is discretized, this term will affect only the bottom layer and the coefficient C must be corrected in way such as the instantaneous discharge must stay the same whatever the number of layer. To ensure the transfer of the bottom roughness to the upper layer, an internal shear stress component must appear at the interface of each layer. This may be expressed by discrete form of the Prandtl internal shear stress formulation; in the x direction we have at the upper interface of a layer k :

$$\tau_{xL_{k-1}} = \frac{\rho \ell^2}{h_k^2} \Delta U_k \sqrt{\Delta U_k^2 + \Delta V_k^2} \tag{18}$$

where ΔU_k and ΔV_k are the differences of the flow velocity between the layer k and k-1 and ℓ is the mixing length. Many authors (Wang 1977, Kokayashi 1980, Leenderstee 1975) use a similar formulation, but the ℓ coefficient is considered as an experimental adjustable coefficient.

3.2 Mixing Length Parameter

As we do not really know what is the value of the mixing length, expect near the bottom where Prandtl assume that it may be proportional to the distance from the bottom, we must find an usable formulation of the mixing length up to the free surface. Montgomery 1943 suggests, from some experiments in particular flow that:

$$\ell = \chi\, z(1 - z/h) \tag{19}$$

χ is the Von-Karman coefficient which is defined by χ = 0,4. As shown on fig. 3, this distribution presents a plane symetry on the z/h = 0,5 plane. This involves the assumption that the distribution of the horizontal stress is similar near the free surface and near the bottom, which is not really satisfactory.

To remedy such situation, Yalin 1977 suggests a distribution of the mixing length based on the assumption of a logarithmic velocity profile:

$$\ell = \chi z \sqrt{(1 - z/h)} \tag{20}$$

As the Montgomery 1943 distribution, this one agrees with the Prandtl assumption near the bottom:

$$\ell = \chi z \tag{21}$$

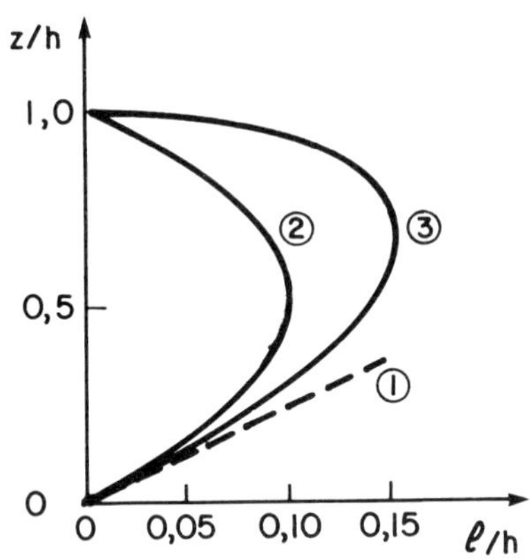

Fig. 3 Mixing length vertical distribution.

The major advantages of the Yalin 1977 distribution are:

a) the mixing length decreases less quickly near the free surface than near the bottom, assuming that the turbulent fluctuation in the neighbourhood of the surface are less damped than by the presence of a rigid boundary at the bottom.

b) the horizontal shear stress becomes zero without the necessity of assuming a zero vertical velocity gradient at the surface. This formulation was retained and adapted to the present model; in a discrete form, for the lower interface of a layer k, equation (20) becomes:

$$\ell_k = \chi h_k (b - k) \sqrt{k/b} \tag{22}$$

and the corrected bottom friction factor can be written:

$$C = C_r - \frac{\sqrt{g}}{\chi b} \sum_{k=1}^{b} \frac{b}{b-k} \tag{23}$$

where C_r is the Chézy reference coefficient for a two dimensional model.

In order to verify the good overall behaviour of this approach, a test was run on the model to simulate a steady state uniform turbulent flow with a five level configuration. The velocity profile obtained is presented at the figure 4.

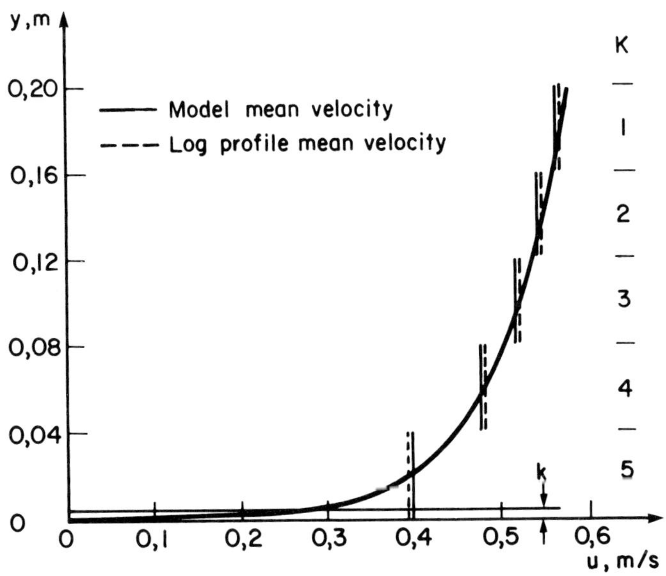

Fig. 4 Velocity profile for a uniform steady flow.

4 SOLUTION METHOD

In steady state configuration, the model of the flow is dominated by a quadratic friction term. The Newton Raphson procedure must be chosen rather than a substitution scheme to solve the non linear system. In fact, it is obvious that the later will not converge because the quadratic term. The incremental form of the method is:

$$[K_T]_{U_{i-1}} \{\Delta\phi_i\} = - [K(\phi)]\{\phi\}_{i-1} \tag{24}$$

where the tangent matrix is:

$$[K_T] = \left[\frac{\partial[K(\phi)]}{\partial\phi}\{\phi\} + [K(\phi)]\right] \tag{25}$$

The non-steady flow is based on an implicit Euler scheme with an iterative Newton-Raphson stabilization procedure:

$$\left[[M] + \Delta t[K]\right]_{i-1}\left\{\Delta\phi^{t+\Delta t}\right\} = [M]\left\{\phi^{t} - \phi^{t+\Delta t}_{i-1}\right\} - \Delta t\,[K(\phi)]\,\{\phi\}_{i-1} \tag{26}$$

These techniques need much CPU time, and some possibilities have been set to reduce it. The tangent matrix can be, with a little loss of accuracy, recomputed at the beginning of each time steps rather than each iteration. In some case, computing time reduction may reach a factor three.

5 APPLICATION

Many numerical tests have been performed with this model. The simulation of the tridimensional structure of the flow in a 180° curved flume in steady state has been compared with a scale model in laboratory. The dimensions used in the mathematical model were the same than for the physical model to prevent any scale effect on the results. The test bring us the confirmation of the rightness of the formulation chosen here. The results of this comparison can be found in Robert 1983.

5.1 Schematic estuary

In the testing process of the model, one goal was to verify that the vertical behaviour of the simulation agrees with the results of a two dimensional vertical models (width averaged) particularly with the Sunderman 1976 formulation based on cubic interpolation function in the vertical direction. The test case can be described as a schematic estuary closed at one end and submitted at the other one to a sinusoidal tidal cycle of 1 meter of amplitude. The flume has a length of 92 km and an width of 4,6 km. The depth is 50 meters except for a central portion where the bottom raise up to a depth of 20 meters. The finite element grid is composed of 10 quadrilateral elements and three layers were specified. The results of some time steps of a complete cycle (fig. 5) show a good agreement with those presented by Sundermann 1976 ones despite the fact that the vertical discretisation of the present model is far less accurate. The time interval of computation was, in this case, 1200 sec. for a full implicit scheme. This case was an occasion to test the validity of the mixing length distribution. The same test was run with a constant interfacial friction factor

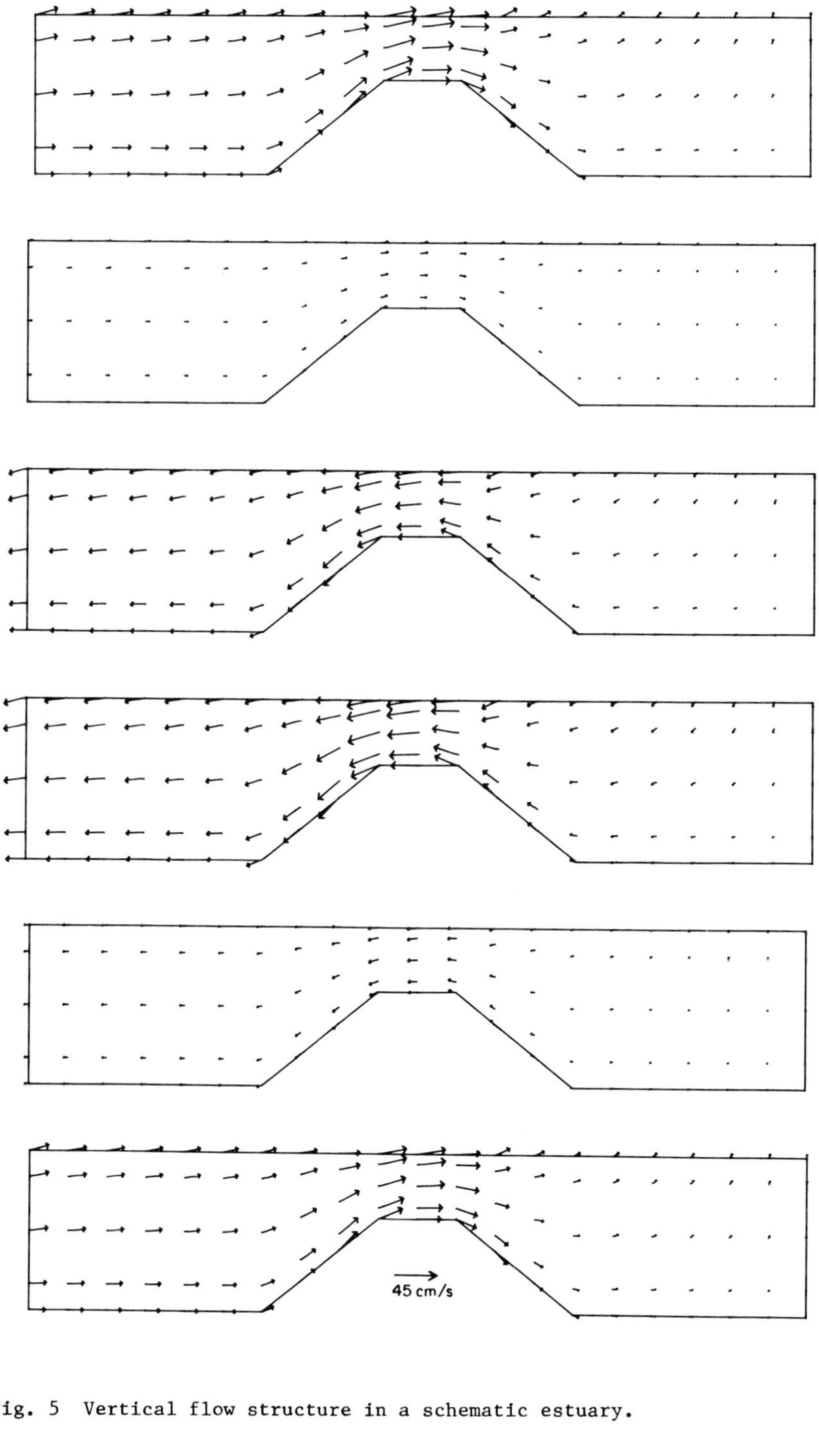

Fig. 5 Vertical flow structure in a schematic estuary.

and shown some irregularities of the vertical velocity profile as an erroneous position of the maximum velocity during the ebb flow. That situation was cured by the use of the distribution presented by eq. 20.

5.2 Wind induced circulation

The second example focuses on wind driven circulation. Some preliminary numerical tests agreed with the Leenderstee 1975 samples in the same condition. As for preceeding example the simulation was performed as a propagation problem; a constant 40 km/h S-W wind blowing over a 10 km x 5 km x 10m rectangular bassin. The result, shown in fig. 6, gives the circulation over the five layers, the water surface elevation and a transversal cross section in the middle of the bassin after six minutes of real time. At this time, the steady state is not totally achieved and the free surface is still moving. The corresponding computing time for this non-linear non-steady 2221 degrees of freedom was about 2h40 on a VAX 850 for a full implicit scheme. Nevertheless, some techniques as the computation of the global matrix only at the beginning of each time step can reduce the computing time to one third and the use of a vectorial computer may be usefull.

6. CONCLUSION

One of the most important concluding remark of this work, is that the automatic adjustment of the interlayer friction factor by the use of a well chosen mixing length distribution gives excellent vertical behaviour of the model. This vertical pattern of the flow is then relatively accurate and the formulation is well adapted to the natural flows. The use of the model is made easy by the fact that the finite element grid is only two dimensions and consequently a coupling with a two dimensional model is provided. This two and three dimensions models should be used to predict the vertical structure of the flow only where it is necessary.

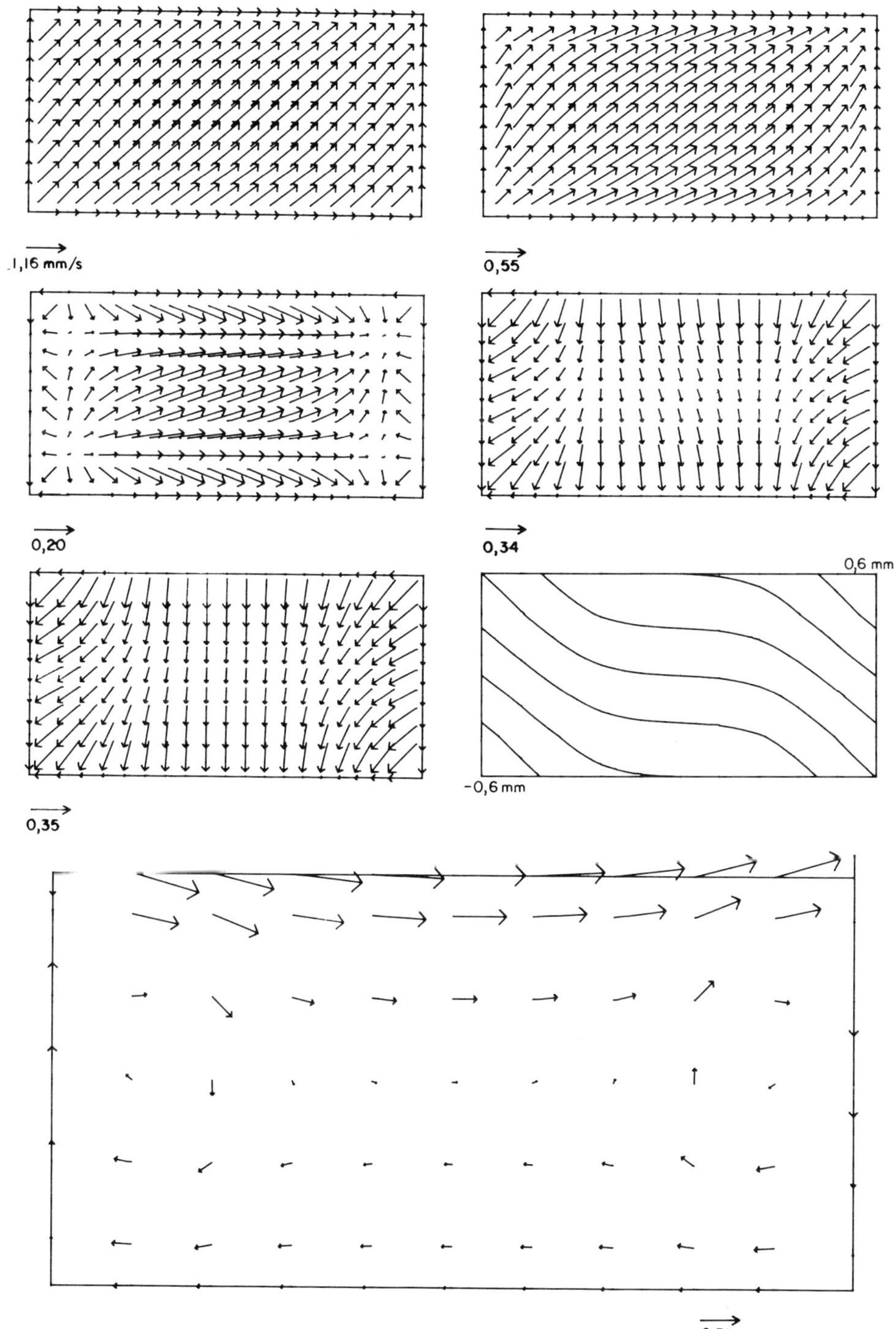

Fig. 6 Wind driven circulation in a rectangular bassin.

7. REFERENCES

Cochet, J.F., 1979. Modélisation d'écoulements stationnaires et non stationnaires par la méthode des éléments finis. Thèse de Docteur-Ingénieur, Université de Technologie de Compiègne, France, 142 p.

Kobayashi, M., Nakota, K. and Kawahara, M., 1980. A Three Dimensional Multi-Leveled Finite Element Model for Density Current Analysis. In: Third Int. Conf. on Finite Element in Flows Problems, Vol. II, Banff, Canada, pp. 80-92.

Leenderstee, J.J., Liu, S.K., Alexander, R.C. and Nelson, A.B., 1975. Three-dimension Model for estuaries and Coastal Seas, Vol II, Aspects of Computation, R-1764-OWRT, The Rand Corporation, California, USA, 165 p.

Montgomery, R.B., 1943. Generalization for Cylinder of Prandtl's Linear Assumption for Mixing Length. Annuals of the New York Academy of Sciences, XLIV: 89-103.

Robert, J.L., 1983. Modélisation tridimensionnelle des écoulements à surface libre, permanents et non permanents, par la méthode des éléments finis. Thèse de doctorat, Université Laval, Québec, Canada, 233 p.

Sunderman, J., 1976. Computation of Barotropic Tides by the Finite Element Method. In: Finite Element in Water Resources, First Int. Conf. Gray, Pinder and Brebbia ed., Princeton University, Pentech Press, London, U.K., pp. 4.31-4.67.

Wang, H.P., 1977. Multi Leveled Finite Element Hydrodynamic Model of Block Island Sound. In: Finite Element in Water Resources, First Int. Conf., Gray Pinder and Brebbia ed., Princeton University, Pentech Press, London, U.K.

Yalin, M.S., 1977. Mechanics of Sediment Transport, Second ed., Pergamon Press, Oxford, U.K., 259 p.

REAL AND SPURIOUS BOUNDARY LAYER EFFECTS IN THREE-DIMENSIONAL HYDRODYNAMICAL MODELS

Bruno M. JAMART and José OZER
Management Unit of the Mathematical Models of the North Sea and the Scheldt Estuary (MUMM). Institut de Mathématique, Avenue des Tilleuls, 15, 4000 Liège, Belgium.

ABSTRACT

On the basis of a simple test problem, three examples of the effect of boundary layers on large scale flows in three-dimensional hydrodynamical models are discussed. In the first case, an artificial, numerically induced residual flow is shown to arise from an inappropriate lateral boundary condition and the staggered nature of the computational grid. The second example demonstrates that the lack of sufficient horizontal resolution along the lateral boundaries can have a deleterious effect on the speed of propagation of long gravity waves. The last example concerns the effects on the speed of propagation of long gravity waves of the parameterization of the bottom stress in 3-D and 2-D models.

1. INTRODUCTION

The present volume bears witness that three-dimensional hydrodynamical models are rapidly coming of age. Among the many reasons why 3-D models sometimes have to be resorted to is the opportunity they offer to better describe boundary layer phenomena.

Particular care is required in modelling and/or interpreting boundary layer effects, both from a numerical and from a theoretical point of view. On the one hand, the numerical procedures need to be accurate and to have "high enough" resolution. On the other hand, the mathematical formulation of the problem needs to be "sufficiently complete".

In this paper, three examples of the effect of boundary layers on large scale flows are discussed on the basis of a simple test problem. The first two examples concern spurious effects and erroneous results. The third case deals with a real, frictional rather than fictional, boundary layer, namely the bottom boundary layer (BBL). There is no implication other than the necessity of carefulness in the relative proportion of incorrect results (2/3) versus correct ones.

In the next two sections, the two 3-D models whose results are used later and the benchmark calculation are briefly described. In sections 4 and 5, two examples are given of spurious effects introduced in the solution by mishandled lateral boundary conditions. In the first case, a spurious boundary layer develops along the lateral solid boundary of the basin, resulting in an artificial, numerically induced residual flow. This case is discussed at length in a recent paper (Jamart and Ozer, 1986) and only a summary is presented here. The second example demonstrates that the lack of sufficient horizontal resolution in the vicinity of a solid vertical boundary can have a deleterious effect on the speed of propagation of long gravity waves, even when the resolution is more than adequate for the wave itself.

In section 6, we investigate the relationship between the speed of propagation of long waves and the parameterization of the bottom stress in mathematical models. In particular, the differences between 2-D and 3-D models, and, in the latter case, between a slip and a no-slip bottom boundary condition are emphasized.

2. MODEL EQUATIONS AND SOLUTION PROCEDURES

The numerical experiments discussed in this paper were performed as part of a model intercomparison exercise described in several earlier publications (Jamart and Ozer, 1985, 1986; Jamart *et al.* , 1986). Hence, only a brief exposé of the "materials and methods" suffices here. Two independently developed three-dimensional models are considered. The models shall be referred to as "FD-model" and "FU-model", respectively, where FD stands for Finite Difference and FU for FUnction. The justification for these denominations will ensue from the description of the numerical procedures.

2.1. The finite difference ("FD") model

The FD-model is a version of the numerical model developed by Paul and co-workers (Paul and Lick, 1981). The governing equations are the 3-D, time-dependent, nonlinear Navier-Stokes equations. It is assumed that the pressure is hydrostatic, the fluid homogeneous, and the Coriolis parameter constant. The turbulent fluxes of momentum are parameterized by means of eddy coefficients, denoted A_h in the horizontal plane and A_v in the vertical direction. The latter may be a function of time, depth, and/or flow variables. In conventional notations, and using the vertically integrated hydrostatic equation, the equations read :

$$\frac{\partial u}{\partial t} + \frac{\partial u^2}{\partial x} + \frac{\partial uv}{\partial y} + \frac{\partial uw}{\partial z} - fv = -g\frac{\partial \eta}{\partial x} + A_h\left(\frac{\partial^2 u}{\partial x^2} + \frac{\partial^2 u}{\partial y^2}\right) + \frac{\partial}{\partial z}\left(A_v\frac{\partial u}{\partial z}\right) \tag{1}$$

$$\frac{\partial v}{\partial t} + \frac{\partial uv}{\partial x} + \frac{\partial v^2}{\partial y} + \frac{\partial vw}{\partial z} + fu = -g\frac{\partial \eta}{\partial y} + A_h\left(\frac{\partial^2 v}{\partial x^2} + \frac{\partial^2 v}{\partial y^2}\right) + \frac{\partial}{\partial z}\left(A_v\frac{\partial v}{\partial z}\right) \tag{2}$$

$$\frac{\partial u}{\partial x} + \frac{\partial v}{\partial y} + \frac{\partial w}{\partial z} = 0 \tag{3}$$

For this model, the boundary conditions are :

(i) along solid, vertical boundaries,

$$u = v = 0 \quad \text{for all t and z;} \tag{4}$$

(ii) at the bottom, $z = -H$,

$$u = v = w = 0 \quad \text{for all t;} \tag{5}$$

(iii) at the surface, defined as the time-mean position of the air-sea interface, and denoted $z = 0$,

$$\frac{\partial \eta}{\partial t} = w \quad \text{for all t;} \tag{6}$$

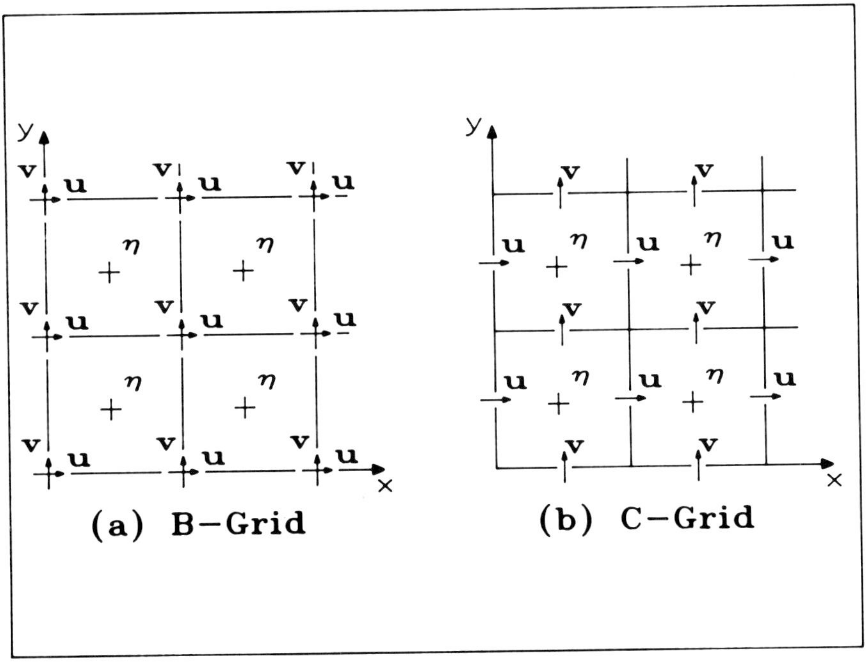

Fig. 1. Computational grids : (a) B-grid used in the FD-model;
(b) C-grid used in the FU-model.

and

$$A_v \frac{\partial u}{\partial z} = \frac{1}{\rho} \tau_w^x \ , \ A_v \frac{\partial v}{\partial z} = \frac{1}{\rho} \tau_w^y \quad \text{for all t,} \tag{7}$$

where τ_w^x and τ_w^y are the components of the prescribed wind stress.

The numerical procedure used to integrate these equations is described in detail by Paul and Lick (1981) and summarized by Jamart *et al.* (1986). For the following discussion, we only need to note that :

– all derivatives, both spatial and temporal, are approximated by finite difference analogs (hence the denomination as "FD-model");

– the spatial (horizontal) distribution of the variables is such that u and v are calculated at the same points (the vertices of a rectangular mesh), whereas w and η are defined at the center of each cell (see Fig. 1(a)); this type of grid arrangement is often called the Arakawa B-grid (Arakawa and Lamb, 1977).

The free surface position is computed by a semi-implicit scheme (Paul, personal communication), thus allowing for a longer time step than that constrained by the C-F-L criterion.

2.2. The spectral ("FU") model

The second 3-D model used in the numerical experiments is, in its present version, limited to the solution of the linearized equations of continuity and motion. Moreover, lateral friction is ignored so that equations (1) and (2) reduce to

$$\frac{\partial u}{\partial t} - fv = -g\frac{\partial \eta}{\partial x} + \frac{\partial}{\partial z}\left(A_v \frac{\partial u}{\partial z} \right) \tag{8}$$

$$\frac{\partial v}{\partial t} + fu = -g\frac{\partial \eta}{\partial y} + \frac{\partial}{\partial z}\left(A_v \frac{\partial v}{\partial z} \right) \tag{9}$$

The continuity equation (3), integrated vertically, and the kinematic conditions at $z = 0$ and $-H$ combine to yield :

$$\frac{\partial \eta}{\partial t} + \frac{\partial}{\partial x}\int_{-H}^{0} u\, dz + \frac{\partial}{\partial y}\int_{-H}^{0} v\, dz = 0 \tag{10}$$

The dynamic boundary condition at the free surface is the same as for the FD-model, *i.e.*, equation (7). At $z = -H$, the bottom stress is assumed proportional to the bottom current

$$A_v \frac{\partial u}{\partial z} = \kappa\, u \ , \quad A_v \frac{\partial v}{\partial z} = \kappa\, v \tag{11}$$

a condition from which the no-slip case of equation (5) can be recovered by taking the limit for $\kappa \rightarrow \infty$.

The specification of an appropriate boundary condition along the coastline is trickier than appears at first sight. A conventional choice for the 3-D linear model consisting of equations (8) to (10) is to set equal to zero the component of the velocity normal to the wall at all depths. Serious difficulties associated with this boundary condition and alternative formulations will be discussed in a later section.

In the FU-model, the depth dependence of the velocity variables is not approximated by the classical, finite difference type of representation but rather by means of a truncated expansion in a series of continuous functions. This approach was first proposed by Heaps (1972). It has since been used and expanded upon by Heaps himself (1981) and by many other researchers (*e.g.*, Nihoul, 1977, Davies, 1977, 1980, and subsequent papers, Owen, 1980, Cooper and Pearce, 1982). Different sets of basis functions have been used in conjunction with this approach. In the FU-model, following Heaps (1972), we choose the eigenfunctions of the vertical diffusion operator.

The discretization of the equations in the horizontal plane is accomplished by a finite difference procedure. The disposition of the variables on the computational grid is different in the FU-model from that used in the FD-model. In the function model, the two components of the velocity vector and the elevation are all calculated at distinct locations. This spatial grid, referred to as a C-grid, is displayed in Fig. 1(b).

Finally, the time-stepping procedure used to integrate the equations of the FU-model is a semi-implicit, alternating-direction method patterned after that proposed by Beckers and Neves (1985) for a vertically integrated model. More details on the numerics of the FU-model are given by Jamart and Ozer (1986).

3. THE TEST PROBLEM

The boundary layer effects discussed in this paper are, we think, of rather general relevance. Yet, we find it useful for demonstration purposes to use one of the benchmark calculations of a model intercomparison experiment currently in progress. All the numerical results to be displayed were obtained by running the models just described to solve the following problem.

The domain is a rectangular basin, closed on all sides, of dimensions 600 x 1200 km, and of uniform depth H = 100 m. The long axis of the basin runs from south to north and is aligned with the y-axis. The latitude is taken as 55° N. The motion starts from rest ($u = v = w = \eta = 0$) and it is forced by a constant, impulsively applied, southward wind stress ($\tau_w^y = -0.1$ N m^{-2}).

To our knowledge, the only complete analytical solution to this problem is that of Ekman (1905) for steady state and linear equations, and that solution is actually valid for flat-bottomed enclosed basins of any shape. At steady state, the two components of the transport are zero everywhere. The surface slope can be calculated from the vertically integrated momentum balance which involves only $\nabla_h \eta$, τ_w , and $\tau_b = \rho \, (A_v \dfrac{\partial \mathbf{u}}{\partial z})_{z=-H}$. The bottom stress can be related to $\nabla_h \eta$ and τ_w by solving the vertical problem. Ekman (1905) assumes that the eddy viscosity A_v is independent of the depth and he uses a no-slip condition at the bottom. A similar solution is shown to also hold for a linear slip condition and any positive, arbitrarily z-dependent A_v by Jamart and Ozer (1987).

For the time-dependent part of the solution, the presence of sharp corners in the domain boundary complicates the mathematical analysis of the problem. Indications as to the type of wave motion to be expected can be found in the studies by Taylor (1921) and Rao (1966) and in the monograph of LeBlond and Mysak (1978). The effects of a constant vertical eddy viscosity on Kelvin waves are discussed by Mofjeld (1980). Precious as are such elements of the system response, we mostly rely on model intercomparison to test the codes of the 3-D models.

4. SPURIOUS LATERAL BOUNDARY LAYER ON A C-GRID

The equations of the FU-model have a well-known steady solution (Ekman, (1905)) for constant depth and wind stress. The residual transport vanishes everywhere. The surface slope and the velocity profiles are horizontally uniform. However, the numerical solution obtained by running to steady state the FU-model as described in Section 2 differs from the exact solution.

In particular, as is shown by Jamart and Ozer (1986, Figs. 4 and 5), the depth-averaged currents are more or less organized in two gyres rotating in opposite directions, with strong flows along the edges of the basin. This pattern is spurious. The solution should be horizontally uniform. The building up of the erroneous residual currents also affects the time-dependent part of the solution.

The source of the error in the numerical procedure is twofold. First, the computational grid used in this model (*i.e.*, the C-grid) requires that averages be performed everywhere to evaluate the Coriolis term in the momentum equations. Second, the boundary condition implemented along the solid walls (*i.e.*, the vanishing of the normal velocity at all depths) is inappropriate for the model's equations.

The inadequacy of the boundary condition $\mathbf{u} \cdot \mathbf{n} = 0$ at all z stems from the fact that the equations of the FU-model do not include the necessary terms (horizontal diffusion) to allow a smooth vanishing of the velocity in a coastal boundary layer. In the absence of rotation, this results in a singularity in the w-field (the vertical velocity becomes infinite at the wall in the continuous solution). With rotation, the conventional evaluation of the Coriolis term on a C–grid (averaging the values at the four closest points) underestimates the deflecting "force" along solid boundaries because at least one of the points entering the mean is located on the boundary. The inadequacy of the boundary condition for the equations of the FU-model, combined with the properties of the C-grid, is responsible for the numerically induced residual current observed in the results of this model.

Different tacks are possible to circumvent the difficulty. The first would be to modify the governing equations so as to model properly the behavior of the fluid in the side-wall friction layers. The requirements for such an approach, and elements of the solution, are discussed by Pedlosky (1979, sections 4.13, 5.12, and 8.3). The second possible tack, which we make here, is to modify the lateral boundary condition so that it is sufficient for the problem to be well-posed and to implement that condition without introducing unwanted effects.

As discussed by Greenspan (1968) for steady motion, by Mofjeld (1980) for Kelvin waves, and by Pedlosky (1979), a natural requirement for the flow near a vertical wall is that the normal transport be zero at some distance from the coast. The thickness of the unresolved boundary layers depends at least on viscosity, depth and latitude, and it may be time-dependent. Nevertheless, those layers are in general quite thin for the scales we are considering in the present discussion. Hence, we set

$$\int_{-H}^{0} \mathbf{u} \cdot \mathbf{n} \, dz = 0 \tag{12}$$

along the closed boundaries of the computational domain.

From a theoretical point of view, equation (12) is a sufficient boundary condition for the equations of the FU-model. Indeed, as is discussed, *e.g.*, by Davies (1987), the "3-D (x,y,z,t)" problem can always be decomposed in a "2-D (x,y,t)" problem plus a "1-D (z,t)" problem. The

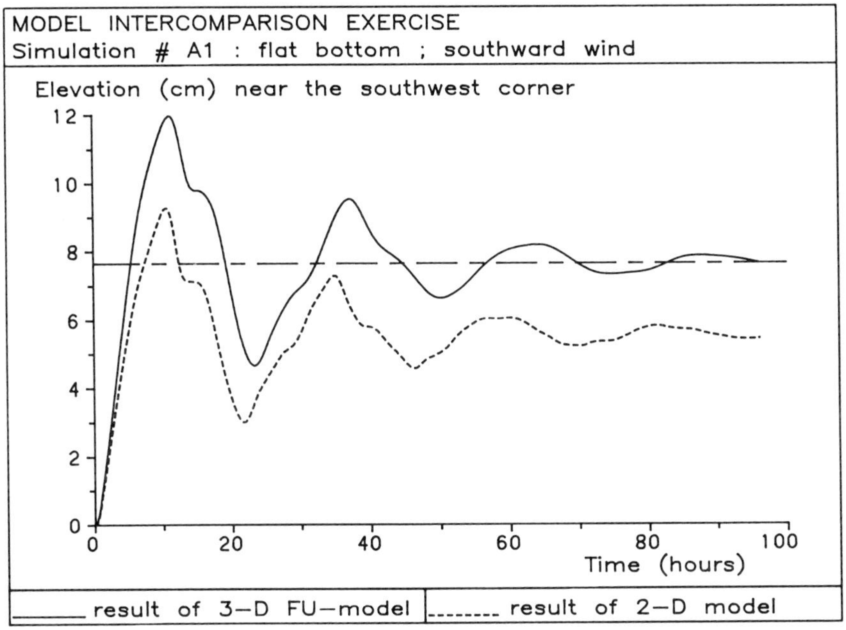

Fig. 2. Time series of sea surface elevation in the southwest corner of the basin. The solid line shows the results of the 3-D FU-model. The horizontal line indicates the elevation of the exact solution at steady state. The dashed line corresponds to the results, discussed in section 6, of a 2-D model.

2-D problem is similar to the classical, vertically-integrated, shallow water wave model for which (12) is an appropriate condition.

From a computational point of view, however, the solution of the 3-D equations does require the evaluation of the Coriolis term at all depths, or, equivalently, for all vertical modes in the function model. In particular, if a C grid is used, some kind of averaging is necessary. Hence, something must be done, in addition to using (12) in the continuity equation. We have shown (Jamart and Ozer, 1986) that a simple stratagem to eliminate the spurious boundary currents observed in the results of the FU-model is to retain in the calculation of the Coriolis term only those points that are not located on the boundary (the "wet-points-only" approach). Other devices to the same effect, which can be referred to as "computational boundary conditions", are presently under study.

When the wet-points-only approach, together with equation (12), is implemented in the FU-model, the solution of the test problem described in Section 3 converges towards the exact steady state solution. Sample results are shown for a particular location (the southwest corner of the basin) in Figs. 2 (elevation) and 3 (depth-averaged speed). For this calculation, A_v is constant and equal to $65 \times 10^{-3}\ m^2\ s^{-1}$; a no-slip condition is used at the bottom; the

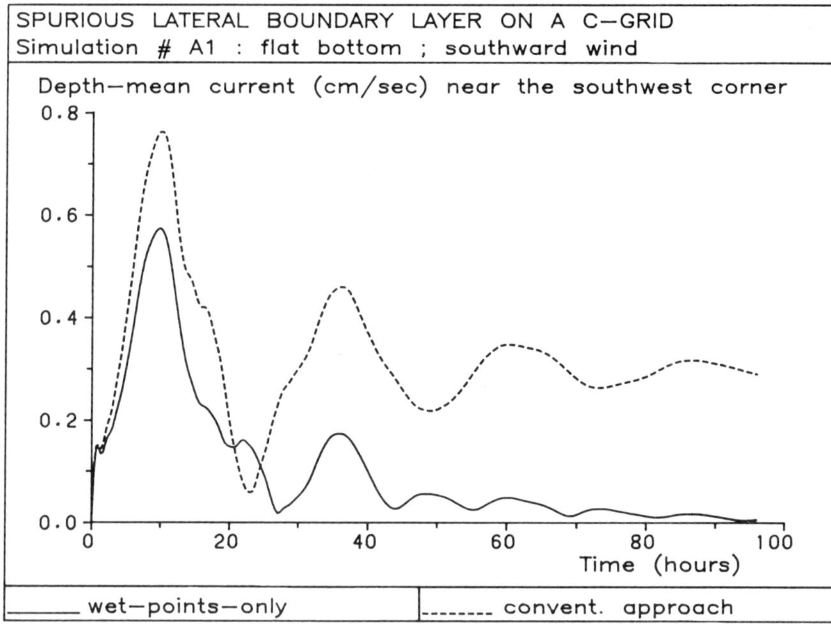

Fig. 3. Time series of the depth-averaged current in the southwest corner of the basin. The solid line shows the result of the 3-D FU-model using the wet-points-only method for the evaluation of the Coriolis term. The dashed line shows the erroneous result obtained when the conventional approach is used.

computational grid is uniform, with $\Delta x = \Delta y = 40$ km; 10 modes are retained in the velocity expansions.

When the Coriolis term is evaluated in the conventional way along the boundary - or, in other words, if the condition $\mathbf{u} \cdot \mathbf{n} = 0$ at all z, which also comprises (12), is used in the horizontal averaging procedure - the depth-averaged velocity does not tend to zero as indicated by the dashed line on Fig. 3. The elevation, however, is hardly affected by the spurious residual current.

5. SPURIOUS LATERAL BOUNDARY LAYER ON A B-GRID

With the aim of verifying the time-dependent part of the results of the FU-model, for which no analytical solution appears available, the test problem was solved with the FD-model using the same parameters as for the FU-model. Previous experiments with a similar model reported by Jamart *et al.* (1986) have indicated that a rather fine vertical spacing is required to obtain a correct evaluation of the bottom stress. Hence, we chose $\Delta z = 2$ m to perform the calculations , which may be somewhat of an overkill considering the unrealistically large value assigned to A_v.

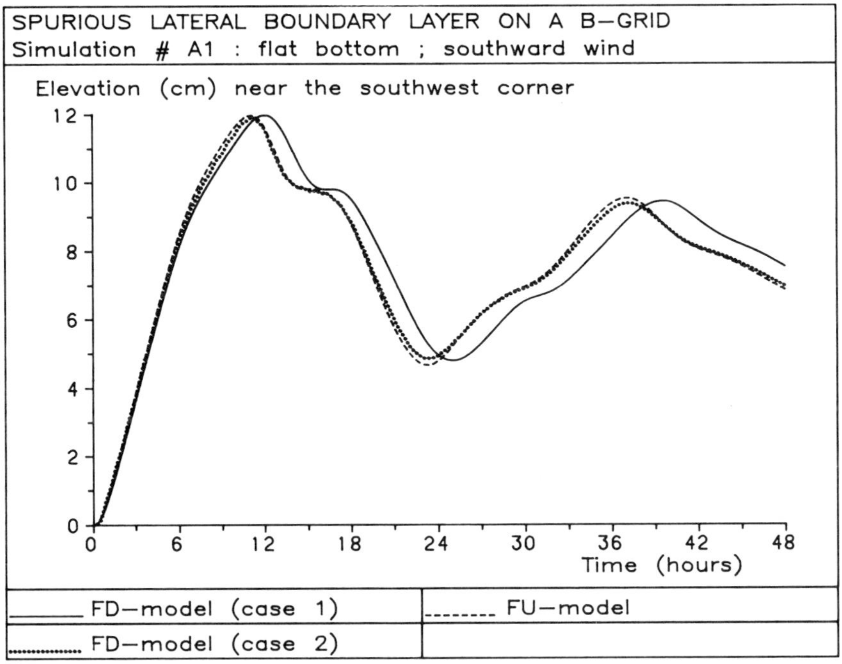

Fig. 4. Time series of sea surface elevation in the southwest corner of the basin. The solid line shows the result of the 3-D FD-model using a uniform grid. The dashed line shows the result of the 3-D FU-model. The dotted line shows the result of the 3-D FD-model using a variable grid.

Preliminary results obtained with the FD-model are shown in Figs. 4 and 5, wherein the corresponding time series calculated with the FU-model are also displayed for comparison. Obviously, there are large discrepancies (mostly phase differences) between the results of the two models.

The analysis of the results of the benchmark calculation, which was also performed with a 2-D (vertically integrated) model, indicates that the transient response of the system can be viewed as a superposition of :

- the building up, in an asymptotic fashion, of the steady state solution for the sea surface slope;
- fairly rapidly damped inertial oscillations of the velocities at the various depth-levels;
- propagating and more slowly damped long gravity waves and the associated quasi-barotropic currents.

Both the shape and the phase speed of the gravity waves are affected by rotation and by bottom friction (see next section). The first mode, which dominates the transient response, is akin to a Kelvin wave along the straight portions of coastline. This wave does not progress at the same

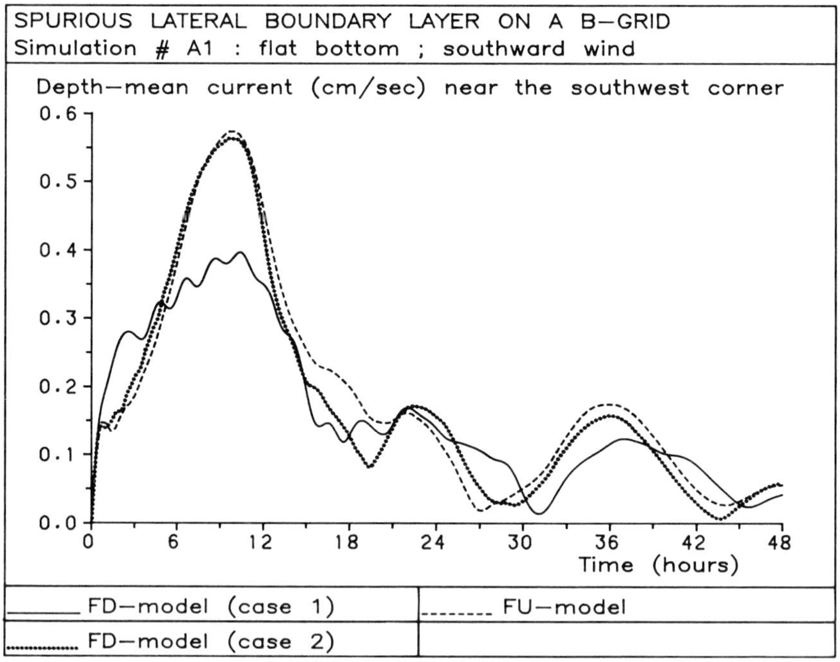

Fig. 5. Time series of the depth-averaged current in the southwest corner of the basin. Solid, dashed and dotted lines are as defined for Figure 4.

speed in the FU- and the FD-model.

Experiments performed with the FD-model have shown that the observed differences cannot be attributed to :

- the presence or absence of the advective terms;
- either the timestep or the degree of implicitness with which various terms are calculated;
- differences in the numerical treatment of variations in the vertical direction; the bottom stresses calculated by the two models in the absence of rotation are the same.

Hence, the only possible explanations of the discrepancy are :

- the effect of the horizontal viscous terms (included in the FD-model only);
- the difference in computational grids (B-grid in the FD-model; C-grid in the FU-model);
- the difference in boundary conditions.

These effects have been studied on the shallow water wave equations by Davey *et al.* (1983) and Hsieh *et al.* (1983), whose results are applicable to the present 3-D calculations.

First, with a Rossby radius $L = c/f \sim 263$ km, the horizontal Ekman number $E_h = A_h / fL^2$ is of order 10^{-4} for $A_h \sim O(10^3\ m^2\ s^{-1})$. Therefore, using the length of the basin perimeter normalized by L to calculate the nondimensional wavenumber l of the dominant mode, the parameter $\varepsilon = E_h / l$ introduced by Davey *et al.* (1983) to estimate the importance of lateral viscosity is found to be $O(10^{-4})$. Horizontal friction should not affect the phase speed. This is confirmed by numerical experiments : the results of the FD-model are not sensitive to variations of A_h.

Second, Hsieh *et al.* (1983) show that, in the absence of lateral viscosity, the phase speed of a free Kelvin wave calculated using a C-grid is independent of the "resolution parameter", defined as $\Delta = \Delta x/L$ assuming $\Delta y = \Delta x$. For the B-grid, however, provided there is any horizontal friction at all, the phase speed is i) always smaller than in the inviscid case; ii) dependent upon Δ, and strongly so if $\varepsilon/\Delta^2 \leq 1$; iii) different whether one uses a free-slip or a no-slip lateral boundary condition. For the two grids, the effect of bottom stress only (assumed to be a linear function of the depth-averaged current in those studies) is basically the same as long as $\Delta \leq 0.3$. Neglecting this bottom stress effect, we find from Fig. 4 of Hsieh *et al.* that for a uniform grid spacing of 40 km yielding $\Delta \sim 0.15$, a free Kelvin wave propagates on a B-grid with no-slip boundary condition at a speed approximately equal to 92% of that calculated on a C-grid. This is precisely the ratio of the times elapsed between the first two maxima in elevation calculated with the FU- and with the FD-model.

In physical terms, the no-slip boundary condition and the B-grid used in the FD-model create a spurious boundary layer wherein the velocities (and hence the transports) are brought to zero over the cell adjacent to the boundary notwithstanding the fact that the viscous layer is not properly resolved with a meshsize of 40 km. In particular, the longshore transport of the Kelvin waves, which should increase exponentially up to a rather short distance from the coast, is significantly underestimated, leading to a decrease in the phase speed.

As was the case in the previous section, there are several ways to eliminate the spurious boundary layer. First, a free-slip lateral boundary condition is shown by Hsieh *et al.* (1983) to affect to a much lesser extent than a no-slip condition the phase speed of a free Kelvin wave calculated on a B-grid. Such an approach is certainly useful for the shallow water wave equations but it could be troublesome for the 3-D calculation of the wind-driven part of the motion. We have not attempted to use a free-slip condition in the FD-model.

A "safe bet" approach is to reduce the uniform meshsize to the point where the lateral viscous layer is well resolved and/or where the long gravity waves are unaffected by the boundary condition. This remedy is computationally expensive.

A third approach is to refine the grid locally, *i.e.*, to use a small Δ only near shore and in the direction normal to the coast. We have repeated the test calculation with the FD-model using a variable grid spacing (Δx = 5, 10, 15, 20, 30, 40 km going from the boundaries towards the center of the basin). The results of that computation are shown in Figs. 4 and 5. They are in excellent agreement with those of the FU-model. We conclude that the effects of the spurious lateral boundary layer, which is still present, have been effectively reduced to noise level.

6. EFFECTS OF THE BOTTOM BOUNDARY LAYER

The modelling of the bottom boundary layer (BBL) is no easy task. The question of how the turbulent processes occuring in the BBL should be parameterized is certainly not yet settled. The choice of appropriate values for whatever parameters are included in a given model is also not trivial and the answer is usually site- and/or model-specific. Nevertheless, it is of interest to study the effects on model results of some parameterizations. This is indeed one of the main goals of the model intercomparison exercise mentionned earlier. This exercise includes i) the FD-model; ii) the FU-model; iii) a conventional 2-D (vertically integrated) storm surge model. In both 3-D models, the turbulent mixing of momentum is represented by means of an eddy coefficient. The bottom stress used in the 2-D model, for the experiments reported here, is a linear function of the depth-averaged current, $\tau_b = \gamma \rho \bar{u}$.

For the test problem described in section 3, and provided that the spurious lateral boundary layers discussed in sections 4 and 5 are taken care of, the results of the FD- and FU-models are almost undistinguishable. The results of the 2-D model, however, are significantly different from those of the 3-D models. For example, the time series of the elevation in the southwest corner calculated with the 2-D model is compared to that of the FU-model in Fig. 2. The friction coefficient γ used for this computation is taken as 2.4×10^{-3} m s^{-1}. The 3-D solution is seen to be always higher than the 2-D solution, to have a first mode oscillation of larger amplitude and longer period, and to have a "smoother" evolution than the elevation calculated with the 2-D model. These three characteristics are even more obvious when the same comparison is done with $f = 0$ (see Figs. 2 and 3 of Jamart and Ozer, 1986).

Heaps (1972) has already observed and commented on such differences, and so have we in several papers. In particular, we have contrasted in some detail the role of the bottom stress in 2-D versus 3-D models for steady state wind-driven circulation in shallow basins (Jamart and Ozer, 1987). This aspect is not discussed here. We have also investigated the effects of bottom stress parameterization on the free oscillations in an enclosed basin (Jamart *et al.*, 1986) using a "semi-analytical" approach. That theory is hereafter improved and expanded upon.

For convenience, and because it is sufficient to demonstrate our main points, we restrict the analysis to the case of a constant A_v. We shall also ignore rotation in the present study of bottom stress effects. The effects of the Coriolis term on waves in a rectangular basin can only be studied numerically. For an inviscid, homogeneous fluid, Rao (1966) has shown that the effect of rotation on the free gravitational oscillations in a 2×1 rectangular basin is to decrease the frequency of the first positive antisymmetric mode and to increase the frequency of all the other modes. The first mode is dominant in the results of our calculations. For the test problem discussed here, the period of the first longitudinal mode (equal to 21.3 h in the absence of friction and of rotation) is lengthened by a factor of about 1.1 ($T_1 \sim 23.4$ h) on the rotating earth. This frequency shift due to rotation only is different from that due to bottom stress only which we discuss hereafter.

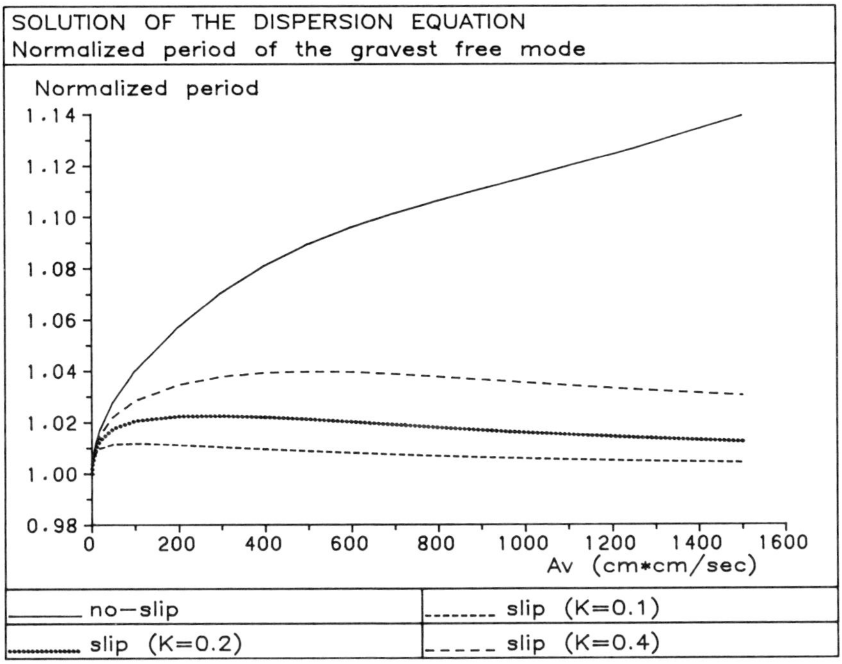

Fig. 6. Period of the first longitudinal free mode, normalized by the period of the frictionless solution, as a function of the vertical eddy coefficient. The solid line corresponds to the case of a no-slip bottom boundary condition. Short dashed, dotted, and long dashed lines show the results in the case of a linear slip condition, with κ = 0.1, 0.2 and 0.4 cm/s, respectively.

For the 2-D, linear model, assuming $\eta \sim \exp\left[i\,(\mathbf{k}\,.\,\mathbf{x} - \omega\, t)\right]$, it is easily shown that the dispersion equation is

$$\omega^2 + i\,\omega\,\frac{\gamma}{H} = |\,\mathbf{k}\,|^2\, g\, H \tag{13}$$

Consider now the 3-D model defined by equations (8) to (10), with f = 0. The vertical problem (a diffusion equation with boundary condition (11) at the bottom and zero shear at the surface) can be solved separately with the pressure gradient as a forcing term. This also holds true for the case f ≠ 0 (see Mofjeld, 1980, for a no-slip bottom boundary condition). The divergence of the bottom stress can then be calculated analytically, and the dispersion equation for plane waves is found to be :

$$\omega^2 = |\,\mathbf{k}\,|^2\, g\, H \left[1 - \frac{\tanh \lambda H}{\lambda H} \left[1 + \frac{\lambda\, A_v}{\kappa} \tanh \lambda H \right]^{-1} \right] \tag{14}$$

where $\lambda = (1 - i) \sqrt{\dfrac{\omega}{2\, A_v}}$. The analysis of equations (13) and (14) is most easily done, in physical terms, by considering separately the effects of the bottom stress on i) the frequency

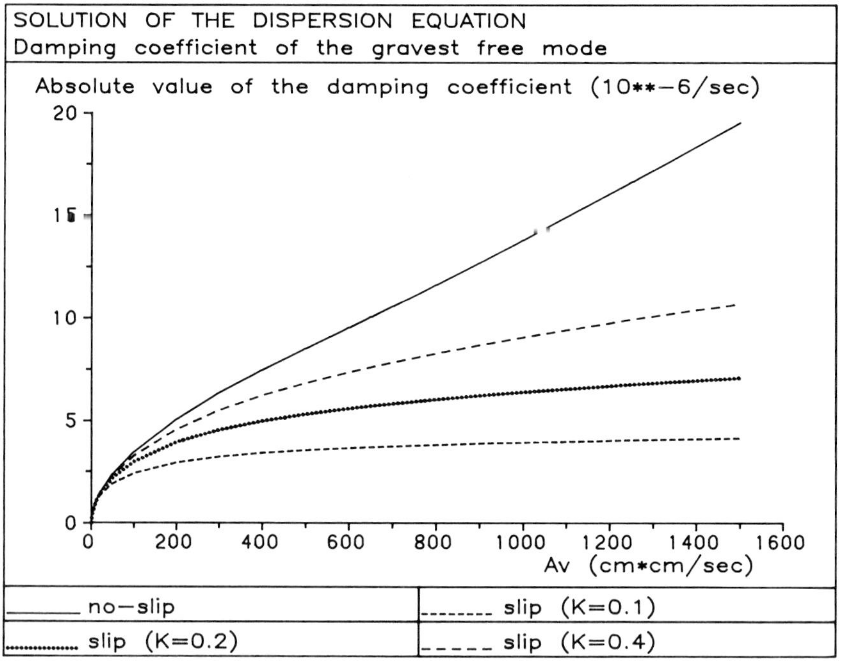

Fig. 7. Bottom stress induced damping coefficient as a function of the vertical eddy coefficient. Line types are the same as in Figure 6

shift with respect to the frictionless case and ii) the damping of the free waves.

For the 2-D model and for reasonable values of γ, (13) indicates that the frequency shift is extremely small for the lowest modes present in our test calculations. For example, with $\gamma = 2.4 \times 10^{-3}$ m s^{-1} , the period of the first longitudinal mode is increased by less than 0.25 hour. The damping coefficient (*i.e.*, the imaginary part, ω_I, of the frequency), given by $-\gamma / 2H$, is independent of the wavenumber.

For the 3-D model, by contrast, the dispersion equation shows that both the frequency shift and the damping coefficient depend on the wavenumber as well as on the depth and the friction parameters (A_v and κ). The solution of (14) for the gravest mode, calculated numerically through an iterative procedure, is shown in Fig. 6 (normalized period as a function of A_v for different values of κ, the no-slip solution being the limit for $\kappa \rightarrow \infty$) and Fig. 7 (damping coefficient for the same parameters). These figures illustrate the sensitivity of the bottom stress effects to the choice of boundary condition.

The phase lag observed in Fig. 2 between 2-D and 3-D results (the latter obtained under the rather unrealistic assumption of no-slip with $A_v = 65 \times 10^{-3}$ m^2 s^{-1}) is fully accounted for by the lengthening effect displayed in Fig. 6. Similarly, the difference in damping coefficients ($\omega_I = -1.2 \times 10^{-5}$ s^{-1} in the 2-D model, versus -0.8×10^{-5} s^{-1} in the 3-D model) explains

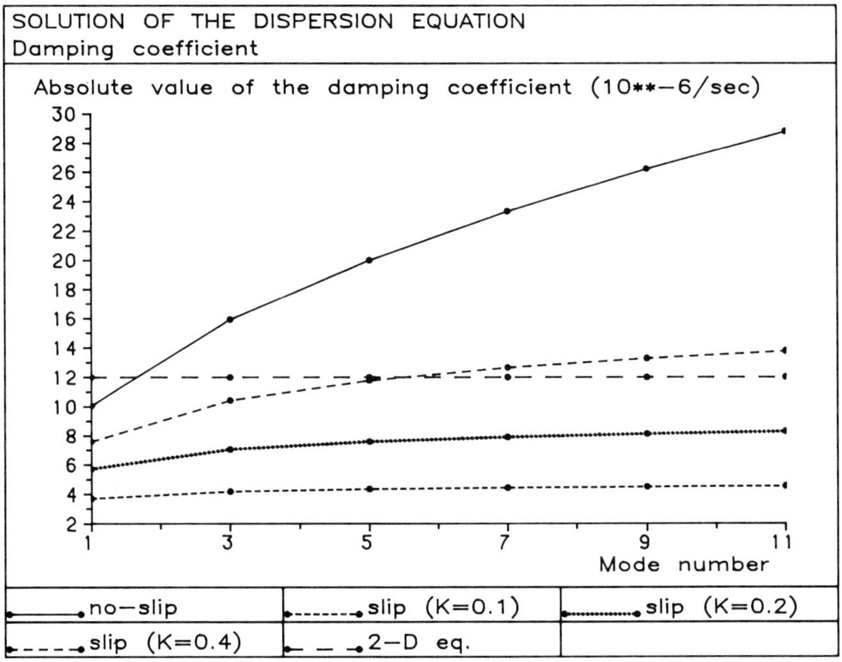

Fig. 8. Bottom stress induced damping coefficient for the first six longitudinal free modes (only odd modes can exist). Line types are the same as in Figure 6. The horizontal line shows the damping coefficient of the 2-D model.

the difference in the amplitude of the dominant oscillation. The relative smoothness of the 3-D time series compared to those of the 2-D results is also consistent with equations (13) and (14). Indeed, as is shown in Fig. 8, the damping coefficient in a model that resolves the vertical structure of the flow is dependent on the wavenumber ($|\omega_I|$ increases with $|\mathbf{k}|$, rapidly so if κ is large). In a 2-D model, on the other hand, at least for the case of linear friction, the higher modes are not damped more rapidly than the lowest ones, resulting in more complicated time evolutions. Our experience is that this last remark also applies to a 2-D model with a quadratic bottom stress.

7. SUMMARY AND CONCLUSIONS

In this paper, three examples of interactions between boundary layer phenomena and large scale flow are discussed. These effects of the boundary layers are either real or spurious.

We have shown how the implementation of an inappropriate lateral boundary condition in a 3-D model using a staggered ("C") grid results in the building up of a spurious boundary layer. We have also shown that the lack of resolution near the solid boundary in a model using a B-grid not only prevents the boundary layer phenomena from being properly modelled but also affects to a significant extent the speed of propagation of long gravity waves. Remedies are proposed to eliminate such spurious boundary layers, or at least, their effects on the large scale flow.

The parameterization of the turbulence in the bottom boundary layer can be a determinant factor not only for the steady state wind-driven circulation in shallow basins (Jamart and Ozer, 1987) but also for the characteristics of long gravity waves. Dispersion equations taking into account the effects of bottom stress in 2-D and 3-D formulations yield a satisfactory qualitative explanation for the different responses observed in numerical experiments. The dispersion equation for models resolving the vertical structure of the flow also indicates the sensitivity of the results to the choice of bottom boundary condition (slip versus no-slip).

Boundary layer phenomena can play an important role in the outcome of three-dimensional hydrodynamical calculations. The complexity of the physics and of the numerical procedures dictates that caution be exercised in interpreting the results of models. Simple numerical experiments - homework type calculations - can be useful in gaining insight and in discriminating physical effects from numerical artefacts.

8. ACKNOWLEDGMENTS

This work was supported by Det norske Veritas within the frame of an industry research and development project called the Bottom Stress Experiment (BSEX) funded by Fina Exploration Norway. We thank Ms Y. Spitz for help with the calculations and comments on the manuscript, Ms C. Coolen for assistance with the computer graphics and Ms A.F. Lucicki for typing the manuscript.

9. REFERENCES

Arakawa, A. and Lamb, V.R., 1977. Computational design of the basic dynamical processes of the UCLA general circulation model. In : Methods of Computational Physics, Vol. 17, Academic Press, New York, 173-265.

Beckers, P.M., and Neves, R.J., 1985. A semi-implicit tidal model of the north European continental shelf. Applied Mathematical Modelling, 9 : 395-402.

Cooper, C.K., and Pearce, B., 1982. Numerical simulations of hurricane-generated currents. Journal of Physical Oceanography, 12 : 1071-1091.

Davey, M.K., Hsieh, W.W., and Wajsowicz, R.C., 1983. The free Kelvin wave with lateral and vertical viscosity. Journal of Physical Oceanography, 13 : 2182-2191.

Davies, A.M., 1977. The numerical solution of the three-dimensional hydrodynamic equations using a B-spline representation of the vertical current profile. In : J.C.J. Nihoul (Editor), Bottom Turbulence, Oceanographic Series, Elsevier, Amsterdam, 19 : 1-25.

Davies, A.M., 1980. Application of the Galerkin method to the formulation of a three-dimensional nonlinear hydrodynamic sea model. Applied Mathematical Modelling, 4 : 245-256.

Davies, A.M., 1987. On extracting current profiles from vertically integrated numerical models. Submitted for publication.

Ekman, V.W., 1905. On the influence of the Earth's rotation on ocean-currents. Arkiv för matematic, astronomi, och fysik, Band 2, No.11.

Greenspan, H.P., 1968. The Theory of Rotating Fluids. Cambridge University Press, Cambridge, 328 pp.

Heaps, N.S., 1972. On the numerical solution of the three-dimensional hydrodynamic equations for tides and storm surges. Mémoires de la Société Royale des Sciences de Liège, 6 (II), 143-180.

Heaps, N.S., 1981. Three-dimensional model for tides and surges with vertical eddy viscosity prescribed in two layers - I. Mathematical formulation. Geophysical Journal of the Royal Astronomical Society, 64 : 291-302.

Hsieh, W.W., Davey, M.K., and Wajsowicz, R.C., 1983. The free Kelvin wave in finite difference numerical models. Journal of Physical Oceanography, 13 : 1385-1397.

Jamart, B.M., and Ozer, J., 1985. Hydrodynamical modelling of time-dependent flows in 2, 2.5, and 3 dimensional space. In : R. Van Grieken and R. Wollast (Editors), Progress in Belgian Oceanographic Research, The University of Antwerpen, 1-14.

Jamart, B.M., Ozer, J., and Spitz, Y., 1986. Bottom stress and free oscillations. In: J.J. O'Brien (Editor), Advanced Physical Oceanographic Numerical Modelling, D. Reidel Publishing Company, 581-598.

Jamart, B.M., and Ozer, J., 1986. Numerical boundary layers and spurious residual flows. Journal of Geophysical Research, 91 (C9) : 10, 621-10, 631.

Jamart, B.M., and Ozer, J., 1987. Comparison of 2-D and 3-D models of the steady wind-driven circulation in shallow waters. Submitted for publication.

LeBlond, P.H., and Mysak, L.A., 1978. Waves in the Ocean. Oceanographic Series, 20, Elsevier, Amsterdam, 602 pp.

Mofjeld, H.O., 1980. Effects of vertical viscosity on Kelvin waves. Journal of Physical Oceanography, 10, 1039-1050.

Nihoul, J.C.J., 1977. Three-dimensional model of tides and storm surges in a shallow well-mixed continental sea. Dynamics of Atmospheres and Oceans, 2 : 29-47.

Owen, A.A., 1980. A three-dimensional model of the Bristol Channel. Journal of Physical Oceanography, 10 : 1290-1302.

Paul, J.F., and Lick, W.J., 1981. A numerical model for three-dimensional variable density hydrodynamic flows. U.S. Environmental Protection Agency Report.

Pedlosky, J., 1979. Geophysical Fluid Dynamics. Springer, New York, 624 pp.

Rao, D.B., 1966. Free gravitational oscillations in rotating rectangular basins. Journal of Fluid Mechanics, 25 : 523-555.

Taylor, G.I., 1921. Tidal oscillations in gulfs and rectangular basins. Proceedings of the London Mathematical Society, 20 : 148-181.

A TROPHIC-DIFFUSION 3D MODEL OF THE VENICE LAGOON

C. DEJAK and G. PECENIK(*)
Dept. of Physical Chemistry, University of Venice, Dorsoduro 2137, Venice.
(*)Dept. of Environmental Protection and Security-PAS, MONTEDIPE S.p.A., via dell'Elettricità 41, Marghera, Venice, Italy.

ABSTRACT

A mathematical 3D diffusion model of the Venice lagoon's central area is presented; its substantial correspondence with a model of tidal agitation enables the estimation of appropriate eddy diffusivity coefficients. The conditions specified at the"open boundaries" are derived from a logarithmic finite differences analysis of a delocalized Gaussian distribution: this approach permits a more realistic representation of continuity and of outward fluxes.

Optimization of the model's program for execution on parallel supercomputers yields marked time and costs savings, and allows for consideration of thermal, chemical and biological state variables and of their interconnections,thus enabling the model to describe the eutrophication phenomenon in this peculiar ecosystem characterized by high turbulent dispersion.

Heat fluxes are analyzed in the bulk and at the air-water interface, taking into account the related processes of insolation, re-irradiation, conduction and evaporation. The thermal stratification in the deeper lagoon channels is analyzed in order to describe the diffusion of heat from the underlying salt wedge in thermal equilibrium with the sea, and to evaluate the sea-lagoon exchanges, although the model does not account for currents at the seaward boundaries.

1. INTRODUCTION

Since the desastrous flooding of the Venice Lagoon in 1966, many efforts have been made to develop theoretical and practical instruments to prevent the deterioration of both the historical town and the lagoon environment. Because of the pressing need for an overall understanding of the lagoon's hydrodynamics, and in order to predict tidal elevations accurately, several finite-difference mathematical hydrodynamical 1D (Di Silvio and D'Alpaos, 1972) and 2D (Ghetti, 1973; Sguazzero et al., 1978) models have been developed, including a still partially operative physical 3D model built at Voltabarozzo (Padova).

Other problems have recently emerged, related mainly to the chemical and thermal pollution and the consequent eutrophication, particularly in the central lagoon, which includes the historical town and the highest italian industrial concentration. To face this situation, a global interdisciplinary project was initiated (Dejak et al., 1985a), with the aim of developing a model suitable not only for studies of pollutants dispersion, but also able to reproduce the spatial and temporal distributions of relevant water quality parameters, up to steady state achievement.

2. CONSTRUCTION OF THE THREE-DIMENSIONAL MODEL

2.1 Modelling approach and preliminary dispersion models

As an alternative to the building of either inadequate advection or impractical advection-diffusion models, the 3D model was conceived on account that the hydrodynamical regime of the lagoon is predominantly governed by tidal motion, and hence the pollutants transport caused by tidal mixing may be considered as equivalent to that essentially induced by a turbulent diffusion phenomenon. This approach is justified for this particular body of water because "residual currents" have not yet been experimentally evidenced, and therefore the tidal movements averaged over more than one tidal cycle have a zero average velocity.

The model was developed through successive stages, involving the building of introductory dispersion models designed to overcome both strictly computational and general methodological and mathematical problems. Initially, a 2D diffusion model was prepared, using a very fine mesh of 100x100 m., in order to conform to the width of the main lagoon channels. This model was subsequently extended to include the advection term, after elaboration of a bathymetric map associating to each of the 17920 grid's squares the average depth (surveyed from a lagoon topographic chart, scale 1:5000) of the corresponding area. Given that the number of measurements of surface and especially of underwater velocities was rather small (Ghetti et al., 1979) the currents were first tentatively computed using a functional relationship, and then by interpolation using an existing hydrodynamical model (D'Alpaos and Di Silvio, 1979) which had been calibrated with the same velocity data and with measured tidal elevations. The integration of the diffusion and of the advection terms, respectively, was performed with an explicit method and a modified second order "upstream" scheme which ensures convergence, consistency and stability. This scheme also presents conservative and transportive properties related to the capability to damp out discontinuities (Roache, 1972). These models constituted the first attempts to include contributions due to the third dimension : in fact, according to the modified numerical scheme, the calculation of the intracell advective-diffusive flux accounts for the difference in depth between two contiguous 3D cells.

2.2 "Open boundary" conditions

The difficulty to describe the rather complex transport processes occuring at the sea entrances and the need to restrict the modelled area require dealing with "open boundaries" across which a continuous flux of pollutant occurs.These situations are commonly handled either by assuming a very far boundary where concentration (or density, temperature) is zero at all times, or by setting equal to zero, at all times, the gradients normal to the boundary.

When applying these almost identical methods, however, an excessive and unrealistic outward mass flux was observed. To avoid this, different procedures were tried, including: a macroscopic balance in addition to the point by point microscopic balance given in the diffusion equation; the solution of a time dependent differential equation at the boundaries, with extrapolations using from three to six eccentric points (Abramowitz and Stegun, 1964). The most satisfactory solution was obtained by approximating the tail of a Gaussian distribution across the boundary, through a finite difference analysis of the logarithm of the concentration, that is using a delocalized Gaussian distribution or a "finite ratio" method.

n=0	n=1	n=2	n=3
$c_0=c_1$	$c_1=\text{const}$	$c_0=c_1R$	$c_0=c_3R^3$

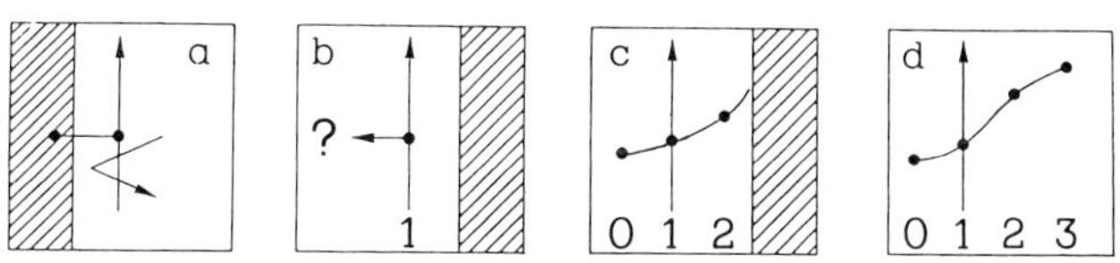

Figure 1. Schematization of possible extrapolations with n points involved : a) no flux, b) impossible extrapolation, c) exponential smoothing, d) Gaussian.

The adopted method has been shown to satisfy continuity of outward mass flux, as checked against commonly proposed conditions (Dejak et al., 1985b). By means of a simple numerical procedure, the method can be used to extrapolate boundary concentrations (c_0, outside the boundary) at all times and at all points, in terms of the values at three inside grid points (c_1, c_2, c_3, within the boundary). Cases appearing in practical application of this method are shown in Fig.1, where $R = c_1/c_2$ (or c_2/c_1 if $c_1 > c_2$).

2.3 Achievement of steady state and related mathematical conditions

With both 2D advection-diffusion models, some of the major features of pollutant dispersion in the lagoon were reproduced. Yet, with the first model steady state was never reached (Dejak et al., 1985c), even after covering a high number of cycles, probably because of the inaccuracies in the velocity field, but mainly as a consequence of the strict bidimensionality of that model. In contrast, by using more realistic tidal velocities, and hence including contributions coming from the third dimension, a steady state was attained with the other model after a reasonable number of semidiurnal tidal cycles, (Dejak et al., 1985d).

The difficulty of attaining steady state with either the 2D diffusion model or the earlier advection-diffusion model with an exactly time averaged tide in two dimensions, challenged the possibility to reach steady state through a purely 3D diffusion model or through its equivalent. Mathematically it may be shown (Dejak et al., 1985d) that steady state cannot be attained through a purely diffusive process either in one or in two dimensions. In fact, considering a n dimensional system (n = 1,2 or 3) and a point-source, assuming that an amount Q of pollutant is released instantaneously, the concentration c, setting $\psi = r/\sqrt{4kt}$, is given by

$$c = \frac{Q}{\sqrt{4\pi kt}^{\,n}}\, e^{-\psi^2}$$

If the point-source emits continuously with polluting strength $\dot{Q}$, the solutions for n equal to 1 and 2, are both diverging and are given respectively by

$$n{=}1 \quad c = \frac{\dot{Q}}{\sqrt{k\pi}}\sqrt{t}\left[e^{-\psi^2} - \psi\sqrt{\pi}\,\mathrm{erfc}\psi\right] = \frac{\dot{Q}}{\sqrt{k\pi}}\sqrt{t}\left[1 - \psi\sqrt{\pi} + \psi^2 - \frac{\psi^4}{6} + \ldots\right]_{t\to\infty} \to \infty$$

$$n{=}2 \quad c = \frac{\dot{Q}}{4\pi k} E_1(\psi^2) = \frac{\dot{Q}}{4\pi k}\left[\ln t - \gamma - \ln\frac{r^2}{4k} + \psi^2 - \frac{\psi^4}{4} + \ldots\right] \quad \underset{t\to\infty}{\to \infty}$$

Standard notations for E_1, erfc(.), $\gamma = \ln\Gamma$, are used (Abramowitz and Stegun, 1964).

Finally, for n=3, the following solution is obtained, which tends to the well known solution of the diffusion equation at steady state:

$$n{=}3 \quad c = \frac{\dot{Q}}{4\pi kr}\,\mathrm{erfc}\psi = \frac{\dot{Q}}{4\pi k}\left[\frac{1}{r} - \frac{1}{\sqrt{k\pi t}}\left(1 - \frac{\psi^2}{3} + \frac{\psi^4}{10} - \ldots\right)\right] \to \frac{\dot{Q}}{4\pi kr}$$

By performing a further integration, it is also easily demonstrated that, within a sphere of radius δr however small centered at the source, in the first two cases, the amount of pollutant grows indefinitely, while for n=3, it assumes the finite value $\dot{Q}\delta r^2/2k$, as time tends to infinity. Hence, contrary to what is sometimes asserted (Diachishin, 1963), the accumulation of a conservative pollutant does not continue indefinitely. This indicates that steady state through a purely diffusive process may be achieved only in three dimensions (Dejak et al., 1985d).

2.4 Choice of the numerical scheme and estimation of 3D eddy diffusivity

Steady state pollutant distributions emerging from the advection-diffusion model were observed to be substantially comparable to those produced by the pure diffusion model in three dimensions (Dejak et al., 1985e). This outcome further supported the approach, whereby the tidal dispersive mechanism might be accounted for by means of a global 3D turbulent diffusion, involving an effective eddy diffusivity larger (by about one order of magnitude) than that which would be needed to reproduce ordinary vorticity at lagoon scale. As a first approximation, space independent eddy diffusivities were included, and to conform with experimental evidence (Kullenberg, 1971; Kullenberg et al., 1973) the vertical diffusivity was assumed to be one order of magnitude smaller than the horizontal diffusivity. Because of the disparity between horizontal (100 m.) and vertical steps (1 m.), and of the resulting large magnitude of the diffusion number, the use of an implicit scheme for vertical diffusion was unavoidable to guarantee stability for reasonably small time increments. Among the many available two and three time levels schemes (Richtmyer and Morton, 1967) that of Laasonen (1949) was selected after comparative 1D analysis with respect to the analytical solution (Baule, 1970). We chose this scheme because, despite discontinuities, it presents a higher stability (as proved by Fourier analysis), and a better convergence with continuous input, than the other implicit methods. Initially, the implicit Laasonen scheme was only adopted for the computation of the vertical diffusion. Later, after the introduction in the program of additional loops required by the nonlinear "open" boundary conditions, the model was made fully implicit, thus achieving a faster computation even for a large diffusion number. The modified model was tested against the original more reliable explicit model, taken as reference; comparable distributions were obtained for a diffusion number up to a value of ten.

The effective eddy diffusion constant embodying tidal agitation was estimated by comparing the steady-state surface distributions obtained respectively with the 3D model and with 2D advection-diffusion model. The optimal value for the horizontal diffusivity determined in this way is equal to 22 m^2/sec (or 7.92 hm^2/h). With this value, both distributions are in reasonable agreement. This result is of great significance as it justifies the utilization of the 3D model, which includes 128x140x20 grid points, as a reliable substitute for the advection-diffusion model for the description of pollutant dispersion in the lagoon.

2.5 Eutrophication model

To account better for the main peculiarities of the lagoon ecosystem, the biological compartment was conceived primarily on the basis of information provided by "in situ" functional investigations. A rather limited study of this type led to the definition of a first "trophic reactor" (Cescon et al., 1976), originally adopted to describe the biological processes. Eight state variables are presently considered : phytoplankton, zooplankton, degradable carbonaceous detritus, dissolved oxygen, nitrogen oxydes, ammonia, orthophosphorus, and temperature.

A further reformulation of the biological compartment was based upon multidisciplinary field experimentations initiated in 1978 under the coordination of a Regional Commission and with the collaboration of the ENEL-DCO, Environmental Dept. (Piacenza). The research, whose main objective was to study the effects of thermal discharges upon the lagoon ecosystem, was soon extended to investigate other related biochemical and physicochemical processes, particularly : phytoplankton growth rate as a function of light, nutrients and temperature, phytoplankton respiration, heterotrophic and nitrification activity, kinetic of release of nitrogen and phosphorus from sediments, benthic oxygen demand, BOD-dissolved oxygen relation. So far, reliable parameter values have been determined for phytoplankton maximum growth rate and half-saturation constants for nitrogen and phosphorus (Bertonati et al., 1985), as well as optimal and maximum growth temperature and light intensity. Parameters not yet experimentally determined were derived by elaboration of literature data (Joergensen, 1979) and from model calibration.

An example of two steady state distributions provided by the model unifying dispersive and trophic-dynamic processes, is shown in Fig.2. The variables displayed are phytoplankton and orthophosphorus.

3. COMPUTATIONAL IMPLEMENTATIONS

The computer program was originally conceived and organized to be executed on traditional architecture computers. Briefly, it is structured in three principal portions: initialization (including parameter definition, bathymetry reading, computation of auxiliary functions); the main iteration(continuous input, computation of 3D diffusion of the eight variables, computation of biological equilibria); output of distribution maps at preassigned time. Of these parts, the central one, pertaining to 3D diffusion, carried out with the fractional step method for each direction, is the most time consuming. On a CDC 7600, one single diffusive step for all eight state variables takes two hundred seconds. With almost forty cycles needed to achieve steady state, such a computational time is unacceptable for routine runs. Furthermore, the modelled water body presents for all layers an extremely irregular shape, with indented boundaries associated with the abruptly changing bathymetry. Also, computational complications arise from the nonlinear "open boundaries" conditions as these imply auxiliary extrapolation loops and from the huge memory requirements which necessitate the programming of laborious swapping of data to and from secondary memories.

The program was also run on the supercomputers HO45480/4 (Hitachi, Tokyo)

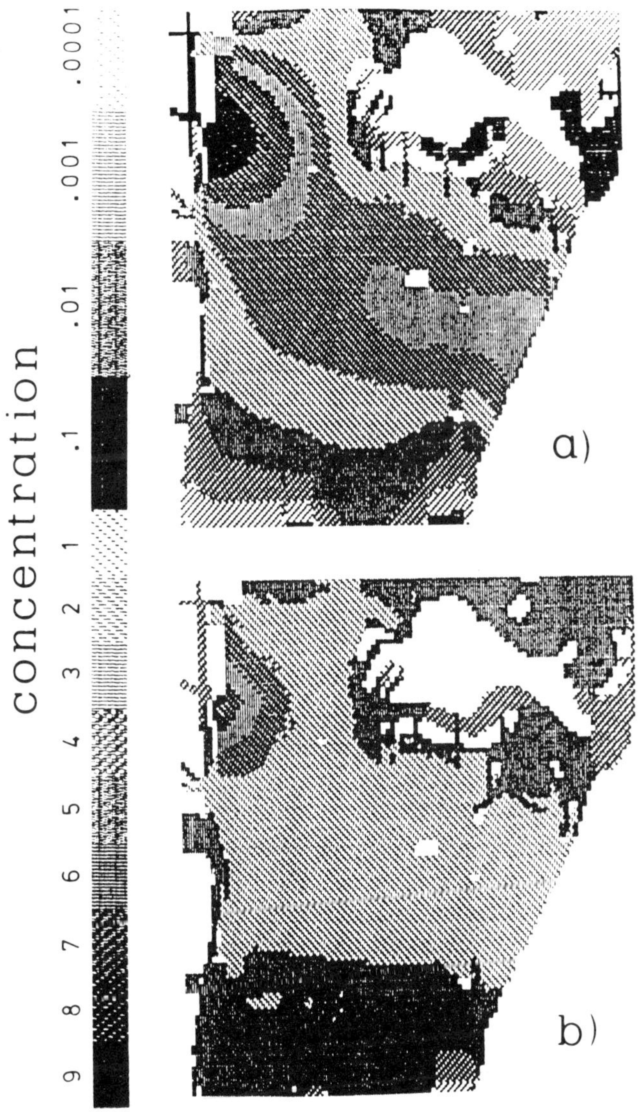

Figure 2. Steady state distributions of a) phytoplankton ,b)orthophosphorus

and CRAY-1 at CISI (Clamart sur Seine) where some attempts to partially vectorize the code were made, yielding only marginal benefits. In order to exploit new parallel architecture computers, only recently accessible in Italy, an extensive reorganization of the program was untertaken to enhance vectorial code generation. Consequently, a number of conditional expressions contained within

"DO-LOOPS" and originally programmed to deal with the many possible situations, were removed and replaced by algebraic expressions involving mutually excluding terms activated by logical coefficients. By analysing and grouping the many different conditional typologies, a limited number of logical coefficients could be identified and packed in one single matrix: the matrix is codified once, at start-up time, and it is decodified, while optimizing the number of auxiliary vectors required, inside each diffusion subroutine.

The storing capacity was still insufficient to handle simultaneously the entire model. This problem was overcome by removing all cells not representing real water volumes, and by packing only the effective elements in eight wide vectors, which replaced the eight 3D matrices originally required. By so doing, one 3D matrix of pointers is now needed for recovering the spatial positions of each element: vectors are scattered, depending on each coordinate, with special procedures to minimize the sizes of the auxiliary arrays.

The reorganized program, as tested on the CRAY X/MP, not only permits a reduction of the computing time to only ten seconds per diffusive step, but it also prevents memory saturation. The possibility of maintaining time and expenses at a reasonable level will allow further important modifications of the model, particularly in the biological compartment, where the inclusion of additional variables is prospected as are more detailed formulations of the interconnections.

4. MODEL APPLICATIONS

Eight years ago, while they were still being developed, the models described above had to be applied to urgent practical problems, mainly connected to the energy sector. Such problems often require a multidisciplinary approach. For the first study, which concerned the dispersion of heat released from the industrial zone (Dejak et al., 1977), a preliminary model was used in which the 3D grid just served to create the equivalence between the real bathymetry and a simple sloping bottom to permit the analytical 3D integration of the diffusion equation (Dejak et al., 1975). From this earlier application it was possible to design rationally the field experiments required to assess the thermal impact on the lagoon of the cooling water discharged from a 1 GW power plant. The five year long research that followed, coordinated by a Scientific-Technical Commission purposely appointed by the Regional Authority, represents even now the only really long term multidisciplinary and systematic investigation of the lagoon ecosystem actually performed (Battaglia et al., 1983). Moreover, based upon indications provided by this work, new integrative directives were tentatively put forward, regulating thermal pollution in the lagoon. The 3D model developed could thus be designed by strictly coupling experimental and theoretical works, as field data allowed for a less abstract modelling, and conversely theoretical work promoted those experimental studies most needed to proceed in the modelling efforts. Indeed, at least one extensive tracer diffusion experiment is required in order to fully implement a combined dispersion eutrophication model. Such an experiment is, at present, firmly impeded by the local Health Authority. To circumvent this obstacle, the model is applied, with the support of the National Research Council, to determine optimal sites for a lagoon water quality continuous monitoring network. The realization of this network is indispensable both for the collection of all data required for a decisive comparison between theory and experience, and for the prediction of the effects of the programmed hydraulic interventions on the lagoon ecosystem.

5. MODEL EXTENSION

The improvement most required of the present model is the introduction of the thermal influence on chemical and biological processes. To achieve this, the current field studies focus on the temperature dependence of rate processes, and the modelling work seeks to make it possible to follow the daily as well as the seasonal evolution of temperature at each point of the modelled area. Even with the seven hundreds or so temperature profiles now available, surveyed in varying depths at selected lagoon sites and for different seasons, attempts to derive temperature values for such a wide model directly form experimental data appear unlikely to succeed, even with the support of sophisticated statistical treatments. The definition of a complete set of temperature values, in fact, may not be derived by correlation of existing historical series of air and water temperatures, both because a large portion of the modelled area is affected by cooling water discharges and because of the highly varying natural heat exchanges between the lagoon and the sea. More precisely, the solution to this problem involves a detailed quantification of the global lagoon heat budget, based on fluxes at its boundaries, and particularly at the air-water interface where the processes of insolation, backscattering, conduction and evaporation occur (Mariotti and Dejak, 1982). Moreover, internal processes affecting the heat dispersion must be analyzed, particularly in the vertical direction where thermocline formation and stratification add further complications.

The modelling approach aims at determining temperature values at all times and at all grid's nodes, and at finding a description, consistent with the pure diffusion model discussed above, of the heat exchange with the atmosphere as well as between the sea and the lagoon. To handle the first problem, a monthly virtual equilibrium temperature, i.e., the temperature associated to a zero net average heat flux, has been calculated. From these, assuming an initial arbitrary punctual temperature value, the equilibrium value is determined by computing heat fluxes, and considering the thermal inertia of water columns of varying depths through an iterative procedure, under the constraint that the yearly sum of the fluxes has to be equal to zero. By successively refining the initial arbitrary temperature, the iteration converges to a unique set of values. The values of the meteoclimatic parameters used in the calculations were the original monthly averaged data collected over a thirty year survey, as well as linearly interplolated; more realistic results were obtained through Fourier interpolation (Fig.3). Tentatively, the same temperature was initially assumed for water columns of different depths (Fig.4), but realistic results were obtained only by accounting for thermal stratification and by adopting depth varying diffusivities. The latter are estimated, assuming a steady vertical heat flux, from the reciprocal of experimental temperature gradients obtained by a best fit with generalized Gamma functions. In this way, the typical evolution of the thermocline for the summer and winter months was obtained (Fig.5), showing a temperature difference of about 4°C between top and bottom layers and an inflection point corresponding to the highest gradient and minimal diffusivity; during intermediate seasons, the temperature profiles are steeper.

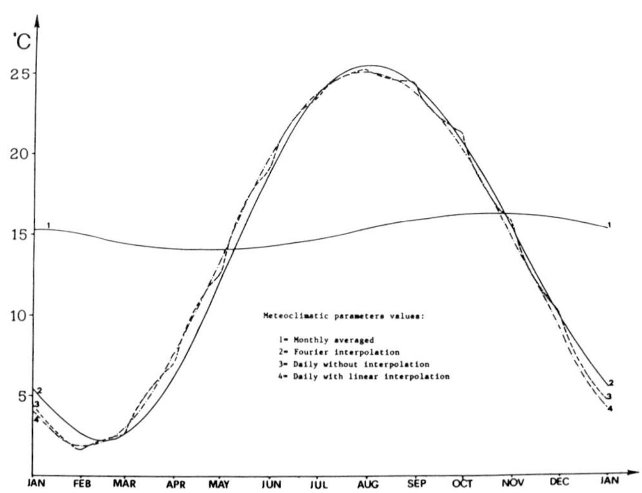

Figure 3. Monthly surface temperature obtained through different interpolation procedures of meteoclimatic parameters.

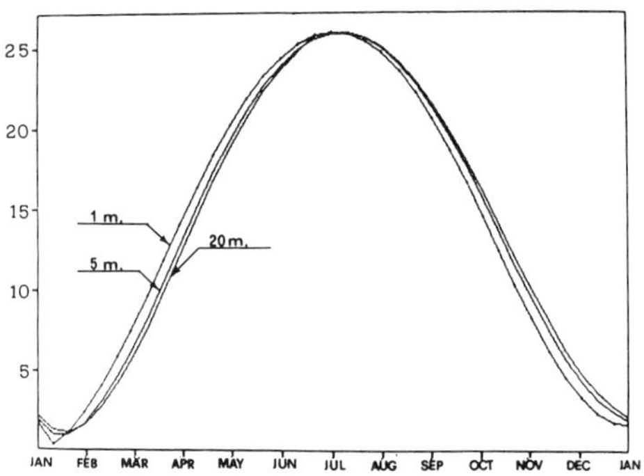

Figure 4. Annual temperature variation with depth.

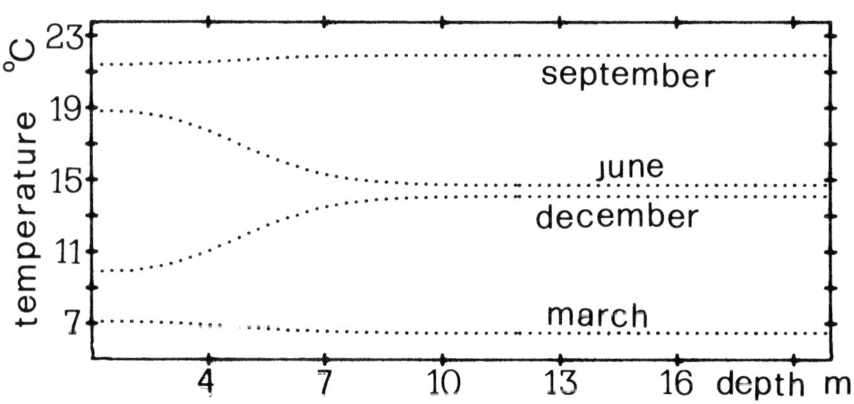

Figure 5. Temperature profiles for typical months.

The distribution of temperature profiles, in fact, suggests the presence in the deeper lagoon channels of an upper layer and of a lower salt wedge in thermal dynamic equilibrium respectively with the atmosphere and the sea. Thus, when the sea water temperature differs greatly from that of the lagoon, consistently with the model's conception, it is assumed that sea-lagoon heat exchanges occur across a thin virtual layer permitting a very slow, yet continuous heat flux and with which the minimal eddy diffusivity is associated. Actually, to define a discontinuity between internal and external temperature, at the open boundaries, would be incompatible with a pure diffusion model because, as the computation proceeds, such a discontinuity would induce inside the modelled area the progressive and uncontrollable formation of unrealistic horizontal temperature gradients. We believe that even a 3D advection model could not adequately reproduce the complex velocity fields required to compute the heat exchanges at the lagoon mouths, not included in this model, but represented by open boundaries. Here, according to the specified conditions, an automatic adjustment occurs between locally computed and sea water temperatures, respectively above and below the virtual separation layer. Based upon the many experimental temperature profiles available, it is now possible to complete the model calibration and thus to estimate more adequate vertical eddy diffusivities. The introduction of seasonally varying diffusivities will increase the model capability to describe both the appearance and the evolution of the eutrophication phenomenon during the most critical periods.

6. CONCLUSIONS

The construction of the 3D trophic-diffusion model described here has been successfuly achieved owing to :

i) the absence in the lagoon of experimentally significant currents having a non zero long term average velocity;

ii) conditions enabling the simulation of tidal agitation by mere second order equations with constant eddy diffusivity;

iii) the capability to calibrate the 3D model, up to the steady state achievement, with a convergent advection-diffusion model including features overcoming its structural two-dimensionality;

iv) the implementation of a simple integration scheme (Laasonen, 1949),insensitive to discontinuities and particularly fitted for use with continuous input;

V) the elaboration of practical and sufficiently accurate "open boundaries" conditions, such as the delocalized Gaussian distribution.

This treatment of transport processes has the following advantages:

1) to account, with a very fine computational grid, for a bathymetry that is not reducible to more regular geometries due to the intermingling of the lagoon's narrow and deep channels;

2) to work with "moderate" computing time and memory requirements and hence at reasonable expenses, notwithstanding the small mesh size and the need to integrate eight interconnected second order differential equations;

3) to enable the simulation of complex ecological processes linked to physical phenomena having 3D charactersitics, and also related to the concomitant transport of the variables involved;

4) to reproduce, with adequate approximations, the seasonal thermal behavior, in spite of complications deriving from heat flux balance, thermal stratification and very irregular sea-lagoon exchanges, avoiding to resort to the awkward and uncommon treatment involving temperature-salinity-density system.

The model, owing to its flexibility, constitutes a practical tool for the investigation of the main factors responsible for the deterioration of the Venice lagoon and a useful instrument to help find a more rational plan for the now unpostponable interventions required to safeguard its ecosystem.

7. ACKNOWLEDGEMENTS

The authors thank Sig. Sergio Manzi for the assiduous collaboration offered in program vectorization. The work has been supported by the National Research Council, Progetto Finalizzato "Energetica 2".

8. REFERENCES

Abramowitz, M. and Stegun, I., 1964. Handbook of mathematical functions. Dover Pub. Inc., N.Y.; p:887,914,991.

Battaglia, B., Datei, C., Gambaretto, G., Guarise, G., Perin, G., Vianello, E. and Zingales, F., 1983. Indagine per la valutazione dei riflessi ambientali del funzionamento a piena potenza della centrale termoelletrica di Fusina. Commissione Tecnico Scientifica per la sperimentazione ed i controlli periodici sulla Centrale Termoelettrica ENEL sita in località Fusina di Porto Marghera, Venezia. Provvedimenti Regione Veneto n.810 and n.811, May, 10, 1979.

Baule, B., 1970. Die Mathematik des Naturfarschers und Ingenieurs., Band IV, Partielle Differentialgleichungen, S. Hirzel Verlag, Leipzig.

Bertonati, M., Dejack, C., Mazzei Lalatta, I. and Pecenik, G., 1985. Eutrophication model of the Venice Lagoon: statistical treatments of phytoplankton growth parameters. (Ecol. Modelling, in press).

Cescon, B., De Angeli, U., Iovenitti, L., Isolati, A., Alfassio Grimaldi, S., 1976. The calibration of a trophic model of the Venice Lagoon. Tecneco S.p.A., Environmental Study Group, S. Ippolito (Pesaro), Italy.

D'Alpaos, L. and Di Silvio, G., 1979. Utilizzazione dei risultati per la taratura dei modelli matematici. In:"Le correnti di marea nella laguna di Venezia". Ministero dei Lavori Pubblici.Comitato per lo studio dei provvedimenti a difesa della citta di Venezia ed a salvaguardia dei suoi caratteri ambientali e monumentali. Eds. Istituto di Idraulica dell'Universita di Padova: pp.82-95.

Dejak, C., Mazzei Lalatta, I., Lavagnini, I. and Saba, G., 1975. Diffusion models for chemical and thermal pollution. In: Proc. of the 3d Int. Congr. Iranian Chemical Society, March 28-April 1.Pahlavi University, Shiraz, Iran. pp.261-266.

Dejak, C., Mazzei Lallata, I. and Meregalli, L., 1977. Diffusion of pollutants in lagoon and in the sea. In: Proc. of the Int. Congr.Louvain-la-Neuve: E4: p.5.

Dejak, C., 1979. Problemi riguardanti gli scarichi di acque di raffredamento di centrali termoelettriche e di altri impianti industriali nella laguna veneta. Report for Ente Zona Industriale Porto Marghera. n.A2: 1-17.

Dejak, C., 1979. Verifica sperimentale sugli scarichi termici della Centrale di Fusina, Ibidem, A3:1-4.

Dejak, C., Mazzei Lalatta, I. and Meregalli, L., 1981. Vertical averages in a three-dimensional diffusion of pollutants. Nuovo Cimento, 4:493-510.

Dejak, C., Mazzei Lalatta, I., Pecenik, G. and Rossi, A., 1985a. Development of a mathematical eutrophication model of the Lagoon of Venice. (Ecol. Modelling: in press).

Dejak, C., Mazzei Lalatta, I., Meregalli, L., Messina, E. and Pecenik, G.,1985b. A two-dimensional diffusion model of the Venice Lagoon and relative open boundaries conditions. (Ecol. Modelling: in press).

Dejak, C., Mazzei Lalatta, I., Messina, E., and Pecenik,G.,1985c. An advection-diffusion pollution model of the lagoon of Venice (Ecol. Modelling:in press).

Dejak, C., Mazzei Lalatta, I., Molin, M., Messina, E. and Pecenik, G., 1985d. Steady state achievement by introduction of true tidal velocities in a pollution model of the Venice Lagoon. (Ecol. Modelling: in press).

Diachishin, A.N., 1963. Waste disposal in tidal waters. J. San. Eng.Div., 89 (SA4):23-43.

Di Silvio, G. and D'Alpaos, L.,1972. Il comportamento della laguna di Venezia esaminato con il metodo propagatorio unidimensionale. Eds. Istituto Veneto di Scienze Lettere ed Arti. Commissione di Studio dei Provvedimenti per la Conservazione e Difesa della Laguna e della città di Venezia.

Ghetti, A., 1973. The hydraulic problems of the Lagoon of Venice. Giornale economico. C.C.I.A., Venezia, 1:34-38.

Ghetti, A., Bartoletti, U., Lippe, E., D'Alpaos, L. and Di Silvio, G., 1979. Le correnti di marea nella laguna di Venezia. Introduzione e parte la. Ministero del Lavori Pubblici. Comitato per lo Studio dei Provvedimenti a Difesa della città di Venezia ed a Salvaguardia dei suoi caratteri ambientali e monumentali. Eds. Istituto di Idraulica; Università di Padova; pp.3-66.

Joergensen, S.E. Handbook of environmental data and ecological parameters. Pergamon Press, New York.

Kullenberg, G.E., 1971. Results of diffusion experiments in the upper region of the sea.Report 12, Kobenhavns Universitet, Institut for Fysisk Oceanografi.

Kullenberg, G.E., Murthy, C.R., and Westerbuy, H. 1973. An experimental study of diffusion characteristics in the thermocline and hypolimnion regions of Lake Ontario. In: Proc. 16th Conf. Great Lakes Res., pp:774-790.

Laasonen, P., 1949.Ueber eine Methode zur Losung der Warmeleitungsgleichung. Acta Math.; pp. 81-309.

Mariotti, M. and Dejak, C., 1982. Valutazione dei Flussi energetici all'interfaccia acqua-aria nella laguna di Venezia. In: Proc. of the 10th Convegno "Ambiente e Risorse", Bressanone Aug. 30-Sept.4.

Richtmyer, D.R. and Morton, K.W., 1967. Difference methods for initial value problems , 2nd Ed.,Interscience Publishers, London.

Roache, P.J., 1972. Computational fluid dynamics. Hermosa Publishers Albuquerque, New Mexico 87108, USA.

Sguazzero, P., Ghignoli, C., Rabagliati, R. and Volpi, G., 1978. Hydrodynamic numerical modelling of the lagoon of Venice. IBM J. Res. Dev., 22 (5): 472-480.

THREE DIMENSIONAL CONTINENTAL SHELF HYDRODYNAMICS MODEL INCLUDING WAVE CURRENT INTERACTION

M.L. SPAULDING AND T. ISAJI
Applied Science Associates, Inc., 70 Dean Knauss Drive, Narragansett, Rhode Island 02882

ABSTRACT

A three dimensional circulation model, suitable for continental shelf and coastal sea regions, has been modified to consider wave current interaction in the bottom boundary layer. The wave current interaction processes are described using a simplified version of the Grant-Madsen approach. The hydrodynamic equations are solved by a weighted residual method in the vertical and by a forward in time, centered in space finite difference technique in the horizontal. The vertical variations in horizontal velocity are represented by an expansion of Legendre polynomials. A split mode computational technique is employed to minimize computational time.

The model has been applied to predict the flow and surface elevation fields for tropical storm Delia (Gulf of Mexico) and compared to available field observations. Sensitivity studies to two and three dimensional representations of the vertical eddy viscosity, constant bottom friction coefficient, and wave current interaction have been performed. Results of these simulations in terms of the velocity and surface elevation fields, local force balances and velocity structure are described.

1 INTRODUCTION

When the wind blows over the water, the applied shear stress at the surface induces a "steady" current in the water column. This steady current then interacts with the sea floor forming a bottom boundary layer which dissipates energy through frictional losses at the sea bed. The wind, at the same time, generates waves at the sea surface which in turn result in a wave boundary layer at the bottom. This wave boundary layer, interacting with the steady current, over a rough bottom, modifies the effective bottom stress and therefore alters the steady current field. While the interaction between the wave induced and steady current fields has been recognized as important in predicting the current fields near the sea floor (Grant and Glenn, 1983; Glenn and Grant, 1983; Grant and Madsen, 1979; Wiberg and Smith, 1983; Cooper and Pearce, 1980, 1982; Pearce et al, 1979) existing continental shelf and coastal sea circulation models (Jaycor, 1979; Gordon and Spaulding, 1974; ASCE, 1980; Swanson and Spaulding, 1978; Hinwood and Wallis, 1975) do not account for the presence of the wave bottom boundary layer.

Grant and Glenn (1983) undertook the extension of earlier work (Grant and Madsen, 1979; Grant and Madsen, 1982) to develop a continental shelf bottom boundary layer model (BBLM) which included the behavior of (1) combined wave and current interaction, (2) moveable bed effects (ripples, near bed transport, bioturbation), (3) stratification due to temperature and salinity, (4) stratification due to suspended sediment and (5) planetary rotation.

This represented a first step toward developing an accurate methodology to estimate wave current interaction effects. The next step, and the goal of this study, is to integrate the BBLM with an existing continental shelf circulation model in order to develop a model system capable of estimating the interaction of currents and waves for a coastal or shelf region.

The specific study objectives within this general context are to: (1) Develop and implement a methodology to predict wave current interaction by integrating the BBLM into a shelf circulation model, and interfacing the resulting model to wind and wave-swell models suitable for storm conditions. (2) Compare the resulting model system predictions against a storm event for which field observations exist.

Section 2 presents an overview of the model system and linkages and a brief description of the model components. In Section 3, the model is applied to simulate tropical storm Delia. A total of thirteen cases are run to explore the sensitivity of model predictions to the vertical representation of the eddy viscosity, to bottom friction coefficient (with and without wave current interaction) and to two and three dimensional descriptions of the flow field. A summary of the study is included in Section 4.

2 THEORETICAL DEVELOPMENT OF THE MODEL SYSTEM

2.1 Model System Overview and Linkages

The design of a model system capable of accurately predicting the three dimensional current structure during a storm must include not only a three dimensional continental shelf hydrodynamics model, modified to incorporate the BBLM, but models of the wind and wave fields as well. A conceptual flow chart of the model system is given in Figure 1.

Given estimates of the storm parameters (in terms of the pressure field, radius to maximum winds, forward velocity, storm track and deflection angle), the storm wind model predicts the spatial and temporal structure of the wind velocity field. This information is subsequently used as input to the storm wave model which estimates the resulting wave environment in the study region. The wave spectra can then be employed to predict the wave induced bottom currents by well known linear or nonlinear wave theory. Using the wind velocity and atmospheric pressure data predicted by the storm wind model, the wave induced bottom currents estimated by the wave model, and the sea bottom sediment characteristics and roughness, a three dimensional continental shelf circulation model is solved iteratively with the BBLM to predict the spatial current structure and surface elevation distributions. The solution is advanced in time by simply updating the storm parameters and repeating the calculation sequence.

In the text to follow, each model system component is described in detail. This presentation gives a brief overview of the theoretical foundation of the approach and its integration into the model system.

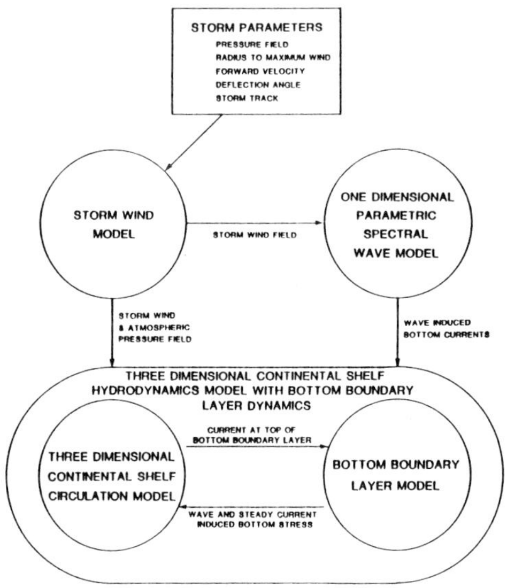

Figure 1. Overview of model system components and linkages.

2.2 Model System Components

(i) Continental Shelf Hydrodynamic Model. The three dimensional numerical hydrodynamic model employed for the present study follows the development originally given in Owen (1980) and reported in our earlier work (Isaji and Spaulding, 1984; Isaji et al, 1982). Since the approach is well documented in the refereed literature, we present only a brief overview here.

The three dimensional conservation equations for water mass and momentum with the Boussinesq and hydrostatic assumptions invoked form the basis for the model. The three dimensional conservation equations suitable for limited shelf waters in spherical polar coordinates are written as:

$$\frac{\partial \zeta}{\partial t} + \frac{1}{R}\frac{\partial}{\partial \phi}(\bar{u}d) + \frac{1}{R\cos\phi}\frac{\partial}{\partial \kappa}(\bar{v}d) - \frac{d\bar{u}}{R}\tan\phi = 0 \qquad (1)$$

Conservation of momentum

$$\frac{\partial u}{\partial t} + \frac{u}{R}\frac{\partial u}{\partial \phi} + \frac{v}{R\cos\phi}\frac{\partial u}{\partial \kappa} + \frac{v^2}{R}\tan\phi + \frac{g}{R}\frac{\partial \zeta}{\partial \phi}$$

$$+ 2v\Omega \sin\phi - \frac{1}{\rho R}\frac{\partial p_a}{\partial \phi} - \frac{\partial}{\partial z}(N_z \frac{\partial u}{\partial z}) = 0 \tag{2}$$

$$\frac{\partial v}{\partial t} + \frac{u}{R}\frac{\partial v}{\partial \phi} + \frac{v}{R\cos\phi}\frac{\partial v}{\partial \kappa} - \frac{uv}{R}\tan\phi + \frac{g}{R\cos\phi}\frac{\partial \zeta}{\partial \kappa}$$
$$- 2u\Omega \sin\phi - \frac{1}{R\cos\phi}\frac{\partial p_a}{\partial \kappa} - \frac{\partial}{\partial z}(N_z \frac{\partial v}{\partial z}) = 0 \tag{3}$$

$$\frac{\partial p}{\partial z} = - \rho g \tag{4}$$

where the following notation has been used:

ϕ,κ,z Spherical polar coordinate system with ϕ and κ measured in the latitudinal and longitudinal direction, respectively. z is measured vertically upward from mean sea level.

u,v,w Components of the current in the ϕ, κ, and z directions, respectively. Overbar indicates vertically averaged values.

p Pressure

p_a Atmospheric pressure

ρ Density

N_z Vertical eddy viscosity

f_c Coriolis parameter ($2\ \Omega\ \sin\phi$), assumed constant, where Ω is the angular speed of the earth's rotation and ϕ the latitude angle.

g Gravitational acceleration

R Radius of the earth

These equations are solved subject to the following boundary conditions. (a) At land boundaries the normal component of velocity is set to zero. (b) At the open boundaries the sea surface elevation is specified as a series of sine waves each with its own amplitude and phase or by appropriate gradients of the local surface elevation or normal velocity field. (c) At the sea surface the applied stress due to the wind is matched to the local stress in the water column and the kinematic boundary condition is satisfied. (d) At the sea floor, a quadratic stress law, based on the local bottom velocity, is used to represent frictional dissipation and a friction coefficient determined by the BBLM parameterizes the loss rate.

Anticipating the use of a weighted residual method, in which vertical variations are represented in terms of a set of basis functions, a new set of independent variables is introduced that transforms both the surface and bottom onto coordinate surfaces. This transformation represents a simple nondimensionalization with the local water column depth. A detailed presentation of the transformed equations can be found in Owen (1980).

The numerical solution methodology follows that of Davies (1977a, b) and Owen (1980). The vertical variations in horizontal velocity are described by an expansion of Legendre polynomials. The resulting equations are then solved by a Galerkin weighted residual method in the vertical and by an explicit finite difference algorithm in the horizontal.

A space staggered grid scheme in the horizontal plane is used to define the study area. Sea surface elevation and vertical velocity are specified in the center of each cell while the horizontal velocities are given on the cell faces. The u and v velocities are defined on the cell faces normal to the ϕ and κ directions, respectively.

To reduce computational costs, a "split-mode" or "two-mode" formulation is used (Simons, 1975; Owen, 1980; Gordon, 1982). In the split-mode approach, the free-surface elevation is treated separately from the internal, three-dimensional flow variables. The free-surface elevation and vertically integrated velocities are calculated using the vertically integrated equations of motion (external mode) for which the Courant-Friedrichs-Levy (CFL) limit must be met. The vertical structure of the horizontal components of the current may then be calculated such that the effects of surface gravity waves are separated from the three-dimensional equations of motion (internal mode). Long period surface gravity waves, therefore, no longer limit the internal mode calculations and much longer time steps are possible.

The external mode equations are approximated by a forward in time, centered in space (FTCS) finite difference scheme. The internal mode is solved using a centered in time technique with vertical diffusive terms centered in time, to ease the vertical time step restriction from the standard FTCS procedure.

(ii) Bottom Boundary Layer Model. Given the extensive computational resources required to perform time dependent three dimensional simulations of storms on a relatively fine mesh and the iterative nature of the full BBLM solution approach (Grant and Madsen, 1979), the hydrodynamic model's computational time would be excessive. A decrease in computational time is clearly required if the proposed approach is to be a viable model. It was therefore decided to simplify the BBLM. The goal of this effort was to maintain the accuracy of the BBLM for conditions present during storm events but drastically reduce the computational time per simulation. The interested reader is referred to Grant and Madsen (1979, 1982), Grant and Glenn (1983), and Spaulding and Isaji (1985) for a thorough presentation of the full and simplified BBLM.

The model is meant to be applied primarily for storm conditions. As a result, there are several assumptions inherent in the model development. These include: (1) Wave velocities at the bed (at a height less than the wave boundary layer thickness) are large compared to the current velocity. Grant and Madsen (1979), indicate that this criteria is formally met for current to wave velocity ratios of 0.25 or less. Practically, however, much larger ratios do not greatly change the predicted drag coefficients (Spaulding and Isaji, 1985). (2) The bed is always rough, either because of bioturbation or because of sediment transport. These simplifying assumptions allow a substantial reduction in the computational effort to calculate the drag coefficient. These limitations are not severe for storm or swell conditions and the model can safely be assumed to be generally applicable. The major limitation for the simplified BBLM case is that stratification is not permitted. Although stratification is not the typical case, it may be important in selected areas and at specific times (e.g., self stratification due to suspended sediment transport).

The drag coefficient calculated by the model is defined in terms of the mean bottom stress

$$\tau_b = \rho C_D \mid U_1 \mid U_1 \tag{5}$$

where C_D is the drag coefficient and U_1 is the mean current velocity at 1 m from the bottom.

As a note on the computational time, the simplified BBLM is approximately 50 times faster than the full BBLM.

(iii) Storm Wind Model. The storm wind model incorporates the symmetric pressure field suggested by Harris (1958).

$$\frac{P - P_o}{P_n - P_o} = e^{-r/R} \tag{6}$$

where

P Pressure at radial location r (mb)

P_o Central pressure (mb)

P Peripheral pressure (mb)

R Radius to maximum winds (km)

The wind field was then modeled using the approach proposed by Jelesnianski (1974). The velocity at any location (r) is given by

$$V(r,\theta) = V_{max}(r) \; [1 - \frac{V_f}{2} (1 - \cos\theta)/\bar{V}_{max}] \tag{7}$$

where V_f is the forward velocity of the storm (knots) and θ is the angle measured clockwise from the locus of maximum winds. The locus of maximum winds is located at an angle of (90+ θ) to the right of the direction of forward motion where the deflection angle is defined as the angle between the true wind direction and a tangent to a circle with the center at the hurricane center. The maximum wind is given by

$$\bar{V}_{max} = .885 \; [10.89 \; (P_n - P_o)^{1/2} - \frac{f_c R}{2} + V_f \tag{8}$$

where f_c is the Coriolis parameter (1/hr.).

The velocity at any distance, r, along the locus of maximum winds is given by

$$\bar{V}_{max} \; (r) = \frac{2Rr}{R^2 + r^2} \bar{V}_{max} \tag{9}$$

where the units of r and R must be consistent.

Although Myers and Malkin (1961) compiled observed data that suggests the deflection angle displays significant temporal and spatial variability for a particular storm, it will be assumed as a constant for the present application. An angle of 5° is normally selected (Cooper and Pearce, 1982).

The wind stress was related to the storm wind using the formulation of Smith and Banke (1975):

$$C_d = (0.63 + 0.66\ U_{10}) \times 10^{-3} \tag{10}$$

where U_{10} is the 10 m wind speed in m/s. This formula gives similar results to those of Garratt (1977), Large and Pond (1981) and Wu (1980) over the wind speed range of interest but is substantially lower than the earlier work of Wu (1969) at wind speeds above 70 m/s.

While the wind field model proposed here is relatively simplistic, it has been selected for the present application because the procedure is well documented (CERC, 1984) and has been extensively used in describing hurricane wind fields for several storms in potential areas of model application (Pearce et al, 1979; Tetra Tech, 1981).

(iv) Storm Wave Model. To hindcast sea states for tropical and extratropical storms, a one dimensional parametric spectral model has been selected based on its simplicity and its successful application to past storm events.

This approach is based on an analysis of the JONSWAP energy spectrum, E(f)

$$E(f) = \alpha\ g^2\ (2\pi)^{-4}\ f^{-5} \exp\left\{-\frac{5}{4}\left(\frac{f}{f_m}\right)^{-4} + \ln\gamma \exp\left[-\frac{(f/f_m - 1)^2}{2\ \sigma^2}\right]\right\} \tag{11}$$

where σ, the width of the spectra, is given as

$$\sigma = \begin{cases} \sigma_a & \text{for } f \le f_m \\ \sigma_b & \text{for } f > f_m \end{cases}$$

f_m is the frequency of the spectral peak, α the Phillips constant, ϕ the peak enhancement factor (ratio of JONSWAP Spectrum peak/Pierson Moskowitz Spectrum peak), g gravity and f wave frequency.

Hasselman et al (1976) proposed that fetch limited wave spectra can be normalized in such a way that there is an approximate invariance of normalized spectral shape with fetch. They proposed that the nonlinear regime wave-wave interactions were responsible for the shape invariance and that it characterized not only idealized fetch limited situations, but also growing wind seas in general.

This parametric approach exploits the shape invariance by representing the spectrum in terms of the five JONSWAP parameters (σ, ϕ, σ_a, σ_b, f_m). Noting evidence of the shape invariance of hurricane generated wave spectra measured by a laser altimeter in hurricane Ava, from wave measurements in hurricane Camille (Hasselman et al, 1976) and from meteorological buoy data during hurricane Elise (Withee and Johnson, 1976), Ross (1976) proposed that the advecting wave field for a hurricane can be determined by the local wind and a fetch given by the local radius of curvature of the wind field.

Following Kitaigorodskii's (1961) procedure that for fetch limited wind seas all wave variables, when nondimensionalized in terms of gravity, g and wind speed, U, should be functions only of the single nondimensional fetch parameter $\xi = gX/U^2$, where X is the fetch, Hasselman et al (1976) represented the nondimensional peak frequency, $\nu = Uf_m/g$ where f_m is the peak frequency, and the nondimensional total energy $\varepsilon = Eg^2/U^4$ as a function of ξ.

Ross (1975) replaced the fetch, X, by r, the radial distance to the eye of the hurricane, to parameterize fetch. The data sets from hurricane Ava, Camille, and Eloise were combined to give the following power law fits to the data

$$\left.\begin{aligned} \nu &= 0.97\ \xi^{-0.21} \\ \varepsilon &= 2.25 \times 10^{-5}\xi^{0.45} \\ \gamma &= 4.7\ \xi^{-0.13} \\ \alpha &= 0.035\ \nu^{0.82} \end{aligned}\right\} \quad \text{if } \xi \leqslant 3 \times 10^{4} \tag{12}$$

where $\xi = gr/U^2$. The spectral shape parameters σ_a and σ_b were found to scatter about an average of 0.1 but indicated no consistent dependence on ξ.

Application of this technique to Ava, Camille, and Eloise showed reasonable comparison to observations. A subsequent comparison of this approach and a full spectral model to Belle showed good agreement to the observed data (Ross and Cardone, 1977).

Using the JONSWAP spectrum E(f) we can calculate the significant wave height, H_s, as

$$H_S = 4\sqrt{m_o} \tag{13}$$

and the mean zero up crossing wave period, T_z, as

$$T_z = \sqrt{\frac{m_o}{m_2}} \tag{14}$$

where the nth order moment of the spectrum is denoted as

$$m_n = \int_o^\infty f^n\, E(f)df \tag{15}$$

Important limitations of this simple one dimensional parametric approach are that swell is not included, fetch limited seas due to coastal boundaries are not possible and the shallow water effects of refraction, percolation, frictional losses and change in nonlinear wave-wave interaction are not represented. Its key advantages are its simplicity and successful experience in representing the wave field in the high wind areas of hurricanes.

3 APPLICATION OF THE MODEL SYSTEM TO TROPICAL STORM DELIA

The model system was applied to tropical storm Delia as a test of the methodology under conditions where wave current interaction is expected to be significant. There also existed wind, wave, and current observations from an offshore platform with which to compare model predictions and to better understand model performance.

A weak cyclonic circulation in the southwestern Caribbean, noted in 31 August 1973 satellite photographs, strengthed to become tropical storm Delia on 2 September 1973 with its center located at approximately 24° N, 88° W. Delia's central pressure deepened during the night of the 2nd and the morning of the 3rd to 987 mb as the storm traveled in a roughly northwesterly course toward Galveston, Texas (Figure 2). Delia continued on its northwesterly course during the afternoon of the 4th. Later on that day the storm crossed the Texas coast near Galveston and became erratic with an ill defined center.

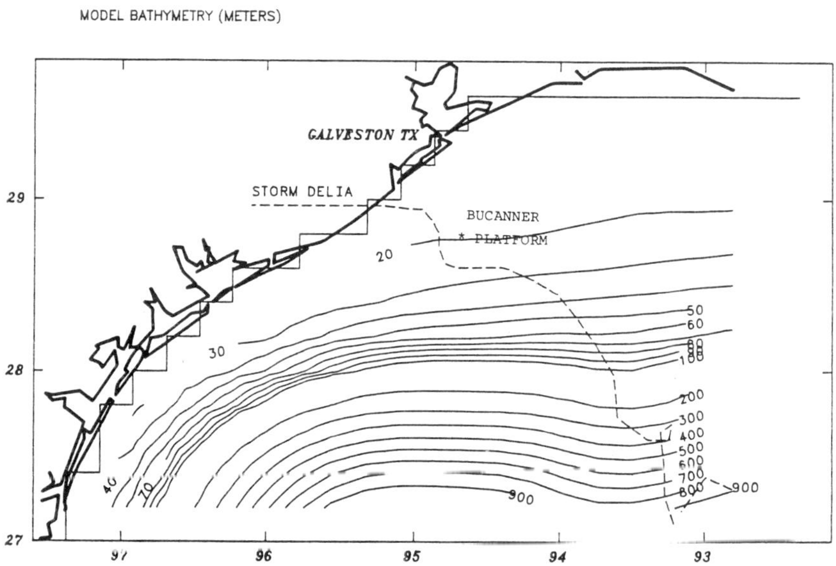

Figure 2. Model grid system, bathymetry and storm track for tropical storm Delia.

The wind field for Delia was calculated using the simple wind model. The radius to maximum winds was assumed at 64 km with a maximum wind speed of 25 m/s. The time dependent central pressure was defined using the curve given in Forristall et al (1977) based on Navy and Air Force dropsonde data. The lowest central pressure was taken as 987 mb with a peripheral pressure of 1013 mb. The storm was assumed to have a forward velocity of 6.1 m/s and a deflection angle of 5°.

During the passage of the storm, the currents were recorded at three depths; 4, 10, and 19 m below the surface in 20 m water depth at the Buccaneer station (Figure 2). The currents were recorded by electromagnetic current meters that were moored on taut wires suspended between the bridge of the platform and a steel anchor. While the instruments functioned throughout the storm period it was noted that a large scour hole had developed in the vicinity of the anchor and caused the meters to rotate about 31°. Since it is impossible to know exactly when this rotation took place, the data is presented here in its original form. It therefore must be recognized that there is, at the minimum, this uncertainty in the reported currents. Forristall et al (1977) give a detailed discussion of the collection and analysis of this data. Wave staff data were also recorded from 3 September 1973 through the storm and provide a measurement of the wave height and period.

The model system was applied to the study area using a 22 km square grid system (Figure 2) where the orientation is along lines of constant latitude and longitude. Longitude is positive to the east and latitude positive to the north. The bathymetric data was derived from the NOAA/NOS charts and the National Geophysical Data Center (NGDC) digitized bathymetric data tapes. The model domain was selected so the offshore boundary was in deep water and the alongshore boundaries were sufficiently removed from the observation site to minimize the problem of accurately specifying the open cross shelf boundary conditions. The boundary conditions for the simulation were an inverted barometric pressure condition on the offshore boundary and no gradient of the vertically averaged velocity normal to the cross shelf boundaries. The procedure in locating the cross shelf boundaries was to perform several simulations where the boundaries were moved progressively closer to the area of interest and determine the smallest domain size where the boundary specifications did not markedly influence the model predictions over the time period of interest. These approximations were similar to those of earlier simulations by Forristall et al (1977) and Cooper and Pearce (1982). Although these boundary conditions are known to reflect waves (Reid, 1975), sensitivity studies showed that this was not a problem for these relatively short duration simulations.

The vertical structure was represented by five Legendre polynomials as a compromise between vertical resolution and computational time. This assessment was made by comparing model predictions with varying numbers of polynomials to an analytic solution for wind driven flow in a rectangular channel with specified surface and bottom stresses.

A total of thirteen cases (Table 1) were investigated using two and three dimensional simulations, with and without the BBLM included. Sensitivity runs were also performed to investigate the impact of alternate bottom drag coefficient values and fixed (constant value) and variable vertical eddy viscosity on model predictions. Drag coefficient values were selected to bracket the typical range for coastal hydrodynamic modeling while the vertical eddy viscosity range was based on earlier simulations of Delia by Forristall et al (1977) and Cooper and Pearce (1982).

TABLE 1

Simulation cases of tropical storm Delia.

CASE NUMBER	3D/2D	VERTICAL EDDY VISCOSITY (cm^2/s)	BOTTOM FRICTION FORMULATION (NON-DIMENSIONAL)
1	3D(A)	CONSTANT 100	BBL(C)
2	3D	CONSTANT 150	BBL
3	3D	CONSTANT 200	BBL
4	3D	VARIABLE IN TIME & SPACE(D)	BBL
5	3D	PARABOLIC IN VERTICAL (E)	BBL
6	3D	CONSTANT 150	0.003
7	3D	CONSTANT 150	0.0005
8	3D	CONSTANT 300	0.0005
9	3D	CONSTANT 300	0.001
10	2D(B)		0.0005
11	2D		0.001
12	2D		0.002
13	2D		0.03

(A) Three dimensional (horizontal and vertical) simulation with five Legendre polynomials
(B) Two dimensional vertically integrated simulation
(C) Simplified Bottom Boundary Layer Model (BBLM)
(D) Visocisty = W*H/RE, where W = surface wind friction velocity,
H = local water depth
RE = 50, the flow Reynolds number
(E) Viscosity = Vm * a1 * (z+a2) * (z+a3)
where Vm = 150, the maximum viscosity,
z = vertical coordinate,
a1,a2,a3 = constants, adjusted to produce the maximum at 10 meters below the surface, and the minimum (50 cm^2/s) at the bottom. For depths shallower than 20 m, the maximum value was used as a constant.

To illustrate the simulations, Case 2 is presented in detail. Other cases will only be summarized.

The surface currents are a combined response to the wind stress and the local pressure gradient. As the storm builds in intensity, the surface induced currents increase reaching values greater than 2 m/s during the height of the storm. The pattern, as expected from the forward motion of the storm, displays an intensification in the right front quadrant.

As the storm induced flows reach the coast they are modified such that the currents are parallel to the coast. As the storm passes southwest of Galveston, the central pressure increases and the storm dissipates. There is a corresponding reduction in the wind induced surface flow. The bottom velocities are, in general, relatively weak throughout the simulation.

Studying the surge response as the storm approaches the shore, water piles up against the coast, east of Galveston, and the southwesterly directed vertically averaged velocities increase in magnitude flowing in a direction parallel to the coast in the nearshore area and parallel to the isobaths in the offshore region. A maximum surge height of 2.25 m is predicted at 37 hours into the simulation. (The simulation starts at 0.0 hrs. on 3 September 1973.) As the storm dissipates, the surge height decreases, although southeast of Galveston Bay it is still present at 49 hours.

To document the vertical structure, Figure 3 shows the east-west and north-south profiles of the current at the Buccaneer measurement site as a function of depth during the storm passage. The response is clearly barotropic with increasing wind stress resulting in increased flow from surface to bottom. The selection of the constant vertical eddy viscosity, in this case, minimizes the near surface vertical current gradient while the BBLM enhanced stress gives significant shear in the water column, particularly near the bottom, during the height of the storm.

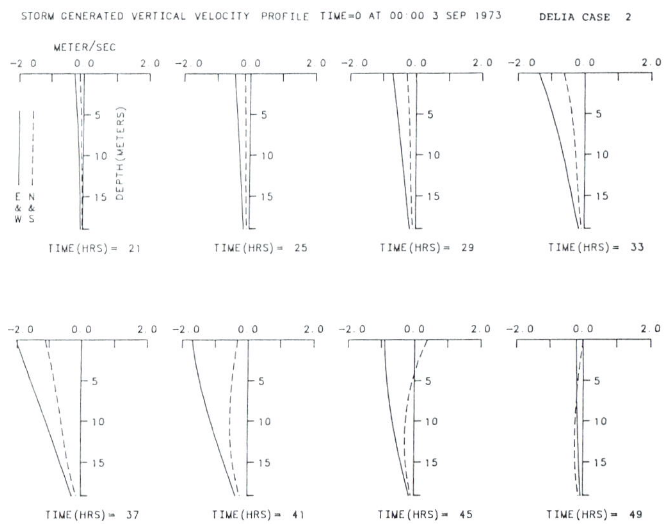

Figure 3. Model predicted vertical current profiles at the Buccaneer observation platform at selected times for Hurricane Delia.

Figure 4 shows a comparison between model predictions and observations (when they exist) for wind velocity, atmospheric pressure, wave height and period, surge height, current velocity at the surface, mid depth and bottom and the bottom drag coefficient (noted as bottom friction in the figure) as a function of time through the entire storm simulation. All are given at the Buccaneer site.

Figure 4 also gives the vertically averaged time dependent momentum balance at the site for the N-S and E-W directions, respectively. The momentum balance includes: local acceleration, free surface elevation gradient, Coriolis force, bottom and surface stress and atmospheric pressure gradient. The nonlinear convective acceleration terms are small in comparison

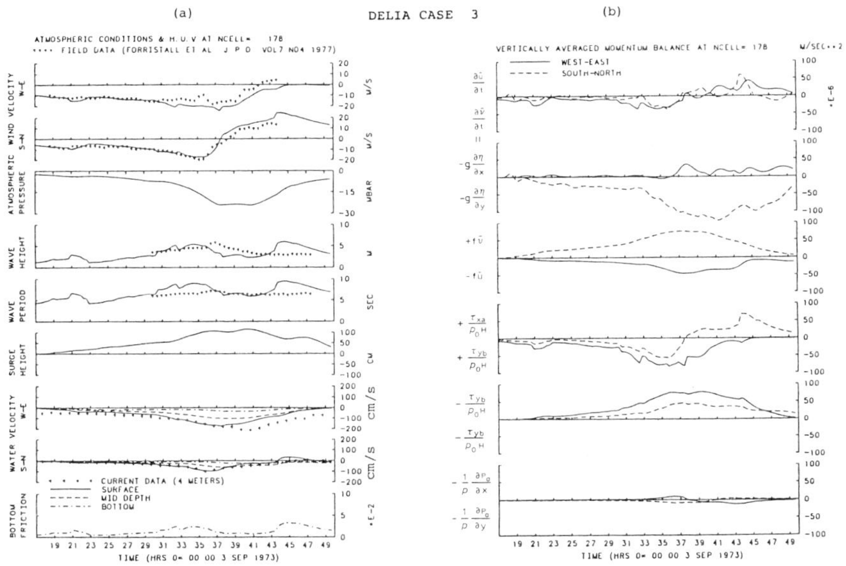

Figure 4. (a) Comparison of observations and model predictions, Delia, for wind, atmospheric pressure, wave height and period, surge height, water velocity (Buccaneer Site) and bottom friction
(b) Vertically averaged momentum at the Buccaneer Platform

to all other terms and have therefore been neglected. This assessment was made by performing simulations with and without the convective acceleration terms included and comparing the results.

Comparing model predictions to observations shows that the simplified storm wind and wave models reasonably reproduce the observed patterns at the Buccaneer platform. The agreement between predicted and observed wind fields is comparable to that of Forristall et al (1977) and Pearce and Cooper (1982). Additional tuning of the wind field description was not appropriate since there was only one point for calibration. The perturbations of the predicted wave field are clearly seen to be in direct response to the changing path and hence fetch of the wind field.

The model predicted currents are shown in comparison to the 4 m, below the surface, current observation. This vertical location was selected as representative of the water column since the observed vertical structure of the horizontal currents was quite uniform. The model predicted surface current is seen to be in good agreement with data both in terms of magnitude and direction. The east-west velocity, however, peaks before and dissipates faster than the observed. The strong westward current observed before the passage of the storm (40-50 cm/s) is typical of the autumn semi-permanent current present in the area (Nowlin, 1971; Estes and Scrudato, 1977; EPA, 1982). EPA (1982) reports currents of 46 cm/s at a heading of 282° for September in the region bounded by 94-95° W longitude and 29-30° N latitude.

The model however is not able to reproduce this feature due to the limited size of the domain and the absence of forcing parameters related to this semi-permanent mean flow. The predicted mid-depth and bottom currents are considerably smaller than observed with the bottom currents 25% of the surface values compared to 95% for the data. The model predicted shear is mainly attributed to the enhanced bottom roughness due to wave-current interaction. The peak bottom friction coefficient is approximately 0.030, an order of magnitude larger than is conventionally used for storm surge modeling. The fluctuations in the bottom friction coefficient are seen as a direct response to changes in the predicted wave environment.

In terms of the force balance in the north-south (approximately cross shelf) direction, the surface slope is roughly balanced by the Coriolis force and the bottom stress. Local acceleration approximately balances the surface wind stress. This balance is maintained throughout the storm with the magnitudes of the forces dependent on the storm strength. The balance in the east-west (along shelf) direction is between the bottom stress, the Coriolis force, and wind stress terms. The surface slope only becomes important after the storm has passed the observation site and is traveling over land. The atmospheric pressure gradient terms play only a minor role in the force balances. If viewed in a coordinate system that is parallel to shore, the primary force balance is geostrophic with a correction for bottom friction. This is increasingly the case as the storm strength builds.

To further illustrate the comparison, Figure 5 shows detailed plots of the currents for each of the three vertical locations for both observations and model predictions. The lower two graphs present the model predicted bottom shear velocity and the roughness height. Although Forristall et al (1977) have derived estimates of the shear velocity and roughness heights for the storm, we have chosen not to use their analysis for comparison to the model because (1) the use of only three points to estimate these values leads to extremely large errors and (2) the assumption of a log velocity profile used to derive their values may not be valid.

The model shows a generally increasing shear velocity peaking at approximately 4 cm/s during the height of the storm and decreasing gradually as the storm dissipates. The roughness height is generally on the order of 1-2 cm but displays substantial peaks of approximately 6-8 cm during the 32-26 and 44-49 hour time periods. The peaks correspond to increases in the wave height and period during these times (Figure 4).

Although not shown here, a detailed analysis of the other cases is presented in Spaulding and Isaji (1985). The principal results will be summarized here. For the three dimensional simulations, with the BBLM included, increasing the vertical eddy viscosity decreases the surface current and increases the mid-depth current. The bottom current remains relatively constant. The surface elevation decreases slightly. The force balance

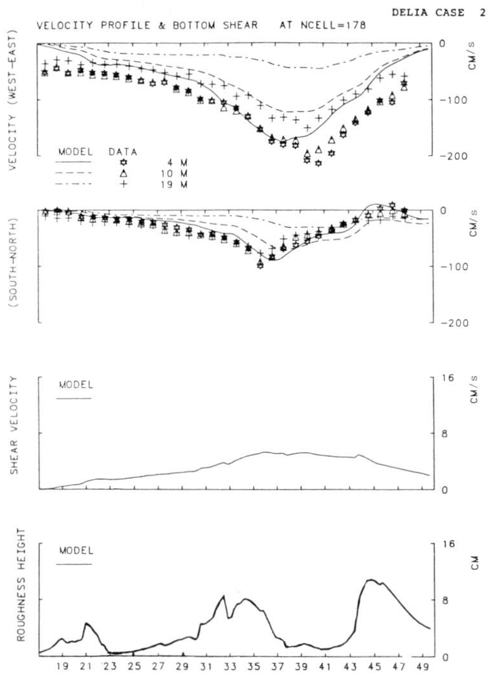

Figure 5. Comparison of model predictions and observations for velocity, bottom shear velocity and bottom roughness height at the Buccaneer Site.

remains basically as discussed earlier with the primary effect of increasing eddy viscosity to lower the surface slope and Coriolis terms and to increase the bottom friction. The variable eddy viscosity formulations, Cases 4 and 5, give solutions much like those for the constant cases with similar mean values.

Cases 6 through 9, respectively, employ the three dimensional model with two values of constant eddy viscosity and selected constant values of bottom friction coefficient. Increasing the eddy viscosity has the same effect as noted earlier, that is to decrease the vertical gradient of the velocity field. Increasing the bottom friction coefficient increases the bottom stress, reduces the surface slopes and reduces the currents at all levels. The larger the friction coefficient, the stronger the vertical current shear.

Comparing a three dimensional simulation, with and without the BBLM, for the same vertical eddy viscosity, 150 cm^2/s (Case 2 and 6), the principal difference is the magnitude of the bottom stress. For the BBLM simulation, the bottom friction coefficient is approximately an order of magnitude larger than for the conventional three dimensional simulation with constant drag coefficient. There is also a reduction in surface elevation and velocity.

The vertical shear rate however remains approximately the same, particularly near the peak of the storm (39 hrs.). The increase in bottom friction coefficient combined with the decreased velocity leads to only a factor of two increase of the BBLM vertically averaged bottom stress compared to the constant drag coefficient case. The force balances are accordingly modified with the constant drag three dimensional case having larger Coriolis and surface slope terms than the BBLM simulation.

Cases 10 through 13 explore model predictions assuming that a two dimensional vertically averaged approach is suitable to describe the flow. This would appear to be a reasonable assumption given the lack of significant water column stratification due to salinity and temperature in the area and the vertical homogeneity of the current observations. Cases 10 through 13 correspond to increasing bottom drag coefficient covering the range (0.0005 to 0.003) that typifies coastal areas. Increasing this friction factor results in a reduction in both the vertically averaged velocity and the surface elevation. The compensating effect of decrease in velocity to increase in drag coefficient results in about constant bottom stress. The reduction in the free surface height and slope is compensated by a reduction in the Coriolis force.

Comparison of model predictions (with the BBLM included) to observations, as shown in Figure 5, indicate that the model under predicts the resulting vertical velocity distribution, especially near the bottom. The model clearly over predicts the bottom friction. The question arises as to why this happens considering that the BBLM gives such good comparisons to observations in other areas, such as the CODE site (Grant et al, 1984).

To investigate this problem, the simplest version of the near bottom stratified (suspended sediment) bottom boundary layer model (Grant and Glenn, 1983) was used to predict the current structure at the Buccaneer site. The reference current was the 3 m depth measurement from Forristall's data (U_{3m}). Wave height (H_s) and period (T) were also extracted from Forristall et al. Table 2 summarizes the input data. The times have been selected to span the most active period during the storm event.

Two cases were selected for simulation. Case 1 assumes a medium sand with a mean diameter of 0.05 cm while Case 2 uses a fine sand, with a mean diameter of 0.01 cm. Both have specific gravities of 2.65. Based on EPA (1982) and Estes and Scrudato (1977), the fine sand is likely to be more representative of the area. Both cases, however, were studied to illustrate the effect of sand grain size on the predictions. For both cases, the model was run with (stratified) and without (neutral) suspended sediment induced stratification.

The results of the simulation, in terms of the friction velocity u_{*c}, roughness height, z_{oc}, and drag coefficient, c_{D3m}, referenced to the 3 m depth are shown in Table 3. Figure 6 shows a comparison between model predictions and observations for current at the three vertical stations as a function of time for the (a) neutral and (b) stratified flows. Only Case 2 is shown since it shows slightly better comparison to observations than Case 1. This seems reasonable since the fine sand approximation is likely to better represent the sediments found at the site.

TABLE 2

Input data for BBLM simulation of tropical storm Delia.

Time (hr)	U_{3m} (cm/s)	H_s (m)	T (s)	h/Lo	A_b (m)	U_b (m/s)	$\frac{Z_r}{A_b}$
30	71.06	3.67	5.51	0.0735	2.280	2.6	1.526
32	87.14	4.19	5.87	0.0647	2.842	3.0416	1.224
35	112.56	4.72	5.94	0.0632	3.245	3.4333	1.072
37	149.38	5.96	6.59	0.0514	4.675	4.457	0.744
38	141.28	5.03	6.25	0.0571	3.691	3.7111	0.943
40	156.54	3.86	5.75	0.0675	2.540	2.776	1.370
42	122.81	3.17	5.74	0.0677	2.083	2.28	1.671
44	92.42	3.13	5.67	0.0694	2.022	2.2407	1.721
47	56.34	3.15	5.81	0.0661	2.102	2.273	1.73

Note

Wave-current relative angle ϕ_c = 0 for all times.

Lo - deep water wave length, h - water depth, A_b - maximum wave excursion, U_b - maximum wave orbital velocity, Z_r - height of current reference velocity.

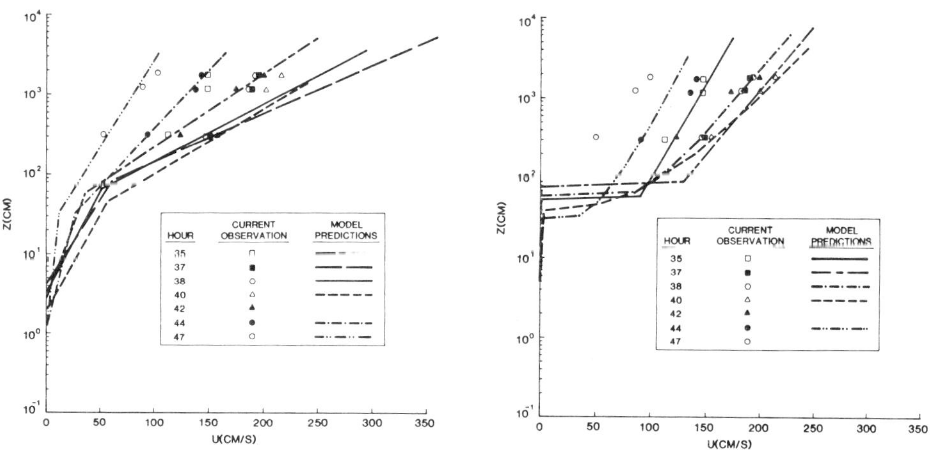

NEUTRAL STRATIFIED

Figure 6. Comparison of BBLM predictions of current versus depth for Case 2, neutral and stratified simulations.

The analysis indicates that during the peak of the storm (hours 35-41), the flow becomes stratified. This stratification results in a reduction in u_{*c} by a factor of 3-5. The bottom drag coefficient referenced to 3 m lowers from 0.027-0.037 to about 0.001. This value of c_D is in good agreement with that empirically determined using the two dimensional vertically averaged simulation. The simple stratified model does well at reproducing the currents at the other levels as well. As the storm dissipates, the flow no longer shows strong stratification, even though the potential exists. The observed velocity profiles at this time are in good agreement with the neutral simulations. This change in response from a stratified to a neutral behavior could be due to bed armoring effects. This explanation appears quite reasonable given the geology of the area where fine sand overlays a silt clay bottom. We cannot however directly confirm this behavior since no suspended sediment concentration data are available during the storm. As the flow returns to a neutral configuration, the friction factor returns to a value on the order of 0.02 and the shear velocity again increases.

TABLE 3

BBLM predictions of friction velocity, roughness and drag coefficient for tropical storm Delia.

TIME (hr)	U_{*c}	Z_{oc}	C_{D3m}	U_{*c}	Z_{oc}	C_{D3m}
	neutral			stratified		
			CASE 1*			
35	20.04	32.23	1.0317	9.93	42.23	0.0078
37	31.37	45.37	0.0441	12.61	4.21	0.0071
38	25.73	33.90	0.0332	11.94	46.44	0.0071
40	22.82	19.60	0.0213	12.89	25.68	0.0068
44	12.75	16.77	0.019	8.34	20.48	0.013
47	8.32	21.24	0.022	5.58	24.64	0.0098
			CASE 2			
35	18.66	27.3	.0275	3.82	45.69	0.0012
37	28.67	37.92	.0368	4.66	68.41	0.00097
38	23.83	28.46	.0285	4.52	51.12	0.001
40	21.24	15.99	.0184	5.08	30.05	0.0011
44	11.95	13.82	.017	3.43	23.23	0.0014
47	7.86	18.09	.0195			

* Case 1 Medium Sand, d = 0.05 cm, C_b = 0.6, Specific gravity - 2.65
Case 2 Fine Sand, d = 0.01 cm, C_b = 0.6, Specific gravity - 2.65

Comparing all simulations to observations suggests that the two dimensional vertically averaged case with a drag coefficient of 0.001 gives an extremely reasonable prediction. If we roughly adjust the reference level to account for the presence of the semi-permanent mean flow (45 cm/s) which has not been modeled, then a slightly larger value of drag coefficient appears more appropriate. The use of the BBLM approach, assuming neutral near bottom stratification significantly underestimates the mean current magnitude in comparison to observations. An independent analysis of the current structure using the BBLM with the observed 3 m currents as input shows that if suspended sediment induced stratification is considered then the predicted and observed current profiles are in good agreement throughout most of the storm.

4 CONCLUSIONS AND RECOMMENDATIONS

Application of the model system was made to tropical storm Delia. The best model predictions, in comparison to the available observations, were obtained using a constant bottom friction coefficient. Incorporation of the wave-current interaction BBLM dynamics in the simulation predicts currents significantly lower than observed. Sensitivity studies for Delia investigating vertical eddy viscosity magnitude and formulation and alternate representations of the vertical structure (from two dimensional vertically averaged to fully three dimensional with increasing numbers of Legendre polynomials) could not reconcile the difference between the model predictions (with the BBLM included) and the observations.

Using observed Delia currents at the surface measurement station and wave conditions as input, simulations were made with the BBLM to describe the current structure and bottom stress levels. Neutral and suspended sediment induced stratification cases were run. Comparison of these simulations to observations suggest that as the storm strength builds the near bottom, water column becomes stratified. When this occurs there is an effective decrease in the bottom stress level and an increase in currents compared to the neutral (non stratified) case. During later stages of the storm bed armoring appears to become active and the water column reverts to the neutral case. While this analysis cannot be confirmed from existing field observations from Delia, it appears highly plausible. It successfully explains the underprediction of the currents by the neutral BBLM included in the hydrodynamic model and the generally good agreement between the two dimensional vertically averaged simulations, using a low value of the bottom friction factor, and the observations.

Simulations of the wind and wave fields for Delia, considering the simplicity of the model formulations and the limited data available for calibration, were generally quite good. The simple one dimensional parametric wave model appears adequate for describing the wind generated wave heights for high intensity small radius storm events in deep and intermediate water depths. In the nearshore area, the use of a shallow water wave model is recommended.

The analyses performed on storm Delia suggest that stratification effects caused by variations in concentrations of resuspended bottom sediments must be carefully analyzed to determine their impact on the current profile. It is clear from the Delia BBLM simulations that stratification can substantially alter the predicted current structure. Based on these observations, it is recommended that the simplified BBLM be modified to incorporate stratification effects. Once this is done, a detailed simulation of a storm for which a substantial amount of data exists, not only on currents but stratification as well, should be performed.

As a note on computational time, the full BBLM requires a factor of 50 more CPU time than the simplified model employed in this study. For a simple storm surge simulation with a square gridded domain of 19 (alongshore) by 5 (onshore - offshore), a resolution of 0.25°, a constant depth of 100 m and five Legendre polynomials requires a 0.0289 s of CPU time per cell per time step. With the simplified BBLM included, the CPU time increases by 0.0092 s per cell per time step. The computational time hence increases by 31.8% with the inclusion of the simplified BBLM. Detailed specifications for this simulation are included as the first test case in a model user's manual (Isaji, 1985). All CPU times are quoted for a PRIME 550 Model II computer with 2 mb of main disk storage. CPU times may change substantially for modifications in the model domain size and configuration depending on paging requirements.

5 ACKNOWLEDGEMENTS

This work was sponsored by the American Gas Association under Contract No. PR169-186 with G.L. Smith and Dr. D.T. Tsahalis serving in the role of technical contract monitors. ASA acknowledges the assistance of Dr. William Grant, Woods Hole Oceanographic Institute (WHOI) who served as a subcontractor. Dr. Grant developed and verified the simplified bottom boundary layer model (BBLM), assisted in the task of interfacing the BBLM with the continental shelf hydrodynamics model, and performed the neutral and stratified BBLM simulations of Delia.

6 REFERENCES

ASCE, Task Committee on Computational Hydraulics, 1980. Sources of Computer Programs in Hydraulics. ASCE Journal of the Hydraulics Division, Vol. 106, No. H45, pp. 915-023.

CERC, 1984. Shore Protection Manual, Volumes I and II. Department of the Army, Coastal Engineering Research Center, Waterways Experiment Station, Vicksburg, Mississippi.

Cooper, C.K. and B.R. Pearce, 1980. On the Forcing Mechanisms Affecting the Bottom Shear Stress in Coastal Waters. J. of Phys. Oceanog., Vol. 10, No. 11, pp. 1870-1876.

Cooper, C. and B. Pearce, 1982. Numerical Simulations of Hurricane Generated Currents. Journal of Physical Oceanography, Vo. 12, p. 1071-1091.

Davis, A.M., 1977a. The Numerical Solution of the Three-Dimensional Hydrodynamical Equations Using a B-Spline Representation of the Vertical Current Profile. In: Bottom Turbulence, Proceedings of the 8th Liege Colloquium on Ocean Hydrodynamics, J.C.J. Nihoul (ed), Elsevier, New York: 27-48.

Davies, A.M., 1977b. Three-Dimensional Model with Depth-Varying Eddy Viscosity. In: Bottom Turbulence, Proceedings of the 8th Liege Colloquium on Ocean Hydrodynamics, J.C.J. Nihoul (ed.), Elsevier, New York: 27-48.

EPA, 1982. Environmental Impact Statement (EIS) for the Galveston, Texas Dredged Material Disposal Site Designation. U.S. EPA Criteria and Standards Division (WH-585).

Estes, E.L. and R.J. Scrudato, 1977. Aquatic Disposal Field Investigations, Galveston, Texas Offshore Disposal Site. Appendix A Investigation of the Hydraulic Regime and Physical Nature of Sedimentation. Texas A&M University, Department of Marine Science, 134 pp.

Forristall, G.Z., R.C. Hamilton, and V.J. Cardone, 1977. Continental Shelf Currents in Tropical Storm Delia: Observations and Theory. J. Phys. Oceanog., Vol. 7, No. 4, pp. 532-546.

Garratt, J., 1977. Review of Drag Coefficients Over Oceans and Continents. Mon. Weather Rev., 105, 915-929.

Glenn, S.M. and W. Grant, 1983. Continental Shelf Bottom Boundary Layer Model, Vol. III-User's Manual. American Gas Association, Project No. PR-153-126.

Gordon, R. and M.L. Spaulding, 1974. A Bibliography of Numerical Models for Tidal Rivers, Estuaries and Coastal Water. Univ. of R.I., Marine Technical Report 32, 55 pp.

Gordon, R., 1982. A Three-Dimensional Numerical Model of Circulation and Salinity Distribution for Estuarine Application. M.S. Thesis, Department of Ocean Engineering, University of Rhode Island, Kingston, R.I. 161 pp.

Grant, W.D. and O.S. Madsen, 1979. Combined Wave and Current Interaction Within a Rough Bottom. Journal of Geophysical Research, 84, (C4), 1797-1808.

Grant, W.D. and O.S. Madsen, 1982. Moveable Bed Roughness in Unsteady Oscillatory Flow. Journal of Geophysical Research, 87, (C1), 469-481.

Grant, W.D. and S.M. Glenn, 1983. A Continental Shelf Bottom Boundary Layer Model, Volume I: Theoretical Development. Technical Report to the American Gas Association, May 1983.

Grant, W., A. Williams, and S. Glenn, 1984. Bottom Estimates and Their Prediction on the Northern California Continental Shelf During CODE 1. The Importance of Wave-Current Interaction, Journal of Physical Oceanography, Vol. 14, p. 506-527.

Harris, D.L., 1958, Meterological Aspects of Storm Surge Generation. Journal of Hydraulics Division, ASCE, paper 1859, 22 pp.

Hasselman, K., D.B. Ross, P. Muller and W. Sell, 1976. A Parametric Wave Prediction Model. J. Phys. Oceanogr., 6, 200-228.

Hinwood, J.B. and I. Wallis, 1975. Review of Models of Tidal Water. ASCE J. of the Hydraulics Division, Vol. 101, No. HY11, pp. 1404-1421.

Isaji, T., 1985. Design Flow Conditions Near Bottom - Phase II Coupling of a Continental Shelf Hydrodynamics Model to a Bottom Boundary Layer Model. Vol. II - Computer Code and User's Manual, PR169-186, American Gas Association, Washington, D.C., pp. 184.

Isaji, T. and M.L. Spaulding, 1984. A Model of the Tidally Induced Residual Circulation in the Gulf of Maine and Georges Bank. J. of Phys. Oceanogr., Vol. 14, No. 6, 1119-1126.

Jaycor, Inc., 1979. Proceedings of the Continental Shelf Physical Oceanographic Model Evaluation Workshop. Document No. J-79-310-001, Research Triangle Park, North Carolina, p. 175.

Jelesnianski, C.P., 1974. SPLASH (Special Program to List Amplitudes of Surges from Hurricanes - II. General Track and Variant Storm Coditions. NOAA Technical Memorandum NWS TDL-52.

Kitaigorodskii, S.A., 1961. Applications of the Theory of Similarity of the Analysis of Wind-Generated Wave Motion as a Stochastic Process. Izv. Akad. Nauk SSSR, Ser. Geofiz., No. 1, 73-80.

Large, W.G. and S. Pond, 1981. Open Ocean Momentum Flux Measurements in Moderate to Strong Winds. J. Phys. Oceanogr., 11, 324-336.

Myers, V.A. and W. Malkin, 1961. Some Properties of Hurricane Wind Fields as Deduced from Trajectories. Report #49, U.S. Dept. of Commerce.

Nowlin, W.D., 1971. Water Masses and General Circulation of the Gulf of Mexico. Oceanology 6(2), p. 28-33.

Owen, A., 1980. A Three Dimensional Model of the Bristol Channel. J. Phys. Oceanogr. 10: 1290-1302.

Pearce, B.R., C. Cooper and E. Doyle, 1979. Hurricane Generated Currents, Civil Engineering in the Oceans IV, ASCE/San Francisco, CA, September 10-12.

Reid, R.O., 1975. Comments on Three-Dimensional Structure of Storm-Generated Currents by G.Z. Forristall. J. Geophys. Res., 8, 1184-1185.
Ross, D.B., 1975. A Comparison of SKYLAB S-193 and Aircraft Views of Surface Roughness and a Look Toward Seasat. Proceedings of the NASA Earth Resources Survey Symposium. Lyndon B. Johnson Space Center, Houston, Texas, June 1975, 1911-1920.
Ross, D.B., 1976. A Simplified Model for Forecasting Hurricane Generated Waves (abstract). Bull. A.M.S. January 113. Presented at American Meteorological Society Conference on Atmospheric and Oceanic Waves, Seattle, Washington, March 29 - April 2, 1976.
Ross, D. and V. Cardone, 1977. A Comparison of Parametric and Spectral Hurricane Wave Prediction Products, Proceedings of NATO Symposium on Turbulent Fluxes Through the Sea Surface, Wave Dynamics and Prediction, Marseille, France, 12-16 September 1977.
Simons, T.J., 1975. Verification of Numerical Models of Lake Ontario, Part II. Stratified Circulation and Temperature Changes. J. Phys. Oceanogr. 5: 98-110.
Smith, S.D. and E.G. Banke, 1975. Variation of the Sea Surface Drag Coefficient with Wind Speed. Q.J.R. Meteorol. Soc., 101, 665-673.
Spaulding, M.L. and T. Isaji, 1985. Design Flow Conditions Near Bottom - Phase II Coupling of a Continental Shelf Hydrodynamics Model to a Bottom Boundary Layer Model. Volume I - Theoretical Development and Application, Contract PR169-186, American Gas Association, Washington, DC, pp. 184.
Swanson, J.C. and M.L. Spaulding, 1978. A Review of Three Dimensional Hydrodynamic Models Suitable for Application to Coastal Sea and Continental Shelf Regions. NASA Langley Research Center, Hampton, VA.
Tetra Tech., Inc., 1981. Coastal Flooding and Storm Surge Model. Federal Emergency Management Agency. Washington, D.C.
Wiberg, P. and J.D. Smith, 1983. A Comparison of Field Data and Theoretical Models for Wave-Current Interactions at the Bed on the Continental Shelf. Continental Shelf Research, Vol. 2, 126-136.
Withee, G.W. and A. Johnson, 1976. Buoy Observations During Hurricane Eloise., Report, Environ. Sci. Div., NOAA Data Buoy Off., NSTL Station, Mississippi.
Wu, J., 1969. Wind Stress and Surface Roughness at Air-Sea Interface. J. Geophys. Res. 74, 444-455.
Wu, J., 1980. Wind-Stress Coefficients Over Sea Surface Near Neutral Conditions: A Revisit, J. Phys. Oceanogr., 10, 727-740.

THREE-DIMENSIONAL MODEL OF CURRENTS IN THE BAY OF SEINE

J.C. SALOMON, B. THOUVENIN and P. LE HIR
IFREMER, B.P. 337, 29273 BREST CEDEX

ABSTRACT

Water movement and mixing processes in the Bay of Seine are essentially induced by tides, winds and density differences due to river inflow and temperature gradients near the coast.

To study tidal currents alone, it is sufficient to use a two dimensional model. However, for wind and density effects, the three spatial dimensions must be taken into account.

For that purpose a three-dimensional model of currents has been developed which solves the nonlinear Navier-Stokes equations in a straightforward manner through a finite difference technique. The model has been tested in various schematic cases for which analytical solutions or results from previous models were available. It was then applied to the Bay of Seine and proved to be an efficient tool for coastal studies.

1. REGIONAL SETTING

The Bay of Seine, as defined by the coastline, is shaped more or less like a regular quadrilateral about one hundred kilometers long and forty to fifty kilometers wide, opening onto the English Channel (Fig. 1). The bay is quite shallow (twenty five meters deep) except for the ancient submerged valley of the Seine which runs diagonally across the bay and reaches depths of forty meters in the northwestern part of the bay.

With an average flow of 400 m^3/s, the Seine is a modest river. Nevertheless, it is the only significant fluvial discharge in the Channel. The estuary, which used to be upstream of the city of Le Havre, has now a smaller area, because of the improvement of the waterway. Over the last hundred and fifty years, the estuarine area has decreased by a factor of two, and at the same time the location has shifted towards the bay. Today the density nodal point is situated downstream of Le Havre.

The characteristics of the bay are quite complicated. Indeed, the present situation includes almost all the possible causes for water movements in the coastal zone :

- The main part of the dynamics is due to the tide. The tidal range can exceed eight meters in spring tides. The currents increase regularly from east to west, from one to three knots. The tidal wave and the associated currents are strongly distorded while propagating in the shallow areas, so that the nonlinear terms in the hydrodynamic equations are important.

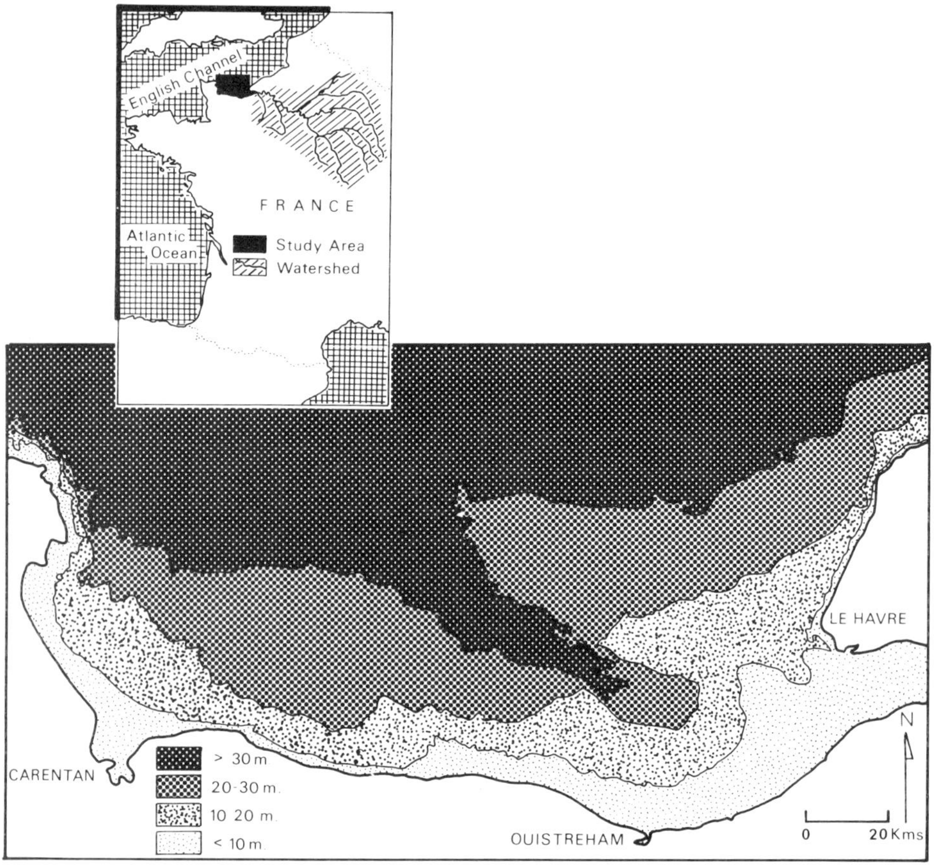

Fig. 1. Location map and bathymetry of the Bay of Seine.

- The wind is another driving force of water movements in the bay, all the more important as it is rather intense and as water depths are small.

- As mentionned before, density circulation does exist too, closely related to the fluvial runoff in the eastern part of the bay. It is almost of estuarine type (converging velocities near the bottom, nodal point) with some geostrophic features (rotation of surface velocities to the right).

- Finally water movements in the bay is partly due to the general circulation in the Channel.

2 MODEL SPECIFICATIONS

The tidal components of the currents in the bay can be studied, for their essential part, with two-dimensional (2D) models, but those which are linked to the density field or to the wind do require a three-dimensional (3D) calculation.

A few years ago we made a first attempt at modelling currents in three dimensions in this area (Salomon and Le Hir, 1980). We proceeded in a quite similar way to that used by Nihoul (1977) at the time, by developing along the vertical coordinate the results of a 2D model at specific points.

The results revealed some interesting apects of the currentology, but such calculations are not really three dimensional:the frictional stress is considered to be in the same direction as the vertically averaged velocity. Moreover, the nonlinear advective terms are difficult to take into account, since the calculations are local.

The technical specifications of the present 3D model are that the model should:

- solve the complete nonlinear equations;
- take into account the precise form of the sea bed (very important in coastal problems), which eliminates "stair step" models;
- have the same behaviour at all times of the computation and over the whole bay.

Due to the depth differences in our zone, and due to the tidal amplitude, we discarded the use of reduced coordinates. As different grids act like different media, waves and dynamic structures are partly reflected and distorted.

On the other hand, the model could have modest performances as far as horizontal turbulent fluxes calculations are concerned because for the time scales we are interested in, these terms are small compared to the advective terms.

Simple considerations on the dynamics of the bay show that due to its restricted dimensions and the intensity of friction forces, its "memory" towards barotropic waves and even baroclinic ones is of a few hours. Its memory towards phenomena of transport of matter (like salt) is much higher. Calculations based on the lack of salt in the bay give a residence time of about a month.

For financial reasons (among others) it was impossible to measure boundary conditions during such a long period (except for the tide) and the project of such simulations was given up. Model indications, in terms of density circulation, will essentially result from the initial salt and temperature distribution, and not from modifications of the boundary conditions during the calculations. Barotropic movements, on the other hand, are driven through the boundary conditions.

3 MODEL CHARACTERISTICS

3.1 Equations

The model solves the Navier-Stokes equations in their usual form, using the hydrostatic approximation, and assuming Fickian diffusion:

$$\frac{\partial u}{\partial t} + u\frac{\partial u}{\partial x} + v\frac{\partial u}{\partial y} - fv + \frac{1}{\rho}\frac{\partial P}{\partial x} - \frac{1}{\rho}\frac{\partial (N_x \frac{\partial u}{\partial z})}{\partial z} - F_x = 0$$

$$\frac{\partial v}{\partial t} + u\frac{\partial v}{\partial x} + v\frac{\partial v}{\partial y} + fu + \frac{1}{\rho}\frac{\partial P}{\partial y} - \frac{1}{\rho}\frac{\partial (N_y \frac{\partial v}{\partial z})}{\partial z} - F_y = o$$

$$\frac{\partial \zeta}{\partial t} + \frac{\partial \int_H udz}{\partial x} + \frac{\partial \int_H udz}{\partial y} = o$$

$$\frac{\partial u}{\partial x} + \frac{\partial v}{\partial y} + \frac{\partial w}{\partial z} = o$$

$$\rho = \rho o \ (1 + aS)$$

$$\frac{\partial S}{\partial t} + u\frac{\partial S}{\partial x} + v\frac{\partial S}{\partial y} + w\frac{\partial S}{\partial z} - \frac{\partial (K_Z \frac{\partial S}{\partial z})}{\partial z} - F_S = o$$

(x, y) : horizontal coordinates
z : vertical coordinate
(u, v, w) : velocity components
ζ : surface elevation
H : total depth of water
ρ : mass per unit volume
f : Coriolis factor
S : salinity
Fx, Fy : horizontal diffusion of momentum
Fs : horizontal diffusion of salt
Nx, Ny : vertical eddy viscosity coefficients
Kz : coefficient for salt diffusion along the vertical

The horizontal diffusion terms Fx, Fy, Fs are physically important in vertically integrated models where they are called dispersion because they include at the same time the vertical velocity gradients and the vertical diffusion. Here they are essentially required to damp small scale computational modes. It would be different if long term simulations were to be done. They are expressed in the following usual way :

$$Fx = Nxx \frac{\partial^2 u}{\partial x^2} + Nxy \frac{\partial^2 u}{\partial y^2}$$

$$Fy = Nyx \frac{\partial^2 v}{\partial x^2} + Nyy \frac{\partial^2 v}{\partial y^2}$$

$$Fs = Kx \frac{\partial^2 S}{\partial x^2} + Ky \frac{\partial^2 S}{\partial y^2}$$

In the present application to the Bay of Seine, which we know to be rather well-mixed, the pressure gradients are expressed in the following simple form:

$$\frac{1}{\rho}\frac{\partial p}{\partial x} = g\frac{\partial \zeta}{\partial x} + ag(\zeta - z)\frac{\partial S}{\partial x}$$

3.2 Coefficients for vertical exchange

The model makes use of the concept of turbulent exchange coefficients parameterized in a simple form. The other, more sophisticated, possibility, which would consists in solving, after closure, the turbulence equations, has been rejected because:

- no initial or boundary conditions are available with enough accuracy,
- it has been established by different investigators that in the case of tidal currents in shallow waters, a relation between the diffusion coefficients and the velocity gradients (or the concentration gradient) through a mixing length, gives reasonable results. No additional complication has been seeked on this point.

The effect of buoyancy forces, which highly modify the mixing length, is introduced by means of a global Richardson number:

$$Ri = \frac{a\,g\,H.\Delta S}{U^2}$$

ΔS: Salinity difference between bottom and surface

U : vertically averaged velocity.

Several expression (generally empirical ones) have been tested in the model. The following relations appear to give the best results:

$$N_z = 7.10^{-4} + 5.10^{-2}U\ \mathrm{Log}(0.1\ z^3 + 1)(1 + 7\ Ri)^{-\frac{1}{4}}$$

$$K_z = 7.10^{-4} + 5.10^{-2}U\ \log(0.1\ z^3 + 1)(1 + Ri)^{-\frac{7}{4}}$$

3.3. Method of resolution

The numerical procedure and the discretization grid are extrapolated from the two dimensional vertical model that we already used in several estuaries: Gironde, St Lawrence (de Borne de Grandpré et al., 1979, 1981), Seine (Salomon, 1981). It is a mixed implicit-explicit finite difference technique.

The originality of the method lies essentially in the fact that the locations of the upper and lower frontiers of the integration area change during the calculations. In order to achieve that, use is made of virtual points located above the surface, and others situated under the bottom (Hamilton, 1975). Details of this technique have been exposed by Thouvenin and Salomon (1984).

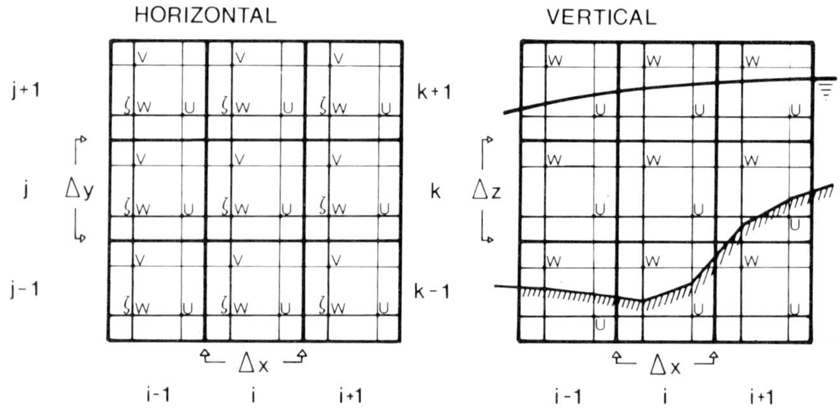

Fig. 2. Computational grids

Briefly stated, at every time step a finite difference method is used to go over the usual calculation sequence inside the area of computations. The results are then extrapolated to external virtual points by analytical functions which take into account the upper and lower boundary conditions :

$$\frac{\partial \vec{V}}{\partial z} = \frac{\vec{\tau}_W}{\rho Nz} \quad \text{and} \quad \frac{\partial S}{\partial z} = o \qquad \text{at the surface}$$

$$Nz\,\frac{\partial \vec{V}}{\partial z} = k\,\vec{V}_{1m}\,/V_{1m}/ \text{ and } \frac{\partial S}{\partial z} = o \qquad \text{at the bottom}$$

$\vec{V}_{1m}$: the velocity vector one meter above the bottom

$\vec{\tau}_W$: wind stress

k : friction coefficient

The functions used are usually polynomials of third degree (splines), except for the velocity near the bottom. To ensure a correct velocity profile just above the bottom where it is known to be logarithmic, we use either a logarithmic curve or an exponential one, obeying the continuity of the velocity curve and of its first derivative.

3.4 Numerical scheme

The numerical scheme is mixed implicit/explicit. All spatial derivatives are centered except advective terms in the salt equation. The pressure gradient term, particularly important for the stability of the scheme, is treated as implicit.

Considering the equations in the sequence described in 3.1., it is possible to solve the complete system in an explicit way, i.e. without solving any matrix.

The complete numerical scheme is described and discussed in Thouvenin (1983).

4. MODEL APPLICATIONS

4.1 Schematic tests

Before applying the model to the Bay of Seine, we tested its performances in some simple and rather schematic cases for which aspects of the solution are known.

In particular, we have studied (Thouvenin, 1983):

- The propagation of a sinusoidal wave in a rectangular basin.
- The free oscillations in a closed basin.
- The velocity field in a channel crossed by a trench.
- The effects of the wind blowing over the open sea.
- Convective movements along isopycnal surfaces.

Here, we shall briefly comment on the last two experiments.

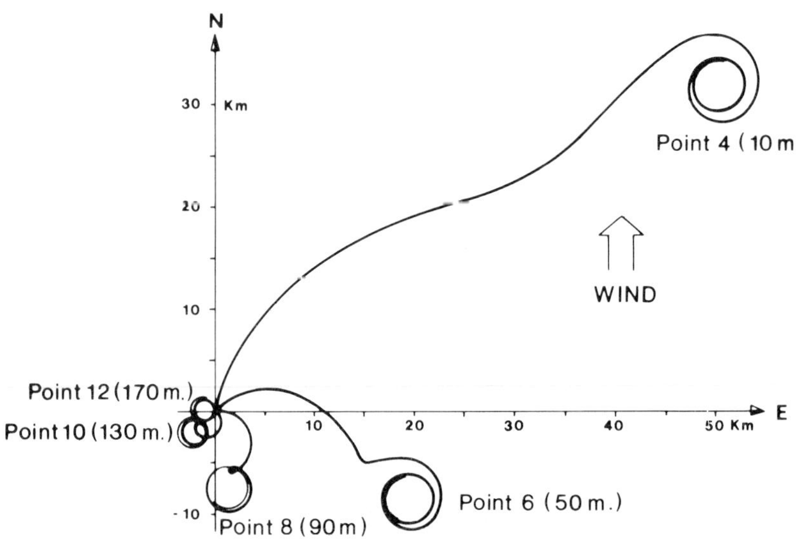

Fig.3. Water particles trajectories under wind action (Distance from the surface indicated in brackets).

Fig. 3 shows the trajectories of water particles induced by a constant wind supposed to blow during 33 hours from the south, and then to stop. According to the known analytical solution, we observe the superposition of the Ekman drift spiral and of inertial oscillations. When the wind stops the inertial rotation, stronger near the surface, diminishes. Some energy is transferred through viscous stress to the lower layers which are thus accelerated.

Comparing the thickness of the Ekman layer produced by the model with the theoretical solution, we find for Nz a difference of 20 %. This difference which sums up several numerical errors seems quite reasonable.

The movement associated to a salt wedge in a rectangular tank is interesting as well (fig.4). Here, we observe the formation of a density driven circulation along the interface and the appearance of a barotropic wave propagating very quickly far from the salt wedge. Also to be noted is the polarisation of vertical movements, more intense on the right side, due to the Coriolis force .

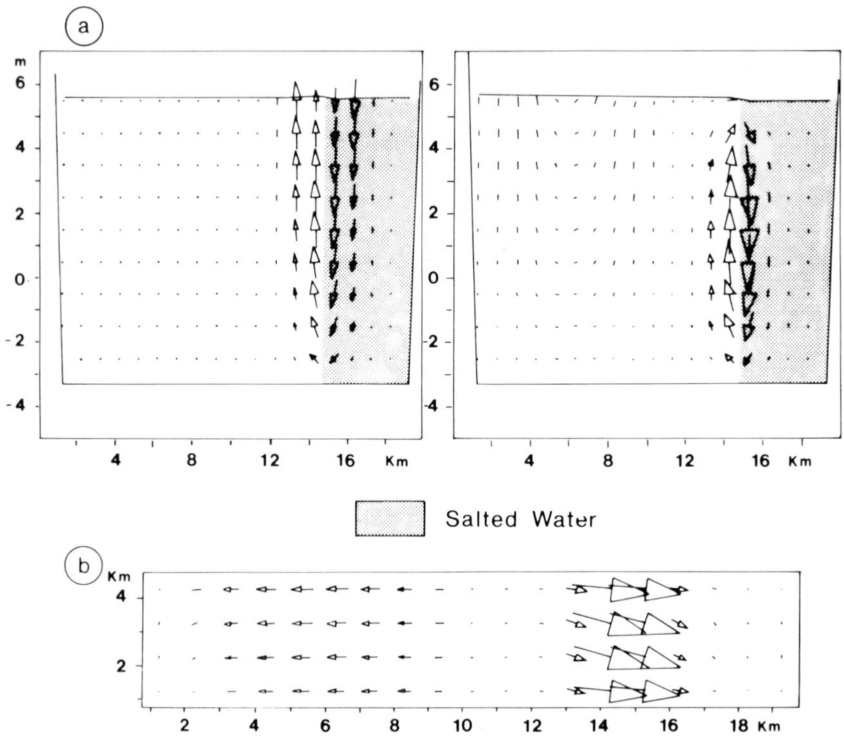

Fig. 4 Salt wedge circulation

a) lateral views after 30 and 300 minutes

b) view from above after 5 hours

4.2 Application to the Bay of Seine

(i) Initial and boundary conditions. Because of computer limitations, the application to the Bay of Seine is done with a relatively coarse grid size (4 km, 4 m). It must be acknowledged that due to the lack of information on boundary conditions, a grid refinement would not have been completely justified.

Initial conditions for salinity are taken from in situ measurements made during the year 78.

Boundary conditions along the open boundary (water levels and velocities) are taken from a 2D model (Salomon, 1985). The vertical structure of the horizontal velocities is supposed to obey a logarithmic profile. The vertical component of the velocity is linearly interpolated between the bottom and the surface where it is calculated through the following relations :

$$w_{bottom} = 0$$

$$W_{surface} = \frac{\partial \zeta}{\partial t} + u \frac{\partial \zeta}{\partial x} + v \frac{\partial \zeta}{\partial y}$$

(ii) Tidal currents. The tidal component of the currents is the essential part of the dynamics but as it seemed already well known through measurements (Le Hir and l'Yavanc, 1985) or 2D modeling (Salomon, 1985), little additional information was expected.

In fact, we observe (see Fig. 5):

- important vertical movements which, of course, could not have been measured. They are the result of the bottom shape and horizontal convergences.

- phase differences and rotations along the vertical which come from the different ratio of the inertial force and the driving force (surface slope) as well as from characteristics of bottom Ekman layer.

Thus we are more aware of the error we make in a 2D model assuming the friction stress on the bottom to be in the same direction as the average velocity.

(iii) Wind induced movements. The wind induced dynamics is typically three-dimensional and cannot be approached by 2D simulations.

This component of the circulation is linked to others and especially to the tide. The results may be very complex and difficult to understand from in situ measurements. The model here is of great assistance, even if its indications are not very accurate.

Theoretically, calculations should take into account the response of the entire Channel to wind forcing and large scale atmospheric pressure gradients. However, some interesting information can be obtained from local simulations about the response of the bay to wind forcing, at least qualitatively.

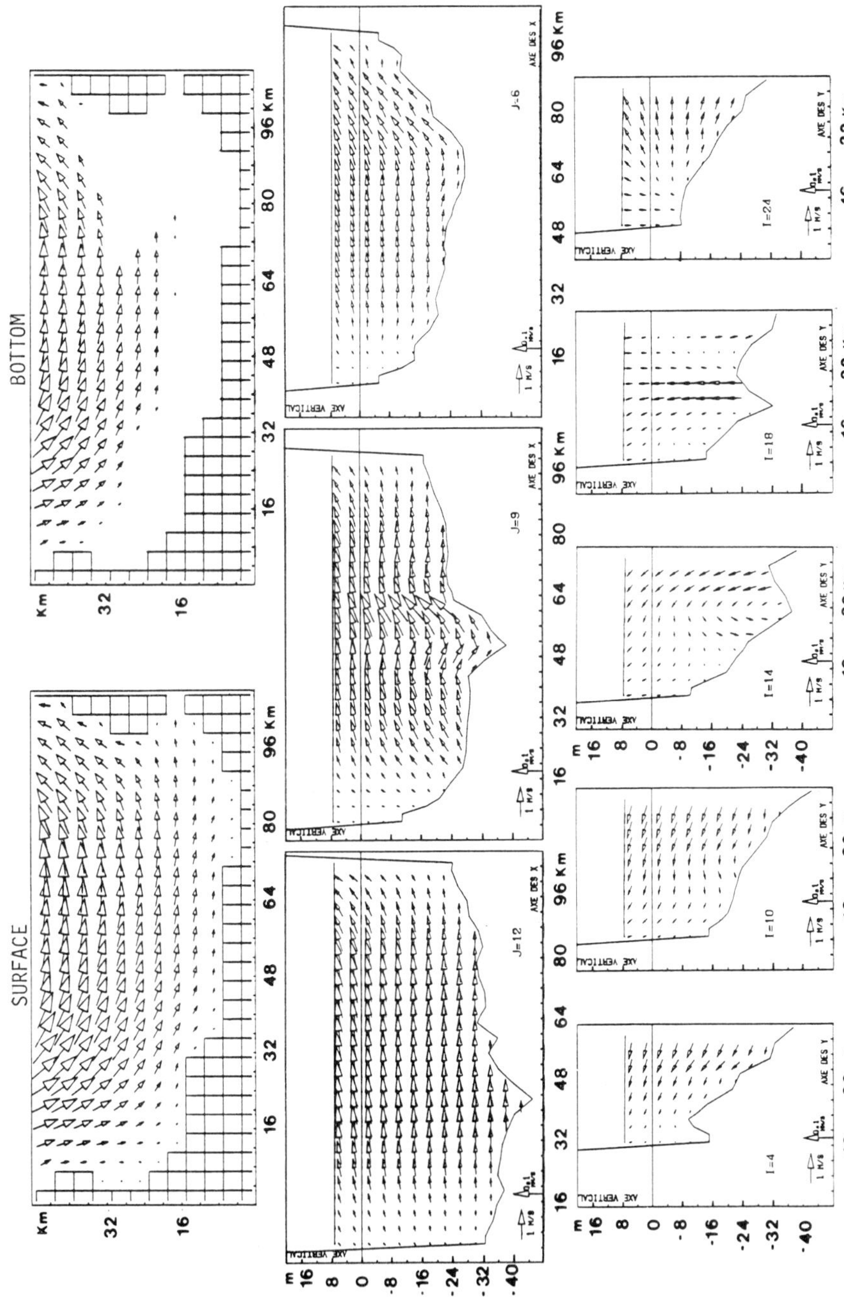

Fig. 5. Three dimensional tidal velocity field during flood (horizontal sections near the surface and the bottom ; E-W vertical sections ; N-S vertical sections).

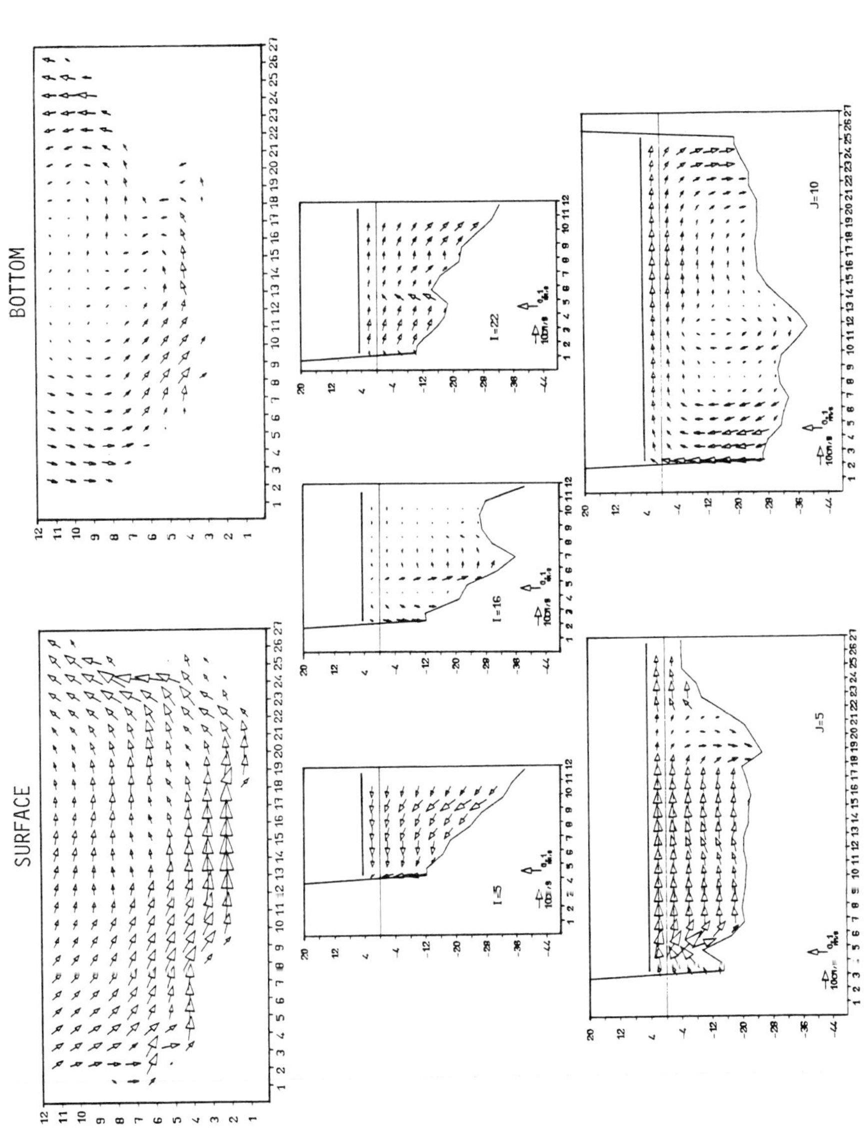

Fig. 6. Wind induced velocity field (horizontal sections near the surface and the bottom ; N-S vertical sections ; E-W vertical sections). Wind blowing from the west.

So we have carried out some simulations of the direct action of the wind on the bay alone.

The boundary condition corresponds to a hypothetical steady regime. In the dynamical equation along the northern boundary a balance is assumed between the surface slope, friction stresses on the bottom and at the surface, and the Coriolis force.

The results presented in Figs. 6 and 7 are for uniform wind fields of 15 m/s. A permanent regime is generally observed after seven hours.

The horizontal sections clearly show the steering of the currents by bathymetric features, particularly near the bottom (fig. 6).

Cross currents are hardly ever observed in the Parfond (ancient Seine valley), even close to the surface, where currents are less related to the bottom shape. Velocities there often rotate a few tens of degrees.

At the surface, the velocities are larger in the shallow zones, a consequence of the fact that the driving force related to the unit of volume or unit of mass is inversely proportional to the water depth. This is the basic reason for the general dissymmetry between the eastern and western parts of the bay.

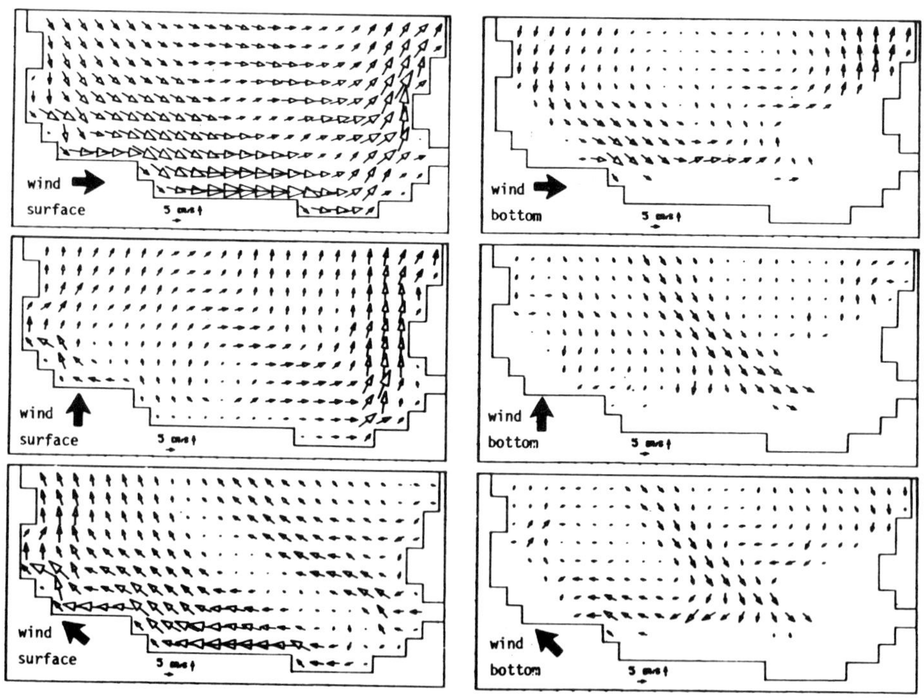

Fig. 7 Horizontal velocity fields for different wind directions.

For the same reason, the wind induced velocities above the Parfond are weaker. When the conditions are such that a deficit of water appears in the southern part of the bay, an adjustment current appears in the Parfond, opposite to the surface velocities. This is a well known feature of wind action in littoral zones and lakes : a back current appears in the deepest parts, generally the waterway. It is also in close agreement with measurements made during this study.

The vertical sections also show very obvious vertical structures corresponding to vertical movements near the banks and helicoidal movements inside the trench of the Parfond, which measurements would not have revealed.

These simulations, as well as the measurements, indicate that the wind induced velocities are often of an order of magnitude of 10 to 30 cm/s, which is much higher than tidal residual currents. This mechanism is thus an essential vector of the long term movements of waters in the bay.

5 CONCLUSION

The model presented in this paper is but one element of a general study of the bay.

The study lasted three years, during which a considerable effort has been made on measurements : three cruises of a few months each have been carried out, involving different measurement locations (up to twelve), sometimes at the surface, at the bottom and at mid-depth. Important human and technical resources have thus been involved.

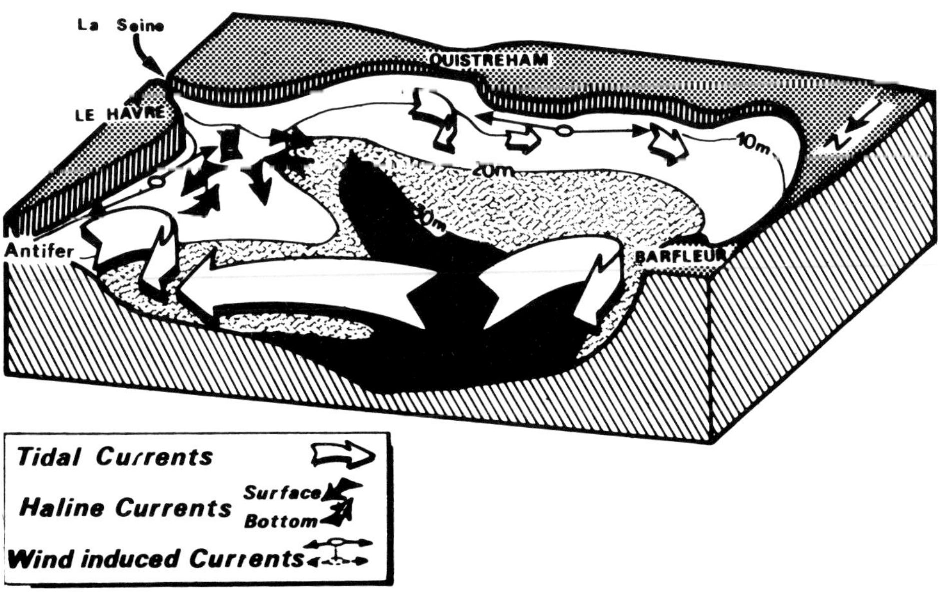

Fig.8 Diagram showing the residual circulation in the Bay of Seine

The measurements produced a lot of interesting indications but,finally,when it came to elaborating a comprehensive synthesis of thedynamical regime of the bay, the main role was devoted to the model. This way we were able to understand what occured and we could separate the proper effects of each of the physical mechanisms involved : wind, tide, density, Coriolis ... This way we understood what measurements where showing.

The diagram for the residualcalculation presented in fig.8, was essentially deduced from model results.

So, it proved to be an efficient tool for conducting coastal studies.

Some advantages of this model have been already mentioned, especially the possibility of conciliating a fixed location of computational points and a continuous representation of the bottom and the surface, in spite of a great tidal range.

It is highly perfectible from the numerical point of view as well as for the turbulent exchange simulations, but before complicating it, it is worth weighting the pros and cons.

Our opinion is that for restricted coastal sites under the influence of the tide and the wind, preventing high stratification to occur, and for which initial and boundary conditions are hardly available, such a model is an interesting compromise and an efficient tool.

6 REFERENCES

De Borne de Grandpré, C., 1979. Modèle bidimensionnel en temps réel de la circulation verticale estuarienne. Application à la Gironde. Oecanologica Acta, 2, 1, 61-68.

De Borne de Grandpré, C., El Sabh, M.I. and Salomon, J.C., 1981. A two-dimensional numerical model of the vertical circulation of tides in the St Lawrence Estuary. Estuarine, Coastal Shelf Sci, 12, 375-387.

Hamilton, P., 1975. A numerical model of the vertical circulation of tidal estuaries and its application to the Rotterdam Waterway, Geophys. J.R. Astron. Soc., 40, 1-21.

Le Hir, P. and l'Yavanc, J., 1985. Observations de courant en baie de Seine. Rapport Scient. CNRS/Greco Manche, 7-14.

Nihoul, J.C.J., 1977. Three-dimensional model of tides and storm surges in a shallow well-mixed continental sea. Dynamics of Atmosphere and Oceans, 2,29-47.

Salomon, J.C. and Le Hir, P., 1980. Etude de l'estuaire de la Seine. Modélisation numérique des phénomènes physiques. Rapport Scient. CNEXO/UBO, 286 p.

Salomon, J.C., 1981. Modelling turbidity maximum in the Seine Estuary. In : J.C.J. Nihoul (editor), Ecohydrodynamics. Elsevier, Amsterdam, 285-317.

Salomon, J.C., 1985. Courantologie calculée en baie de Seine. Rapport Scient. CNRS/Greco Manche, 15-22.

Thouvenin, B., 1983. Modèle tridimensionnel de circulation et de dispersion pour des régions côtières à marée. Thèse 3e cycle U.B.O., 269 p.

Thouvenin, B. and Salomon, J.C., 1984. Modèle tridimensionnel de circulation et de dispersion en zone côtière à marée. Premiers essais : cas schématique et baie de Seine. Oceanologica Acta, 7,4, 417-429.

TIDAL STREAMS IN SHALLOW WATER

P.P.G. DYKE
Department of Mathematics and Statistics, Plymouth Polytechnic, Drake Circus, Plymouth, Devon, PL4 8AA, U.K.

ABSTRACT

The analysis of tides is particularly suited to perturbation methods, since there is normally a dominant oscillatory flow. In shallow water, a variety of non-linear effects can be justifiably considered. The advection terms, free surface terms and variations in vertical and horizontal momentum transfer are all important. This contribution examines, analytically, some of these effects, solves some simple problems and provides insight into how the steady circulation and higher harmonics can be thus generated.

1 INTRODUCTION

In the shallow continental shelf regions that surround most continents and which are particularly wide adjacent to North West Europe, barotropic tides are a very important source of water movements. Tides are predominantly oscillatory in character; however, non-linear interactions cause an asymmetry and give rise to tidal harmonics. These non-linear interactions occur through the advection terms, the free surface and in the representation of turbulent momontum oxohango. For oxamplo, if the primary tidal oscillations are plane waves

$$u(x,y,t) = U\cos(kx + ly - \sigma t) \tag{1}$$

$$v(x,y,t) = V\sin(kx + ly - \sigma t) \tag{2}$$

with frequency σ rad s^{-1}, wave number vector (k,l) and tidal ellipse having semi-major axis U, semi-minor axis V; it is easily seen that a typical advective term that occurs in the equations of motion, say

$$u\frac{\partial v}{\partial x} = \tfrac{1}{2}kUV\ (1 + \cos 2(kx + ly - \sigma t)) \tag{3}$$

gives rise to a steady term and a second harmonic. This is simple. However, the solution of the full three dimensional partial differential equations which govern tidal dynamics is far more of a challenge. Oceanographers are still arguing about how best to represent turbulent momentum exchange. Quite early on in a typical computation, it is beneficial to concentrate the attention on a specific phenomenum. For example, in a series of papers Tee (1976, 1977, 1979, 1980, 1981, 1982, 1985) and Tang and Tee (1986) have addressed the steady problem which requires elimination of time-dependent effects. This is most conveniently done by assuming all variables proportional to $e^{i\sigma t}$ so that $\partial/\partial t$ is replaced by $i\sigma$ ($i = \sqrt{-1}$) wherever it occurs. The fact that this can still remain a lifetime study is an indication of the complexity of the tidal problem. We shall follow this scheme in some of the later sections of this paper.

This paper is solely concerned with tides. However, let us not lose sight of the many and various other causes of coastal currents. If an estimation of the strength and direction of a current was required, then although tides may well provide the main input, from an engineering viewpoint it would be foolhardy indeed to ignore wind induced current, storm induced flows (including surges), density currents, and currents that arise from sea surface slopes outside the shelf region. If such an estimate is required, the analytical methods of this paper have to be abandoned and numerical techniques such as those of Davies (1986) come into their own. Davies favours a spectral approach. One could use finite differences in the vertical, this is done by, for example, Backhaus (1983). This paper is concerned more with fundamentals and less with final applications, however it is hoped that some of the ideas herein will later find their way into these applications.

Tides are phenomena consisting of a dominant mode added to which are higher order but smaller effects. An obvious candidate to choose from the limited number of analytical techniques available is perturbation. This requires all variables to be expanded in power series of some small parameter (eg. wave slope or ratio of boundary layer depth to total depth). The non-linear problem then transforms into a series or progressively more complicated, but linear, problems. Perturbation methods are old fashioned and have received a bad press recently due to their overuse outside regions of applicability and validity. The tidal problem is tailor made for perturbation methods, so no excuse is required for using them here.

2 THE GOVERNING EQUATIONS

For a shallow, coastal sea region it is reasonable to assume that the Coriolis acceleration is constant and that variation in density can be ignored, at least in a preliminary investigation. To include density stratification would be to include internal tides which, admittedly, cause baroclinic residuals that can be important, but which would only confuse at this early stage. A simple (the simplest) treatment of baroclinic residuals can be found in Dyke (1980). The conservation of momentum and mass equations take the form

$$\frac{\partial \underline{u}}{\partial t} + (\underline{u}.\underline{\nabla}_H)\underline{u} + w\frac{\partial \underline{u}}{\partial z} + \underline{f} \times \underline{u} = -g\,\underline{\nabla}_H\zeta + \frac{\partial}{\partial z}\left[N\frac{\partial \underline{u}}{\partial z}\right] + A_H\nabla_H^2\underline{u}\,, \tag{4}$$

and

$$\frac{\partial\zeta}{\partial t} + \underline{\nabla}_H \cdot \int_{-h}^{\zeta} \underline{u}\,dz = 0\,. \tag{5}$$

In these equations, z, the coordinate axis which points vertically upward, has a zero value at mean sea level, and $\zeta(x,y,t)$ describes the elevation of the sea surface. The hydrostatic and Boussinesq approximations have been applied, $\underline{u} = \underline{u}(x,y,z) = (u,v)$ is the horizontal fluid velocity, $w = w(x,y,z)$ the vertical fluid velocity. $\underline{\nabla}_H$ is the horizontal gradient operator with x east and y north, N is a coefficient of eddy viscosity which may vary with position and/or velocity, $z = -h(x,y)$ is the sea bed. $\underline{f}$ is twice the angular velocity of the earth and A_H is a constant horizontal diffusion coefficient. Since it is our wish to model tides which are mainly the result of gravitational forces dictated by the frequency of the earth-moon system, it seems a reasonable assumption to permit only oscillations of frequency σ, the frequency of the dominant M_2 tide. This means that we adopt the scheme mentioned in the introduction of replacing $\partial/\partial t$ by $i\sigma$ where σ is the aforementioned frequency. As mentioned in section one, analytical progress is facilitated by adopting a perturbation technique based upon expanding ζ, $\underline{u}$ and w as power series of a small parameter. The parameter we choose is ϵ the wave slope which obeys

$$\epsilon = O(|\zeta|/|\zeta+h|)\,. \tag{6}$$

Accordingly

$$\zeta = \epsilon\zeta_0 + \epsilon^2\zeta_1 + \ldots \tag{7}$$

$$\underline{u} = \epsilon\underline{u}_0 + \epsilon^2\underline{u}_1 + \ldots$$

$$w = \epsilon w_0 + \epsilon^2 w_1 + \ldots$$

The zero order equations are

$$i\sigma\underline{u}_0 + \underline{f} \times \underline{u}_0 = -g\underline{\nabla}_H\zeta_0 + (N\underline{u}_{0z})_z + A_H\nabla_H{}^2\underline{u}_0 \ , \tag{8}$$

and $i\sigma\zeta_0 + (hu_0)_x + (hv_0)_y = 0$. (9)

Implicit in deriving these last two equations is that the total depth is much larger than the tidal amplitude. (Otherwise ϵ would cease to be small). Tee (1986) has referred to this as the weakly non-linear case. We later relax this condition, ie. abandon the perturbation technique, temporarily. The only remaining non-constant yet to be expanded is the eddy viscosity N. To be as general as possible, let us write

$$N = \delta N_0 + \delta^2 N_1 + \ldots \tag{10}$$

where δ is another expansion parameter, the definition of which will depend on how N changes in various regimes. However, until section four of this article, N will not be expanded.

The first order equations will be

$$i\sigma\underline{u}_1 + (\underline{u}_0 \cdot \underline{\nabla}_H)\underline{u}_0 + w_0 \frac{\partial \underline{u}_0}{\partial z} + \underline{f} \times \underline{u}_1 = -g\underline{\nabla}_H\zeta_1 + (N\underline{u}_{1z})_z + A_H\nabla_H^2\underline{u}_1 \tag{11}$$

$$i\sigma\zeta_1 + (hu_1)_x + (hv_1)_y + (\zeta_0 u_0)_x + (\zeta_0 v_0)_y = 0 \ . \tag{12}$$

With adroit use of Taylor's series, subsequent order equations may be found, with the cross terms from previous order solutions providing the forcing.

The boundary conditions are quite straightforward since we are concerned with tides which are very long waves (hydrostatic). Thus

$\underline{u} = \underline{0}$ at coastlines (13)

modified to

$\underline{u}.\underline{n} = 0$ at coastlines (14)

if A_H, the horizontal diffusion coefficient of momentum is set to zero. At the sea bed, there must be either no flow, or zero normal flow, again dependent upon whether friction is allowed, ie.

$\underline{u} = \underline{0}$ at $z = -h$, if N is non-zero at the sea bed (15)

$$\underline{u}.\nabla_H h = 0 \text{ at } z = -h \qquad \text{if } N = 0 \text{ close to the sea bed.} \tag{16}$$

The sea surface boundary condition reduces to $p = P$ atmospheric at $z = \zeta(x,y,t)$ as a consequence of having assumed hydrostatic balance. (This has already been subsumed in eliminating pressure from (4)).

Now we shall make a few simplifying assumptions. First of all, set $A_H = 0$. This means we neglect horizontal boundary layers and demand only that the component on velocity normal to the coast is zero.

Next we suppose that a quadratic friction law applies at the sea bed

$$\underline{\tau}_b = \rho\gamma|\underline{u}|\underline{u} \ . \tag{17}$$

This is hardly a simplifying assumption, except that in tidal regimes it is possible to use Fourier Analysis to linearise (17) (see, for example Marchuk and Kagan (1984), p244), without significant loss of information. Finally, we will ignore surface stress, which rules out consideration of wind waves and tide-wind wave interaction.

If we vertically integrate the zero order momentum equation (8), we obtain, in component form

$$i\sigma\hat{u} - f\hat{v} = -g\zeta_x - \tau_b{}^x/\rho h \ , \tag{18}$$

$$i\sigma\hat{v} + f\hat{u} = -g\zeta_y - \tau_b{}^y/\rho h \ , \tag{19}$$

where the bottom stress terms arise from undoing the stress/rate of strain (Newtonian) term in (8) and assuming that the bottom stress is the same as the stress in the sea itself at the sea bed. The notation is as follows:

$$\underline{u}_0 = (u_0, v_0) = \text{Re}\{(U,V)e^{i\sigma t}\} \ , \tag{20}$$

and

$$(\hat{u},\hat{v}) = \frac{1}{h}\int_{-h}^{0} (U,V)\ dz \ . \tag{21}$$

The continuity equation is

$$i\sigma\zeta + (h\hat{u})_x + (h\hat{v})_y = 0 \ , \tag{22}$$

and elimination of $\hat{u}$ and $\hat{v}$ from (18) and (19) with substitution into (22) leads to a second order elliptic partial differential equations for $\hat{u}$ and $\hat{v}$, namely,

$$(\sigma^2-f^2)\ \hat{u} = i\sigma g\zeta_x + fg\zeta_y + i\sigma\tau_b{}^x/\rho h - f\tau b^y/\rho h \ , \tag{23}$$

and $(\sigma^2-f^2)\ \hat{v} = -fg\zeta_x + i\sigma g\zeta_y - f\tau_b{}^x/\rho h + i\sigma\tau_b{}^y/\rho h$, (24)

whence $i\sigma(\sigma^2-f^2)\zeta + i\sigma g[(h\zeta_x)_x + (h\zeta_y)_y] + fg[(h\zeta_y)_x - (h\zeta_x)_y]$

$$+ \frac{i\sigma}{\rho h}\ [(h\tau_b{}^x)_x + (h\tau_b{}^y)_y] + \frac{f}{\rho h}\ [(h\tau_b{}^y)_x - (h\tau_b{}^x)_y] = 0 \ . \tag{25}$$

Equation (25) reduces to the Helmholtz wave equation if h and the bottom stress are considered constant. The boundary conditions on ζ are

$$(\hat{u}.\hat{v}).\underline{n} = 0 \text{ on coasts} \ , \tag{26}$$

$\underline{n}$ being a vector normal to the coast, and $\hat{u}$ and $\hat{v}$ being expressed in terms of ζ through (23) and (24). At open sea boundaries ζ is specified.

Some special solutions to this problem are examined in section three. It is instructive to look at an alternative derivation of the zero order equations, this time retaining z dependence, and more importantly not explicitly demanding a small ϵ. All that is demanded is that non-linear terms are in some sense small.

2.1 Direct Approach

Consider again equation (4) with no non-linear terms, namely

$$\frac{\partial \underline{u}}{\partial t} + \underline{f} \times \underline{u} = -g\underline{\nabla}_H\zeta + \frac{\partial}{\partial z}\left\{ N\frac{\partial \underline{u}}{\partial z}\right\} + A_H\nabla_H^2\underline{u} \ . \tag{27}$$

Differentiate this with respect to z, which eliminates the first term on the right hand side. (Assuming A_H constant). Using the suffix derivative notation

$$\underline{u}_{zt} + \underline{f}x\underline{u}_z = (N\underline{u}_z)_{zz} + A_H\nabla_H^2\underline{u}_z \tag{28}$$

Writing $\underline{u} = (u,v)$ and $\Phi = u_z + iv_z$, equation (28) is written in complex form

$$\Phi_t + if\Phi = (N\Phi)_{zz} + A_H\nabla_H^2\Phi \tag{29}$$

If A_H can be ignored, the transformation

$$R = N\Phi e^{ift} \tag{30}$$

turns equation (29) into the standard diffusion equation

$$R_t = NR_{zz} \tag{31}$$

where N can be a general function of x,y,z, but not t.

Alternatively, if N is not a function of x or y (or t) A_H can be retained in equation (29), and the equation for R is

$$R_t = NR_{zz} + A_H\nabla_H^2 R \ . \tag{32}$$

If N _is_ time dependent, this can be catered for by the addition of a term $-N_tR/N$ on the left hand side of equation (31) or (32), or more naturally by dealing directly with equation (29). However, the beauty of deriving (31) and (32) is that it opens up all those analytical methods (Crank (1975)) or numerical methods for solving the diffusion equation.

The quantity Φ and its complex conjugate $\bar{\Phi}$ are derivatives of the rotary tidal current. These were first introduced in a slightly different form by Thorade in 1928 (Defant (1961), Soulsby (1983)). If N is constant, these rotary tidal currents resemble the Ekman spiral but with f replaced by $\sigma + f$ and $\sigma - f$ and with the two spirals spiralling very differently since $\sigma + f >> \sigma - f$ (and strange things happen at $\sigma = f$).

The equations that have been derived in this section are general for a linear or quasi-linear formulation. If, however, the advection terms cannot be neglected even at zero order, then the problem needs re-examination. It has recently been suggested Davies (1985) that it is physically reasonable to let eddy viscosity vary as the current speed or its square. In either case, the previous formulation again fails due to the term

$$\frac{\partial}{\partial z} N \frac{\partial \underline{u}}{\partial z}$$

becoming fully non-linear. It is still justifiable to ignore the advection terms if horizontal variation of velocity is significantly slower than vertical variation. This will be the case far away from coasts and especially where there are vertical shear layers. Analytical progress is very difficult unless some kind of perturbation method is employed. Certainly, if eddy viscosity depends upon the square of velocity, then the vertical transfer of momentum becomes a term cubic in velocity in some general sense. If it is still assumed that the primary oscillation is of frequency σ, that is

$$(u,v,\zeta) \text{ proportional to } e^{i\sigma t} \tag{33}$$

then a cubic non-linearity will generate terms of time dependence $e^{3i\sigma t}$. In the language of tides, an M_2 tide can in this way lead to the generation of M_6 tides, but with no M_4 tides. Alan Davies (personal communciation) has indicated that M_6 is present in some strength near North Sea coasts whereas M_4 is almost entirely absent. This would support the above eddy viscosity dependence and suggest that the energy of the tide is dissipated more through vertical transfer of momentum and less via horizontal advection (which would give rise to M_4 since these terms are velocity squared (or $e^{2i\sigma t}$) non-linearities).

3 RESIDUAL FLOW

Most analytical progress is possible when attention is restricted to time independent flow so that by time averaging over several tidal cycles, the complications caused by non linear generation of harmonics is eliminated. Referring back to the perturbation expansions of the last section; equations (11) and (12) (with $A_H = 0$) have to be solved. The complimentary function is the same as in the zero order problem, with the extra terms consisting of (known) products of zero order terms. The whole solution, complimentary function plus particular integral, is then time averaged over several tidal periods. Obviously, since there are non-linear interactions, the assumption equivalent to (33) is technically false and $\partial/\partial t$ has to be carried explicitly in the equations. This is of no consequence here since time averaging later takes place. It is worth noting that no steady currents arise out of the cubic non-linearities mentioned in the last section, but they would arise out of a friction term containing quadratic terms in velocity (or elevation). The solution for $\underline{u}_1$ will be of the form

$$\underline{u}_1 = \underline{u}_E + \underline{u}_{11}e^{i\sigma t} + \underline{u}_{12}e^{2i\sigma t} \tag{34}$$

where $\underline{u}_E$ is independent of time, provided N is not of a form that generates even higher harmonics. Upon time averaging, denoted by the time honoured overbar,

$$\overline{\underline{u}_1} = \underline{u}_E \tag{35}$$

and $\underline{u}_E$ is the Eulerian Residual Current, or Eulerian Tidal Residual. The Eulerian residual is what is observed by a stationary observer or current meter. Particles which are moving with the fluid, or very nearly with the fluid in the cases of pollution, fish eggs and the like, move with Lagrangian residual velocity. Because the flow is primarily oscillatory, we can use the well known, but often misused analysis in the introduction of Longuet-Higgins

(1953) which shows that

$$\bar{\underline{u}}_L = \underline{u}_E + \bar{\underline{u}}_S \tag{36}$$

where $\bar{\underline{u}}_L$ is the Lagrangian residual flow, and $\bar{\underline{u}}_S$ is the Stokes' Drift, given entirely in terms of zero order flow:

$$\bar{\underline{u}}_S = \mathrm{Re}\left\{ \frac{1}{i\sigma} \left[\overline{u_o \frac{\partial u_o}{\partial x}} + \overline{v_o \frac{\partial u_o}{\partial y}} \, , \, \overline{u_o \frac{\partial v_o}{\partial x}} + \overline{v_o \frac{\partial v_o}{\partial y}} \right] \right\} . \tag{37}$$

The question now arises as to the relative importance of $\underline{u}_E$ and $\bar{\underline{u}}_S$. Johns and Dyke (1972) assumed that $\underline{u}_E$ was zero outside a boundary layer. Moore (1970) showed that, in the absence of friction if there are no closed geostrophic contours (depth contours for constant Coriolis parameter), then the total Lagrangian current $\bar{\underline{u}}_L$ is zero, ie. $\underline{u}_E = -\bar{\underline{u}}_S$ and is certainly not zero (unless $\bar{\underline{u}}_S = 0$, not the case here). Examination of the Stokes' drift in a frictionless sea thus gives information about Eulerian residuals provided there are no closed geostrophic contours.

Tee (1980) calculates $\bar{\underline{u}}_L$ via a numerical procedure and decides that both $\underline{u}_E$ and $\bar{\underline{u}}_S$, in general, contribute to $\bar{\underline{u}}_L$ with neither dominating. It is interesting that Moore (1970) shows that $\bar{\underline{u}}_L$ is, in general, indeterminate in a frictionless sea and follows geostrophic contours. The results of Tee (1980) must, therefore, be heavily dependent on how frictional dissipation at the sea bed is parameterised.

3.1 Some Stokes' Drift Results

From the results of the previous section, it is well worthwhile calculating Stokes' drift because this is opposite in direction and equal in magnitude to the Eulerian residual current when there are no closed geostrophic contours.

If it is assumed that

$$\frac{\partial}{\partial y} = 0 \tag{38}$$

and the depth is allowed to vary, then geostrophic contours run north-south and are not closed. The particularisation (38) is applicable to eastward travelling long crested tidal waves. With zero friction, equation (25) reduces to the manageable

$$(\sigma^2 - f^2)\zeta + g(h\zeta_x)_x = 0 \, . \tag{39}$$

if we let

$$\zeta = A(x) \sin \sigma t \ , \tag{40}$$

$$u_o = U(x) \cos \sigma t \ , \tag{41}$$

$$v_o = V(x) \sin \sigma t \ , \tag{42}$$

$$\text{then } U = \frac{\sigma g}{\sigma^2 - f^2} A'(x) \tag{43}$$

$$\text{and } V = \frac{-fg}{\sigma^2 - f^2} A'(x) \ . \tag{44}$$

This analysis then becomes appropriate for standing waves with a straight north-south coastline. The Stokes' drift, from equation (36) is

$$\underline{\bar{u}}_S = (U_S, V_S)$$

$$\text{where } U_S = 0 \tag{45}$$

$$V_S = \frac{-fg^2}{2(\sigma^2 - f^2)^2} A'(x) \ A''(x) \tag{46}$$

where, from (39), A(x) satisfies

$$(hA')' + \frac{\sigma^2 - f^2}{g} A = 0 \ . \tag{47}$$

At the coast, say $x = x_1$,

$$\frac{dA}{dx} = 0 \ . \tag{48}$$

Now in this configuration, $\underline{u}_E = -\underline{\bar{u}}_S$, and a zero of $\underline{\bar{u}}_S$ would give a zero of Eulerian residual. We can perform some interesting analysis to obtain a criterion for zeros of V_S.

By Rolle's theorem, provided (U,V) is not zero other than at the coast, the zero of V_S nearest the coast occurs when

$$\frac{d^2A}{dx^2} = 0 \ . \tag{49}$$

At this location, say $x = \xi$, from equation (47)

$$h'A' + h_o \lambda^2 A = 0 \qquad \text{at } x = \xi \tag{50}$$

where h_0 is some reference depth and

$$\lambda^2 = \frac{\sigma^2-f^2}{gh_0} . \tag{51}$$

Integrating (47) with respect to x and using the mean value theorem, one obtains

$$hA' = h_0\lambda^2 \ (x_1-x)A(\xi_2) \tag{52}$$

$0 \leqslant \xi_2 \leqslant x_1$ (all x).

Hence, eliminating A' from (50) and (52),

$$\frac{dh}{dx} (x_1-\xi \ ) = -h \frac{A(\xi_2)}{A(\xi \ )} \tag{53}$$

where dh/dx is evaluated at $x = \xi$, the zero of A"(x).

So far, (53) is exact. However, if we assume that the ratio of the amplitude on the right hand side is approximately unity, as is the case for tides, then

$$-\frac{1}{h}\frac{dh}{dx} \simeq \frac{1}{(x_1-x)} \tag{54}$$

at a location where V_S is zero.

Equation (54) is satisfied identically for a constant slope sea bed. Dyke (1973) examines particular cases and finds that reversals in V_S occur when the topography is convex upwards, and no reversals occur in V_S when the topography is concave upwards.

4 FRICTIONAL EFFECTS

There are many ways of introducing friction. Through the perturbation scheme, equations (7), (8), (9) and (10), through methods such as those of section (2.1) and through dimensional arguments favoured by those working in turbulence (Monin and Yaglom (1971)). The recent review by Soulsby (1983) is recommended reading. Unfortunately, there is no best way to represent friction.

Returning to perturbation methods, if $\delta = \epsilon$ in equation (10), no friction enters the equations at zero order, and the flow can be considered of boundary layer type with frictional effects restricted to sea-bed and sea-surface. Perhaps on continental shelves it is more reasonable to assume

$$N = N_0 + \epsilon N_1 + \ldots \tag{55}$$

whence similar equations to Tee (1979, 1980) are obtained but with additional frictional terms at higher order. Now, what are N_0, N_1, etc? A good question; let us see the results of assuming an expansion such as (55). Equation (8) remains unchanged save N is replaced by N_0. If N_0 is constant or linear in z (Soulsby (1983)), the result is Ekman type spirals or distorted versions thereof involving Kelvin functions (see Soulsby's article for details). If N_0 is a function of velocity, an idea increasing in popularity, then the linear perturbation scheme will give rise to non-linear equations in u_0, v_0 and ζ_0. If time dependence is retained, then the occurrence of non-linear terms as forcing terms on the right hand side of equation (11) enables one to see more easily how higher harmonics are generated.

ACKNOWLEGMENT

The author is indebted to Kim Tee for access to unpublished material and to Alan Davies for useful discussions.

REFERENCES

Backhaus, J.O., 1985. A semi-implicit with an explicit scheme in a three-dimensional hydrodynamic model. Cont. Shelf Res. 2: 243-254.

Crank, J., 1975. The Mathematics of Diffusion, Second Edition, Oxford University Press, 414 pp.

Davies, A.M., 1985. On determining current profiles in oscillatory flows. Applied Math. Mod. 9.

Davies, A.M., 1986. Numerical modelling of marine systems. In Numerical Modelling - Applications to Marine systems. J. Noye (ed) Elsevier North Holland (to appear).

Defant, A., 1961. Physical Oceanography, Volume 2, Pergamon Press, Oxford, 590 pp.

Dyke, P.P.G., 1973. Zeros of tidal mass transport under standing waves. Riv. Ital. Geof. 22: 353-358.

Dyke, P.P.G., 1980. On the Stokes' drift induced by tidal motions in a wide estuary. Est. Coast. Mar. Sci 11: 17-25.

Johns, B. and Dyke, P.P.G., 1972. The structure of the residual flow in an offshore tidal stream. J. Phys. Oceanog. 2: 73-79.

Longuet-Higgins, M.S., 1953. Mass transport in water waves. Phil. Trans. Roy. Soc. Lond. A245: 535-581.

Marchuk, G.I. and Kagan, B.A., 1984. Ocean Tides: Mathematical Models and Numerical Experiments. Pergamon Press, Oxford, 292 pp.

Monin, A.S. and Yaglom, A.M., 1971. Statistical Fluid Mechanics. Vol. I MIT Press, Cambridge Mass, 769 pp.

Moore, D., 1970. The mass transport velocity induced by free oscillations at a single frequency. Geoph. Fluid Dyn. 1: 237-247.

Soulsby, R.L., 1983. The bottom boundary layer of shelf seas. In Physical Oceanography of Coastal and Shelf Seas. B. Johns (ed), Ch.5, 189-266. Elsevier Science Publishers, Amsterdam.

Tang, Y. and Tee, K.T., 1986. Effects of the mean and tidal current interaction on the tidally induced residual current. J. Physical Oceanography (submitted).

Tee, K.T., 1976. Tide-induced residual current, a 2-D nonlinear numerical mode. J. Mar. Res. 34: 603-628.

Tee, K.T., 1977. Tide-induced residual current - verification of a numerical model. J. Phys. Oceanog. 7: 396-402.

Tee, K.T., 1979. The structure of three-dimensional tide-generating currents. Part I Oscillating Currents. J. Phys. Oceanogr. 9: 930-944.
Tee, K.T., 1980. The structure of three-dimensional tide-induced currents. Part II Residual Currents.
Tee, K.T., 1981. A three dimensional model for tidal and residual currents in bays. In Transport Models for Inland and Coastal Waters, 284-309.
Tee, K.T., 1982. The structure of three-dimensional tide-generating currents: experimental verification of a theoretical model. Est. Coast and Shelf Sci. 14: 27-48.
Tee, K.T., 1985. Depth-dependent studies of tidally induced residual currents on the sides of Georges Bank. J. Phys. Oceanogr. 15: 1818-1846.

MODELLING AND OBSERVATIONS OF THE RESIDUAL CURRENT OFF SOUTHWEST NOVA SCOTIA

K.T. TEE and P.C. SMITH
Bedford Institute of Oceanography, Department of Fisheries and Oceans, Dartmouth, Nova Scotia, B2Y 4A2 (Canada)

D. LEFAIVRE
Institut Maurice Lamontagne, 850 Route De La Mer, Ste-Flavie, Quebec, G5H 3Z4 (Canada)

ABSTRACT

Using a depth-dependent weakly nonlinear tidal model, the three-dimensional structure of the tidally induced residual current near Cape Sable off the southwestern tip of Nova Scotia, is computed. The result is compared to 17 current meter observations located at seven stations. Examples of the depth-averaged residual current, and of the vertical structure of the current at all the experimental stations are presented.

Using a multiple regression technique, the tidally induced residual currents are estimated from the current meter records. The estimated residual currents at two stations are well reproduced by the model. At one station, the predicted residual currents are smaller than the observed values. For the rest of the stations, the comparison is not conclusive because of large uncertainties in the estimated residual currents.

1 INTRODUCTION

Three-dimensional tidally induced residual currents have been computed in several nonlinear depth-dependent numerical tidal models (e.g., Leendertse and Liu, 1975). However, the results have not been examined carefully, and the structures of the current have not always been verified by observations. One of the reasons may be the lack of current meter data in the modelled area that indicate the signals (spring-neap oscillation) of the tidally induced residual currents. Another reason may be that there are problems in separating the tidally induced residual current from those induced by wind and freshwater runoff, especially when the latter forcings are dominant. The objectives of this paper are: (1) to compute the three-dimensional tidally induced residual current with a numerical model for the Cape Sable area, off southwestern Nova Scotia (Fig. 1); (2) to estimate the tidally induced residual current from 17 current meter data located at seven stations; and (3) to compare the model residual currents with those estimated from the data.

The coastal waters off the southwestern tip of Nova Scotia (Cape Sable

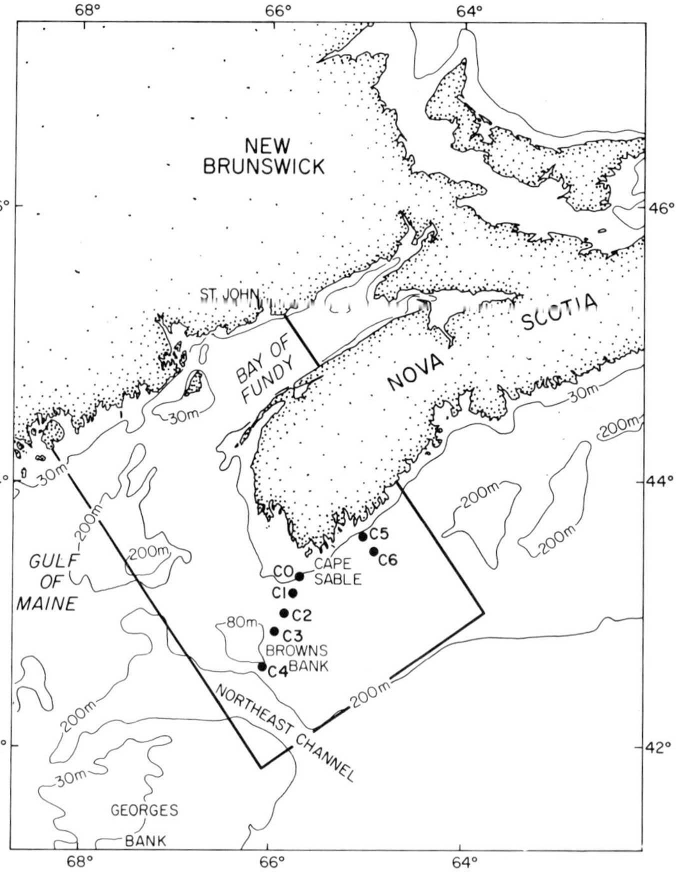

Fig. 1. Location map for the Bay of Fundy and Gulf of Maine.

area, Fig. 1) are among the most productive areas on the eastern coast of North America. It has been suggested that this is an area of upwelling (Lauzier, 1967; Garrett and Loucks, 1976; Smith, 1983). The modelled area is indicated by solid lines which extend across the Bay of Fundy and the Scotian Shelf in the north-eastern boundary, along the edge of the Scotian Shelf (~200 m isobath) at the south-eastern boundary, across the Gulf of Maine at the south-western boundary, and along the New Brunswick coast at the north-western boundary. There are 36 grid points at the north-western and south-eastern boundaries, and 47 grid points at the north-eastern and south-western boundaries. The horizontal grid spacing is 7.047 km. There are two important topographic features near Cape Sable: one is the convex shape of isobaths around the Cape, and another is the Browns Bank, a shallow submarine bank approximatley 50 km long and 30 km wide. The crest of the Bank is ≈50 m below the surface.

The mooring array consists of a line of five sites (C0 to C4) from Cape Sable across Browns Bank, and two more sites (C5, C6) on Scotian Shelf (Fig. 1). The depths of the stations are: 30 m for C0, 60 m for C1 and C5, and 110 m for the rest of the stations. The current measurements are available from C0 between October 1981 and April 1983, and from C1 and C2 between April 1979 and April 1983, and from C3 to C6 between April 1979 and August 1980.

A three-dimensional nonlinear numerical tidal model is described in section 2 and an analysis of the current meter data in section 3. In section 4 we discuss the computed residual current in the Cape Sable area and the comparison of the currents with the observations at stations C0-C6.

2 NUMERICAL MODEL

A simple method of calculating the three-dimensional tidal and residual currents has been described in detail by Tee (1979, 1980, 1986). Only those aspects that are related directly to the following solution and discussion are presented here.

2.1 The governing equations

For a Cartesian system with coordinates x, y, and z, where x and y are the horizontal coordinates and z is the vertical coordinate measured vertically upward from the mean sea level, the equations describing the tidal motion of a homogeneous fluid in shallow water are

$$\frac{\partial \underset{\sim}{u}}{\partial t} + \underset{\sim}{u} \cdot \nabla \underset{\sim}{u} + w \frac{\partial \underset{\sim}{u}}{\partial z} + \underset{\sim}{f} \times \underset{\sim}{u} = -g\nabla\zeta = \frac{\partial}{\partial z} N \frac{\partial \underset{\sim}{u}}{\partial z} + A_h \nabla^2 \underset{\sim}{u} \tag{1a}$$

$$\frac{\partial \zeta}{\partial t} + \nabla \cdot \int_{-D}^{\zeta} \underset{\sim}{u} dz = 0 \tag{1b}$$

where $\underset{\sim}{u}$ is the horizontal velocity vector with components u and v in coordinates x and y; w is the vertical velocity component; D is the depth of the bottom below mean sea level; ζ is the height of the water surface above mean sea level; f is the Coriolis parameter; N = N(x,y,z) is a vertical eddy viscosity coefficient; A_h is a horizontal eddy viscosity coefficient; g is gravity; ∇ is the horizontal gradient operator; and ∇^2 is the horizontal Laplacian operator.

By assuming that the advective term is small compared to the other terms in the momentum equation, a case we call the weakly nonlinear case, we obtain (Tee, 1979) the first order equations for the oscillating current ($\underset{\sim}{u}_1$)

$$\frac{\partial \underset{\sim}{u}_1}{\partial t} + \underset{\sim}{f} \times \underset{\sim}{u}_1 = -g\nabla\zeta_1 + \frac{\partial}{\partial z} N \frac{\partial \underset{\sim}{u}_1}{\partial z} + A_h \nabla^2 \underset{\sim}{u}_1 \tag{2a}$$

$$\frac{\partial \zeta_1}{\partial t} + \nabla \cdot D\underset{\sim}{U}_1 = 0 \tag{2b}$$

and the second-order equations for the residual current ($\underset{\sim}{u}_2$)

$$\overline{\underset{\sim}{u}_1 \cdot \nabla \underset{\sim}{u}_1} + \overline{w_1 \frac{\partial \underset{\sim}{u}_1}{\partial z}} + \underset{\sim}{f} \times \underset{\sim}{u}_2 = -g\nabla\zeta_2 + \frac{\partial}{\partial z} N \frac{\partial \underset{\sim}{u}_2}{\partial z} + A_h \nabla^2 \underset{\sim}{u}_2 \tag{3a}$$

$$\nabla \cdot (D\underset{\sim}{U}_2 + \overline{\zeta_1 \underset{\sim}{u}_{1s}}) = 0 \tag{3b}$$

where the overbar denotes the time average over a tidal period, $\underset{\sim}{u}_{1s}$ is the value of $\underset{\sim}{u}_1$ at surface (z=0), and $\underset{\sim}{U}_1$ and $\underset{\sim}{U}_2$ are the depth averaged values of $\underset{\sim}{u}_1$ and $\underset{\sim}{u}_2$.

2.2 The boundary conditions

The boundary conditions for a semi-enclosed basin are

(a) $\frac{\partial \underset{\sim}{u}_1}{\partial z} = 0$ and $\frac{\partial \underset{\sim}{u}_2}{\partial z} = \frac{\underset{\sim}{\tau}_{2s}}{\rho N}$ at z = 0, where (Tee, 1980)

$$\underset{\sim}{\tau}_{2s} = -\rho \frac{\partial}{\partial z} N \frac{\partial}{\partial z} \overline{\zeta_1 \underset{\sim}{u}_1} \Bigg|_{z=0} , \tag{4}$$

and

(b) $\underset{\sim}{u}_1 = \underset{\sim}{u}_2 = 0$ at z = -D and at the coastal boundary.

The open boundary conditions require the specification of the tidal elevation (ζ_1) and the depth-averaged residual current ($\underset{\sim}{U}_2$). The tidal elevation is taken from Greenberg's (1983) large-scale model. Two boundary conditions on $\underset{\sim}{U}_2$ are tested: (1) $\underset{\sim}{U}_2 = 0$ and (2) $\underset{\sim}{U}_2$ specified from Greenberg's model. The result shows that the residual currents near Cape Sable, including the waters off the tip of the Cape and around Browns Bank, are generally the same for the two boundary conditions. The sensitivity of the residual circulation to the open boundary conditions is further examined by modelling the same area with a depth-averaged non-linear numerical tidal model. The linear version of this model is used to produce the depth-averaged tidal current for the 3-D tidal model. By assuming zero advection at the open boundary, we again find that the depth-averaged residual currents off the tip of Cape Sable, and around Browns Bank for the non-linear depth-averaged tidal model are consistent with those for the 3-D tidal model. The reason for the insensitivity of the residual current to the open boundary condition is that strong residual current near the Cape Sable area is mostly developed locally in the shallow water region, and not affected significantly by the generally weak current at the open boundary.

2.3 The vertical eddy viscosity

The vertical eddy viscosity (N) is specified as:

$$N = \begin{cases} \nu_0 \, (1 + R_2 \, (D + z))^2 & \text{for } z + D \leq \delta_z \\ N_m = \nu_0 \, (1 + R_2 \delta_z)^2 & \text{for } z + D \geq \delta_z \end{cases} \tag{5}$$

where $\nu_0 = 1.4 \times 10^{-6}$ m^2s^{-1} is the molecular eddy viscosity, δ_z is the thickness of the transition layer between laminar and turbulent flow, and R_2 is a parameter. This form of N has been applied by Johns and Dyke (1971, 1972), and Tee (1979, 1980, 1982, 1985). There are two adjustable parameters in (5): (1) the vertical eddy viscosity above the bottom transition layer, N_m; and (2) the thickness of the transition layer δ_z. The value of N_m can be estimated from the relationship between vertical eddy viscosity (N) and linear bottom friction coefficient (λ) (Tee, 1980),

$$\lambda = -\left\{\frac{1}{D} \int_{-D}^{o} \int_{-D}^{z'} \frac{z}{N} \, dz dz'\right\}^{-1} \tag{6}$$

The bottom friction in the depth-dependent model will approximate that in the depth-independent model if (6) is satisfied.

The value of λ can be estimated from

$$\lambda(x,y) = K_I \frac{\gamma U_m}{D} \tag{7}$$

where K_I is a constant taken to be one, γ is the bottom friction coefficient taken to be 0.0021, and U_m is the current amplitude. For given values of λ, (6) is used to compute N_m. The value of δ_z is chosen by comparing the vertical variation of the observed and computed tidal current at station C1 (Fig. 1). By using the observed tidal current at mid-depth (30 m), and estimating N_m from (6), the vertical variation of the tidal current can be computed for various values of δ_z. Fig. 2 shows the comparison for the winter period. The observed tidal current is obtained by a 29-day tidal analysis centered at day 56, 1980 (February 25). The water column at this period is homogeneous. The best fit is obtained with δ_z = 0.15 m, and is shown in Fig. 2. The comparison is also performed for the fall period when the water column is stratified: σ_t increased from 23.83 kg m^{-3} near the surface to 25.13 near the bottom. For the same δ_z as that used in the winter period (= 0.15 m), the best fit between the computed and observed tidal current is found by reducing N_m to 20% of its value in winter period (0.0077 m^2s^{-1}). Figure 3 shows the comparison for the period centered at day 280, 1980 (October 7). Although the observed semi-major axis and the orientation

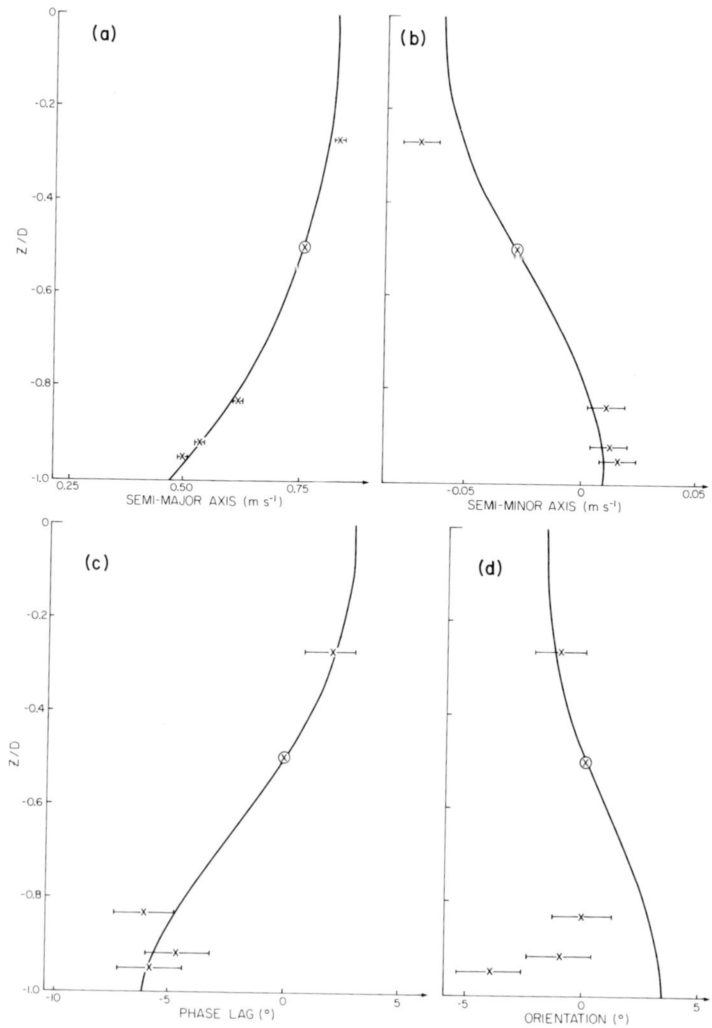

Fig. 2. Vertical variation of the tidal current in homogeneous water - comparison between theoretical and experimental results at station C1: (a) semi-major axis; (b) semi-minor axis (negative value indicates clockwise rotation); (c) phase lag; and (d) orientation of the semi-major axis. The best fit between the computed and observed results is for $\delta_z = 0.15$ m and N_m computed from (6). The observation was taken between February and March, 1980.

of this axis are well simulated by the model, the comparison of the semi-minor axis and the phase lag are less satisfactory. More study is required to improve the comparison. In the following, only the parameters for the homogeneous case are applied.

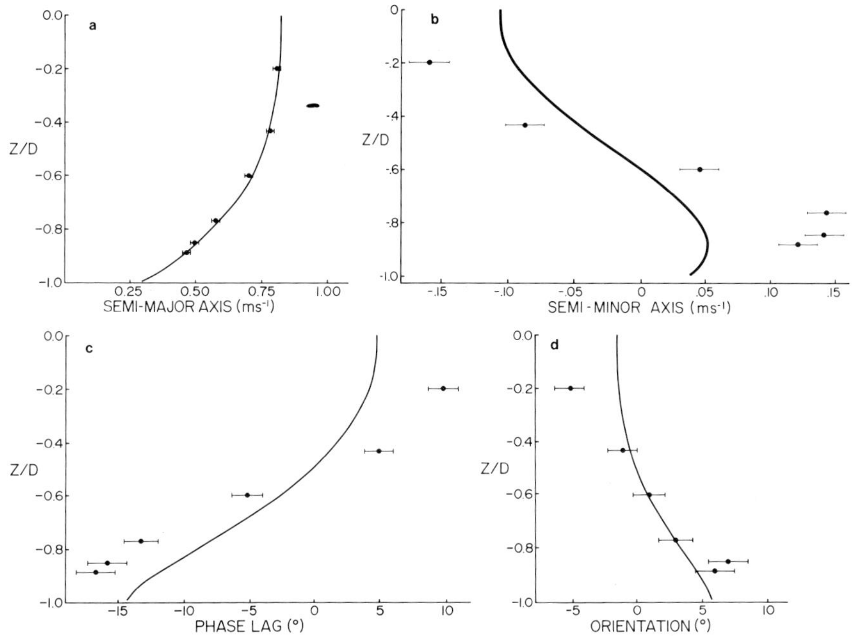

Fig. 3. Vertical variation of the tidal current in stratified water - comparison between theoretical and experimental results at station C1: (a) semi-major axis; (b) semi-minor axis (negative value indicates clockwise rotation); (c) phase lag; and (d) orientation of the semi-major axis. The best fit between the computed and observed results is for $\delta_z = 0.15$ m and $N_m = 0.2 \times$ (value of N_m for homogeneous water). The observation was taken between September and October, 1980.

2.4 Computational procedure

The two-dimensional depth-averaged nonlinear tidal model described by Tee (1976) is first used to compute the depth-averaged tidal current. The linear friction coefficient, λ, is then estimated from (7), where U_m is the semi-major axis of the computed depth-averaged tidal current. The value of N_m can be computed from (6) using the estimated λ. The depth-averaged tidal current is recomputed using the linearized model, which neglects the nonlinear terms and sets the bottom stress of the tidal current ($\underset{\sim}{\tau}_{b1}$) to be $\rho\lambda \underset{\sim}{U}_1$, where ρ is density and $\underset{\sim}{U}_1$ is the depth-averaged tidal current. Knowing the values of $\underset{\sim}{U}_1$ and N, and using the boundary condition (4), the vertical variation of the tidal current can be computed analytically from (2) (Tee, 1979). The accuracy of the computation can be improved by updating the bottom stress in the depth-averaged tidal model to be that computed from the vertical variation of the tidal current ($\underset{\sim}{\tau}_{b1} = \rho N\, \partial \underset{\sim}{u}_1/\partial z$ at $z = -D$). However, it was found that the improvement was only 1-2%, and was neglected here. The same result was found by Tee (1979, 1980, 1985).

Using the computed $\underset{\sim}{u}_1$, and setting the bottom stress of the residual current ($\underset{\sim}{\tau}_{b2}$) equal to $\rho\lambda\underset{\sim}{U}_2$, where $\underset{\sim}{U}_2$ is the depth-averaged residual current, we compute $\underset{\sim}{U}_2$ by transforming the depth-averaged equations for residual current (depth-average of (3a)) into the vorticity equation, which is then solved numerically using a relaxation method (Tee, 1986). From the known values of $\underset{\sim}{U}_2$, the vertical variation of $\underset{\sim}{u}_2$ can be solved numerically from (3a) (Tee, 1980). The accuracy of $\underset{\sim}{U}_2$ is then improved by updating the bottom stress ($\underset{\sim}{\tau}_{b2}$) in the depth-averaged momentum equation from $\underset{\sim}{\tau}_{b2} = \rho N \partial \underset{\sim}{u}_2/\partial z|_{z} = -D$. The computation is iterated until $\underset{\sim}{\tau}_{b2}$ is not significantly improved from the updated $\underset{\sim}{u}_2$. Here, it was found that five iterations were sufficient to obtain an accurate solution (i.e. to within .01%).

2.5 Spring-Neap computation

The spring-neap variation of the tidal and residual currents is simulated by changing the tidal forcing at the open boundary by ±25%, which corresponds approximately to the $M_2 \pm N_2$ or $M_2 \pm S_2$ forcing on the continental shelf (see Table 1 of Butman et al., 1983). It will be shown in section 3 that this spring-neap computation is important for comparing the computed and observed residual currents.

Because the inertial acceleration of $\underset{\sim}{u}_2$ ($\partial \underset{\sim}{u}_2/\partial t$) is not included in the computation, the above simulation of the spring-neap variation is valid only if $\lambda \gg \Delta\sigma$, where λ is the linear friction coefficient (7), and $\Delta\sigma = |\sigma(M_2) - \sigma(N_2,S_2)|$ is the difference between M_2 frequency and N_2 or S_2 frequency. In the Cape Sable area, the condition of $\lambda \gg \Delta\sigma$ is valid in most of the area except in the deep water region near the south-western and north-eastern open boundaries. The values of $\lambda/\Delta\sigma$ for all the experimental stations are shown in Table 1. For the difference between M_2 and N_2 frequencies, the values of $\lambda/\Delta\sigma$ are greater than three at most of the stations except C6 where $\lambda/\Delta\sigma$ = 1.5. The values of $\lambda/|\Delta\sigma|$ are also greater than three for $\Delta\sigma = \{\sigma(S_2) - \sigma(M_2)\}$ at most of the stations except C5 and C6 where $\lambda/\Delta\sigma$ = 1.6 for C5 and 0.8 for C6. Thus, neglecting the inertial acceleration in the computation of spring-neap variation of $\underset{\sim}{u}_2$ is reasonable at most of the stations except the weak dissipative areas of C5 and C6. Note that because strong residual currents near Cape Sable are generated locally in the shallow water region, the problem of $\lambda/\Delta\sigma$ not much larger than one in the deep water (weak residual current) area near the open boundaries is not expected to affect significantly the results of spring-neap variation in the shallow water region.

TABLE 1

Depths and ratios of the linear friction coefficient (λ) to the monthly ($\sigma(M_2-N_2)$) and fortnightly ($\sigma(S_2-M_2)$) frequencies for stations C0-C6.

Stations	$\lambda/\sigma(M_2-N_2)$	$\lambda/\sigma(M_2-S_2)$	Depth (m)
C0	38.3	20.3	30
C1	19.2	10.2	60
C2	6.8	3.6	110
C3	6.2	3.3	110
C4	6.2	3.3	110
C5	3.1	1.6	60
C6	1.5	0.8	110

3 ANALYSIS OF THE EXPERIMENTAL DATA

The experimental data, measured by the Aanderaa RCM-5 current meters, were sampled at 30-min intervals. The depths of the instruments, and the lengths of the record used for analyses, are given in Table 2. The time series of the current components were low-passed with a Cartwright filter (half-power at 31 h) and subsampled at 6 h intervals. The low-passed data

TABLE 2

Depths of the current meters, periods of data used in the spectral analyses, values of α computed from (12), and comparison between the mean tidally induced residual current ($\bar{u}_2$) estimated from current meter observations and that computed from the numerical model. Estimated residual currents larger than 1.5 times the confidence limits are underlined.

Station Name	Depth (m)	Period	α	Observed $\bar{u}_2$ Magnitude (m sec^{-1})	Observed $\bar{u}_2$ Angle (°)	Computed $\bar{u}_2$ Magnitude (m sec^{-1})	Computed $\bar{u}_2$ Angle (°)
C0	15	Oct.81 - Apr.83	2.2	<u>0.129 ±0.025</u>	257 ±10	0.099	257
	22	Mar.82 - Apr.83	2.1	<u>0.131 ±0.028</u>	247 ±11	0.093	257
	25	Mar.82 - Apr.83	2.1	<u>0.100 ±0.026</u>	247 ±13	0.091	257
C1	15	Oct.79 - Aug.80	3.0	0.030 ±0.030	250 ±90	0.045	278
	30	Oct.81 - Apr.83	2.7	0.024 ±0.021	246 ±40	0.042	281
	40	Aug.80 - Apr.83	2.4	<u>0.040 ±0.024</u>	239 ±31	0.037	283
	50	Oct.81 - Apr.83	2.1	<u>0.045 ±0.026</u>	226 ±30	0.035	285
	55	Aug.80 - Oct.81	2.0	0.027 ±0.022	228 ±40	0.031	286
	57	Aug.80 - Oct.82	2.0	<u>0.026 ±0.015</u>	217 ±28	0.030	286
C2	50	Nov.78 - Aug.80	2.1	0.021 ±0.021	229 ±86	0.012	343
	90	Aug.80 - Oct.81	2.0	0.049 ±0.033	138 ±49	0.015	344
	100	Oct.79 - Mar.82	1.9	0.027 ±0.018	44 ±33	0.014	345
C3	15	Aug.79 - Aug.80	1.6	<u>0.181 ±0.068</u>	77 ±22	0.042	123
C4	50	Apr.79 - Mar.80	1.7	0.071 ±0.053	240 ±45	0.028	74
C5	15	Apr.79 - Mar.80	1.0	0.152 ±0.123	128 ±38	0.012	270
	50	Apr.79 - Aug.80	1.1	0.036 ±0.030	93 ±47	0.010	249
C6	100	Apr.79 - Aug.80	2.9	0.014 ±0.013	149 ±56	0.005	279

($\underset{\sim}{u}$) can be written as the sum of mean ($\bar{\underset{\sim}{u}}$) and modulating ($\hat{\underset{\sim}{u}}$) components,

$$\underset{\sim}{u} = \bar{\underset{\sim}{u}} + \hat{\underset{\sim}{u}} \tag{8}$$

For the tidally induced residual current, the modulating component contains monthly $\{\approx 28$ day, $\Delta\sigma = \sigma(M_2) - \sigma(N_2)\}$ and fortnightly $\{\approx 14$ day, $\Delta\sigma = \sigma(S_2) - \sigma(M_2)\}$ periods. Evidence of the tidally rectified residual current can be obtained from the coherent modulation in these two frequency bands between the low-passed data and the amplitude of the tidal signal (Tee, 1977; Butman et al., 1983; Smith, 1983).

Besides the tidal forcing, the wind forcing can also induce the low frequency modulations in the current meter data. Using these two forcings, the modulation at each frequency can be described by an optimum model

$$\underset{\sim}{u} = \underset{\sim}{a}F + \underset{\sim}{b}E + \underset{\sim}{c}N + \text{noise} \tag{9}$$

where $\underset{\sim}{u}$ is the low-passed current meter data with components u and v in the x- and y-directions, F, E and N are the modulation of the tidal forcing, and the east and north components of the wind velocity, and $\underset{\sim}{a}$, $\underset{\sim}{b}$ and $\underset{\sim}{c}$ are complex vector coefficients with x and y components corresponding to u and v currents. The tidal forcing, F, is written as

$$F = U_a^{\alpha} \tag{10}$$

where U_a is the time-varying amplitude of the semi-major axis (Smith, 1983) and α is a spatially variable coefficient.

Using moored current meters in the Minas Basin at the head of the Bay of Fundy, where only the tidal forcing is important for low frequency oscillations in the residual current, Tee (1977) found that the value of α varies between 1.0 and 1.4. The value of 1.0 was used by Smith (1983) to estimate the tidally induced residual current at stations C1 (50 m depth) and C3 (44 m depth). In Butman et al.'s (1983) study of the tidally induced residual current on the southern side of Georges Bank, the tidal forcing is taken to be the square of the low-passed tidal elevation. In these previous studies, wind forcing was not included in the spectral analysis of the tidally induced residual current. Here, α is estimated from the spring-neap computation described in section 2.5. Using (9) and (10), and neglecting the wind forcing, which is not included in the numerical model, α can be estimated from

$$\alpha = \frac{\{\underset{\sim}{u_2}(\text{spring}) - \underset{\sim}{u_2}(\text{neap})\}/\underset{\sim}{u_2}(\text{mean})}{\{U_a(\text{spring}) - U_a(\text{neap})\}/U_a(\text{mean})} \tag{11}$$

Theoretically, α is generally different for the x- and y-components of the residual currents. However, it was found that, unless the direction of the observed residual current is close to one of the axes, α in both the x- and y-components can be estimated from the magnitude of the residual current ($|\underset{\sim}{u}_2|$),

$$\alpha = \frac{\{|\underset{\sim}{u}_2(\text{spring})|-|\underset{\sim}{u}_2(\text{neap})|\}/|\underset{\sim}{u}_2(\text{mean})|}{\{U_a(\text{spring}) - U_a(\text{neap})\}/U_a(\text{mean})} \qquad (12)$$

The values of α, estimated from (12), for all the experimental stations are shown in Table 2. The values vary from a minimum of 1.0 at 15 m for station C5 to a maximum of 3.0 at the same depth for station C1.

For given values of α, and the time series of $\underset{\sim}{u}$, F, E and N, the values of $\underset{\sim}{a}$, $\underset{\sim}{b}$ and $\underset{\sim}{c}$ can be estimated from (9) using the spectral method described by Garrett and Toulany (1982). The least squares determination of the x component of $\underset{\sim}{a}$, $\underset{\sim}{b}$ and $\underset{\sim}{c}$ for each frequency $\Delta\sigma$ is given by

$$\begin{aligned} a_x\Phi_{FF} + b_x\,\Phi_{EF} + c_x\,\Phi_{NF} &= \Phi_{UF} \\ a_x\,\Phi_{FE} + b_x\,\Phi_{EE} + c_x\,\Phi_{NE} &= \Phi_{UE} \qquad (13) \\ a_x\,\Phi_{FN} + b_x\,\Phi_{EN} + c_x\,\Phi_{NN} &= \Phi_{UN} \end{aligned}$$

where a_x, b_x and c_x are the complex x-component of $\underset{\sim}{a}$, $\underset{\sim}{b}$ and $\underset{\sim}{c}$, $\Phi_{ij}(\Delta\sigma)$ is the cross-spectrum between variables i and j at frequency $\Delta\sigma$. Similar formulae can be used to obtain a_y, b_y and c_y, the y-component of the coefficients $\underset{\sim}{a}$, $\underset{\sim}{b}$ and $\underset{\sim}{c}$.

The tidally induced residual current averaged over the period of observation, $\overline{\underset{\sim}{u}_2}$, is estimated from

$$\overline{\underset{\sim}{u}}_2 = \underset{\sim}{a}\,\overline{U}_a^{\alpha} \qquad (14)$$

where $\overline{U}_a$ is the mean value of U_a.

4 RESULTS AND DISCUSSIONS

4.1 Results of the numerical simulation

Figure 4 shows the depth-averaged residual current induced by the mean (M_2) tide. The basic features in the residual circulation near the Cape Sable area are: (1) the westward jet near the coast, which passes through stations C0 and C1; (2) the clockwise circulation around Browns Bank, which passes through station C3; and (3) the eastward flow east of Browns Bank, which passes through station C4. Stations C2, C5 and C6 are situated in the weak current areas. Near the southern corner of the open boundary, there are strong eastward flows on the northern side of Georges Bank and north-western inflow in the deep channel. At the entrance to the Bay of Fundy (north-western of Yarmouth, Fig. 4), there are irregular residual flows. These irregularities, attributed to a resolution problem, have previously been observed in two depth-averaged tidal models which have similar horizontal

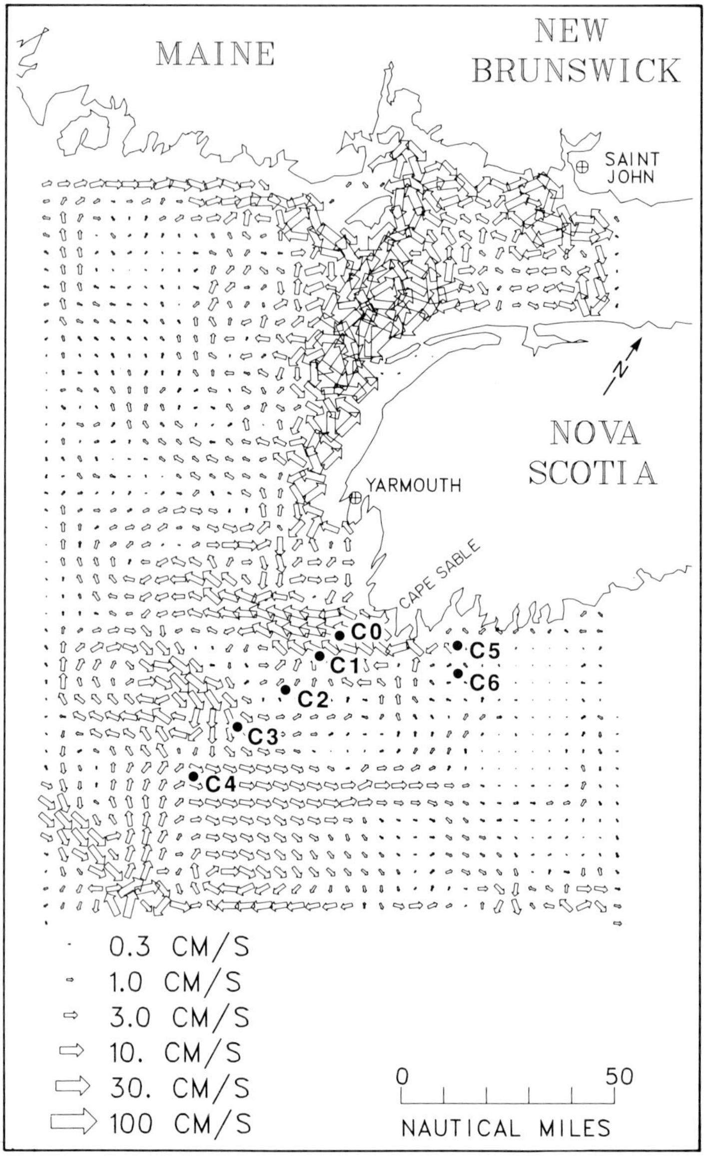

Fig. 4. The computed depth-averaged residual current induced by the mean (M_2) tide.

grid spacings to those applied in this study (≈7 km), but which includes the whole Gulf of Maine - Bay of Fundy system (Greenberg, 1983; Isaji and

Spaulding, 1984). Here, we study only the residual current near Cape Sable and around Browns Bank.

Examples of the vertical structure of the residual current are shown in Fig. 5 for the results at all the experimental stations. The magnitudes of the residual current decrease with depth for all stations except C2 and C6. The directions of the current, measured clockwise from true North, increase with depth at stations C0, C1 and C2, and decrease at C3, C4 and C5. At C6, the direction first increases from the surface to a maximum at 30 m, and then

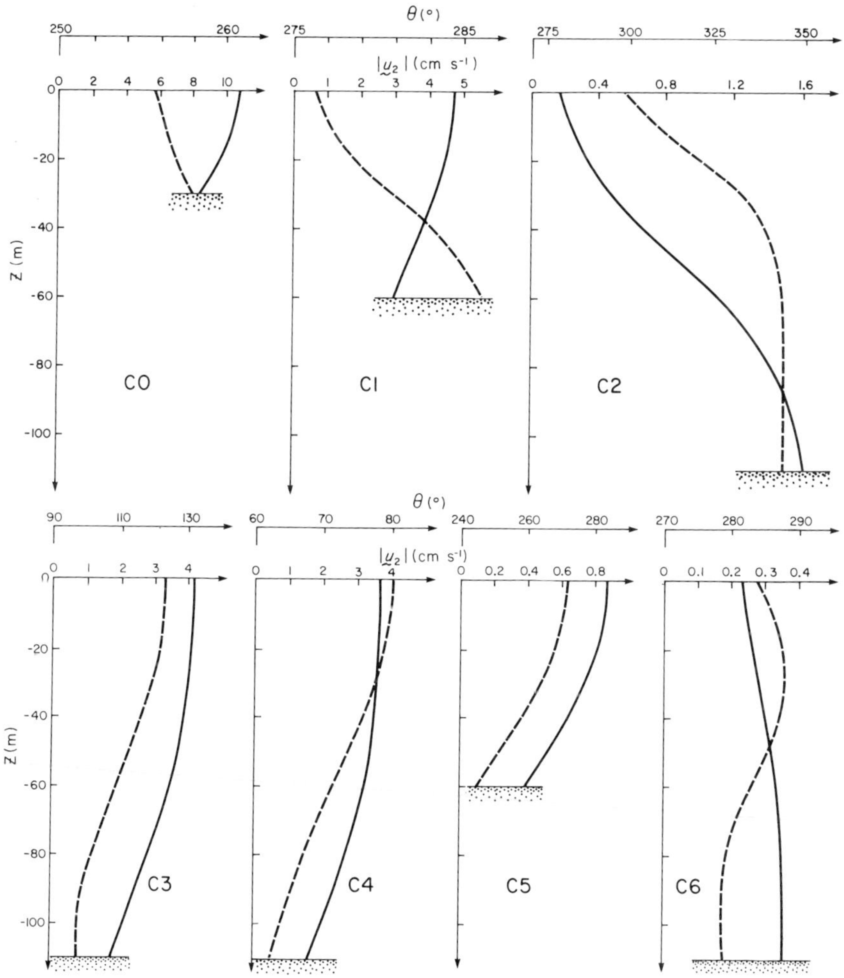

Fig. 5. The vertical structure of the magnitudes ($|\underset{\sim}{u}_2|$, solid line) and directions (θ, dashed line) of the tidally induced residual current at stations C0-C6, computed from the numerical model.

decreases toward the bottom. The vertical variations of the direction are generally between 10 and 26°, except at C0 and C2 where the direction changes by 2.3° and 45°, respectively.

4.2 Comparison between computed and observed residual currents

The predicted mean tidally induced residual current for the current meter stations are shown in columns 7 and 8 of Table 2. The detailed vertical structure of the current at these stations has been shown in Fig. 5. The estimated $\bar{\underset{\sim}{u}}_2$ (from 9 and 14) is shown in columns 5 and 6 of the same Table. The value of $\bar{\underset{\sim}{u}}_2$, estimated by using the monthly frequency, is more or less the same as that using the fortnightly frequency. The band-averaged result is shown in Table 2. The 95% confidence limits for the currents are also shown in the Table. Estimated residual currents that are larger than 1.5 times the confidence limits are underlined. The values of α used in the estimation are shown in column 4.

The estimated $\bar{\underset{\sim}{u}}_2$ is most reliable at station C0 where the estimated $\bar{\underset{\sim}{u}}_2$ is much larger than the confidence limit. There is also some reliability in estimating $\bar{\underset{\sim}{u}}_2$ at 40 m, 50 m and 57 m for station C1, and at 15 m for station C3, where the estimated $\bar{\underset{\sim}{u}}_2$ is larger than 1.5 times the confidence limits. For the rest of the current meters, the values of $\bar{\underset{\sim}{u}}_2$ are comparable to the confidence limit, and the reliability of the estimation is poor.

From Table 2, we can see that the observed $\bar{\underset{\sim}{u}}_2$ at station C0 is reproduced very well by the numerical model. At station C1, the magnitudes of $\bar{\underset{\sim}{u}}_2$ is also reproduced well, but the directions of the current are off by about 40 to 50 degrees. However, the uncertainties in the direction of the estimated $\bar{\underset{\sim}{u}}_2$ at this station are large, and the directions of the computed $\bar{\underset{\sim}{u}}_2$ change rapidly at nearby model grid points (Fig. 4). Thus, the disagreement in the computed and observed directions may not be significant. At station C3, the predicted $\bar{\underset{\sim}{u}}_2$ is much lower than the observed value. One possible explanation of this disagreement is that, because the clockwise circulation around the bank is largely generated by topographic rectification of the tidal current {similar to the generating mechanism on Georges Bank and Norfolk Sandbank (Huthnance, 1973; Loder, 1980; Tee, 1985)}, the horizontal grid spacing of about 7 km used in the model may be too coarse to resolve the detailed topography on the sides of the Bank.

The computed $\bar{\underset{\sim}{u}}_2$ at stations C2 and C4 are about half of the observed values, and those at C5 and C6 are much smaller than the observed values. However, because the 95% confidence limits for the observed residual current at these stations are very large ($|\bar{\underset{\sim}{u}}_2|$ less that 1.5 times the confidence limits), the disagreement may not be conclusive.

5 CONCLUSION

The aim of this study is to compute the three-dimensional tidally induced residual current in the Cape Sable area, and to compare the model current with current meter observations at stations C0 to C6 (Fig. 1). By using Tee's (1979, 1980, 1986) three-dimensional tidal model, the depth-averaged residual current and the vertical structure of the current at stations C0 to C6 are shown in Figs. 2 and 3.

Using a multiple regression technique, described by Garrett and Toulany (1982), and including both the tidal and wind forcings, the tidally induced residual currents are estimated for all the current meter records (Table 2).

The estimated $\overline{\underset{\sim}{u}}_2$ at station C0 is reproduced very well by the model. At station C1, the magnitude of $\overline{\underset{\sim}{u}}_2$ is also reproduced well. Although the directions of the current at this station are off by about 40 to 50 degrees, this may not be significant because of the large uncertainty in the estimation and the rapid variation of direction in the computed currents. At C3, on the northern side of Browns Bank, the estimated $\overline{\underset{\sim}{u}}_2$ are much lower than the observed values, which may be because the horizontal grid spacing used in the model ($\simeq$7 km) is not sufficient to resolve the topography in the area. For the rest of the stations (C2, C4-C6), the predicted $\overline{\underset{\sim}{u}}_2$ is smaller than the estimated values. However, the result may not be conclusive because of large uncertainties in the estimated $\overline{\underset{\sim}{u}}_2$.

6 REFERENCES

Butman, B., Noble, M., Chapman, D.C. and Beardsley, R.C., 1983. An upper bound for the tidally rectified current at one location on the southern flank of Georges Bank. J. Phys. Oceanogr., 13: 1452-1460.

Garrett, C.J.R. and Loucks, R.M., 1976. Upwelling along the Yarmouth Shore of Nova Scotia. J. Fish. Res. Board Can., 33: 116-117.

Garrett, C. and Toulany, B., 1982. Sea level variability due to meteorological forcing in the northeast Gulf of St. Lawrence. J. Geophys. Res., 87: 1968-1978.

Greenberg, D.A., 1983. Modelling the mean barotropic circulation in the Bay of Fundy and Gulf of Maine. J. Phys. Oceanogr., 13: 886-904.

Huthnance, J.M., 1973. Tidal current asymmetries over Norfolk sandbanks. Estuarine Coast. Mar. Sci., 1: 89-99.

Isaji, T. and Spaulding, M.L., 1984. A model of the tidally induced residual circulation in the Gulf of Maine and Georges Bank, J. Phys. Oceanogr., 14: 1119-1126.

Johns, B., and Dyke, P., 1971. On the determination of the structure of an offshore tidal stream. Geophys. J.R. Astr. Soc. 23: 287-297.

Johns, B. and Dyke, P., 1972. The structure of the residual flow in an offshore tidal stream. J. Phys. Oceanogr., 2: 73-79.

Lauzier, L.M., 1967. Bottom residual drift on the continental shelf area of the Canadian Atlantic Coast, J. Fish. Res. Board Can., 24: 1845-1858.

Leendertse, J.J. and Liu, S.K., 1975. Modelling of three-dimensional flows in estuaries, P-5461, The Rand Corporation Publication, 18 pp.

Loder, J.W., 1980. Topographic rectification of tidal currents on the sides of Georges Bank. J. Phys. Oceanogr., 10: 1399-1416.

Smith, P.C., 1983. The mean and seasonal circulation off southwest Nova Scotia, J. Phys. Oceanogr., 13: 1034-1054.

Tee, K.T., 1976. Tide-induced residual current, a 2-D non-linear numerical tidal model. J. Mar. Res., 34: 603-628.

Tee, K.T., 1977. Tide-induced residual current-verification of a numerical model. J. Phys. Oceanogr., 1: 396-402.

Tee, K.T., 1979. The structure of three-dimensional tide-generating currents. Part I: Oscillating currents. J. Phys. Oceanogr., 9: 930-944.

Tee, K.T., 1980. The structure of three-dimensional tide-induced current. Part II: Residual current. J. Phys. Oceanogr., 10: 2035-2057.

Tee, K.T., 1982. The structure of three-dimensional tide-generating currents: Experimental verification of a theoretical model. Estuarine, Coastal and Shelf Sci., 14: 27-48.

Tee, K.T., 1985. Depth-dependent studies of tidally induced residual currents on the sides of Georges Bank. J. Phys. Oceanogr., 15: 1818-1846.

Tee, K.T., 1986. Simple models to simulate three-dimensional tidal and residual currents. In: Three-dimensional shelf models, Meteor. Monogr., No. 5, Amer. Geophys. Union (in press).

A THREE-DIMENSIONAL WEAKLY NONLINEAR MODEL OF TIDE-INDUCED LAGRANGIAN RESIDUAL CURRENT AND MASS-TRANSPORT, WITH AN APPLICATION TO THE BOHAI SEA

SHIZUO FENG
Institute of Physical Oceanography, Shandong College of Oceanology, 5 Yushan Rd., Qingdao, The People's Republic of China.

ABSTRACT

A three-dimensional theory of the tide-induced Lagrangian residual currents and of the inter-tidal transport processes is proposed, based on a weakly nonlinear dynamical model of tides and intra-tidal transport processes. The first order Lagrangian residual velocity is the mass-transport velocity, that is the sum of the Eulerian residual velocity and the Stokes' drift velocity. The second order dynamics shows that the next contribution is a "Lagrangian drift velocity" which is a tidally periodic function of the tidal phase at which the marked water parcel is released. A set of field equations governing the mass-transport velocity is derived, which shows that the mean Lagrangian residual circulation is more relevant than the Eulerian residual circulation to the description of inter-tidal flow processes. A new inter-tidal transport equation for the tidal cycle average of the concentration of any conservative and passive tracer is derived. That equation correctly describes the Lagrangian nature of convective transport. The importance of the Lagrangian residual drifts for the dispersion or spreading of tracers in long-term transport processes is emphasized. Finally, a comparison between the present theory and observations in the Bohai Sea, China, shows good qualitative agreement.

1. INTRODUCTION

In recent years, research on the hydrodynamics of coastal seas and tidal estuaries has focused on the inter-tidal processes which determine the long-term transport and distribution of important water properties such as salinity and temperature, and the concentration of pollutants or any tracer. While the dominant observable motion in coastal seas, such as the Bohai Sea and the East China Sea, is the rotary circulation associated with tides, the inter-tidal transport processes are mainly dependent on the residual circulation, but not on the tidal currents themselves, at least not in a direct way. It is becoming increasingly clear that the residual circulation should be described and studied by means of the Lagrangian residual velocity, since it is more relevant to use a Lagrangian mean velocity rather than a Eulerian mean velocity to determine the origin of water masses. The Lagrangian mean velocity of a marked water parcel leads to the concept of Lagrangian residual velocity, or Lagrangian residual current. It should be pointed out that the study of Lagrangian residual currents is a relatively recent topic, and thus any further investigation of Lagrangian residual currents may be a valuable contribution both from a theoretical and from a practical point of view.

As a first approximation of the Lagrangian residual velocity, the mass-transport velocity has been shown to be the sum of the Eulerian residual velocity and of the Stokes' drift velocity (Longuet-Higgins, 1969). As is well known, the Eulerian residual velocity has been defined as the Eulerian mean velocity, *i.e.*, as the velocity averaged at a fixed spatial point over one or several tidal cycles. The Lagrangian residual velocity, however, is related to the Lagrangian mean velocity of a marked water parcel, *i.e.*, to the velocity averaged by following the marked water parcel over the tidal cycles. It is almost self-evident that, in contrast to the Eulerian residual velocity, the Lagrangian residual velocity is not only a function of the location where the parcel is released but also of the tidal phase at which that parcel is released. Indeed, the Lagrangian mean velocity depends upon the trajectory that the marked water parcel follows in the tidal field (Zimmerman, 1979; Cheng and Casulli, 1982; Cheng, 1983). By considering second order dynamics, this dependence of the Lagrangian residual velocity on the tidal phase has been revealed in a depth-averaged two-dimensional model (Feng *et al.*, 1986a). It should be pointed out that a depth-averaged two-dimensional analysis of the Lagrangian residual velocity is, in general, questionable from the viewpoint of hydrodynamics because the interpretation of averaging a Lagrangian velocity over depth is not very clear unless the vertical water column is assumed to move in the tidal field as a rigid body (Stern, 1975; Alfrink and Vreugdenhil, 1981; Feng, 1986a). The dynamics of the Lagrangian residual current should be rigorously treated in a three-dimensional space. Therefore, a study of a three-dimensional theory of Lagrangian residual circulation in coastal seas is of much importance not only for practical applications, but also to satisfy the requirements of Lagrangian dynamics. In this paper, a weakly nonlinear three-dimensional model of Lagrangian residual current is introduced. The residual circulation in continental shelf seas, semi-enclosed shallow seas, gulfs or bays and tidal estuaries is, generally speaking, driven by tides, winds acting on the sea surface, variations of water density, and open boundary forces. While much effort has been applied to the study of residual circulation driven by winds and density variations, only recently has attention been drawn to the tide-induced residual circulation (Nihoul and Ronday, 1975; Heaps, 1978; Tee, 1976). In this paper, to avoid confusing the main issues, the effects due to surface wind and baroclinic variations are not included; only the dynamics of tide-induced Lagrangian residual circulation is discussed. The effects of wind and density variations on the Lagrangian residual circulation have been examined in previous papers (Feng, 1986b; Feng and Cheng, 1986).

Here, the concept and the definition of Lagrangian residual velocity are reexamined. As stated above, the Eulerian residual velocity has been defined as the Eulerian mean velocity. Nevertheless, it would not seem proper to maintain that the Lagrangian residual velocity is synonymous with the Lagrangian mean velocity of a marked water parcel. Undoubtedly, the Lagrangian mean velocity of a marked water parcel in the tidal field is a Lagrangian velocity as far as inter-tidal processes are concerned, since it is proportional to the net displacement of the marked water parcel over the averaging period (several tidal cycles). On the other hand, it is natural to expect that the Lagrangian residual velocity could be used as a Eulerian field variable, and the aggregate of such local velocities may be specified as a Eulerian field of flow for the inter-tidal processes. Thus it can be seen that the Lagrangian residual velocity cannot be unconditionally defined as the Lagrangian mean velocity of a marked water parcel unless the latter satisfies the continuity equation for an incompressible fluid. These conditions are discussed in sections 3 and 6 of this paper.

Field equations governing the Eulerian residual circulation have been presented by Nihoul and Ronday (1975), who show that the introduction of nonlinear coupling between tides into the equations leads to the generation of a "tidal stress". Furthermore, a three-dimensional baroclinic model for the Eulerian residual circulation has also been proposed (Feng *et al.*, 1984), in which there exists not only a "tidal stress" but also a "tidal surface source" due to the nonlinear interaction between tides. However, it is more relevant to the present problem to derive a set of field equations governing the mean circulation of the Lagrangian residual current. In this paper, a set of field equations for the tide-induced mean circulation is introduced. These equations govern the mean Lagrangian residual circulation and are the most appropriate for the description of inter-tidal processes.

A long-term transport equation, namely, a convection-diffusion equation for the inter-tidal processes describes the balance between convective and dispersive transports for the tidally averaged concentration of any conservative and passive tracer. As is well known, in the conventional long-term transport equation, the convection velocity is the Eulerian residual velocity and an assumption of "tidal dispersion" has to be made (Fischer *et al.*, 1979). As stated above, however, for the inter-tidal transport processes the convective transport should be characterized by the Lagrangian residual velocity. Furthermore, the physics and dynamics of the "tidal dispersion" are not well understood (Pritchard, 1954; Bowden, 1967; Fischer *et al.*, 1979). A new inter-tidal transport equation has already been proposed, but it is a two-dimensional depth-averaged equation (Feng *et al.*, 1986b). In this paper, we derive a three-dimensional inter-tidal transport equation which describes correctly the Lagrangian nature of convective transport without introducing the "tidal dispersion" terms.

In the last two sections of this paper, the Lagrangian dynamics of tracer spreading on long-term scales and a preliminary application of the proposed model to the Bohai Sea, China, are briefly described.

2. FORMULATION OF THE MODEL

Based on a nonlinear three-dimensional tidal model (Feng, 1977), the generalized nondimensional dynamic problem proposed by Feng (1986a) is as follows :

i) Field equations

$$\nabla \cdot \mathbf{u} = 0 \,, \tag{1}$$

$$\frac{\partial u}{\partial \theta} + \kappa \, \mathbf{u} \cdot \nabla u - fv = - \frac{\partial Z}{\partial x} + \frac{\partial}{\partial z} \left(\nu \frac{\partial u}{\partial z}\right) , \tag{2}$$

$$\frac{\partial v}{\partial \theta} + \kappa \, \mathbf{u} \cdot \nabla v + fu = - \frac{\partial Z}{\partial y} + \frac{\partial}{\partial z} \left(\nu \frac{\partial v}{\partial z}\right) , \tag{3}$$

$$\frac{d}{d\theta}\,(N\,\boldsymbol{\xi}) = \mathbf{u}\ , \tag{4}$$

$$\frac{\partial S}{\partial \theta} + \kappa\,\mathbf{u}\,.\,\nabla S = \varepsilon\,\frac{\partial}{\partial z}\,(k\,\frac{\partial S}{\partial z})\ ; \tag{5}$$

ii) Boundary conditions :

$$\text{at } z = \kappa\,Z\ ,\quad w = \frac{\partial Z}{\partial \theta} + \kappa\,(u\,\frac{\partial Z}{\partial x} + v\,\frac{\partial Z}{\partial y})\ , \tag{6}$$

$$\text{and}\qquad \frac{\partial u}{\partial z} = \frac{\partial v}{\partial z} = \frac{\partial S}{\partial z} = 0\ ; \tag{7}$$

$$\text{at } z = -\,h\ ,\quad \mathbf{u} = 0\ , \tag{8}$$

$$\text{and}\qquad \frac{\partial S}{\partial z} = 0\ , \tag{9}$$

where

$$\mathbf{x} = (x,y,z) = \left(\frac{x_*}{L}\ ,\ \frac{y_*}{L}\ ,\ \frac{z_*}{h_c}\right),$$

$$\theta = t_*\ \omega\ ,$$

$$h = \frac{h_*}{h_c}\ ,$$

$$S = \frac{S_*}{S_c}\ ,$$

$$Z = \frac{Z_*}{Z_c}\ ,$$

$$\mathbf{u} = (u,v,w) = \left(\frac{u_*}{u_c}\ ,\ \frac{v_*}{u_c}\ ,\ \frac{w_*}{w_c}\right),$$

$$\boldsymbol{\xi} = (\xi,\eta,\zeta) = \left(\frac{\xi_*}{\xi_c}\ ,\ \frac{\eta_*}{\xi_c}\ ,\ \frac{\zeta_*}{\zeta_c}\right),$$

$$\nabla = \mathbf{e}_1\,\frac{\partial}{\partial x} + \mathbf{e}_2\,\frac{\partial}{\partial y} + \mathbf{e}_3\,\frac{\partial}{\partial z}\ ,$$

$$u_c = \kappa\,\sqrt{g\ h_c}\ ,$$

$$w_c = \frac{h_c}{L}\ u_c\ ,$$

$$\xi_c = \kappa\,N\,L\ ,$$

$$\zeta_c = \kappa\,N\,h_c\ ,$$

$$f = \frac{f_*}{\omega}\ ,$$

$$\nu = \frac{\nu_*}{\nu_c}\ ,$$

$$k = \frac{k_*}{k_c}\ ,$$

$$\nu_c = \omega\ h_c^2\ ,$$

$$\omega = \frac{\sqrt{g\ h_c}}{L}\ .$$

Three nondimensional parameters, κ, N, and ε have been introduced. They are defined by

$$\kappa = \frac{Z_c}{h_c} , \tag{10}$$

$$N = 1 + (n-1)\ \kappa , \tag{11}$$

$$\varepsilon = \frac{k_c}{\omega\ h_c^2} . \tag{12}$$

In these equations, t_* denotes time; (x_*,y_*,z_*) form a Cartesian coordinate system on an f-plane with corresponding unit vectors $(\mathbf{e}_1,\mathbf{e}_2,\mathbf{e}_3)$; (u_*,v_*,w_*) are the velocity components in the $(\mathbf{e}_1,\mathbf{e}_2,\mathbf{e}_3)$ directions; Z_* is the displacement of the free surface from the undisturbed sea surface; h_* is the water depth measured from the undisturbed sea surface; g is the gravitational acceleration; f_* is the Coriolis parameter; ν_* is the eddy viscosity; k_* is the eddy diffusivity; S_* is the concentration of any conservative and passive tracer; (ξ_*,η_*,ζ_*) are the Lagrangian displacements in the $(\mathbf{e}_1,\mathbf{e}_2,\mathbf{e}_3)$ directions of a marked water parcel at time t_*, for which the initial condition is $(\xi_*,\eta_*,\zeta_*) = 0$ when $t_* = t_{0*}$, and thus the position of the marked water parcel can be expressed as $(x_*,y_*,z_*) = (x_{0*},y_{0*},z_{0*}) + (\xi_*,\eta_*,\zeta_*)$; L and h_c denote the horizontal and vertical scales, respectively; ω^{-1} is the time scale; the quantities with the subscript "c" indicate the characteristic value of the corresponding dimensional quantities; n is the number of tidal cycles over which the tidal current velocity is averaged to filter the periodic current velocity and to derive the residual velocity; κN is a measure of the relative scale of tidal excursion to L ; ε/κ is a measure of the relative effect on a concentration of the eddy diffusion compared to convection.

Let us assume (Feng, 1986a) :

$$O(\kappa N) = O(\kappa) \tag{13}$$

$$O(\kappa) < 1 \tag{14}$$

$$O(\varepsilon/\kappa) = O(\kappa). \tag{15}$$

These conditions imply weakly nonlinear dynamics : the nonlinearity due to the eddy viscosity and diffusion are excluded, and the transport mechanism is dominated by convection.

Given the scale of tidal excursion, $\xi_c = \kappa\ N\ L$, and conditions (13) and (14), the velocity of a marked water parcel can be expanded in a Taylor series expansion about $\mathbf{x}_0$, or

$$\mathbf{u}\,(\mathbf{x}_0 + \kappa\, N\, \boldsymbol{\xi}\,,\, \theta) = \mathbf{u}\,(\mathbf{x}_0,\theta) + \kappa\, N\, \boldsymbol{\xi}\, .\, (\nabla \mathbf{u})_0$$

$$+ \frac{\kappa^2 N^2}{2} \left\{ \xi^2 \left[\frac{\partial^2 \mathbf{u}}{\partial x^2}\right]_0 + \eta^2 \left[\frac{\partial^2 \mathbf{u}}{\partial y^2}\right]_0 + \zeta^2 \left[\frac{\partial^2 \mathbf{u}}{\partial z^2}\right]_0 \right.$$

$$+ 2\,\xi\,\eta \left[\frac{\partial^2 \mathbf{u}}{\partial x\, \partial y}\right]_0 + 2\,\xi\,\zeta \left[\frac{\partial^2 \mathbf{u}}{\partial x\, \partial z}\right]_0$$

$$\left. + 2\,\eta\,\zeta \left[\frac{\partial^2 \mathbf{u}}{\partial y\, \partial z}\right]_0 \right\} + O\,(\kappa^3) \qquad (16)$$

where the notation $(\ )_0$ indicates that the term is evaluated at $\mathbf{x}_0$.

The displacement of the marked water parcel can be expressed by

$$N\, \boldsymbol{\xi} = \int_{\theta_0}^{\theta} \mathbf{u}\,(\mathbf{x}_0 + \kappa\, N\, \boldsymbol{\xi}\,,\, \theta')\, d\theta' \qquad (17)$$

where $\theta_0 = \omega\, t_0$

A perturbation technique with the small parameter κ has been used to derive and solve the jth-order model (j = 0, 1, ...) of the dynamic problem proposed, (1)-(9) (Feng, 1977 and 1986a). In the present paper, we suppose that the solution for the tidal currents, and in particular the Eulerian residual currents and the Lagrangian displacements of marked water parcels, have been obtained. Thus, the basis to calculate the Lagrangian residual velocity and the concentration of any conservative and passive tracer has been laid.

For clarity, a nonlinear M_2 tidal system is used, instead of a complicated tidal system including several astronomical tides and associated shallow water constituents, to examine the Lagrangian residual current. It is natural to select the circular frequency of M_2 as the characteristic circular frequency, and thus the nondimensional circular frequency and period of M_2 are 1 and 2π respectively. The zero-th order model, or M_2 tide, can be expressed, in a nondimensional form, as $u_0 = u_0' \cos\theta + u_0'' \sin\theta$, where the superscripts $'$ and $''$ indicate the harmonic coefficients. As is well known, the first order model contains the M_4 tide and the (first order) Eulerian residual velocity. The second order model contains the M_6 tide and other harmonics of frequency equal to that of the M_2 tide. Here, the harmonics of the order of $O(\kappa^j)$ (j = 3, 4, ...) are not considered.

3. LAGRANGIAN RESIDUAL CURRENT INDUCED BY AN M_2-TIDAL SYSTEM

Let us use angular brackets < > to denote a tidal cycle average of any quantity A, or $\langle A \rangle = \frac{1}{2\pi n} \int_{\theta_0}^{\theta_0 + 2\pi n} A\, d\theta$, where n = 1,2,... . The Eulerian residual velocity, $\mathbf{u}_{ER}$, defined as the Eulerian mean velocity (*i.e.*, the velocity averaged at a fixed spatial point, say $\mathbf{x}_0$, over n tidal cycles) is expressed as $\mathbf{u}_{ER} = \langle\, \mathbf{u}\,(\mathbf{x}_0\,,\,\theta)\, \rangle$. The Lagrangian mean velocity, $\mathbf{u}_L$, of a marked water parcel released from the point $\mathbf{x}_0$ at time θ_0 and moving over the n tidal cycles is defined as $\mathbf{u}_L = \langle \mathbf{u}(\mathbf{x}(\mathbf{x}_0,\theta),\theta)\rangle$. However, as pointed out in section 1, the Lagrangian residual velocity $\mathbf{u}_{LR}$ cannot be unconditionally defined as the Lagrangian mean velocity $\mathbf{u}_L$ unless $\mathbf{u}_L$ satisfies the continuity equation for an incompressible fluid. In fact, in the proposed weakly nonlinear model, assumption (13) ensures that $\mathbf{u}_L$ satisfies the continuity equation for an incompressible fluid. We shall prove that point at the end of this section. Thus, here the Lagrangian residual velocity can be expressed as $\mathbf{u}_L$, or

$$\mathbf{u}_{LR} = \frac{1}{2\pi n} \int_{\theta_0}^{\theta_0 + 2\pi n} \mathbf{u}\,(\mathbf{x}(\mathbf{x}_0,\theta),\theta)\; d\theta \tag{18}$$

The substitution of (16) and (17) into (18) yields the Lagrangian residual velocity induced by an M_2-tidal system, to be correct to the second order harmonics and normalized by $u_{rc} = \kappa u_c$, as follows

$$\mathbf{u}_{LR} = \mathbf{u}_{LM} + \kappa\, \mathbf{u}_{LD} \tag{19}$$

where

$$\mathbf{u}_{LM} = \mathbf{u}_{ER} + \mathbf{u}_{SD} \tag{20}$$

and

$$\mathbf{u}_{LD} = \mathbf{u}'_{LD} \cos\theta_0 + \mathbf{u}''_{LD} \sin\theta_0 \tag{21}$$

$$\mathbf{u}'_{LD} = \mathbf{u}''_0 \cdot \nabla \mathbf{u}_{LM} - \mathbf{u}_{LM} \cdot \nabla \mathbf{u}''_0$$
$$\mathbf{u}''_{LD} = -\,\mathbf{u}'_0 \cdot \nabla \mathbf{u}_{LM} + \mathbf{u}_{LM} \cdot \nabla \mathbf{u}'_0$$

The first order Lagrangian residual velocity expressed as the sum of the Eulerian residual velocity $\mathbf{u}_{ER}$ and the Stokes' drift velocity $\mathbf{u}_{SD} = \langle N\boldsymbol{\xi}_0 \cdot (\nabla \mathbf{u}_0)_0 \rangle$ is the mass- transport velocity $\mathbf{u}_{LM}$, which was first introduced by Longuet-Higgins (1969). Equation (20) is called Stokes' formula. The Lagrangian drift velocity, $\mathbf{u}_{LD}$, was only recently revealed and named (Feng *et al.*, 1986a; Feng, 1986a). The second order dynamics shows the dependence of the Lagrangian residual velocity on the tidal phase θ_0 at which the marked water parcel is released from the fixed point $\mathbf{x}_0$.

In the previous depth-averaged model, the "two-dimensional" Lagrangian drift velocity traces out an ellipse on a hodograph plane as the initial tidal phase θ_0 varies from 0 to 2π; in other words, when the "marked water columns" are released from a fixed position continuously over a tidal period, the ensemble of the terminal positions of the "marked water columns" after a tidal cycle form an ellipse in the "two-dimensional space", the Lagrangian residual ellipse (Cheng *et al.*, 1986; Feng *et al.*, 1986a). The Lagrangian residual velocity derived in a three-dimensional space, equation (19), or the Lagrangian drift velocity, (21), has a similar but three-dimensional structure (Feng, 1986a). Here it should be pointed out that this unique property of the Lagrangian residual velocity reflects its Lagrangian nature since the Lagrangian residual velocity is born of the Lagrangian mean velocity of a marked water parcel in the tidal field but the latter depends on the trajectory that such a parcel follows and the parcels follow different trajectories depending upon the time, θ_0, of their release at $\mathbf{x}_0$.

Noting that $(\mathbf{x}_0,\theta_0)$ is to be selected arbitrarily, and then using $(\mathbf{x},\theta)$ instead of $(\mathbf{x}_0,\theta_0)$, the Lagrangian residual velocity can be viewed as a Eulerian field variable and the aggregate of such local velocities may be specified as a Eulerian field of flow provided that the Lagrangian residual velocity expressed by (19)-(21) satisfies the continuity equation for an incompressible fluid. As a matter of fact, by taking the divergence of (19)-(21) and going through some algebraic manipulations, we have

$$\nabla \cdot \mathbf{u}_{LR} = 0 . \tag{22}$$

Hence, the definition used here of the Lagrangian residual velocity as the Lagrangian mean velocity of a marked water parcel is valid. The case in which $\mathbf{u}_{LR}$ cannot be defined by (18) will be examined and discussed in section 6.

Thus, (19)-(21) show that the Lagrangian residual velocity field is similar to the tidal current velocity field in the sense that it is a sum of a tidally periodic fluctuation plus the tidal cycle average since $\mathbf{u}_{LM} = \langle\mathbf{u}_{LR}\rangle$. However, given that $O\left(|\kappa\, \mathbf{u}_{LD}| / |\mathbf{u}_{LM}|\right) = \kappa$, they are different because the tidally periodic part of the tidal current velocity field is typically greater than the residual part by one or more orders of magnitude. In contrast to the Eulerian residual velocity field which is steady, the Lagrangian residual velocity field is obviously a time-dependent field of flow.

4. A SET OF FIELD EQUATIONS FOR THE MEAN LAGRANGIAN RESIDUAL CIRCULATION INDUCED BY AN M_2-TIDAL SYSTEM

As stated in section 1, while much effort has been applied to the study of residual circulation driven by the wind on the sea surface and the variation of water density, only recently has attention been drawn to the tide-induced residual circulation. However, in coastal seas, where the dominant observable motions are tides, the residual circulation is induced not only by the wind on the sea surface and the horizontal gradient of water density but also by the nonlinear coupling of tides, as pointed out by Nihoul and Ronday (1975). A scale analysis on the general circulation in the Bohai Sea and the East China Sea has also shown that the tide-induced residual circulation is, in general, a component of the general residual circulation (Feng *et al.*, 1986). The residual circulation is conventionally derived from current-meter records using filter techniques or time averages of time series records to remove tidal variations, *i.e.*, the residual circulation is conventionally defined as the Eulerian residual circulation. However, it is becoming increasingly clear that the residual circulation should be related to the Lagrangian residual velocity since the problem of residual circulation is to describe and understand the inter-tidal transport processes (Csanady, 1982; Feng, 1986b). Thus, it might be appropriate to define the (tide-induced) residual circulation as the (tide-induced) mean Lagrangian residual circulation. To describe the tide-induced mean Lagrangian residual circulation and to study some of its characteristics, a set of field equations governing the tidal cycle average of the Lagrangian residual velocity, or the mass-transport velocity, $\mathbf{u}_{LM}$, is derived as follows :

i) Field equations

$$\nabla \cdot \mathbf{u}_{LM} = 0 , \tag{23}$$

$$- f\, v_{LM} = - \frac{\partial \langle Z_1 \rangle}{\partial x} + \frac{\partial}{\partial z} (\nu \frac{\partial u_{LM}}{\partial z}) + \pi_1 , \tag{24}$$

$$f\, u_{LM} = - \frac{\partial \langle Z_1 \rangle}{\partial y} + \frac{\partial}{\partial z} (\nu \frac{\partial v_{LM}}{\partial z}) + \pi_2 ; \tag{25}$$

ii) Boundary conditions :

$$\text{at } z = 0 , \quad w_{LM} = 0 , \tag{26}$$

$$\text{and} \quad \frac{\partial u_{LM}}{\partial z} = \frac{\partial v_{LM}}{\partial z} = 0 ; \tag{27}$$

$$\text{at } z = -h , \quad \mathbf{u}_{LM} = 0 ; \tag{28}$$

where

$$\begin{aligned}
\pi_1 = & - \frac{\partial}{\partial x} \langle \frac{1}{2} \boldsymbol{\xi}_0 \cdot \nabla Z_0 \rangle \\
& + \frac{\partial}{\partial z} \langle (\frac{5}{2} \frac{\partial \xi_0}{\partial x} + \frac{\partial \eta_0}{\partial y}) \nu \frac{\partial u_0}{\partial z} + (\frac{\partial \xi_0}{\partial y} + \frac{1}{2} \frac{\partial \eta_0}{\partial x}) \nu \frac{\partial v_0}{\partial z} \rangle \\
& + \frac{\partial}{\partial z} \langle \nabla \nu \cdot \boldsymbol{\xi}_0 \frac{\partial u_0}{\partial z} \rangle - \langle \frac{\partial \boldsymbol{\xi}_0}{\partial z} \cdot \nabla (\nu \frac{\partial u_0}{\partial z}) \rangle \\
& + \frac{1}{2} \langle \frac{\partial^2 \xi_0}{\partial z \partial x} (\nu \frac{\partial u_0}{\partial z}) + \frac{\partial^2 \eta_0}{\partial z \partial x} (\nu \frac{\partial v_0}{\partial z}) + \xi_0 \frac{\partial^2}{\partial z \partial x} (\nu \frac{\partial u_0}{\partial z}) + \eta_0 \frac{\partial^2}{\partial z \partial x} (\nu \frac{\partial v_0}{\partial z}) \rangle
\end{aligned}$$

$$\begin{aligned}
\pi_2 = & - \frac{\partial}{\partial y} \langle \frac{1}{2} \boldsymbol{\xi}_0 \cdot \nabla Z_0 \rangle \\
& + \frac{\partial}{\partial z} \langle (\frac{5}{2} \frac{\partial \eta_0}{\partial y} + \frac{\partial \xi_0}{\partial x}) \nu \frac{\partial v_0}{\partial z} + (\frac{\partial \eta_0}{\partial x} + \frac{1}{2} \frac{\partial \xi_0}{\partial y}) \nu \frac{\partial u_0}{\partial z} \rangle \\
& + \frac{\partial}{\partial z} \langle \nabla \nu \cdot \boldsymbol{\xi}_0 \frac{\partial v_0}{\partial z} \rangle - \langle \frac{\partial \boldsymbol{\xi}_0}{\partial z} \cdot \nabla (\nu \frac{\partial v_0}{\partial z}) \rangle \\
& + \frac{1}{2} \langle \frac{\partial^2 \xi_0}{\partial z \partial y} (\nu \frac{\partial u_0}{\partial z}) + \frac{\partial^2 \eta_0}{\partial z \partial y} (\nu \frac{\partial v_0}{\partial z}) + \xi_0 \frac{\partial^2}{\partial z \partial y} (\nu \frac{\partial u_0}{\partial z}) + \eta_0 \frac{\partial^2}{\partial z \partial y} (\nu \frac{\partial v_0}{\partial z}) \rangle
\end{aligned}$$

In these equations, (π_1,π_2) represent the nonlinear coupling of astronomical tides and can be naturally named "tide-induced body force". The tidal force contains two parts. The first term characterizes the nonlinear interaction between the tidal displacement and the tidal elevation, and it is horizontally irrotational. The second part represent the effect of eddy viscosity; this term is rotational. (u_{LM},v_{LM}) and w_{LM} are the horizontal and vertical components of $\mathbf{u}_{LM}$, respectively, and $\langle Z_1 \rangle$ is the residual elevation.

The conceptual difference between the set of field equations for the Lagrangian residual circulation derived here, (23)-(28), and that for the Eulerian residual circulation (Nihoul and

Ronday, 1975; Feng *et al.*, 1984) is revealed by the differences in the kinematic and kinetic boundary conditions at $z = 0$. Equation (26) shows that there is no "tidal surface source", or $\frac{\partial}{\partial x} \langle u_0 Z_0 \rangle + \frac{\partial}{\partial y} \langle v_0 Z_0 \rangle$, at $z = 0$ and equation (27) shows that there is no "tidal stress", or $- \langle Z_0 \frac{\partial^2}{\partial z^2}(u_0 , v_0) \rangle$, at $z = 0$. Thus, the handling of the continuity equation is simplified when the mean Lagrangian residual circulation is used to describe inter-tidal processes. There are other attractive features to this formulation.

If a material surface in the water is specified geometrically by the equation $F(\mathbf{x},\theta) = \text{const.}$, F is a quantity which is invariant for a water parcel on the surface, so that :

$$\frac{DF}{D\theta} = \frac{\partial F}{\partial \theta} + \kappa\, \mathbf{u} \,.\, \nabla F = 0 \tag{29}$$

at all points on the surface. In particular, the equation of any surface bounding the sea water must satisfy (29). $F(\mathbf{x},\theta)$ can also be written as $F_0 + \kappa F_1 + \kappa^2 F_2 + O(\kappa^3)$, like the other variables. Substituting this expansion into equation (29), and taking a tidal cycle average of the latter, we have

$$\mathbf{u}_{LM} \,.\, \nabla \langle F \rangle = 0 \tag{30}$$

where F_0 has been approximately expressed by $\langle F \rangle$. The kinematic boundary condition at the sea surface (26) can be derived as a special case of equation (30) (Feng, 1986b).

Equation (30) shows that in the individual derivative of the tidal cycle average of a material surface, $\frac{D}{D\theta} \langle F \rangle$, the convective velocity is the mean Lagrangian residual velocity rather than the Eulerian residual velocity. This further confirms that the mean Lagrangian residual circulation is appropriate to the description of inter-tidal flow processes. In fact, equation (30) is also valid when F is any conservative quantity in the flow field.

The set of field equations governing the mass-transport velocity, (23)-(28), can in principle be solved if the forcing function (π_1,π_2), *i.e.*, the tide-induced body force, is given and if the eddy viscosity $\nu = \nu(\mathbf{x})$ is assumed to be known. A numerical model to solve the set of equations (23)-(28) was proposed by Song (1986). If the sea water is assumed to be an inviscid fluid, the tide-induced body force (π_1,π_2) is reduced to an irrotational force and the set of equations (23)-(28) describes some geostrophic motions. This implies that the careful and correct selection of the eddy viscosity, $\nu(\mathbf{x})$, is of much importance for the numerical simulation of the mean Lagrangian residual circulation in coastal seas.

5. INTER-TIDAL TRANSPORT EQUATION

The convection-diffusion equation for the tidal cycle average of the concentration of any conservative and passive tracer is also called the long-term transport equation, or the inter-tidal transport equation (Feng et al., 1986b). It can be derived by means of equation (5). Substituting $S = S_0 + \kappa S_1 + \kappa^2 S_2 + O(\kappa^3)$ into equation (5), the zero-th order equation is $\partial S_0 / \partial \theta = 0$, which says that the tidal cycle average of the concentration, <S>, can be approximately evaluated by S_0. The first order equation is $\partial S_1 / \partial\theta + \mathbf{u}_0 \cdot \nabla S_0 = 0$, and the second order equation is as follows

$$\frac{\partial S_2}{\partial\theta} + \mathbf{u}_0 \cdot \nabla S_1 + \mathbf{u}_1 \cdot \nabla S_0 = E \frac{\partial}{\partial z} (k \frac{\partial S_0}{\partial z}) \tag{31}$$

where $E = \varepsilon / \kappa^2$, and, in view of (15), we have O(E) = 1.

Substituting the first order equation into equation (31) and using the zero-th order equation, a tidal cycle average of equation (31) yields the inter-tidal transport equation, or

$$\mathbf{u}_{LM} \cdot \nabla \langle S \rangle = E \frac{\partial}{\partial z} (\langle k \rangle \frac{\partial \langle S \rangle}{\partial z}) \tag{32}$$

to the zero-th order approximation, in which S_0 has been approximated by <S>.

The equation derived here, (32), is different from the traditional long-term transport equation (Fischer *et al.*, 1979). On the one hand, in the latter, the convection has been unreasonably represented by the Eulerian residual velocity, but in equation (32) the convective transport is reasonably expressed by the Eulerian mean of the Lagrangian residual velocity, *i.e.*, by the mass-transport velocity. On the other hand, an assumption on the so-called "tidal dispersion" has to be introduced into the conventional long-term transport equation (Fischer *et al.*, 1979), whereas equation (32) can describe correctly the Lagrangian nature of the convection affecting inter-tidal transport processes without introducing any hypothesis for tidal dispersion.

Let $\overline{A}$ denote the depth-averaged quantity of A, or $\overline{A} = \frac{1}{h}\int_{-h}^{0} A\, dz$. An inter-tidal transport equation satisfied by $\overline{\langle S \rangle}$ is derived as follows :

$$\overline{u_{LM}} \frac{\partial \overline{\langle S \rangle}}{\partial x} + \overline{v_{LM}} \frac{\partial \overline{\langle S \rangle}}{\partial y} = Pe^{-1} \frac{1}{h} \nabla \cdot (h \underline{\mathbf{D}} \cdot \nabla \overline{\langle S \rangle})$$

where P_e denotes the Peclet number for the residual motion, $Pe = \kappa u_c L / D_c$; $\underline{\mathbf{D}}$ denotes the dispersion coefficient tensor due to the vertical shear of the horizontal component of the mass-transport velocity (Bowden, 1965), and D_c represents its scale. Since Pe is usually a large parameter (Bowden, 1965; Feng et al., 1986b), the convective transport effect is greater than the dispersive one. Thus the equation just derived is reduced to

$$\overline{u_{LM}} \frac{\partial \overline{\langle S \rangle}}{\partial x} + \overline{v_{LM}} \frac{\partial \overline{\langle S \rangle}}{\partial y} = 0 \tag{33}$$

Equation (33) suggests that, at the zero-th order approximation, the transport process is determined purely by the convection, so that the concentration isolines for $\overline{\langle S \rangle}$ coincide with the streamlines of the depth-averaged mass-transport velocity. As a matter of fact, an equation similar to (33) can be derived under similar conditions for the horizontally two-dimensional tidal problem (Feng *et al.*, 1986b). Of course, equation (33), derived in a three-dimensional space, behaves as the depth-average of a three-dimensional mass-transport velocity. It should be pointed out that this equation is valid in the "interior" of a basin because the diffusion or dispersion becomes more pronounced in the "boundary region" (Feng *et al.*, 1986b).

6. LONG-TERM TRANSPORT PROCESSES

The conclusions derived above concerning the Lagrangian residual velocity and the inter-tidal transport processes are based upon a weakly nonlinear dynamical model of tidal flow and intra-tidal transport processes. Several hypotheses have been made, including that $O(\kappa N) = O(\kappa)$, or $O(N) = 1$, in which $O(\kappa) < 1$. Given that $N = 1 + (n-1)\kappa$, the condition $O(N) = 1$ is valid in the cases of $O(n) < \kappa^{-2}$. If we assume $O(\kappa) = 10^{-1}$, then $O(n) = \kappa^{-2} \sim 100$, which implies that the theory and model given above could not be directly applied to the cases of long averaging period of time such as a season or a year. It is worth giving the following approach to these problems of long-term transport processes.

Instead of the Lagrangian residual velocity, the original Lagrangian mean velocity of a marked water parcel has to be used here, or

$$\mathbf{u}_L = \frac{1}{2\pi n} \int_{\theta_0}^{\theta_0 + 2\pi n} \mathbf{u}\,(\mathbf{x}(\mathbf{x}_0,\theta),\theta)\, d\theta \tag{34}$$

The definition (34) can be reduced to (18) only when $O(n) < \kappa^{-2}$, of course, including $n = 1$. Nevertheless, the Lagrangian mean velocity expressed by (34) can be formulated by means of the Lagrangian residual velocity. As a matter of fact, noting that

$$\mathbf{u}_L \equiv \mathbf{u}_L\,(\mathbf{x}_0,\theta_0;n) = \frac{1}{2\pi n} \int_{\theta_0}^{\theta_0 + 2\pi n} \mathbf{u}\,(\mathbf{x}(\mathbf{x}_0,\theta),\theta)\, d\theta$$

$$= \frac{1}{n} \sum_{j=1}^{n} \frac{1}{2\pi} \int_{\theta_0+(j-1)2\pi}^{\theta_0 + 2\pi j} \mathbf{u}\,(\mathbf{x}(\mathbf{x}_0,\theta),\theta)\, d\theta = \frac{1}{n} \sum_{j=1}^{n} \mathbf{u}_L\,(\mathbf{x}_{j-1},\theta_0+(j-1)2\pi;1)\ ,$$

where $\mathbf{x}_j = \mathbf{x}_0 + \kappa^2 2\pi \sum_{i=0}^{j-1} \mathbf{u}_L\,(\mathbf{x}_i,\ \theta_0 + 2\pi i\ ;1)$ $(j=1,2,...n)$, using (18), and substituting (19)-(21) into the expression just derived, we have

$$\mathbf{u}_L\,(\mathbf{x}_0,\theta_0;n) = {}^{M}\mathbf{u}_L\,(\mathbf{x}_0;n) + \kappa\ {}^{D}\mathbf{u}_L\,(\mathbf{x}_0,\theta_0;n) \tag{35}$$

where

$${}^{M}\mathbf{u}_L\,(\mathbf{x}_0;n) = \frac{1}{n} \sum_{j=1}^{n} \mathbf{u}_{LM}\,(\mathbf{x}_{M,j-1}) \tag{36}$$

$${}^{D}\mathbf{u}_L\,(\mathbf{x}_0,\theta_0;n) = {}^{D}\mathbf{u}_L'\,(\mathbf{x}_0;n)\cos\theta_0 + {}^{D}\mathbf{u}_L''\,(\mathbf{x}_0;n)\sin\theta_0 \tag{37}$$

$${}^{D}\mathbf{u}_L'\,(\mathbf{x}_0;n) = \frac{1}{n}\left\{ \sum_{j=1}^{n} \mathbf{u}_{LD}'\,(\mathbf{x}_{M,j-1}) + \kappa^2\, 2\,\pi \sum_{j=1}^{n-1} \left[\mathbf{u}_{LD}'\,(\mathbf{x}_{M,j-1}) \cdot \nabla \sum_{i=j}^{n-1} \mathbf{u}_{LM}\,(\mathbf{x}_{M,i}) \right]\right\}$$

$${}^{D}\mathbf{u}_L''\,(\mathbf{x}_0;n) = \frac{1}{n}\left\{ \sum_{j=1}^{n} \mathbf{u}_{LD}''\,(\mathbf{x}_{M,j-1}) + \kappa^2\, 2\,\pi \sum_{j=1}^{n-1} \left[\mathbf{u}_{LD}''\,(\mathbf{x}_{M,j-1}) \cdot \nabla \sum_{i=j}^{n-1} \mathbf{u}_{LM}\,(\mathbf{x}_{M,i}) \right]\right\}$$

$$\mathbf{x}_{M,j} = \mathbf{x}_0 + \kappa^2\, 2\,\pi \sum_{i=0}^{j-1} \mathbf{u}_{LM}\,(\mathbf{x}_{M,i}) \tag{38}$$

$$\mathbf{x}_{M,0} = \mathbf{x}_0 \quad \text{and} \quad \sum_{a}^{b} = 0 \quad \text{if} \quad b < a$$

Obviously, the formulae (35)-(38) will be exactly reduced to the formulae (19)-(21), using the Lagrangian residual velocity $\mathbf{u}_{LR}$ instead of the Lagrangian mean velocity $\mathbf{u}_L\,(\mathbf{x}_0,\theta_0;1)$, if $n = 1$. It is certainly expected that the formulae (35)-(38) can be approximately reduced to the formulae (19)-(21), using the Lagrangian residual velocity $\mathbf{u}_{LR}$ instead of the Lagrangian mean velocity $\mathbf{u}_L\,(\mathbf{x}_0,\theta_0;n)$, if $O(n) < \kappa^{-2}$ even though $n > 1$. In fact, we have $\mathbf{x}_{M,j} = \mathbf{x}_0$ as $O(n) < \kappa^{-2}$ in view of the terms $(\kappa^2\, 2\,\pi \sum_{j}^{n}) = O(\kappa^2\, n) < 1$ contained in the formulae (35)-(38); and then,

$${}^{M}\mathbf{u}_L\,(\mathbf{x}_0;n) = \frac{1}{n}\sum_{j=1}^{n} \mathbf{u}_{LM}\,(\mathbf{x}_0) = \mathbf{u}_{LM}\,(\mathbf{x}_0)$$

and

$({}^{D}\mathbf{u}_L'\,(\mathbf{x}_0;n),\ {}^{D}\mathbf{u}_L''\,(\mathbf{x}_0;n)) = (\mathbf{u}_{LD}'\,(\mathbf{x}_0),\ \mathbf{u}_{LD}''\,(\mathbf{x}_0))$; and finally, $\mathbf{u}_L\,(\mathbf{x}_0,\theta_0;n) = \mathbf{u}_{LR}\,(\mathbf{x}_0,\theta_0)$, approximately.

When n is of an order of magnitude equal to or greater than κ^{-2}, the Lagrangian mean velocity $\mathbf{u}_L\,(\mathbf{x}_0,\theta_0;n)$ cannot be proven to satisfy the continuity equation for incompressible fluids. Thus, the Lagrangian residual velocity cannot be defined as the Lagrangian mean velocity, or $\mathbf{u}_L\,(\mathbf{x}_0,\theta_0;n)$, when $O(n) \geq \kappa^{-2}$. It is evident that the Lagrangian mean velocity $\mathbf{u}_L\,(\mathbf{x}_0,\theta_0;n)$ is, in general, not only a function of $(\mathbf{x}_0,\theta_0)$ but also dependent on n. Explicitly containing n reflects rationally the experience of the Lagrangian motion of the marked water parcel moving from the initial position $\mathbf{x}_0$ at time θ_0 to the terminal position $\mathbf{x}_n$ at time $\theta_0 + 2\pi n$. Because the Lagrangian mean velocity is directly proportional to the net Lagrangian displacement of the marked water parcel, with a proportionality coefficient $(2\pi n)^{-1}$, in view of (35)-(38), the latter behaves well like the former, and the horizontal projection of the net Lagrangian displacement will trace out an ellipse over a tidal cycle. The equations also show that the net Lagrangian displacements, or $2\,\pi\, n\, \mathbf{u}_L\,(\mathbf{x}_0,\theta_0,n)$, of marked water parcels released from the position $\mathbf{x}_0$ at tidal phase θ_0 over such a long time as a season can be of an order of magnitude greater than those over one or a few of tidal cycles. Thus, the Lagrangian residual drifts, or $2\,\pi\, n\, \kappa\, {}^{D}\mathbf{u}_L\,(\mathbf{x}_0,\theta_0;n)$, of which the horizontal components trace out an ellipse over a tidal cycle, play a more pronounced role in the dispersion, for example, of pollutants on "long" time scales than on "short" scales, although the Lagrangian drift velocities, or $\kappa\, {}^{D}\mathbf{u}_L\,(\mathbf{x}_0,\theta_0;n)$, are of the same order of magnitude in both cases. For example, the ratio of the Lagrangian residual drift for $O(n) = \kappa^{-2}$ to that for $O(n) = 1$ is $O(\kappa^{-2})$, which suggests the importance, at least the potential importance, of the Lagrangian residual drifts for long-term dispersion phenomena though $O(\kappa\,|\,{}^{D}\mathbf{u}_L\,(\mathbf{x}_0,\theta_0;n)\,|\,/\,|\,{}^{M}\mathbf{u}_L\,(\mathbf{x}_0;n)\,|\,) = O(\kappa) < 1$. Detailed descriptions of the phenomena and the dynamics mentioned in this section can be found in another paper

(Feng, 1986c).

7. AN APPLICATION TO THE BOHAI SEA, CHINA

The dynamic model proposed in the present paper should be verified through a test of the theory against field data in a realistic situation. This three-dimensional model can be conveniently applied to the summer residual circulation and the inter-tidal transport processes in the Bohai Sea, China (Fig. 1). A scale analysis has shown that the tide-induced nonlinear effect and the Huanghai Sea Warm Current which enters the Bohai Sea through the Bohai Strait may be the principal factors contributing to the formation of the summer circulation in the Bohai Sea. The wind stress on the sea surface and the baroclinic effects are negligible and less important, respectively (Feng *et al.*, 1986). Thus, the total transport through the Bohai Strait has to be prescribed; that transport can be obtained from the calculation of the current speed based upon field data. For the tides, we can use an existing three-dimensional nonlinear numerical model of the M_2-tide in the Bohai Sea. The calculated depth-averaged mass-transport velocity field as the mean residual circulation of the summer in the Bohai Sea has recently been obtained (Sun *et al.*, 1986), and is exhibited in Fig. 2. According to equation (33), the concentration isolines for the depth-averaged tidal cycle mean of the concentration of any conservative and passive tracer must approximately coincide with the streamlines of the depth-averaged mass-transport velocity. The salinity can be conveniently used as such a tracer in the Bohai Sea for the summer. The salinity distribution at the depth of 10 m in the Bohai Sea for June, 1958, as derived from observations, is shown in Fig. 3. This picture can be used qualitatively to represent a typical summer distribution of the depth-averaged salinity in the Bohai Sea. Even without including other factors likely to affect the distribution of salinity, such as stratification, the Huanghe River runoff and the coastal dispersion boundary layer, the patterns of isohaline contours of Fig. 3 and the mass-transport velocity field of Fig. 2 seem to be in surprisingly good qualitative agreement. Furthermore, the pattern of the mean Lagrangian residual circulation in the Bohai Sea for the summer can be used to explain dynamically the existence of an area, situated in the northeastern part of Laizhou Bay, where the water is of relatively high salinity and transparency and of relatively low temperature and concentration of suspended sediments. These conditions explain the presence, in that area, of such bottom fauna as spatangia (Su *et al.*, 1986). It is of interest to note that the pattern of the summer circulation in the Bohai Sea is only one big counterclockwise gyre if the coupled mean tide-induced Lagrangian residual circulation is not considered (Feng *et al.*, 1986). Such a circulation cannot be used to interpret the existence of the area stated above. The present theory for predicting long-term transport processes and studying their dynamics in coastal seas seems to be satisfactory, at least qualitatively.

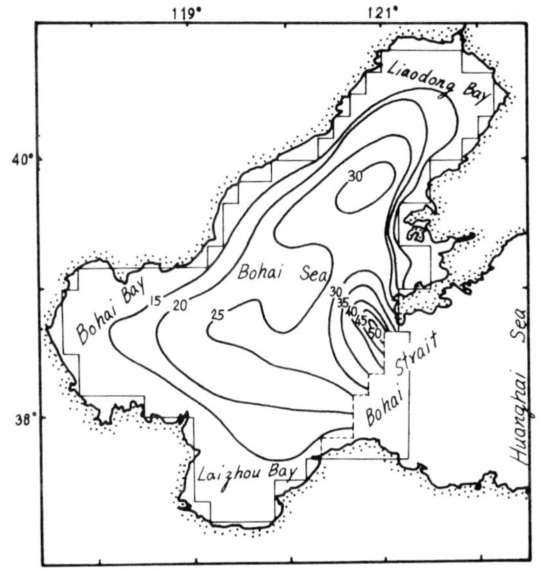

Fig. 1. Depth distribution of the Bohai Sea, China, (m).

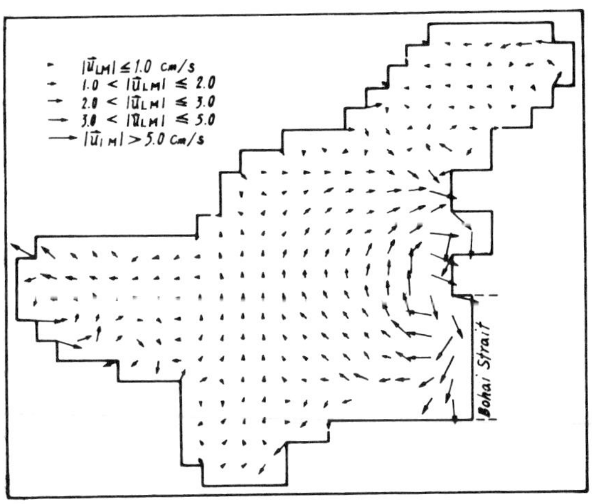

Fig. 2. Computed depth-averaged mass-transport velocity field (cm/sec).

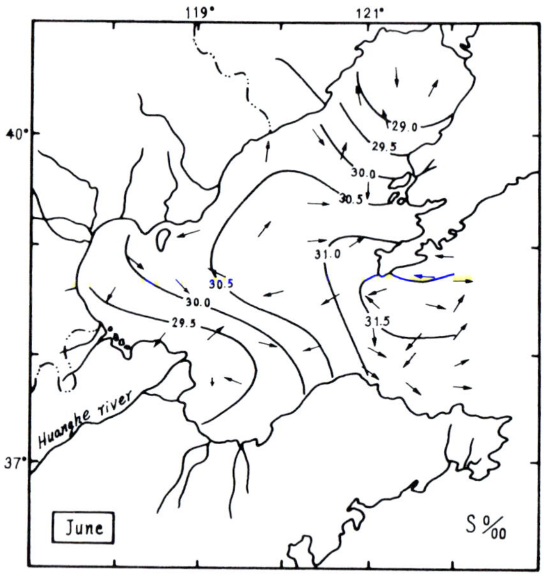

Fig. 3. Salinity distribution (‰) measured at the depth of 10 m in the Bohai Sea in June 1958. Arrows denote observed Eulerian residual currents.

8. CONCLUSION

Based upon a three-dimensional weakly nonlinear theory, the Lagrangian residual velocity is approximately expressed as a sum of a tidally periodic fluctuation, which we call the Lagrangian drift velocity, and of the tidal cycle average, *i.e.*, the mass-transport velocity. The Lagrangian residual velocity is different from the Eulerian residual velocity, which is steady, and it is similar to the tidal current velocity. A set of field equations governing the mass-transport velocity is derived. These equations show that the mean Lagrangian residual circulation is more relevant than the Eulerian one to the description of inter-tidal flow processes. In particular, a new inter-tidal transport equation for the tidal cycle average of the concentration of any conservative and passive tracer is proposed, in which the convective transport is characterized by the mass-transport velocity and there is no need for an ad hoc hypothesis on "tidal dispersion". It is also shown that the Lagrangian residual drift, which is related to the Lagrangian drift velocity, is potentially the most important factor controlling the dispersion or spreading of tracers in the long-term transport processes. And finally, when the theory is preliminarily applied to the Bohai Sea, China, a good qualitative agreement between the present theory and field data is revealed.

9. REFERENCES

Alfrink, B.J. and Vreugdenhil, C.B., 1981. Residual currents. Delft Hydraul. Lab., Delft, The Netherlands, Rep. R 1469-11, 42 pp.

Bowden, K.F., 1965. Horizontal mixing in the sea due to a shearing current. J. Fluid Mech., 21 : 83-95.

Bowden, K.F., 1967. Circulation and diffusion. In: G.H. Lauff (Editor). Estuaries. Publ., No. 83, AAAS, Washington, D.C., pp.15-36

Cheng, R.T., 1983. Euler-Lagrangian computations in estuarine hydrodynamics. In : C. Taylor, J.A. Johnson and R. Smith (Editors), Proc. of the Third Intern. Conf. on Num. Meth. in Laminar and Turbulent Flow. Pineridge Press, pp. 341-352.

Cheng, R.T. and Casulli, V., 1982. On Lagrangian residual currents with applications in South San Francisco Bay, California. Water Resour. Res., vol. 18, 6 : 1652-1662.

Cheng, R.T., Feng, S. and Xi, P., 1986. On Lagrangian residual ellipse. In : J. van de Kreeke (Editor), Intern. Conf. on Physics of Shallow Estuaries and Bays, Lecture Notes on Coastal and Estuarine Studies. Springer-Verlag, pp. 102-113

Csanady, G.T., 1982. Circulation in the Coastal Ocean. D. Reidel Publ. Comp., Dordrecht/Boston/London, 279 pp.

Feng, S., 1977. A three-dimensional nonlinear model of tides. Scientia Sinica, vol. 20, 4 : 436-446.

Feng, S., 1986a. A three-dimensional weakly nonlinear dynamics on tide-induced Lagrangian residual current and mass-transport. Chinese J. of Oceanology and Limnology, vol. 4, 2 : 139-158.

Feng, S., 1986b. On the fundamental dynamics of barotropic circulation in shallow seas. Acta Oceanologica Sinica (submitted for publication).

Feng, S., 1986c. On tracer spreading in a long-term transport process (in preparation).

Feng, S., and Cheng, R.T., 1986. Formulation of the governing equations for Lagrangian residual current and residual transport. Intern. Symposium on Physics of Shallow Bays, Estuaries and Continental Shelves. Shandong College of Oceanography, Qingdao, China, November 1986 (in preparation).

Feng, S., Cheng, R.T. and Xi, P., 1986a. On tide-induced Lagrangian residual current and residual transport, Part I : Residual current. Water Resour. Res., vol. 22, 12 : 1623-1634.

Feng, S., Cheng, R.T. and Xi, P., 1986b. On tide-induced Lagrangian residual current and residual transport, Part II : Residual transport with application in South San Francisco Bay, California. Water Resour. Res., vol. 22, 12 : 1635-1646.

Feng, S., Xi, P. and Zhang, S., 1984. The baroclinic residual circulation in shallow seas. Chinese J. of Oceanology and Limnology, vol. 2, 1 : 49-60.

Feng, S., Xi, P. and Zhang, S., 1986. Numerical modeling of the general circulation. In : C.K. Tseng and M. Tomczak (Editors), Oceanography of East China Sea, Huanghai Sea and Bohai Sea, Chapter 6. Lecture Notes on Coastal and Estuarine Studies. Springer-Verlag (submitted for publication).

Fischer, H.B., List, E.J., Koh, R.C.Y., Imberger, J. and Brooks, N.H., 1979. Mixing in Inland and Coastal Waters. Academic Press, New York, 483 pp.

Heaps, N.S., 1978. Linearized vertically-integrated equations for residual circulation in coastal seas. Deut. Hydrog. Z., 31 : 147-169.

Longuet-Higgins, M.S., 1969. On the transport of mass by time-varying ocean currents. Deep Sea Res., 16 : 431-447.

Nihoul, J.C.J. and Ronday, F.C., 1975. The influence of the "tidal stress" on the residual circulation. Tellus, 27 : 484-489.

Pritchard, D.W., 1954. A study on salt balance in a coastal plain estuary. J. Mar. Res., 13 : 133-144.

Song, L., 1986. A hydrodynamic velocity-splitting model with a depth-varying eddy viscosity in shallow seas. Acta Oceanologica Sinica (accepted for publication).

Stern, M.E., 1975. Ocean Circulation Physics. Academic Press, New York, 246 pp.

Su, Z., Wiseman, W.T., Fan, Y., Gao, S., Qian, Q. and Yang, Z., 1986. Analyses of hydrologic characteristics of the area adjacent to the Huanghe Estuary (personal communication).

Sun, W., Xi, P. and Song, L., 1986. Numerical calculation of the three-dimensional tide-induced Lagrangian residual circulation in the Bohai Sea. Acta Oceanologica Sinica (submitted for publication).

Tee, T.K., 1976. Tide-induced residual current : A 2-D nonlinear numerical model. J. Mar. Res., 31 : 603-628.

Zimmerman, J.T.F., 1979. On the Euler-Lagrangian transformation and the Stokes' drift in the presence of oscillatory and residual currents. Deep Sea Res., 26A : 505-520.

THREE DIMENSIONAL NUMERICAL MODEL FOR THERMAL IMPACT STUDIES

M. DARRAS, P. DONNARS and P. PECHON
Research Engineers - Electricité de France, Laboratoire National d'Hydraulique
Chatou, France

ABSTRACT

The three dimensional numerical model ODYSSEE was developed to study the near-field thermal impact of coastal nuclear projects. A finite difference scheme is used to solve the governing equations. To have a good reproduction of vertical mixing, an accurate turbulence modelling is used involving the effect of vertical thermal gradient. The application of the model to the test case of a flume with a stratified flow shows that the model does fairly well in spite of small discrepancies which occur when the advection term and the diffusion term are not negligible in the equations of turbulent fluxes of momentum. Then the model is applied to the case of the nuclear power station of Gravelines. The measured and computed temperature profiles are very similar ; the vertical thermal front and the horizontal stratification are well represented. Nevertheless the horizontal mixing modelling will be improved in the future because it is not accurate enough in some particular cases when velocity shears are strong.

1 - INTRODUCTION

The thermal studies of coastal nuclear power stations using a once-through cooling system require a good knowledge of the heated water dilution. The marine currents ensure the mixing of temperature so that the dispersion in the ambient flow far from the outlet can be estimated with a two-dimensional numerical model assuming that current and temperature are homogeneous over the depth. In the near-field of a buoyant discharge outlet, the vertical mixing is not strong enough to force the vertical homogenization of water. Then a three-dimensional numerical model or a physical model is needed in order to represent the buoyancy effects in the fluid.

The purpose of the paper is to show the recent developments and applications of the three-dimensional model ODYSSEE, which computes simultaneously curent repartition and heat dilution.

2 - THE NUMERICAL MODEL ODYSSEE

2.1 The basic equations

A three-dimensional modelisation needs a simultaneous solution of the heat dilution and of the velocity field, because the heat repartition creates buoyancy effects which influence the current, which in turn influences the heat repartition.

The flow is governed by the unsteady Navier-Stokes equations and the mass conservation equation. Simultaneously the temperature repartition is obtained by solving the advection - dispersion equation, the exchange of heat between sea and atmosphere being negligible here :

$$\frac{dT}{dt} = \text{div}\ (K\ \text{grad}\ T)$$

where T : temperature

K : coefficient of turbulent diffusion.

The temperature distribution act on the velocities through the local water density by the following equation :

$$-\frac{1}{\rho}\ \frac{\partial \rho}{\partial T} = \beta$$

where ρ : local water density

β : coefficient of thermal expansion

2.2 The assumptions

The equations are simplified using the following assumptions :

- In the case of tidal or wind induced currents and when the bottom topography is gentle, the flow pattern is almost horizontal and the vertical acceleration is small compared with the gravity. The pressure is thus assumed to be hydrostatic in the Navier-Stokes equations.
- The variation of the water density in the Navier Stokes equations is linearized using the Boussinesq approximation.

2.3 The turbulence modelling

The turbulent fluxes of momentum and of temperature are modelled with the Prandtl's mixing length hypothesis by horizontal and vertical eddy viscosities.

Horizontally, the velocity gradients are generally small, so the advection transfers prevail over the turbulent exchanges. Therefore an arbitrary horizontal viscosity coefficient is taken (from a reasonable range of values).

On the other hand, the vertical turbulent exchanges are not negligible compared with the vertical advection because the latter is very weak. Consequently, the vertical turbulent transfers require an accurate model, as a result of the substantial interference with thermal phenomena.

Moreover, it is not desirable to increase the complexity of the system to be solved by introducing additional multidimensional differential equations.

Consequently, a compromise is chosen, as follows.

The equations used to describe the correlations between the temperature and velocity fluctuations ($\overline{u'_i u'_j}$ and $\overline{u'_i T'}$, i and j = 1 to 3 for x, y, z), as well as the turbulent energy, $k = \overline{(u'^2 + v'^2 + w'^2)}/2$, and $\overline{T'^2}$, are described through the Launder (1975) model for unknown terms. Dissipation, ε , is then modelled as follows :

$$\varepsilon = C_E \frac{k^{1.5}}{l},$$

where $C_E = 0.47$ and l is a length scale. The length scale is assumed to be constant in the fluid except near the bottom and the sea-surface where it varies linearly with the distance from the boundary.

The horizontal gradients are assumed to be negligible, compared with the vertical gradients. Moreover, the classical local equilibrium assumption is adopted : the energy produced is assumed to be instantaneously absorbed or dissipated at the same place, and therefore the diffusion, transport and unsteady terms are not taken into account. These approximations will be discussed in the first application.

It can be shown (Dewagenaere 1979) that the eddy viscosity can be determined from the resulting system, which can be solved algebraically. The resulting turbulent kinetic energy depends on the Richardson number (fig 1) :

$$Ri = \frac{\dfrac{g}{\overline{T}} \dfrac{\partial \overline{T}}{\partial z}}{\left(\dfrac{\partial \overline{u}}{\partial z}\right)^2 + \left(\dfrac{\partial \overline{v}}{\partial z}\right)^2}$$

In a stable medium (Ri < 0), the energy decreases rapidly, and falls to zero when Ri = 0,4.

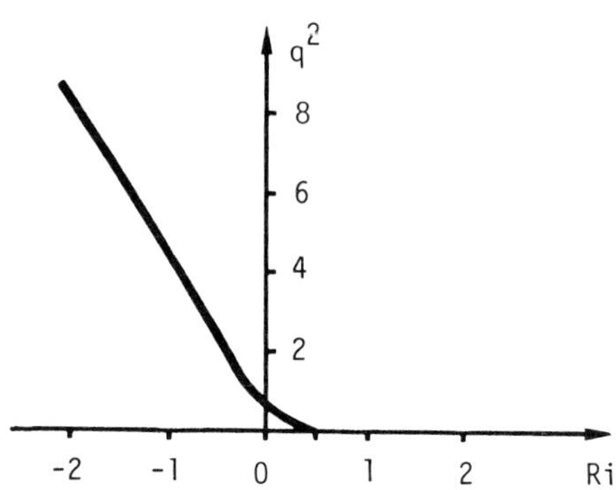

Fig. 1 - Non-dimensional turbulent kinetic energy

2.4 The numerical solution

2.4.1 Use of curvilinear coordinates

In order to avoid evolution of the calculation field because of the free surface variations, a curvilinear coordinate z* is used to get a transformed domain independant of time (fig 2) :

$$z^* = \bar{S}\,(z - z_F) \,/\, (S - z_F)$$

where $\bar{S}$ is a horizontal reference level

S (x, y, t) water surface level

z_F (x,y) bed level

z (x, y) vertical coordinate

The equations written with the new coordinate are of the same form (see Benque et Al. 1982).

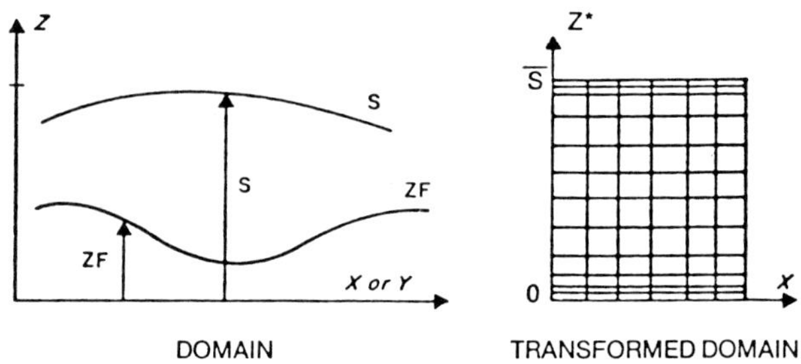

Fig. 2 - Curvilinear coordinates

2.4.2 The boundary conditions

a - The currents

- open boundaries

Generally, at open sea boundaries, the mass-currents integrated over depth can be obtained from field measurements or from a two-dimensional computation at a larger scale.

To deduce a velocity profile at the boundaries, the following Ekman type relations are integrated (Benque et Al. 1982). The open boundaries must be located in areas with regular bottom in order to allow to neglect horizontal gradients

$$\frac{\partial u}{\partial t} = \frac{\partial}{\partial z}\left(\nu_z \frac{\partial u}{\partial z}\right) + fv - \frac{1}{\rho}\frac{\partial P}{\partial x}$$

$$\frac{\partial v}{\partial t} = \frac{\partial}{\partial z}\left(\nu_z \frac{\partial v}{\partial z}\right) - fu - \frac{1}{\rho}\frac{\partial P}{\partial y}$$

where ν_z : turbulent viscosity

f : coriolis parameter

- coastal boundaries

Velocity components are zero or a slip condition is taken.

- free surface

. without wind, we assume :

$$\frac{\partial u}{\partial z} = 0, \quad \frac{\partial v}{\partial z} = 0 \text{ and } w = 0$$

. with wind, a surface shear tress $\vec{\tau}_w$ is generated, depending on the wind speed and related to the flow velocity by the equation :

$$\nu_z \frac{\partial \vec{u}}{\partial z} = \vec{\tau}_w$$

- sea bed

Two types of conditions have been tested :

. the velocity is zero. This condition imposes refined mesh grid near the bottom, in order to have a good description of the gradients.

. a slip velocity condition is imposed by assuming that shear stress is a quadratic function of the bed velocity (see Coëffe et al. 1982).

b - The temperature

At the open boundaries or at the outlet, when the fluid flows into the domain the temperature repartition is imposed for the advection step, whereas when it flows out no condition is needed for this step (free outing condition). For the diffusion step the variation of temperature along the normal is set to zero at all the boundaries .

2.4.3 The numerical scheme

The governing equations are solved using a finite differences scheme based on a splitting of operators, with a rectilinear computational grid in the direction x, y, z*.

A completely implicit solution of the equations would lead to a prohibitive computation time. On the other hand, a completely explicit solution has some restriction upon the time step. It can be shown (Coëffe et Al. 1982) that the most restrictive conditions affect the determination of the free-surface wave and the vertical diffusion. Consequently these two steps are solved implicitly, the others being treated explicitly.

The advection terms are solved by the characteristics method, and the vertical diffusion is treated with a double-sweep algorithm.

The pressure term and the continuity equation are solved together. The remaining equations averaged over the depth lead to 2D equations which relate the sea surface elevation and the two components of the fluxes. They are treated with a 2D iterative method using an alternating direction operator with coordination.

Then the horizontal velocity profiles are modified with the new sea surface gradient and the vertical velocity is calculated by integrating the local continuity equation.

3 - APPLICATIONS

Two applications were performed to test the validity of the assumptions and to evaluate the predictive capability of the model ODYSSEE.

The first chosen case is a quasi-two dimensional experimental flume with a stratified flow, for which extensive measurements were available (Gartrell 1979). The turbulence modelling could be discussed regarding the comparison of the measured temperature and velocity profiles and the numerical results.

In the second application, the dilution of the heated water plume of the nuclear power station of Gravelines was simulated and the results were compared with field surveys to test the reliability of the model.

3.1 Stratified flows in a flume

3.1.1 Description of the flume (fig 3)

The data were taken from measurements made in a 40.m recirculating flume, with an average flow depth of 45 cm. Two pumps and a temperature control system allowed the production of a two-layered density-stratified flow. The lower layer had an initial depth of 30 cm while the upper layer had an initial depth of 15 cm.

Profiles of the two velocity components and of the temperature were made at various points along the flume centerline.

3.1.2 Tests with smooth bottom

Two similar tests were investigated ; only the initial conditions at the entrance of the model differed :

	mean velocity cm/s		ΔU =	mean temperature °C		ΔT =
	upper layer U1	lower layer U2	U1 - U2	upper layer T1	lower layer T2	T1 - T2
test 1	14,0	7,5	6,5	34,1	32,4	1,7
test 2	9,8	9,0	0,8	30,9	30,2	0,7

In these two cases, the bottom of the physical model was smooth so that little turbulence was generated from the bottom. A slip condition was imposed at the bottom and on the sides of the flume. The experimental temperature and velocity profiles were taken at the upstream boundary of the numerical flume.

The computed results (fig 4) were very close to the measurements in the test 1 whereas some discrepancies could be noticed in the test 2 : although the shape of the velocity profile was somewhat preserved, it is staggered compared to measurements and the temperature gradient is too strong at the interface.

The observed differences between measured and computed results in the second test could be explained by considering the equation of kinetic turbulent energy. In the test 2, the Richardson number $Ri = g \frac{\partial \overline{T}}{\partial z}/\overline{T} \left(\frac{\partial \overline{u}}{\partial z}\right)^2$ was relatively large because $|\partial \overline{u}/\partial z|$ was small ; thus the local production of turbulent kinetic energy (that is $-\overline{u'w'}\,\partial \overline{u}/\partial z$) was small and the transport equation was dominated by diffusive terms such as $\partial u'^2 w'/\partial z$, terms neglected in the turbulence model. In fact the shape of the velocity profile was not modelled badly because the dominant term in the Reynolds equation was the pressure gradient term, while the transport equation for T had no analogue for a heat source :

$$\overline{u}\frac{\partial \overline{u}}{\partial x} + \overline{w}\frac{\partial \overline{u}}{\partial z} = -\frac{1}{\rho}\frac{\partial \overline{P}}{\partial x} - \frac{\partial \overline{u'w'}}{\partial z} + \frac{\partial^2 \overline{u}}{\partial z^2}$$

$$\overline{u}\frac{\partial \overline{T}}{\partial x} + \overline{w}\frac{\partial \overline{T}}{\partial z} = - \frac{\partial \overline{T'w'}}{\partial z}$$

For these reasons, the term $\partial \overline{T'w'}/\partial z$ was modelled poorly in the test 2 ; this induced a bad mixing of temperature at the interface of the two layers.

On the contrary, in the test 1 the term of local production of kinetic energy prevailed with respect to the transport and the diffusion terms because the velocity and the temperature gradients are large at the interface. Therefore the turbulent mixing of temperature was well reproduced in that case.

note : The shape of the velocity profile was satisfying in the two tests but the value of the integrated velocity profile was smaller than the measured one. This was due to the fact that the experimental model was not really bidimensional because of friction on the flume sides. This induced greater velocities along the centerline in the physical model.

3.1.3 Effect of changes of the bottom roughness

This experiment was different from the others in the way that it was isothermal and the roughness of the bed changed at x = 4,5 m from rough

(Darcy-Weisbach friction factor f = 0,05) to smooth (f = 0,017). Thus, one was able to measure the reaction of the model to sudden changes of the bed roughness.

Figure 5 shows that the model behaved well in the first part of the smooth section but the agreement was not so good after x = 20 meters. This was due to the following reason : in the experiment, turbulent bursts generated in the roughened section continued to propagate in the smooth section, continuing the turbulent mixing. In the model, the "memory" of the various high turbulence levels was maintained only in the smooth section in the gradient of $\overline{u}$. Hence, the model reduced $-\overline{u'w'}$ too rapidly after the change in bed roughness and the mixing did not persist enough.

3.1.4 Summary of the results

Though some small discrepancies could be displayed when the advection and the diffusion terms in the turbulence model were not negligible, ODYSSEE did fairly well. So it was applied to a real case which is described in the following paragraph.

3.2 Dilution of a heated water Plume under the action of tidal currents

3.2.1 Description

The model ODYSSEE was used to evaluate the temperature field near the coastal nuclear power plant of Gravelines located in the North of France, near Dunkirk.

The power station is built on a sand mound encroaching upon the sea. To minimize recirculation between the intake and the outlet, cool waters are withdrawn inside the outer harbour of Dunkirk while heated waters are discharged on the shore out of the harbour by mean of a free surface channel (fig 6). For the 2 units of 900 MWe considered in this application, the mean value of the discharge is 120 m^3/s ; it induces velocities of 0.20 m/s at high water and 1.5 m/s at low water approximatively, at the end of the outlet channel. For these conditions, the heated temperature elevation is 7°C.

During mean spring tide, the tidal amplitude reaches 5 meters and the velocities exceed 1 m/s. The drift during a tidal period goes eastward and its value fluctuates between 2 km and 3 km.

The bottom slope is gentle and the mean water depth (under lowest tide level) does not exceed 13 meters in the studied area except in the navigation channel of the harbour where it reaches 20 meters.

3.2.2 The numerical model

To test the predictive capability of the model, the thermal dilution of the discharged flume was computed. The studied domain covered an area of

5. x 3. km^2 including the outlet works while the intake was not modelled ; only the mouth of the harbour was represented (fig 6).

The computation grid had 32 640 nodes, the mesh-size was 75 meters long horizontaly and varied vertically, from 0.10 meter at the bottom and the surface, to 2.0 meters at mid-depth. The depth-integrated flows imposed at the open boundaries and at the harbour entrance were provided by interpolation of the results of a larger two-dimensional numerical model for mean spring tide conditions. Moreover the discharged flow imposed at the channel end is balanced by an additional flow distribution along the harbour mouth corresponding to the intake withdrawal.

At the open boundaries, the temperature was zero when the water flowed into the domain and was derived from the advection step when the water flowed out. This means that we assumed that water which went out of the domain and came back after the slack period had a negligible temperature. The reliability of this assumption depended of course on the domain dimensions and a compromise had to be done here to avoid a prohibitive number of computational nodes.

3.2.3 The results

During the flood period (fig 7a) the outlet discharge was advected eastwards against the harbour dyke and a strong dilution was observed at the work end because of the mixing in front of the harbour mouth (effects of the currents acceleration and the bottom irregularities). A horizontal thermal front could be noticed on the western limit of the plume. Moreover a vertical stratification due to the buoyancy effects could be noticed.

When the tide was lowering (fig 7b), the current flowing out of the harbour deviated the plume northwards. During ebb-tide (fig 7c) the current flowed eastwards and the heated plume was transported along the coast. The plume was still stratified but the dilution in the ambient flow was not as effective as during the flood period : at mid ebb-tide the heat area exceeding 6°C at the surface was around 0.22 km^2 while it was only 0.09 km^2 at mid flood.

First the computed thermal impact at the water surface was compared with infrared teledetection (surface aerial thermographies). This comparison was in fact qualitative because the power of the station was not the same during the teledetection survey and in the computed case (6 MWe instead of 2 MWe).

The thermal front observed by teledetection (fig 8) is well reproduced by the model and the predicted distribution of temperature is similar. Nervertheless the figure 9 illustrating the numerical and the measured distribution of temperature at mid flow shows that the model underestimated the dilution in the strong velocity shears occuring at the harbour pass, and that the computed plume was attached on the dyke while it was detached in nature. That means that the modelling of the horizontal mixing is too simple when the velocity shear is strong.

The vertical profiles of temperature were compared for the same conditions of outlet discharge. The figure 10 representing the profiles at various moments shows that the model reproduced correctly the vertical stratification. The measured temperature was generally higher than the computed one because it includes the far field heating, but the shapes are similar.

4 - CONCLUSIONS

The three dimensional model ODYSSEE was succesfully tested in two cases where stratification phenomena were important. It was shown that the modelling of horizontal mixing has to be improved in some particular situations where velocity shears are strong. Nevertheless the model ODYSSEE is a very interesting tool for the prediction of thermal impact. It can also be of use for other type of studies like dilution of all kind of pollutant or evaluation of sediment movement involving three-dimensional effects, and the knowledge of the three-dimensional circulation is also often needed for simulating ecological processes.

5 - ACKNOWLEDGEMENT

The authors are very much indebted to the people who developed the model, P. Dewagenaere, J.C. Soliva and M.C. Burg, under the supervision of J.P. Benque. They appreciated the work done in cooperation with G. Gartrell, during a 9 months stay at the LNH.

6 - REFERENCES

Benque J.P., Hauguel A. and Viollet P.L., 1982. Engineering Application of Computational Hydraulics. Volume 2. Pitman.

Coëffe Y., Warluzel A. and Burg M.C., 1982. Three-dimensional Numerical model for tidal and wind generated flow. Coastal Engineering, Cape Town.

Darras M., Allen H., Cavelier D. and Caudron L. 1984. Synthése de mesures de température dans le rejet chaud d'une centrale en bord de mer. Rapport I.5. 18e journées de l'hydraulique, Marseille, Société hydraulique de France.

Dewagenaere P., 1979. Thesis, University Pierre et Marie Curie, Paris VI.

Gartrell G. , 1979. Report KH-R-39 Keck Hydraulics Lab. California Institute of Technology, Pasadena CA.

Gartrell G. and Pearson H.J., 1982. "On the scaling of the vertical diffusivity in stably stratified flows". Submitted for publication.

Gartrell G., 1983. Modélisation d'écoulements stratifiés à l'aide du modèle ODYSSEE Rapport de stage LNH.

Launder B.E., 1975. Journal of Fluid Mechanic 67. pp 569-581

Launder B.E., 1972. Heat and mass transport topics in applied physics, 12 Springer-Verlag.

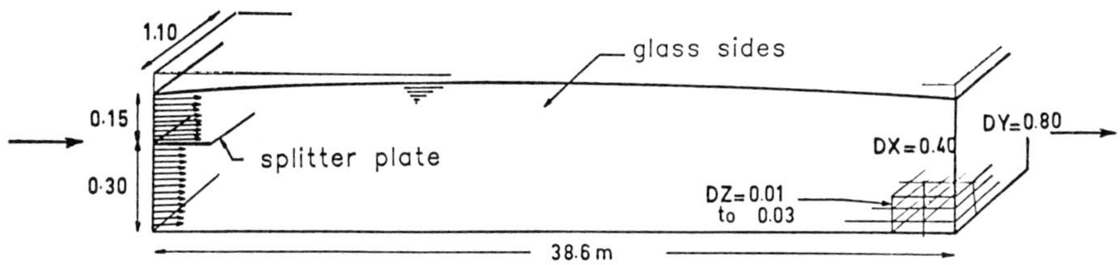

Fig. 3 - The experimental flume and the mesh grid

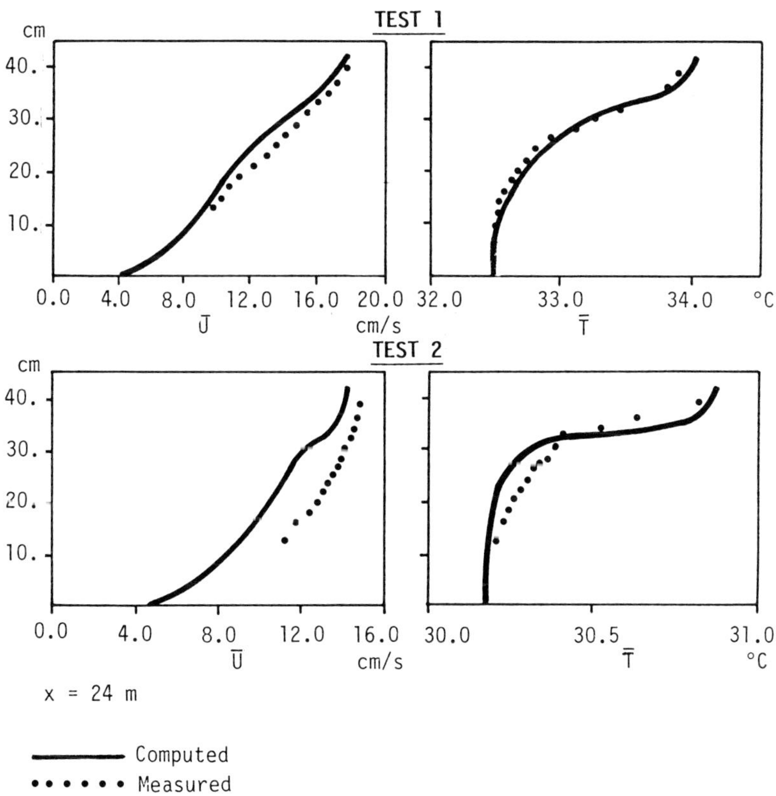

Fig. 4 : Tests in a flume with smooth bottom

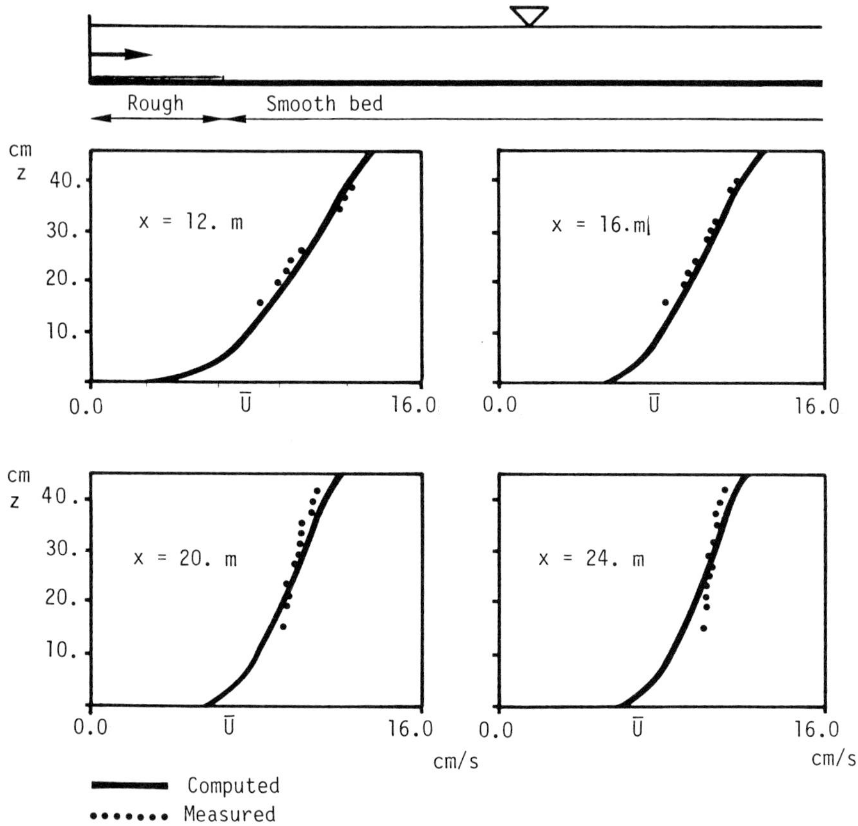

Fig 5 : Test in a flume with varying bottom roughness
Velocity profiles in the smooth part.

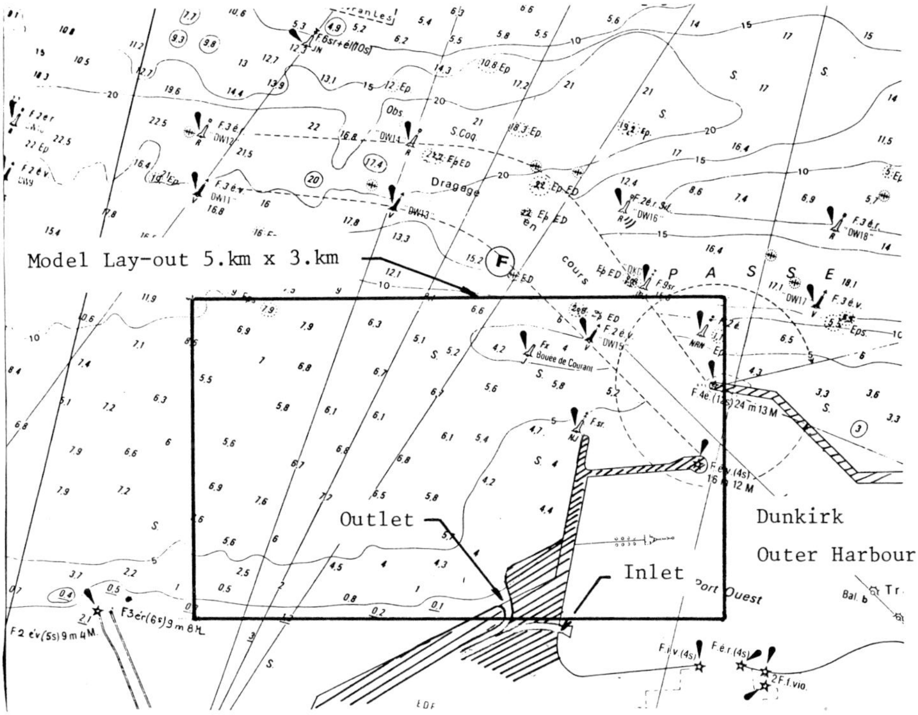

Fig. 6 - The model lay-out for the thermal study of Gravelines nuclear power plant

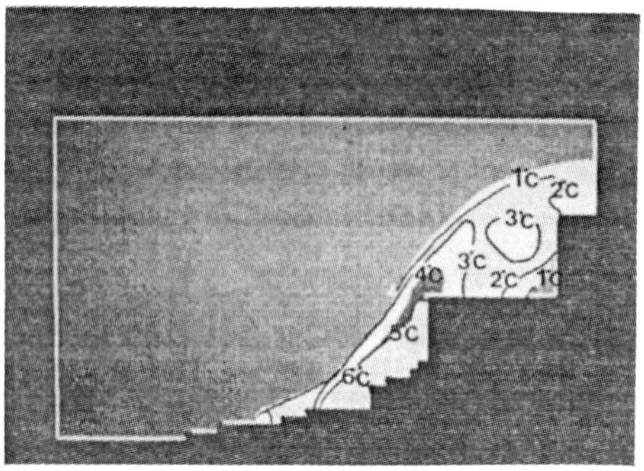

a - mid flood

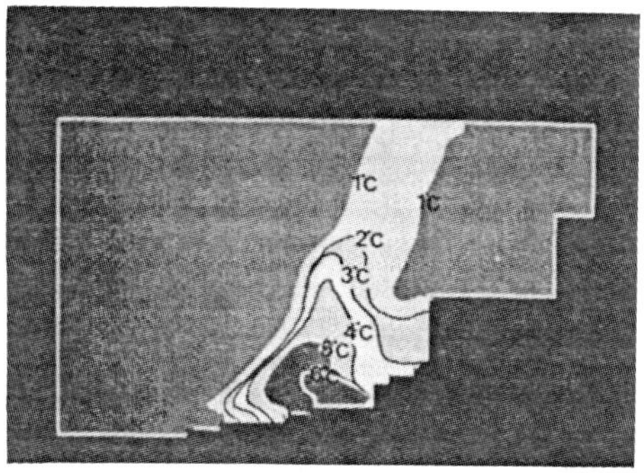

b - slack water

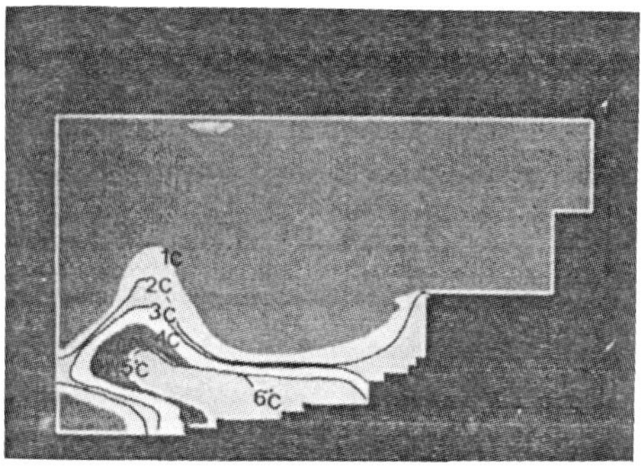

c - beginning of ebb tide

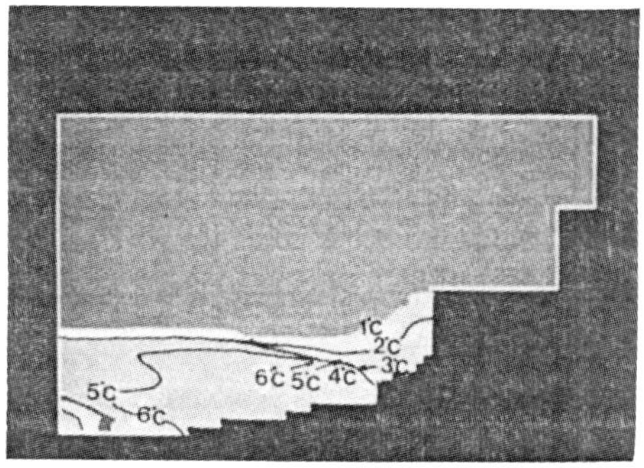

d - mid ebb tide

Stratification

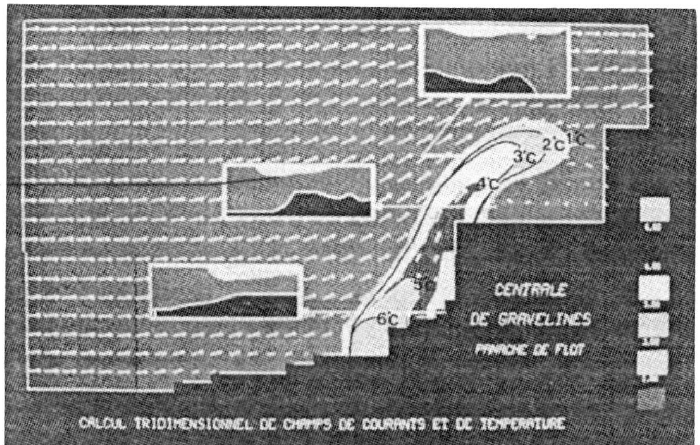

e - vertical sections of the plume (flood period)

Fig. 7 - The calculated plume evolution during tidal cycle

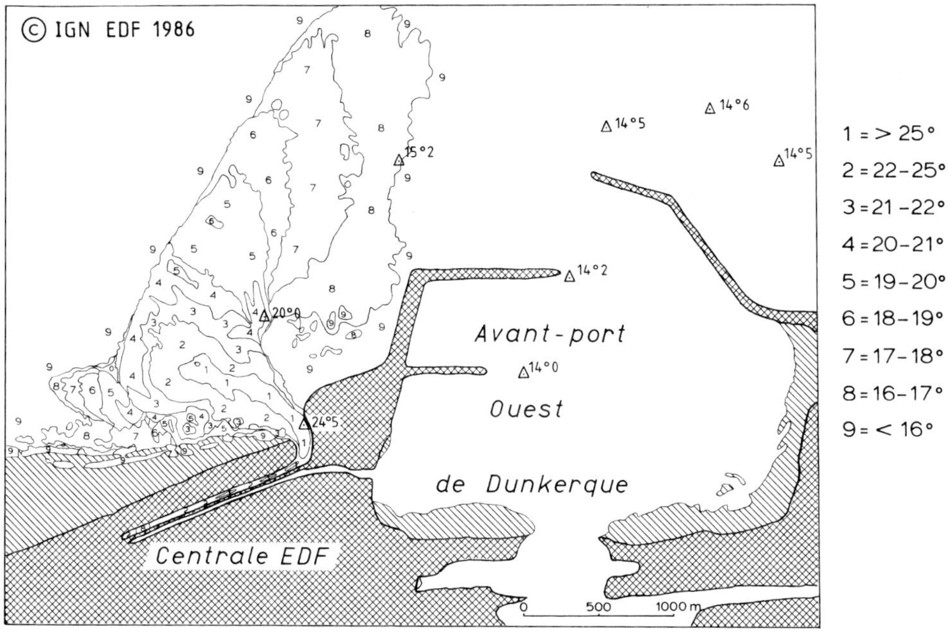

Fig. 8 - Observed temperature distribution drawn from an infrared teledetection of the plume HW + 3 h (to compare with fig. 7-b).

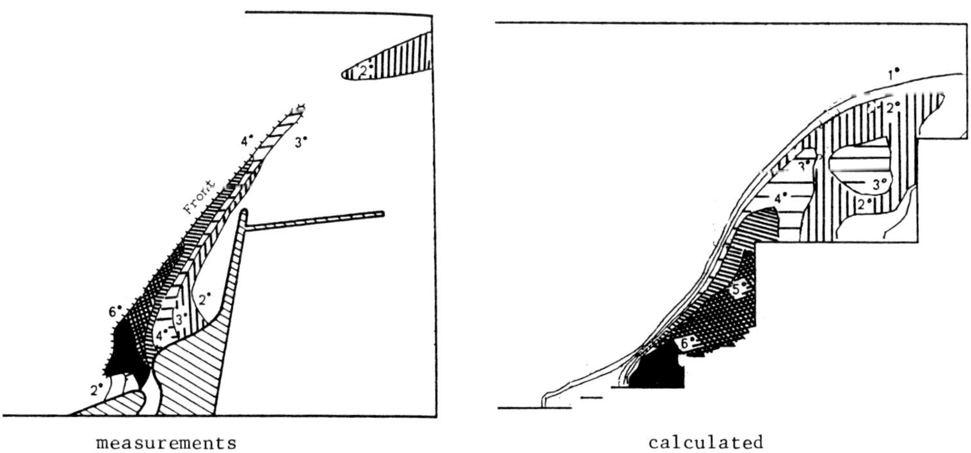

Fig. 9 - Temperature elevation at the surface - mid flood

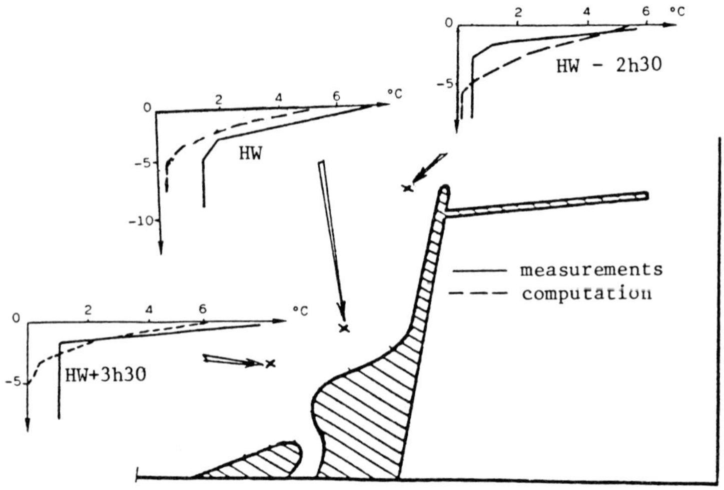

Fig.10-Temperature profiles over the depth

ESTIMATION OF STORM-GENERATED CURRENTS

N. S. HEAPS and J. E. JONES
Institute of Oceanographic Sciences, Bidston Observatory, Birkenhead, Merseyside, U.K.

ABSTRACT

Some relatively simple calculations are proposed for estimating the vertical structure of storm-generated currents at any particular location on a continental shelf. The water is supposed to be homogeneous. Horizontal gradients of storm-surge elevation from a two-dimensional vertically-integrated numerical model are introduced into a one-dimensional model for the vertical current structure at the location, along with wind stress and atmospheric pressure gradients. The changing distribution of currents through the depth is determined as the dynamic response to those forces.

1 INTRODUCTION

It is a fact of experience that the changing distribution of the elevation of the sea surface due to storm surges can be well predicted using two-dimensional vertically-integrated numerical models (Heaps 1983). This assumes that good meteorological forecasts, pertaining to the surge generation, are available. Thus, the elevation field may be obtained satisfactorily without considering the vertical distribution of current: studies by Proudman (1954), Roed (1979) and Johns et al (1983) all reached this conclusion. Specifically, the investigation by Johns and co-workers found little difference between the performance of a complex three-dimensional model and that of a much simpler two-dimensional vertically-integrated model in simulating storm surges in the Bay of Bengal. Given that a model is suitably designed and properly adjusted with an appropriate level of frictional dissipation, whether it is two-dimensional or three-dimensional appears to be largely irrelevant to the satisfactory reproduction of surge elevations.

Taking the above facts into account, the present paper employs horizontal gradients of elevation derived from a well-tested vertically-integrated numerical storm-surge model to drive a one-dimensional model for the vertical distribution of surge current at any particular location (figure 1). Wind stress and atmospheric pressure gradients at that location, derived from

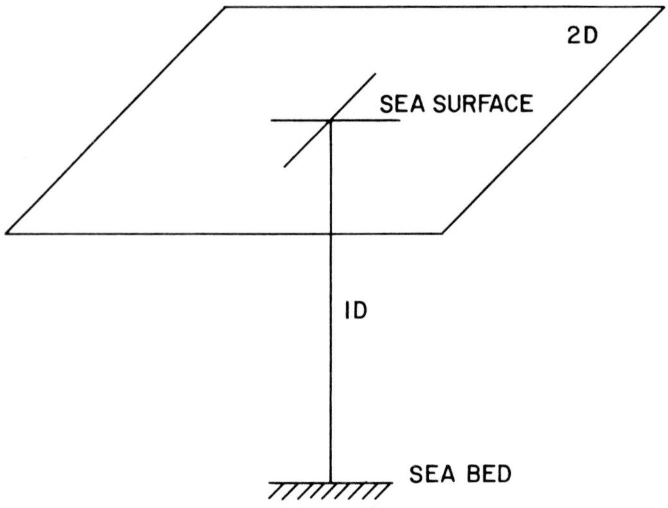

Fig. 1. Combination of a one-dimensional (1D) model through the vertical, sea surface to sea bed, and a two-dimensional (2D) vertically-integrated model in the horizontal.

meteorology, are also involved in the driving. The three-dimensional problem of determining current structure through the vertical is thus resolved into separate two-dimensional and one-dimensional computations, the former preceding the latter.

Similar earlier work on the determination of the distribution of storm-surge currents through the vertical at a point, notably that done by Forristall (1974, 1980), connects the one-dimensional and two-dimensional models of figure 1 through bottom friction. The latter is evaluated at the sea bed in the one-dimensional model and is expressed in terms of certain convolution integrals involving time histories of the wind stress and the storm-surge gradient at the point location. Surge elevation gradients from the two-dimensional model again drive the one-dimensional model to obtain the changing vertical structure of current, but their derivation in the two-dimensional scheme is based on a difficult integro-differential equation (arising from the three-dimensional form of bottom friction assumed) which leads to a complicated analysis. In the present paper it is suggested that such recourse to the three-dimensionality of the problem in determining surge-elevation gradients is largely unnecessary, asserting rather that a well-adjusted two-dimensional vertically-integrated model of conventional design is sufficient to provide them. The simplification resulting from this assertion is considerable.

2 BASIC EQUATIONS

The hydrodynamical equations of motion are taken in the form

$$\frac{\partial u}{\partial t} - \gamma v = -gP + \frac{\partial}{\partial z}(\mu\frac{\partial u}{\partial z}) \tag{1}$$

$$\frac{\partial v}{\partial t} - \gamma u = -gQ + \frac{\partial}{\partial z}(\mu\frac{\partial v}{\partial z}) \tag{2}$$

where

$$P = \frac{\partial \zeta}{\partial x} + \frac{1}{\rho g}\frac{\partial p_a}{\partial x}\ , \qquad Q = \frac{\partial \zeta}{\partial y} + \frac{1}{\rho g}\frac{\partial p_a}{\partial y} \tag{3}$$

Boundary conditions are, at the sea surface,

$$-\rho(\mu\frac{\partial u}{\partial z})_0 = F\ , \qquad -\rho(\mu\frac{\partial v}{\partial z})_0 = G \tag{4}$$

and, at the sea bed,

$$-\rho(\mu\frac{\partial u}{\partial z})_h = k\rho u_h\ , \qquad -\rho(\mu\frac{\partial v}{\partial z})_h = k\rho v_h \tag{5}$$

The above formulation and subsequent analysis follows that of Heaps (1974) and the notation is

t	time
x,y,z	Cartesian coordinates forming a left-handed set in which x, y are measured in the horizontal plane of the undisturbed sea surface and z is depth below that surface,
h	undisturbed depth of water,
ζ	elevation of the water surface,
u,v	components of current at depth z in the directions of increasing x, y respectively,
p_a	atmospheric pressure on the sea surface
F,G	components of wind stress on the sea surface in the x, y directions,
μ	coefficient of vertical eddy viscosity,
k	coefficient of friction in a linear law of bottom stress,
ρ	density of the water, assumed constant (homogeneous sea)
γ	geostrophic coefficient, also assumed constant,
g	acceleration of the Earth's gravity.

Suffix 0 indicates evaluation at the surface z=0 and suffix h evaluation at the sea bed z=h.

By a vertical integration, Heaps (1974) showed that equations (1) and (2), subject to conditions (4) and (5) may be transformed into:

$$\frac{\partial u_r}{\partial t} + \lambda_r u_r - \gamma v_r = - g a_r P + \frac{F}{\rho h} \tag{6}$$

$$\frac{\partial v_r}{\partial t} + \lambda_r v_r + \gamma u_r = - g a_r Q + \frac{G}{\rho h} \tag{7}$$

for $r = 1,2,3,\ldots,$ where

$$u_r = \frac{1}{h}\int_o^h u f_r dz \;, \quad v_r = \frac{1}{h}\int_o^h v f_r dz \;, \quad a_r = \frac{1}{h}\int_o^h f_r dz \tag{8}$$

The λ_r $(r=1,2,3,\ldots)$ are the ascending eigenvalues and the f_r $(r=1,2,3,\ldots)$ the corresponding eigenfunctions through depth satisfying

$$\frac{\partial}{\partial z}\left(\mu\frac{\partial f}{\partial z}\right) = - \lambda f \;, \qquad 0 \leqq z \leqq h \tag{9}$$

$$\left(\frac{\partial f}{\partial z}\right)_o = 0 \tag{10}$$

$$\left(\mu\frac{\partial f}{\partial z}\right)_h + k f_h = 0 \tag{11}$$

$$f_o = 1 \tag{12}$$

Thus, solutions of (9)-(12) yield

$$\lambda = \lambda_r \;, \; f = f_r \; (r = 1,2,3,\ldots.) \tag{13}$$

Eddy viscosity μ and frictional coefficient k are assumed to be independent of the time t and hence so also are λ_r and f_r. In general, we have $\mu=\mu(x,y,z)$, $k=k(x,y)$ and consequently $\lambda_r = \lambda_r(x,y)$, $f_r=f_r(x,y,z)$. Apart from the positional dependency, therefore, f_r is a function of z defined in the depth range $0 \leqq z \leqq h$. The functions f_r form an orthogonal set in that range which enables u and v to be expressed in terms of u_r and v_r in the series forms:

$$u = \sum_{r=1}^{\infty} \Phi_r u_r f_r \quad , \quad v = \sum_{r=1}^{\infty} \Phi_r v_r f_r \tag{14}$$

where

$$\Phi_r = \frac{h}{\int_o^h f_r^2 dz} \tag{15}$$

Equation (14) shows that u and v may each be expressed, at any time, as a linear combination of a set of current modes through the vertical, namely $f_r(z)$, $r=1,2,3\ldots$. Since u_r and v_r vary with time, so also does the proportion of each mode in the total sum giving either u or v.

Considering motion at any particular position, the method of the present work is to solve (6), (7) through time for u_r and v_r with F, G, P, Q as prescribed forcing functions varying with time. The Laplace transformation method is employed to do this. Then (14) is used to deduce the variations in horizontal current through the depth at the position.

Eddy viscosity μ is chosen to be independent of the depth z for the purposes of the immediate analysis; it is related however to wind and tide conditions (see section 4). This choice of invariability through depth provides a base-line solution of equations (9)-(12). It may be justified if attention is mainly directed towards the vertical distribution of current in the main body of the vertical water column, away from the surface and bottom boundary layers. In any case, further elaboration to account for a depth-dependent μ is speculative at the present time.

In the above circumstances of a depth dependent μ, Heaps (1974) found

$$\lambda_r = \mu\alpha_r^2/h^2 \tag{16}$$

$$f_r = \cos\alpha_r\xi \quad (\xi = z/h) \tag{17}$$

$$a_r = \sin\alpha_r/\alpha_r \tag{18}$$

$$\Phi_r = 2/(1 + a_r \cos\alpha_r) \tag{19}$$

where

$$\alpha_r \ (r = 1,2,3 \ldots) \tag{20}$$

denote the ascending non-negative roots of the equation

$$\alpha\tan\alpha = c \ (c = kh/\mu) \tag{21}$$

Further, taking the series in (14) to say M terms (a practical necessity) and adding a correction term to allow for the truncation, leads to working formulae:

$$u = \sum_{r=1}^{M} \Phi_r u_r f_r + \frac{2hF}{\rho\mu\pi^2} R_M(\xi) \tag{22}$$

$$v = \sum_{r=1}^{M} \Phi_r v_r f_r + \frac{2hG}{\rho\mu\pi^2} R_M(\xi) \tag{23}$$

where

$$R_M(\xi) = (3\xi^2 - 6\xi + 2)\frac{\pi^2}{12} - \sum_{r=1}^{M} \frac{\cos r\pi\xi}{r^2} \tag{24}$$

3 VERTICAL DISTRIBUTION OF CURRENT AT ANY PARTICULAR LOCATION

It is assumed that F, G, P, Q are prescribed at regular intervals δt, from some origin of time t=0, at the position where the vertical distribution of current is to be estimated. Wind stress components F, G and atmospheric pressure gradients $\partial p_a/\partial x$, $\partial p_a/\partial y$ come from meteorological analyses while residual sea level gradients $\partial\zeta/\partial x$, $\partial\zeta/\partial y$ are derived from a two-dimensional vertically-integrated model. The P and Q are evaluated from (3) using model grid-point values of ζ and p_a to approximate the spatial derivatives (see section 5). Values are denoted by:

$$F = F_j,\ G = G_j,\ P = P_j,\ Q = Q_j;\ \text{at } t = j\delta t \quad (j = 0,1,2,3,\ldots.) \tag{25}$$

Over a typical time interval $[j\delta t, (j+1)\delta t]$, assuming a linear variation of each variable with time, we have:

$$\left.\begin{aligned} F &= F_j + (F_{j+1} - F_j)(t - j\delta t)/\delta t \\ G &= G_j + (G_{j+1} - G_j)(t - j\delta t)/\delta t \\ P &= P_j + (P_{j+1} - P_j)(t - j\delta t)/\delta t \\ Q &= Q_j + (Q_{j+1} - Q_j)(t - j\delta t)/\delta t \end{aligned}\right\} \tag{26}$$

Figure 2 indicates the type of variation considered. Average components of wind stress over $[j\delta t, (j+1)\delta t]$ are:

$$\hat{F}_j = \tfrac{1}{2}(F_j + F_{j+1}), \quad \hat{G}_j = \tfrac{1}{2}(G_j + G_{j+1}) \tag{27}$$

and the associated average resultant wind stress is therefore

$$\hat{E}_j = (\hat{F}_j^2 + \hat{G}_j^2)^{\frac{1}{2}} \tag{28}$$

On the basis of $\hat{E}_j$, the eddy viscosity μ for the interval may be estimated: $\mu=\mu_j$ (see section 4). It is assumed that μ is independent of the depth coordinate, remains invariant over each interval, but generally changes from one interval to the next according to the change in wind stress.

Taking Laplace transforms in the equations (6) and (7), defining (Jaeger 1961) the transform of a variable $y(\tau)$ as

$$\bar{y}(p) = \int_0^\infty e^{-p\tau} y(\tau)\,d\tau\ , \tag{29}$$

yields

$$p\bar{u}_r - u_r^o + \lambda_r\bar{u}_r - \gamma\bar{v}_r = -ga_r\bar{P} + \bar{F}/\rho h \tag{30}$$

$$p\bar{v}_r - v_r^o + \lambda_r\bar{v}_r + \gamma\bar{u}_r = -ga_r\bar{Q} + \bar{G}/\rho h \tag{31}$$

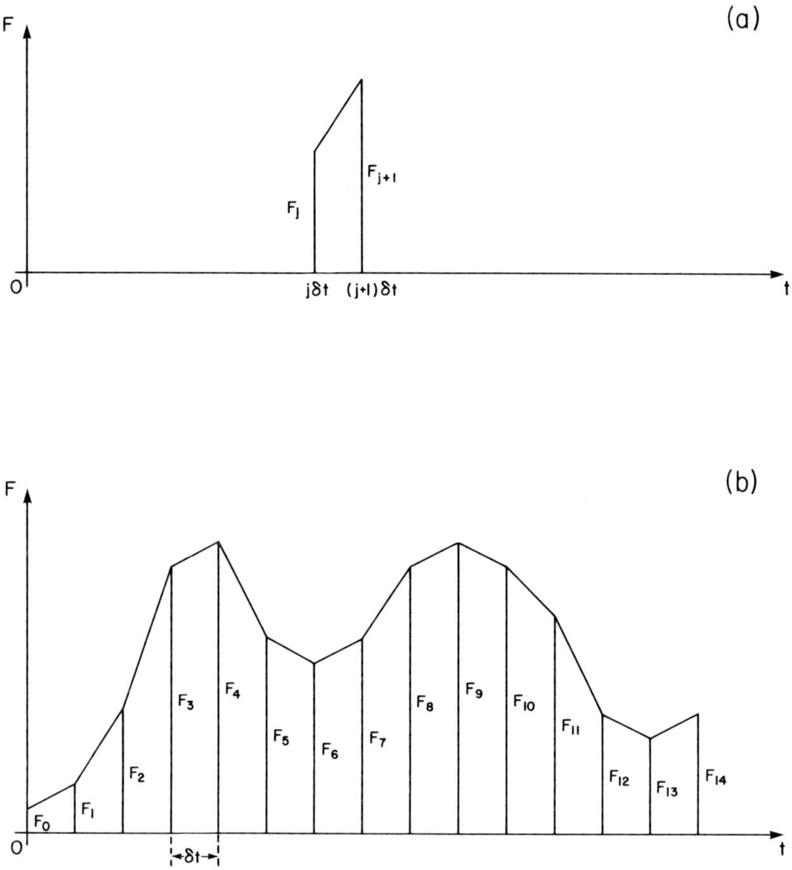

Fig. 2. Diagrams illustrating the variation of F with time t, (a) over interval $[j\delta t,(j+1)\delta t]$ and (b) for all t.

Here, τ is the time variable while u_r^o and v_r^o denote the initial ($\tau=0$) values of u_r and v_r respectively. Using (30) and (31) to solve for $\overline{u}_r$, $\overline{v}_r$ gives:

$$\overline{u}_r = \frac{\gamma(v_r^o - ga_r\overline{Q} + \overline{G}/\rho h) + (p + \lambda_r)(u_r^o - ga_r\overline{P} + \overline{F}/\rho h)}{(p + \lambda_r)^2 + \gamma^2} \tag{32}$$

$$\overline{v}_r = \frac{(p + \lambda_r)(v_r^o - ga_r\overline{Q} + \overline{G}/\rho h) - \gamma(u_r^o - ga_r\overline{P} + \overline{F}/\rho h)}{(p + \lambda_r)^2 + \gamma^2} \tag{33}$$

Laplace transform inversion for the interval $[j\delta t,(j+1)\delta t]$ with

$$\tau = t - j\delta t \tag{34}$$

may be performed on (32) and (33) using standard methods. In this, note from (26) that

$$\left.\begin{aligned}
\bar{F} &= \frac{F_j}{p} + \frac{(F_{j+1} - F_j)}{\delta t}\frac{1}{p^2} \\
\bar{G} &= \frac{G_j}{p} + \frac{(G_{j+1} - G_j)}{\delta t}\frac{1}{p^2} \\
\bar{P} &= \frac{P_j}{p} + \frac{(P_{j+1} - P_j)}{\delta t}\frac{1}{p^2} \\
\bar{Q} &= \frac{Q_j}{p} + \frac{(Q_{j+1} - Q_j)}{\delta t}\frac{1}{p^2}
\end{aligned}\right\} \quad (35)$$

The result of the inversion is

$$\begin{aligned}
u_r = {} & (u_r^o \cos\gamma\tau + v_r^o \sin\gamma\tau)e^{-\lambda_r\tau} \\
& + V_j\delta t(\gamma - \gamma e^{-\lambda_r\tau}\cos\gamma\tau - \lambda_r e^{-\lambda_r\tau}\sin\gamma\tau) \\
& + U_j\delta t(\lambda_r - \lambda_r e^{-\lambda_r\tau}\cos\gamma\tau + \gamma e^{-\lambda_r\tau}\sin\gamma\tau) \\
& + (V_{j+1} - V_j)(\gamma\tau - K + Ke^{-\lambda_r\tau}\cos\gamma\tau + \nu e^{-\lambda_r\tau}\sin\gamma\tau) \\
& + (U_{j+1} - U_j)(\lambda_r\tau - \nu + \nu e^{-\lambda_r\tau}\cos\gamma\tau - Ke^{-\lambda_r\tau}\sin\gamma\tau)
\end{aligned} \quad (36)$$

$$\begin{aligned}
v_r = {} & (v_r^o \cos\gamma\tau - u_r^o \sin\gamma\tau)e^{-\lambda_r\tau} \\
& + V_j\delta t(\lambda_r - \lambda_r e^{-\lambda_r\tau}\cos\gamma\tau + \gamma e^{-\lambda_r\tau}\sin\gamma\tau) \\
& - U_j\delta t(\gamma - \gamma e^{-\lambda_r\tau}\cos\gamma\tau - \lambda_r e^{-\lambda_r\tau}\sin\gamma\tau) \\
& + (V_{j+1} - V_j)(\lambda_r\tau - \nu + \nu e^{-\lambda_r\tau}\cos\gamma\tau - Ke^{-\lambda_r\tau}\sin\gamma\tau) \\
& - (U_{j+1} - U_j)(\gamma\tau - K + Ke^{-\lambda_r\tau}\cos\gamma\tau + \nu e^{-\lambda_r\tau}\sin\gamma\tau)
\end{aligned} \quad (37)$$

where

$$\left.\begin{aligned}
U_j &= \frac{1}{\delta t}\left(\frac{F_j}{\rho h} - g a_r P_j\right)\frac{1}{\lambda_r^2+\gamma^2} \\
V_j &= \frac{1}{\delta t}\left(\frac{G_j}{\rho h} - g a_r Q_j\right)\frac{1}{\lambda_r^2+\gamma^2} \\
K &= \frac{2\lambda_r\gamma}{\lambda_r^2+\gamma^2} \\
\nu &= \frac{\lambda_r^2-\gamma^2}{\lambda_r^2+\gamma^2}
\end{aligned}\right\} \quad (38)$$

with u_r and v_r given by (36) and (37), the vertical structure of horizontal current at the position under consideration, for $j\delta t \leq t < (j+1)\delta t$, is determined (see equations (16)-(24)) from

$$\left.\begin{aligned} u &= \sum_{r=1}^{M} \Phi_r u_r f_r + \frac{2hF}{\rho\mu_j \pi^2} R_M(\xi) \\ v &= \sum_{r=1}^{M} \Phi_r v_r f_r + \frac{2hG}{\rho\mu_j \pi^2} R_M(\xi) \end{aligned}\right\} \quad (39)$$

in which

$$\left.\begin{aligned} \lambda_r &= \mu_j \alpha_r^2/h^2 \\ f_r &= \cos\alpha_r\xi \quad (\xi = z/h) \\ a_r &= \sin\alpha_r/\alpha_r \\ \Phi_r &= 2/(1 + a_r\cos\alpha_r) \\ R_M(\xi) &= (3\xi^2 - 6\xi + 2)\frac{\pi^2}{12} - \sum_{r=1}^{M} \frac{\cos r\pi\xi}{r^2} \end{aligned}\right\} \quad (40)$$

where

$$\alpha_r \ (r = 1,2,3, \dots M) \tag{41}$$

denote, in ascending order, the first M non-negative roots of

$$\alpha \tan \alpha = c \quad (c = kh/\mu_j) \tag{42}$$

As in previous work (Heaps 1974), c may be taken as a known constant, implying $k=c\mu_j/h$ (variable with μ_j). Then α_r, f_r, a_r, Φ_r are unchanging but λ_r differs from one interval to the next because of its dependence on μ_j. Alternatively, if c varies between intervals, as would obtain with k a known constant, then so also would α_r, f_r, a_r, Φ_r leading changes through time in the structure of the vertical modes.

Currents u, v through the depth z are evaluated over each interval $[j\delta t,(j+1)\delta t]$ using (39) and (40) with u_r and v_r from (36) and (37). The calculations for interval j+1 follow directly after those for interval j. Thus the computations advance in a series of successive steps through time, corresponding respectively to j = 0,1,2,3.... . The finishing values of u_r, v_r for one interval are taken as the starting values u_r^o, v_r^o for the next. Initially, it is convenient to take $u_r^o = v_r^o = 0$ corresponding to a state of zero initial motion.

4 EDDY VISCOSITY

Using a result due to Svensson (1979), and putting together arguments given by Csanady (1976) and Heaps (1984), for wind-driven flow during the interval $[j\delta t,(j+1)\delta t]$ we take

$$\mu = \mu_W \tag{43}$$

where

$$\left.\begin{aligned} \mu_W &= 0.065hu_* && h \leqq h_W \\ &= 0.026u_*^2/\gamma && h > h_W \end{aligned}\right\} \quad (44)$$

and

$$h_w = 0.4u_*/\gamma \quad ; \tag{45}$$

u_* is the friction velocity given by

$$u_* = (\hat{E}_j/\rho)^{\frac{1}{2}} \tag{46}$$

and $\hat{E}_j$ the average wind stress over the interval, defined by (28).

Turbulence associated with the tides is generated by the action of frictional stress, S_b say, at the sea bed. It is therefore supposed that the associated eddy viscosity, μ_T say, may be estimated from equations (44)-(46) replacing u_* by u_{*b} where

$$u_{*b} = (\hat{S}_b/\rho)^{\frac{1}{2}} \tag{47}$$

Here, $\hat{S}_b$ represents the bottom stress S_b averaged over a tidal cycle. Assume a quadratic law:

$$S_b = K\rho(u_m{}^2 + v_m{}^2) \tag{48}$$

in which K is a frictional coefficient and u_m, v_m depth-mean components of tidal current for the mean tide given by

$$u_m = a\cos\omega t, \quad v_m = b\sin\omega t \tag{49}$$

where a,b are the semi-major and semi-minor axes of the tidal current ellipse and ω is the tidal frequency. Then integration over a tidal cycle leads to the result:

$$\hat{S}_b = \tfrac{1}{2}K\rho(a^2 + b^2) \tag{50}$$

Hence

$$u_{*b} = (K/2)^{\frac{1}{2}} (a^2 + b^2)^{\frac{1}{2}} \tag{51}$$

and substitution in (44), (45) yields

$$\left.\begin{aligned} \mu_T &= 0.065h(K/2)^{\frac{1}{2}} (a^2 + b^2)^{\frac{1}{2}} && h \leq h_T \\ &= 0.013K (a^2 + b^2)/\gamma && h > h_T \end{aligned}\right\} \tag{52}$$

where

$$h_T = 0.4(K/2)^{\frac{1}{2}} (a^2 + b^2)^{\frac{1}{2}}/\gamma \tag{53}$$

Typically,

$$K = 0.0026 \tag{54}$$

in which case

$$\left.\begin{aligned} \mu_T &= 0.00234h(a^2 + b^2)^{\frac{1}{2}} && h \leq h_t \\ &= 0.338 \times 10^{-4} (a^2 + b^2)/\gamma && h > h_T \end{aligned}\right\} \tag{55}$$

where

$$h_T = 0.0144 (a^2 + b^2)^{\frac{1}{2}}/\gamma \tag{56}$$

The form for μ_T, $h \leq h_T$, is comparable to that suggested by Bowden (1953).

Using values obtained from (44) and (55), for combined wind and tide we take

$$\left.\begin{aligned} \mu_j &= \mu_T \text{ for } \mu_T > \mu_W \\ &= \mu_W \text{ for } \mu_W > \mu_T \end{aligned}\right\} \quad (57)$$

This equation determines μ_j, i.e. μ in the interval $[j\delta t,(j+1)\delta t]$, at the particular position under consideration, in terms of $\hat{E}_j$, a, b, at that position.

5 HORIZONTAL GRADIENTS

Horizontal gradients of ζ and p_a are required for the determination of P and Q from (3). These may be evaluated as follows.

Suppose the point at which currents are to be estimated coincides with grid point 0 of a two-dimensional numerical model (figure 3). Then values of ζ from the model, at neighbouring grid points may be employed to compute $\partial\zeta/\partial x$ and $\partial\zeta/\partial y$ at 0. Thus, at any particular time, let ζ_{-3}, ζ_{-2}, ζ_{-1}, ζ_1, ζ_2, ζ_3 denote the ζ-values at the centres of the grid boxes located symmetrically on either side of 0 in the x-direction. These values are marked in figure 3. Then standard formulae for numerical differentiation (Whittaker and Robinson 1944; Comrie 1949, p.544) give, going to first order differences,

$$\left(\frac{\partial\zeta}{\partial x}\right)_o = \frac{1}{2\Delta x}\left(\zeta_1 - \zeta_{-1}\right) \quad (58)$$

or going to third order,

$$\left(\frac{\partial\zeta}{\partial x}\right)_o = \frac{1}{12\Delta x}\left(-\zeta_2 + 8\zeta_1 - 8\zeta_{-1} + \zeta_{-2}\right) \quad (59)$$

or, to fifth order,

$$\left(\frac{\partial\zeta}{\partial x}\right)_o = \frac{1}{60\Delta x}\left(\zeta_3 - 9\zeta_2 + 45\zeta_1 - 45\zeta_1 + 9\zeta_{-2} - \zeta_{-3}\right) \quad (60)$$

Here, Δx denotes the grid spacing in the x-direction, supposed to be a constant. Suffix 0 in the above formulae indicates evaluation at grid point 0.

The same expressions may be used in the y-direction to give $(\partial\zeta/\partial y)_o$, with Δx replaced by Δy - the grid spacing in that direction. Values of ζ associated with appropriate neighbouring grid points, lying symmetrically on either side of point 0 in the y-direction, are then employed.

Knowing values of p_a at the same grid points enables $(\partial p_a/\partial x)_o$ and $(\partial p_a/\partial y)_o$ to be evaluated similarly.

Alternatively, for convenience, P and Q may be determined from

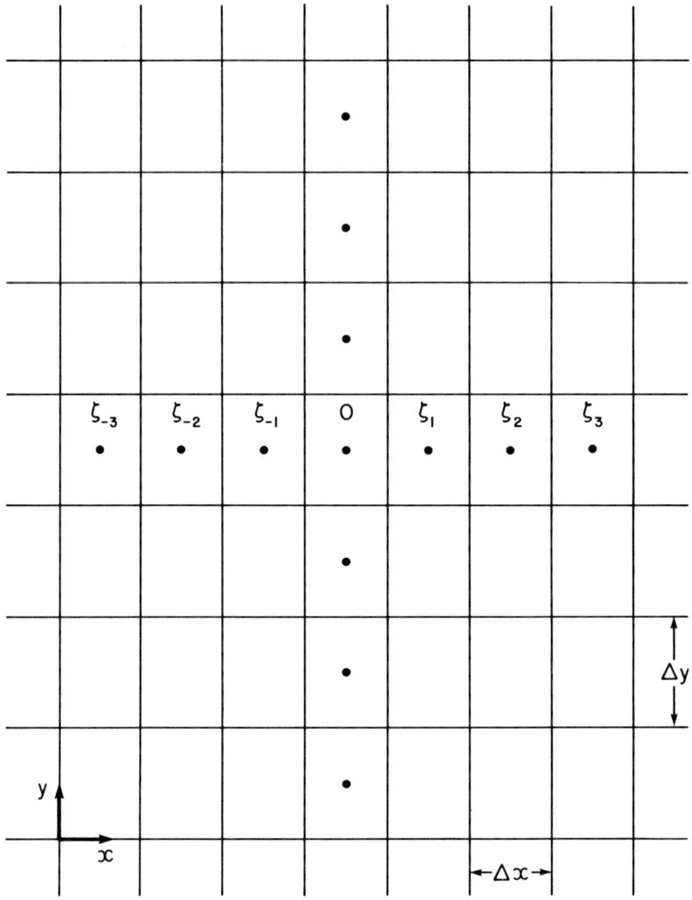

Fig. 3. Grid points at which ζ is required for the evaluation of $\partial\zeta/\partial x$ and $\partial\zeta/\partial y$ at point 0. Values of in the x-direction through 0 are marked. Each point lies at the centre of a grid square.

$$P = \frac{\partial}{\partial x}(\zeta-\zeta_H) \quad , \quad Q = \frac{\partial}{\partial y}(\zeta-\zeta_H) \tag{61}$$

where ζ_H denotes the hydrostatic elevation associated with the atmospheric pressure p_a

$$\frac{\partial p_a}{\partial x} = -\rho g\frac{\partial \zeta_H}{\partial x} \quad , \quad \frac{\partial p_a}{\partial y} = -\rho g\frac{\partial \zeta_H}{\partial y} \tag{62}$$

Then, grid-point values of $\zeta-\zeta_H$ may be used to obtain the required horizontal gradients at 0 for the evaluation of P and Q there.

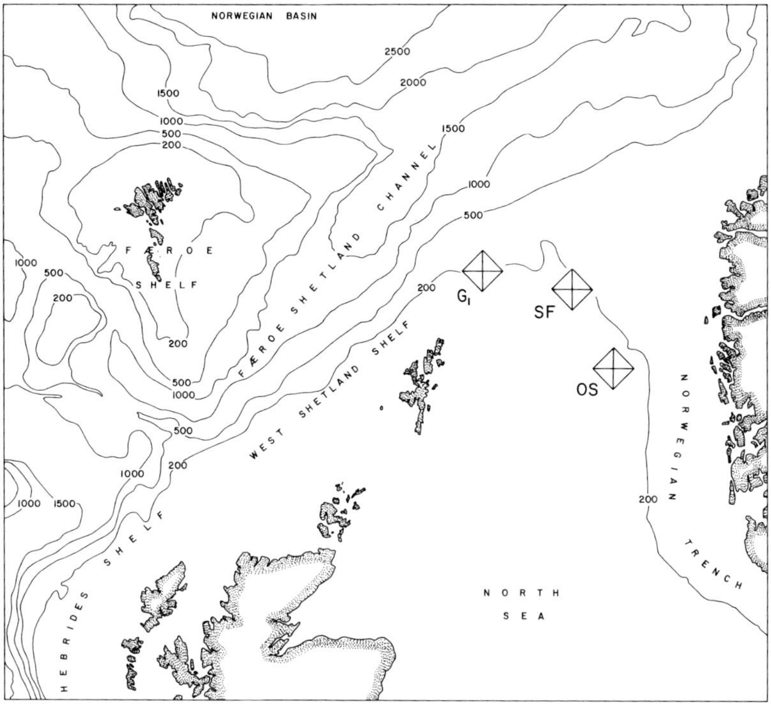

Fig. 4. Geographical positions of Stations G1, Statfjord (SF) and Oseberg (OS).

6 APPLICATIONS OF THE THEORY

The theory of this paper has been applied to predict storm-generated currents measured at three stations located near the shelf edge to the north of the North Sea. The stations, marked in figure 4, are:

	G_1	(61°30.7'N, 0°2.5'E)
Statfjord	SF	(61°19.54'N, 1°55.12'E)
Oseberg	OS	(60°33.9'N, 2°47.7'E)

The current meter positions through the vertical at each station are illustrated in figure 5. Thus at G_1 (total depth 191m), records come from 41m below the sea surface and 25m above the sea bed; at SF (total depth 150m), from 30, 70m below the surface and 3m above the bed; and at OS (total depth 102m), from 2, 12, 25, 50m below the surface and 3m above the bed.

For every meter, residuals were extracted from the raw data by removal of the tidal signal, this being determined on the basis of a full harmonic tidal analysis. However the sampling interval of the raw data varied from meter to meter. At G_1 measurements were half hourly; at Oseberg these were hourly for the top three meters and every 10 minutes for the others. The Statfjord

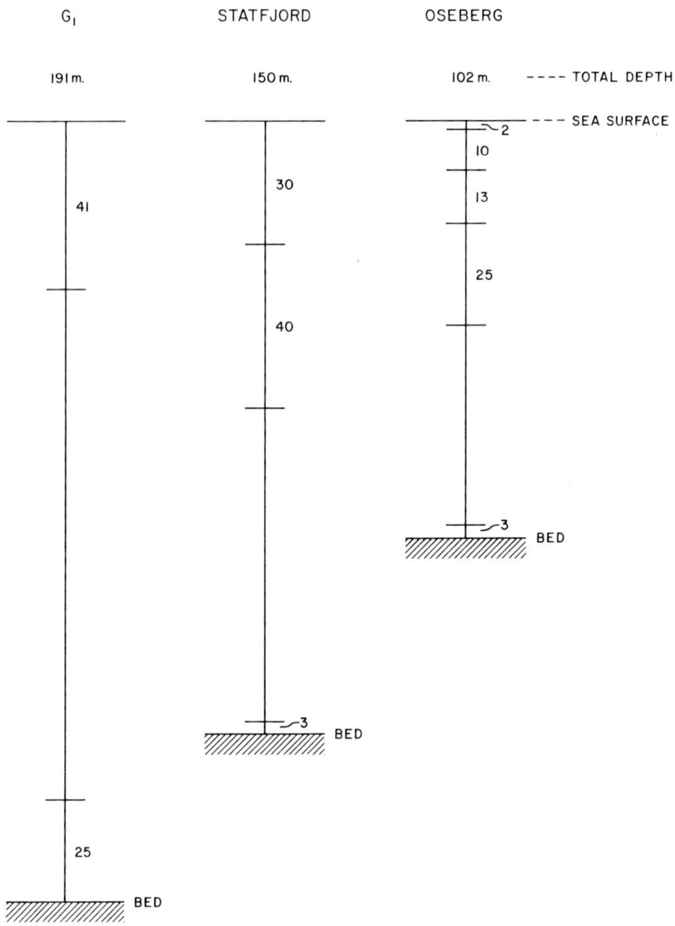

Fig. 5. Current meter positions through the vertical at Stations G_1, Statfjord and Oseberg during January 1983. Distances through the vertical are marked in metres.

measurements were hourly. For convenience, by a combination of a low pass filter and cubic spline interpolation the extracted residuals were converted to hourly values at integer clock hours.

The measurements at G_1 were part of the Continental Slope Experiment (CONSLEX) during the winter of 1982-3. The Oseberg data (raw measurements + residuals) were provided by Norsk Hydro. Det Norske Veritas supplied the Statfjord data which was collected under the auspices of the Norwegian Meteorological Institute.

Comparisons between theory and observation have been made throughout the whole of January 1983. Thus for each meter, hourly time series of east-going

TABLE 1

Parametric values used in applying the theory

Parameter	Units	G1	SF	OS
h	m	191	150	102
γ	s^{-1}	$1.278x10^{-4}$	$1.278x10^{-4}$	$1.270x10^{-4}$
g	ms^{-2}	9.81	9.81	9.81
a	ms^{-1}	10.0	4.5	8.1
b	ms^{-1}	5.0	4.5	3.6
δt	hr	1	1	1
Δx	km	26.56	26.56	26.56
Δy	km	37.11	37.11	37.11
ρ	$g\ cm^{-3}$	1.025	1.025	1.025
M		30	30	30
C		2	2	2

(u) and north-going (v) current components have been plotted directly against corresponding values obtained from the theory. Power spectra involving the various series have elucidated properties of coherence and phase spanning a range of periods from the Nyquist frequency of two hours, to beyond the inertial period (approximately 14 hours).

A relentless pattern of strong winds from a westerly quarter was a key feature of the weather in the area to the north of the North Sea during January 1983, making that area and period suitable for a study of storm-induced currents.

Parametric values used in our present application of the theory are listed in Table 1. At each station, current components u, v through depth were evaluated at hourly intervals (δt = 1 hour) using the expressions (36)-(42). In the latter, we took c=2 as in previous work (Heaps 1974) and M=30 (this being an over-prescription of the number of nodes, to ensure accuracy within the limits of the theory).

For each station the computations of current required hourly values of P and Q, [see equations (3) and (61)]. To this end, hourly surge elevations ζ were obtained from runs with a two-dimensional vertically-integrated model (CXX) of the North-West European continental shelf which extended beyond the shelf edge. This model was developed by Dr. R. A. Flather at the Institute of Oceanographic Sciences, Bidston. The model had a grid size of 1/3° latitude by 1/2° longitude (figure 6). In the vicinity of stations G1, SF and OS this effectively gave hourly ζ at the grid points of a rectangular network such as that in figure 3, with Δx = 26.56km and Δy = 37.11km as entered in Table 1. Meteorological data yielded 3-hourly values of hydrostatic elevation ζ_H at coincidental points and times; these were converted to hourly values by a

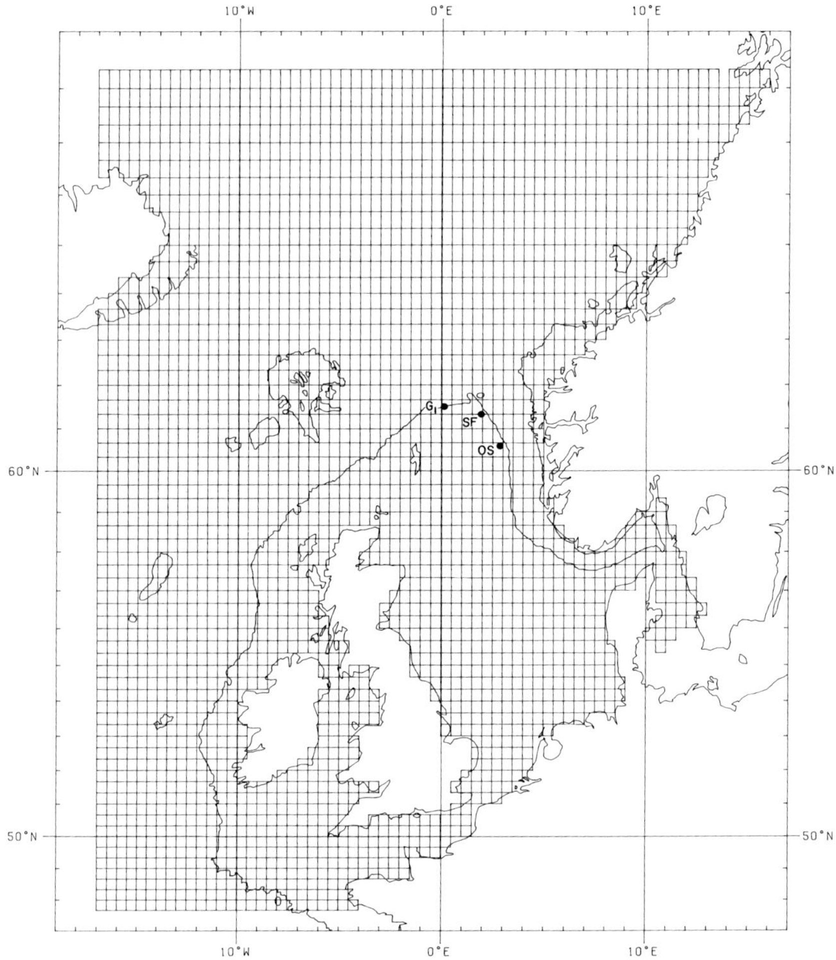

Fig. 6. Two-dimensional model CXX, showing grid network and the positions G_1, SF, OS.

linear interpolation. Then, hourly values of P and Q were evaluated at each station using difference formula (59) with each term on the right hand side of the equation replaced by the appropriate $(\zeta - \zeta_H)$.

Also at each station the computations of current required hourly values of the wind stress components F, G [see equation (21)]. Meteorological data provided 3-hourly values, which were converted to hourly values by linear interpolation. Eddy viscosity μ (depth uniform and changing from one hourly interval to the next), also required in computing the currents, was deduced from the hourly wind stress components employing the formulae of section 4. Semi-major and semi-minor axes of the tidal current ellipse at each station

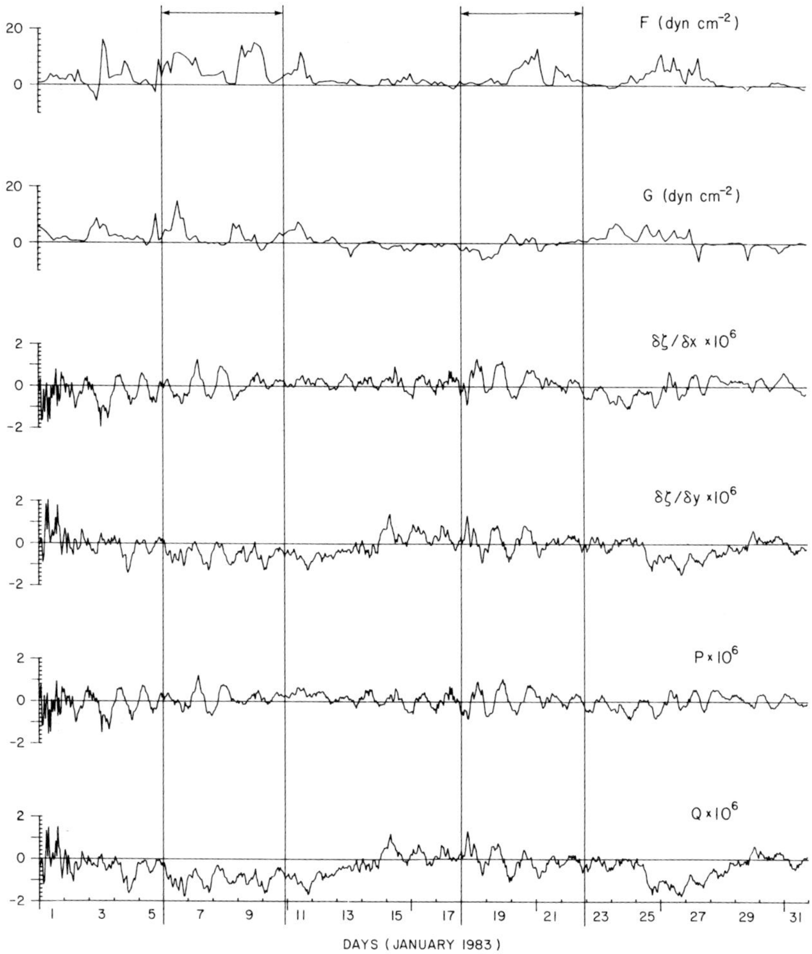

Fig. 7. Variations of F, G, $\partial\zeta/\partial x$, $\partial\zeta/\partial y$, P and Q at G_1 throughout January 1983.

were involved in this deduction but tidal effects on μ were insignificant in the present calculations (a, b small, see Table 1).

7 PRESENTATION OF RESULTS

The forcing functions of wind stress F, G and gradient P, Q used to drive the model at the point G_1 are depicted in figure 7. It may be noted that the 6th-10th January was a period of high wind stress followed by a quieter period during the middle of the month. Subsequently the 18th-22nd January was period of fairly high regular gradient forcing.

Throughout the whole month the $\partial\zeta/\partial x$ and $\partial\zeta/\partial y$ values closely follow the P and Q curves. Thus by equations (61) and (62) we may deduce that the

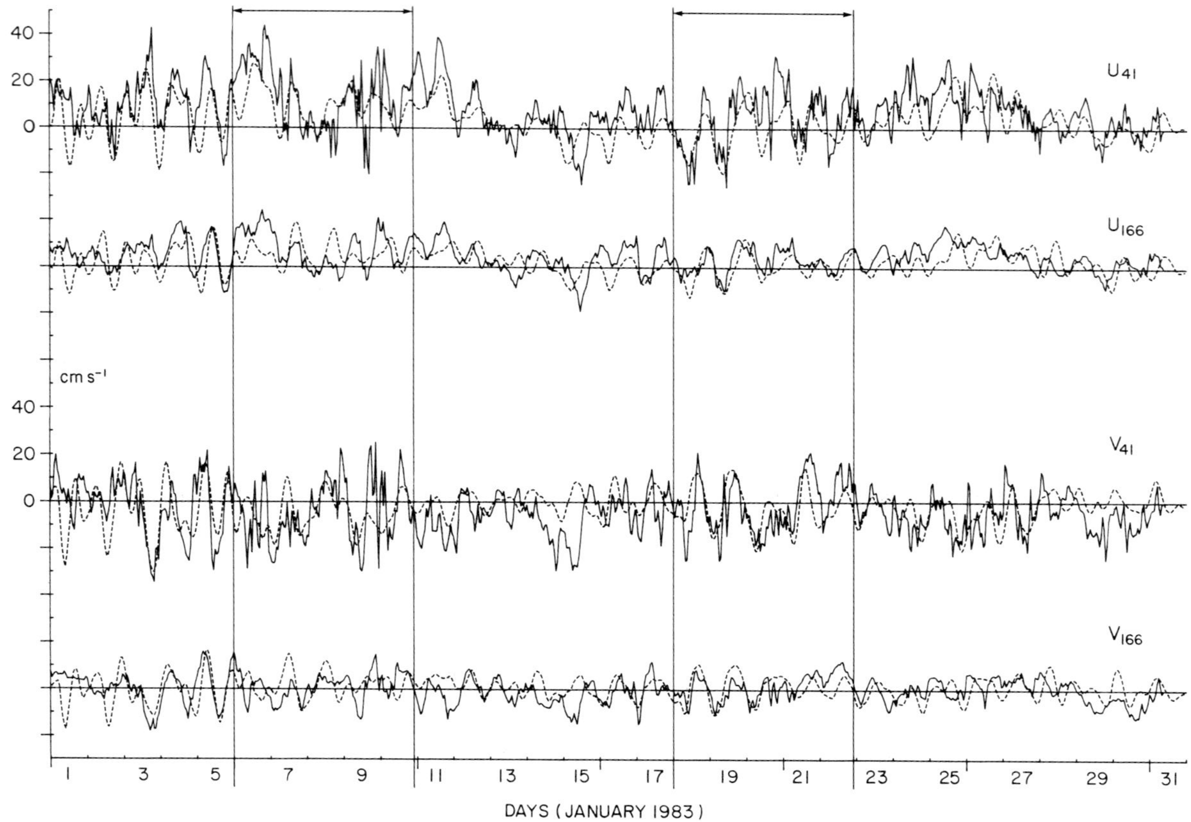

Fig. 8. Theoretical predictions of u- and v-currents at G_1, at 41 and 166 metres below the sea surface, compared with the corresponding currents measured at those depths: for the whole of January 1983; ——— observation, ----- model prediction. For the two five-day periods delineated, vertical profiles of current are drawn in figures (12a) and (12b).

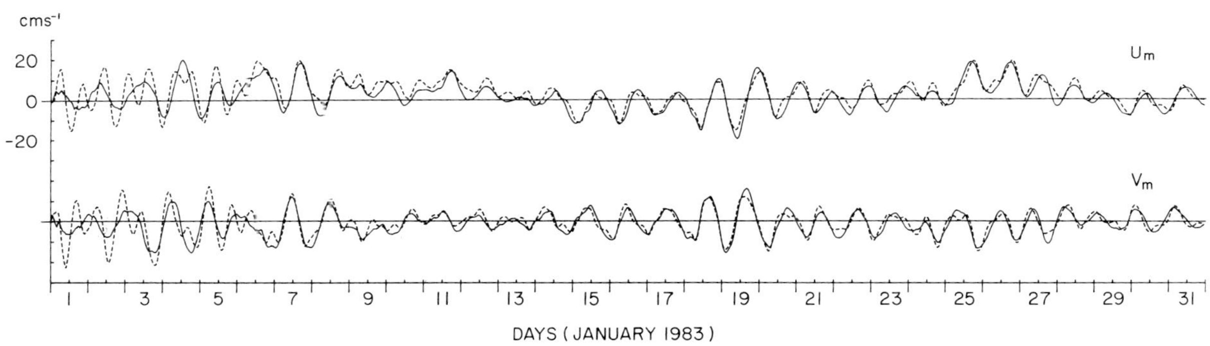

Fig. 9. Components of depth-mean current u_m, v_m at G_1, as derived from the one-dimensional model through the vertical (-----) and the two-dimensional model in the horizontal (———).

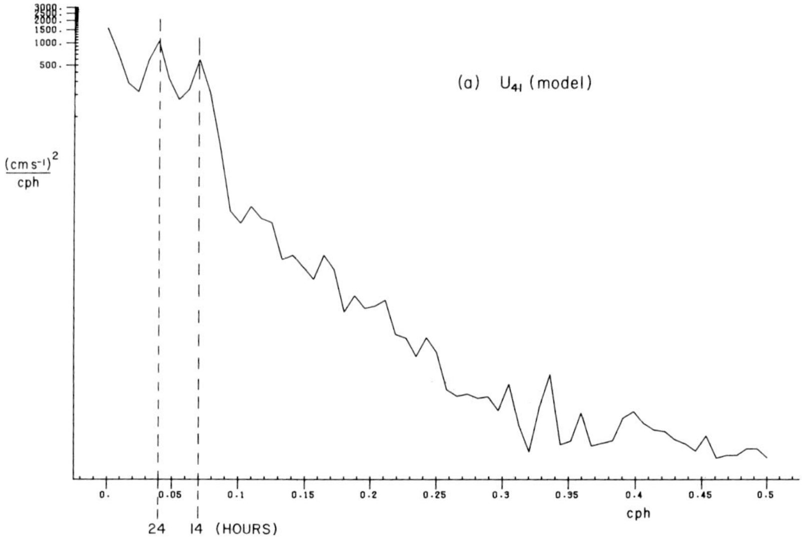
3000.
2500.
2000.
1500.
1000.
500.
(a) U_{41} (model)
$\frac{(cm\,s^{-1})^2}{cph}$
0.
0.05
0.1
0.15
0.2
0.25
0.3
0.35
0.4
0.45
0.5
cph
24 14 (HOURS)

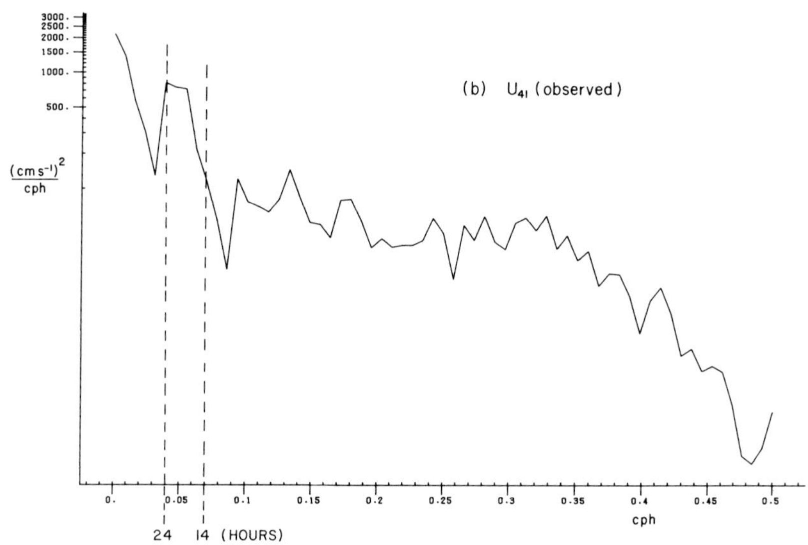
3000.
2500.
2000.
1500.
1000.
500.
(b) U_{41} (observed)
$\frac{(cm\,s^{-1})^2}{cph}$
0.
0.05
0.1
0.15
0.2
0.25
0.3
0.35
0.4
0.45
0.5
cph
24 14 (HOURS)

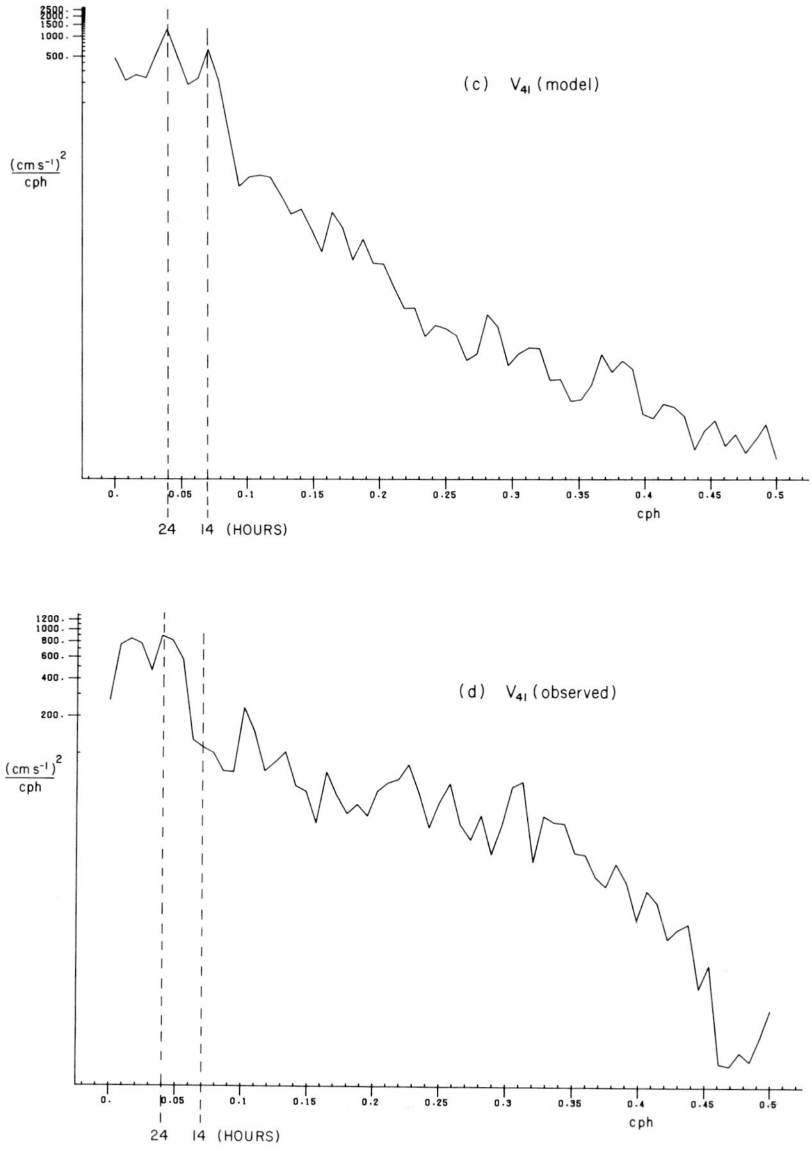

Fig. 10. Power spectra comparisons for the east-directed currents [(a) model and (b) observed] and north-directed currents [(c) model and (d) observed] for the top meter at G_1.

contribution to the forcing from the atmospheric pressure gradients must be small.

Figure 8 is a comparison between the observed currents at G_1 with those predicted by the model. There is an obvious measure of correlation between the two sets of data, particularly so for oscillations with a near-diurnal period or greater. However there are higher frequency components in the observations not reproduced in the model. In both cases current amplitudes at the top meter position are greater than at the lower meter.

Figure 9 illustrates that the overall depth-mean current response of the one-dimensional model is in very good agreement with the depth-mean current calculated by the driving two-dimensional model. This indicates that the frictional dissipation in both models is comparable. There are however some initial oscillations in the one-dimensional model due to the start-up of the model from rest. This generates some transient inertial oscillations.

The power spectra displayed in figure 10 highlight the features already noted in figures 8 and 9. There is a distinct diurnal peak in both the observed and model spectra which are probably due to continental shelf waves. Cross spectra between the u- and v-components of current yield a 90° phase lag for both observed and model data, as might be expected for shelf waves. Also the amplitudes of the u- and v-components are approximately equal and in general reflect the properties of shelf waves as have been observed at a nearby station (Stevenson 61°30'N, 0°0'E).

The model spectra, 10(a) and 10(c), show a distinct peak at 14 hours due to inertial oscillations triggered by the initial shock applied to the model. The spectra from observations, 10(b) and 10(d), show no such peak, however they do exhibit a high frequency contribution absent in the model spectra. Part of this may be traceable to the meteorological input to the model which derives from 3-hourly values, ie less frequent than the hourly observations.

To compare more closely the observed current profile through depth, with that calculated by the model; a cross correlation, between the respective observed and model spectra was performed. For the low frequency regime this showed that at G_1 the observations exhibited a greater shear through depth than the model, for both the u- and v-currents, ie the model behaviour was somewhat more barotropic.

It is of interest to compare the surface wind speed and the surface current calculated by the model. Figure 11 shows that at G_1 the maximum wind speed just exceeded 20m s^{-1} and the calculated surface current reached a maximum just in excess of 40cm s^{-1}. The ratio between surface current and wind speed has a mean of about 0.016 which is low compared with the usually accepted 0.03 or 3% value. As can be seen the value of 0.03 is reached only infrequently. Omission of surface boundary layer effects ie using an eddy viscosity which is

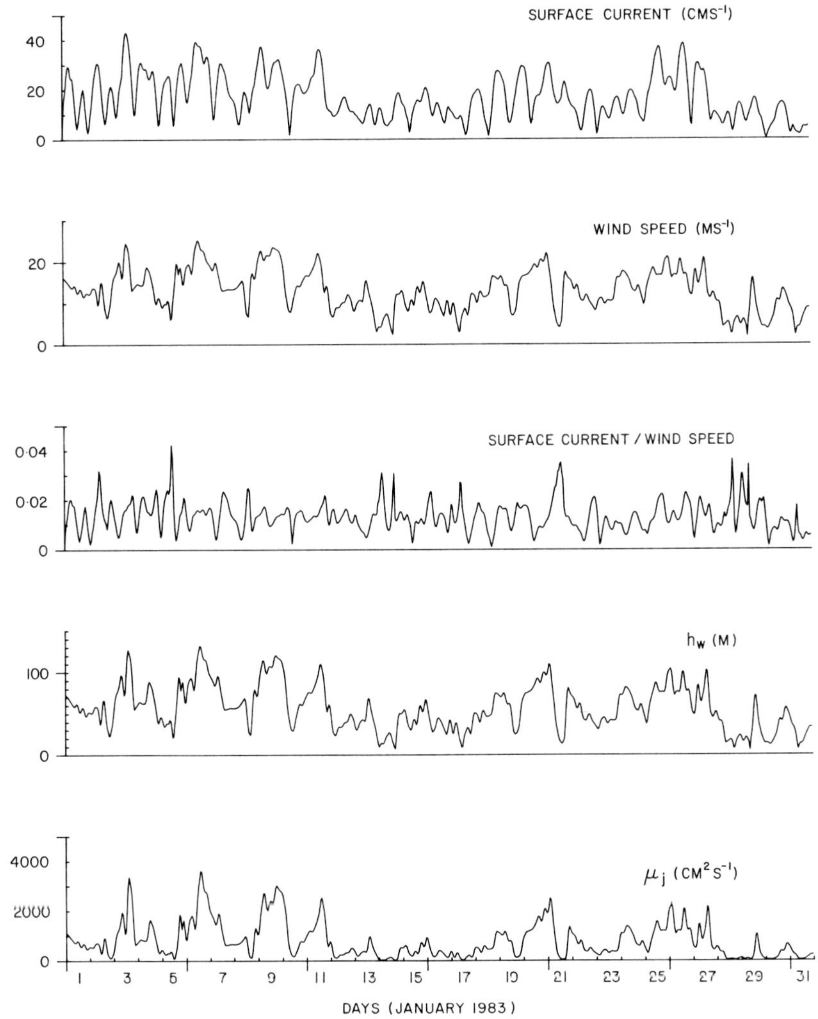

Fig. 11. Variations of surface current, wind speed, the ratio of surface current and wind speed, h_w and μ_j at G_1 throughout January 1983.

constant through the vertical, may be the cause of this discrepancy.

Figure 11 also depicts the depth of frictional influence h_w which is generally less than about 100m, well within the total depth of 191m. The coefficient of eddy viscosity μ_j reaches values of 3000 to 4000 $cm^2 s^{-1}$ during high wind episodes but drops to about 400 $cm^2 s^{-1}$ during the quiet period in the middle of the month.

Figures 12(a) and 12(b) provide a comparison between current profiles through depth for a period of high wind stress, and a period when gradient forcing predominates. Due to the prevailing westerly wind during January 1983, the east component of stress F has been selected as representative of the total wind stress and heads the current profiles at six-hourly intervals.

In both figures, a barotropic response with a 24-hour oscillation may be noted. This is indicative of shelf wave behaviour. Overall the high wind stress current profiles exhibit a greater current amplitude throughout depth with somewhat more structure than in the lower wind case.

We can now turn to a comparison between model results and observations at the Oseberg position, figures 13(a) and 13(b). Here, in shallower water observations are available at five current meter positions through depth. However data from the top two meters are unavailable from January 14th onwards and there are also short breaks in the data from the other three meters about this time.

In figure 13(a), considering the initial 13 days, it can be seen that the model predictions diminish less rapidly than the observations after the peak currents on the 7th January. Generally the fit between model and observations is poorer than for G_1 and once again inertial oscillations are clearly visible in the model output near the start of computations. There also appears to be a mean easterly current missing from the model predictions for the bottom two meters. It is possible that the cause of the difference in the mean currents is due to the 'slope current' which can be traced in many observations along the shelf edge and may be driven by density effects not included in these models. However in this context it is also perhaps worth noting that the lower two meters at Oseberg were of a different design (Aanderaa RCM-4) to the top three meters (SIMRAD UCM VI).

In the model results, current amplitudes decrease from the sea surface downwards. In the observations there is a corresponding decrease for the top three meters but increased current amplitudes are evident for the lower two meters.

Spectral analysis shows a clear peak around a period of 50 hours in the model data, but this is not particularly discernable in the observations. The coherence between the observed and model spectral is greater than 0.60 for periods greater than 50 hours.

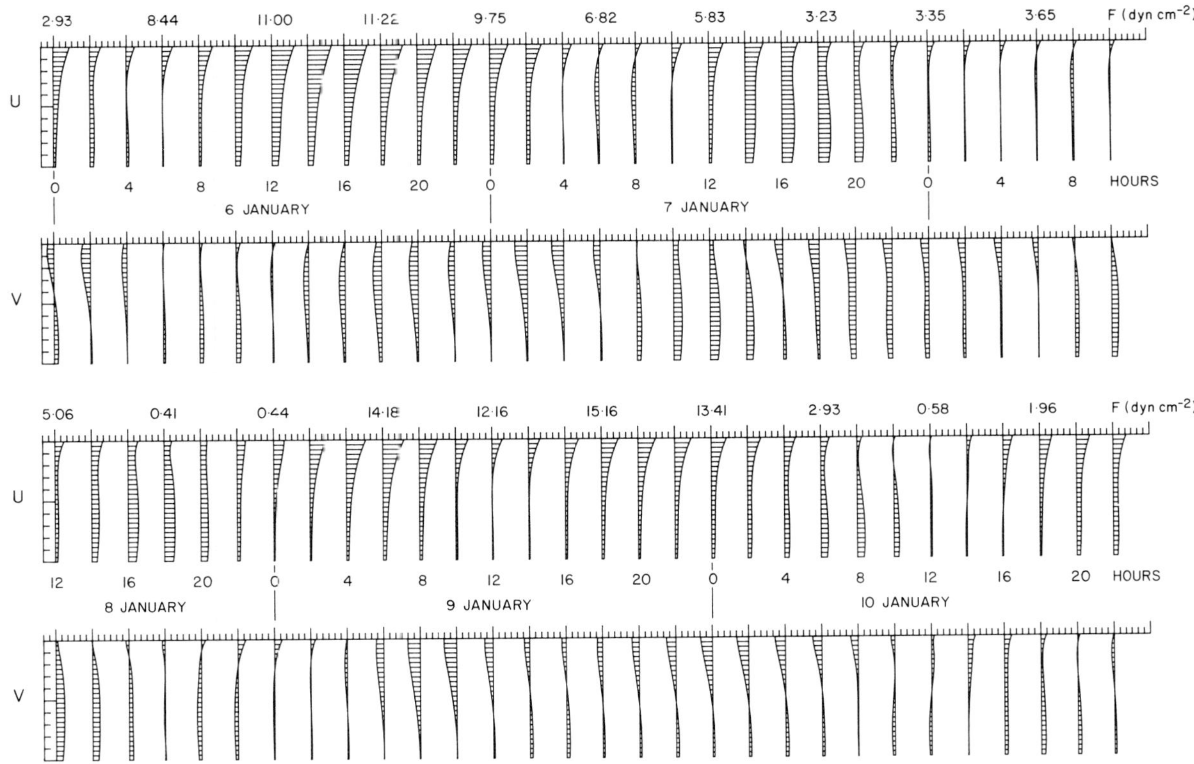

Fig. 12a. Vertical profiles of u- and v-current at G_1, 6-10 January 1983, shown at two-hourly intervals. Westerly stresses, F dyn cm^{-2}, at six-hourly intervals, head the various profiles. Units of current: 10cm s^{-1} (horizontal). Depths (vertical) normalized to fractions of unity.

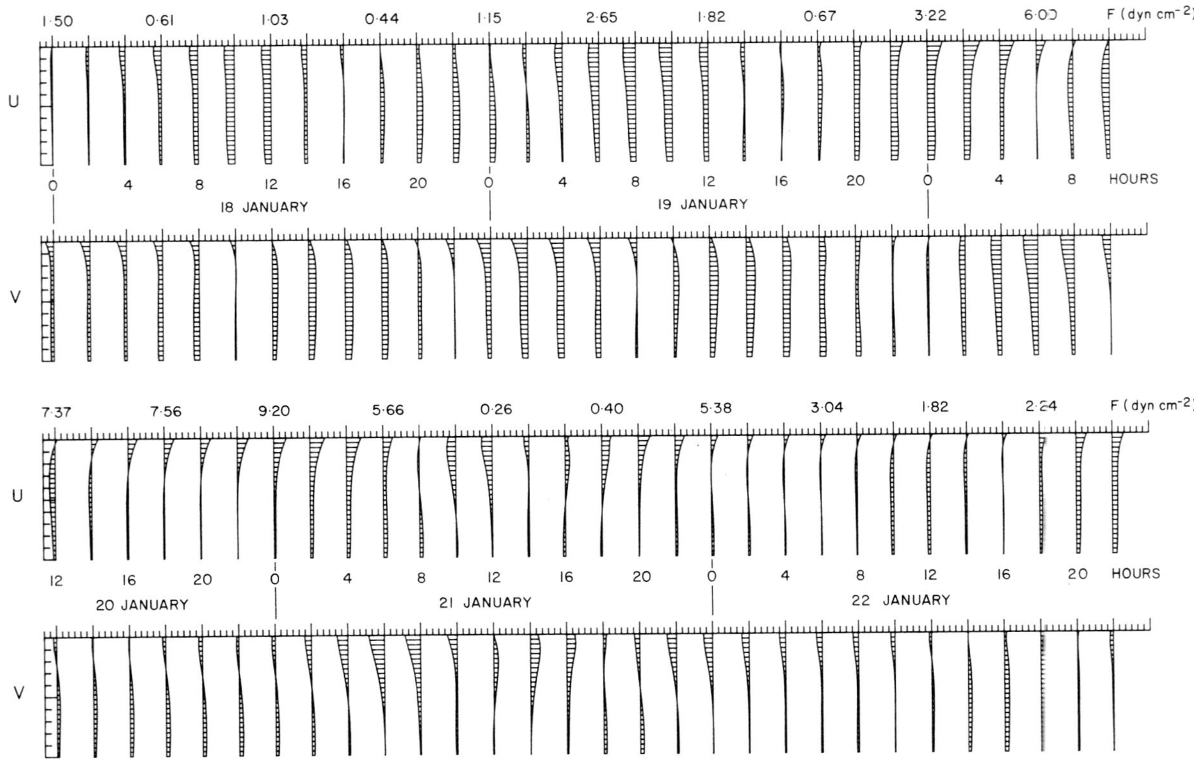

Fig. 12b. Vertical profiles of u- and v-current at G_1, 18-22 January 1983, shown at two-hourly intervals. Westerly stresses, F dyn cm^{-2}, at six-hourly intervals, head the various profiles. Units of current: 10cm s^{-1} (horizontal). Depths (vertical) normalized to fractions of unity.

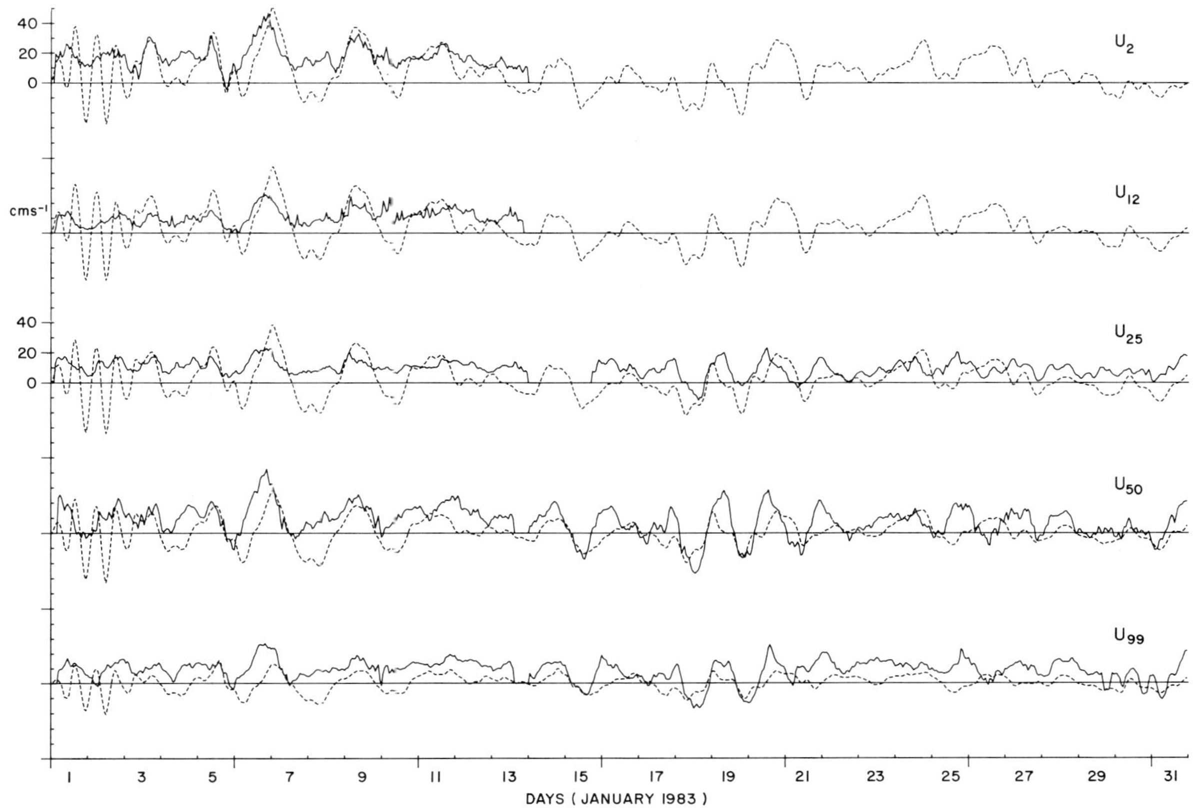

Fig. 13a. Theoretical predictions of u-currents at OS, at 2, 12, 25, 50, 99 metres below the sea surface, compared with the corresponding currents measured at these depths: for the whole of January 1983; ——— observation, ----- model prediction.

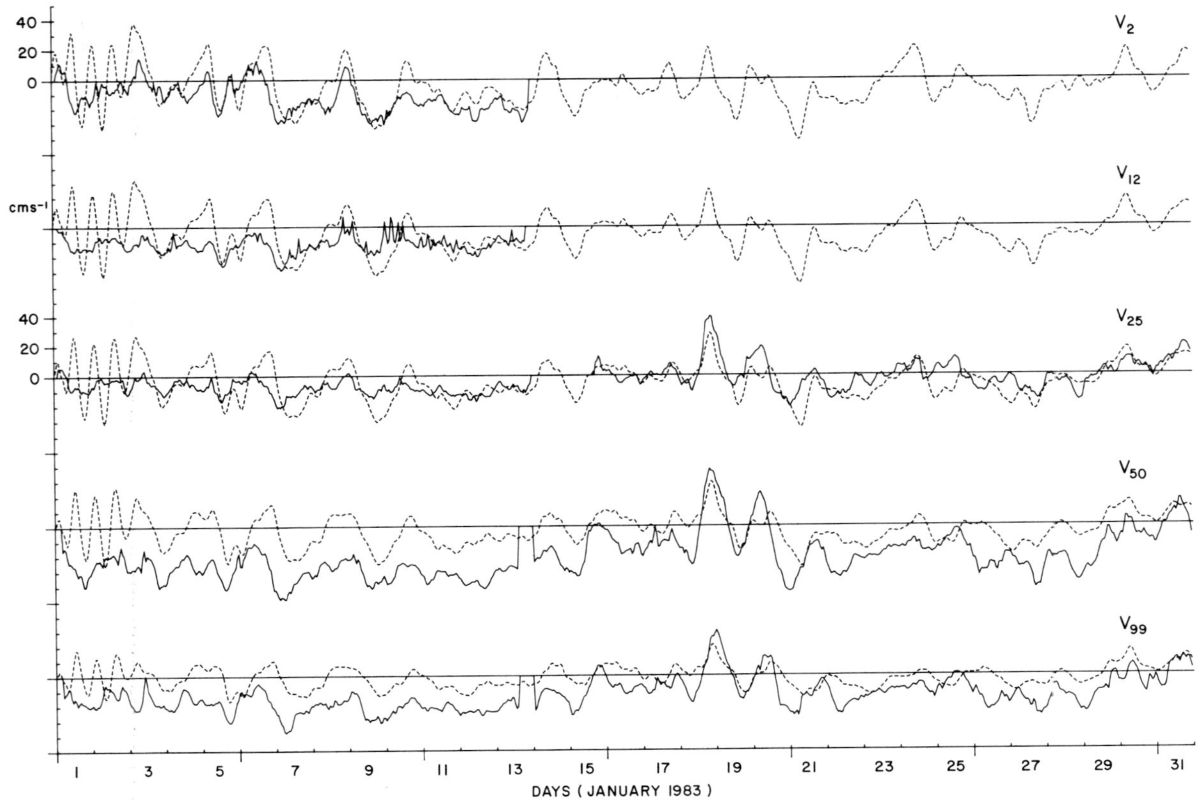

Fig. 13b. Theoretical predictions of v-currents at OS, at 2, 12, 25, 50, 99 metres below the sea surface, compared with the corresponding currents measured at these depths; for the whole of January 1983; ——— observation, ----- model prediction.

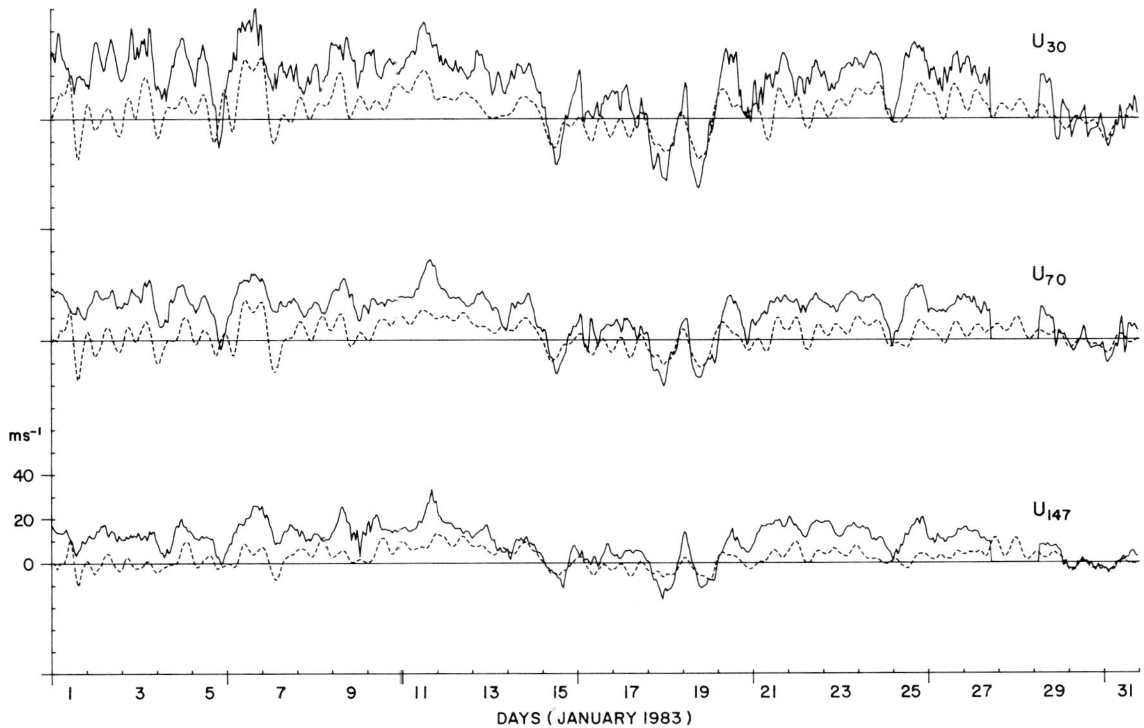

Fig. 14a. Theoretical predictions of u-currents at SF, at 30, 70 and 147 metres below the sea surface, compared with the corresponding currents measured at those depths: for the whole of January 1983; ——— observation, ----- model prediction.

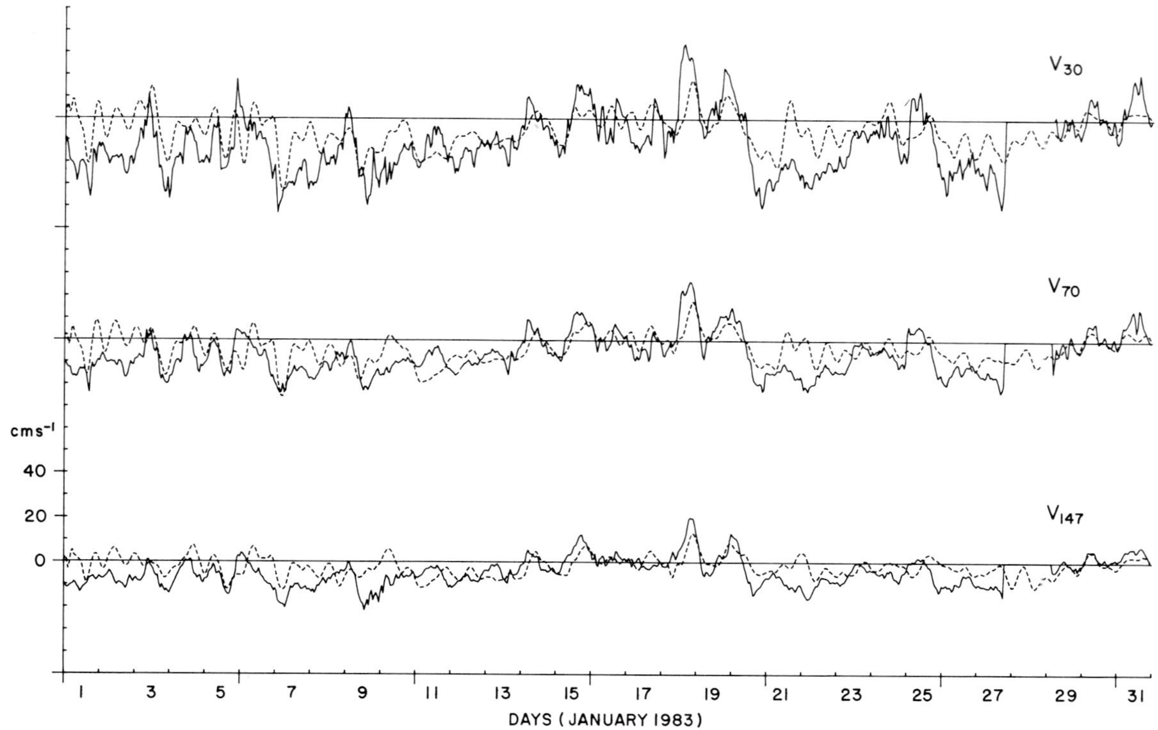

Fig. 14b. Theoretical predictions of v-currents at SF, at 30, 70 and 147 metres below the sea surface, compared with the corresponding currents measured at those depths: for the whole of January 1983; ——— otservation, ----- model prediction.

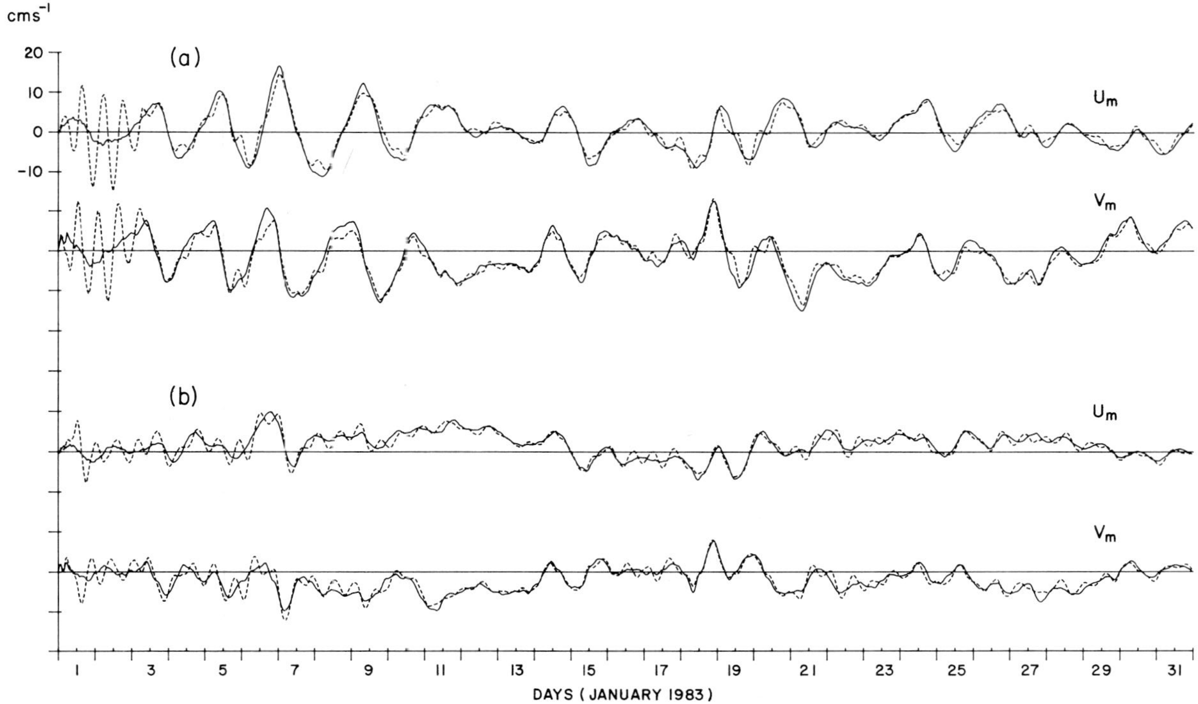

Fig. 15. Components of depth-mean current u_m and v_m at (a) Oseberg and (b) Statfjord, as derived from the one-dimensional model through the vertical (-----) and the two-dimensional model in the horizontal (———).

Figure 13(b) displays generally the same features, again there is a discrepancy at the two lowest meter positions. Here there appears to be a mean southerly current missing from the model predictions. Also in the model spectra the 50 hour peak appears only in the top two positions.

When a cross spectral calculation is carried out at Oseberg, the top three meters exhibit similar behaviour as noted at G_1 For the low frequency regime the observations show a greater shear through depth than the model results in both the u- and v-components. For the lowest two meters however, this trend is not so clear. The model and observed shear for both u- and v-components appear comparable.

Considering the Statfjord position, figure 14(a) shows that the fit between model and observed data is again poorer than at G_1. In particular an easterly current appears to be absent from the model predictions. In both sets of data, the current amplitudes diminish through depth. Spectral analysis gives a clear indication of a 36 hour period in both model and observed data. The spectral coherence between the model and observations is greater than about 0.75 for periods longer than this.

Figure 14(b) displays similar features, here, there seems to be just a small southerly current absent from the model data. Spectra of both model and observations still exhibit a peak around 36 hours but it is less clear.

At Statfjord the cross-spectral calculation reveals that, in the low frequency regime both observed and model u-currents exhibit a comparable shear through depth. However for the v-currents the observed shear appears to be somewhat greater than in the model.

Finally figure 15 compares depth-mean currents derived from the one-dimensional model with those from the driving two-dimensional model at (a) Oseberg and (b) Statfjord. Again there is good agreement apart from some inertial oscillations at the beginning of the model record.

8 CONCLUDING REMARKS

The work presented here shows that a relatively simple one-dimensional model, driven by meteorological forcing and sea-surface gradients from a two-dimensional model can give an overall reasonable representation of low frequency current behaviour through depth at any selected point.

However the present model is inapplicable for current structure in the near-surface and near-bottom boundary layers where a constant eddy viscosity is unrealistic. Nevertheless as can be seen in the Oseberg and Statfjord data there is still reasonable agreement at a position only 3m above the sea bed.

It is clear from the observations that at the locations chosen the current profiles through depth were near-barotropic. Although the model did show the barotropic-type shelf waves this was not a test of the model under high shear conditions as might be expected under very strong wind conditions. Of course the model is inapplicable to stratified conditions when high shears can occur.

Nevertheless the observations did show some current structure. Cross-spectral calculations showed that the present model tended to underestimate the current shear through depth, though in certain cases the observed and model shear were comparable.

The depth-mean current comparison between the one-dimensional and two-dimensional models shows that the frictional dissipation is well represented.

It is obvious that a more complex one-dimensional model may be needed to deal with some or all of the limitations of the present scheme. However the measure of success so far achieved indicates that for certain applications at least, the complexity of a full three-dimensional model for current structure may not be necessary.

ACKNOWLEDGEMENTS

The authors are grateful to Norsk Hydro, especially L. I. Eide for supplying data from the Oseberg position. We are also very grateful to Dr. B. Aune of Det Norske Meteorologiske Institutt for permission to use their Statfjord data and to F. E. Dahl of Det Norske Veritas for supplying it.

Dr. Judith Wolf kindly provided data from the CXX two-dimensional model. She also presented this paper at Liege on behalf of one of the authors (N.S.H.) and her help in advising the other author (J.E.J.) on completing the manuscript is much appreciated.

Thanks are also due to Dr. J. M. Vassie for providing spectral analysis programs, Mr. R. A. Smith for preparing and photographing the diagrams and to Mrs. J. Huxley for typing the manuscript.

REFERENCES

Bowden, K.F., 1953. Note on wind drift in a channel in the presence of tidal currents. Proc. R. Soc., A, 219: 426-446.

Comrie, L.J., 1949. Chambers's Six-Figure Mathematical Tables, Vol. 2: Natural Values. W. & R. Chambers Ltd., Edinburgh.

Csanady, G.T., 1976. Mean circulation in shallow seas. J. Geophys. Res., 81: 5389-5399.

Forristall, G.Z., 1974. Three-dimensional structure of storm-generated currents. J. Geophys. Res., 79: 2721-2729.

Forristall, G.Z., 1980. A two-layer model for hurricane-driven currents on an irregular grid. J. Phys. Oceanogr., 10: 1417-1438.

Heaps, N.S., 1974. Development of a three-dimensional model of the Irish Sea. Rapp. P.-v. Reun. Cons. perm. int. Explor. Mer., 167: 147-162.

Heaps, N.S., 1983. Storm surges, 1967-1982. Geophys. J.R. astr. Soc., 74: 331-376.
Heaps, N.S., 1984. Vertical structure of current in homogeneous and stratified waters. pp. 153-207 in, Hydrodynamics of Lakes, ed. K. Hutter, Springer-Verlag, 341 pp.
Jaeger, J.C., 1961. An Introduction to the Laplace Transformation. Methuen & Co. Ltd., London.
Johns, B., Sinha, P.C., Dube, S.K., Mohanty, U.C. and Rao, A.D., 1983. Simulation of storm surges using a three-dimensional numerical model: an application to the 1977 Andhra cyclone. Q. Jl. R. met. Soc., 109: 211-224.
Proudman, J., 1954. Note on the dynamical theory of storm surges. Arch. Met. Geophys. Bioklim, A, 7: 344-351.
Roed, L.P., 1979. Storm surges in stratified seas. Tellus, 31: 330-339.
Svensson, V., 1979. The structure of the turbulent Ekman layer. Tellus, 31: 340-350.
Whittaker, E.T. and Robinson, G., 1944. The Calculus of Observations. Blackie and Son Ltd., London.

A COUPLED 2-D/3-D MODELLING SYSTEM FOR COMPUTATION OF TIDAL AND WIND-INDUCED CURRENTS

J.M. USSEGLIO-POLATERA AND P. SAUVAGET
SOGREAH Consulting Engineers, BP 172 X, 38042 GRENOBLE Cedex (FRANCE)

ABSTRACT

Although computer costs are rapidly decreasing, it is not yet common to carry out 3-D modelling all over large areas where detailed results are required. The authors present the development of a promising code allowing the definition of a 3-D block in any 2-D computational field. 3-D computation is limited to the area where it is absolutely necessary while the feed back of 3-D behaviour towards the outer area is accurately simulated. The algorithm based upon fractional steps and the solution provided for the coupling problems are described in detail.

1 INTRODUCTION

Mathematical modelling for coastal engineering or coastal environment problems displays typical features that may significantly influence the trend of code developments.

Since most of the time water is very shallow and there is no significant stratification, two-dimensional (2-D) models are well adapted, but sometimes the conventional 2-D hypotheses are locally inadequate (sharp depth variations or strong wind forcing) and three-dimensional (3-D) models are recommended. Indeed, it is exactly where the flow field displays 3-D behaviour that reliable results are often required (sediment transport, ship manoeuvring).

On the other hand, detailed results are needed locally (for instance, around new structures) and the grid size must be very fine (sometimes lower than 20 m) whereas the tidal excursions of water particles may be very long (several kilometers), hence the area involved in the flow computation may be very large. Consequently, a large number of computational points is needed despite the most advanced techniques of grid refinement (Chenin-Mordojovich et al, 1985).

Advanced and reliable 3-D codes exist but up to now, only a few people are ready to pay for 3-D simulations on such a dense grid for any local coastal engineering project. This is mainly a cost efficiency concern but the investigation of any method able to provide an effective, even partial solution to this problem, does involve a series of interesting theoritical problems.

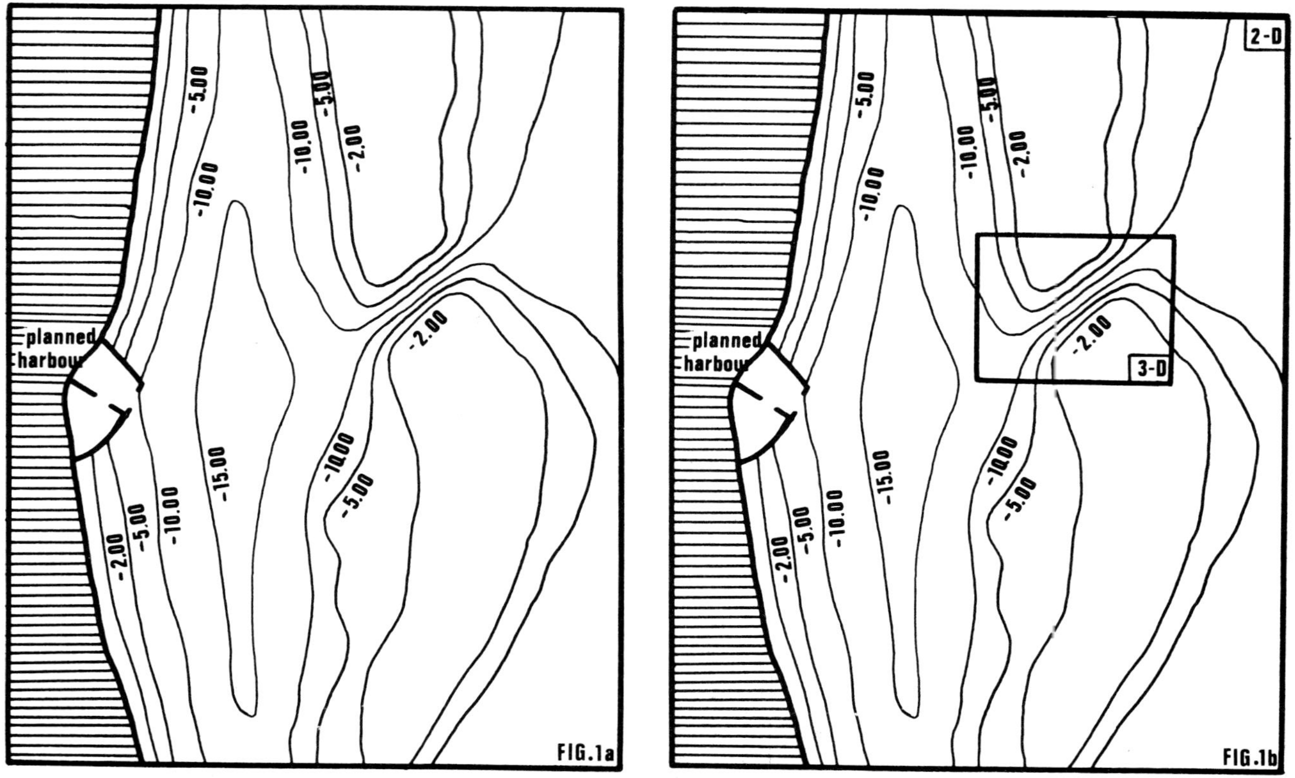

—— EQUAL DEPTH CURVES

FIG.1 : SCHEMATIC EXAMPLE OF A REAL APPLICATION REQUIRING A COUPLED 2-D / 3-D MODELLING SYSTEM

Starting from a schematic but typical example, the authors will show the content and the first applications of a new code coupling 2-D and 3-D equations.

2 PRESENTATION OF THE PROBLEM USING A SCHEMATIC EXAMPLE

Consider the schematic example shown on Fig.1a. A new harbour is planned in a coastal area. The shoreline is nearly linear, the bathymetry very uneven, (sudden depth but extensive shallows off-shore). The main access to the harbour is a natural channel crossing the shallows but the channel position and direction vary significantly in time due to major sedimentological changes. A significant tidal range induces alternating tidal currents parallel to the shoreline.

It is assumed that the problem concerns simulation of the tidal flow in order to study the sedimentological processes in the channel area.

3-D modelling is particularly recommended because the velocities near the channel vary considerably in direction and intensity throughout the water layer.

Since the tidal range is wide, the tidal currents are strong. Thus, there is a large area involved in the generation of the flow field around the channel. 3-D modelling over the whole area is possible but it would be very expensive. On the other hand, in most of the area, 2-D modelling is adequate and even recommended near the shoreline because flooding and drying of tidal flats has not yet been reliably simulated in 3-D models (as far as we know).

A solution with nested (imbedded) models would be possible : a 2-D model would give boundary conditions for a local 3-D model (fig.1b) provided a reasonable assumption for the vertical distribution of horizontal velocities on the 2-D/3-D boundary line. But the main drawback of the nested models is obvious : the simulation will provide no feed back of the internal information towards the surrounding area and if the inner structure changes significantly, for instance because of civil engineering works, feed back will be needed.

Clearly, a coupling method is necessary.

3 A NEW COUPLED 2-D/3-D MODELLING SYSTEM

A new modelling system has been developed in order to provide an effective solution to the above-mentioned problem. This new code couples as far as possible 2-D and 3-D equations so that, in any given computational area, a 3-D block may be defined. Therefore, the computation is 3-D where it is absolutely necessary, and 2-D elsewhere. For the case of the above example, the boundaries of the 3-D block could be the same as the boundaries of the internal model.

3.1 Preamble

For clarity,the local velocity and the depth-averaged velocity are first defined. The local (horizontal) velocity which varies throughout the water layer is a typical 3-D modelling parameter. The depth averaged (horizontal) velocity is obtained by integrating the local velocity over the water layer.

3.2 Governing 2-D and 3-D equations

Consider now both the 2-D and 3-D governing shallow water wave equations. In the momentum equations, the different terms may be related to the various physical processes occurring simultaneously, namely advection, diffusion, wave propagation, wind forcing, Coriolis deviation and bed friction. In table 1, these equations are presented using this split operator approach which is the basis of fractional step resolutions.

Now consider the formulation of each of these processes.

For advection, horizontal momentum diffusion and wave propagation, integration over the water layer of the 3-D formulation gives the 2-D formulation. Therefore, 2-D and 3-D formulations are consistent. This is the same for Coriolis deviation although this is responsible for most of the typical 3-D behaviour when wind is involved.

Owing to the assumption of hydrostatic pressure distribution, the vertical velocity remains very low and vertical advection is of very little importance with respect to horizontal advection.

Wind forcing and bed friction are surface stresses. In two dimensions, their formulation depends on typical 2-D parameters: bed friction depends on the depth-averaged velocity according to Manning-Strickler formulation and wind effects directly influence the depth-averaged velocity as well. The 3-D formulation is very different because bed and wind stress can be held as actual surface stresses and their propagation throughout the water layer is modelled through the formulation of vertical diffusion processes.

3.3 Coupling problems

This analysis of the governing equations emphasizes obvious inconsistencies in the formulation inducing the main coupling problems. Consequently, it seems necessary to find a coupling algorithm displaying the following properties :

. the physical processes having a similar formulation in 2-D and 3-D must be isolated in order to use similar algorithms for these 2-D and 3-D operators ;

the algorithm must provide some solution to maintain the coherence in the results for the processes which are a priori inconsistent.

In addition, at each time step, boundary conditions are needed for the local velocities on the 2-D/3-D boundary line where only the depth-averaged velocities are known (σ - coordinates are used in the 3-D block).

3.4 Algorithm of the coupled code

The coupling algorithm is based on the splitting-up techniques developed by Yanenko (1968). The governing equations are split into several fractional steps corresponding to the various physical processes involved. For each time increment, the various steps are solved successively using a numerical approach adapted to their mathematical properties. The theory of fractional step techniques shows that the governing equations are approximated to first-order accuracy in time. This technique is already applied in existing 2-D and 3-D codes developed by Sogreah Consulting Engineers and the French Electricity Board (EDF/LNH) (Benque et al, 1982, Burg et al, 1982).

The coupling algorithm in which nine fractional steps are involved is presented first. The solution to the above-mentioned problem of surface stress representation will be described subsequently (3.5).

The nine successive fractional steps are the following :

. 1st step (2-D advection): advection of the depth-averaged velocities is computed using the 2-D characteristics method throughout the computational field except in the 3-D block. The spatial interpolation at the foot of the characteristics line is calculated using linear or cubic interpolation.

. 2nd step (horizontal 3-D advection): horizontal advection of the local velocities is calculated using the same method of characteristics as mentioned for the 3-D block.

. 3rd step (vertical 3-D advection): vertical advection of the local velocities is computed within the 3-D block, using a 1-D characteristic method, a vertical integration of the horizontal local velocities then yields the advected depth-averaged velocities in the 3-D block.

. 4th step (2-D diffusion): diffusion of the depth-averaged velocities is performed throughout the computational area by splitting the operator in each horizontal direction and using the implicit double sweep method; the computation is not necessary within the 3-D block but simplifies the programming effort while involving a negligible increase in cost. Moreover, it allows for continuity across the 2-D/3-D boundary line since the advected depth-averaged velocities have been evaluated consistently in the 1st and 3rd steps.

TABLE 1.

A/ **2D SHALLOW WATER WAVE EQUATIONS SOLVED IN CYTHERE ES1 USING THE SPLIT-OPERATOR APPROACH**

ADVECTION

$$\left\{\begin{array}{l}\frac{\partial U}{\partial t} + \frac{\partial(\bar{u}U)}{\partial x} + \frac{\partial(\bar{v}U)}{\partial y} = 0 \\ \frac{\partial V}{\partial t} + \frac{\partial(\bar{u}V)}{\partial x} + \frac{\partial(\bar{v}V)}{\partial y} = 0\end{array}\right.$$

DIFFUSION

$$\left\{\begin{array}{l}\frac{\partial U}{\partial t} - \frac{\partial}{\partial x}(K\frac{\partial U}{\partial x}) - \frac{\partial}{\partial y}(K\frac{\partial U}{\partial y}) - FV = 0 \\ \frac{\partial V}{\partial t} - \frac{\partial}{\partial x}(K\frac{\partial V}{\partial x}) - \frac{\partial}{\partial y}(K\frac{\partial V}{\partial y}) + FU = 0\end{array}\right.$$

PROPAGATION

$$\left\{\begin{array}{l}\frac{\partial U}{\partial t} + gh\frac{\partial Z}{\partial x} + \frac{\tau_{SX}}{\rho} + \frac{\tau_{BX}}{\rho} = 0 \\ \frac{\partial V}{\partial t} + gh\frac{\partial Z}{\partial y} + \frac{\tau_{SY}}{\rho} + \frac{\tau_{BY}}{\rho} + 0 \\ \frac{\partial Z}{\partial t} + \frac{\partial U}{\partial x} + \frac{\partial V}{\partial y} = 0 \quad \text{(mass conservation)}\end{array}\right.$$

B/ **3D SHALLOW WATER WAVE EQUATIONS SOLVED IN ODYSSEE USING THE SPLIT-OPERATOR APPROACH**

HORIZONTAL ADVECTION

$$\left\{\begin{array}{l}\frac{\partial u}{\partial t} + u\frac{\partial u}{\partial x} + v\frac{\partial u}{\partial y} = 0 \\ \frac{\partial v}{\partial t} + u\frac{\partial v}{\partial x} + v\frac{\partial u}{\partial y} = 0\end{array}\right.$$

VERTICAL ADVECTION

$$\left\{\begin{array}{l}\frac{\partial u}{\partial t} + w\frac{\partial u}{\partial z} = 0 \\ \frac{\partial v}{\partial t} + w\frac{\partial v}{\partial z} = 0\end{array}\right.$$

HORIZONTAL DIFFUSION

$$\left\{\begin{array}{l}\frac{\partial u}{\partial t} - \frac{\partial}{\partial x}(K\frac{\partial u}{\partial x}) - \frac{\partial}{\partial y}(K\frac{\partial u}{\partial y}) - Fv = 0 \\ \frac{\partial v}{\partial t} - \frac{\partial}{\partial x}(K\frac{\partial v}{\partial x}) - \frac{\partial}{\partial y}(K\frac{\partial v}{\partial y}) + Fu = 0\end{array}\right.$$

VERTICAL DIFFUSION

$$\left\{\begin{array}{l}\frac{\partial u}{\partial t} - \frac{\partial}{\partial z}(\nu_Z\frac{\partial u}{\partial z}) = 0 \\ \frac{\partial v}{\partial t} - \frac{\partial}{\partial z}(\nu_Z\frac{\partial v}{\partial z}) = 0\end{array}\right.$$

PROPAGATION

$$\left\{\begin{array}{l}\frac{\partial u}{\partial t} + g\frac{\partial Z}{\partial x} = 0 \\ \frac{\partial v}{\partial t} + g\frac{\partial Z}{\partial y} = 0 \\ \frac{\partial u}{\partial x} + \frac{\partial v}{\partial y} + \frac{\partial w}{\partial z} = 0 \quad \text{(mass conservation)}\end{array}\right.$$

with:

x, y, z and t = the independant space and time variables

U, V = the unit-width discharges through the depth in the x- and y- directions; $\bar{u}$, $\bar{v}$ = the corresponding depth-averaged velocities

u, v, w = the local velocities in the x-, y- and z- directions respectively

Z = the free surface elevation / h = the flow depth

K, ν_Z = the coefficients of horizontal and vertical diffusion

τ_{BX}, τ_{BY} = the bed friction shear stresses / τ_{SX}, τ_{SY} = the surface wind shear stresses /

F = the Coriolis acceleration parameter / g = the acceleration due to gravity ρ = the density of the water

. 5th step (horizontal 3-D diffusion): horizontal diffusion of the local velocities is computed within the 3-D block using the same implicit double sweep method; a vertical integration of the horizontal local velocities then yields the new depth-averaged velocities in the 3-D block.

. 6th step (vertical 3-D diffusion): vertical diffusion of the local velocities is computed inside the 3-D block using the implicit double-sweep method. Up to now, turbulence effects have been modelled according to Prandtl's mixing length hypothesis and the mixing length is assumed to be constant in the fluid except near the bottom and the sea-surface where it varies linearly with the distance from the boundary (Burg et al, 1982). A vertical integration of the new local velocities enables an evaluation of the momentum sink due to wind action, bed friction and turbulent diffusion to be made (see 3.5 for details).

. 7th step (free surface elevations): the wave propagation step consists in an equation relating depth averaged velocities and surface elevations which are typical 2-D parameters. Accordingly, in the coupled code, throughout the computational field, free surface elevations are computed using a 2-D algorithm, applied to the equations of the 2-D propagation step of table 1: an iterative method based on a special variant of the conjugate gradient approach is carried out through an alternating direction operator with co-ordination in space (Benqué et al, 1982). Inside the 3-D block, new friction coefficients are evaluated in order to obtain, with 2-D formulation of bed shear stress, the head loss computed in the previous step (see 3.5 for details). Once surface elevations are known, depth averaged velocities can be evaluated over the whole computational field.

. 8th step (boundary velocity profile): at this time, the vertical distribution of horizontal velocities on the 2-D/3-D boundary line is needed. An Ekman type relation is currently used (Burg et al, 1982). This relation is certainly not the most appropriate relation for all problems but it is particularly suitable when wind is involved, work is at present under way to improve this formulation.

. 9th step (local velocities after propagation): the horizontal local velocities are corrected according to the variations of the free surface elevations. They do not need to be integrated since the actual depth averaged velocities have already been computed with the same numerical scheme as in the 7th step. Finally, the vertical local velocities are evaluated.

3.5 A solution to the problems of surface stress

(i) bed friction

In 2-D, the momentum sink due to bed friction is expressed as a bed stress $\vec{\tau}_B$, related to the Chezy or Strickler head loss formulae, as follows (for instance, with Chezy's approach) :

$$\vec{\tau}_B = \rho g \frac{\vec{U}\,\|\vec{U}\|}{Ch^2\,h^2} \tag{1}$$

where: ρ = the density of the water,

g = the acceleration due to gravity,

$\vec{U}$ = the unit-width discharge vector,

$\|\vec{U}\|$ = the modulus of the unit-width discharge vector,

h = the flow depth,

Ch = Chezy's coefficient.

The friction head loss therefore only depends on the depth averaged velocity.

In 3-D, a slip velocity condition is imposed on the sea bed, assuming that the shear stress is a quadratic function of the bed velocity (Burg et al., 1982) :

$$\nu_Z \frac{\delta \vec{u}_B}{\delta Z} = \frac{\vec{u}_B\,\|\vec{u}_B\|}{c^2} \tag{2}$$

$$\nu_Z = k^2\,.\,\Delta Z^2\,.\left\|\frac{\delta \vec{u}}{\delta Z}\right\| \tag{3}$$

where :

$\vec{u}_B$ = the bed velocity vector (no vertical component),

c = a constant,

ν_Z = the eddy viscosity (or vertical diffusion coefficient),

k = the Von Karman constant, i.e. 0.41,

ΔZ = the distance between the bed and the point where the condition is applied, which must be within the logarithmic boundary layer.

In this case, the friction head loss depends on the local velocity and is transferred to the depth averaged velocity by mean of vertical turbulent diffusion.

Clearly, the friction formulation in the two systems is very different and a relation between Ch, the Chezy coefficient, and c is not immediately apparent. Assuming unidimensional flow, a real relationship exists; consistency is excellent as far as steady flow is concerned but becomes much less so in the case of transient flows.

In the coupled algorithm, the free surface elevations are calculated all over the computational field using the 2-D algorithm (7th step). However, the momentum sink due to bed friction is already known in the 3-D block (6th step). Therefore, a new coefficient Ch_e is evaluated for each computational point within the 3-D block in order to obtain with the 2-D formulation the head loss computed by the 3-D formulation. As things stand at present, Ch_e is given by the following equation :

$$\frac{\Delta \vec{u}}{\Delta t} = g \frac{\vec{u} \, \| \vec{u} \|}{Ch_e^2 \, h} \tag{4}$$

where:

$\dfrac{\Delta \vec{u}}{\Delta t}$ = momentum sink computed in the vertical diffusion step,

$\vec{u}$ = depth-averaged velocity at the end of the 2-D diffusion step.

Studies are at present in progress to investigate a similar formulation for an implicit expression of bed friction. This formulation is, however, far more complex and considerable programming investments will be required.

(ii) <u>wind</u>

In order to allow for wind effects correctly, the 2-D bed friction formula has been somewhat modified in order to take into account certain 3-D aspects. With Chezy's approach :

$$\vec{\tau}_B = \rho g \frac{(\vec{U} + \vec{q}) \, . \, \{ \| \vec{U} \| + \| \vec{q} \| \}}{Ch^2 \, h^2} \tag{5}$$

This formulation corresponds to (1) but an additional discharge, called "q", has been introduced. "q" depends on the Ekman depth, the wind speed, Coriolis deviation, the water depth and on the current velocity. More details on this formulation are available (see Hamm et al. (1985)). This formulation eliminates some well-known drawbacks of the conventional 2-D formulation of wind effects: the poor representation of the vertical profile of horizontal velocities on the one hand, and the zero depth averaged velocities leading to zero bed friction, on the other hand. (This is the case when a steady wind is blowing over a flat bottom basin). Consequently, good consistency is obtained for wind effects.

4 APPLICATION EXAMPLE

In order to illustrate the present state of these developments, a simple but interesting test case is described below.

The computational area is a rectangular basin closed on three sides, 3.800 m long and 1.100 m wide (Fig.2). Its average depth is 4 m. An S-shaped channel crosses the basin in its middle portion and the average water depth in this channel is 12 m. Chezy's friction coefficient is 40. The boundary condition on the open side is a significant tidal range (6 m).

The computation was performed over half a tidal cycle, starting from high water (ebb). Three runs were performed. The first one with a 2-D code, CYTHERE ES1 (Benque et al., 1982), and the last one with the coupled 2-D/3-D code, CYTHERE 3-D. The computational grid size was 100 m hence the number of computational points in the horizontal plane is 468 (39 x 12). A 3-D block was defined covering an area of 1.600 x 900 m. The smallest distance between the 2-D/3-D boundary line and the channel is 300 m.

The following results have been sketched : the depth averaged velocities obtained with the two codes (Fig.3) and the horizontal velocities in the 3-D block at various fractions of the depth (Fig.4 and Fig.5).

It is clear that the coupling scheme introduces no singularity near the 2-D/3-D boundary lines. The particular flow field observed with the results of the 2-D/3-D calculation near the western entrance to the channel exists as well as with the results of the 2-D calculation. Accordingly, it emphasizes the efficiency of the 2-D code itself and the reliability of such a tool with regard to coupling purposes.

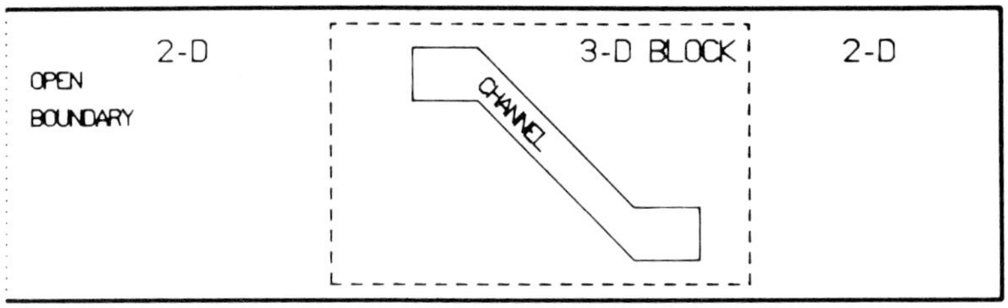

200 M (LENGTH SCALE)

MEAN DEPTH : 4.00 M

TIDAL RANGE : 6.00 M

CHANNEL DEPTH : 12.00 M

BED FRICTION : 32 (STRICKLER)

NUMBER OF COMPUTATIONAL POINTS :

2-D : 39 x 12 3-D BLOCK : 17 x 10 x 18

HORIZONTAL GRID SIZE : 100 M

TIME STEP = 120. SEC. (CFL > 10)

FIG.2 : APPLICATION EXAMPLE : RECTANGULAR BASIN CLOSED ON THREE SIDES

Going into detail, discrepancies are apparent and further investigations have shown that this is because bed friction is slightly under-estimated in the channel in 2-D, due to the significant depth variations (4 m versus 12 m). Clearly, the feed back of the internal information to the surrounding area is satisfying.

Finally, the coupled 2-D/3-D code provides an interesting estimate of the variations in direction and intensity of the horizontal velocities around and within the channel. These results are directly usable as far as sedimentation processes or ship manoeuvring problems are concerned.

The cost efficiency of the coupled code has been evaluated, and is approximately proportional to the number of computational points, i.e.

(a) 468 (= 39 x 12) for full 2-D computation,

(b) 8 424 (= 39 x 12 x 18) for full 3-D computation,

(c) 3 528 (= (17 x 10 x 18) + (39 x 12)) for the coupled code.

The ratio between (b) and (c) will be significantly higher in most real applications because in this schematic case the 2-D area is limited to the minimum necessary around the 3-D block.

5 CONCLUSIONS

An original tool has been developed. The results obtained on regular geometries and reasonably even bathymetries are very promising. Work is at present under way to test the code on real applications displaying complex natural layouts provided that the basic hypotheses are respected.

6 REFERENCES

Benque, J.P., Cunge, J.A., Feuillet, J., Hauguel, A. and Holly, F.M., 1982. New method for tidal current computation. Journal of the Waterway, Port, Coastal and Ocean Division, ASCE, 108 WW3: 396-417.

Burg, M.C., Coeffe, Y. and Warluzel, A., 1982. Tridimensional numerical model for tidal and wind generated flow. Proc. 18th Coastal Engineering Conf., ASCE, Capetown, South-Africa, 635-651.

Chenin-Mordojovich, M.I. and Hauguel, A., 1985. The internal refined grid in particular areas inside a 2-D mathematical model. Proc. 21st Congress. IAHR, Melbourne, Australia.

Hamm, L., Quetin, B. and Usseglio-Polatera, J.M., 1985. Two dimensional modelling of wind-induced currents in coastal and harbour areas. Proc. Int. Conf. on Numerical and Hydraulic Modelling of Ports and Harbours, Birmingham, U.K., 13-22.

Usseglio-Polatera, J.M. and Chenin-Mordojovich, M.I., 1985. Numerical problems in coupling two- and three-dimensional models: the CYTHERE 3-D system. Proc. Int. Symp. on Refined Flow Modelling and Turbulence Measurements, Iowa City, USA.

Yanenko, N.N, 1968. Méthodes à pas fractionnaires. Armand Colin, Paris, 203 pp.

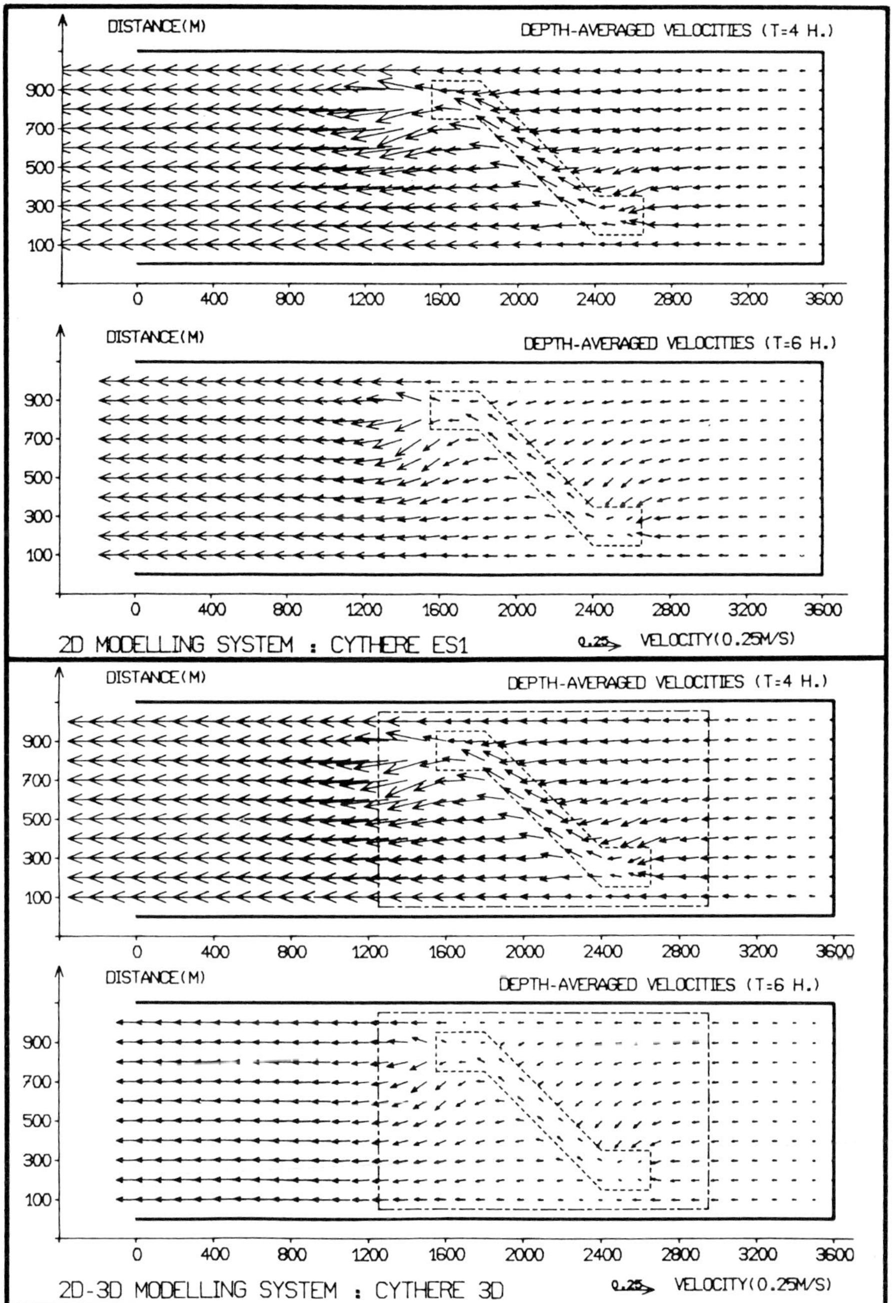

FIG.3 : DEPTH AVERAGED VELOCITIES COMPUTED WITH THE 2-D CODE AND THE COUPLED 2-D / 3-D CODE, 4 HOURS AFTER HIGH WATER AND AT LOW WATER (APPLICATION EXAMPLE OF FIG.2)

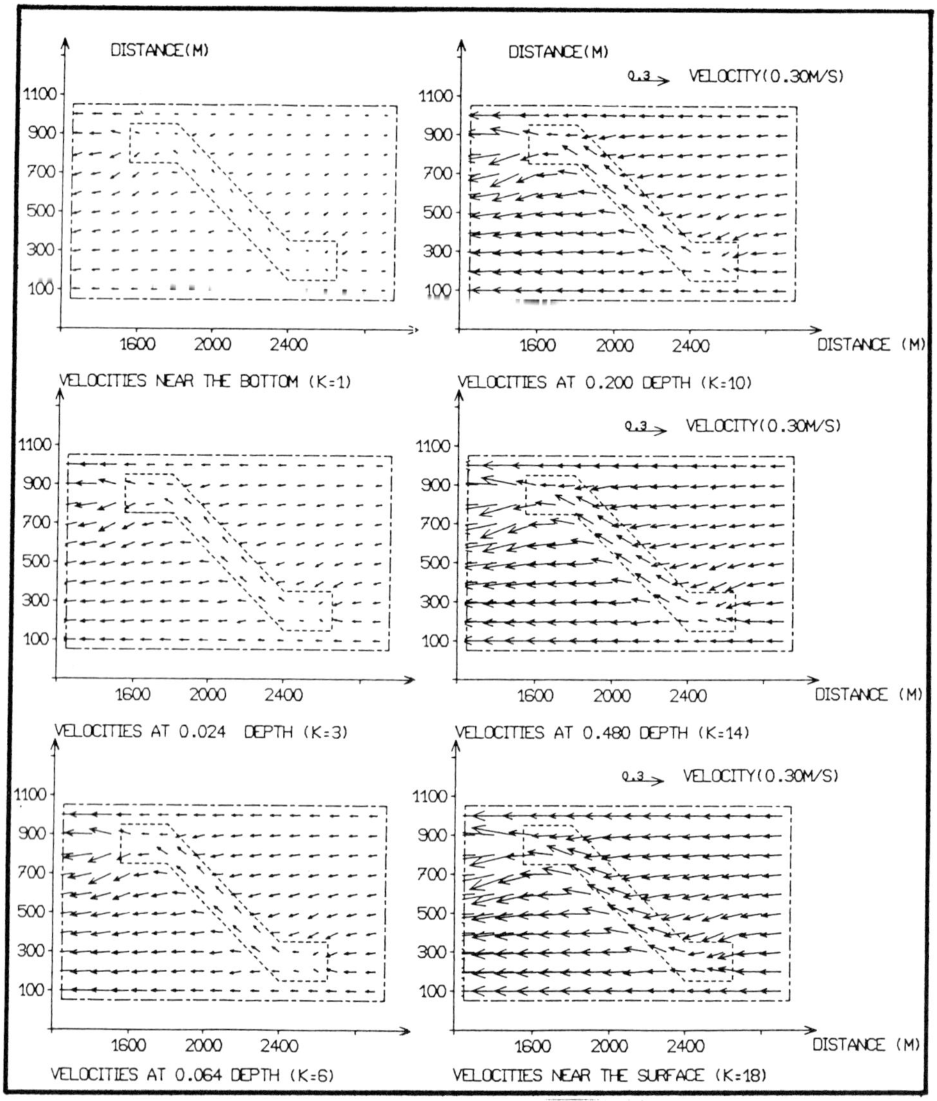

FIG.4 : HORIZONTAL VELOCITIES IN THE 3-D BLOCK AT VARIOUS FRACTIONS OF THE DEPTH (SIGMA COORDINATES) 4 HOURS AFTER HIGH WATER (APPLICATION EXAMPLE OF FIG.2)

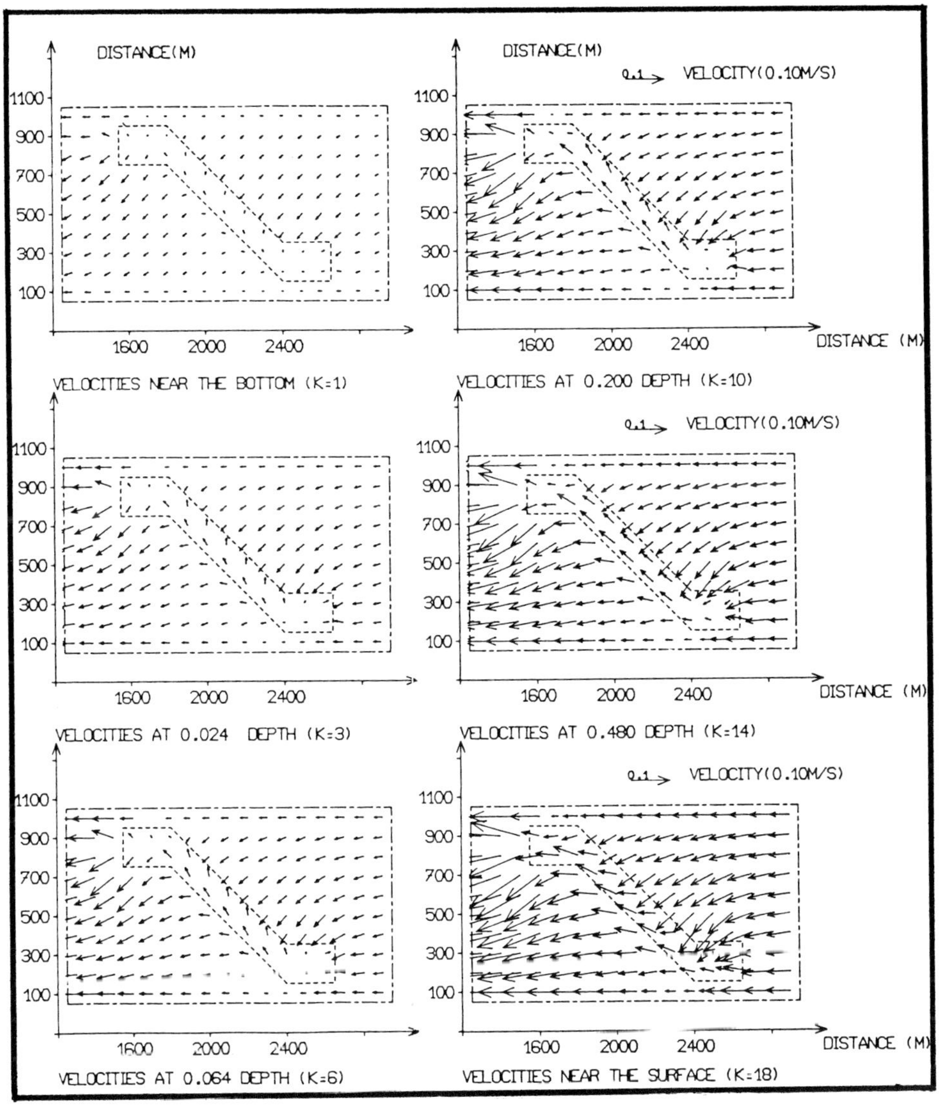

FIG.5 : HORIZONTAL VELOCITIES IN THE 3-D BLOCK AT VARIOUS FRACTIONS OF THE DEPTH (SIGMA COORDINATES) AT LOW WATER (APPLICATION EXAMPLE OF FIG.2)

A HIGH RESOLUTION THREE-DIMENSIONAL MODEL SYSTEM FOR BAROCLINIC ESTUARINE DYNAMICS AND PASSIVE POLLUTANT DISPERSION

J. KROHN and K. DUWE
GKSS Forschungszentrum Geesthacht, P.O. Box 1160, 2054 Geesthacht
K.D. PFEIFFER
Institut für Meereskunde, Universität Hamburg, Troplowitzstr.7, 2000 Hamburg 54
(Federal Republic of Germany)

ABSTRACT

In order to examine the dynamics of the Elbe estuary different three-dimensional models have been developed. In its lower part the Elbe estuary is characterized by a significant influence of baroclinicity connected with fresh water discharge and strong small scale variations of bottom topography, in the upper part pure fresh water conditions prevail. The system of models therefore consists of high resolution baroclinic and barotropic versions, a basic version of 250 m resolution covering the total estuary and finer segments (50 - 100 m resolution) being applied to areas of specific interest. The shallow water equations are formulated semi-implicitly to yield a suitable discretization of the space and time domain; the drying of tidal flats is handled without giving rise to shockwave-like disturbances in the velocity field.

The model has been extended to treat the transport of passive tracers by applying a 'Monte-Carlo'-technique. The transformation into the Lagrangian formulation uses linear interpolation and some modifications taking into account the high variability of the flow field and the complex morphology.

Two areas have been chosen - the brackish water zone and the fresh water zone just seaward of Hamburg - to simulate the dispersion of a number of released passive tracers under tidal influence. Results are obtained for release near the surface and release evenly distributed over the water depth. The calculated particle paths and particle distributions generally show a large variability, depending on the location and the tidal phase of insertion.

1 INTRODUCTION

The river Elbe, flowing into the North Sea at the southeastern corner of the German Bight is one of the largest rivers in Europe (about 1100 km total length). Its drainage area includes large towns (e.g. Prague, Dresden, Berlin, Hamburg) and industrial areas. Consequently the water is considerably polluted by urban sewage and industrial effluents. The hazardous effect of pollutants becomes even more dangerous in the tidally influenced lower portion of the river where the flow reverses within the semidiurnal tidal cycle. This means that the same volume of water passes a fixed point (e.g. a sewage introduction) several times. The relative effect of fresh water discharge on the velocity field decreases with increasing river cross-sectional areas. Therefore a water particle needs a longer time to proceed downstream, a typical value for mean discharge conditions is 20 days travel time from Hamburg to the North Sea (125 km) compared to two days for the same distance within the non-tidal part of the river.

Originating from storm-surge prediction models these environmental problems motivated the development of more sophisticated models to examine, i.e. to understand and predict, the dynamical features of the whole Elbe estuary.

This paper deals with a description of the present state of essentially three-dimensional models for the circulation and transport of passive tracers and gives some results obtained so far.

2 HYDROGRAPHY

The tidal wave progresses into the Elbe estuary up to the weir at Geesthacht, i.e. about 140 km. In its lower part the estuary is characterized by a significant influence of baroclinicity. North Sea water and fresh water are mixed forming a pronounced brackish water zone that may be defined by its salinity, varying between $1 \cdot 10^{-3}$ (nearly fresh water) and $34 \cdot 10^{-3}$ (North Sea water). The horizontal extent of this zone (up to 50 km) and its position depend upon meteorological conditions and on the fresh water discharge which varies between 400 m^3/s in September and 1200 m^3/s in April. A 15 m deep navigational channel giving access to the port of Hamburg is adjoining extensive tidal flat areas leading to strong topographic gradients. Forced by a tidal range of up to 5 m this results in complex dynamics including flooding and falling dry with strong currents exceeding 2 m/s in the deeper parts of the river. The dynamical flow pattern with strong horizontal and vertical current shear is accompanied by vertical stratification due to the salt wedge intrusion into the estuary. An example is illustrated in Fig. 1 for a station in a narrow channel within the tidal flats, showing vertical salinity gradients of 10 10^{-3} over 5 m. These mesoscale phenomena in contrast to the large scale flow in the deep waterways show the high variability of the entire flow regime that itself causes high rates and a high variability of erosion, sedimentation and redistribution of suspended matter and water properties. The water levels and currents depend nonlinearly on the river discharge (seasonal variation), the tidal range in the German Bight (spring/neap cycle, stormsurges) and the meteorological forcing. So it seems impossible to define a mean tide as being representative for the estuary.

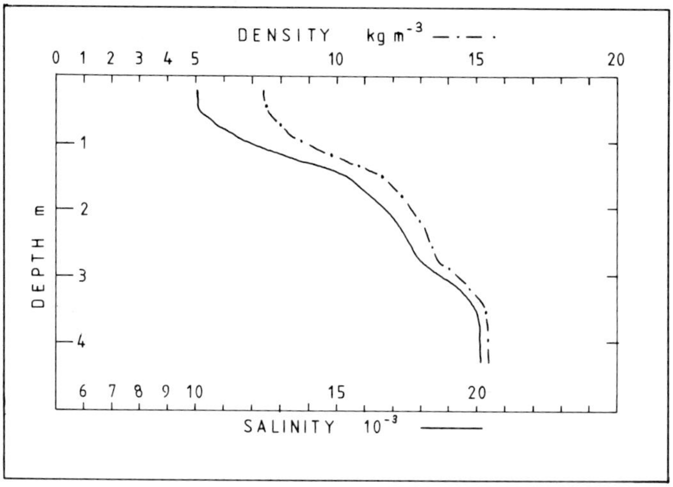

Fig. 1. Observed vertical profiles of salinity and density (calculated from conductivity) in a narrow channel, 10 km west of Brunsbüttel

3 THE MODEL SYSTEM

There are a number of motivations to develop an estuarine model system of different components. As a consequence of the descriptions given in chapter 2 it is obvious that only a three-dimensional model can take into account the observed horizontal and vertical variabilities. Due to present technical limitations it is impossible to handle an all-purpose model for the entire estuary with sufficient resolution. In order to overcome this problem the strategy has been different. If one is interested in a general study of the entire estuary as, e.g. long-term or storm surge studies, a relatively coarse resolution would be sufficient, whereas for processes linked to the velocity field like the formation of a turbidity zone, salt wedge formation and erosion/sedimentation of sand, a regionally limited area approach is appropriate. So the idea has been to combine large scale (250 - 500 m grid resolution) and smaller scale (50 - 100 m resolution) models to form a three-dimensional model system.

3.1 The estuarine circulation model system

The model system is based on a semi-implicit, two-time-level formulation of the shallow water equations for barotropic shelf seas (Backhaus, 1983). An extension for vertically integrated flow upon tidal flats (very thin layer of water, flooding, falling dry) has been given by Duwe and Hewer (1982), whereas Duwe et al. (1983) introduced the corresponding three-dimensional baroclinic version including some verifications. An application of this model to the fresh water zone together with a comparison between the three- and two-dimensional (vertically integrated) versions will be reported by Krohn and Lobmeyr (1986).

Pfeiffer and Sündermann (1986) extended the model in order to overcome the technical restriction that only the uppermost computational layer was allowed to fall dry. Presently this specific model is formulated explicitly in a classical FTCS scheme (FTCS - finite difference approximation forward-in-time and centered-in-space).

According to the different problems treated, four model components are available (Table 1).

TABLE 1
Overview over the four model components. Model numbers refer to Fig. 2

Model area Model number	Function
Entire estuary 1	Overview/Production of boundary values Study of the oscillation system
Brackish water zone 2	Study of velocity field in brackish water zone Formation of turbidity zone/salt wedge
Neufeld wadden area 3	Verification of 3D velocity field with vertically high resolving model
Fresh water zone 4	Verification of 3D velocity field (barotropic) with extensive measurements

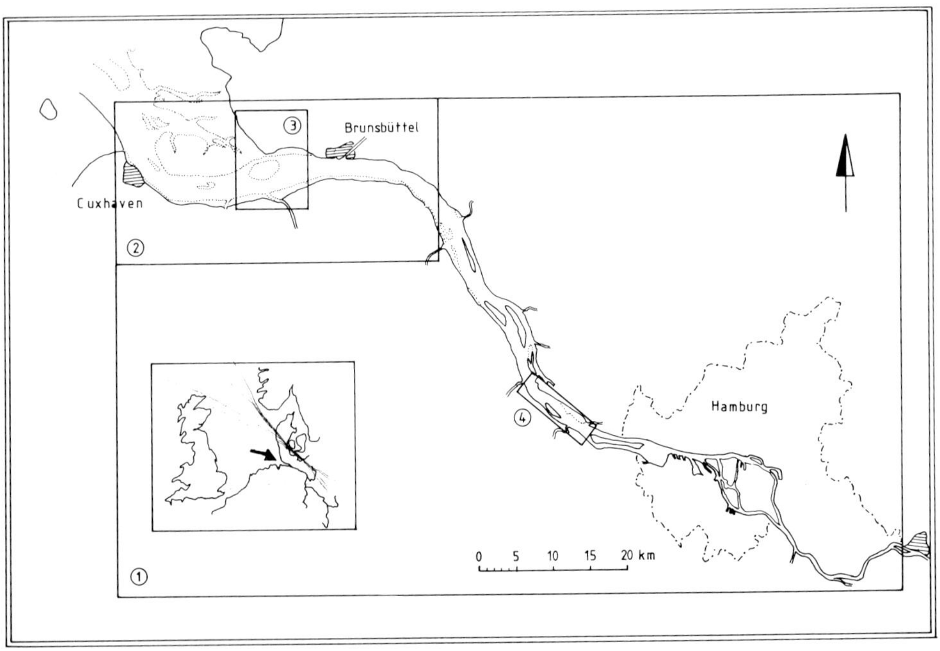

Fig. 2. Schematic map of the Elbe estuary. Areas of the different models are indicated by rectangles, the encircled numbers refer to Table 1.

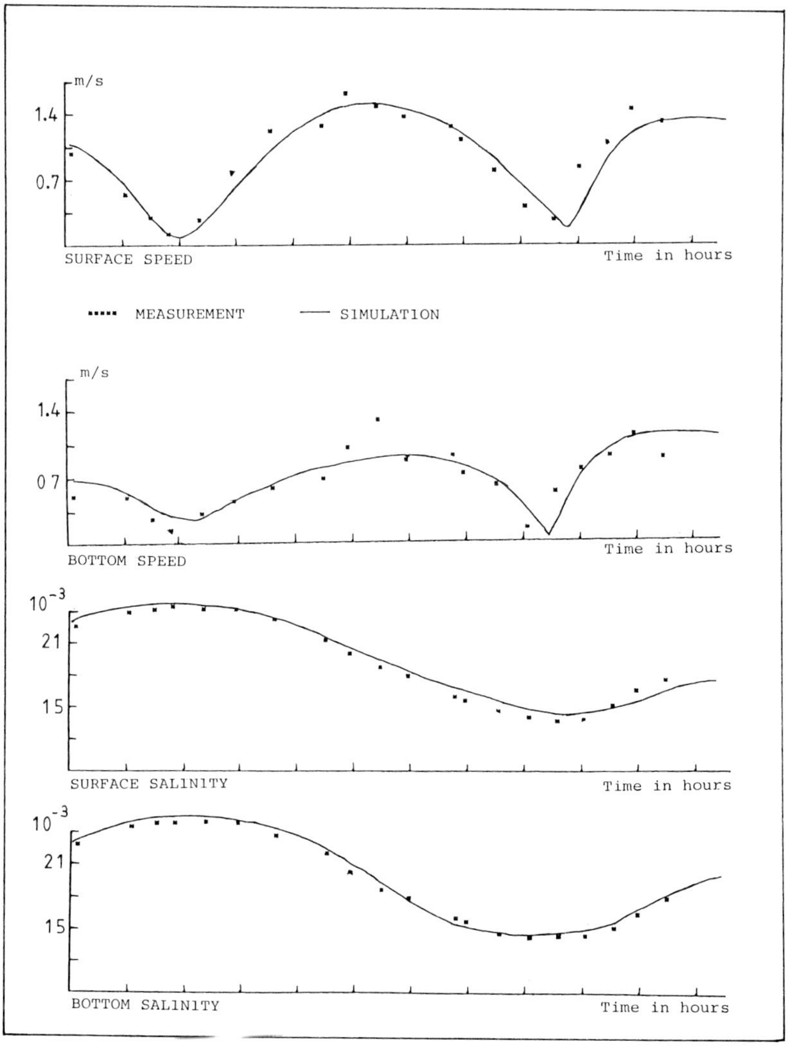

Fig. 3. Model no. 2: Observed versus computed surface velocity, bottom velocity, surface salinity and bottom salinity (from top to bottom) at a station in the navigational channel between Cuxhaven and Brunsbüttel.

In order to illustrate the horizontal variability of the flow and salinity fields, the results of a simulation for a typical autumn situation (september, low fresh water discharge) and spring tide are given in Figs. 4 to 6. The density distribution near low water (Fig. 4) shows a pronounced variation between the deeper and shallower regions. This topographic effect can be observed near high water, too. Here strong gradients are generated especially in areas where relatively dense North Sea water is enabled to flow over flooded tidal flats to

The areas covered by the different models are sketched in Fig. 2. In Table 2 some data are given on horizontal (Δs) and mean vertical (Δz) grid resolution, the timestep (Δt) and the Courant number, i.e. the factor by which the maximum timestep allowed for an explicitly formulated numerical scheme, is exceeded due to the use of an implicit formulation. The last two columns indicate the number of grid points handled in the calculations and the computational time needed to simulate one semidiurnal cycle on a Siemens 7.882 computer.

TABLE 2
Details of the four segment models

Model	Δs (m)	Δz (m)	Δt (s)	CFL factor	No. of wet points	CPU (h)
Entire estuary	250	4	150	10	18 000	2
Brackish water zone	250	2	150	10	16 000	1.5
Neufeld wadden area	250	1	10	10	9 000	2.5
Fresh water zone	100	1.5	60	5	9 000	1

The basic model (no. 1) extends over the entire estuary from the seaward boundary to the weir at Geesthacht where the tidal wave is artificially stopped. This model gives an overview over the oscillation system, i.e. the water level movement especially for storm surge calculations. Secondly it provides boundary values for the limited area "segment models" of higher resolution.

The segment model no. 2 has been designed to study the 3D velocity field in the most variable part of the estuary, the brackish water zone. It is essentially baroclinic so that the formation and movement of a strong vertical density gradient ('salt wedge') can be simulated. Therefore it has the same horizontal resolution as a model no. 1 but a higher vertical resolution (2 m). At the boundaries, the time series of water levels is prescribed, measured values at the seaward boundary and model data of the basic model at the eastern boundary. For the density cyclic boundary conditions at the seaward boundary and modelled data at the other one are used.

There have been some field experiments in order to verify the velocity field. For example Fig. 3 shows the computed and observed time series of near surface and bottom velocities at a station in the deep navigational channel. The model also reproduces the observed vertical time lag of flow reversal between near bottom and surface layers. Near low water slack time, at low salinity (see lower part of Fig. 3), the flow reversal starts near the bottom, due to friction. At high water slack time, baroclinicity predominates and the near surface flow changes its direction before the bottom water.

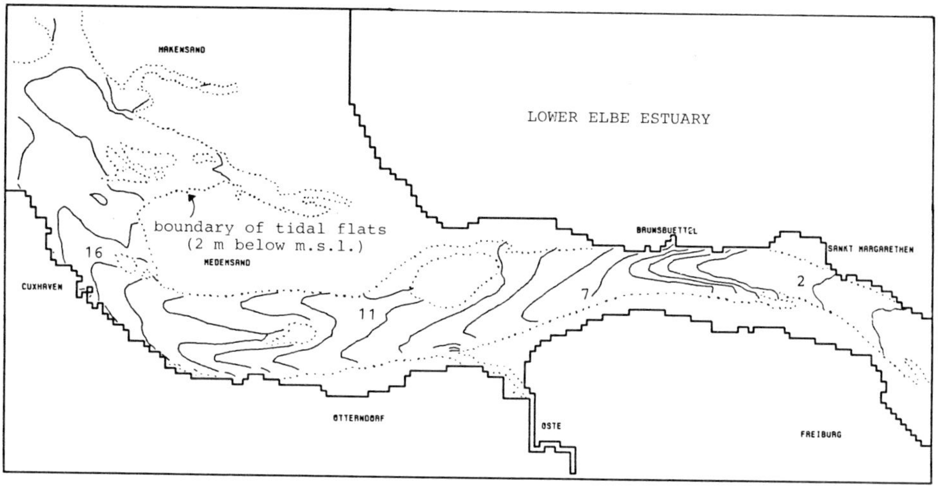

Fig. 4. Model no. 2: Density distribution in sigma-t at low water in Cuxhaven for mean September discharge and spring tide (surface layer).

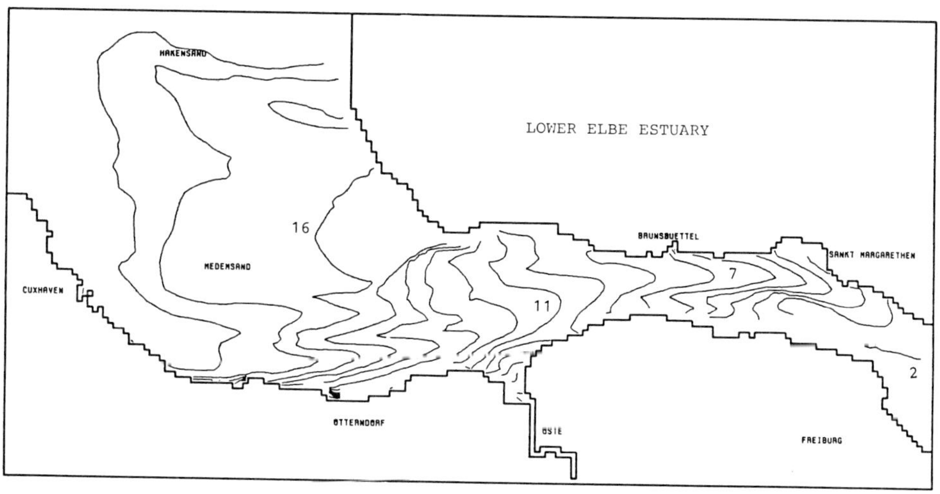

Fig. 5. Same as Fig. 4 but for high water.

meet relatively light water transported by the ebb current. For the same season and astronomical situation the vertically integrated Eulerian residual transports are depicted in Fig. 6 to illustrate the variable flow pattern. The highest values are linked to the deeper parts and directed towards the sea, whereas on tidal flats and in the narrow channels the tidal mean transport may be directed up-estuary. A longitudinal section of density shows that during one

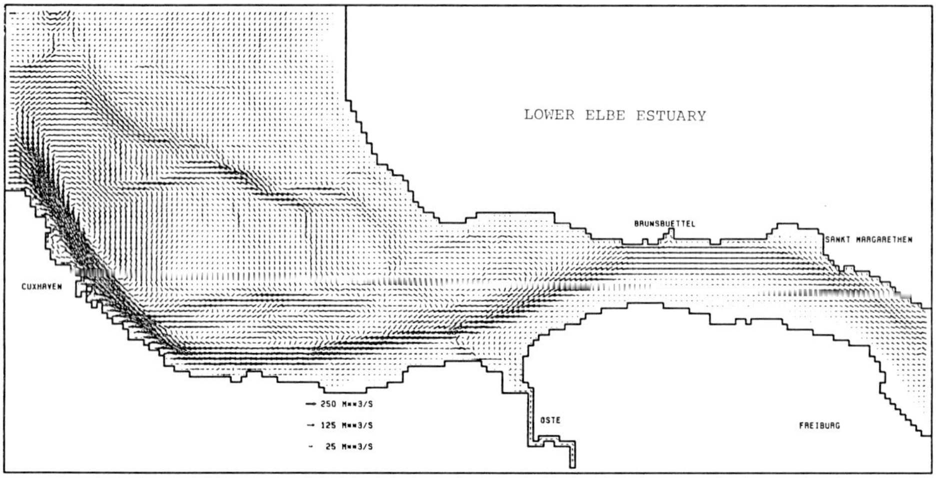

Fig. 6. Model no. 2: Eulerian residual transports (vertically integrated) for mean September discharge and spring tide.

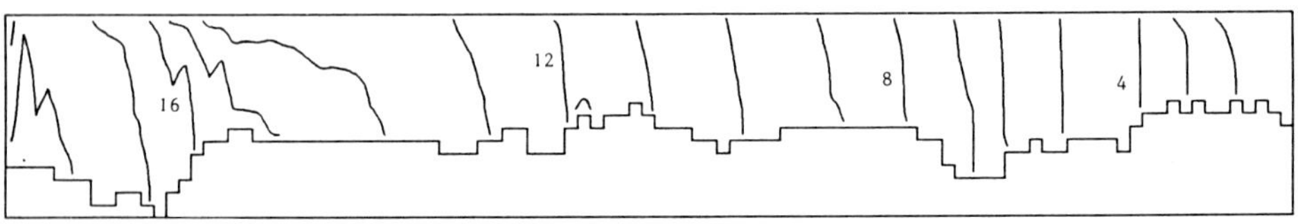

Fig. 7. Model no. 2: Longitudinal section of density in sigma-t at low water in Cuxhaven.

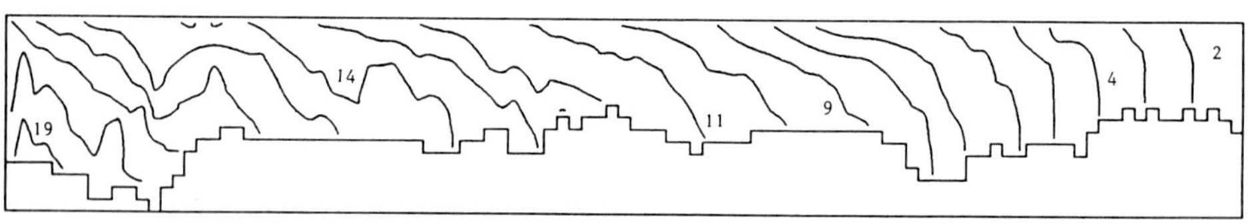

Fig. 8. Same as Fig. 7 but for high water.

tidal cycle the estuary changes from well-mixed conditions (Fig. 7) near low water to partially mixed and even stratified conditions (Fig. 8) near high water in its upper and lower parts, respectively.

Segment model no. 3 incorporates the two most significant topographic features, the deep navigational channel and an adjacent tidal flat, surrounded by a narrow channel up to 5 m deep. This model is an extension of the previous ones as it is capable of treating more than one computational layer to take part in the process of falling dry and flooding. This feature has been motivated by the strong current shear and density gradients in the uppermost meters of the brackish water zone water column. The model itself has been described in detail recently (Pfeiffer and Sündermann, 1986). Here some additional results will be demonstrated. The ability of simulating the flow field in very shallow areas is demonstrated in Fig. 9, where a vertical cross-section from north to south across the model area is shown during high and low water. For comparison with measurements, data have been selected from stations whose flooding duration matches those of neighbouring grid cells. The results are given in Fig. 10. The qualitative agreement is good, especially the reproduction of the non-harmonic behaviour and the typical peak after flood begins.

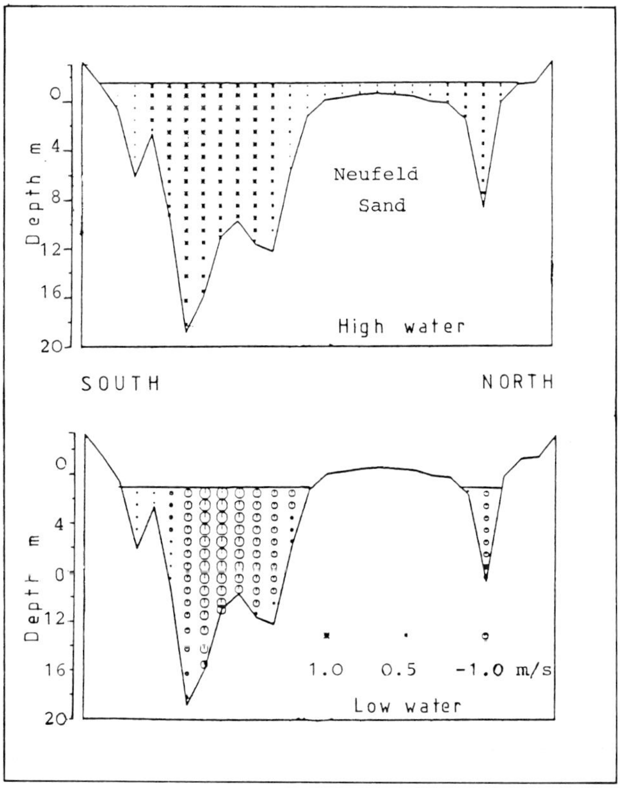

Fig. 9. Model no. 3: Vertical cross section from north to south showing calculated velocities at high water (top) and low water (bottom). Positive (negative) values indicate flood (ebb) currents. The diameter of the symbols is linearly varying with the magnitude of the velocity.

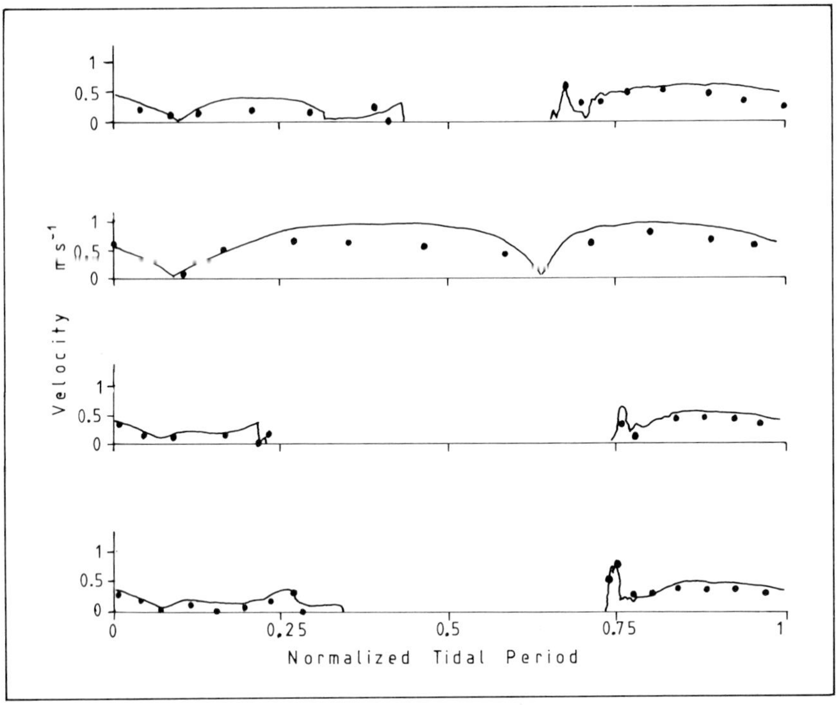

Fig. 10. Model no. 3: Observed versus calculated velocities at points on tidal flats. Full dots indicate observation, solid line denotes calculation. First two measurements (from top) were taken on Nov. 6, 1979, second pair on July 6, 1981. (Courtesy of Wasser- und Schiffahrtsamt Cuxhaven.)

Segment model no. 4 constitutes an application to the fresh water zone of the Elbe estuary. In the area considered extensive measurements have been performed (Michaelis and Knauth, 1985) used for model testing (Krohn and Lobmeyr, 1986). The topography (Fig. 11) clearly shows the deep navigational channel (up to 15 m) connected with strong lateral gradients. The model is forced by prescribed water levels at its two open boundaries, taken from official tide gauge records. These, however, had to be corrected in order to reproduce another tide gauge station data within the model area. In the future this problem will be overcome by using water level boundary values produced by the basic model of the total estuary. A more thorough discussion of this topic is given in the above mentioned paper.

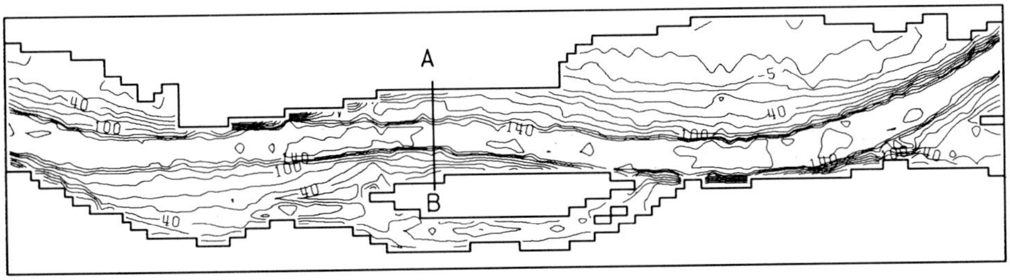

Fig. 11. Model no. 4: Topography, depths in decimeters referred to local chart datum (mean sea level minus 1.1 m). Section A-B indicates profile where model results are compared with data (see Fig. 13).

After this correction, the observed and computed cross-sectional averages of the velocity match quite satisfactorily (Fig. 12). The measured and calculated near surface velocity distributions across the river are shown in Fig. 13 at four different instants during one tidal cycle. The decrease towards the river banks are reproduced. Existing differences might be due to the still unsufficiently resolved bottom topography. A typical velocity field near low water is shown in Fig. 14. Apart from the deep navigational channel in the eastern part, the flow has already reversed its direction, especially in the shallow southern areas.

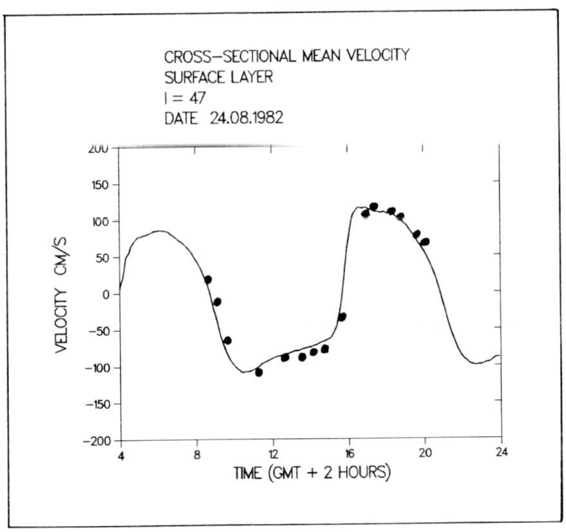

Fig. 12. Model no. 4: Comparison between observed (dots) and calculated velocities of the uppermost computational layer. Values are integrated over the river width.

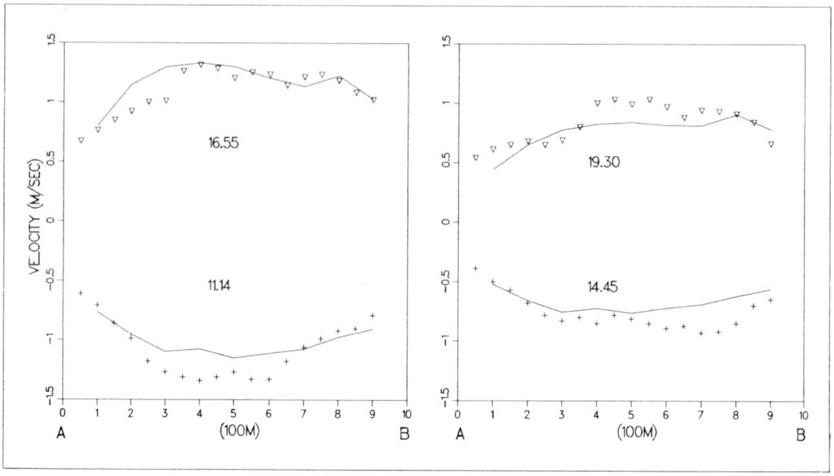

Fig. 13. Model no. 4: Across river profile of observed (symbols) and calculated velocities in the uppermost layer at different times (UTC, 24.8.1982). Profile runs from A to B, see Fig. 11.

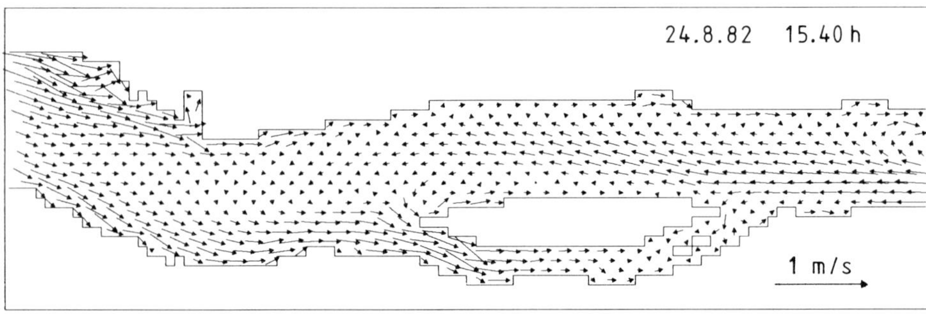

Fig. 14. Model no. 4: Computed velocity vectors in surface layer, low water slack time. Vectors plotted every second grid point. Ebb current to the left.

3.2 The passive tracer transport model

As a first step towards the simulation of suspended matter transport and the formation of a turbidity zone the transport of a passive tracer is computed, i.e. a completely dissolved conservative substance. Due to the computational effort tracer methods are mainly suitable for tracking a limited number of particles released into the water as may be the case for ship accidents or waste water discharge.

Consider a given concentration C of a passive substance in a three-dimensional flow field u = (u,v,w). Its temporal and spatial change is described by the well-known advection-diffusion equation

$$\frac{\partial C}{\partial t} + u\frac{\partial C}{\partial x} + v\frac{\partial C}{\partial y} + w\frac{\partial C}{\partial z} + k_h \nabla^2 C + k_v \frac{\partial^2 C}{\partial z^2} = 0 \qquad (1)$$

(t - time; x,y,z - cartesian coordinates; k_h, k_v - horizontal, vertical diffusion coefficients.) In order to overcome problems arising from numerical diffusion caused by formulations of the nonlinear advection terms, a Lagrangian approach has been used. This method (see, e.g. Bork and Maier-Reimer, 1978) essentially avoids numerical diffusion as the nonlinear terms vanish. The velocity field is split into an advective part (calculated velocities) and a stochastic part (fluctuations defined below). It should be admitted that averaging of velocities may principally lead to small diffusion-like effects. Turbulent diffusion is simulated by a statistical random process ('Monte-Carlo method') by adding for each spatial coordinate a random number to the velocity component. Due to special properties of the estuarine flow field some modifications to the classical method have been necessary.

Numerical experiments carried out in order to test different formulations of the turbulent diffusion led to

$$u = \bar{u} + u', \quad v = \bar{v} + v', \quad w = \bar{w} + w', \qquad (2)$$

the fluctuations being represented by

$$u' = \alpha\, p_h \qquad \alpha \in (-1,1) \qquad (3)$$

$$v' = \beta\, p_h \qquad \beta \in (-1,1) \qquad (4)$$

$$w' = \gamma\, p_v \qquad \gamma \in (-1,1) \qquad (5)$$

where α, β and γ are random numbers out of a "top hat" distribution over the given interval, and p_h and p_v define the bandwidth of the horizontal and vertical fluctuations. Their values are determined after

$$k_i = \frac{1}{6}(p_i)^2 \Delta t, \qquad i = (h, v) \qquad (6)$$

(Δt - timestep), see Maier-Reimer and Sündermann (1982) so that k_h is of the order of 1 m^2/s and k_v of the order of 0.05 m^2/s, both values representing observed means. In the vicinity of solid boundaries a 'no-slip' condition for the velocity component parallel to the boundary is applied, giving a more realistic approach with respect to conservation of transports computed in the Eulerian grid. If a particle tries to cross a solid boundary or the free surface, which can only be caused by the random component, the particle is transported with another random component satisfying the no-flux boundary condition.

The strong variability of the flow requires that the original timestep (depending on the relationship between grid size and time step of the current model) be decreased by a factor of 4 in the transport calculation. This may lead to a temporal interpolation of the velocities.

Fig. 15 shows the evolution of particle distributions 1, 6 and 10 hours after release near the surface at a point near the navigational channel in the brackish water zone. Due to longitudinal current shear the cloud is stretched considerably whereas lateral diffusion seems to be negligible. After flow reversal the concentration moves back and - in the case of hazardous substances - could affect the tidal flats and coastal zone.

An application to the fresh water zone segment model is depicted in Fig. 16. All calculations have been performed with 10 000 particles. A detailed description can be found in Duwe et al. (1986).

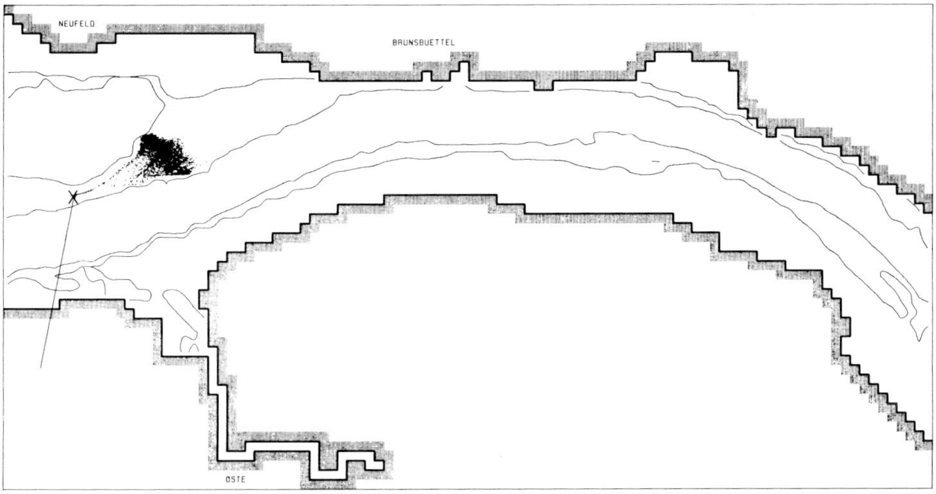

Fig. 15 a. Model no. 2: Particle distribution 1 hour after release at marked point (X)

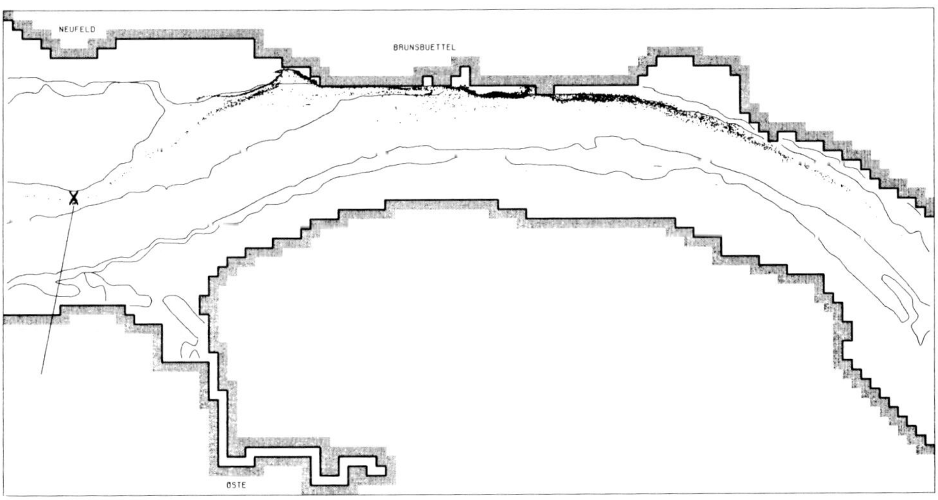

Fig. 15 b. Model no. 2: Particle distribution 6 hours after release at marked point (X)

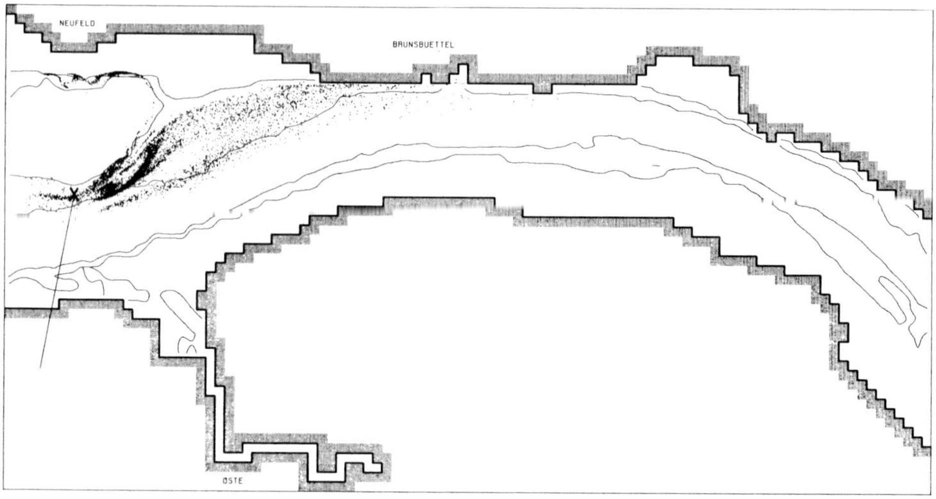

Fig. 15 c. Model no. 2: Particle distribution 10 hours after release at marked point (X)

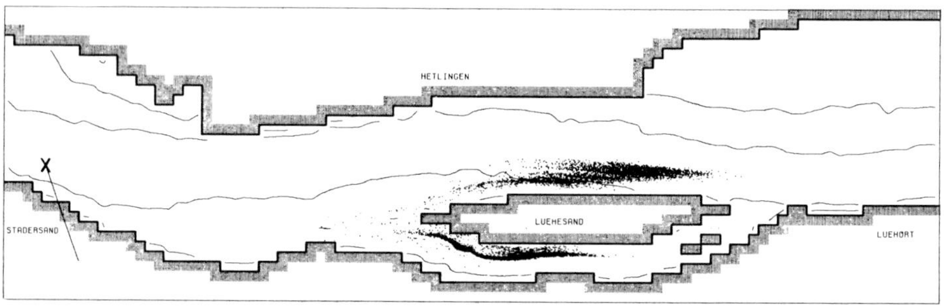

Fig. 16. Model no. 4: Particle distribution 2 hours after release at marked point (X)

4 CONCLUSIONS

On the basis of a semi-implicit formulation of the shallow water equations a set of numerical models for estuarine applications has been developed. It combines high spatial resolution and computational efficiency as the courant number of one is considerably exceeded. With respect to the grid size and available observations the velocity field is well reproduced. It is obvious, however, that more detailed observations are necessary for further verification, especially of the vertical structure of the flow in the vicinity of the bottom. Together with improved formulations of transport phenomena, e.g. resolution of front-like structures, this will lead to a model system that is capable of calculating the transport of dissolved as well as suspended matter.

5 REFERENCES

Backhaus, J., 1983. A semi-implicit scheme for the shallow water equations for application to shelf sea modelling. Cont. Shelf Res., 2(4): 243-254.

Bork, I. and Maier-Reimer E., 1978. On the spreading of power plant cooling water in a tidal river applied to the river Elbe. Adv. Water Res., 1.

Duwe, K. and Hewer, R., 1982. Ein semi-implizites Gezeitenmodell für Wattgebiete. D. Hydrogr. Zeitschr., 35: 223-238.

Duwe, K.C., Hewer, R.R. and Backhaus, J.O., 1983. Results of a semi-implicit two-step method for the simulation of markedly nonlinear flow in coastal seas. Cont. Shelf Res., 2(4): 255-274.

Duwe, K., Krohn, J., Pfeiffer, K., Riedel-Lorjé, J.C. and Soetje, K.C., 1986. Ausbreitung von wassergefährdenden Stoffen in der südlichen Deutschen Bucht und im Elbe-Ästuar nach Freisetzung durch Schiffe. Subm. to D. Hydrogr. Zeitschr.

Krohn, J. and Lobmeyr, M., 1986. Comparison of two- and three-dimensional high resolving estuarine models with observations in the Elbe river. In preparation.

Maier-Reimer, E. and Sündermann, J., 1982. On tracer methods in computational hydrodynamics. In: M.B. Abbott and J.A. Cunge (Editors). Engineering applications of computational hydraulics, Volume 1. Pitman, Boston/London/Melbourne, pp. 198-217.

Michaelis, W. and Knauth, H.-D., 1985. Das Bilanzierungsexperiment 1982 (BILEX '82) auf der Unterelbe. GKSS Forschungszentrum Geesthacht, Report GKSS 85/E/3 (unpubl.), 212 pp.

Pfeiffer, K.D., Sündermann, J., 1986. Ein dreidimensionales Flachwassermodell mit vertikaler Auflösung im Tidehubbereich: Entwicklung und erste Anwendungen. Die Küste, 43: 149-165.

A THREE DIMENSIONAL NUMERICAL MODEL OF SEMI-DIURNAL TIDES ON THE EUROPEAN CONTINENTAL SHELF

A. M. DAVIES
Institute of Oceanographic Sciences, Bidston Observatory, Birkenhead, Merseyside L43 7RA (England)

ABSTRACT

A three dimensional hydrodynamic tidal model of the North west European Continental shelf is developed using a staggered finite difference grid in the horizontal and a spectral method through the vertical.

Some preliminary calculations of the M_2 and S_2 tides performed with the model are used to demonstrate the sensitivity of computed tides in the North Sea to tidal input at the continental shelf edge.

Histograms of errors between observed and computed M_2 and S_2 tides, at one hundred and twenty nine offshore and coastal tide gauges, show that the model can adequately reproduce the tides in the North Sea, provided an accurate specification of the tides along the shelf edge is available. Some indication of the accuracy of the computed spring tidal currents is obtained by comparing computed surface currents with observations.

1 INTRODUCTION

In earlier papers (Davies and Furnes 1980, Heaps and Jones 1981) three dimensional numerical models were used to simulate only the M_2 component of the tide, with Davies (1986) computing both the M_2 and M_4 tides. However on the North-West European shelf, the S_2 component of the tide is also important. In this paper the M_2 and S_2 tidal currents are computed by integrating the tidal model with both M_2 and S_2 input on the open boundary, and separating the two components by harmonic analysis of the resulting computed time series. By this means bottom stress and turbulence in the model are determined by the combination of these tidal constituents. To the author's knowledge this is the first time that a three dimensional tidal model has been used to simulate these tidal constituents in combination.

The model uses a finite-difference grid in the horizontal, and a Galerkin-spectral method in the vertical. The functions used through the vertical are eigenfunctions of an eigenvalue problem involving the vertical variation of eddy viscosity. By this means an accurate and computationally efficient (Davies and Stephens 1983) model can be developed. Unlike early spectral models (Davies and Furnes 1980, Heaps and Jones 1981) which were restricted to certain vertical variations of eddy viscosity, in the present model eddy viscosity can vary through the vertical and with horizontal position and time, in an arbitrary manner.

The three-dimensional numerical model has a grid resolution of 1/3° latitude by 1/2° longitude and covers the continental shelf (see Fig. 1, showing the grid and the location of some tide gauges). The open boundary of the model coincides with the shelf edge. This is a natural boundary between the deep ocean and shallower shelf, and has been chosen as an open boundary because it coincides with off-shore tide gauge measurements made by Cartwright (1976), which form the basis of the boundary input to the model together with data from a two dimensional N.E. Atlantic model of Flather, Proctor and Wolf (personal communication). Along these open boundaries M_2 and S_2 tidal input are determined, and a radiation condition is employed to allow disturbances from the interior of the model to propagate outwards.

Co-tidal charts of M_2 and S_2 over the region of the shelf are constructed from the model output, and the accuracy of the model is determined by comparing amplitude and phase of tidal elevations with observations taken at a number of tide gauges. In most cases the model can satisfactorily reproduce the observed tides.

Mean spring tidal currents at sea surface and sea bed over the whole region of the shelf are determined from the model. Sea surface values of these currents are in good agreement with similar distributions derived from observations (Howarth 1982, Howarth and Pugh 1983).

2. THREE-DIMENSIONAL SPECTRAL MODEL

In this section we briefly describe the major steps in the formulation of the three dimensional Galerkin-spectral model. The interested reader is referred to Davies and Owen (1979), Davies (1980, 1983a,b), Owen (1980) for more details.

2.1 Hydrodynamic equations

The equations for continuity and motion for a homogeneous fluid, neglecting the advective terms and shear in the horizontal, in polar coordinates are given by

$$\frac{\partial \zeta}{\partial t} + \frac{1}{R\cos\varphi}\left\{\frac{\partial}{\partial \chi}\int_0^h u\,dz + \frac{\partial}{\partial \varphi}\int_0^h v\cos\varphi\,dz\right\} = 0 \tag{1}$$

$$\frac{\partial u}{\partial t} - \gamma v = \frac{-g}{R\cos\varphi}\frac{\partial \zeta}{\partial \chi} + \frac{\partial}{\partial z}\left\{\mu\frac{\partial u}{\partial z}\right\} \tag{2}$$

$$\frac{\partial v}{\partial t} + \gamma u = \frac{-g}{R}\frac{\partial \zeta}{\partial \varphi} + \frac{\partial}{\partial z}\{\mu\frac{\partial v}{\partial z}\} \tag{3}$$

where χ , φ are east-longitude and north-latitude respectively, with z depth below the undisturbed surface. In these equations, t denotes time, ζ elevation of the sea surface above the undisturbed level, h undisturbed depth of water, R radius of earth, γ Coriolis parameter, u,v east-going and north-going components of current at depth z, and g acceleration due to gravity. The coefficient of eddy viscosity μ varies with x, y, z and t.

For tides a zero stress surface boundary condition is required, giving

$$-\rho\{\mu\frac{\partial u}{\partial z}\}_0 = 0 \quad , \quad -\rho\{\mu\frac{\partial v}{\partial z}\}_0 = 0 \tag{4a,b}$$

with subscript zero denoting evaluation at z=0.

Applying a slip condition at the sea bed and using a quadratic law of bottom friction, gives

$$-\{\mu\frac{\partial u}{\partial z}\}_h = ku_h\{u_h^2 + v_h^2\}^{\frac{1}{2}} \qquad -\{\mu\frac{\partial v}{\partial z}\}_h = kv_h\{u_h^2 + v_h^2\}^{\frac{1}{2}} \tag{5a,b}$$

where k is the coefficient of bottom friction, and subscript h denotes evaluation at z=h.

2.2 Application of the Galerkin method

We now briefly consider the solution of equations (1) to (3) subject to boundary conditions (4a,b) and (5a,b) in terms of the Galerkin method with a basis set of functions f_r (r=1,2,...m) through the vertical. In general the choice of basis functions f_r is arbitrary and functions such as B-splines, Chebyshev or Legendre polynomials (Davies and Owen 1979) have proved very successful. In the formulation developed here, these basis functions are taken as modes of an eigenvalue problem involving the vertical variation of eddy viscosity μ. For convenience they are defined in terms of sigma coordinates σ, over the interval $0\leqslant\sigma\leqslant1$, rather than over the interval $0\leqslant z\leqslant h$. The transformation to sigma coordinates is accomplished using

$$\sigma = z/h \tag{6}$$

Expanding the two components of velocity u,v in terms of m functions $f_r(\sigma)$ and horizontal-space and time dependent coefficients $A_r(\chi,\varphi,t)$ and $B_r(\chi,\varphi,t)$ gives

$$u(\chi,\varphi,\sigma,t) = \sum_{r=1}^{m} A_r(\chi,\varphi,t)f_r(\sigma) \tag{7}$$

$$v(\chi,\varphi,\sigma,t) = \sum_{r=1}^{m} B_r(\chi,\varphi,t)f_r(\sigma) \tag{8}$$

Using the Galerkin method in the vertical space domain, equations (1), (2) and (3) are first transformed to sigma coordinates using equation (6). The sigma form of these equations is then multiplied by each of the basis functions $f_k(\sigma)$ and integrated with respect to σ over the interval 0 to 1. By integrating the term involving the vertical eddy viscosity, boundary conditions (4a,b) and (5a,b) in sigma coordinates can be included, (see Davies 1980, 1983a,b for details). Finally, expansions (7) and (8) are substituted into the resulting equations, giving,

$$\frac{\partial \zeta}{\partial t} + \frac{1}{R\cos\varphi} \sum_{r=1}^{m} \left[\frac{\partial}{\partial \chi}\{hA_r\int_0^1 f_r d\sigma\} + \frac{\partial}{\partial \varphi}\{hB_r\cos\varphi\int_0^1 f_r d\sigma\}\right] = 0 \tag{9}$$

$$\sum_{r=1}^{m} \frac{\partial A_r}{\partial t}\int_0^1 f_r f_k d\sigma - \gamma \sum_{r=1}^{m} B_r\int_0^1 f_r f_k d\sigma = \frac{-g}{R\cos\varphi}\frac{\partial \zeta}{\partial \chi}\int_0^1 f_r d\sigma$$

$$-\frac{f_k(1)ku_h(u_h^2+v_h^2)^{\frac{1}{2}}}{h} - \frac{1}{h^2}\sum_{r=1}^{m} A_r\int_0^1 \mu\frac{df_r}{d\sigma}\frac{df_k}{d\sigma}d\sigma \tag{10}$$

where $k=1,2,\ldots,m$

With a similar equation to (10) derived from the V equation of motion (3).

2.3 Calculation of vertical modes

For an arbitrary basis set of functions, the m equations in (10) and the corresponding V equations, are coupled together through bottom friction and the viscosity term. However the coupling through the viscosity term can be removed, with an associated significant saving in computer time (Davies and Stephens 1983), by using a basis set of modes which are eigenfunctions of an eigenvalue problem involving the eddy viscosity, of the form

$$\frac{d}{d\sigma}\left[\mu\frac{df}{d\sigma}\right] = -\tilde{\varepsilon}f \tag{11}$$

with $\tilde{\varepsilon}$ and f denoting respectively eigenvalue and eigenfunction.

Writing μ in terms of a time-dependent and horizontally varying part $\alpha(\chi,\varphi,t)$ and vertically varying function $\Phi(\sigma)$, gives

$$\alpha(\chi,\varphi,t)\frac{d}{d\sigma}\left[\Phi\frac{df}{d\sigma}\right] = -\tilde{\varepsilon}f \tag{12}$$

Since α is a function of χ,φ,t, it only affects the eigenvalues, and the eigenfunctions are determined from

$$\frac{d}{d\sigma}\left[\Phi\frac{df}{d\sigma}\right] = -\tilde{\varepsilon}f \tag{13}$$

with $\tilde{\varepsilon}=\alpha\varepsilon$

For an arbitrary vertical variation of Φ, the eigenvalues of (13) can be readily computed using either the Galerkin method (Davies 1983a,b) or the Runge-Kutta approach (Davies and Furnes 1984). Extensive details of the method are given in these papers and will not be repeated here.

In order to solve equation (13), surface and sea bed boundary conditions, consistent with those given in equations (4a,b) and (5a,b) must be specified. It can be readily shown (Davies 1983a,b) that boundary conditions,

$$\Phi\frac{df}{d\sigma}\bigg|_0 = 0 \quad , \quad \Phi\frac{df}{d\sigma}\bigg|_1 = 0 \tag{14}$$

are appropriate and give rise to real eigenvalues and eigenvectors of (13).

For convenience the eigenfunctions are normalized by requiring that

$$f_r(0) = 1 \qquad r=1,2,\ldots,m \tag{15}$$

The eigenfunctions of (13) are orthogonal, and consequently

$$\int_0^1 f_r f_k d\sigma = 0 \qquad r\neq k \tag{16}$$

Also with boundary condition (14), we have,

$$\int_0^1 f_r d\sigma = 0 \qquad r=2,3,\ldots,m \tag{17}$$

2.4 Spectral form of the hydrodynamic equations

With a basis set of eigenfunctions (modes), equations (9), (10), and the corresponding V equation can be simplified (see Davies 1983a,b for details) by using (13), (15), (16) and (17), to give,

$$\frac{\partial\zeta}{\partial t} + \frac{1}{R\cos\varphi}\left\{\frac{\partial}{\partial\chi}\left[hA_1\int_0^1 f_1 d\sigma\right] + \frac{\partial}{\partial\varphi}\left[hB_1\int_0^1 f_1 d\sigma\right]\right\} = 0 \tag{18}$$

$$\sum_{r=1}^{m}\frac{\partial A_r}{\partial t}\int_0^1 f_r f_r d\sigma - \gamma\sum_{r=1}^{m} B_r\int_0^1 f_r f_r d\sigma = \frac{-g}{R\cos\varphi}\frac{\partial\zeta}{\partial\chi}\int_0^1 f_r d\sigma$$

$$-f_r(1)\,\frac{ku_h(u_h^2+v_h^2)^{\frac{1}{2}}}{h} - \frac{A_r\varepsilon_r\alpha}{h^2}\int_0^1 f_r f_r d\sigma \tag{19}$$

$$\sum_{r=1}^{m} \frac{\partial B_r}{\partial t} \int_0^1 f_r f_r d\sigma + \gamma \sum_{r=1}^{m} A_r \int_0^1 f_r f_r d\sigma = \frac{-g}{R} \frac{\partial \zeta}{\partial \varphi} \int_0^1 f_r d\sigma \tag{20}$$

$$-f_r(1) \frac{k v_h (u_h^2+v_h^2)^{\frac{1}{2}}}{h} - \frac{B_r \varepsilon_r \alpha}{h^2} \int_0^1 f_r f_r d\sigma$$

where

$$u_h = \sum_{r=1}^{m} A_r f_r(1) \quad , \quad v_h = \sum_{r=1}^{m} B_r f_r(1) \tag{21a,b}$$

Equations (18), (19) and (20) are the working equations which have to be solved to find the variation of ζ, A_r and B_r over the sea area, subject to initial and boundary conditions. Currents at any depth can then be calculated from the A_r and B_r using expansions (7) and (8).

Any arbitrary vertical variation of eddy viscosity, varying with horizontal position and time can be incorporated in the model.

In order to solve equations (18), (19) and (20) it is necessary to discretize in the horizontal and with time. Horizontal discretization is accomplished using a staggered finite different grid, in which ζ, U, V are evaluated at different grid points. The time discretization is accomplished using a forward time-stepping method. Details of these techniques are standard and will not be presented here.

Solutions are generated from a state of zero displacement and motion. Along a closed boundary the normal component of current is set to zero. At the open boundaries of the model, M_2 and S_2 tidal input determined from a three dimensional model (Davies and Furnes 1980), and observations (Cartwright 1976) together with data from a two dimensional model of the N.E. Atlantic (Flather, Proctor and Wolf personal communication) were specified. A three dimensional radiation condition was employed along the open boundary to allow disturbances from the interior of the model to propagate outward. Details of these conditions can be found in Davies and Furnes (1980) and will not be given here.

An appropriate parameterization of vertical eddy viscosity to use in a three-dimensional model is difficult to determine. Davies and Furnes (1980), related eddy viscosity to depth mean currents $\bar{u}$ and $\bar{v}$, by,

$$\mu = K(\bar{u}^2+\bar{v}^2)/\sigma$$

with σ a frequency of $1.0\times10^{-4}s^{-1}$, and K a dimensionless coefficient having a value $K=2.0\times10^{-5}$. This formulation and these parameters are used in the present calculation.

The coefficient of bottom friction k, used in (5a,b), should have a value comparable with observed values of C_{100}, where C_{100} is the coefficient which

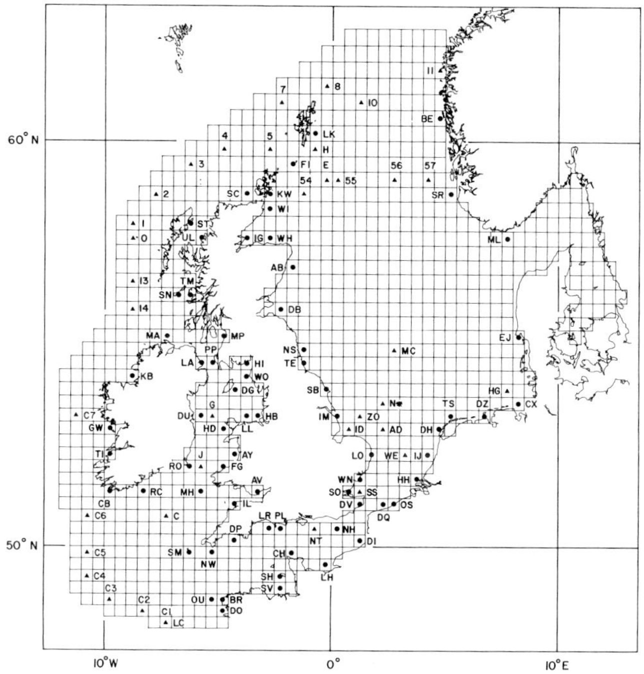

Fig. 1. Finite-difference grid of the three-dimensional continental shelf model, showing position of some tide gauges used in comparisons.

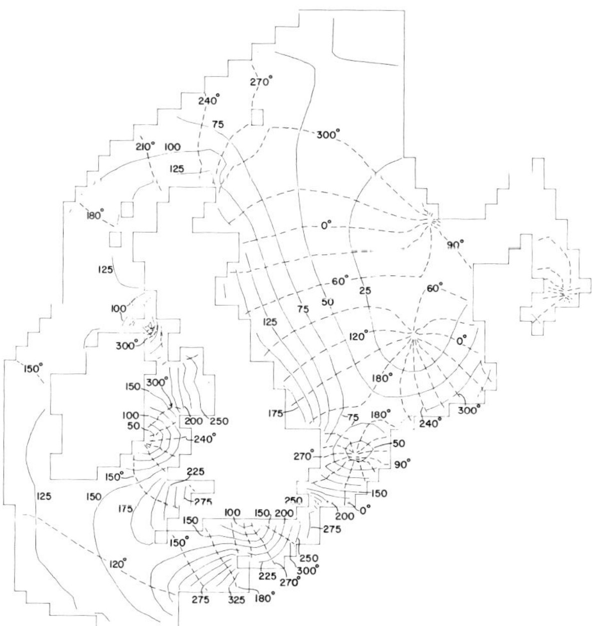

Fig. 2. M_2 co-tidal chart showing amplitude in cm (——) and phase in degrees (----)

relates bottom stress to the current 100cm above the sea bed. Values of C_{100} range from 4.30×10^{-3} Wolf (1980) to $5.29 \pm 1.7 \times 10^{-3}$ (Bowden and Ferguson 1980). In the present calculations a value of 5.0×10^{-3} was employed; a value consistent with this range of observed values.

3 COMPARISON OF OBSERVED AND COMPUTED ELEVATIONS AND CURRENTS

3.1 Elevations

Co-tidal charts of the M_2 and S_2 components of the tide computed by harmonically analysing output from the model, using standard methods, are given in Figures 2 and 3a,b.

Referring to Figure 2, it is apparent that the model reproduces the known (Lennon 1961) three M_2 amphidromes in the North Sea. The position of these three amphidromes and the distribution of co-amplitude and co-phase lines shown in Figure 2 is in excellent agreement with M_2 co-tidal charts derived from higher resolution North sea models (Davies 1976, Prandle 1980, Mathisen and Johansen 1983, Davies, Sauvel and Evans 1984) and also co-tidal charts derived from observations (Lennon 1961).

In order to examine the sensitivity of the S_2 tide to changes in the S_2 tidal input along the open boundary, results from two calculations (namely calculations A and B) will now be described. In the first calculation (Calcn. A), the S_2 tidal input (ζ_T, q_T) to the radiation condition (Davies and Furnes 1980) was interpolated from a two dimensional N.E. Atlantic model of Flather, Proctor and Wolf (personal communication). The M_2 tidal input to the radiation conditions was identical to that used by Davies and Furnes (1980). The M_2 and S_2 tides were separated by a harmonic analysis of the tidal time series computed by the model with this boundary forcing. It was found that although the M_2 tidal elevation, amplitude and phase was in good agreement with observations derived from off-shore tide gauges, in the vicinity of the open boundary (see Table 1), the S_2 amplitude was of the order of up to 10cm too high (approximately 20%), (see gauges 5 and 6 in Table 1). This error gave rise to an amplitude error of the order of 6cm, at off shore tide gauges H, 54, 55, 56, 57 which in essence mark the northern entrance to the North Sea (see Table 2). (These tide gauges lie approximately along a line from Wick to Bergen, see Davies, Sauvel and Evans (1984)).

Ideally an iterative process should then be used (Flather 1976) to derive a new ζ_T , q_T for the S_2 tidal input to the radiation condition. However such an iterative process is computationally expensive with a three dimensional model since each iteration involves a full 3D tidal simulation with the associated harmonic analysis to separate M_2 and S_2. In practice a number of calculations were performed, with ζ_T , q_T for the S_2 tide, reduced by the order of approximately 15% in the region of the open boundary where the S_2 tidal amplitude exceeded the observed. By this means a second solution (Calcn. B) was obtained in which the S_2 amplitude and phase was in better agreement with observations at off-shore tide gauges near the model open boundary (see Table 1); in the Celtic Sea and particularly at rigs H, 54, 55, 56, 57 in the northern North Sea (see Table 2). The M_2 tidal input to the model in both Calcns. A and B was identical to that used by Davies and Furnes (1980). It is evident from Tables 1 and 2, that with this input the M_2 tide is in good agreement with observations at offshore tide gauges. It is also apparent that the M_2 tide at these gauges is not significantly influenced (of order 1cm amplitude and 1 degree phase) by changes in S_2 tidal input.

TABLE 1

OSTG Boundary	M_2 Obs		Calcn. A		Calcn. B		S_2 Obs		Calcn. A		Calcn. B	
	h (cm)	g°	h (cm)	g°	h (cm)	g°	h (cm)	g°	h (cm)	g°	h (cm)	g°
C1	129	102	129	103	129	103	45	137	50	140	41	137
C2	111	108	128	104	128	104	36	138	45	142	43	138
C3	116	107	120	103	120	103	39	140	44	144	42	135
C4	109	109	121	108	121	108	37	142	44	146	43	139
C5	112	116	124	113	124	113	38	150	45	148	44	146
C6	112	121	123	121	123	121	37	154	45	150	44	153
C7	117	144	113	144	113	144	41	179	46	179	42	178
West Coast/Northern North Sea Boundary												
14	112	163	111	164	111	164	42	196	46	189	42	202
13	110	167	109	168	109	168	41	202	46	191	40	207
1	108	179	103	179	103	179	40	212	42	212	37	216
2	104	190	101	190	101	190	39	224	43	225	36	227
3	93	205	91	206	91	206	35	238	39	246	33	243
4	84	228	83	225	83	225	31	261	38	270	30	261
5	70	252	74	259	74	259	26	287	37	305	27	297
6	54	264	58	256	58	256	20	298	30	302	21	294
7	56	256	53	257	53	257	21	293	28	299	19	294
8	55	282	53	282	53	282	20	316	25	323	19	320
9	66	302	59	291	59	291	23	333	29	334	21	329
10	51	282	50	289	50	289	19	328	23	335	17	327
11	52	285	52	279	52	279	18	320	22	328	18	317

Table 1. Comparison of computed and observed amplitude h(cm) and phase g(degrees) for the M_2 and S_2 components of the tide at off-shore gauges in the vicinity of the shelf edge.

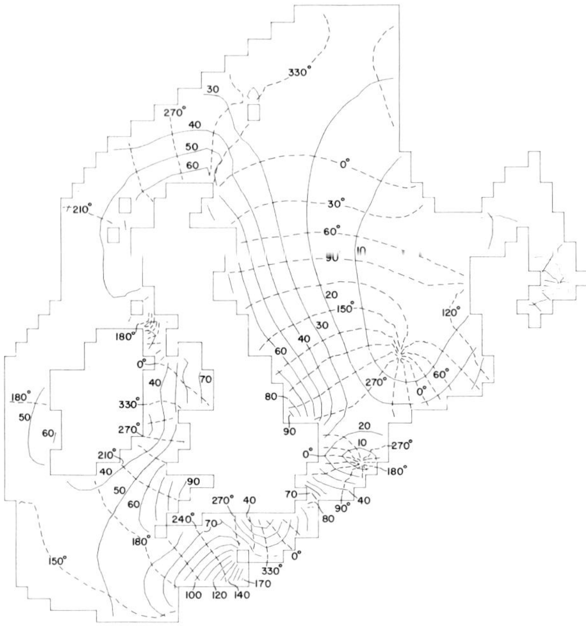

Fig. 3a. S_2 co-tidal chart showing amplitude in cm (——) and phase in degrees (----), derived from calculation A.

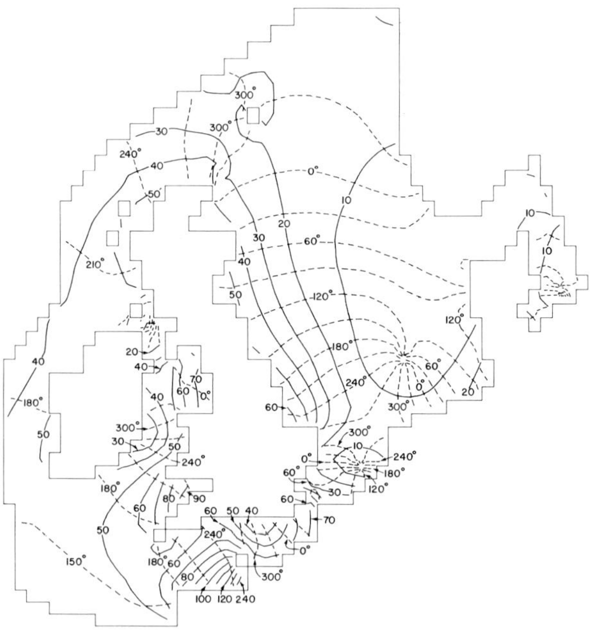

Fig. 3b. S_2 co-tidal chart showing amplitude in cm (——) and phase in degrees (----), derived from calculation B.

Cotidal charts for the S_2 tide, derived from calculations A and B are shown in Figs 3a,b. It is apparent from these Figures, that the amplitude of the S_2 tide in the northern North Sea, computed in calculation A (see Fig 3a and Tables 2 and 3), is significantly higher than that in calculation B, which is in reasonable agreement with observations. The spatial distribution of the S_2 tide derived from calculation B is in good agreement with that given by Mathisen and Johansen (1983), from a high resolution two-dimensional North Sea model.

TABLE 2

Celtic Sea/ Irish Sea	M_2 Obs h (cm)	g°	Calcn. A h (cm)	g°	Calcn. B h (cm)	g°	S_2 Obs h (cm)	g°	Calcn. A h (cm)	g°	Calcn. B h (cm)	g°
H	123	112	128	109	128	108	41	145	46	148	44	140
G	138	122	139	116	139	116	45	158	48	156	47	151
F	156	136	155	129	155	129	52	178	49	174	50	168
C	163	144	160	134	160	134	55	190	49	184	51	175
D	189	142	181	134	182	134	66	187	59	187	59	177
B	144	154	134	144	134	144	49	202	41	210	42	196
J	112	184	100	172	101	173	44	230	44	237	41	224
G	138	309	134	305	135	305	44	344	39	327	39	328
North Sea												
H	50	309	54	301	54	301	19	338	25	348	19	341
54	67	323	71	316	72	315	24	360	33	6	25	358
55	50	318	51	314	52	314	18	354	24	2	18	355
56	32	309	33	307	33	307	13	345	16	357	12	349
57	25	286	25	291	25	290	10	328	12	346	9	338

Table 2. Comparison of computed and observed amplitude h (cm) and phase g (degrees) for the M_2 and S_2 components of the tide at off-shore gauges.

Considering initially the distribution of the S_2 tide over the North Sea. It is evident from Tables 2 and 3, that in the northern North Sea, in calculation A, the S_2 tide computed with the model has a significant error, in that the computed amplitude is the order of +10cm to +15cm above the observed (see for example Wick and Aberdeen). This reflects an intensification of the boundary error (of order 5cm) as the tide propagates into the model. On the other hand, solution B, which was optimised to simulate the S_2 tide at off-shore tide gauges in northern North Sea (see Table 2), contains only a small error, of order -4cm at Aberdeen; -5cm at North Shields. However an examination of the S_2 tidal solution in the Southern Bight of the North Sea particularly along the Dutch, German and Danish coasts, revealed that on average both solutions underestimate the amplitude of the S_2 tide; with calculation B, giving a significantly lower amplitude than calculation A (see Table 3). This suggests that within the model, the dissipation of tidal energy within the shallow regions of the North Sea is larger than that which would occur in nature.

It is particularly interesting to note the change in amplitude and phase of the M_2 tidal elevation at coastal gauges, due to the differences in S_2 input to the model. Although Tables 1 and 2, showed this to be small at off-shore gauges, it is evident from Table 3, that changes in S_2 do significantly affect the amplitude, although not the phase of the M_2 tide. For example the change of the order of 10cm in the amplitude of M_2 at Shivering Sands between calcns. A and B (see Table 3).

TABLE 3

	M_2 Obs		Calcn. A		Calcn. B		S_2 Obs		Calcn. A		Calcn. B	
	h (cm)	g°	h (cm)	g°	h (cm)	g°	h (cm)	g°	h (cm)	g°	h (cm)	g°
Wick	101	322	109	309	112	308	35	360	48	3	35	354
Invergordon	136	337	151	326	154	326	48	15	68	20	50	11
Aberdeen	130	23	127	13	129	13	45	61	54	68	41	59
North Shields	158	89	154	81	158	82	53	130	65	142	48	133
Scarborough	171	111	175	117	181	117	58	153	73	181	54	172
Immingham	225	163	191	154	202	155	74	214	82	232	60	222
Inner Dowsing	197	162	183	175	194	175	68	209	80	256	57	246
Shell AD	75	194	74	194	77	194	29	242	34	275	25	265
Lowestoft	71	258	66	246	69	248	22	297	26	316	19	303
Shivering Sands	182	346	130	336	140	337	54	40	42	71	29	62
Ostende	176	5	158	4	169	5	52	58	45	96	33	86
Helgoland	109	312	76	318	81	319	29	20	23	86	16	81
Cuxhaven	134	344	101	316	109	318	34	54	32	85	23	80
Esbjerg	66	36	51	26	55	27	16	98	15	142	10	136

Table 3. Comparison of computed and observed amplitude h(cm) and phase g(degrees) for the M_2 and S_2 components of the tide at some coastal gauges in the North Sea.

The effect of the S_2 tide upon the M_2 tide, is through the non-linear terms in the equations. In particular its influence upon energy dissipation arises through changes in bottom stress through two mechanisms. The first mechanism is direct, in that bottom stress depends in a non linear manner upon the total current at the sea bed, which is increased when the S_2 current is increased. The second is indirect, in that vertical eddy viscosity is related to the square of the total current, and hence increases with increasing S_2 current. A consequence of an increase in eddy viscosity; is a decrease in current variation through the vertical, and in the limit as the eddy viscosity goes to infinity the current will have no vertical structure. In this case bottom

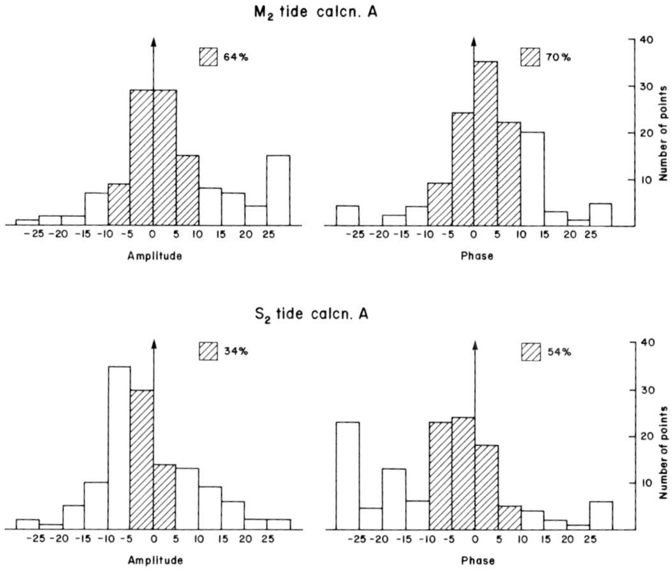

Fig. 4a. Histogram of the distribution of errors in calculation A, for the M_2 and S_2 components of the tide.

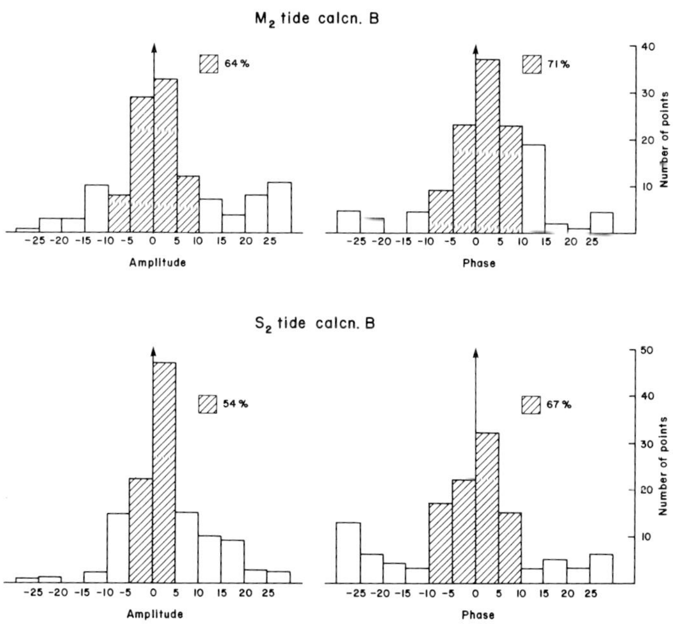

Fig. 4b. Histogram of the distribution of errors in calculation B, for the M_2 and S_2 components of the tide.

current and depth mean current will have the same value. Hence the bottom stress will be related to the depth mean current (the situation in a vertically integrated model). In such circumstances the use of a frictional coefficient $k = 0.005$ will be inappropriate (a value of $k = 0.0025$ is usual in a vertically integrated model Flather (1976)) and will produce excessive damping in the model.

The high level of damping of the M_2 and S_2 tide in the Southern Bight of the North Sea, suggests that the value of bottom friction coefficient k, and viscosity coefficient K (equation 22) although appropriate for a simulation of the M_2 tide alone, (Davies and Furnes 1980) may require some slight reduction when the M_2 and S_2 tide are simulated together.

Histograms of errors, based upon comparisons at one hundred and twenty nine coastal and off-shore tide gauges are given in Figs 4a,b. It is evident from these figures that both the amplitude and phase of the S_2 tide is significantly over-estimated in calculation A. In calculation B however, the amplitude of the S_2 component agrees to within 5cm of the observed at over 54% of the tide gauges, with phase agreeing to within 10 degrees at 67% of the gauges. A slight bias towards an underestimation of amplitude and phase is evident in calculation B, see Fig 4b. The grid of the model however is very coarse in the Southern Bight of the North Sea, and with such a grid it may be unrealistic to expect a high level of accuracy from the model, due to limitations of grid resolution.

3.2 Currents

The crucial test of any three dimensional model is its ability to reproduce the vertical variation of tidal current from sea surface to sea bed. Comparisons of the form presented by Davies and Furnes (1980), for the M_2 tide, are presently in progress for the S_2 tide, and will be reported subsequently.

However, some indication of the spatial and vertical variation of the M_2 and S_2 tidal currents at spring tide can be obtained from Figs 5a,b. The maximum tidal current on the shelf will occur at spring tide when the M_2 and S_2 tidal currents are in phase. Figs 5a,b, show contours of spring tidal current at sea surface and sea bed. Comparing Fig 5a with near surface tidal current amplitudes derived from observations (see Fig 4.15 in Howarth and Pugh 1983) it is evident that the model reproduces the spatial distribution of the spring tidal currents to within an acceptable level of accuracy. In general surface spring tidal currents computed with the model are the order of a few percent higher than those given by Howarth and Pugh (1983). This may be due

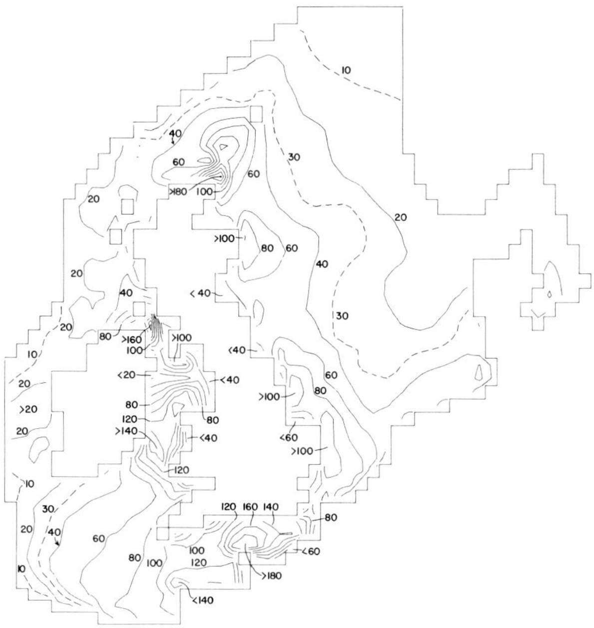

Fig. 5a. Computed contours of the major axis (M_2 + S_2) at the sea surface.

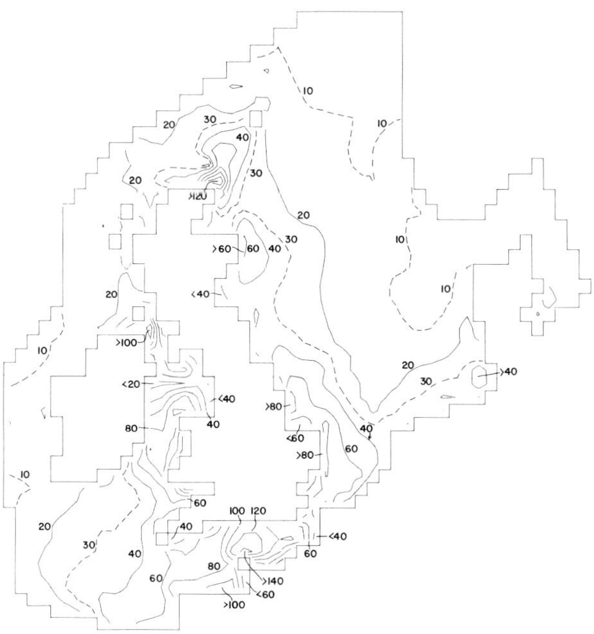

Fig. 5b. Computed contours of the major axis (M_2 + S_2) at the sea bed.

to the fact that Figures 5a,b are based upon Calculation A, in which, as we have seen, the S_2 input to the model was of the order of 10% higher than the observed. Although in general, both calculations yielded a lower amplitude of S_2 in the shallow regions. The other possibility is the difficulty of drawing accurate distribution of surface tidal currents from point current observations taken at depth (Howarth 1982, Howarth and Pugh 1983).

It is evident from Figures 5a,b, that spring currents near the sea bed are significantly lower than the surface currents.

4 CONCLUDING REMARKS

A three dimensional numerical tidal model of the North West European Shelf has been developed in this paper. The model incorporates a quadratic law of bottom friction, and eddy viscosity can vary through the vertical and with horizontal position and time in an arbitrary manner. By using the Galerkin method with a basis set of eigenfunctions, a computationally economic and accurate method (Davies and Stephens 1983) of solving the hydrodynamic equations has been presented. This method unlike the use of grid boxes through the vertical (see for example Davies and Stephens 1983), yields a continuous current profile from sea surface to sea bed.

Computed M_2 and S_2 tidal elevations and currents over the region of the shelf are in general in reasonably good agreement with observations taken in the area. A rigorous comparison of observed and computed tidal currents at different points in the vertical water column is presently in progress and results will be reported subsequently.

The use of a quadratic law of bottom friction, rather than the application of a no-slip condition at the sea bed does however mean that the height above the sea bed at which the bottom currents are computed cannot be determined. Consequently important information to the off-shore oil industry, on the spatial and temporal variations of near bed currents and stresses (of particular importance in determining sediment transport and possible erosion of the sea bed in the vicinity of off-shore structures) cannot be determined.

The method developed here has recently been extended to deal with a no-slip bottom boundary condition. In this case near bed currents and bed stresses can be determined. Results from these calculations are presently being analysed. This extension appears scientifically worth pursuing and is of significant practical importance to the oil industry, where near bed measurement programmes are already in progress.

The sensitivity of tidal currents to changes in the vertical profile of eddy viscosity is also of scientific importance, and could yield an improvement in accuracy of computed tidal current profiles. The use of a constant eddy viscosity from sea surface to sea bed is questionable following results from recent field measurements (Wolf 1980) which show a near parabolic variation from surface to bed, with viscosity reaching a maximum at a point below mid-depth. The model developed here could incorporate such a variation and it would be of significant scientific and practical importance to examine the effect of such a profile upon computed tidal currents.

It is apparent that the model developed in this paper can be readily extended to examine a number of important scientific problems in tidal physics, and sediment transport, with the associated benefits to the off-shore oil industry, and applications to various pollution problems.

5 ACKNOWLEDGEMENTS

The author is indebted to Dr. N. S. Heaps for a number of useful comments and suggestions in connection with this paper.

Data made available by participants in the JONSDAP experiment and by Mr. M. J. Howarth from a Celtic sea experiment, without which the comparisons made in this paper would not be possible, are very much appreciated.

Numerical model results supplied by Drs. Flather, Proctor and Wolf from a two-dimensional North-East Atlantic model are gratefully acknowledged.

The numerical modelling work described in this paper was funded by a Consortium consisting of the Natural Environment Research Council and the Department of Energy.

6 REFERENCES

Bowdon, K.F. and Ferguson, S.R., 1980. Variations with height of the turbulence in a tidally-induced bottom boundary layer. Marine Turbulence. Proceedings of the 11th Liege Colloquium on Ocean Hydrodynamics, J.C.J. Nihoul, Ed. Elsevier, Oceanography Series Vol. 28, 259-286.

Cartwright, D.E., 1976. Shelf boundary tidal measurements between Ireland and Norway. Mem.Soc.R.Sci.Liege, 10, 133-139.

Davies, A.M., 1976. A numerical model of the North Sea and its use in choosing locations for the deployment of offshore tide gauges in the JONSDAP '76 oceanographic experiment. Dtsch. Hydrogr. Z. 29, 11-24.

Davies, A.M., 1980. Application of the Galerkin method to the formulation of a three-dimensional non-linear hydrodynamic numerical sea model, Applied Mathematical Modelling, 4, 245-256.

Davies, A.M., 1983a. Formulation of a linear three-dimensional hydrodynamic sea model using a Galerkin-Eigenfunction method. International Journal of Numerical methods in Fluids, 3, 33-60.

Davies, A.M., 1983b. Application of a Galerkin-Eigenfunction method to computing currents in homogeneous and stratified seas. pg. 287-301 in Numerical Methods for Fluid Dynamics ed K.W. Morton and M.J. Baines, Academic Press (London).

Davies, A.M., 1986. A three-dimensional model of the Northwest European Continental Shelf, with application to the M_4 tide. J. Phys. Oceanogr., 16, 797-813.

Davies, A.M. and Owen, A., 1979. Three-dimensional numerical sea model using the Galerkin method with a polynomial basis set. Appl. Math. Modelling, 3, 421-428.

Davies, A.M. and Furnes, G.K., 1980. Observed and Computed M_2 tidal currents in the North Sea. J. Phys. Oceanogr., 10, 237-257.

Davies, A.M. and Furnes G.K., 1984. On the determination of vertical structure functions for time dependent flow problems (to appear in Tellus).

Davies, A.M. and Stephens, C.V., 1983. Comparison of the finite difference and Galerkin methods as applied to the solution of the hydrodynamic equations. Applied Mathematical Modelling, 7, 226-240.

Davies, A.M., Sauvel, J., and Evans, J., 1984. Computing near coastal tidal dynamics from observations and a numerical model. Continental Shelf Research, 4, 341-366.

Flather, R.A., 1976. A tidal model of the north west European continental shelf. Mem.Soc.Roy.Sci. Liege, 10, 141-164.

Heaps, N.S. and Jones, J.E., 1981. Three-dimensional model for tides and surges with vertical eddy viscosity prescribed in two layers - II. Irish Sea with bed friction layer. Geophysical Journal Royal Astronomical Society, 64, 303-320.

Howarth, M.J., 1982. Tidal currents of the continental shelf. In : A.H. Stride (Editor) Offshore Tidal Sands. Chapman and Hall, London pp. 10-26.

Howarth, M.J. and Pugh, D.T., 1983. Observation of tides over the Continental Shelf of North-West Europe, pg. 135-188 in Physical Oceanography of Coastal and Shelf Seas ed B. Johns, published Elsevier Oceanography Series No. 35.

Lennon, G.W., 1961. The deviation of the vertical at Bidston in response to the attraction of ocean tides. Geophys. J.R. Astron.soc., 6, 64-84.

Mathisen, J.P., Johansen, O., 1983. A numerical Tidal and Storm Surge Model of the North Sea. Marine Geodesy, 6, 267-291.

Owen, A., 1980. A three-dimensional model of the Bristol Channel. Journal of Physical Oceanography, 10, 1290-1302.

Prandle, D., 1980. Co-tidal charts for the Southern North Sea. Deutsche Hydrographische Zeitschrift, 33, 68-81.

Wolf, J., 1980. Estimation of shearing stresses in a tidal current with application to the Irish sea, in, Marine Turbulence, ed, J.C.J. Nihoul, Elsevier Scientific Publishing Company, Amsterdam, pp 319-344.

A GENERAL THREE-DIMENSIONAL EDDY-RESOLVING MODEL FOR STRATIFIED SEAS

I.D. JAMES
Institute of Oceanographic Sciences, Bidston Observatory,
Birkenhead, L43 7RA, England.

ABSTRACT

Models which predict the development of baroclinic features, such as fronts and eddies, in limited sea areas, particularly in the shelf seas, are required for several purposes. For example, the circulation in these features affects biological activity and can determine the paths of pollutants. Also, the enhanced currents in eddies can affect offshore engineering, as in the Norwegian Coastal Current, where near-surface velocities of over 1.5 m s^{-1} have been recorded.

Fundamental problems with models of this kind are the inclusion of both barotropic and baroclinic flows, with their different time scales, the need to retain sharp frontal gradients without rippling or excessive smearing and the treatment of open boundary conditions. The model described here uses a finite-difference sigma coordinate grid box scheme, a split into barotropic and baroclinic modes with a semi-implicit scheme for the barotropic part, a hybrid advection scheme to retain positivity and sharpness at fronts and partly prescribed, partly radiation boundary conditions. Arbitrary formulations of vertical eddy viscosity and diffusivity, varying in space and time, may be used. The coding is designed for efficient use on the Cray-1s computer. The results show the ability of the model to simulate the evolution of eddies in a coastal current front.

1 INTRODUCTION

A fundamental length scale for motions in stratified seas is the baroclinic Rossby radius of deformation $(g'h')^{\frac{1}{2}}/f$, where g' is the reduced gravity ($= g\Delta\rho/\rho_0$ for a two layer fluid, where $\Delta\rho$ is the density difference and ρ_0 the reference density), h' is a depth scale ($= h_1(H - h_1)/H$ for a two-layer fluid, upper layer thickness h_1, total depth H), and f is the Coriolis parameter. This length scale is the ratio of the (first mode) long internal wave phase speed to the Coriolis parameter, and is the typical scale of the most energetic baroclinic eddies seen in satellite images of the shelf seas, implying diameters from 5 to 50 km. This determines the resolution necessary for an eddy-resolving model. Sharp gradients are formed as the eddies advect and stir the water : this process is analogous to frontogenesis in the atmosphere. It can enhance already large horizontal gradients produced in the shelf seas by variations in tidal mixing and fresh water outflow at the coasts. Therefore the ideal model not only resolves the Rossby radius scale but also allows sharp gradients to develop and propagate realistically. This problem is shared with weather prediction models, which, although easily resolving the major depressions and anticyclones, benefit in the neighbourhood of fronts from increased resolution. The requirement of high resolution limits the feasibility of the models, with present-day computers, to sub-regions of the shelf, so the solution will depend on the choice of open boundary conditions. The relative lack of monitoring of the sea (compared with the atmosphere) limits the development of a forecasting model, but it is possible to compare the model results qualitatively with what is known from observations, aided particularly by remote sensing, and laboratory models.

There are many potential applications of eddy-resolving models of shelf seas. Transport and dispersal of pollutants are likely to be dependent on stirring by the eddies in many regions, while fronts are probable areas of concentration by convergent flow. There is rapid transport in frontal jets, while the instability of fronts provides a mechanism for cross-frontal transfer, which may determine how quickly pollutants are dispersed out of coastal regions into the open sea. It is not possible to parameterize all these effects in a coarse-grid model. Frontal zones are known to be biologically productive as here nutrients can be mixed or upwelled into regions with adequate light for phytoplankton growth. Pingree et al. (1979) show an example of enhanced production in a frontal eddy. The eddies may cause cross-frontal transfer of nutrients, and may be expected to increase vertical motions. Enhanced horizontal velocities in eddies can cause problems to offshore engineers : near-surface flows have been measured at over 1.5ms^{-1} in eddies in the Norwegian Coastal Current in areas of oil and gas production (James and McClimans 1983, Carstens et al. 1984). The current increases rapidly (over a few hours) to these values as the edge of the eddy moves past the measuring site, so here there is a clear application for a forecasting model.

2 MODEL EQUATIONS

The model to be described here is a development of that of James (1986). It is based on the primitive equations with the Boussinesq, incompressible, hydrostatic and f-plane approximations. They include a prognostic equation for density ρ in the form of bucyancy $b = g(\rho_o - \rho)/\rho_o$. Arbitrary vertical eddy viscosity A and diffusivity K are assumed, but there is no imposed horizontal mixing. As Mellor and Blumberg (1985) point out, horizontal diffusion should not be required for models that resolve the dominant horizontal eddy scale dictated by the Rossby radius (or variable bottom topography), as is the case here. An imposed horizontal diffusion would smear out fronts. There do remain possible numerical problems such as a tendency for nonlinear terms to drive energy towards the grid scale and rippling near sharp gradients, which can be masked by introducing a horizontal diffusion term. However, the aim here is to use a numerical scheme which takes care of these problems with, if possible, a minimum of numerical diffusion.

The equations in a rectangular (x,y,z,t) coordinate system, right-handed with z upwards, are then

$$\partial u/\partial t = -L(u) + fv - \partial\phi/\partial x + \partial(A\partial u/\partial z)/\partial z, \tag{1}$$

$$\partial v/\partial t = -L(v) - fu - \partial\phi/\partial y + \partial(A\partial v/\partial z)/\partial z, \tag{2}$$

$$\partial b/\partial t = -L(b) + \partial(K\partial b/\partial z)/\partial z, \tag{3}$$

$$b = \partial\phi/\partial z, \tag{4}$$

$$L(1) = 0, \tag{5}$$

where the operator L is defined by

$$L(a) = \partial(u\, a)/\partial x + \partial(v\, a)/\partial y + \partial(w\, a)/\partial z \tag{6}$$

$$\text{and} \quad \phi = (\text{total pressure})/\rho_o + gz \tag{7}$$

A transformation to sigma coordinates allows the inclusion of a free surface $z = \zeta(x,y,t)$ and bathymetry $z = -h(x,y)$ with the same number of grid points in tho vertical at each position, arbitrarily spaced. The varying depth becomes a constant interval in σ, with $\sigma = 0$ on the surface and $\sigma = -1$ on the sea bed.

The transformation is to $(\alpha,\beta,\sigma,\tau)$ coordinates, with

$$\alpha = x,\ \beta = y,\ \sigma = (z - \zeta)/H,\ \tau = t \tag{8}$$

where $H(x,y,t) = h + \zeta$ is the total depth.

The transformed equations are:

$$\partial(Hu)/\partial\tau = -\Lambda(u) + Hfv - H\partial\phi/\partial\alpha + Hb(\partial\zeta/\partial\alpha + \sigma\partial H/\partial\alpha) + H^{-1}\partial(A\partial u/\partial\sigma)/\partial\sigma \tag{9}$$

$$\partial(Hv)/\partial\tau = -\Lambda(v) - Hfu - H\partial\phi/\partial\beta + Hb(\partial\zeta/\partial\beta + \sigma\partial H/\partial\beta) + H^{-1}\partial(A\partial v/\partial\sigma)/\partial\sigma \tag{10}$$

$$\partial(Hb)/\partial\tau = -\Lambda(b) + H^{-1}\partial(K\partial b/\partial\sigma)/\partial\sigma \tag{11}$$

$$\partial H/\partial\tau = -\int_{-1}^{0}[\partial(Hu)/\partial\alpha + \partial(Hv)/\partial\beta]d\sigma \tag{12}$$

$$\Omega = H^{-1}\int_{\sigma}^{0}[\partial(Hu)/\partial\alpha + \partial(Hv)/\partial\beta]d\sigma + (\sigma/H)\int_{-1}^{0}[\partial(Hu)/\partial\alpha + \partial(Hv)/\partial\beta]d\sigma \tag{13}$$

$$\phi = P_a/\rho_o + g\zeta - H\int_{\sigma}^{0} b\, d\sigma \tag{14}$$

where P_a is atmospheric pressure and the operator Λ is defined by

$$\Lambda(a) = \partial(Hua)/\partial\alpha + \partial(Hva)/\partial\beta + H\partial(\Omega a)/\partial\sigma \tag{15}$$

The vertical velocity w in the original coordinates is

$$w = H\Omega + (1 + \sigma)(D\zeta/D\tau) + \sigma(Dh/D\tau). \tag{16}$$

3 THE GRID SCHEME AND TIME STEPPING

The Arakawa B-grid, which has u and v defined at common points at the corners of a grid square, with b, h and ζ defined at the centre (see James (1986) for a diagram) is used to define the finite-difference equations in this model. This was found to be much more satisfactory than the C-grid (which has u and v defined at the centres of adjacent sides of the square) for maintaining a sharp front in approximate geostrophic balance. Jamart and Ozer (1987, this colloquium) found a similar problem of spurious flows with the C-grid, in their case next to a rigid boundary in a rectangular sea model. This trouble near boundaries, whether rigid or frontal, arises from the averaging necessary in the Coriolis term on the C-grid.

The variables have positions denoted by a single index running through the three-dimensional grid. This provides long inner do-loops in the time-stepping part of the calculation which are vectorizable on the Cray-1s. Points outside the calculation area are then discarded (put to zero) and boundary points modified before the next time step.

A leapfrog scheme is used for the time differencing. This allows use of the unconditionally stable DuFort-Frankel method for the vertical mixing terms. The separation of solutions on odd and even time steps found in leapfrog methods is suppressed by the time filter of Asselin (1972), which replaces value a^n at time level n by the value

$$a_1^n = a^n + \tfrac{1}{2}\nu(a_1^{n-1} - 2a^n + a^{n+1}) \tag{17}$$

after the new value a^{n+1} has been calculated. Asselin (1972) showed that any $\nu > 0$ damps the unwanted computational mode and that ν should be taken less than 1. The higher values of ν damp the higher frequencies more. A comparison of the results with $\nu = 0.1$ and $\nu = 0.8$ for the problem shown here confirms this, the changes being very slight.

4 SEMI-IMPLICIT TREATMENT OF THE BAROTROPIC TERMS

The Courant-Friedrichs-Levy (CFL) condition on the time step for explicit methods is extremely restrictive for fine grids and barotropic waves. In James (1986) the equations were split into a gravity-wave (or barotropic) part, and the remainder, which were solved using different time steps. For the problem shown, the barotropic time steps had to be as little as ten seconds to satisfy the CFL condition, while the CFL condition based on the baroclinic (internal) wave phase speed allowed the longer time step to be six minutes. A significant improvement in computation time has been achieved by solving the barotropic part by a semi-implicit method. This method also has the advantage that the scheme can be chosen to suppress two-grid-interval noise, which can be troublesome on the B-grid.

The velocity is divided as follows:

$$u = \bar{u}_g + u_r,\ v = \bar{v}_g + v_r, \tag{18}$$

where
$$\partial(H\bar{u}_g)/\partial\tau = -(H/\rho_o)(\partial P_a/\partial\alpha) - Hg\partial\zeta/\partial\alpha + Hf\bar{v}_g + F_S - F_B \tag{19}$$

and
$$\partial(Hu_r)/\partial\tau = Hfv_r + A_M + H^{-1}\,\partial(A\,\partial u_r/\partial\sigma)/\partial\sigma - F_S + F_B, \tag{20}$$

where A_M contains nonlinear and baroclinic terms and F_S and F_B are surface and bottom stress. There are similar equations for $\bar{v}_g$ and v_r.

The semi-implicit scheme is similar to that of Kwizak and Robert (1971). The variables $\bar{u}_g$, $\bar{v}_g$ and ζ are found at time level n+1 from

$$(H\bar{u}_g)^{n+1} = (H\bar{u}_g)^{n-1} - 2\delta t g H^n \tfrac{1}{2}((\partial\zeta/\partial\alpha)^{n+1} + (\partial\zeta/\partial\alpha)^{n-1}) + 2\delta t\,(H^n f\bar{v}_g{}^n + F_S{}^n - F_B{}^n) \tag{21}$$

$$(H\bar{v}_g)^{n+1} = \ldots \tag{22}$$

$$\zeta^{n+1} = \zeta^{n-1} - 2\delta t[\tfrac{1}{2}(\partial(H\bar{u}_g)^{n+1}/\partial\alpha + \partial(H\bar{u}_g)^{n-1}/\partial\alpha) + \tfrac{1}{2}(\partial(H\bar{v}_g)^{n+1}/\partial\beta + \partial(H\bar{v}_g)^{n-1}/\partial\beta) + \partial(H\bar{u}_r)^n/\partial\alpha + \partial(H\bar{v}_r)^n/\partial\beta]. \tag{23}$$

Here, variations in P_a have been neglected, δt is the barotropic time step and $\bar{u}_r$, $\bar{v}_r$ are integrals of u_r and v_r over the depth with respect to σ. Substitution of (21) and (22) into (23) leads to an implicit equation for ζ^{n+1}, which can be solved by relaxation. To allow vectorization, simultaneous relaxation is used, with ζ^n as a first guess for ζ^{n+1}. Boundary values of ζ^{n+1} are required: an explicit boundary condition which fixes this value throughout the relaxation process speeds convergence. Over relaxation is allowed, but the optimum parameter is not obvious.

The equation for ζ involves terms of the form

$$\partial(H\partial\zeta/\partial\alpha)/\partial\alpha + \partial(H\partial\zeta/\partial\beta)/\partial\beta \qquad (24)$$

which can be finite-differenced in several ways. The value at a particular ζ-point involves values of ζ at some or all of the eight surrounding ζ-points. The false two-grid interval noise, with its chess-board appearance, arises from a separation of solutions on two subsets of the grid. If a five-point representation of (24) is used (the centre point and the furthest corner points) these two solutions remain unconnected. If, however, a nine-point representation is used the two solutions are connected and the two-grid interval wave is suppressed. This is similar to the scheme of Mesinger (1973).

This semi-implicit scheme may still involve a different time step from that of the baroclinic equation (20), but for the results shown here the same time step (six minutes) was used. This necessitated a change in the arrangement of time levels shown in James (1986) so that the barotropic and baroclinic levels 3 were concurrent.

5 HYBRID ADVECTION SCHEME

The hybrid advection scheme used here was described by James (1986). It is intended to give some of the advantages of the flux-correction scheme of Boris and Book (1976) with much less computational effort. The ideal advection scheme would combine positivity (so there would be no unrealistic rippling, leading to obviously wrong density values, near a front) with low numerical diffusion. The upwind scheme has positivity but high numerical diffusion. Both the flux-correction scheme and the hybrid scheme calculate the advective flux as a weighted average of that from a low-order, positive scheme such as upwind differencing and a higher order scheme, here centred differencing. Hence,

$$\text{Flux} = C \times (\text{flux from upwind scheme}) + (1-C) \times (\text{flux from centred scheme}) \qquad (25)$$

The difference between the flux-correction and the hybrid scheme lies in the method of choosing C. Here, C is simply defined from the b field from a test, due to Harten (1978), for the edge of fronts, where the density gradient changes. The value of C defined at the b-point labelled i, for the α-directed flux, is given by

$$C_{\alpha i} = | (|\Delta_{i+\frac{1}{2}} b| - |\Delta_{i-\frac{1}{2}} b|)/S |^{p} \quad (26)$$

for $S > \varepsilon$ and $C_{\alpha i} = 0$ for $S \leqq \varepsilon$, where

$$\Delta_{i+\frac{1}{2}} b = b_{i+1} - b_{i} \text{ and } S = |\Delta_{i+\frac{1}{2}} b| + |\Delta_{i-\frac{1}{2}} b|,$$

i+1 being the label of the point next to i in the $+\alpha$ direction. The parameters p and ε allow some further control of the sharpness of the switch factor C. Values of C at points required for the flux calculations can be defined as maxima of those at the appropriate neighbouring b-points. This definition of C ensures that the upwind scheme is used near the edge of a front, where the gradient of b suddenly changes from a low value, so preventing ripples, while the centred scheme is used when density gradients are low. James (1986) showed that this hybrid scheme has significantly less numerical diffusion than the upwind scheme, while later it is shown that adjustment of p can control the amount of numerical diffusion. The hybrid scheme is used in this model for the advective terms for all variables, in all three directions.

6 OPEN BOUNDARY CONDITIONS

Open boundary conditions for a limited area model should allow waves and eddies generated within the model to pass out through the boundary without reflection, while preserving the influence of known external values, say those obtained from a coarse-grid model of the surrounding area. This ideal boundary condition is not known for a general three-dimensional stratified model. Radiation conditions, which allow influences to pass out through the boundary, were proposed by Orlanski (1976). They are based on the Sommerfeld radiation condition, which for variable a at an α = constant boundary is

$$\partial a/\partial t + c\,\partial a/\partial \alpha = 0 \quad (27)$$

The phase speed (or advection velocity) c is calculated from interior points next to the boundary, using (27). In effect, this is looking for a pattern approaching the boundary and assuming that it continues across the boundary. If the problem implies a phase velocity known from theory (for example, $(gh)^{\frac{1}{2}}$ for barotropic Kelvin waves) this may be substituted for c, but this is not the case here.

Hence if the unknown boundary value at time level n+1 has label i, it is related to interior points i-1 and i-2 (for the example of an α = constant boundary with α increasing out of the model area) by

$$a_i^{n+1} = [(1-\mu)a_i^{n-1} + 2\mu\, a^n_{i-1}] \,/\, (1+\mu) \qquad (28)$$

$$\text{where } \mu = (a_{i-1}^{n-1} - a_{i-1}^{n+1}) \,/\, (a_{i-1}^{n+1} + a_{i-1}^{n-1} - 2a^n_{i-2}),$$

unless this μ is negative, in which case there is inflow and μ is put equal to zero, or if this $\mu > 1$, in which case μ is put equal to 1. This is the implicit version of the Orlanski boundary condition, which is usable if a_{i-1}^{n+1} is already calculated before a_i^{n+1} is needed. The explicit form calculates μ from one time level earlier. Whatever the value of μ, (28) takes the form of an interpolation formula involving values at earlier times and the neighbouring point. Chapman (1985) has tested various open boundary conditions for a barotropic model and found this Orlanski condition the most satisfactory overall (with a sponge added for a waves only case). Blumberg and Kantha (1985) used a radiation condition with $c = (gh)^{\frac{1}{2}}$ and found a steady emptying of the model basin. To prevent this, they introduced an extra term $(a_o-a)/T_f$ to the right hand side of (27), to bring the boundary value of a (in their case, surface elevation) back to a given value a_o, with time constant T_f. This may be termed a "loosely clamped" boundary, where "clamped" refers to the case $T_f = 0$ (so $a = a_o$) and "radiational" to the case T_f infinite, when equation (27) holds. Chapman (1985) presumed the basin emptying to be the result of $(gh)^{\frac{1}{2}}$ being an overestimate of c.

A further development of the Orlanski radiational boundary condition was made by Raymond and Kuo (1984), who allowed for radiation in a general direction rather than simply normal to the boundary as in (27). Another possibility, for the B-grid, is to linearise the u_r and v_r equations next to the boundary, determine ζ and b on the boundary from the boundary conditions and then calculate the linearised u and v. However, it has been found more satisfactory here to apply the open boundary conditions to velocity as well as ζ and b, as this avoids unrealistic spurious velocities running along the boundary. Clearly, there are numerous possible open boundary conditions, the best of which is likely to depend on the problem in hand, and the question is still an active topic of research. For the results shown here, for a coastal current with a given inflow, several different outflow conditions were taken to test the sensitivity of the model to them.

7 RESULTS OF A COASTAL CURRENT MODEL

This model was applied to the problem of the release of a patch of less dense fluid into a coastal current, and its subsequent development into an

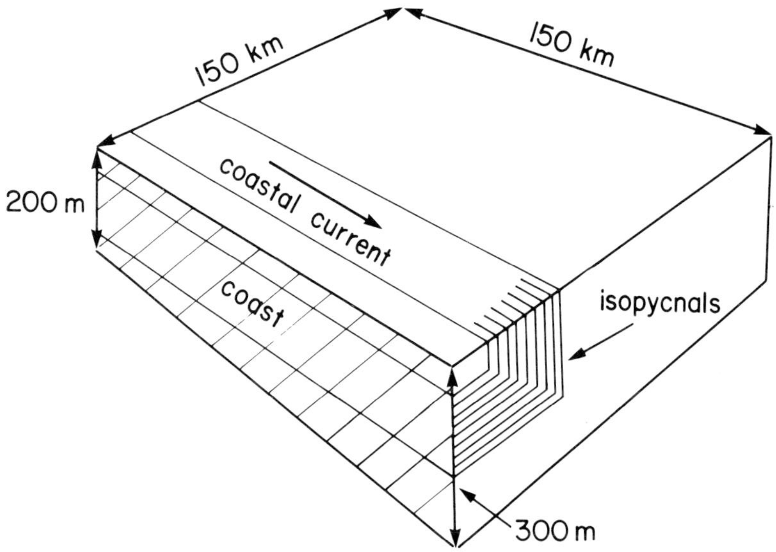

Fig.1. Basic configuration of the model region.

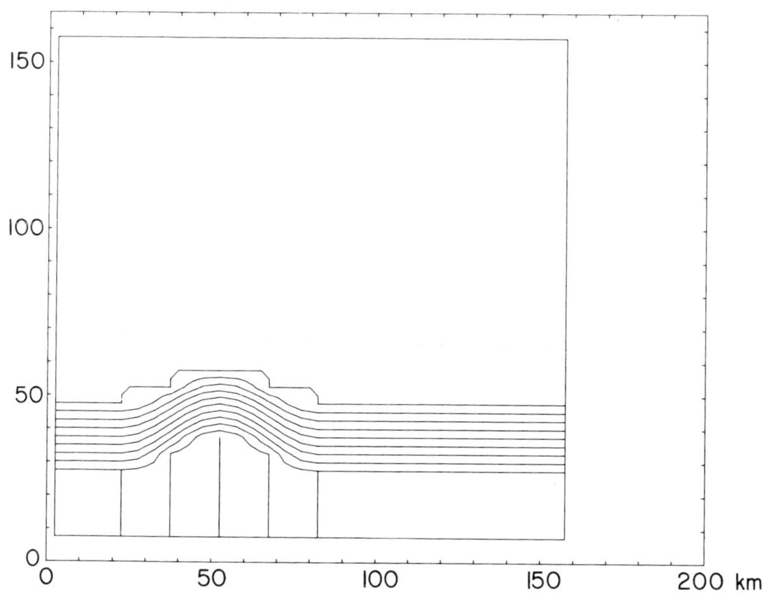

Fig.2. Initial conditions. (a) Surface density contours.

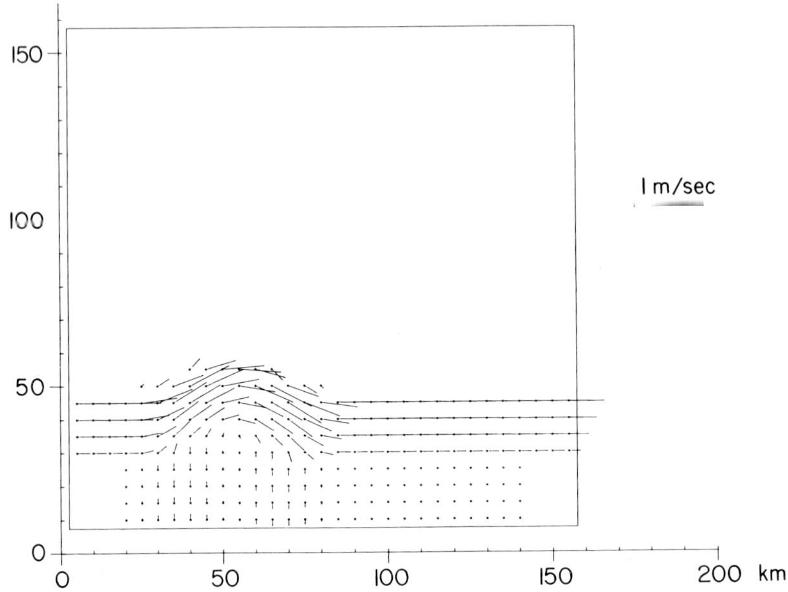

Fig.2(b). Surface current vectors.

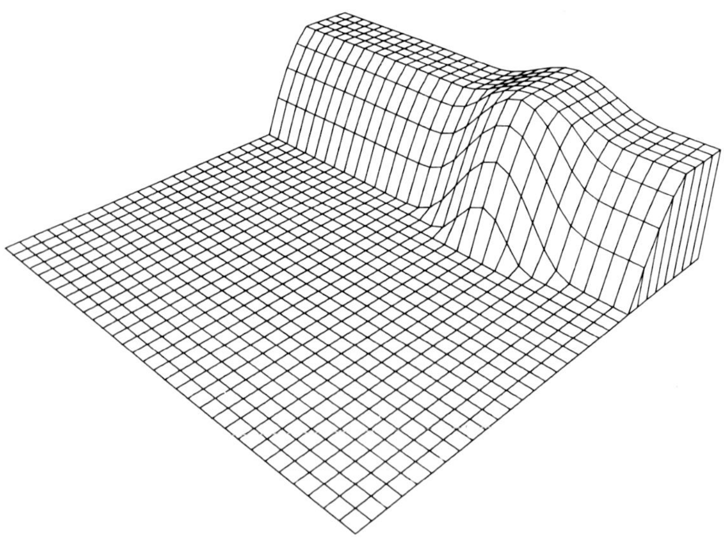

Fig.2(c). Perspective plot of the surface elevation, viewed from offshore.

eddy. The basic conditions, shown in Fig. 1, are similar to those of the Norwegian Coastal Current in the Norwegian Trench, with a bottom sloping from 200 to 300 m and a density change across the current of one σ_t unit. Fig. 2(a) shows the surface density contours with an introduction of less dense water in a patch near the coast. This is taken as the initial density for the model run. The initial currents (the surface values are shown in Fig. 2(b)) and the initial surface elevation (Fig. 2(c)) are calculated on the assumption of geostrophic flow with zero velocity at the sea bed. This provides a smooth start to the calculation. The initial change in level across the front at the inflow (or outflow) boundary was 10.3 cm.

The boundary conditions used in obtaining the results shown in Fig. 3 (the surface density, surface and bottom currents and surface elevation after 6 days) were appropriate to a coastal current controlled at the inflow: inflow velocity and density prescribed, zero gradient of elevation at the inflow and Orlanski radiation at the two remaining open boundaries. These runs had a little vertical mixing ($A=K=20$ cm^2 s^{-1}), a quadratic bottom friction coefficient of 0.0025, Asselin time filter with $\nu = 0.8$, and hybrid switch factor parameters $p = 1$ and $\varepsilon = 10^{-4}$. The grid had 32 x 32 (horizontal) x 12 (vertical) points and 5 km horizontal resolution.

These results show a progression of the original disturbance along the coast at a rate of the order of 10 km d^{-1}, as observed for Norwegian Coastal Current eddies (f is taken to be that for 60°N). It has developed into an anticyclonic eddy together with a cyclonic eddy which is more evident in the bottom velocities. This agrees with the surface prominence of anticyclonic "whirls" (Carstens et al. 1984) and the laboratory observations of Griffiths and Linden (1981) that while motion in the upper layer is primarily anticyclonic, cyclonic motion in the lower layer entrains upper layer fluid and gives the characteristic "backwards-breaking" appearance. This vortex pair structure was also seen in the model of James (1984), which uses a different numerical scheme based on a rigid-lid assumption and periodic boundary conditions, and a sinusoidal initial disturbance. Other eddies are present in these results, left behind at the position of the original disturbance, and ahead of the main vortex pair.

The latter eddy may be thought to be rather dependent on the outflow boundary condition. However, with the outflow boundaries clamped at the initial value (Fig. 4) there is very little difference from the radiational solution except very near the boundary. This result, of near-identical solutions, was found also for the loosely-clamped condition, with T_f between 1 and 5 hours. The conclusion is that the model is insensitive to which of these boundary conditions is taken, at least up to the times shown. This conclusion may change as the main disturbance reaches the boundary.

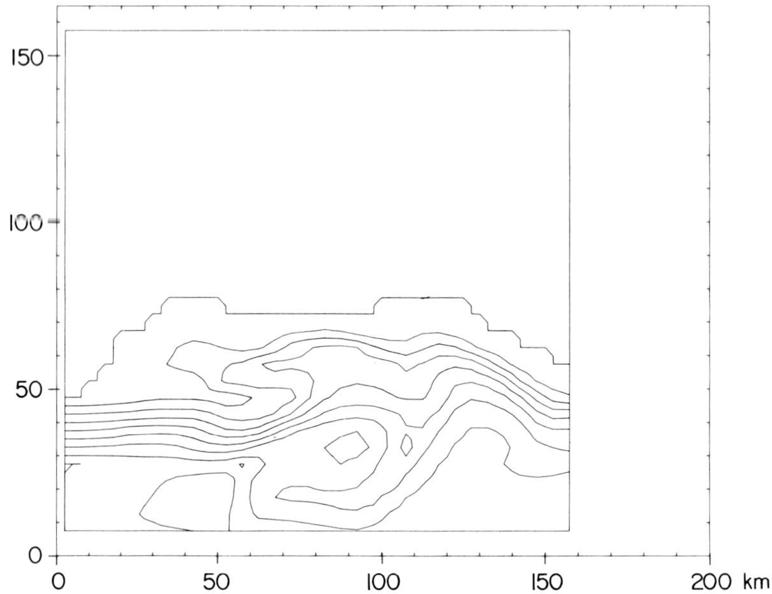

Fig.3. After 6 days. (a) Surface density.

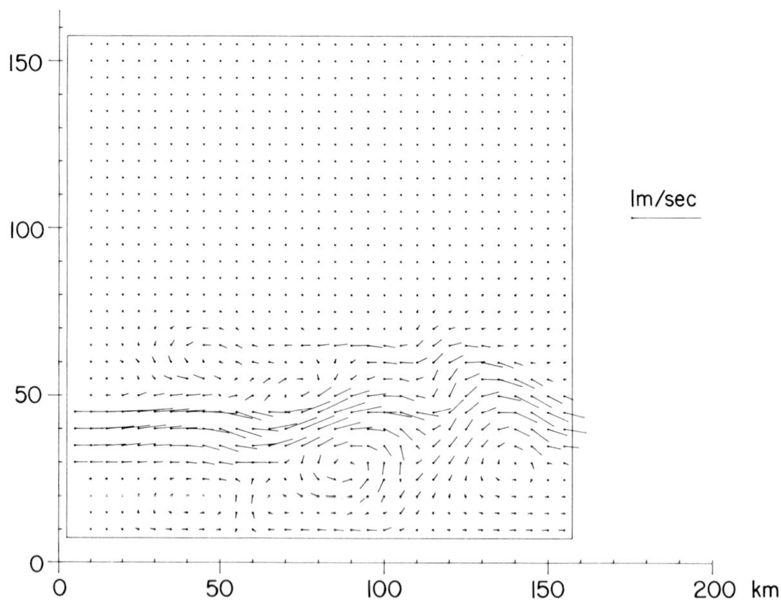

Fig.3(b). Surface currents.

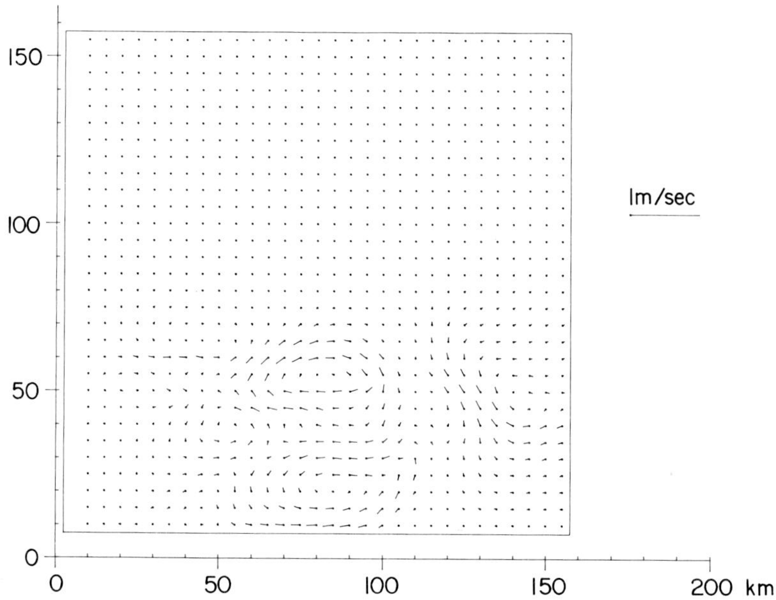

Fig.3(c). Bottom currents.

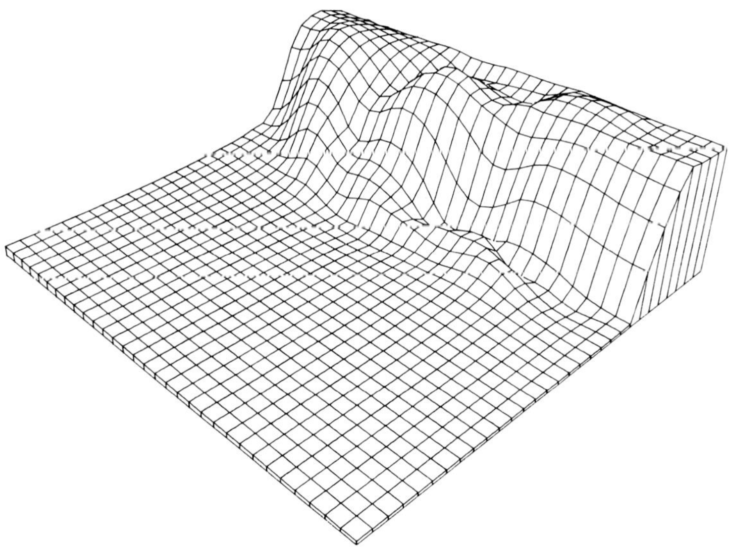

Fig.3(d). Surface elevation.

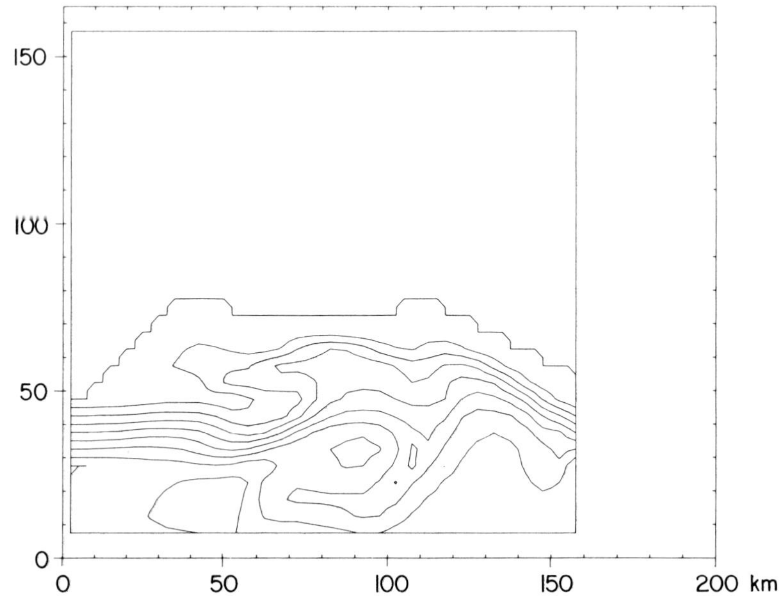

Fig.4. After 6 days with the clamped boundary condition. (a) Surface density.

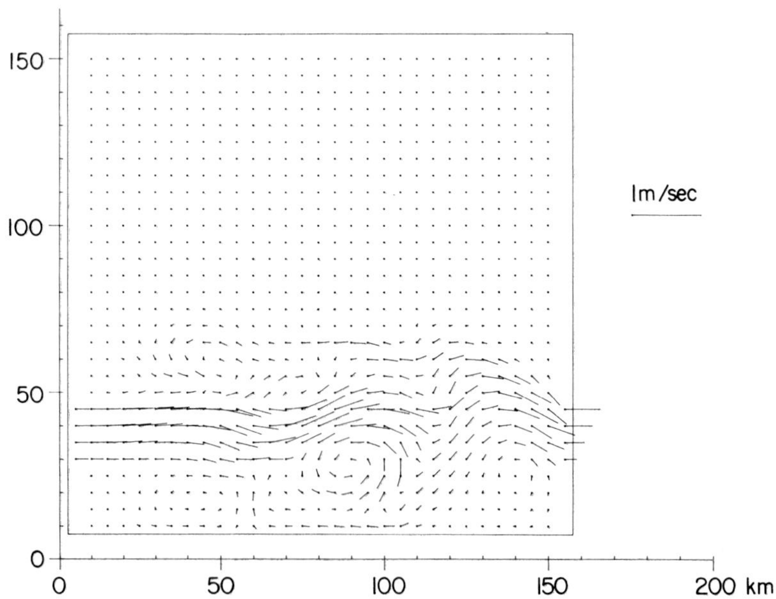

Fig.4(b). Surface currents.

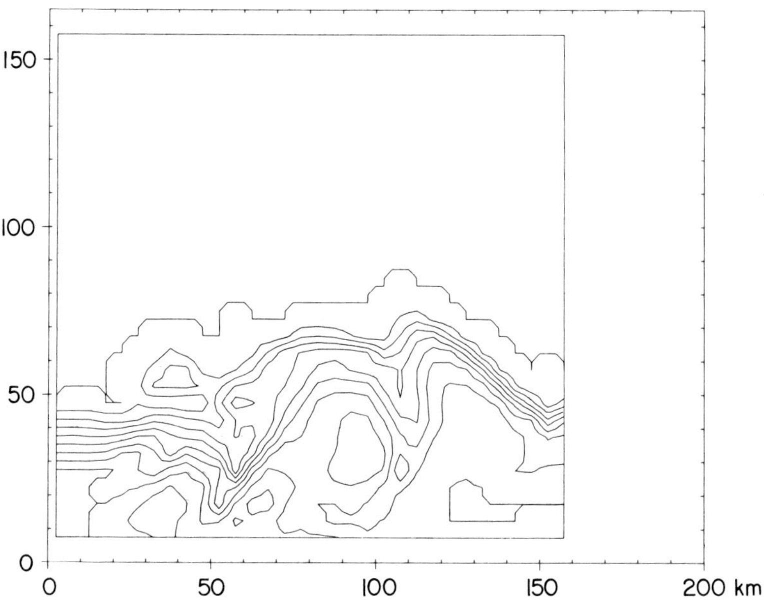

Fig.5. After 6 days, for the frictionless case with p=4. (a) Surface density.

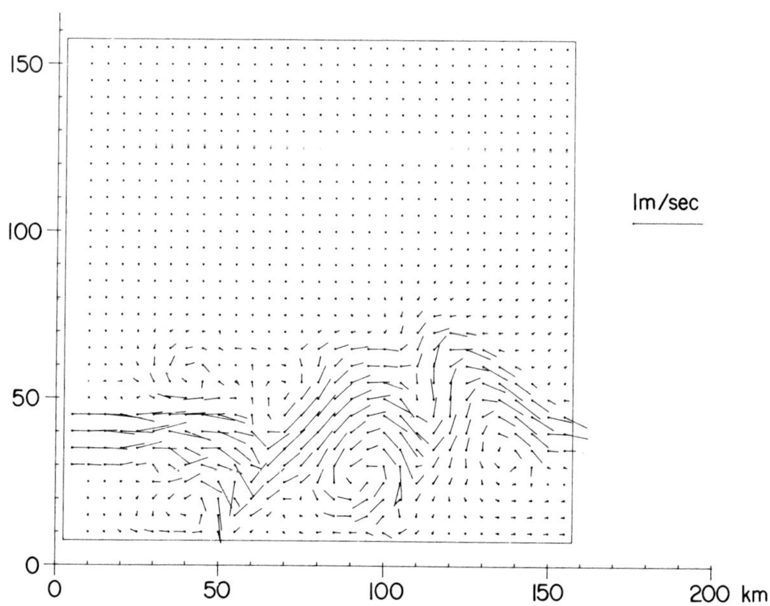

Fig.5(b). Surface currents.

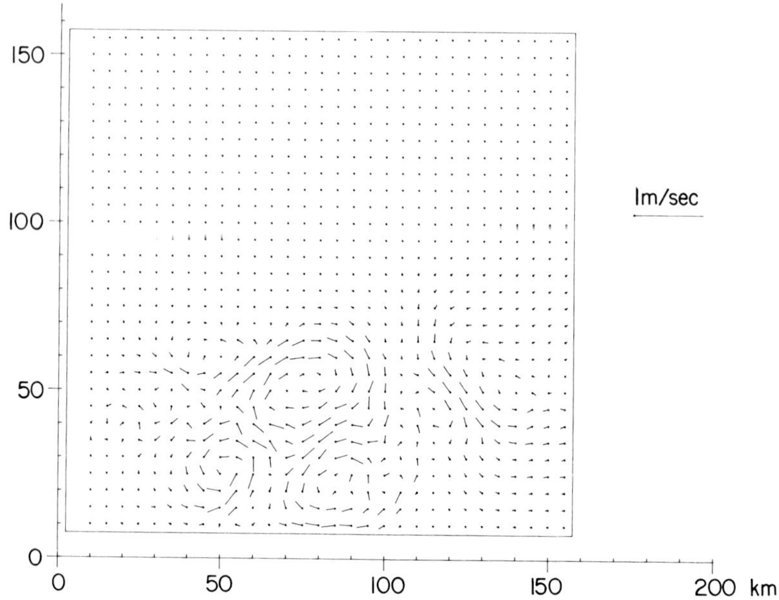

Fig.5(c). Bottom currents.

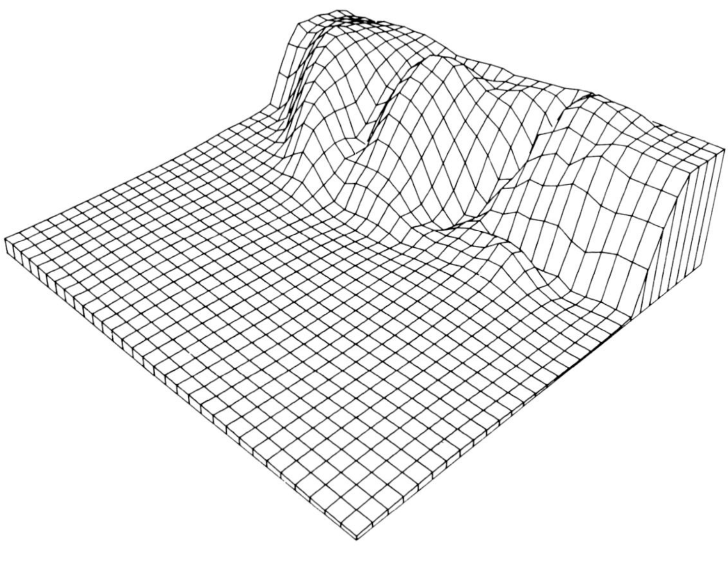

Fig.5(d). Surface elevation.

The different outflow boundary conditions do produce small differences in the mean level (of the order of a centimetre, much smaller than those of Blumberg and Kantha (1985)). Such differences in level can be produced by small differences in the currents at the boundaries.

The region away from the coast where density is uniform has barotropic motion only. This remains much smaller than the flow in the coastal current.

This model is able to run happily with zero imposed friction (A,K and surface and bottom stresses zero). The remaining damping is due to numerical diffusion and the Asselin time filtering. This can then be adjusted to as low a level as possible. Such a frictionless case, with $\nu = 0.1$ and $p = 4$, is shown in Fig. 5, with radiational open boundaries. Compared with Fig. 3, this shows a much sharper result, with larger bottom currents, and a vortex pair directed at a greater angle to the coast. A significant part of this change is due to the increase in the hybrid switch parameter p. This underlines the importance of the advective terms in baroclinic eddy-resolving models.

8 CONCLUSIONS

For this problem, the model ran in about 40 seconds per day on a Cray-1s. This shows the feasibility, with present and forthcoming computer facilities, of eddy-resolving primitive equation baroclinic models for regions of the shelf and for time periods from several days up to weeks. A vortex pair has again been found to result from the disturbance, and subsequent instability, of a front.

Open boundary conditions and advection schemes remain areas for further investigation: here, radiational open boundaries and a hybrid advection scheme have been found useful.

9 REFERENCES

Asselin, R., 1972. Frequency filter for time integrations. Mon.Weath.Rev., 100: 487-490.

Blumberg, A.F. and Kantha, L.H., 1985. Open boundary condition for circulation models. J.Hydraul.Engng., ASCE, 111: 237-255.

Boris, J.P. and Book, D.L., 1976. Solution of continuity equations by the method of flux-corrected transport. Methods Comput.Phys., 16: 85-129.

Carstens, T., McClimans, T.A. and Nilsen, J.H., 1984. Satellite imagery of boundary currents. In: J.C.J. Nihoul (Editor), Remote Sensing of Shelf Sea Hydrodynamics. Elsevier Oceanography Series, 38. Elsevier, Amsterdam/Oxford/New York/Tokyo, pp. 235-256.

Chapman, D.C., 1985. Numerical treatment of cross-shelf open boundaries in a barotropic coastal ocean model. J.Phys.Oceanogr., 15: 1060-1075.

Griffiths, R.W. and Linden, P.F., 1981. The stability of buoyancy driven coastal currents. Dynam.Atmos.Oceans, 5: 281-306.

Harten, A., 1978. The artificial compression method for computation of shocks and contact discontinuities: III - Self-adjusting hybrid schemes. Math.Comput., 32: 363-389.

Jamart, B.M. and Ozer, J., 1987. Real and spurious boundary layer effects in three-dimensional hydrodynamic models. In: J.C.J. Nihoul (Editor), Three-dimensional Models of Marine and Estuarine Dynamics. Elsevier Oceanography Series. Elsevier, Amsterdam/Oxford/New York/Tokyo (this volume).

James, I.D., 1984. A three-dimensional numerical shelf-sea front model with variable eddy viscosity and diffusivity. Cont.Shelf Res., 3: 69-98.

James, I.D., 1986. A front-resolving sigma coordinate sea model with a simple hybrid advection scheme. Appl.Math.Modelling, 10: 87-92.

James, I.D. and McClimans, T.A., 1983. Coastal current whirls in laboratory and numerical models. Ocean Modelling, 53: 1-3 + figs. (unpublished manuscript).

Kwizak, M. and Robert, A.J., 1971. A semi-implicit scheme for grid point atmospheric models of the primitive equations. Mon.Weath.Rev., 99: 32-36.

Mellor, G.L. and Blumberg, A.F., 1985. Modelling vertical and horizontal diffusivities with the sigma coordinate system. Mon.Weath.Rev., 113: 1379-1383.

Mesinger, F., 1973. A method for construction of second-order accuracy difference schemes permitting no false two-grid-interval wave in the height field. Tellus, 25: 444-458.

Orlanski, I., 1976. A simple boundary condition for unbounded hyperbolic flows. J.Comput.Phys., 21: 251-269.

Pingree, R.D., Holligan, P.M. and Mardell, G.T., 1979. Phytoplankton growth and cyclonic eddies. Nature, 278: 245-247.

Raymond, W.H. and Kuo, H.L., 1984. A radiation boundary condition for multi-dimensional flows. Quart.J.R.Met.Soc., 110: 535-551.

A 3-D MODEL OF THE SEVERN ESTUARY

J WOLF
Institute of Oceanographic Sciences, Bidston Observatory,
Birkenhead, Merseyside L43 7RA, U.K.

ABSTRACT

The Severn estuary is a high-energy tidal regime with strong residuals generated by nonlinear effects. The present knowledge of the tidal and residual flow is reviewed. Some measurements of current profiles were used to calculate the M_2 tidal parameters and a residual flow and also the friction stress through depth at a particular location. A 3-dimensional (3-D) model of the inner Bristol Channel and Severn Estuary was used to generate tidal and residual currents in the area and these are compared with the available observations.

1 INTRODUCTION

The Severn Estuary lies enclosed between the coasts of S. Wales and S.W. England, see Fig. 1. In this paper, discussion will be limited mainly to (i) the lower Severn Estuary, between the Shoots (upstream of Avonmouth) and the Holms Islands, and (ii) the inner Bristol Channel between Swansea Bay and the Holms Islands. The mean water depth ranges from over 50m at the seaward end to less than 10m near Avonmouth. In the Severn Estuary the deep water is confined to rather narrow channels with extensive shallow flats.

The normal maximum tidal range in this area is the third largest in the world and the largest in the British Isles. The tide is predominantly semi-diurnal and the tidal wave enters from the Celtic Sea and is amplified all the way up the Bristol Channel (Heaps, 1967). The amplification stems partly from energy propagating into an increasingly narrow area (Prandle and Rahman, 1980) and partly from a broad-band resonance with central period between 11 and 12 hours between the shelf edge and the tidal limit of the Severn (Fong and Heaps, 1978). These high tidal amplitudes lead to very strong tidal currents of over 4 knots in the deep channels of the inner Bristol Channel and Severn Estuary. There is also a large spring-neap cycle in the tidal range. The mean spring range at Avonmouth is 12.3m while the mean neap range is 6.5m, which represents a variation in energy levels of 4:1,

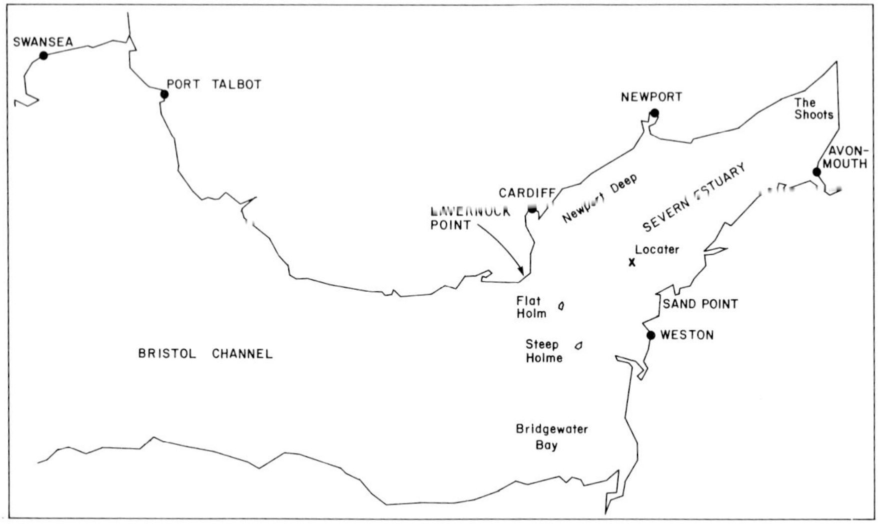

Fig. 1. The Bristol Channel and Severn Estuary, showing the location of observations from M.V. Gardline Locater, September 1978.

since the energy is proportional to the amplitude squared. The tide is also quite nonlinear with the ebb flow sometimes lasting 2 hours longer than the flood.

Due to bottom friction the tidal wave is partially progressive i.e. currents are in advance of elevations by less than 90° (Hunt, 1964). There is a net energy flux into the Channel which is dissipated in the shallow regions where friction is strongest. Owen (1980) calculates the energy flux and shows that the energy dissipation reaches a maximum in the area between Swansea Bay and Lavernock Point. The progressive nature of the M_2 tide also results in a net residual transport of water or Stokes' drift, into the estuary (Longuet-Higgins, 1969; Prandle, 1978). By continuity of mass this is compensated by a seaward-flowing Eulerian residual current (Uncles and Jordan, 1980). Other components of residual flow may be produced by winds, density gradients, fresh-water flow, topography and other tidal nonlinearities e.g. the advection of momentum and nonlinear friction. Uncles (1982a) examined the relative importance of these effects in a depth-averaged numerical model and compared his findings with observations. He found that advection is the most important mechanism driving many gyres, especially off headlands (Pingree, 1978; Tee, 1976), followed by the seaward-flowing continuity residual balancing the Stokes' drift. Depth-averaged density currents were found to be important in the eastern Bristol Channel but small in the Severn Estuary. His observations and model currents gave reasonable qualitative agreement, but observations of residuals are quite sparse. Observations show the predominantly down-channel residual flow with some indications of gyres near the coast. However the observed residuals are not uniform through the water column. The direction of the residuals tends to exhibit veering with increasing depth below the surface, which is explained partly by differential flow due to density gradients and possibly also by friction. Neither of the latter effects can be properly modelled in a depth-averaged model.

The Severn Estuary was once considered to be a typical well-mixed estuary with no significant vertical differences of salinity. However Bowden (1967) points out that vertical salinity differences of up to 0.5% have been observed and that a longitudinal density gradient requires some vertical current structure to balance it since it produces a depth-varying pressure gradient. Also he quotes values of the effective longitudinal dispersion from observation that are 10-100 times larger than those predicted from well-mixed theory. This suggests that the dispersion due to advection of these small vertical and transverse gradients is not negligible. However, the longitudinal and transverse salinity gradients in the Bristol Channel are not large, and at most produce residuals of a few centimetres per second. No thermal stratification occurs except locally in calm sunny weather (Parker and Kirby, 1981).

A potentially important factor in residual flow which may also affect the oscillatory tidal flow is the density stratification caused by suspended sediment. To date this has not been included in numerical models. The dynamics of the sediment balance are complicated and not yet fully understood. There is fairly recent evidence that the suspended sediment can produce density stratification at least of the same order as the salinity distribution and locally perhaps much larger. Parker and Kirby (1981) have carried out extensive surveys of suspended sediment and in particular found a marked suspended sediment front aligned longitudinally in the Severn Estuary (Kirby and Parker, 1982).

Sediments in the Bristol Channel/Severn Estuary do not show the classical relationship of grain size with energy levels (Hamilton, 1979). Normally the higher energy regimes coincide with a sediment population of coarser grain size than that of low energy regimes but the reverse is true in the upper Severn Estuary. Hamilton suggests that this is due to the complexity and intensity of sorting in the highly energetic oscillatory tidal currents. The Severn Estuary has the highest energy levels of any British estuary, but there is no source of supply of coarse sands and gravels hence the only mobile sediments are generally fine sands and muds. These are very mobile due to the exceptionally strong currents. The fine clay sediment has the property of flocculating into larger particles which then have enhanced settling rates. The sediment moves in and out of suspension between maximum tidal stream to slack water. It is estimated that 70% of the sediment which is mobile at springs is settled at neaps (Parker and Kirby, 1981). The main bodies of settled mud are in Bridgewater Bay and Newport Deep. The Bristol Channel and the deep channels of the Severn Estuary have mainly rocky floors with perhaps a thin layer of sediment. At maximum flood and ebb the suspended sediment tends to be well-mixed through the depth. Towards slack water this sediment settles out, forming characteristic layers with locally large vertical gradients of density (Parker and Kirby, 1981). These features of suspended sediment may have a strong influence on the density gradients which in turn affect the residual flow and possibly even the tidal flow. Horizontal density gradients can produce residuals while vertical density gradients inhibit vertical mixing and consequently the exchange of both mass and momentum.

This high energy tidal regime has the effect of enhancing various nonlinear effects which may elsewhere be insignificant. These should not be omitted without careful consideration. The next section discusses some observed current profiles from a station in the Severn Estuary and the calculation of frictional shear stresses. Section 3 describes the results of a 3-D model for the tidal and residual currents in this area. These are compared with the observed current structure.

2 OBSERVED CURRENT PROFILES

The measurements were made from M.V. Gardline Locater in September 1978 by I.O.S. Taunton as part of their cohesive sediments project. Observations were obtained on the 12-13/9/78 and 22-24/9/78. The longest continuous series was from 07.41 GMT on 12/9/78 until 05.00 GMT on 13/9/78. The location of the observations is shown in Fig. 1. The ship was anchored on station using 4 anchors to reduce any movement which might contaminate the measurements. Two arrays of current meters were deployed, both continuously monitored from the ship. The first array of 5 current meters was lowered until it just touched bottom i.e. the instruments were at fixed heights above the sea-bed. The second array of 3 current meters was positioned a fixed distance below the sea surface. During the tidal cycle the two arrays overlapped by a varying amount. Some of the current meters measured flow only, while others measured

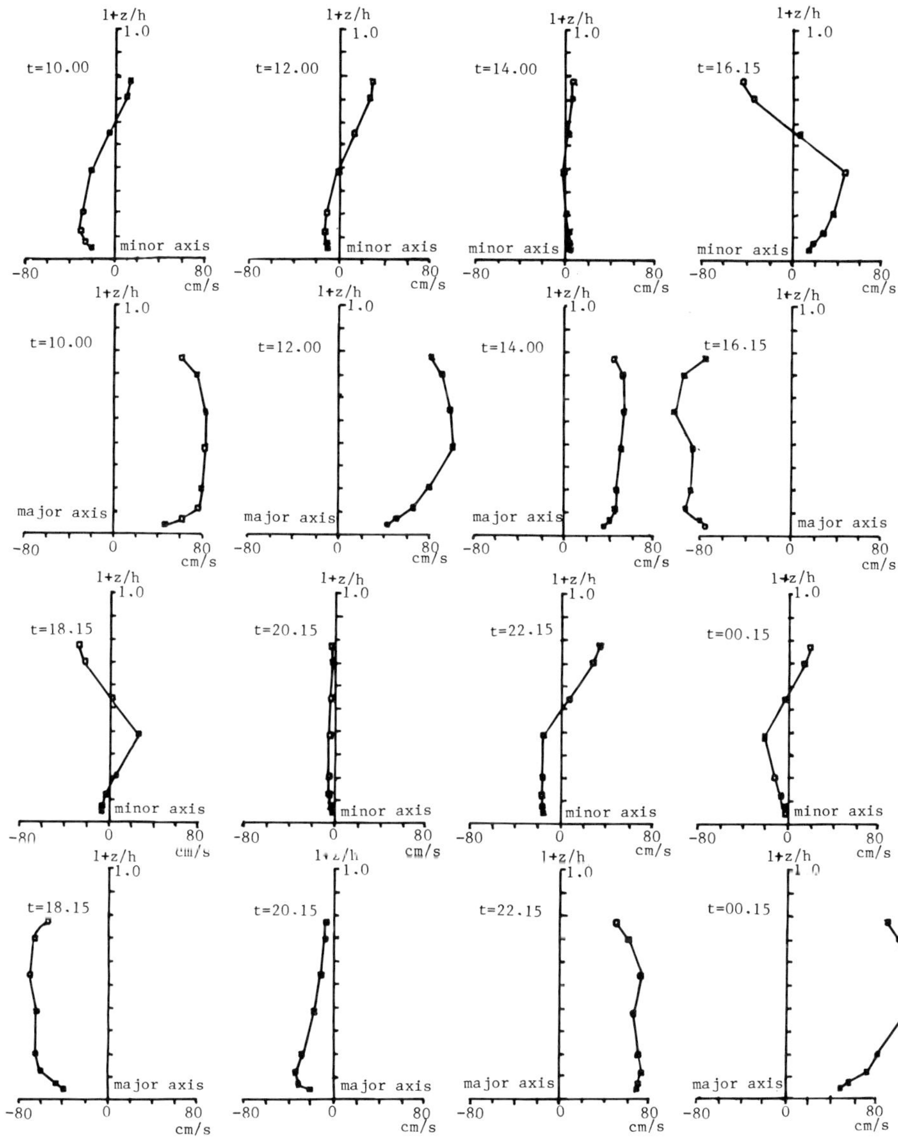

Fig. 2. Major and minor axis current profiles through a tidal cycle. 12-13/9/78. Time, t, is in G.M.T. Mean depth = 21.4m. Tidal range = 8.75m. Time of HW = 14.52. LW = 21.15.

direction as well as speed. Pressure sensors were included to check the depths of the meters.

The direction of the current for each meter was obtained by linear interpolation and extrapolation from the three directional observations made at each time. The speed and direction time series from each level were converted into north and east components. A Fourier analysis was used to obtain the M_2 amplitude and phase of each component. Then an M_2 current ellipse was calculated and the orientation of the major axis was determined for each height above the sea-bed. The depth-averaged value of the major axis orientation was chosen as a more representative direction along which to resolve the current components. This direction was about 027° (Magnetic). The north and east components of the total current were then converted to components along the major (now labelled x) and minor (y) axis directions of the M_2 tidal current. Some current profiles are shown at 2-hourly intervals through one tidal cycle in Fig. 2.

It can be seen that the v-component of current (minor axis) changes sign through depth. This reflects the rotation of the major axis of the ellipse clockwise with depth below the surface. In fact the M_2 currents are almost rectilinear at every depth. The u-component of current (major axis) is more interesting. It shows a marked reduction in current speed near the sea surface which shows up in the top three current meter records and therefore is not just an instrument malfunction. It does not seem to be confined to the floating array, since at low tide the two arrays overlap and the top meter of the bottom array is above the lowest meter of the surface array. It is also not a function of the choice of the directions for resolution of the current components as it can be seen in the original current speed profiles.

The current time series were used to calculate frictional stress time series using linearised equations as in a previous paper, Wolf (1980). The surface and bottom stress must be prescribed. The surface stress was set to zero. Winds were generally light (less than 16 knots, giving a maximum surface stress of less than 1 dyne/cm^2), so this approximation was reasonable. The bottom stress was related to the depth-mean current by a quadratic friction law. The amplitude and phase of the M_2 tidal current components and of the M_2 stress components through depth are plotted in Fig. 3. The major axis is dominant and the calculations more reliable. The current phase is almost constant through depth as is the stress phase. The stress amplitude decreases almost linearly with height above the sea-bed. Obtaining a classical eddy viscosity by relating stress to the velocity gradient, would produce negative values near the surface, since the velocity gradient becomes negative there. This suggests that the simple model used does not fully

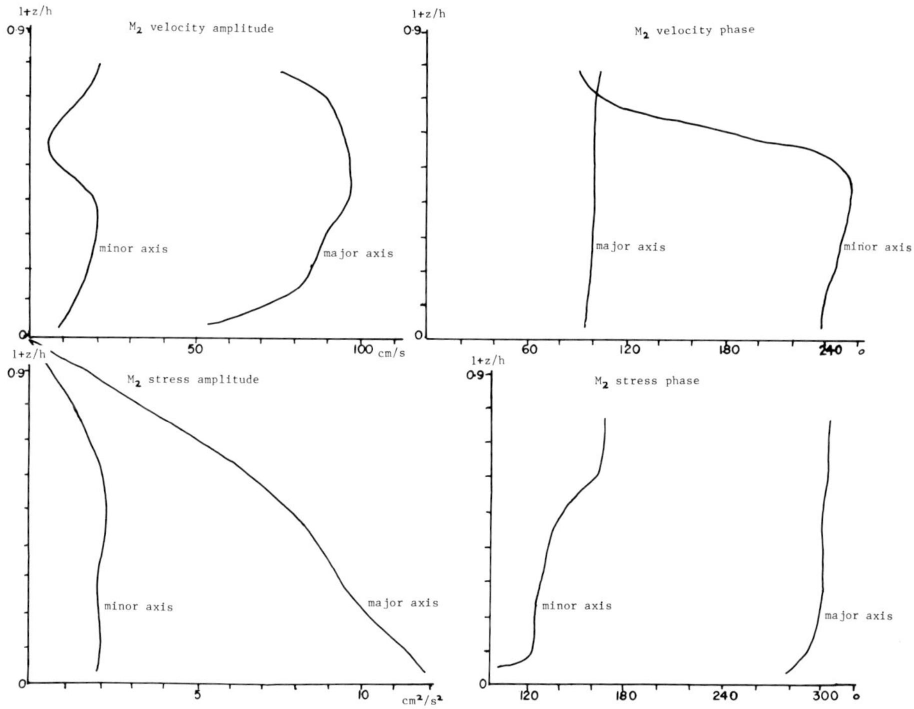

Fig. 3. M_2 current and stress amplitude and phase as a function of depth.

predict the frictional stresses. It seems likely that nonlinear effects such as advection are important in this area. The next section describes the results of a fully nonlinear 3-D model which shows the complexity of the flow in the area.

3 THE 3-D MODEL

In order to examine the structure of tidal and residual flow in the Severn Estuary a three-dimensional model was required which should be of sufficiently fine resolution to resolve the complex topography of the area in some detail. This includes deep channels, sand banks and inter-tidal flats. The Bristol Channel has been modelled in three dimensions on a grid of approximately 4km square by Owen (1980b) and Stephens (1983). This resolution does not give sufficient detail in the Severn Estuary amounting to only about three grid-boxes across the channel. In order to use a finer grid without reducing the time step as in an explicit scheme a semi-implicit alternating direction finite difference scheme was applied to a model of the Bristol Channel/Severn Estuary which had previously only been modelled in 2-D (Owen, 1980a; Proctor,

1981b) with a grid-size of about 1km. The finite difference scheme is described in Wolf (1983), with the results for the M_2 tidal elevation in the Bristol Channel and Severn Estuary. The M_2 cotidal chart is well-reproduced.

The vertical eddy viscosity used in the model was taken from Owen (1980b). This means that the eddy viscosity is parabolic through depth with a maximum at mid-depth and varies in space and time depending on the magnitude of the local velocity and depth. This form of viscosity is suggested in part by the results of Wolf (1980) and also by the work of Bowden and Hamilton (1975).

The model was used to obtain three-dimensional tidal residuals. Figs. 4 (a) and (b) shows the residuals obtained by averaging over an M_2 tidal cycle for the surface and near-bottom currents in the model. The advective terms have been included and these provide the main source of the tidal residuals. In particular, they drive the gyres which may be seen at all depths near headlands and all corner points (the latter produce grid-scale eddies which are unrealistic). Two very pronounced gyres may be seen on either side of Lavernock Point. The clockwise gyre is the result of a lee eddy on the ebb tide and the anticlockwise one on the flood. The predominantly seaward flow due to compensation for Stokes' drift is also evident in the central channel. Other eddies corresponding to sandbanks appear outside Swansea Bay (Owen, 1980a). The overall pattern is very similar to that obtained by Owen (1980a) and confirmed by Uncles (1982a), who shows the main causes of residual flow to be advection and continuity.

This is the first attempt to examine residuals in this area in three dimensions on such a fine scale. In general, the surface and bottom residuals are very similar except that the surface currents are much stronger, but in certain areas there is a marked change in direction with depth. There is a tendency for a veering (i.e. clockwise rotation) of 10° to 20° in the current direction from the surface to the bottom. However, in some areas e.g. the SW corner near the open boundary and east of Lavernock Point, it is in the opposite sense i.e. anticlockwise. This characteristic is qualitatively in agreement with the observations shown by Uncles (1982a) for the few stations available in the inner Bristol Channel.

In the Severn Estuary, the residuals show more variability, being in opposite directions at surface and bottom at some grid points and changing direction markedly from one grid point to the next. It would therefore be difficult to compare quantitatively with observations as the resolution of the model may be insufficient here. Some of the variability may be unrealistic, due to grid-scale eddies.

Uncles and Jordan (1979) look in detail at residual currents at two stations in the Severn Estuary, station A being just seaward of Steep Holme

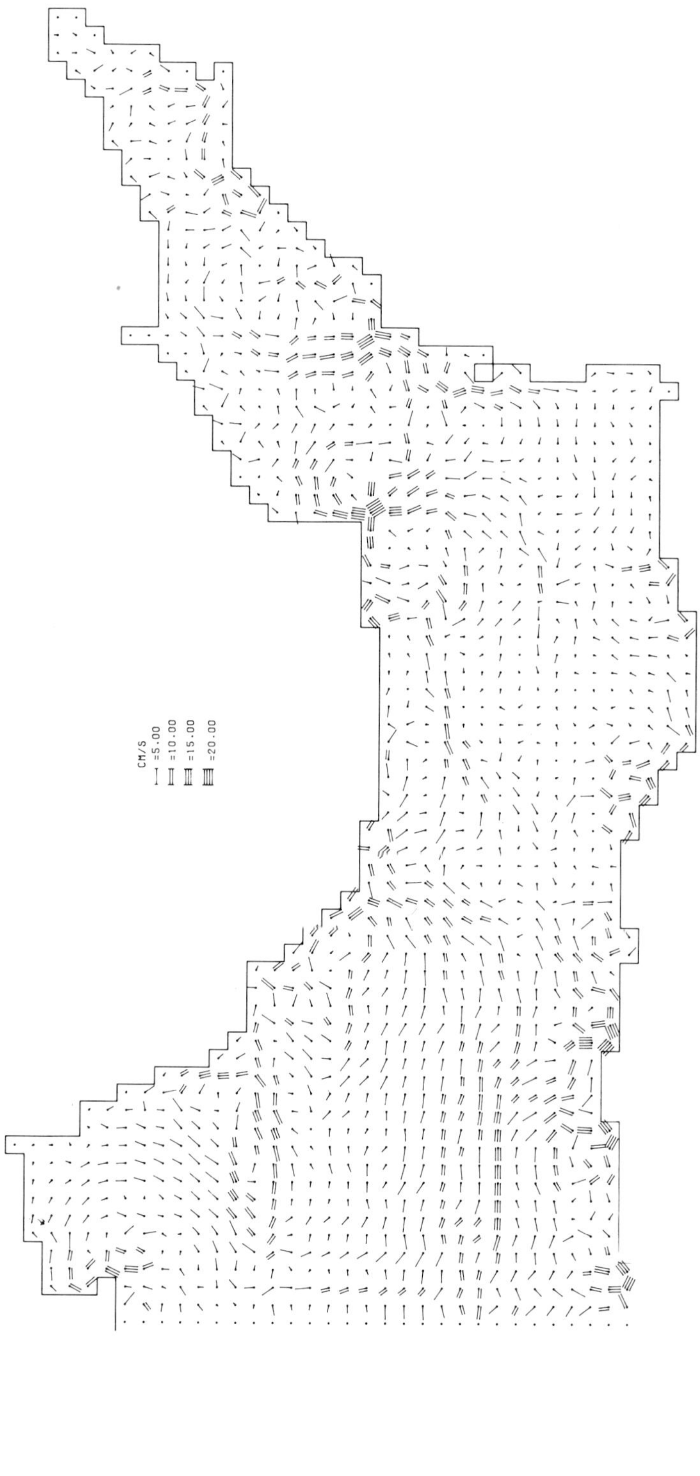

Fig. 4(a). Surface residuals.

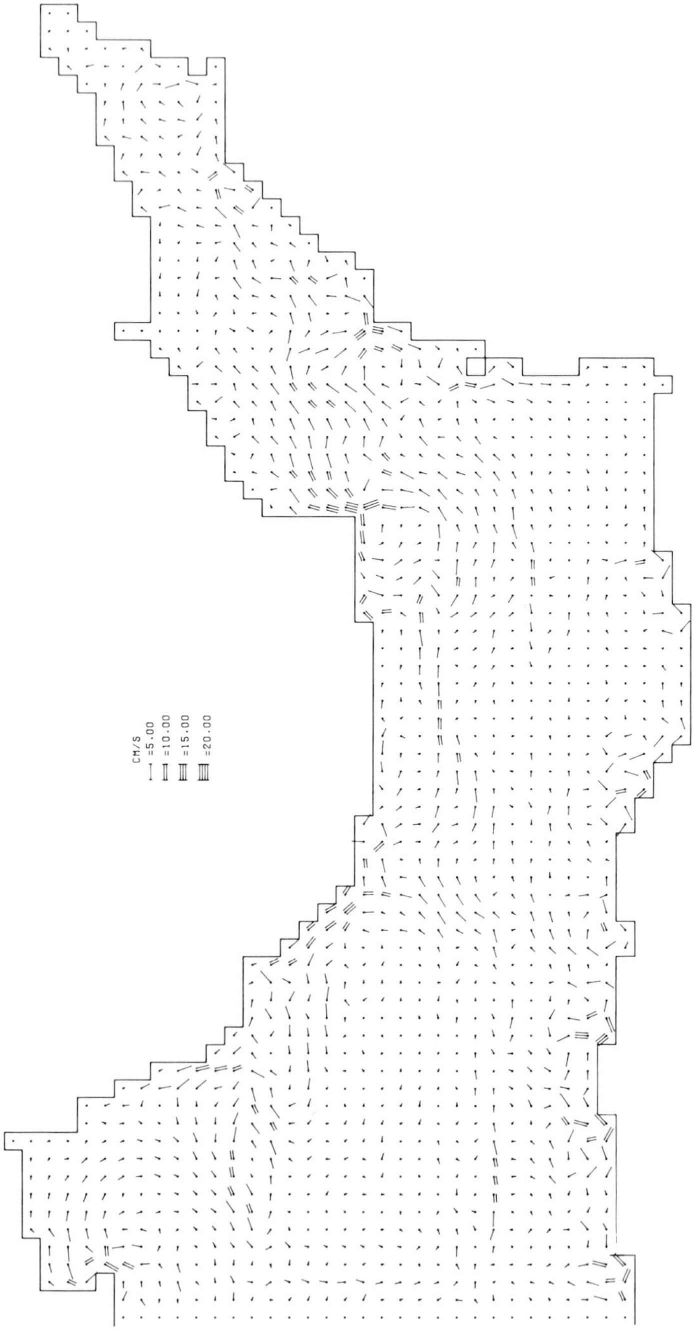

Fig. 4(b). Bottom residuals.

and station B being about 17km upstream. Both stations lie approximately in mid-channel. At station A the model gives a surface residual seawards of about 5cm/s (half the observed residual but in the right direction), the bottom residuals have the right magnitude of 4cm/s but in the opposite direction to the observed residual, the model residuals rotate clockwise with increasing depth below the surface whereas the observed residuals rotate anticlockwise. At station B the observed rotation is again anticlockwise with increasing depth except for a reversal in direction very near to the surface. The model gives an anticlockwise rotation and the near-bed residual current direction is in good agreement but the surface residual is in the opposite direction to that observed. The speeds of 5-10cm/s are the correct order of magnitude. These disagreements between the model and observations may be due to flow driven by the longitudinal salinity gradient although this has been shown to be small in the Severn Estuary (Stephens, 1983) and generally in the same direction from surface to bottom. The discrepancies may also be due to insufficient resolution of the model in the Severn Estuary. Figs. 4(a) and (b) illustrate how much variability there is in the residuals from one grid box to the next.

The tidal currents and residuals from the model are examined in more detail over a cross-section through the Locater position, shown in Fig. 1. In this section there is a deep channel with a complex system of banks around it. The M_2 and residual currents were resolved in the NE- and NW-going directions, approximately aligned along and across the main channel. The M_2 current amplitudes are shown in Figs. 5(a) and (b). The tidal currents are almost rectilinear. The maximum currents of about 100cm/s are found in the deepest part of the section, flowing NE or SW. There is no sign of a reduction in current amplitude near the surface. The residuals, shown in Figs. 6(a) and (b), seem to flow NE on the south (English) side of the section and SW on the north (Welsh) side, which agrees with the general picture described by Parker and Kirby (1981). The transverse residuals are directed NW almost everywhere and are small except near the English coast where there is quite a strong offshore current fed by alongshore currents converging on the headland of Sand Point. Continuity is satisfied in the horizontal by return flow at other sections. These features cannot be checked against observations unfortunately.

The tidal and residual currents at the Locater position are compared in Fig. 7. The tidal current amplitude is in good agreement except for the near-surface reduction in amplitude of the longitudinal component in the observations. The phase is in good agreement for the longitudinal component, however this is not the case for the transverse component. There is a

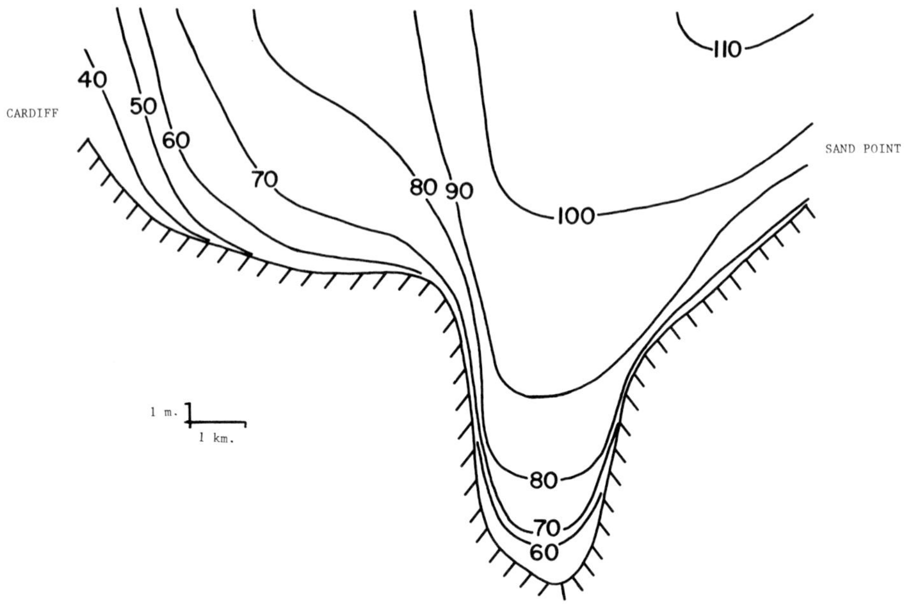

Fig. 5(a). M_2 current amplitude. NE-going component. cm/s.

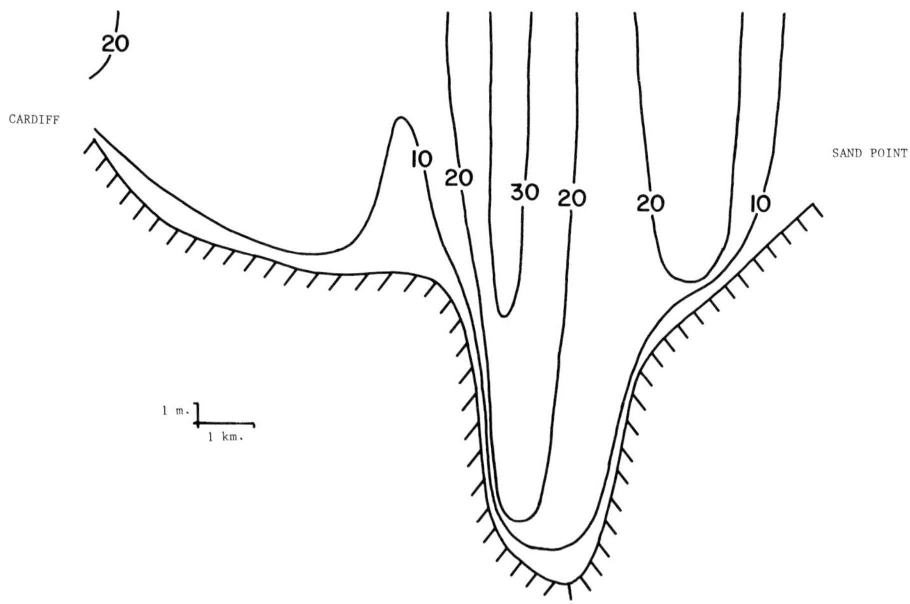

Fig. 5(b). M_2 current amplitude. NW-going component. cm/s.

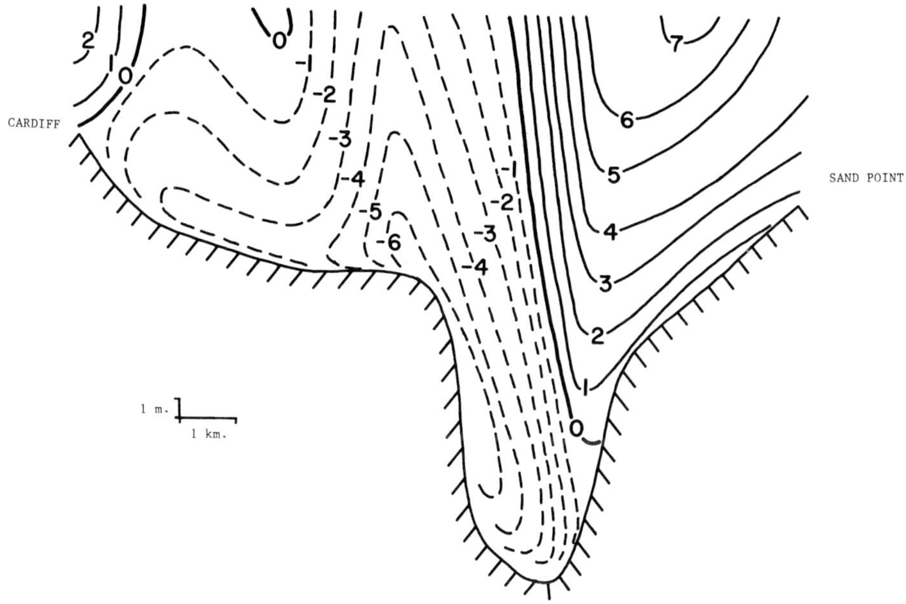

Fig. 6(a). NE-going residual current. cm/s.

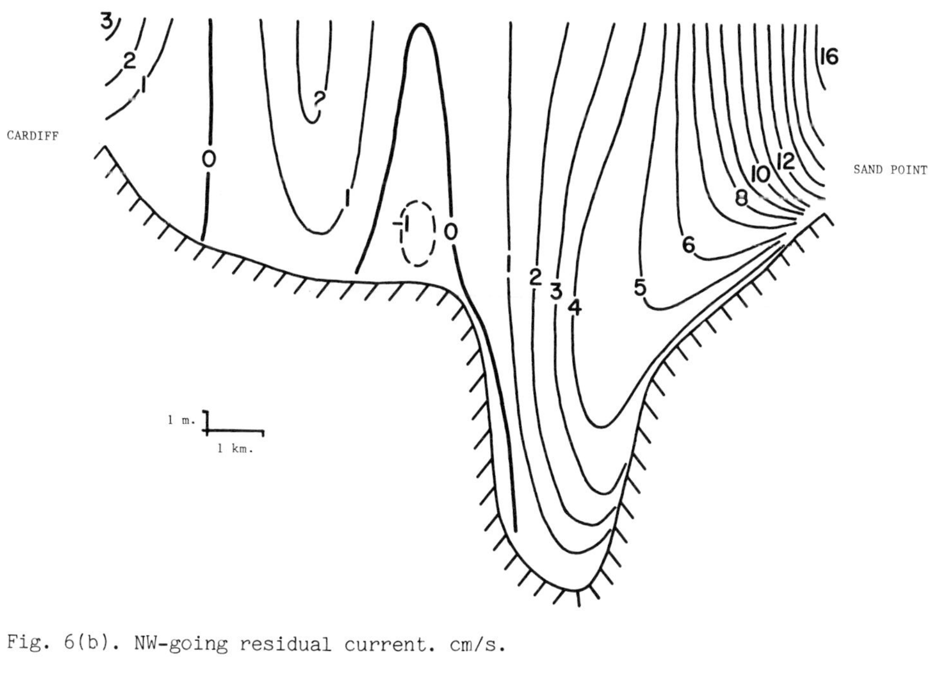

Fig. 6(b). NW-going residual current. cm/s.

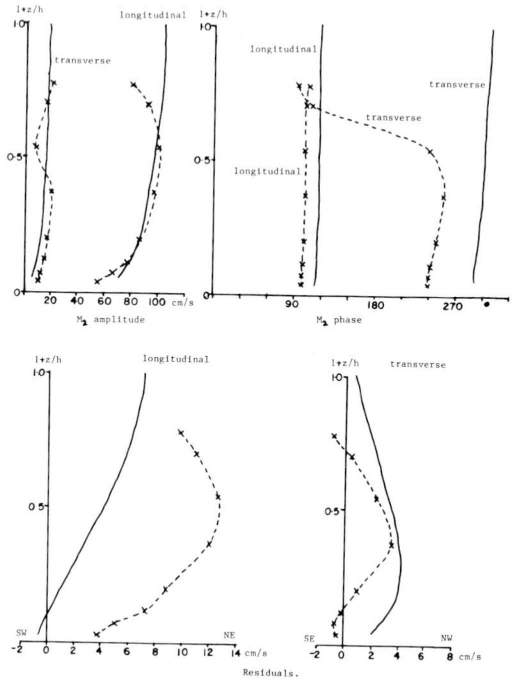

Fig. 7. Comparison of observed and model current profiles. Longitudinal axis is aligned 027°/207°M. Transverse axis is aligned 117°/297°M. —— model ---- observed.

discrepancy of about 45° near the bed and this becomes about 150° near the surface, as the observed transverse current phase changes through 180° at mid-depth. This phase reversal corresponds to a change in the direction of alignment of the current ellipse major axis. The direction of major axis given by the model is 35° near the bed, rotating clockwise to 37° near the surface, which is very close to the observed direction up to mid-depth, where the observed major axis starts to rotate in the opposite direction, being aligned almost N-S near the surface. The direction of rotation of the currents is anticlockwise near the bed in both model and observations but the model then exhibits clockwise rotation throughout the upper part of the water column while the observed currents remain anticlockwise until above mid-depth when they become clockwise.

The residuals show some qualitative agreement. The longitudinal residual is directed NE throughout the depth but the model gives less than half the observed magnitude and does not exhibit a maximum at mid-depth. The transverse residuals are directed NW through the main part of the water column, model and observations being fairly close, but the model does not give the reverse flow at surface and bottom seen in the observations.

4 DISCUSSION

The model gives quite good agreement with the observed direction and magnitude of tidal and residual flow but does not reproduce exactly the details of the vertical profiles of current. It demonstrates the importance of the advective terms but it does not include density effects which may be important. The grid-size of approximately 1km may not be sufficiently small to resolve all the topographic features and the smallest eddies.

The dominant forces in this area are a balance between the surface elevation pressure gradient and the bottom friction, on a tidal time scale, since this is a very energetic tidal regime in rather shallow water. Nonlinearities are important in generating residual flows. It may be that a more sophisticated model for friction is necessary in such an area.

From previous studies the salinity is very well-mixed in the Severn Estuary and unlikely to make much difference to the current structure. However the density stratification induced by suspended sediment may be relatively important. A strong transverse gradient of suspended sediment has been observed in the Severn Estuary which may have effects on the residual currents such as occur at other fronts due to temperature and salinity. It may also affect the tidal currents through nonlinear effects, i.e. the advective terms, and by the reduction in turbulence and viscosity due to the stratification. The pattern of tidal residuals suggest that the density front may be partly explained by the suspended sediment being trapped in the gyres near its origins in Bridgewater Bay.

5 REFERENCES

Bowden, K.F, 1967. Circulation and diffusion. In: G.H. Lauff (Editor), Estuaries. A.A.A.S. Publication no. 83, Washington, pp. 13-36.

Bowden, K.F. and Hamilton, P., 1975. Some experiments with a numerical model of circulation and mixing in a tidal estuary. Est. Cstl. Mar. Sci, 3: 281-301.

Fong, W.W. and Heaps, N.S., 1978. A note on quarter-wave tidal resonance in the Bristol Channel. I.O.S. Report no. 63 (unpubl.), 11 pp + figs.

Hamilton, D., 1979. The high energy, sand and mud regime of the Severn Estuary, S.W. Britain. In: R.T. Severn, D. Dineley, L.E. Hawker (Editors), Tidal power and estuary management. Colston papers no. 30. Scientechnica, Bristol, pp. 162-172.

Heaps, N.S., 1967. Storm Surges. In: H. Barnes (Editor), Oceanography and marine biology, annual review. Allen and Unwin, London, pp. 11-47.

Hunt, J.N., 1964. Tidal oscillations in Estuaries. Geophys. J. R. astr. Soc., 8: 440-455.

Kirby, R. and Parker, W.R., 1982. A suspended sediment front in the Severn Estuary. Nature, 195: 396-399.

Longuet-Higgins, M.S., 1969. On the transport of mass by time varying ocean currents. Deep-Sea Res., 16: 431-447.
Owen, A., 1980a. The tidal regime of the Bristol Channel: a numerical modelling approach. Geophys. J.R. astr. Soc., 61: 59-75.
Owen, A., 1980b. A three-dimensional model of the Bristol Channel. J.P.O., 10: 1290-1302.
Parker, W.R. and Kirby, R., 1981. The behaviour of cohesive sediment in the inner Bristol Channel and Severn Estuary in relation to construction of the Severn Estuary. I.O.S. Report no. 117 (unpubl.), 54 pp.
Pingree, R.D., 1978. The formation of the Shambles and other banks by tidal stirring of the seas. J. Mar. Biol. Assoc. U.K., 58: 211-226.
Prandle, D., 1978. Residual flows and elevations in the southern North Sea. Proc. R. Soc. Lond., A, 359: 189-228.
Prandle, D. and Rahman, M., 1979. Tidal response in estuaries. J.P.O., 10: 1552-1573.
Proctor, R., 1981. Mathematical modelling of tidal power schemes in the Bristol Channel. In: Proc. 2nd Int. Symp. on Wave and Tidal Energy, Cambridge, September 1981. BHRA Fluid Engineering, pp. 33-51.
Stephens, C.V., 1983. Hydrodynamic modelling developments for the west coast of the British Isles. Liverpool University, Ph.D thesis, 213 pp.
Tee, K.-T., 1976. Tide-induced residual current, a two-dimensional non-linear numerical tidal model. J. Mar. Res., 34: 603-629.
Uncles, R.J., 1982. Computed and observed residual currents in the Bristol Channel. Ocean. Acta, 5: 11-20.
Uncles, R.J. and Jordan, M.B., 1979. Residual fluxes of water and salt at two stations in the Severn Estuary. Est. Cstl. Mar. Sci., 9: 287-302.
Uncles, R.J. and Jordan, M.B., 1980. A one-dimensional representation of residual currents in the Severn Estuary and associated observations. Est. Cstl. Mar. Sci., 10: 39-60.
Wolf, J., 1980. Estimation of shearing stresses in a tidal current with application to the Irish Sea. In: J.C.J. Nihoul (Editor) Marine Turbulence. Proc. 11th Int. Liege Colloq. on Ocean Hydrodynamics, 1979. Elsevier Amsterdam, pp. 319-344.
Wolf, J., 1983. A comparison of a semi-implicit with an explicit scheme for a three-dimensional hydrodynamic model. Cont. Shelf Res., 2: 215-229.

THE VARIATIONAL INVERSE METHOD REVISITED*

C. PROVOST
L.O.D.Y.C., Université P. et M. Curie, Tour 14,2E, 4 place Jussieu, 75252 Paris Cedex 05, France.

ABSTRACT

The variational inverse method addresses the classical problem of estimating the three dimensional field of velocity from hydrographic data. It seeks the smoothest velocity field (in the sense of an arbitrarily defined norm) wich is consistent with the data and selected prescribed misfits. These misfits should not be zero: they represent errors in the observations and in the approximate dynamical constraints. The admission of errors necessitates the choice of weights. The choice is made by prior estimation of the relative error variances. The solutions are produced in one single step by a global optimization by demanding overall consistency in the whole domain. By varying the misfits relatively to one another in their respective admissible range, we explore the full envelope of physically plausible estimates of the average geostrophic flow. We discuss the choice of weights, the choice of smoothing criteria, and problems such as resolution of the solutions, compatibility between constraints and parameter estimations.

* Paper presented at the Colloquium and submitted too late for publication in this volume.

THE BRANCHING OF THE GULF STREAM REVISITED USING THE VARIATIONAL INVERSE METHOD*

F. MARTEL and C. PROVOST
L.O.D.Y.C., Université P. et M. Curie, Tour 14,2E, 4 place Jussieu, 75252 Paris Cedex 05, France.

ABSTRACT

We investigate the circulation, and the mass and heat transports in the region between the Grand Banks and the Mid-Atlantic Ridge. We use the variational inverse method developed by Provost and Salmon (J. Mar. Res., 44, 1-34, 1986) to analyse the dense data set described by Clarke *et al.* (J. Phys. Oceanogr., 10, 25-65, 1980). Using only two constraints (geostrophy and total transport dynamics), we can conclude that the data does not support Worthington's hypothesis of two independent gyres. All the flows obtained from our solutions show a branching of the Gulf Stream. However, the solutions are not well resolved in areas with steep bottom topography, and in regions of strong shear. More constraints (data and dynamics) are needed to understand clearly the behavior of the flow, and to establish firmly the "average dynamical balances" in this complex region.

* Paper presented at the Colloquium and submitted too late for publication in this volume.

ABOUT A DIAGNOSTIC ANALYSIS OF THE HISTORICAL HYDROGRAPHIC DATA IN THE TROPICAL ATLANTIC*

C. PROVOST and M. S. SUK
L.O.D.Y.C., Université P. et M. Curie, Tour 14,2E, 4 place Jussieu, 75252 Paris Cedex 05, France.

ABSTRACT

We present an application of the variational inverse method (Provost and Salmon, 1986, J. Mar. Res., 44, 1-34) to the historical hydrographic data of the tropical Atlantic. The idea is to develop a method for analyzing hydrographic data in the tropics and to evaluate the present data set. This paper discusses only the use of temperature and salinity.

We had to face three main difficulties. First, the geostrophic constraint that controls efficiently the vertical shear of the horizontal velocity in mid-latitudes is not valid in the equatorial band. We modify the traditional geostrophic relation with the explicit inclusion of vertical eddy diffusion near the equator. Second, the tropical Atlantic exhibits a strong seasonal variability. We invert the data season by season. Finally, the area is large (20°N-20°S), and the original method has to be adapted to handle large domains and large data sets.

We define the averaging spatial and time scales and estimate the aliasing errors in the data with respect to that definition. The misfits on the data constraint are adjusted in order to filter the aliasing present in the data, and the misfits on the dynamical constraints are adjusted in relation to the estimated accuracy of the corresponding dynamical equation. We explore the domain of solutions for each season with a special attention to where, how, and at which rates water masses cross the equator and to heat transport estimates. Only spring season has a sufficient data coverage for a small domain of solutions. However, even in spring, uncertainty on longitudinally averaged heat transport estimates is large. The inadequacy of the historical data is mainly due to the lack of deep hydrographic measurements and of meridionally closely spaced stations across the equator. Estimated heat transport at depth accounts for a large part of the estimated total transport.

* Paper presented at the Colloquium and submitted too late for publication in this volume.